EX LIBRIS
UNIVERSITATIS SANCTI JOANNIS

Spectroscopic Properties of Inorganic and Organometallic Compounds

Volume 19

A Specialist Periodical Report

Spectroscopic Properties of Inorganic and Organometallic Compounds

Volume 19

A Review of the Recent Literature Published up to Late 1985

Senior Reporters
G. Davidson, *Department of Chemistry, University of Nottingham*
E. A. V. Ebsworth, F.R.S.E., *Department of Chemistry, University of Edinburgh*

Reporters
S. J. Clark, *City University, London*
S. Cradock, *University of Edinburgh*
K. B. Dillon, *University of Durham*
J. D. Donaldson, *City University, London*
S. M. Grimes, *City University, London*
B. E. Mann, *University of Sheffield*
D. W. H. Rankin, *University of Edinburgh*
H. E. Robertson, *University of Edinburgh*

The Royal Society of Chemistry
Burlington House, London W1V 0BN

ISBN 0-85186-173-3
ISSN 0584-8555

Printed in Great Britain by
Whitstable Litho Ltd., Whitstable, Kent

Foreword

The coverage of this volume is essentially the same as last year's; production by camera-ready copy has presented no insurmountable difficulties, though the result is of course less uniform and somewhat less pleasing to the eye than are books produced by traditional methods.

I am as ever grateful to the authors for their excellent work, and I deeply regret that purchasers will have to pay so much to benefit from it.

E. A. V. Ebsworth

Contents

Conversion Factors

1 kJ mol^{-1}		1 kcal mol^{-1}	
2.3901×10^{-1}	kcal mol^{-1}	4.1840	kJ mol^{-1}
1.0364×10^{-2}	eV $atom^{-1}$	4.3364×10^{-2}	eV $atom^{-1}$
8.3593×10	cm^{-1}	3.4976×10^{2}	cm^{-1}
2.5061×10^{6}	MHz	1.0486×10^{7}	MHz

1 cm^{-1}		1 MHz	
1.1963×10^{-2}	kJ mol^{-1}	3.9903×10^{-7}	kJ mol^{-1}
2.8592×10^{-3}	kcal mol^{-1}	9.5370×10^{-8}	kcal mol^{-1}
1.2399×10^{-4}	eV $atom^{-1}$	4.1357×10^{-9}	eV $atom^{-1}$
2.9979×10^{4}	MHz	3.3356×10^{-5}	cm^{-1}

1 eV $atom^{-1}$	
9.6485×10	kJ mol^{-1}
2.3060×10	kcal mol^{-1}
8.0655×10^{3}	cm^{-1}
2.4180×10^{8}	MHz

Mössbauer spectra: $E_{\gamma}(^{57}Fe) = 14.413$ keV

1 mm s^{-1}	
4.639×10^{-6}	kJ mol^{-1}
1.109×10^{-6}	kcal mol^{-1}
4.808×10^{-8}	eV $atom^{-1}$
3.878×10^{-4}	cm^{-1}
1.162×10	MHz

For other Mössbauer nuclides, multiply the above conversion factors by E_{γ}(keV)/14.413

1
Nuclear Magnetic Resonance Spectroscopy

BY B. E. MANN

1 Introduction

Following the criteria established in earlier volumes, only books and reviews directly relevant to this chapter are included, and the reader who requires a complete list is referred to the Specialist Periodical Reports 'Nuclear Magnetic Resonance',[1] where a complete list of books and reviews is given. Reviews which are of direct relevance to a section of this Report are included in the beginning of that section rather than here. Papers where only ^{1}H n.m.r. spectroscopy is used are only included when the ^{1}H n.m.r. spectra make a non-routine contribution, but complete coverage of relevant papers is still attempted where nuclei other than proton are involved. In view of the greater restrictions on space, and the ever growing numbers of publications, many more papers in marginal areas have been omitted. This is especially the case in the sections on solid-state n.m.r. spectroscopy, silicon and phosphorus.

A number of reviews have appeared including 'N.m.r. and inorganic chemistry',[2] 'N.m.r. of metal nuclides. Part II: the transition metals',[3] 'Scalar spin-spin interactions of nuclei in diamagnetic coordination compounds',[4] 'Steady state techniques for low sensitivity and slowly relaxing nuclei',[5] 'Effect of substituents in R_nM organometallic compounds and changes in direct constants of the $^{1}\underline{J}(M,^{13}C)$ spin-spin interaction',[6] 'Nuclear magnetic resonance spectroscopy of organic analytical reagents and their metal complexes',[7] 'Nuclear magnetic resonance spectroscopy of chlorophylls and corrins',[8] 'Structural properties of calmodulin, an intracellular calcium ion-modulator protein, as revealed by different n.m.r. techniques',[9] and 'Elucidation of the structure and metal sequestering properties of metallothionein by nuclear magnetic resonance'.[10]

A number of papers have been published which are too broadly based to fit into a later section and are included here. The general magnitudes of n.m.r. isotope shifts have been discussed.[11] Recycled flow n.m.r. spectroscopy has been used to investigate ^{13}C,

^{15}N, ^{29}Si, ^{31}P, and ^{113}Cd n.m.r. spectra of model compounds.[12] Scalar relaxation of heteronuclear multiple quantum coherences and relative signs of nuclear spin-spin coupling constants have been examined and applied to $\underline{J}(^{35}Cl,^{1}H)$ and $\underline{J}(^{35}Cl,^{29}Si)$, which have opposite signs in $SiHCl_3$.[13] The ^{13}C magnetic shielding in cyclopentadienyl complexes has been calculated by eliminating effects due to charge.[14] The results for the first overlapping sphere Z_α-SW chemical shift calculations reproduce the trends in ^{13}C n.m.r. measurements for the CO, CS, CN, and C_5H_5 ligands in $Ni(CO)_4$, $Cr(CO)_5CS$, and $Fe(C_5H_5)(CO)_2CN$.[15] ^{13}C n.m.r. spectra of RN=CHCH=NR in various coordination modes have been discussed with particular reference to the $Ru_3(CO)_{12}$/RN=CHCH=NR system.[16] Metal-allyl bonding has been studied by using $^{1}\underline{J}(^{13}C,^{13}C)$. $^{1}\underline{J}(^{13}C,^{13}C)$ varies between 59 Hz for Li or $K(C_3H_5)$ down to 40 Hz for some transition-metal complexes.[17] N.m.r. data have also been reported for complexes of ethylenediphosphinetetraacetic acid (^{31}P),[18] 1,3-bis(2-hydroxyphenyl)-1,3-propanedione (^{13}C),[19] and 7-methylguanosine (^{13}C).[20]

2 Stereochemistry

This section is subdivided into ten parts which contain n.m.r. information about Groups IA and IIA and transition-metal complexes presented by Groups according to the Periodic Table. Within each Group, classification is by ligand type.

Complexes of Groups IA and IIA.——^{1}H, ^{7}Li, and ^{13}C n.m.r. spectroscopy has been used to show that $Li[Ph_2CCLiCPh]$ has this structure.[21] A similar investigation has been carried out using $\underline{J}(^{13}C,^{13}C)$ and $\underline{J}(^{13}C,^{6}Li)$ on isotopically enriched materials.[22] Solutions of alkylaryllithium salts chelated by $Me_2NCH_2CH_2NMe_2$ or $Me_2N(CH_2CH_2NMe)_3Me$ have been studied by ^{7}Li n.m.r. spectroscopy. Good correlations between the chemical shift and the pK of the parent hydrocarbon were found.[23] The ^{13}C n.m.r. spectrum of ^{6}Li-benzovalene shows $^{1}\underline{J}(^{13}C,^{6}Li)$.[24] ^{1}H, ^{7}Li, and ^{13}C n.m.r. spectra of $MSiMe_{\underline{n}}Ph_{3-\underline{n}}$, M = Li, K, have been discussed in terms of π-polarization of the phenyl rings.[25] At low temperatures $\underline{J}(^{29}Si,^{6}Li)$ and $\underline{J}(^{29}Si,^{7}Li)$ in $Ph_{3-\underline{n}}Me_{\underline{n}}SiLi$ are observed.[26] A wide range of unsaturated hydrocarbons have been reduced by alkali metals in liquid ammonia, and detected <u>in situ</u> by ^{13}C n.m.r. spectroscopy.[27] K^- in $[K(15\text{-crown-}5)_2]^+K^-$ in Me_2O has been observed by ^{39}K n.m.r. spectroscopy.[28] The ^7Li and ^{31}P n.m.r. spectra of $LiPR_2$ show $^1\underline{J}(^{31}P, ^7Li)$.[29] The relaxation times $\underline{T}_1$ of free Na^+ and Na^+ bound to macromolecules have been determined simultaneously by the two-dimensional n.m.r. method by using ^{23}Na n.m.r. spectroscopy.[30] The ^{39}K n.m.r. spectrum of K/Cs alloy dissolved in 12-crown-4 or 15-crown-5 shows the presence of K^-.[31] N.m.r. data have also been reported for $EtCMeCHCH_2Li$ (^{13}C),[32] 1-(Me_2N)-3-lithiopropane (^{13}C),[33] $Bu^tMC{=}C{=}C{=}CMBu^t$ (M = Li, $SiMe_3$; ^{13}C),[34] $\{Li(tmed)\}_2$-$C_6H_4(CHSiMe_3)_2$ (^{13}C),[35] $Li\{OC(=CH_2)C_6H_4CH_2NMe_2\text{-}2\}_4$ (^{13}C),[36] $(C_5Me_5)Li(\mu\text{-}EBu^t)_2(EBu^t)Li(tmed)$ (E = O, S; ^{13}C),[37] $[Bu^t_2C{=}NLi(hmpa)]_{\underline{n}}$ (^7Li),[38] and $Li\{(R^1O)_2P(O)CH{=}COR^2\}$ (^{31}P).[39]

A $^{13}\bar{C}$ n.m.r. study of the biosynthesis of bacteriochlorophyll using ^{18}O has shown that ^{18}O is incorporated into all possible sites.[40] Similarly the incorporation of 1-[1-^{13}C]glutamate and [2-^{13}C]-glycine into bacteriochlorophyll has been studied by ^{13}C n.m.r. spectroscopy.[41] The structure of a bacteriochlorophyllide dimer in solution has been determined by ^1H n.m.r. spectroscopy.[42] ^{43}Ca and ^{113}Cd n.m.r. spectroscopy has been used to investigate structural differences in the two calcium-binding sites of the porcine intestinal calcium-binding protein.[43] Shift reagents have been employed for ^{43}Ca n.m.r. studies of calcium-binding proteins.[44] Quadrupole coupling constants have been determined for ^{43}Ca and ^{25}Mg in $M(acac)_2$, M = Ca, Mg. ^{13}C $\underline{T}_1$ measurements were also performed.[45] N.m.r. data have also been reported for $Mg(C_5H_5)(CH_2Bu^t)$ (^{13}C),[46] $Mg(C_5H_3R^1R^2)_2$ (^{13}C),[47] $[Mg(anthracene)]_{\underline{n}}$ (^{13}C),[48] 1,2-dimethylanthracenemagnesium (^{13}C),[49] Mg-inosine-5'-monophosphate (^{13}C),[50] $Be(R^1COCHCOR^2)_2$ (^{13}C, ^{19}F),[51] and $MUO_2(OAc)_4$ (M = Mg, Ba, Co, Zn; ^{19}F).[52]

Complexes of Groups IIIA and IVA, the Lanthanides, and Actinides. The ^{89}Y n.m.r. spectra of some organoyttrium complexes such as $(MeC_5H_4)_3Y(thf)$ have been reported, and vary over a range of 400 p.p.m. For $(MeC_5H_4)_3Y(thf)$, $\underline{J}(^{89}Y,^1H)$ = 27 Hz.[53] The ^{17}O chemical shifts are related to the lowest electronic transition energies of UO_2 complexes.[54] ^1H, ^{17}O, and ^{79}Br n.m.r. spectra have been recorded for aqueous-organic solutions of the perbromates of uranyl and neptunyl and the coordination of water and anion investigated.[55, 56] The ^{19}F relaxation in gaseous UF_6, WF_6, and MoF_6 is consistent with relaxation dominated by spin-rotation interaction.[57] The possibility of the use of ^{19}F n.m.r. spectra in ^{235}U enrichment determination in UF_6 (gas) has been investigated. The linewidth modification induced by various ^{235}U enrichments is related to the ^{19}F-^{235}U indirect scalar interaction, modulated by rapid ^{235}U quadrupole relaxation.[58, 59] N.m.r. data have also been reported for $(C_5Me_5)LaCl_3Na(OEt_2)$ (^{13}C),[60] Ln($\underline{o}$-$C_6H_4CH_2NMe_2)_3$ (^{13}C),[61] $(C_5Me_5)_2NdCH(SiMe_3)_2$ (^{13}C),[62] $(C_5Me_5)_2Sm(thf)_2$ (^{13}C),[63] $[(C_5Me_5)Sm(\mu$-I)(thf)$_2]_2$ (^{13}C),[64] $(C_5H_5)LuC_6H_4PPh_2CH_2$ (^{13}C),[65, 66] $(C_5H_5)_2Lu(\mu$-$CH_3)_2Li$(tmed) (^{13}C),[67] Yb$\{N(SiMe_3)_2\}_2(AlMe_3)_2$ (^{13}C),[68] $[(C_5H_5)_3U(MW_5O_{19})_2]^{5-}$ (M = Nb, Ta; ^{17}O, ^{93}Nb),[69] $[Th(BH_3Me)_4]_2$ (^{11}B, ^{13}C),[70] $(C_5Me_5)_2Th(\mu$-$PPh_2)_2Ni(CO)_2$ (^{31}P),[71] $[UO_2\{C_5H_3N(CR{=}NCH_2CH_2$-$N{=}CR)_2C_5H_3N\}]^{2+}$ (^{13}C),[72] and $UO_2(R^1COCHCOR^2)_2$ (^{13}C, ^{19}F).[73]

The ^1H n.m.r. spectrum of $(\eta^5{:}\eta^5$-$C_{10}H_8)(C_5H_5)_2M_2(\mu$-$H)_2$, M = Ti, Nb, has been reported. The titanium analogue has a thermally accessible paramagnetic state, and hence the shifts are temperature dependent.[74] ^1H n.m.r. spectroscopy has been used to show exchange in $[(C_5Me_5)_2Hf(N_2)]_2(\mu$-$N_2)$, and the ^{13}C n.m.r. spectrum of $(C_5Me_5)_2HfH_2(CO)$ was recorded.[75] In $(C_5H_5)_2Ti(\mu$-$CH_2)_2M(C_5H_5)_2$, M = Ti, Zr, Hf, both ^1H and ^{13}C n.m.r. spectra show high-frequency shifts for the methylene groups, which were discussed in relation to metal-carbene or metal-metal bonding in those complexes.[76] ^{47}Ti and ^{49}Ti n.m.r. chemical shifts, recently reported for $(C_5H_5)_2TiX_2$ have been reassigned.[77] In $R_2Si(C_5H_4)_2MCl_2$, M = Ti, Zr, the ^{13}C n.m.r. signal of the bridgehead carbon is substantially to low-frequency from the other ring carbon atoms.[78] N.m.r. data have also been reported for $(C_5Me_5)_2ZrH(CH{=}CHBu^t)$ (^{13}C),[79] $(C_5H_5)_2M(SiMe_3)(BH_4)$ (M = Zr, Hf; ^{13}C),[80] $(C_5H_5)_2ZrR^1R^2$ (R^1, R^2 = H, Cl; ^{13}C),[81] M(1-norbornyl)$_4$ (M = Ti, Zr, Hf; ^{13}C),[82] $(C_5Me_5)\{C_5Me_3(CH_2)_2\}Ti$ (^{13}C),[83] $[(C_5H_5)_2M(CH_2PPh_2)_2Rh(CO)Cl]_2$ (M = Ti, Zr; ^{31}P),[84] $Ti(OC_6H_3Bu^tCMe_2CH_2)(OAr)_2$ (^{13}C),[85] $Zr(CH_2Ph)_4$ (^{13}C),[86] $(Bu^t_3CO)_2ZrMe(OCMe_2CHCH_2O)$ (^{13}C),[87] $Zr(CH_2Ph)(\eta^2$-$PhCH_2C{=}NC_6H_3Me_2)$ (^{13}C),[88] $(C_5H_5)_{\underline{n}}Ti$(1-norbornyl)$_{4-\underline{n}}$ (^{13}C),[89] $TiCH_2CH_2C(CH_2)CH_2CHR$-$(C_5Me_5)_2$ (^{13}C),[90] $(C_5Me_5)_2TiCMe{=}CMeCH_2CH_2$ (^{13}C),[91] $(C_5H_4CPh_2)_2Ti$ (^{13}C),[92] $(C_5H_5)_2TiCH{=}CHCH{=}CH$ (^{13}C),[93] $(C_5H_5)_2ZrPh\{CH(SMe)(SiMe_3)\}$ (^{13}C),[94] $(C_5H_5)_2ZrAr(CH{=}PPh_2)$ (^{13}C, ^{31}P),[95] $(C_5H_4Me)_2ZrCPh{=}CPh$-$CPh{=}CPh$ (^{13}C),[96] $(C_5H_5)_2Zr(\eta^1,\eta^5$-$C_5H_5)(\mu$-CO)Ru(CO)(PMe$_3$) ($^{13}C$),[97] $(C_5H_5)_2HfEt(CH{=}PPh_3)$ (^{13}C, ^{31}P),[98] $(C_5H_5)_2TiCl(CH_2)_{\underline{n}}CH{=}CH_2$ (^{13}C),[99] $(C_5H_5)_2TiNBu^t{=}BBu^tCH_2$ (^{11}B, ^{13}C),[100] $(C_5H_5)_2Ti(PMe_3)(CO)$ (^{13}C),[101] $Ti_2(C_5Me_5)_2(CH_2C_5Me_4)(\mu$-$O)_2$ (^{13}C),[102] $(C_5H_5)_2ZrCH(CH_2Bu^t)(SnR_3)Cl$ (^{13}C),[103] $(C_5Me_5)_2ZrCl(CH_2PPh_2)$ (^{31}P),[104] $(C_5H_5)_2HWCHPhOZrPh(C_5H_5)_2$ (^{13}C, ^{31}P),[105] $(C_5H_5)_2ZrCl(PhC{=}CPh)ClZr(C_5H_5)_2$ (^{13}C),[106] $(C_5H_5)_2$-$ZrC_6H_4C\{{=}W(CO)_5\}O$ (^{13}C),[107] $(C_5H_5)_2Zr(\eta$-$CH{=}NR)Cl$ (^{13}C),[108]

$(C_5H_5)_2Zr(\eta^2\text{-}COSiMe_3)(OSO_2CF_3)$ (^{13}C),[109] $(C_5Me_5)HfCl_2(\eta^2\text{-}COPBu^t{}_2)$ (^{13}C, ^{31}P),[110] $(C_5H_5)_2M(\eta^4\text{-diene})$ (M = Zr, Hf; ^{13}C),[111] (C_5Me_5)-$(C_7H_7)M$ (M = Ti, Zr, Hf; ^{13}C),[112] $TiW\{\mu\text{-}C(tol)=CH_2\}(\mu\text{-}CO)(CO)$-$(C_5H_5)_3$ (^{13}C),[113] $(C_5H_5)TiCl_2(NHR)$ (^{13}C),[114] $(C_5H_5)_2Ti(PMe_3)_2$ (^{13}C, ^{31}P),[115] $[(C_5H_5)TiM_5O_{18}]^{3-}$ (M = Mo, W; ^{17}O),[116] $[(C_5H_4R)Ti$-$P_2W_{17}O_{61}]^{7-}$ (^{31}P, ^{183}W),[117] $[(C_5H_5)_2Ti(SCH_2CH_2PPh_2Me)_2]^{2+}$ (^{31}P),[118] $(C_5H_5)_2Zr(N=CHR)Cl$ (^{13}C),[119] $(C_5H_5)ZrXNBu^tBHNC_5H_6Me_4$ (^{11}B, ^{13}C),[120] $(C_5H_5)_2Zr(\mu\text{-}PPh_2)Mo(CO)_4$ (^{31}P),[121] $(C_5H_5)_2Zr(\mu\text{-}PR_2)_2W(CO)_4$ (^{31}P),[122] $(C_5H_5)_2Hf(\mu\text{-}PEt_2)_2Mo(CO)_4$ (^{31}P),[123] $(C_5Me_5)Zr(\mu\text{-}OCH_2$-$PPh_2)_3NiL$ (^{31}P),[124] $(C_5H_5)(C_5H_4PPh_2)Zr(OBu^t)Cl$ (^{13}C, ^{31}P),[125] $[(C_5H_5)_2ZrCl]_2$ (^{13}C),[126] $Ti\{PrB(NBu^t)_2BPr\}Cl_4$ (^{11}B, ^{13}C),[127] $MCl_3\{N(SiMe_2CH_2PMe_2)_2\}$ (M = Zr, Hf; ^{31}P),[128] $[ZrW_5O_{19}H_2]^{2-}$ (^{183}W),[129] and $Zr(OAr)_2(ArNCR)_2$ (^{13}C).[130]

Complexes of V, Nb, and Ta.—^{51}V chemical shifts vary over a range of 1400 p.p.m. with a temperature coefficient of up to +1.2 p.p.m. deg^{-1} for $[(C_5H_5)V(SnCl_3)(CO)_3]^-$. $^1\underline{J}(^{117,119}Sn,^{51}V)$ and $^1\underline{J}(^{51}V,$-$^{13}C)$ were determined.[131] For $[V(CO)_5L]^-$ the shielding of the ^{51}V nucleus decreases in the order alkene > alkyne ~ N_2 > SO_2 > CS_2 > {O}. Compared to η^1-coordination, η^2-coordination gives rise to a deshielding contribution of 100-280 p.p.m.[132] In $(C_5H_4Me)Nb(\eta^2$-$RC_2R)Cl_2$ the ^{13}C n.m.r. chemical shifts of the acetylenic carbon atoms suggest that the alkyne donates four electrons to the niobium atoms.[133] The ^{13}C isotope effect in the ^{51}V n.m.r. spectra of $[V(CO)_6]^-$ has been determined. The ^{13}C n.m.r. spectrum was also recorded.[134] In $[V(CO)_5PR_3]^-$ the ^{51}V shift trends correlate with the integral ligand strength as quantified by Graham's σ- and π-parameters. The ^{31}P n.m.r. spectra were also recorded.[135] ^{51}V and ^{55}Mn chemical shifts of analogous compounds have been compared for $[V(CO)_5L]^-$, $V(NO)(CO)_4L$, and $Mn(NO)_3L$.[136] N.m.r. data have also been reported for $Ta(BH_4)(CO)_2(PMe_3)_3$ (^{13}C, ^{31}P),[137] $[M(2\text{-}CH_2$-$C_6H_4)_2(C_5H_5)_2]^+$ (M = Nb, Ta; ^{13}C),[138] $(C_5H_5)_2M(CO)C(CF_3)=CH_2$ (M = Nb, Ta; ^{19}F),[139] $Ta\{(Bu^tCH_2)_4C_5CH_2CMe_2CH_2\}Cl_3$ (^{13}C),[140] $(Bu^t{}_3SiO)_2$-$Ta(CHPh)(CH_2Ph)$ (^{13}C),[141] $(Bu^tC_5H_4)_2M(CO)Fe_3(CO)_{10}(\mu\text{-}H)(\mu\text{-}CO)$ (M = Nb, Ta; ^{13}C),[142] $(C_5Me_5)Ta(\eta^4\text{-diene})Cl_2$ (^{13}C),[143] $(C_5H_5)V(CO)_{4-\underline{n}}$-$(CNR)_{\underline{n}}$ (^{51}V),[144] $(C_5H_5)_2Nb(SC=CR_2)(SCH=CR_2)$ (^{13}C),[145] (C_5Me_5)-$TaCl_3\{OCH(SiMe_3)C(O)OEt\}$ (^{13}C),[146] $VX(CNR)_3(NO)_2$ (^{51}V),[147] and $[V(CO)_5L]^-$ (^{51}V).[148]

A decrease of ^{51}V shielding in the order $[VO_2Cl_2]^-$ > $[VOCl_4]^-$ > $[VX_3(N_3S_2)]_2{}^{2-}$ (X = N_3 > Cl) has been reported.[149] For $V(OR^1)_3$-

(=NR^2) the ^{51}V n.m.r. spectrum shows $^{1}\underline{J}(^{51}V,^{14}N)$.[150] $[VO_4H]^{2-}$ has been investigated by ^{51}V n.m.r. spectroscopy in a mixed lyotropic mesophase.[151] For a series of oxovanadium(V) compounds, linear relations exist between $\delta(^{51}V)$ and substituent parameters such as the electronegativity, Pearson's hardness parameter, and Taft's electronic and steric constants.[152] High-field ^{51}V and ^{17}O n.m.r. spectra have been determined for peroxyvanadates in aqueous solution; five new species, including four which are dimeric, were identified.[153] $[HPV_{14}O_{42}]^{8-}$ shows a pH-dependent oxygen exchange, which correlates with ^{17}O, ^{31}P, and ^{51}V chemical shifts.[154] Mixed vanadium-tungsten polyoxo complexes in aqueous solutions have been identified by ^{17}O and ^{51}V n.m.r. spectra.[155] ^{31}P and ^{51}V n.m.r. spectroscopy has been used to study vanadotungstophosphate heteropoly acids in solution.[156] The influence of ^{51}V, ^{93}Nb, ^{181}Ta, ^{95}Mo, ^{97}Mo, and ^{183}W coupling on ^{17}O line widths has been determined. Only ^{51}V and ^{93}Nb coupling affects ^{17}O linewidths.[157] N.m.r. data have also been reported for $VCl_2(NBu^t)(NHBu^t)(NH_2Bu^t)$ (^{51}V),[158] $[Cl_2VS_2N_3]_{\underline{x}}$, $[Cl_4MS_2N_3]^-$ (^{51}V, ^{95}Mo),[159] $M(NBu^t)Cl_3(NH_2Bu^t)$ (M = Nb, Ta; ^{13}C),[160] TaF_5L (^{19}F),[161] $TaBr_3(PMe_2Ph)_2$ (^{31}P),[162] $[VOF_3(NO_3)]^-$ (^{19}F, ^{51}V),[163] $[PV_{\underline{n}}W_{12-\underline{n}}O_{40}]^{(3+\underline{n})-}$ (^{17}O, ^{31}P, ^{51}V),[164] $[MW_5O_{18}S]^{3-}$ (M = Nb, Ta; ^{17}O),[165] $[(OC)_3M(\underline{cis}\text{-}Nb_2W_4O_{19})]^{3-}$ (M = Mn, Re; ^{17}O),[166] $Ta_2(OC_6H_4Me\text{-}\underline{p})_{10}$ (^{13}C),[167] and $[Nb_6S_{17}]^{4-}$ (^{93}Nb).[168]

Complexes of Cr, Mo, and W.—The ^{95}Mo $\underline{T}_1$ and $\underline{T}_2$ have been measured for a number of compounds. Quadrupolar relaxation is the only significant mechanism involved, allowing $\underline{T}_1$ values to be interpreted in terms of the electric field gradient, molecular size and shape, solution viscosity, temperature, and solvent-solute interactions. ^{97}Mo relaxation times were also reported.[169] N.m.r. data have also been reported for $CrH_4(dmpe)_2$ (^{13}C, ^{31}P),[170] $(C_5Me_5)Cr(CO)_3H$ (^{13}C),[171] $[DFeCr(CO)_9]^-$ (^{2}H),[172] $[MoH(CN)_7]^{4-}$ (^{13}C),[173] $MoH(SR)(dppe)_2$ (^{31}P),[174] $(C_5H_5)MoH\{C(CN)=CHCN\}$, $(C_5H_5)\overline{Mo\{C_5H_4C(CF_3)=C}H\}\{C(CF_3)=CH_2\}$ (^{13}C, ^{19}F),[175] $\{Mo(C_5H_5)(CO)_2\}_2(\mu\text{-}H)(\mu\text{-}ER_2)$ (E = P, As; ^{31}P),[176] $[(C_5H_5)M(CO)_2(\mu\text{-}H)(\mu\text{-}PPh_2)Pt(PPh_3)(CO)]^+$ (M = Mo, W; ^{31}P),[177] $WH_6(PMe_3)_3$ (^{31}P),[178] $WH_4(\eta^4\text{-}C_6H_8)(PMe_3)_2$ (^{31}P),[179] $W(\eta^4\text{-}C_5H_6)(PMe_3)_3H_2$ (^{13}C, ^{31}P),[180] $(C_6H_8)WH_2(PMe_3)_3$ (^{13}C, ^{31}P),[181] $WCl_2H_2(PMe_3)_4$ (^{31}P),[182] $WH(\eta^2\text{-}CH_2PMe_2)(PMe_3)_4$ (^{13}C, ^{31}P),[183] $(C_5H_4CH_2)W(C_5Me_5)H$ (^{13}C),[184] $(C_5H_5)WHI(NO)(PR_3)$ (^{31}P),[185] $(C_5H_5)W_2Os(CO)_4(\mu\text{-}Ctol)(\mu\text{-}H)H$ (^{13}C),[186] $W_2H(OPr^i)_4(\mu\text{-}C_4Me_4)(\mu\text{-}CPh)$ (^{13}C),[187]

$(OC)_4CrCH_2PPh_2CR^1R^2PPh_2$ (^{31}P),[188-190] $(C_5H_5)(OC)_2Cr\{HC{=}CHC(O)Me\}$ (^{13}C),[191] $(OC)_5MC(OEt){=}N{=}CR^1R^2$ (M = Cr, W; ^{13}C),[192] $[(OC)_5CrCCNR_2]^+$ (^{13}C),[193] $(OC)_5M{=}CROMe$ (M = Cr, W; ^{13}C),[194] $(OC)_5M{=}C(OR)C(OMe){=}CH$-OMe (M = Cr, W; ^{13}C),[195] $(OC)_5Cr{=}CNHC({=}PPh_3)C(O)NMe$ (^{13}C, ^{31}P),[196] $(OC)_4Cr{=}C(OMe)C_{10}H_5(OMe)(OMe)$ (^{13}C),[197] $(OC)_5Cr{=}C{=}C{=}CPh(NMe_2)$ (^{13}C),[198] $(OC)_5M{=}CPhSR$ (M = Cr, W; ^{13}C),[199] $Co_2MRe(\mu_3\text{-}Ctol)(CO)_{15}$ (M = Cr, Mo, W; ^{13}C),[200] $(C_5Me_5)ClGeM(CO)_5$ (M = Cr, W; ^{13}C),[201] $(C_3H_3BMeNBu^t)M^1(M^2Me_3)(CO)_3$ (M^1 = Cr, Mo, W; M^2 = Ge, Sn, Pb; ^{11}B, ^{13}C, ^{119}Sn, ^{207}Pb),[202] $M_2(CH_2)_4(NMe_2)_4$ (M = Mo, W; ^{13}C),[203] (C_5H_5)-$Mo(O_2CCF_3)(CH_2CH_2CF_3)$ (^{13}C, ^{19}F),[204] $(C_5H_5)M(CO)_3CH_2C_3H_4Me$ (M = Mo, W; ^{13}C),[205] $(C_5H_5)Mo(CO)_3\{CHROC(O)Me\}$ (^{13}C, ^{31}P),[206] $(C_5H_5)Mo\{C_4$-$(CF_3)_4\}(SC_6F_5)(CNBu^t)_2$ (^{19}F),[207] $[Mo(C{=}CHPh)(dppe)(\eta^7\text{-}C_7H_7)]^+$ (^{13}C),[208] $Mo_2Br_2({=}CHSiMe_3)_2(PMe_3)_4$ (^{31}P),[209] $(OC)_5MC(NR^1R^2)N{=}CPhR_3$ (M = Cr, Mo, W; ^{13}C),[210] $(C_5H_5)MoI\{C(NMe)(CH_2)_2CH_2\}(CO)_2$ (^{13}C),[211] (mesityl)$(O_2)Mo{=}C$(mesityl)$(PBu^n{}_3)$ (^{13}C, ^{31}P),[212] $(C_5H_5)Mo(CO)_2$-$(SnPh_3){=}C(CH_2)_3NMe$ (^{13}C),[213] $(C_5H_5)Mo\{P(OMe)_3\}_2CCHRBu^t$ (^{13}C, ^{31}P),[214] $[(C_5H_5)_2Mo_2(CO)_4(CH_2CCCH_2OH)]^+$ (^{13}C),[215] $(C_5H_5)_3MoCo_2$-$(\mu_3\text{-}CMe)(\mu\text{-}CO)(CO)_2$ (^{13}C),[216] $[W(CO)_5R]^-$ (^{13}C),[217] $[WO_3(CH_2SiMe_3)]^-$ (^{13}C),[218] $(C_5H_5)W(O)(CHSiMe_3)(CH_2SiMe_3)$ (^{13}C),[219] $(C_5H_5)W(NO)(CH_2$-$SiMe_3)_2$ (^{13}C),[220] $W_2Cl_3(NMe_2)_2(CH_2NMe)(CHCH_2)(PMe_2Ph)_2$ (^{13}C, ^{31}P),[221] $W(C_5H_5)(CH{=}CH_2)I$ (^{13}C),[222] $W\{C(OEt)CHPhCHPhC(OEt)(PMe_3)\}$-$(CO)_4$ (^{13}C, ^{31}P),[223] $(C_5H_5)W(CO)(C_2H_2)(COR)$ (^{13}C),[224] $W\{C(CHO)$-${=}C{=}O\}(CO)_2(PPh_3)_2(NO)$ (^{13}C, ^{31}P),[225] $(C_5H_5)W(CO)_3C\{CPh{=}C(CN)_2\}{=}C$-$(CN)_2$ (^{13}C),[226] $[(OC)_3(dppe)W(C_2R)]^-$ (^{13}C),[227] $(Me_3SiCH_2)_4W_2(\mu\text{-}C$-$SiMe_3)\{\mu\text{-}MeCH(CH_2)CCSiMe_3\}$ (^{13}C),[228] $[(C_5H_5)(CO)_2W(\mu\text{-}CHR)M(CO)_4L]^-$ (M = Cr, Mo, W; ^{13}C),[229] $Fe_2(CO)_6W(C_5H_5)(CO)(\mu\text{-}PPh_2)(\mu_3\text{-}CR)$ (^{13}C, ^{31}P),[230] $RuW(\mu\text{-}Cl)(\mu\text{-}CMe)Cl(CO)_2(PPh_3)(C_5H_5)$ (^{13}C, ^{31}P),[231] $W_2(CO)_8(\mu\text{-}PPh_2)(\mu\text{-}PPh_2C{=}C{=}CHCH{=}CMe_2)$ (^{13}C, ^{31}P),[232] $(C_5H_5)(CO)_2Mn$-$W(CO)_4(\mu\text{-}C{=}CHCO_2Me)$ (^{13}C),[233] $Fe_2W(\mu_3\text{-}CMe)(\mu\text{-}CO)(CO)_8(C_5H_5)$ (^{13}C),[234] $[W(CO)_4C(O)(CH_2)_3PPh_2]^-$ (^{13}C),[235] $W\{{=}CH(\underline{p}\text{-}tol)\}I(CO)_2$-$(C_5H_5)$ (^{13}C),[236] $(OC)_5W{=}C(\underline{p}\text{-}tol)O(CH_2)_2CH{=}CH_2$ (^{13}C),[237] $(OC)_4WC(OEt)CHPhCHPhCOEt$ (^{13}C),[238] $(OC)_5W{=}CCPh(OEt)XC(Y)NCy$ (^{13}C),[239] $Fe(C_5H_4PPh_2)C_5H_4C(OMe){=}W(CO)_4$ (^{13}C),[240] $[(SCN)_3(OC)_2$-$W{=}C{=}NEt_2]^{2-}$ (^{13}C),[241] $W(CHBu^t)(OCH_2Bu^t)_2X_2$ (^{13}C),[242, 243] (C_5H_5)-$W_2(CO)_4({=}CNBu^t)$ (^{13}C),[244] $(C_5H_5)_2W_2Ir_2(CO)_7(CHCO_2Me)_2$ (^{13}C),[245] $Cl(OC)_2(py)_2WCR$ (^{13}C),[246] $[W(CO)Cl_2(CR)(py)(alkene)]^-$ (^{13}C),[247] $W(CR)(OR)_3$ (^{13}C),[248] $[(dppe)(OC)_3WCCH_2Ph]^+$ (^{13}C, ^{31}P),[249] (Pr^iO)-$(RN{=})W(\mu\text{-}\eta^2,\eta^1\text{-}C_2SiMe_3)(\mu\text{-}CSiMe_3)W(OPr^i)_3$ (^{13}C),[250] $W_2\{(\underline{p}\text{-}tol)NCN$-$(\underline{p}\text{-}tol)\}(OBu^t)_6$ (^{13}C),[251] and $[Et_2NCW(CO)_2(\mu\text{-}PPh_2)_2M(CO)_4]^-$ (M = Mo, W; ^{13}C, ^{31}P).[252]

^{13}C and ^{31}P n.m.r. spectra have been used to show that $W(CO)_3$-$\{P(OMe)_3\}(\eta^4-C_8H_8)$ is a mixture of fac- and mer-isomers.[253] The effect of slippage of the cyclopentadienyl ring on n.m.r. parameters has been investigated for (indenyl)$Mo(CO)_2$(1-Me-allyl) and [(indenyl)$IrH(PPh_3)_2]^+$.[254] For a number of molybdenum compounds such as $[(C_5H_5)Mo(CO)_2(NR^1{=}CR^2C_5H_4N)]^+$, the ^{95}Mo signal shows the presence of diastereomers.[255] A study of the aryl ^{13}C chemical shifts and $^1J(^{13}C,^1H)$ for (3-8-η-[2.2]paracyclophane)chromium and some related complexes has shown that they correlate with chromium-carbon distances.[256] The geometric configuration of $(\eta^6-9-RC1_3H_9)$-$Cr(CO)_3$ has been determined using ^1H ASIS.[257] It has been shown that the enantiomeric purity of (arene)$Cr(CO)_3$ complexes bearing an aldehyde group can be determined using $Eu(hfc)_3$.[258] A comparison of the ^1H n.m.r. spectra of veratrole and (veratrole)$Cr(CO)_3$ shows that protons ortho to MeO are less shielded upon complexation and this was discussed in terms of relative regioselectivity.[259] ^1H and ^{13}C n.m.r. spectroscopy, including two-dimensional techniques, has been used to distinguish between the two isomers of $Cr(\eta^6$-estradiol)$(CO)_3$.[260] The ^{31}P chemical shift of the η^6-bonded ligand in $Mo(\eta^6-PhPR_2)(PPhR_2)$(dppe) correlates with the sum of the cone angles of the three π-bonded ligands in each complex.[261] N.m.r. data have also been reported for $[M(C_2H_4)_2(CO_2)(PMe_3)_2]_2$ (M = Mo, W; ^{13}C, ^{31}P),[262] $(C_5Me_5)Mo(CO)_2(C_3HPh_2CO_2)$ (^{13}C),[263] $(C_5H_5)WMe$-$(C_2H_4)(O)$ (^{13}C),[264] $(C_5H_5)Mo(CO)(NO)$(olefin) (^{13}C),[265] $Mo(CO)_4$-$(PMe_2OCHPhCH{=}CH_2)$ (^{13}C, ^{31}P),[266] $Mo(Me_3CS)_2(Bu^tNC)_2(R^1C_2R^2)$ (^{13}C),[267] $Mo(S-2,4,6-C_6H_2Pr^i{}_3)_4$(alkyne) ($^{13}C$),[268] $W_2Cl_4(NMe_2)_2$-$(C_2Me_2)(py)_2$ (^{13}C),[269] $MX_2(S_2CNR_2)_2(C_8H_{12})$ (M = Mo, W; ^{13}C),[270] $W(C_2H_4)_2(CO)(PMe_3)_3$ (^{13}C, ^{31}P),[271] $W(HC_2Ph)$(maleic anhydride)-$(CNEt_2)_2$ (^{13}C),[272] $[(C_5H_5)W(PMe_3)_2(R^1C_2OR^2)]^+$ (^{13}C, ^{31}P),[273] $MoW\{\mu-C_2(tol)_2\}(CO)_4(C_5H_5)\{\eta^5-C_9H_6(COMe)\}$ (^{13}C),[274] $Mo_2(CO)_4$-$(\mu-\eta^1,\eta^2-CCH_2)(C_5Me_5)_2$ (^{13}C),[275] $(C_5H_5)_2Mo(\eta^2-O{=}CH_2)$ (^{13}C),[276] $[(OC)_5W]_2\{\mu_2;\eta^2-Se{=}CH(tol)\}$ (^{13}C),[277] $(OC)_4Mn(\mu-PPh_2)Mo(C_5H_5)(CO)$-$(C_4H_7)$ (^{13}C),[278] $(C_5H_5)Mo(CO)_2(\eta^3-CH_2PPhNMe)$ (^{13}C),[279] $M(CO)_3(PR_3)$-$\{\eta^3-MeC(S)SMe\}$ (M = Cr, Mo, W; ^{13}C, ^{31}P),[280] $(C_5H_5)M(CO)_2(C_3R_3CO)$ (M = Cr, Mo, W; ^{13}C),[281] $Mo(C_3Bu^t{}_2)\{OC(CF_3)_2\}_2L_2$ (^{13}C, ^{19}F, ^{31}P),[282] $W(CO)(C_3H_3R^1CO)(S_2CNR^2{}_2)_2$ (^{13}C),[283] $Zn[W(CO)_2(\eta-MeCHC_6H_4$-$Me)(C_5H_5)]_2$ (^{13}C),[284] $\{W(OBu^t)_3\}_2(\mu-C_3H_7NCNC_3H_7)$ (^{13}C),[285] $Cr(\eta^4$-$C_4Ph_4)(C_2Ph_2)(CO)_2$ (^{13}C),[286] $(C_5H_5)Mo(NO)(\eta^4$-diene) (^{13}C),[287] Mo(norbornadiene)$_2(CO)_2$ (^{13}C),[288] $M(S_2CNR_2)_2(CO)\{(C_8H_{12})_2CO\}$ (M = Mo, W; ^{13}C),[289] $[W(\eta^4-C_4Ph_4H)(S_2CNEt_2)_2]^+$ (^{13}C),[290] $[Cr\{\eta^5-C_5H_5$-$C{=}C(CH{=}CH_2)_2\}(CO)_3]^-$ (^{13}C),[291] $[(C_5H_5)(OC)_2M]_2(\mu-PPhH)_2$ (M = Cr,

Mo, W; ^{31}P),[292] $(C_5H_5)M(CO)_2P(CF_3)_2$ (M = Cr, Mo, W; ^{19}F, ^{31}P),[293] $(C_5H_5)_2Cr_2(CO)_4Se_n$ (n = 1, 2; ^{13}C),[294] $(C_5Me_5)Cr(NO)_2Br$ (^{13}C),[295] $(C_5Me_5)_2M_2O_4$ (M = Cr, Mo; ^{13}C),[296] $(C_5H_5)Mo(CO)_2\{PPh(OCH_2CH_2)_2N\}$ (^{31}P),[297] (fulvene)$Mo_2(CO)_5L_2$ (^{13}C, ^{31}P),[298] $(C_5H_5)M(CO)_3P{=}C(SiMe_3)_2$ (M = Mo, W; ^{13}C, ^{31}P),[299] $\{(C_5Me_5)Mo\}_2(\mu,\eta^6-P_6)$ (^{13}C, ^{31}P),[300] $[(C_5H_5)(OC)_2(Ph_3P)Mo(OCMe_2)]^+$ (^{31}P),[301] $(C_5H_5)Mo(dppe)Cl$ (^{31}P),[302] $(C_5H_5)_2Mo_2(CO)_4(\mu-PC_6H_2Bu^t{}_3)$ (^{31}P),[303] $(C_5H_5)_2Mo_2(CO)_4(Bu^tCP)RhCl(PF_2NMe_2)CO$ (^{19}F, ^{31}P),[304] $(C_5H_5)Mo(CO)_2Co_2(CO)_6C(CO_2R)$ (^{13}C),[305] $(C_5H_5)_2Mo_2FeTe_2(CO)_7$ (^{13}C, ^{125}Te),[306] $[(C_5H_4Me)MoFeTe_2IrCl(PPh_3)_2(CO)_6]^+$ (^{31}P, ^{125}Te),[307] $(C_5H_5)W(CO)_3(PPh_2BH_3)$ (^{11}B, ^{31}P, ^{183}W),[308] $(C_4S_4)\{W(C_5H_5)(CO)_3\}_2$ (^{13}C),[309] $[(C_5H_5)W(CO)_2(PMe_3)SbBu^tMeCl]^+$ (^{31}P),[310] $(Bu^tO)O_2W(\eta^5,\eta^5-Et_4C_5CH_2CH_2C_5Me_4)Rh(CO)_2$ (^{13}C),[311] $Fe_2W(\mu_3-OCCH_2R)(\mu-PPh_2)_2(CO)_5(C_5H_5)$ (^{13}C, ^{31}P),[312] [(benzocycloheptatrienyl)Cr(CO)$_3$]$^-$ (^{13}C),[313] $Cr(\eta^6-C_6H_5CHO)(CO)_3$ (^{13}C),[314] Cr(η^6-arene)(CO)$_3$ (^{13}C),[315] Cr(η^6-naphthalene)$_2$ (^{13}C),[316] $Cr(CO)_3(C_{14}H_{12}R)$ (^{13}C),[317] $Cr(CO)_3(C_{18}H_{22})$ (^{13}C),[318] $Mo(C_4H_4C{=}CPh_2)(CO)_3$ (^{13}C),[319] and $[Mo(\eta-C_7H_7)(dppe)_2]^+$ (^{31}P).[320]

Spin-lattice relaxation times of ^{95}Mo, ^{97}Mo, ^{17}O, and ^{13}C for $Mo(CO)_6$ have been reported. The rotational correlation time of the molecule was obtained from the c.s.a. relaxation of ^{13}C. Quadrupolar coupling constants were calculated for ^{17}O, ^{95}Mo, and ^{97}Mo.[321] The ^{95}Mo n.m.r. parameters have been measured for $Mo(CO)_{6-n}(py)_n$. T_1 was determined for ^{95}Mo and ^{13}C for some complexes.[322] The ^{95}Mo chemical shift range of some seven coordinate molybdenum(II) isocyanide complexes is about 1100 p.p.m. ^{13}C, ^{14}N, and ^{31}P n.m.r. spectra were also recorded where relevant.[323] ^{19}F and ^{31}P chemical shift data support the σ-donor/π-acceptor model of the Cr-P bond in $Cr(CO)_5L$ and *cis*-$Cr(CO)_4L_2$, L = $RPSCH_2CH_2S$.[324] ^{95}Mo chemical shifts have been reported for a range of complexes, $[Mo(CO)_5X]^-$. The chemical shift values extend over 1000 p.p.m. and reflect the importance of ligand field strength, polarizability, and electronegativity factors in determining the ^{95}Mo chemical shifts. The decreasing shielding effect is in the order H^- > $[CN]^-$ > I^- > $[NCO]^-$ > $[NCS]^-$ > $[NCSe]^-$ > Br^- > $[N_3]^-$ > $[NO_2]^-$ > Cl^- > $[O_2CH]^-$ > $[O_2N]^-$ > $[OMe]^-$ > F^-.[325] The ^{13}C, ^{17}O, ^{29}Si, and ^{31}P n.m.r. spectra of $(OC)_4Mo(Ph_2PO)_2SiMeR$ have been studied. The chemical shift ranges of the carbonyl ^{13}C and ^{17}O, the phenyl C(1) ^{13}C and ^{31}P resonance are relatively large, and, with the exception of the *cis* CO ^{17}O chemical shifts, the correlations between the chemical shifts of the various resonances are excellent.[326] For

$Mo(CO)_4$(substituted bipy) the influence of the solvent on the chemical shifts of the bipyridine ring increases significantly as a result of coordination to the metal.[327] Complexes of the type $Mo(CO)_4(Ph_2PO)SiMeR$ exhibit two ^{13}C resonances for the cis-CO and phenyl carbon atoms. The differences between the chemical shifts of the two resonances can be correlated with the Taft steric parameters of the R groups.[328] The ^{31}P n.m.r. spectrum of $Mo(CO)_3\{\overline{P(OCMe_2CH_2)_2N}\}_3$ shows $^{1}J(^{95}Mo,^{31}P)$ = 210 Hz.[329] N.m.r. data have also been reported for $[Cr(CO)_5NCCPh(BH_3)(OMe)]^-$ (^{13}C),[330] $(OC)_5MNH{=}CPh_2$ (M = Cr, W; ^{13}C),[331] $(OC)_5Cr(PHPhPXPh)$-$Cr(CO)_5$ (^{31}P),[332] $(C_6H_2Bu^t{}_3P{=}CHPh)Cr(CO)_5$ (^{13}C, ^{31}P),[333] $M(CO)_5$-(R_3PTe) (M = Cr, Mo, W; ^{13}C, ^{31}P, ^{125}Te),[334] $\{(OC)_5M\}_2RP{=}PR$ (M = Cr, Mo, W; ^{31}P),[335] $\{R^1PC(O)CR^2{=}CR^2\}M(CO)_5$ (M = Cr, Mo, W; ^{13}C, ^{31}P),[336] $(C_6H_2Bu^t{}_3P{=}PC_6H_2Bu^t)M(CO)_5$ (M = Cr, Mo, W; ^{31}P),[337] $R_2NPHM(CO)_5$ (M = Cr, W; ^{13}C, ^{31}P),[338] $M(CO)_5(CNNPPh_3)$ (M = Cr, Mo, W; ^{13}C, ^{31}P),[339] $M(CO)_5(PC_4H_2Me_2)_2Fe$ (M = Cr, Mo, W; ^{31}P),[340] $(C_5H_5)Fe(CO)_2PH(NPr^i{}_2)M(CO)_5$ (M = Cr, Mo, W; ^{31}P),[341] $(OC)_5M\{Ph_2P$-$OS(O)R\}$ (M = Cr, Mo, W; ^{31}P),[342] $[OPh_2PM(CO)_5]^-$ (M = Cr, Mo, W; ^{31}P),[343] $Ph_2PN_3M(CO)_5$ (M = Cr, W; ^{31}P),[344] $M_2(CO)_{10}(SbRPR_3)$ (M = Cr, Mo, W; ^{31}P),[345] $\{C_4H_3(COMe)Ph\}Mo(CO)_5$ (^{13}C, ^{31}P),[346] $[W(CO)_5]^{2-}$ (^{13}C),[347] $(OC)_5WNH{=}CHR$ (^{13}C),[348] $\{C_2Ph_2(CO_2Et)P\}W(CO)_5$ (^{13}C, ^{31}P),[349] $(OC)_5WPPh_2OsH(CO)_2(PPh_2H)(PPh_2Me)$ (^{31}P),[350] $\{C_4Me_2H_2$-$(CH_2Cl)P\}W(CO)_5$ (^{13}C, ^{31}P),[351] $(Et_2N)_2P(CH{=}CH_2)W(CO)_5$ (^{31}P),[352] $\{C_2PhH(CH_2CH_2Cl)\}W(CO)_5$ (^{13}C, ^{31}P),[353] $\{\overline{CH_2CH_2CH{=}CMeCMe{=}CHP}(OMe)\}W$-$(CO)_5$ (^{13}C, ^{31}P),[354] $(OC)_5\overline{W(\mu\text{-}PPh_2)Re}(CO)_4$ (^{13}C, ^{31}P),[355] $(OC)_4Cr$-$(R^1N{=}CR^2CR^3NR^1)$ (^{13}C),[356, 357] $Cr(CO)_4(Ph_2P)_3N$ (^{31}P),[358] $(OC)_4Cr$-$(R^1{}_2P)_2C{=}PR^2{}_3$ (^{31}P),[359] $\{(OC)_4M\}_2ER_2$ (E = S, Se; M = Cr, Mo, W; ^{13}C),[360] $(OC)_4M\{EMe_2P(CF_3)_2\}_2M(CO)_4$ (E = P, As; M = Cr, Mo, W; ^{19}F, ^{31}P),[361] $Mo(CO)_4Br_2(Ph_2PCH_2SiMe_2CH{=}CH_2)$ (^{31}P),[362] $Pr^iN\{PPh(NH$-$Pr^i)\}_2Mo(CO)_4$ (^{31}P),[363] $(OC)_4W(py)L$ (^{31}P),[364] $(OC)_4W(R_2P)_2Te$ (^{13}C, ^{31}P, ^{125}Te),[365] $Cr_3(CO)_{10}(Bu^tP{=}PBu^t)(PBu^t)$ (^{31}P),[366] $(PhCO)(OC)_3$-$W(\mu\text{-}PPh_2)_2Re(CO)Et$ (^{13}C, ^{31}P),[367] (1,5,9-triazacyclododecane)-$Mo(CO)_3$ (^{95}Mo),[368] $HB(pz)_3Mo(CO)_3X$ (^{13}C),[369] $Cr(CO)_3(dppm)_2$ (^{31}P),[370] $M(CO)_3\{RSi(OCH_2EMe_2)_3\}$ (M = Cr, Mo, W; ^{31}P),[371] $Mo(CO)_3$-(dppe)$\{MeN(PF_2)_2\}$ (^{31}P),[372] $(MeNPMe)_4Mo(CO)_3$ (^{31}P),[373] $M(CO)_3$-$\{Ph\overline{PC_6H_4PPh(CH_2)_3X(CH_2)_2C}H_2\}$ (M = Mo, W; X = PPh, NMe; ^{13}C, ^{31}P),[374] $(NC)_2Pd(dppm)_2Mo(CO)_3$ (^{31}P),[375] $MoRu(CO)_6(dppm)_2$ (^{13}C, ^{31}P),[376] $(OC)_3W(\mu\text{-}dppm)_2MX$ (M = Cu, Ag; ^{31}P),[377] $[Cr(CO)_2(CS)$-$\{P(OMe)_3\}]_2$ (^{13}C, ^{31}P),[378] $\{HB(Me_2pz)_3Mo(CO)_2\}_2S$ (^{95}Mo),[379] $[W_2(OPr^i)_6(\mu\text{-}CO)]_2$ (^{13}C),[380] and $W(O)Cl_3(CNR)$ (^{13}C).[381]

^{15}N T_1 and ^{15}N-$\{^1H\}$ n.O.e. measurements have been reported for

^{15}N in N_2 complexes of Mo, W, Re, and Os, together with ^{13}C and ^{31}P n.m.r. data. It was found that the relaxation is predominantly due to c.s.a. and the rotational correlation times were determined. The N_α in the rhenium complex is also relaxed by $^{185,187}Re$. The ^{14}N linewidths were also determined. The ^{31}P relaxation is fully dipolar.[382] ^{95}Mo and ^{14}N n.m.r. studies of seven coordinate molybdenum(VI) monoxo, nitride, and phenylimido complexes have been reported. The chemical shifts of the complexes increase in the order L = NPh < NO < N < O < NS, while the linewidths of these compounds increase in the order L = O < N << NPh < NS < NO.[383] N.m.r. data have also been reported for $MoBr(NNCHMeC_6H_{13})(Ph_2P\text{-}CHMeCHMePPh_2)_2$ (^{13}C),[384] $Mo(NNPh)(SC_6H_2Pr^i{}_3)_3(NCMe)$ (^{15}N),[385] $MoO(C_5H_{10}NO)_2(C_6H_4O_2)$ (^{13}C, ^{95}Mo),[386] $MCl_2(NMe_2)_4(Me_2PCH_2PMe_2)$ (M = Mo, W; ^{31}P),[387] $Mo(N_2)\{(Ph_2PCH_2CH_2)_2PPh\}(PMe_2Ph)_2$ (^{31}P),[388] $\{Mo(Ntol)(S_2CNEt_2)_2\}_2O$ (^{17}O),[389] $F_4M{=}NTeF_5$ (M = Mo, W; ^{19}F),[390] $Mo\{N_2C(O)Ph\}\{NHNHC(O)Ph\}\{PhC(S)NMeO\}_2$ (^{13}C),[391] $Mo_2O_5(C_{13}H_9N_2O)_2\text{-}(C_3H_7NO)(H_2O)_{1.5}$ (^{95}Mo),[392, 393] $MoO_2(C_{13}H_9N_2O)_2(EtOH)_2$ (^{95}Mo),[394] $MoO_2Cl_2\{(Pr^iO)_2P(O)CH_2C(O)NEt_2\}$ (^{13}C, ^{31}P),[395] $W(N_2)_2(dppm)_2$ (^{15}N, ^{31}P),[396] $[WOF_4(MeNNMePh)]^-$ (^{19}F),[397] $[(WCl_5)_2\{\mu\text{-}NC(CF_3)_2N\}]^{2-}$ (^{19}F),[398] $[W(NPh)(\mu\text{-}OMe)(OMe)_3]_2$ (^{13}C),[399] $W(Schiff\ base)_2$ (^{13}C),[400] $W(pinacol)_3$ (^{13}C),[401] $M\{P(OEt)_3\}_6$ (Cr, Mo, W; ^{31}P),[402] $Mo_2(OC_6H_3Me_2)_4(PMe_3)_4$ (^{31}P),[403] $Mo_2(O_2CCF_3)_4(PR_3)_2$ (^{19}F, ^{31}P),[404] $Mo_2F_2(OBu^t)_4(PMe_3)_2$ (^{19}F, ^{31}P),[405] $Mo\{PhP(CH_2CH_2S)_2\}_2$ (^{31}P),[406] $Mo_2Cl_4(PR_3)_4$ (^{31}P),[407] $Mo_2Br_4(dppe)_2$ (^{31}P),[408] $W_2(O_2CBu^t)_4(PPh_3)_2$ (^{31}P, ^{183}W),[409] and $WOCl_4(PR_3)_2$ (^{31}P).[410]

The ^{17}O n.m.r. spectrum of $(hmpa)CrO(O_2)_2$ shows two distinct signals.[411] A special probe for ^{95}Mo n.m.r. spectroscopy has been constructed and used to study some binuclear Mo(II) compounds.[412] A total of 38 representative oxo-Mo(IV) compounds have been studied by ^{95}Mo n.m.r. spectroscopy.[413] The ^{31}P chemical shift anisotropy of $Na_8HW_4PO_{34}$ shows that it has an A-type structure. The ^{13}C and ^{183}W n.m.r. spectra of $[\{(C_5H_5)Fe(CO)_2Sn(OH_2)\}_3W_{18}P_2O_{68}]^{9-}$ were also reported.[414] ^{17}O and ^{183}W n.m.r. spectroscopy has been used to study paratungstate anions in aqueous solutions.[415] ^{17}O, ^{31}P, and ^{183}W n.m.r. spectra of $H_3PW_{12}O_{40}.6H_2O$ show that during catalytic photochemical dehydrogenation of organic substrates it remains unchanged.[416] For $[NaP_5W_{30}O_{110}]^{14-}$ the ^{23}Na n.m.r. spectrum shows two signals for inside and outside ions. The ^{31}P n.m.r. spectra were also recorded.[417] The binding of $[MoO_4]^{2-}$ and $[MoS_4]^{2-}$ to bovine serum albumin has been studied by ^{95}Mo n.m.r. spectroscopy.[418] ^{33}S n.m.r. spectroscopy has been applied to a wide range of compounds including $[MS_nO_{4-n}]^{2-}$, M = Mo, W.[419] The ^{19}F T_1 for WF_6, MoF_6, and UF_6 is dominated by spin rotation interactions.[420] The modification of a RYa-2305 spectrometer to observe ^{19}F n.m.r. spectra from MoF_6, WF_6, and UF_6 has been described.[421] N.m.r. data have also been reported for $[Ph_3SiOMoO_3]^-$ (^{13}C, ^{29}Si),[422] $Mo_2O_2(C_2O_4)_2(OH_2)_4$ (^{13}C),[423] complexes containing the $[Mo_2O_5]^{2+}$ core (^{95}Mo),[424] molybdate complexes of aldoses (^{13}C),[425] $[Ph_3PCH_2CO_2Et]_2[Mo_6O_{19}]$ (^{13}C, ^{31}P, ^{95}Mo),[426] $[Mo_8O_{26}]^{4-}$ (^{17}O, ^{95}Mo),[427] $[PW_{11}O_{39}]^{7-}$ (^{17}O, ^{31}P, ^{183}W),[428] $K_{28}Li_5H_7P_8W_{48}O_{184}\cdot 92H_2O$ (^{31}P, ^{183}W),[429] $[MoS_3(OSiMe_3)]^-$ (^{17}O, ^{95}Mo),[430] MoO_2L_2 (^{13}C),[431] $Mo_3S_4(C_2O_4)_3(OH_2)_3H_2$ (^{13}C),[432] $\{C_4H_8NCS_2MoO\}_2(\mu\text{-}S)_2$ (^{95}Mo),[433] MOF_4 (M = Mo, W; ^{19}F),[434] $Mo_3Cl_9\{S(CH_2CH_2S)_5CH{=}CH_2\}_2(thf)$ (^{13}C),[435] $Mo_2(S_2CNEt_2)_3(S_3C_2NEt_2)(S_3C_2)$ (^{13}C),[436] $MoS_4(AuPEt_3)_2$ (^{31}P),[437] $W_2S_4(S_2CNBu^i{}_2)_2$ (^{13}C, ^{183}W),[438] and NF_4CrF_6 (^{19}F).[439]

Complexes of Mn, Tc, and Re.—^{13}C T_1 values in $[Re_3(\mu\text{-}H)_4(CO)_{10}]^-$ have been determined. It was found that scalar coupling to rhenium is the main pathway. This has permitted the estimation of $^1J(^{187}Re,^{13}C)$ of 360 and 397 Hz.[440] N.m.r. data have also been reported for $[Mn(AlH_4)(dmpe)_2]_2$ (^{13}C, ^{27}Al, ^{31}P),[441] $Mn_2H_2(CO)_6\{(EtO)_2P\}_2O$ (^{31}P),[442] $[Mn_2H(CNH)(CO)_5(dppm)_2]^+$ (^{13}C, ^{15}N),[443] $\{(C_5Me_5)Mn(CO)_2H\}_2SiH_2$ (^{13}C),[444] $(\mu\text{-}H)FeMn(CO)_6\{\mu,\mu'\text{-}N(p\text{-}tol)CH_2CH_2N(p\text{-}tol)\}$ (^{13}C),[445] $[Mn_2H(\mu_2\text{-}\eta^2\text{-}CO)(CO)_4(dppm)_2]^+$ (^{31}P),[446] $(\eta^4\text{-}C_{12}H_{10})ReH_3(PCy_3)_2$ (^{31}P),[447] $(C_4H_4N)ReH_2(PPh_3)_2$ (^{31}P),[448] $[ReH(CO)(NO)(PPh_3)_3]^+$ (^{31}P),[449] $(\mu\text{-}H)FeRe(CO)_6(Pr^iDAB\{H,Me\})$ (^{13}C),[450] $[(C_5H_5)(ON)Re(\mu\text{-}H)(PCy_2)Pt(PPh_3)_2]^+$ (^{31}P),[451] $HRe(CO)_3\{P(OPh)_3\}_2$ (^{13}C, ^{31}P),[452] $[Re_3(\mu\text{-}H)_4(CO)_{10}]^-$ (^{13}C, ^{17}O),[453] and $[(OC)_4Re(PMe_2Ph)Os_3H(CO)_{10}(\mu\text{-}C{=}CPhC_2Ph)]^-$ (^{13}C, ^{31}P).[454]

The ^{31}P n.m.r. spectrum of $[Mn(dmpe)_3]^+$ is a sextet due to manganese coupling of 242 Hz.[455] In $Re(O)I(R^1C_2R^2)_2$ the acetylenes do not rotate, according to ^{13}C n.m.r. data.[456] ^1H and ^{13}C 2D n.m.r. spectroscopy has been used to characterise $Re(\eta^6\text{-}C_6H_6)(\eta^5\text{-}C_7H_7)$ and $Re(\eta^6\text{-}C_6H_6)(\eta^5\text{-}C_7H_9)$.[457] N.m.r. data have also been reported for $Mn\{(CH_2)_3OPPh_2\}(CO)_3L$ (^{31}P),[458] $(OC)_5MC(O)C(O)M(CO)_5$ (M = Mn, Re; ^{13}C),[459] $Ph_2PCH_2CH_2C(O)CMn(CO)_2(dppm)$ (^{13}C, ^{31}P),[460] $Mn(CO)_2(dppe)\{P(OPh)_3\}(C_2Ph)$ (^{31}P),[461] (tetraphenylporphyrin)$SnMn(CO)_4HgMn(CO)_5$ (^{13}C),[462] $(OC)_4MnPMe_2CH_2CH_2Si(OMe)_2$ (^{31}P),[463] $(C_5H_5)Mn(CO)_2{=}Pb{=}Mn(CO)_2(C_5H_5)$ (^{13}C),[464] $(C_5H_4Me)(OC)Mn(\mu\text{-}CO)(\mu\text{-}CR{=}C{=}O)Mn(CO)_4$ (^{13}C),[465] $(C_5H_5)(OC)_2MnPt(\mu\text{-}C{=}CHPh)L_2$ (^{13}C, ^{31}P),[466, 467] $(C_5H_5)(OC)Fe(\mu\text{-}COMe)(\mu\text{-}CO)Mn(C_5H_4Me)(CO)$ (^{13}C),[468] $M_2(CO)_9({=}COCH_2CH_2O)$ (M = Mn, Re; ^{13}C),[469] $(Me_4ReO)_2Mg(thf)_4$

(^{13}C),[470] $(C_5H_4CHO)ReMe(NO)(PPh_3)$ (^{13}C, ^{31}P),[471] $(C_5H_5)(OC)_2$-$\overline{ReCH_2CH_2CH_2CH_2}$ (^{13}C),[472] $[(OC)_4\overline{Re\{C(=CPh_2)NRCO}\}]^-$ (^{13}C),[473] $(C_5H_5)Re(NO)(PPh_3)C(OMe)=CHCH_2Ph$ (^{31}P, and ^1H n.O.e),[474] $(C_5Me_5)Re(CO)(NO)C(OSiMe_3)=PBu^t$ (^{13}C, ^{31}P),[475] $\{(C_5Me_5)Re(CO)_2\}_2$-$(\mu-\eta^2,\eta^{2'}-C_6H_6)$ (^{13}C),[476] $Re(\eta^2-O=C=CC_4H_4)(PMe_3)_3(NO)$ (^{13}C, ^{31}P),[477] $[(C_5H_5)Mn(CO)_2(allyl)]^+$ (^{13}C),[478] $(\eta^3$-dienyl$)Mn(CO)_4$ (^{13}C),[479] $[(C_4Me_4)Mn(CO)_4]^+$ (^{13}C),[480] $(C_4H_4N)Mn(CO)_2(PPh_3)$ (^{13}C, ^{31}P),[481] $RP\{Mn(C_5H_5)(CO)_2\}_2$ (^{31}P),[482] $(C_5H_4Me)(CO)(PR^1{}_3)(PR^2{}_3)Mn$ (^{31}P),[483] $(C_5Me_5)(OC)_2MnAs(CHR^1R^2)Mn(CO)_2(C_5Me_5)$ (^{13}C),[484] (C_5Me_5)-$Mn(CO)_2(S_2)$ (^{13}C),[485] $(C_5H_5)Re(PPh_3)(NO)(PPh_2O)$ (^{13}C),[486] $[(C_5H_5)$-$Re(\underline{p}-N_2C_6H_4OMe)(CO)_2]^-$ (^{13}C),[487] $(\eta^5-C_5H_2Ph_2OR)Re(CO)_3$ (^{13}C),[488] $[\{(C_5Me_5)Re\}_3O_6]^{2+}$ (^{13}C),[489] $Mn(\eta^5$-methylenecyclohexadienyl$)(CO)_3$ (^{13}C),[490] Mn(cyclohexadienyl)$(CO)_2L$ (^{13}C),[491] and $Mn(\eta^5-C_6H_6Ph)$-$(CO)_2(PPh_3)$ (^{13}C).[492]

^1H, ^{13}C, ^{14}N, and ^{55}Mn n.m.r. data for $[Mn(CNR)_6]^{2+}$ have been recorded as a function of solvent and temperature. The ^{55}Mn signals are sharp.[493] For $MnSC(NR)PPh_2(CO)_{4-\underline{n}}(PR_3)_{\underline{n}}$ the deshielding effects in the ^{55}Mn n.m.r. spectrum follow the same trends as in analogous molybdenum and vanadium complexes.[494] The ^{31}P chemical shift and $^2\underline{J}(^{31}P,^{31}P)$ have been reported for various $Mn(C_5H_5)$-$(CO)_{3-\underline{n}}(PR_3)_{\underline{n}}$ complexes and discussed.[495] ^{31}P, ^{119}Sn, and ^{99}Tc n.m.r. spectroscopy has been used to study mixed aqueous solutions of EHDP, tin, and pertechnetate.[496] The ^{187}Re n.m.r. spectra of $[ReO_4]^-$ and $[ReS_4]^-$ have been measured.[497] N.m.r. data have also been reported for $[Mn(CO)(dppe)_2(NCR)]^+$ (^{31}P),[498] $Mn(CO)_5(O_3SCF_3)$ (^{13}C, ^{55}Mn),[499] $[Mn(CO)_5(PR_3)]^+$, $Co(CO)_2L(NO)$ (^{55}Mn, ^{59}Co),[500] $Mn(CO)_5Re(CO)_4(PPh_3)$, $EtC(O)OsEt(NCMe)(CO)_3$ (^{13}C, ^{31}P),[501] $(OC)_4$-$\overline{M(CH_2)_{\underline{n}}PPh_2}$ (M = Mn, Re; ^{13}C, ^{31}P),[502] $(C_6H_4O_2)_2PMn(CO)_4PR_3$ (^{13}C, ^{31}P),[503] $(OC)_4(Me_3SiPH_2)MnI$ (^{31}P),[504] $Mn_2(CO)_8(P_4N_4Bu^t{}_4)$ (^{13}C, ^{31}P),[505] $[Mn(CO)_3(bipy)(PBu^n{}_3)]^+$ (^{31}P),[506] $(OC)_3(R_3P)\overline{MSC(NR)PPh_2}$ (^{13}C, ^{31}P),[507] $[Mn_2H(CO)_5(dppm)_2]^+$ (^{31}P),[508] $MnRu(\mu-PPh_2)(CO)_6$-$(PPh_3)_2$ (^{13}C, ^{31}P),[509] $[(OC)_3Mn(\mu-dppm)_2PtHBr]^+$ (^{31}P),[510] $[Re(CNR)_2(dppe)_2]^+$ (^{31}P),[511] $ReCl(N_2)(CNR)\{P(OMe)_3\}_3$ (^{31}P),[512] $[Re_2Cl_3(dppm)_2(CNBu^t)_3]^+$ (^{31}P),[513] $Re_2Cl_4(CNR)(dppm)_2$ (^{31}P),[514] $Re(CO)_5(NCSe)$ (^{14}N),[515] $Re_4Cl_2(CO)_{15}(MePPMePMe)$ (^{31}P),[516] $[Re_2Cl_3(dppm)_2(CO)_2L]^+$ (^{31}P),[517] $ReCl(N_2)(R_2PCH_2CH_2PR_2)_2$ (^{31}P),[518] and $Re_2Cl_4(OEt)_2(PPh_3)_2$ (^{31}P).[519]

Complexes of Fe, Ru, and Os.—For a series of $M(CO)_4(SiMe_{3-\underline{n}}Cl_{\underline{n}})H$, M = Fe, Ru, Os, the $SiMeCl_2$ derivative usually gives the ^{29}Si resonance at highest frequency. For (arene)Ru(CO)$(SiCl_3)_2$ there is a linear correlation between $\delta(^{29}Si)$ and ν(CO).[520] The quadrupolar effects of ^{59}Co on linewidths in $HMCo_3(CO)_{12}$, M = Fe, Ru, have been discussed and ^{31}P and ^{59}Co $\underline{T}_1$ values were reported.[521] The ^{31}P n.m.r. spectrum of $[RuH_5(PPh_3)_2]^-$ is a sextet, and hence the compound is a pentahydride.[522] A ^{31}P n.m.r. study of $[RuX(PMe_2Ph)_4]^+$, X = H, C_2Ph, has provided direct evidence for square-pyramidal geometry.[523] ^2H n.m.r. spectroscopy has been used to show that RuHCl-$(PPh_3)_3$ deuterates specifically in the o-position.[524] ^1H n.O.e. and ^{31}P n.m.r. measurements have been used to demonstrate the structure of $(C_5H_5)Ru(Ph_2PCHMeCH_2PPh_2)H$.[525] ^{13}C n.m.r. spectroscopy has been used to demonstrate the formation of $[H_3Ru_4(CO)_{12}]^-$ from $H_4Ru_4(CO)_{12}$.[526] In OsH(CO)(η^3-CH_2CRCH_2)$(PPh_3)_2$ the orientation of the η^3-allyl ligand has been determined by ^1H n.O.e. difference spectra. ^{13}C and ^{31}P n.m.r. spectra were also recorded.[527] The ^{187}Os chemical shift of $\{(\underline{p}\text{-cymene})_2Os_2(\mu\text{-}H)_3\}^+$ has been determined as -2526 p.p.m. relative to OsO_4 by two-dimensional n.m.r. spectroscopy.[528] The use of $^1\underline{J}(^{187}Os,^1H)$ in determining the structure of $H_3Os_3(CO)_9CX$ has been discussed.[529] N.m.r. data have also been reported for $Fe_3(\mu\text{-}H)(\mu\text{-}PR_2CH_2PR)(CO_9)$ (^{13}C, ^{31}P),[530] Fe_3-$(CO)_9(\mu\text{-}H)(\mu\text{-}CH_2)$ (^{13}C),[531] $HFe_3Rh(\mu_4\text{-}\eta^2\text{-}C{=}CHR)(CO)_{11}$ (^{13}C),[532] $Ru(C_8H_{12})(C_8H_{10})P(OMe)_3$ (^{13}C, ^{31}P),[533] $Ru(C_5H_5)(PPh_3)H_3$ (^{13}C, ^{31}P),[534] $[H_3Ru(PPh_3)_3]^-$ (^{31}P),[535] $Ru(C_5H_5)$(dppe)H (^{31}P),[536] $RuRhH_2$-Cl(cod)$(dppm)_2$ (^{13}C, ^{31}P),[537] $Ru_2H_2(PPh_3)_4$ (^{31}P),[538] $Ru_2(pz)_2Cl$-(pzH)$(cod)_2(\mu\text{-}H)$ (^{13}C),[539] $RuRhH_3(dppm)_2$ (^{31}P),[540] $H_2Ru_2(NH)(CO)_9$ (^{15}N),[541] $HRu_3(\mu_3\text{-}\eta^3\text{-}MeOCCHCH)(CO)_9$ (^{13}C),[542] $HRu_3(\mu\text{-}CE)(CO)_9L$ (^{13}C),[543] $Ru_3(\mu_2\text{-}D)(\mu_2\text{-}\eta^2\text{-}OCEt)(CO)_{10}$ (^2H),[544] $[OsH_3(PMe_2Ph)_3]^-$ (^{31}P),[545] $[(C_6H_6)OsH(CO)(PPr^i{}_3)]^+$ (^{31}P),[546] $(C_6H_6)OsH(CH{=}CHPh)$-$(PPr^i{}_3)$ (^{13}C, ^{31}P),[547] $OsH(OH)(PMe_3)_4$ (^{31}P),[548] $Os_3(\mu\text{-}H)_2(\mu\text{-}dppm)$-$(CO)_8$ (^{31}P),[549] $\{Os(C_6H_6)(\mu\text{-}H)\}_3\{\mu_3\text{-}(CH_2)_3CH\}$ (^{13}C),[550] $(\mu\text{-}H)Os_3$-$(CO)_9\{\eta^1\text{-}C(OMe)_2\}(\mu_3\text{-}CPh)$ (^{13}C),[551] $(\mu\text{-}H)Os_3(CO)_8(\mu_3\text{-}AsBu^t)$-$[\mu(N,As)NSNAsBu^t{}_2]$ (^{13}C),[552] $Os_3Pt(\mu\text{-}H)_2(\mu_4\text{-}C)(CO)_{10}(PCy_3)$ (^{13}C, ^{31}P),[553] $HOs_3(CO)_{10}(C_5H_8N)$ (^{13}C),[554] $Os_4H_2(CO)_{14}L$ (^{13}C, ^{31}P),[555] $Os_3Pt(CO)_{10}(\mu\text{-}H)(\mu\text{-}CH_2)(PCy_3)$ (^{31}P, ^{195}Pt),[556] and $(C_5H_5)NiOs_3$-$(\mu\text{-}H)_3(CO)_8(Ph_2PC_2Pr^i)Co_2(CO)_6$ (^{13}C).[557]

A polemic on the criticism of their earlier work on conformational analysis of $(C_5H_5)Fe(CO)(PPh_3)(CH_2R)$ using ^1H n.m.r. spectroscopy has appeared, and been answered.[558] ^1H n.O.e difference n.m.r. spectroscopy on $(C_5H_5)Fe(NO)(PPh_3)CH_2SiMe_3$ indicates that the Me_3SiCH_2 group preferentially oscillates between the C_5H_5 and PPh_3 ligands.[559] Two-dimensional ^1H n.m.r. spectroscopy has been used to assign all 23 aromatic protons in [Ru{2-(4-nitrophenyl)-pyridine}(bipy)$_2$]$^+$.[560] N.m.r. data have also been reported for FeMe$_2$(dmpe)$_2$ (^{13}C, ^{31}P),[561] [FeMe(CO)$_2${P(OMe)$_3$}{(Me$_2$P)$_2$C$_6$H$_4$}]$^+$ (^{31}P),[562] [MR(CO)$_2$L(PMe$_3$)$_2$]$^+$ (M = Fe, Ru; ^{31}P),[563] Fe(CH$_2$PPh$_3$)(CO)$_4$ (^{13}C, ^{31}P),[564] MR$_2$(dmpm)$_2$ (M = Fe, Ru; ^{31}P),[565] Fe(CH$_2$CX$_2$CX$_2$CH$_2$)L-(CO)$_3$ (^{13}C, ^{31}P),[566] Fe[CONPr^{i}CH{C(CO$_2$Me)=CCO$_2$Me}CH=NPr^i](CO)$_2$L (^{13}C, ^{31}P),[567] [(ArO)$_3$PFe(CHO)(CO)$_3$]$^-$ (^{13}C, ^{31}P),[568] (C$_3$F$_7$)Fe(CO)$_3$-I(S=PR$_2$H) (^{13}C, ^{31}P),[569] (C$_5$H$_5$)Fe(CO)$_2$(norbornyl) (^{13}C),[570] (C$_5$H$_5$)-Fe(CO)$_2$R (^{13}C),[571, 572] [(C$_5$H$_5$)Fe(=CHCH=CMe$_2$)(CO)$_2$]$^+$ (^{13}C),[573] (C$_5$H$_5$)Fe(CO)$_2$CHMeSPh (^{13}C),[574] [(C$_5$H$_5$)Fe(CO)$_2$(=CHR)]$^+$ (^{13}C),[575] [(C$_5$H$_5$)Fe(CO)$_2${C(OMe)=CMePEt$_3$}]$^+$ (^{13}C, ^{31}P),[576] (C$_5$Me$_5$)Fe(CO)$_2$CO-PR^{1}R^2 (^{13}C, ^{31}P),[577] (C$_5$Me$_5$)Fe(CO)(PPh$_3$)(CH$_2$OMe) (^{13}C, ^{31}P),[578] [(C$_5$H$_5$)(OC)(Ph$_3$P)FeCMeOCH$_2$Mo(CO)$_3$(C$_5$H$_5$)]$^+$ (^{13}C),[579] (C$_5$H$_5$)Fe(CO)-{P(OPh)$_3$}C(CH$_2$L)=CH$_2$ (^{13}C),[580] (C$_5$H$_5$)Fe(CO)(PPh$_3$)(COCH$_2$CHOHMe) (^{13}C),[581] (C$_5$H$_5$)Fe(CO){C$_5$H$_2$(OMe)(O)Me$_2$} (^{13}C),[582] (C$_5$H$_5$)FeR{Ph$_2$P-(CH$_2$)$_n$PPh$_2$} (^{31}P),[583] [Fe$_2$(CO)$_6$(CHO)(μ-PR$_2$)$_2$]$^-$ (^{2}H, ^{31}P),[584] Fe$_2$(CO)$_6$(μ-OCC$_9$H$_{19}$)$_2$ (^{13}C),[585] (C$_5$H$_5$)$_2$Fe$_2$(μ-CO)(μ-CCH=CHR) (^{13}C),[586] Fe$_3$(CO)$_9$(μ$_3$-CF)$_2$ (^{13}C, ^{19}F),[587] Fe$_3$(CO)$_8${C(OEt)CMeCPhCPh} (^{13}C),[588] (C$_5$H$_5$)Fe$_2$Co(CO)$_2$(μ-CO)$_2$(PR$_3$)(μ-CSMe) (^{13}C, ^{31}P),[589] [RuMe(CO)(CNBu^t)(Ph$_2$PCH$_2$)$_3$CMe]$^+$ (^{31}P),[590] (C$_5$Me$_5$)RuMe(η^3-C$_3$H$_5$)Br (^{13}C),[591] [(C$_6$Me$_6$)Ru(PPh$_3$)Me(C$_2$H$_4$)]$^+$ (^{13}C, ^{31}P),[592] C$_{10}$H$_6$(CH$_2$)$_2$Ru-(PMePh$_2$)$_4$ (^{13}C, ^{31}P),[593] Ru(O$_2$CCF$_3$)R(CO)(PPh$_3$)$_2$ (^{19}F, ^{31}P),[594] Ru(CO)(PPh$_3$)$_2$(CHMeCN)(bzac) (^{31}P),[595] [Os(2,3,5-η-C$_{10}$H$_{13}$)(η-C$_6$H$_3$Me$_3$)]$^+$ (^{13}C),[596] (Ph$_3$P)$_2$(CO)ClRuCH$_2$CH$_2$CX=O (^{13}C),[597] [(p-MeC$_6$H$_4$O)$_3$P]$_3$(p-MeC$_6$H$_4$O)$_2$POC$_6$H$_3$MeRuCl (^{13}C, ^{31}P),[598] Ru(C$_6$H$_4$-PPh$_2$)(PMe$_3$)$_3$(S$_2$CH) (^{31}P),[599] (C$_5$H$_5$)Ru{C(OMe)=CHPh}(PPh$_3$)$_2$ (^{13}C),[600] Ru(CO)$_2$(OH)(COEt)(PPh$_3$) (^{31}P),[601] (C$_5$Me$_5$)(CO)$_2$Ru(COBu^t)$_2$ (^{13}C),[602, 603] [Ru(CHO)(CO)(dppe)$_2$]$^+$ (^{31}P),[604] [(C$_5$H$_5$)Ru(dppe)CCPh-(C$_7$H$_7$)]$^+$ (^{13}C),[605] (C$_5$H$_5$)Ru(PPh$_2$CHRCHRPPh$_2$)CH$_2$CH$_2$Ph (^{13}C, ^{31}P),[606] (μ-CH$_2$)Ru$_3$(CO)$_8$(RN=CHCH=NR) (^{13}C),[607] Ru$_3$(CO)$_9$(μ$_3$-η^3-PhPCH$_2$PPhC$_6$H$_4$) (^{31}P),[608] (C$_6$H$_4$MePr^i)Os(C$_6$H$_4$PPh$_2$)X (^{13}C, ^{31}P),[609] and [Os$_3$(CO)$_{10}$(μ-CH$_2$)(μ-NCO)] (^{13}C).[610]

The structure of Fe(η-C$_4$H$_4$)(CO)$_3$ has been determined from the partially oriented ^{1}H and ^{13}C n.m.r. spectra in nematic solvents.[611] Substituent effects of electron distribution in (C$_5$H$_5$)Fe-(C$_5$H$_4$CH=CHC$_6$H$_4$R) have been compared on the basis of ^{13}C n.m.r. chemical shifts.[612, 613] ^{13}C n.m.r. spectroscopy has been used to study polarization of charge in (C$_5$H$_5$)Fe(C$_5$H$_4$COCH=CHC$_6$H$_4$X).[614] The interaction between ferrocene and cyclodextrin has been studied by ^{1}H n.m.r. spectroscopy.[615] Cyclopentadienyl ^{1}H anisotropic shifts of substituents have been discussed for some ferrocenes.[616] The ^{13}C n.m.r. spectrum of $Fe\{C_5H_4S(CH_2)_5SC_5H_4\}$ shows large shifts due to interaction between the sulphur and iron.[617] The ^{13}C chemical shift data of the α-ferrocenyl-β-t-butylvinyl cation have been compared with model compounds to give proof of hybridisation and structure.[618] ^{57}Fe n.m.r. data for a number of ferrocenes have been correlated with structure.[619] N.m.r. data have also been reported for $[(C_5H_5)Fe(CO)_2\{(CH_2{=}CHCH_2N(SO_2Me)SNHSO_2Me\}]^+$ (^{13}C),[620] $[(MeCH{=}C{=}CHMe)Fe(C_5H_5)(CO)_2]^+$ (^{13}C),[621] $[M(\eta^2{:}\eta^3\text{-}C_{11}H_8Ph_3)(CO)_3]^+$ (M = Fe, Ru; ^{13}C),[622] $(Et_2C_2B_2Me_2S)_2Fe(CO)$ (^{11}B, ^{13}C),[623] $Fe_2(CO)_6\text{-}S(CC_6H_6Me_4)$ (^{13}C),[624] $Fe_2(CO)_5\{\mu\text{-}C(OEt)C(CF_3)C(CF_3)\}\{\mu\text{-}CPhCPhH\}$ (^{13}C, ^{19}F),[625] $[(OC)_2Fe_2(\mu\text{-}CO)\{\mu\text{-}C{=}CHCH(PMe_3)Et\}$ (^{13}C, ^{31}P),[626] $(C_5H_5)_2Ni_2Fe_2(CO)_5(\mu\text{-}PPh_2)(\mu_4\text{-}\eta^2\text{-}C_2Ph)$ (^{13}C, ^{31}P),[627] $[(\eta^6\text{-}C_6H_6)\text{-}(\eta^2\text{-}CH_2{=}CHR)M(PMe_3)_2]^+$ (M = Ru, Os; ^{13}C, ^{31}P),[628] (*p*-cymene)Os-$(PMe_3)(C_2H_4)$ (^{31}P),[629] $[Os_3(CO)_{10}(\mu\text{-}COCH_2)]^-$ (^{13}C),[630] $Fe(C_3H_5)\text{-}(C_3H_5.tcne)(PPh_3)(CO)_2$ (^{31}P),[631] $(C_5H_5)Fe_2(CPhCPhC{=}CH_2)(CO)(\mu\text{-}CO)$ (^{13}C),[632] $Fe_2(CH_2CHCH_2)(CO)_6(SEt)$ (^{13}C),[633] $[\{(C_5H_5)_2(CO)_2Fe_2(\mu\text{-}CO)\}_2(\mu\text{-}C_3H_5)]^+$ (^{13}C),[634] $[Fe_3(CO)_9\{CCOC(O)Me\}]^-$ (^{13}C),[635] $Fe_3(CO)_9(PPh_3)_2(\mu_3\text{-}\eta^2\text{-}MeC_2OEt)$ (^{13}C),[636] $Ru(CO)_2(\eta^3{:}\eta^3\text{-}C_{12}H_{20})$ (^{13}C),[637] $MCl(PPh_3)(NO)\{C(CH_2)_3\}$ (M = Ru, Os; ^{13}C),[638] $Ru_4(CO)_{10}\text{-}(\mu\text{-}CO)_2(PhPCPhCPh)$ (^{31}P),[639] $[Os(\eta^3\text{-}PhC_3CHPh)(PMe_3)_4]^+$ (^{13}C, ^{31}P),[640] $[Fe\{C_6H_6(OMe)(PR_3)\}(CO)_3]^+$ (^{13}C, ^{31}P),[641] $Fe\{C_6H_5Me_2\text{-}(OMe)\}(CO)_3$ (^{13}C),[642] $Fe(diene)(CO)_3$ (^{13}C),[643-648] $Fe\{C_7H_9(COMe)\}\text{-}(CO)_3$ (^{13}C),[649] $\{H_2C{=}C(CR{=}CH)_2C{=}CH_2\}\{Fe(CO)_3\}_2$ (^{13}C),[650] $Fe\{\eta^4\text{-}Fe(CO)_3PBu^tCR^1CR^2\}(CO)_3$ (^{31}P),[651] $(R^1N{=}CR^2CR^3{=}NR^4)Fe_2(CO)_5\{P(OMe)_3\}$ (^{31}P),[652] $Fe_2(CO)_6\{\mu\text{-}CPhCPhC(CF_3)CHCF_3\}$ (^{13}C),[653] $[Ru(C_5H_5)(\eta^4\text{-}diene)(CO)]^+$ (^{13}C),[654] $[(C_5H_5)Fe(CO)_{3\text{-}\underline{n}}\{P(OR)_3\}_{\underline{n}}]^+$ (^{13}C, ^{31}P),[655] $(C_5H_5)Fe(CO)_2SP(S)(OEt)_2$ (^{13}C),[656] $(C_5Me_5)Fe(CO)_2PPh_2BH_3$ (^{11}B, ^{31}P),[657] $[Fe(C_5H_5)(CO)(CS)]_2$ (^{13}C),[658] $[(C_5H_5)Fe\{(Ph_2P)_3CH\}]^+$ (^{31}P),[659] $(C_5H_5)Fe(CO)_2P{=}C(OSiMe_3)Bu^t$ (^{13}C),[660] $(C_5Me_5)Fe(CO)_2\text{-}PClBu^t$ (^{31}P),[661] $[(C_5Me_5)Fe(NO)]_2$ (^{13}C),[662] $(C_5Me_5)(CO)_2MP{=}P(C_6H_2\text{-}Bu^t{}_3)$ (M = Fe, Ru; ^{13}C, ^{31}P),[663] $(C_5H_5)Fe(CO)_2OC(O)R$ (^{13}C),[664] $\{(C_5H_5)Fe(CO)_2\}_2SeMe$ (^{13}C),[665] $[(C_5H_5)Fe(CO)_2(Me_3Sb{=}Se)]^+$ (^{13}C),[666] $[Fe_2(CO)_6(\eta^5{:}\eta'^5\text{-}C_{16}H_{16})]^+$ (^{13}C),[667] $MM'_2(\mu_3\text{-}CO)_2(C_5Me_5)_3$ (M, M' = Co, Rh, Ir; ^{13}C),[668] $(C_5H_5)_2Ni_2Fe(CO)_3(C_2Ph_2)$ (^{13}C),[669] $\{(C_5Me_5)\text{-}Rh\}_3\{Ru(CO)_3\}(\mu_3\text{-}CO)_2$ (^{13}C),[670] $[Ru(\eta^5\text{-}C_5H_4R)(CO)(PPh_3)L]^{\underline{n}+}$ (^{13}C, ^{31}P),[671] $[(C_4H_3SCH_2C_5H_4)Ru(PPh_3)_2]^+$ (^{13}C, ^{31}P),[672] $(C_5H_5)RuL(PPh_3)\text{-}SO_2R$ (^{13}C, ^{31}P),[673] $Ru(indenyl)(PPh_3)_2Cl$ (^{31}P),[674] $[(1\text{-}3,5,6\text{-}\eta^5\text{-}C_8H_{11})RuL_3]^+$ (^{13}C),[675] $[(\eta^5\text{-}C_7H_9)Ru(CNR)_3]^+$ (^{13}C),[676] $Fe(C_5H_4S\text{-}CH_2)_2C_6H_4$ (^{13}C),[677] [*m*]-ferrocenophane-1,*n*-diones (^{13}C),[678] (C_5H_5)-

$Fe(C_5H_4CHXCH_2OMe)$ (^{13}C),[679] $Fe(C_5H_4SPr^i)_2PtCl_2$ (^{13}C, ^{195}Pt),[680] α-naphthylferrocene (^{13}C),[681] $[(C_6Me_6)Fe(C_5H_4C_5H_4)Fe(C_5H_5)]^+$ (^{13}C),[682] $(C_5Me_5)_2M$ (M = Ru, Os; ^{13}C),[683] $Fe(C_5H_4E)_2MCl_2$ (E = S, Se; M = Ge, Sn; ^{13}C),[684] $Fe(C_5H_4E^1)_2E^2Ph$ (E^1 = S, Se; E^2 = P, As),[685] $(C_5H_4PR^1R^2)(C_5H_4PR^3R^4)Fe$ (^{31}P),[686] $\{1,3\text{-}C_5H_3(PBu^t{}_2)_2\}Fe\text{-}\{C_5H_3(PBu^t{}_2)CHMeNMe_2\text{-}1,2\}$ (^{31}P),[687] $[(C_5H_5)RuCl(PPh_3)Ph_2PC_6H_5Ru\text{-}(C_5H_5)]^+$ (^{31}P),[688] $[(SC_5H_4)Ru\{C_5H_4S(CH_2CH_2O)_{\underline{n}}CH_2CH_2\}Ag]^+$ (^{13}C),[689] $[(C_5Me_5)(C_5Me_4CH_2)Ru]^+$ (^{13}C),[690] $\{C_3H_3BMeN(SiMe_3)\}_2Ru$ (^{11}B),[691] $[(FC_6H_5)M(C_5H_5)]^+$ (M = Fe, Ru; ^{19}F),[692] $[(arene)Fe(C_5H_5)]^+$ (^{13}C),[693] $[(C_5H_5)Fe(\eta^6\text{-}C_6Me_5CH_2PPh_2)Rh(diene)Cl]^+$ (^{13}C, ^{31}P),[694] $[(C_5H_5)Fe(\eta^5\text{-}C_6H_7)]^+$ (^{13}C),[695] $[Ru(1\text{-}3,5,6\text{-}\eta\text{-}C_8H_{11})(\eta^6\text{-}C_8H_{10})]^+$ (^{13}C),[696] $Fe(C_6H_6)(\eta^4\text{-}C_6H_7X)$ (^{13}C),[697] $[Ru(dppb)(\eta^6\text{-toluene})]^+$ (^{31}P),[698] and $Ru(CO)\{C_6H_3MeC(O)C_6H_4Me\}XL_2$ (^{13}C, ^{31}P).[699]

A combination of ^{13}C and ^{15}N n.m.r. spectra obtained at high frequency with 90% ^{13}C and 99% ^{15}N enrichment reveals two mechanisms for the conversion of $[Fe(CN)_5(NO)]^{2-}$ to $[Fe(CN)_5(NO_2)]^{4-}$. ^{15}N and ^{77}Se n.m.r. spectra were reported for $[Fe_4Se_3(NO)_7]^-$.[700] A correlation between $\delta(^{31}P)$ and Mossbauer isomeric shifts in $[Fe(CN)_5L]^{3-}$ has been investigated and discussed.[701] The electronic effects induced by the ligands in $Fe(CO)_2L(PhCH{=}CHCOMe)$ have been studied by ^{13}C n.m.r. spectroscopy.[702] The ^{57}Fe n.m.r. spectrum of carbonmonoxymyoglobin has been reported. Analysis of $\underline{T}_1$ and $\underline{T}_2$ data has yielded the chemical shift anisotropy and protein rotational correlation time.[703] The ^{57}Fe chemical shifts of sperm whale carbonylmyoglobin and two model complexes have been determined by double-resonance methods in doubly enriched ^{57}Fe,^{13}C-labelled samples. Deprotonation of the axial imidazole in the model complex caused a large low-frequency ^{57}Fe chemical shift.[704] The ^{57}Fe n.m.r. spectrum of carbonylmyoglobin has been determined, including $\underline{T}_1$ and chemical shift anisotropy.[705] N.m.r. data have also been reported for $[Fe(CNR)_6]^{2+}$ (^{13}C),[706] $[M(CNR)_6]^{2+}$ (M = Ru, Os; ^{13}C),[707] $Fe(CO)_4\{PH(C_6H_2Bu^t{}_2CMe_2CH_2)\}$ (^{13}C),[708] $Fe(CO)_4(PRCl_2)$ (^{31}P),[709] $Fe(CO)_4\{P(NEt_2)(C_8H_8)\}$ (^{31}P),[710] $PhP\{PPhFe(CO)_4\}_2C_6H_4$ (^{13}C, ^{31}P),[711] $\{(OC)_4Fe\}_2(C_6H_2Bu^t{}_3P{=}CH_2)$ (^{13}C, ^{31}P),[712] $(Me_3Si)_2N\text{-}P({=}CHSiMe_3)Fe(CO)_4$ (^{13}C, ^{31}P),[713] $(Pr^i{}_2NP)_2COFe_2(CO)_6$ (^{13}C, ^{31}P),[714] $Fe_2(CO)_6(P_3N_3Cl_4)$ (^{31}P),[715] $(Me_2PCH_2PMeCH_2PMe_2)Fe_3(CO)_{11}$ (^{13}C, ^{31}P),[716] $Fe_2(CO)_6(\mu\text{-}CO)(\mu\text{-dppm})$ (^{31}P),[717] $Pt\{Fe(CO)_3NO\}_2\text{-}(NCPh)_2$ (^{195}Pt),[718] $M_3(CO)_{10}(HPPh_2)_2$ (M = Ru, Os; ^{31}P),[719] $Os_4(CO)_{13}(AuPEt_3)_2$ (^{31}P),[720] $[Cu_3Fe_3(CO)_{12}]^{3-}$ (^{13}C, ^{17}O, ^{63}Cu),[721] $RP\{Fe(CO)_3L\}_3$ (^{31}P),[722] $Fe_2(CO)_6(Bu^tPNBu^t)$ (^{31}P),[723] $\{\mu_2\text{-PC-}$

$(SiMe_3)_2\}_2Fe_2(CO)_6$ (^{31}P),[724] $\{CyCSFe_2(CO)_6\}_2Hg$ (^{13}C),[725] $Fe_3(CO)_9$-(μ_3-PPh)(μ_2-PPh_2CPhOEt) (^{31}P),[726] $[Fe_4(AuPR_3)(CO)_{13}]^-$ (^{13}C),[727] $[Ru_3FeC(CO)_{12}]^{2-}$ (^{13}C),[728] M(1,3,4-thiadiazole-2,5-dithiolato)$_2$-$(CO)_2(PPh_3)_2$ (M = Ru, Os; ^{31}P),[729] $Ru_3(CO)_9\{HC(PPh_2)_3\}$ (^{13}C, ^{31}P),[730] M(phthalocyaninato)$(CNR)_2$ (M = Fe, Ru; ^{13}C),[731] $[Fe(CO)_2$-$\{(Me_2As)_2C_6H_4\}Me(PMe_3)]^+$ (^{31}P),[732] $FePdPt(CO)_4(dppm)_2$ (^{31}P),[733] $Ru(C_5H_4NS)_2(CO)_2(PPh_3)$ (^{31}P),[734] $[Ru(CO)_2(\mu\text{-}O_2CEt)]_2$ (^2H, ^{13}C),[735] $(Ph_3P)AuRuCo_{3-\underline{n}}Rh_{\underline{n}}(CO)_{12}$ (^{31}P),[736] $[Os(CO)_2(dppm)_2]^{2+}$ (^{31}P),[737] $Cl_2Os(\mu\text{-}CO)_2(\mu\text{-}dppm)_2RhCl$ (^{31}P),[738] and $[Ru(CO)(PPh_3)_2L(NCR)]^+$ (^{31}P).[739]

A detailed ^1H n.m.r. investigation, including n.O.e. measurements, has been used to determine the conformation of [Ru(5,5'-diphenyl-2,2',2"-terpyridyl)$_2$]$^{2+}$.[740] N.m.r. data have also been reported for $[Fe(bipy)_3][C(CN)_2]_2$ (^{13}C, ^{14}N),[741] lipid-haem (^{31}P),[742] Fe{(cyclohexanedione dioxime)$_3$(BH)$_2$} (^{11}B),[743, 744] [(benzotriazole)Ru$(NH_3)_5]^+$ (^{13}C),[745] [Ru(bipyrimidine)$_3$]$^{2+}$ (^{13}C),[746] $[Ru\{HC(CH_2NHCH_2CH_2NHCH_2)_3CH\}]^{2+}$ (^{13}C),[747] $[M(bipy)_2$-$(SMe_2)_2]^{2+}$ (^{13}C),[748] Ru(H_2edta)(o-benzoquinone diimine) (^{13}C),[749] $[Ru\{(Ph_2PCH_2CH_2)_2NH\}_2ClO_4]^+$ (^{14}N, ^{31}P),[750] [Ru(6-mercaptopurine)$_2$-$(PPh_3)_2]^{2+}$ (^{31}P),[751] $RuCl_3(NS)(PPh_3)_2$ (^{31}P),[752] Ru_2Cl_4\{1,1'-2,2'-$(Ph_2P)_2$-naphthalene\}$_2(NEt_3)$ (^{31}P),[753] Ru(dppe)(tos)$_2$ (^{13}C, ^{31}P),[754] $RuCl_2\{Ph_2P(CH_2CH_2PPh)_2CH_2CH_2PPh_2\}$ (^{31}P),[755] $Ru_2Cl_5(PEtPh_2)_4Ag$-$(PEtPh_2)$ (^{31}P),[756] and $OsCl_2(PMe_2Ph)_3$ (^{31}P).[757]

Complexes of Co, Rh, and Ir.—The effect of nephelauxetism has been incorporated empirically into a theoretical relationship between ^{59}Co chemical shifts and the optical transition energies for cobalt(III) complexes, allowing an empirically found correlation between the chemical shift and the internal ligand-field strength parameter to be rationalised.[758]

The ^{31}P n.m.r. spectrum of $[\{(C_5H_5)Rh(PR_3)\}_2(\mu\text{-}H)_3]^+$ has been analysed to determine $^1\underline{J}(^{103}Rh,^{103}Rh)$ = 11.4 Hz.[759] The first observation of $\underline{J}(Li,^1H)$ in $(C_5Me_5)IrH_2SiMe_3Li$(pmdeta) and $(C_5Me_5)IrH_3$-Li(pmdeta) has been reported. The ^7Li and ^{13}C n.m.r. spectra were reported.[760] ^{19}F and ^{31}P n.m.r. spectra have been used to determine $^1\underline{J}(^{31}P,^{19}F)$ for $[PF_6]^-$ to indicate its coordination in IrHCl-$(PPh_3)_2(CO)PF_6$.[761] In $[Ir_2Cu_3H_6(MeCN)_3(PMe_2Ph)_6]^{2+}$, the $[AX]_3$ patterns in the ^1H and ^{31}P n.m.r. spectra suggest the hydride atoms are trans to the Ir-P vectors.[762] N.m.r. data have also been reported for CoH(dppe) (^{31}P),[763] $[CoHCl(dppm)_2]^+$ (^{31}P),[764] (dppe)RhH_4-

$(\eta^3\text{-}C_3H_4Pr^i)$ (^{31}P),[765] $Rh(C_5H_5)(PPh_3)(C_2H_4)$, $(C_5H_5)RhH(C_2H_4)SiR_3$ (^{13}C, ^{31}P, ^{29}Si),[766] $(C_5Me_5)RhH(SiEt_3)(C_2H_4)$ (^{13}C, ^{29}Si, ^{103}Rh),[767] $HM\{(R_2PO)_2H\}_2PR_2OH$ (M = Rh, Ir; ^{31}P),[768] $RhH(CO)(Ph_2PCH_2)_3CMe$ (^{31}P),[769] $[MH(SnCl_3)_5]^{3-}$ (M = Rh, Ir; ^{119}Sn),[770] (M = Rh; ^{119}Sn),[771] $[RhH(SnCl_3)_4Cl]^{3-}$ (^{119}Sn),[772] $[RhH(SnCl_3)_4(PBu^n{}_3)]$ (^{31}P, ^{119}Sn),[773] $[Rh_2(C_5H_5)_2(\mu\text{-}CO)(\mu\text{-}H)(\mu\text{-}dppm)]^+$ (^{31}P),[774] $[Rh_2(\mu\text{-}H)(\mu\text{-}CO)(CO)_2(\mu\text{-}dppm)_2]^+$ (^{31}P),[775] $(dppe)Rh(\mu\text{-}H)(\mu\text{-}Cl)IrH_2(PEt_3)_2$ (^{31}P),[776] $[Rh_3(\mu\text{-}dpmp)_2(H)_2(CO)_2(\mu\text{-}Cl)_2]^+$ (^{13}C, ^{31}P),[777] $[Rh_3Ag_3H_9\{(Ph_2PCH_2)_3CMe\}_3]^{3+}$ (^{31}P),[778] $[IrH_4(PMe_2Ph)_3]^+$ (^{31}P),[779] $IrH_2(SnCl_3)(cod)PR_3$ (^{31}P, ^{119}Sn),[780] $IrH(CO)(PPh_3)_2(ol)$ (^{31}P),[781] $HIr(CO)(PPh_3)_3$ (^{31}P),[782] $PhP\{(CH_2)_3PPh_2\}_3\overline{IrH\{C(O)C_6H_{4\text{-}\underline{n}}Me_{\underline{n}}C}H_2\}$ (^{31}P),[783] $(Bu^t{}_2PCMe_2CH_2)IrHCl(CH_2CMe_2CH_2Bu^t)$ (^{31}P),[784] $\overline{Ir(CO)}XH(PEt_3)_2(PH_2)$ (^{31}P),[785] $(C_5Me_5)(PMe_3)Ir(CH{=}CH_2)H$ (^{13}C),[786] $(C_5Me_5)(Ph_3P)Ir(OR)H$ (^{13}C, ^{31}P),[787] $[IrH\{Pr^i{}_2P(CH_2)_3PPr^i\}\{OCHPhC(O)NHCH_2Ph\}]^+$ (^{13}C, ^{31}P),[788] $Ir_2H_2Cl_2(CO)_2(dppm)_2$ (^{31}P),[789] $\{(C_5Me_5)Ir\}_2H_3(BH_4)$ (^{11}B, ^{13}C),[790] and $HIr_4(CO)_7(Ph_2PCH{=}CHPPh_2)(PhC_6H_4PCH{=}CHPPh_2)$ (^{31}P).[791]

A fairly linear relationship between log k for $P(OMe)_3$ dissociation and $\delta(^{31}P)$ in the series $(MeO)_3PCo(dmgH)_2R$ has been found. The ^{13}C n.m.r. data were also reported.[792] The ^1H and ^{31}P n.m.r. spectra of a series of cobalamins have been reported. The dependence of chemical shift on axial ligand follows the expected order.[793] The analysis of the ^{13}C and ^{31}P n.m.r. spectra of $(C_5Me_5)_2Rh_2(\mu\text{-}CH_2)_2\{\mu\text{-}CH_2CR(CH_2CR{=}CH_2)CH_2\}$ and $[(C_5Me_5)_2Rh_2(\mu\text{-}CH_2)_2(dppm)]^{2+}$ has yielded $^1\underline{J}(^{103}Rh,^{103}Rh)$ = ca. 13Hz.[794] N.m.r. data have also been reported for $[MeCo(7\text{-}Me\text{-}salen)(en)]^+$ (^{13}C),[795] $RCo(dmgH)_2(PPh_3)$ (^{31}P),[796] $RCo(dmgH)_2(py)$ (^{13}C),[797] $(C_5H_5)_2Co_2X_2(\mu\text{-}PMe_2)(\mu\text{-}CH_2PMe_2)$ (^{31}P),[798] $[(C_5H_5)Co(PMe_3)(NHCMeNMeC{=}CH_2)]^+$ (^{13}C),[799] $[(C_5H_5)(PMe_3)Co(RNHNMeC{=}CH_2)]^+$ (^{13}C),[800] $[M(\overline{COCHRCH_2N}H)_6]^{3+}$ (M = Co, Rh; ^{13}C),[801] $(C_5H_5)Co(CO)(\overline{CNMeCH_2CH_2N}Me)$ (^{13}C),[802] $(\mu_3\text{-}CMe)Co_3(CO)_7(\mu\text{-}dppm)$ (^{13}C),[803] $RhMeBr(PMe_3)_3(CH_2OMe)$ (^{13}C),[804] $(C_5H_5)Rh(PMe_3)_2CH_2CH_2X$ (^{13}C, ^{31}P),[805] $(C_5Me_5)Rh(CO)I(CH_2I)$ (^{13}C, ^{31}P),[806] $(C_5H_5)Rh(CH_2X^1)X^2(PMe_3)$ (^{13}C, ^{31}P),[807] $(C_5Me_5)MR^1R^2(CO)$ (M = Rh, Ir; ^{13}C),[808] $[(C_5H_5)Rh(CH_2I)\{P(OEt)_3\}_2]^+$ (^{13}C, ^{31}P),[809] $Rh(CH_2CMe_2PPhBut)(MeO_2CCH{=}CHCO_2Me)$ (^{13}C),[810] $(C_5Me_5)\overline{MC_6H_4CO}_2$ (Rh, Ir; ^{13}C),[811] $(C_5Me_5)_2Rh_2(CO)_2(\mu\text{-}C{=}CR_2)$ (^{13}C),[812] $Rh_2(CO)_2(dppm)_2(C_2R)$ (^{31}P),[813, 814] $Rh_2(CO)_2(Ph_2PCH_2C_3HN_2CH_2PPh_2)(COMe)(\mu\text{-}PPh_2)$ (^{31}P),[815] $RhEt\{C(O)C_9NH_6\}(py)_2Cl$ (^{13}C, ^{31}P),[816] Rh(octaethylporphyrin)$\{C(O)CH_2OH\}$ (^{13}C),[817] $M(PPh_2CH_2CH_2SiMe_2)_2XL$ (M = Rh, Ir; ^{31}P),[818] $Rh_2Cl_3(\mu\text{-}HgCl)(CO)_2(\mu\text{-}dppm)_2$ (^{13}C, ^{31}P),[819] RhCl(cod)(carbene) (^{13}C, ^{31}P),[820] $M(PPr^i{}_3)_2$-

(=C=CHPh)Cl (M = Rh, Ir; ^{13}C),[821] $(C_5Me_5)_2Rh_2(CO)_2(\mu\text{-}C{=}CR_2)$ (^{13}C),[822] $(C_5H_5)Rh(PR_3)(\mu\text{-}C{=}CH_2)CuCl$ (^{13}C),[823] $Rh_2W(\mu\text{-}CO)(\mu_3\text{-}CMe)$-$(CO)_2(C_5H_5)(C_5Me_5)_2$ (^{13}C),[824] $(C_5H_5)_2Rh_2(\mu\text{-}CO)(\mu\text{-}CH_2)(\mu\text{-}F_3CC{=}CCF_3)$ (^{19}F),[825] $Rh_2(\mu\text{-}pz)_2(\mu\text{-}C_6F_4PPh_2)Br(CO)(BrC_6F_4PPh_2)$ (^{31}P),[826] $(C_5Me_5)_2Rh_2(\mu\text{-}CO)\{\mu\text{-}C(O)CH_2CMe_2CH\}(O_2CCF_3)$ (^{2}H, ^{13}C, ^{19}F),[827] $(C_5Me_5)IrR^1R^2$(dmso) (^{13}C),[828] $Ir(C_6H_4C_5H_4N)_3$ (^{13}C),[829] [IrCl(CHO)-$(dppe)_2]^+$ (^{31}P),[830] $IrCl_2(CO_2H)(CO)(PMe_2Ph)_2$ (^{31}P),[831] $Ir(=CH_2)\{N(SiMe_2CH_2PPh_2)_2\}$ (^{13}C, ^{31}P),[832] $Ir(Ph_2PCH_2CH_2SiMe_2)$-$(PPh_3)(CO)_2$ (^{31}P),[833] and $[(dppe)_2IrAu(PPh_3)]^+$ (^{31}P).[834]

The ^{31}P n.m.r. spectrum of $[N(CH_2CH_2PPh_2)_3Rh(\eta^2\text{-}CS_2)]^+$ is AB_2X.[835] The ^{59}Co chemical shift of $(C_5H_5)Co(cod)$ and related compounds correlate linearly with catalytic activity and regio-selectivity.[836] ^{1}H two-dimensional and ^{13}C n.m.r. spectroscopy have been used to determine the structure of $(C_5H_5)Co(diene)$.[837] The ^{103}Rh $\underline{T}_1$ has been measured for $(cod)Rh\{B(pz)_4\}$ at 3 different magnetic fields. ^{13}C n.m.r. spectroscopy was used to determine the molecular correlation time, and then the ^{103}Rh chemical shift anis-otropy was calculated as 6 500 p.p.m.[838] ^{1}H n.O.e and ^{13}C n.m.r. spectroscopy has been used to determine the structure of compounds such as $(cod)Rh(NC_4H_3CH{=}NCHMePh)$.[839] ^{1}H n.O.e. difference n.m.r. spectroscopy has been used to determine the structure of (cod)-$Rh(OC_6H_4NCHMePh)$.[840] N.m.r. data have also been reported for $Co_3(C_5H_5)_3(SMe)_2(F_3CC_2CF_3)_2$ (^{13}C, ^{19}F),[841] $(C_5H_5)_2(C_5Me_5)Rh_2Co$-$(CO)_3(F_3CC_2CF_3)$ (^{13}C, ^{19}F),[842] $RhCl(PPh_3)_2$(alkene) (^{13}C),[843] (indenyl)Rh(diolefin) (^{13}C),[844] $(C_5Me_5)Ir(CO)(C_2H_4)$ (^{13}C),[845] $IrCl(CH_2{=}CHC_6H_{13})(PPh_3)_2$ (^{31}P),[846] $Ir_2(cod)_2(\mu\text{-}pz)_2(\mu\text{-}MeO_2CC_2CO_2Me)$ (^{13}C),[847] $Co(\eta^3\text{-}RPCHPR)(CO)_3$ (^{13}C, ^{31}P),[848] $(\eta\text{-}1\text{-}Me\text{-}C_3H_4)Rh(Pr^i{}_2P$-$CH_2)_2CH_2$ (^{13}C, ^{31}P),[849] $[(C_5H_5)Rh(C_8H_8OR)]^+$ (^{13}C),[850] $[Rh(\eta^3\text{-}CH_2$-$C_6H_4Me\text{-}\underline{o})(PPh_3)_2]^+$ (^{31}P),[851] $(C_5H_5)_2Rh_2\{C(CF_3)C(CF_3)C(O)NCOPh\}$ (^{13}C, ^{19}F),[852] $(C_5H_5)Co(OCCMeCMeCO)$ (^{13}C),[853] $[CH_2(CHMePPh_2)_2Rh$-$(nbd)]^+$ (^{13}C, ^{31}P),[854] $[Rh\{C_6H_4(PPhCH_2CH_2CH_2)_2PH\}(nbd)]^+$ (^{13}C, ^{31}P),[855] $(C_5H_5)Rh(C_8F_8)$ (^{19}F),[856] $[Rh(cod)\{Ph_2P(CH_2CH_2O)_{\underline{n}}CH_2CH_2$-$PPh_2\}]^+$ (^{13}C, ^{31}P),[857] $[(cod)Rh\{Ph_2P(OCH_2CH_2)_{\underline{n}}OC_6H_4PPh_2\}]^+$ (^{13}C, ^{31}P),[858] Rh(cod)(4-Me-quinoline-$\underline{N}$-oxide)(PPh_3) (^{31}P),[859] [(cod)Rh-$(Ph_2PC_5H_4NC_5H_4NPPh_2)_2Rh(cod)]^{2+}$ (^{31}P),[860] $[Pt_2Rh(\mu_3\text{-}S)_2(PPh_3)_4$-$(cod)]^+$ (^{13}C, ^{31}P, ^{195}Pt),[861] $[Ir(cod)(\mu\text{-}pz)]_2$ (^{13}C),[862] $[(C_5H_5)Co$-$\{Ph_2P(CH_2)_{\underline{n}}PPh_2\}I]^+$ (^{31}P),[863] $[(C_5H_5)Co(CO)(CN)]^-$ (^{13}C),[864] $[(C_5H_5)_2Co_2(\mu\text{-}PMe_2)_2(\mu\text{-}CH{=}NMe)]^+$ (^{31}P),[865] $\{(C_5H_5)Co\}_2(\mu\text{-}PMe_2)_2$-$(\mu\text{-}SO_2)$ (^{31}P),[866] $(C_5H_5)_2Co_2(\mu\text{-}PR_2)_2$ (^{31}P),[867] $[\{(C_5H_5)(Me_3P)Co$-$(S)(CMeNMe)\}_2C]^{2+}$ (^{13}C),[868] $Co_2Rh_2(C_5Me_5)_2(\mu_3\text{-}CO)(\mu\text{-}CO)(CO)_4$

(^{13}C),[869] $(C_5Me_5)CoI_2(CNPh)$ (^{13}C),[870] $(C_4H_4BNPr^i)M(C_4H_4BNPr^i)M$-$(C_4H_4BNPr^i)$ (M = Co, Rh; ^{11}B),[871] $(C_5H_5)Rh\{S_2C_2(CO_2Me)_2\}$ (^{13}C),[872] $Rh_2(C_5H_5)_2(\mu$-$CO)(\mu$-dppm$)(\mu$-$AgOPF_2O)$ (^{31}P),[873] $[(C_5Me_5)(C_5H_3Me_2)Rh]^+$ (^{13}C),[874] $[(C_5Me_5)Rh(P_3O_9)]^-$ (^{13}C, ^{31}P),[875] and $[Ir(C_6H_6)(PPh_3)_2]^+$ (^{13}C).[876]

The ^1H, ^{19}F, and ^{31}P n.m.r. spectra of Rh(ß-diketone)$(CO)_{\underline{n}}L_{2-\underline{n}}$ indicate considerable delocalisation of the electron density.[877] The ^{31}P n.m.r. spectrum of $[Rh(PPh_3)_2$(4-hydroxypyridine)$(CO)]^+$ shows the presence of two isomers,[878] while for $[MX\{Ph_2P(CH_2)_{\underline{n}}$-$PPh_2\}(CO)]_2$ there are two conformers observed by ^{31}P n.m.r. spectroscopy at low temperature.[879] The ^{31}P $\underline{T}_1$ values for $P(C_6H_4X)_3$, $IrCl(CO)\{P(C_6H_4X)_3\}_2$, and $AuCl\{P(C_6H_4X)_3\}$ vary between 28 and 4.4 s. The n.O.e values were also determined.[880] N.m.r. data have also been reported for $(C_6H_4O_2)_2PCo(CO)_3(PPh_3)$ (^{13}C, ^{31}P),[881] $Co_2(Bu^tBNBu^t)(CO)_3$ (^{11}B, ^{13}C),[882] $Co_2(\mu$-$PPh_2)_2(CO)_4(PPh_2H)_2$ (^{31}P),[883] $[Co_2(CO)_4\{(Et_2PCH_2CH_2)_2P\}_2CH_2]^{2+}$ (^{31}P),[884] $Co_4(CO)_{10}\{\mu$-$(RO)_2P\}_2NEt$ (^{31}P),[885] $Co_4(CO)_{10}(\mu_4$-$PPh)_2$ (^{31}P),[886] $Co_4(CO)_7$(dppm)-$(Ph_2P)_3CH$ (^{13}C, ^{31}P),[887] $RhCl(CO)_2(PPh_3)_2$ (^{13}C, ^{31}P),[888] $[Rh\{(Ph_2P)_2NR\}_2][Rh(CO)_2Cl_2]$ (^{31}P),[889] $Rh(CO)Cl(\mu$-$Ph_2AsCH_2PPh_2)_2$-$PtCl_2$ (^{31}P),[890, 891] $(Ph_3P)Rh(CO)$(1-Ph-3-Me-4-benzoyl-5-pyrazolone) (^{31}P),[892] $Rh(CO)(O_2CH)(PCy_3)_2$ (^{13}C, ^{31}P),[893] $Rh_2(\mu$-$Cl)_2(CO)_2\{O(CH_2$-$CH_2)_2NP(OCH_2CH_2)_2CH_2\}$ (^{31}P),[894] $Rh_2(CO)_2Cl_2\{u$-$(Ph_2P)_2C_5H_3N\}_2$ (^{31}P),[895] $[Rh_2(Ph_2PNPPh_2)_2(CO)(\mu$-$CO)_2]^+$ (^{31}P),[896] $[Rh(\mu$-$Bu^t{}_2P)$-$(CO)_2]_2{}^{2-}$ (^{13}C, ^{31}P),[897] $Rh_2(\mu$-$pz)_2I_2(CO)_2(\mu$-dppm) (^{31}P),[898] $Rh_2(\mu$-$N_2CHCO_2Et)(CO)_2(dppm)_2$ (^{31}P),[899] $[Rh_2(\mu$-$OR)(CO)_2(\mu$-dppm$)_2]^+$ (^{31}P),[900] $Rh_2(\mu$-$Cl)(CO)_2(Ph_2PCH_2C_3HN_2CH_2PPh_2)$ (^{31}P),[901] $[PdRh_2$-$(CO)_2Cl_3\{(Ph_2PCH_2)_2AsPh\}_2]^+$ (^{13}C, ^{31}P),[902] $Rh_2(\mu$-$CO)[\mu$-$(Ph_2P)_2$-$NMe]_2X_2$ (^{31}P),[903] $[Rh_3\{\mu$-$(Ph_2PCH_2)_2PPh\}_2(\mu$-$CO)(CO)(\mu$-$X)X]^+$ (^{13}C, ^{31}P),[904] $[Rh_4(\mu$-$Me_3C_3N_2)(CO)_4(\mu$-dppm$)_4]^{2+}$ (^{31}P),[905] $Rh_4(CO)_{10}$-(-diop) (^{31}P),[906] $IrX(CO)_{\underline{n}}$(dppe) (^{31}P),[907] $[Ir_2(CO)_2(\mu$-$OH)(dppm)_2]^+$ (^{31}P),[908] and $[Ir_6(CO)_{15}(\overline{N}O)]^-$ (^{15}N).[909]

Replacement of NH_3 protons by deuterium in a series of 10 Co(III) complexes leads to large changes in the $\underline{T}_1$ values of imidazole ligand protons adjacent to the NH_3 ligands. N.O.e. measurements between the NH_3 protons show that large $\underline{T}_1$ changes occur on NH_3 deuteriation. ^{13}C n.m.r. spectra were also recorded.[910] ^{15}N chemical shifts of Co(III) complexes of $[^{15}NO_2]^-$ have been determined and coordination shifts of between 135.0 and 161.5 p.p.m. found.[911] Inequivalent methyl groups in $SC_4H_3CH_2O_2Co(dmgH)_2(py)$ have been attributed to changes in hydrogen bonding.[912] A complete ^{13}C n.m.r. signal assignment has been made for some corrinoids using two-dimensional INADEQUATE spectroscopy.[913] ^{1}H n.O.e and ^{13}C n.m.r. spectroscopy have been used to determine the structure of $[\{(MeC_6H_4)PhPCH_2CH_2PPh(C_6H_4OMe)\}Ir\{O{=}CR^1NHC(CO_2Me){=}CHR^2\}]^+$.[914] ^{59}Co Chemical shifts of a variety of dithiocarbamate complexes cover a wide range, but do not correlate simply with electrochemical data.[915, 916] N.m.r. data have also been reported for $[Co(NH_3)_5Cl[lasalocid A]_2$ (^{13}C),[917] $[Co(NH_3)_5(threonine)]^{3+}$ (^{13}C),[918] $[(H_3N)_4Co(P_2O_7)]^-$ (^{31}P),[919] $[Co(en)_2(hfac)]^{2+}$ (^{13}C, ^{19}F),[920] Co(glycylglycine)X(en) (^{13}C),[921] $[(en)_2Co(SO_3)(S_2O_3)]^-$ (^{13}C),[922] $[(en)_2Co(methylmalonato)]^+$ (^{13}C),[923] $[CoCl(NH_2CH_2CMe_2NH_2)(NH_2CH_2CMe_2N{=}CHCMe_2NH_2)]^{2+}$ (^{13}C),[924] $[CoCl_2(NH_2CN_2CMe_2NH_2)_2]^+$ (^{13}C),[925] $[CoCl(en)(NH_2CH_2CH{=}NCH_2CH_2NH_2)]^+$ (^{13}C),[926] $[\{Co_2(H_2NCH_2CH_2SH)_6\}M]^{n+}$ {M = Co(III), Cd, Pb(II), Ni(II), Zn, Hg(II); ^{13}C},[927] $[Co\{(1R,2S)$-2-(aminomethyl)cyclohexylamine$\}_3]^{3+}$ (^{13}C),[928] $[CoCl_2\{(1R,2R)$-dimethyl-1,2-cyclohexanediamine$)_2]^+$ (^{13}C),[929] $[Co\{H_2N(CH_2)_2NH(CH_2)_2NH_2\}(salicylaldehydo)]^{2+}$ (^{13}C),[930] $[Co(NH_3)_2(trien)]^{3+}$ (^{13}C),[931] $[Co(CN)_2$(trimethylenediamine-N,N'-diacetato)$]^-$ (^{13}C),[932] $[CoCl_2$(1,7-bis{(S)-2-pyrrolidinyl}-2,6-diazaheptane)$]^+$ (^{13}C),[933] $[Co$(3-aminomethyl-3-methylazetidine)$(OH_2)Cl]^{2+}$ (^{13}C),[934] $[Co(acac)(Ph_2PCH_2CH_2NHCH_2CH_2NHCH_2CH_2PPh_2)]^{2+}$ (^{13}C),[935] [Co(ox)(5-S-Me-2,3,2-tet)]$^+$ (^{13}C),[936] $[CoCl_2$(3,7-diazanonane-1,9-diamine)$]^+$ (^{13}C),[937] $[(9,17\text{-}Me_2\text{-}13\text{-}NO_2$-1,3,5,7,11,15-hexaaza-tetracyclo[11.5.1.1^{3,9}.1^{5,17}]henicosane)Co$]^{3+}$ (^{13}C),[938] $[Co\{OC_6H_4CH{=}N(CH_2)_mNH(CH_2)_nNH(CH_2)_pN{=}CHC_6H_4O\}]^+$ (^{13}C),[939] $Co(dmgH)_2(N_3)$(imidazole) (^{13}C),[940] $Co(dmgH)_2(SCN)$(1-Et-imidazole) (^{13}C),[941] $Co(dmgH)_2Cl$(imidazole) (^{13}C),[942] vitamin B_{12} derivatives (^{13}C),[943, 944] $Co(acac)_2(N_3)$(imidazole) (^{13}C),[945] bis{tris(L-cysteinato-N,S)cobaltate(III)-μ-S,μ-S',μ-S"}cobaltate(III) (^{13}C),[946] Co(III) complexes of ATP (^{31}P),[947] Co(III) complexes of S-(carboxymethyl)-L-cysteine (^{13}C),[948] $[Co(bipy)_2(NCS)_2]^+$ (^{59}Co),[949] $[Rh\{(+)\text{-}binap\}(MeOH)_2]^+$ (^{31}P),[950] $Rh(NHNC_6H_4R)Cl_3(PEtPh_2)_2$ (^{31}P),[951] $[Rh(oxalato)_2(Rpy)_2]^-$ (^{13}C),[952] $MCl(PF_2NMe_2)_3$ (M = Rh, Ir; ^{19}F, ^{31}P),[953] $[Rh(PPh_3)_3(MeCN)][BF_4]$ (^{13}C, ^{19}F, ^{31}P),[954] $[RhAu\{NC_5H_3(CH_2PPh_3)_2\}_2]^+$ (^{31}P),[955] $Rh(NO)(O_2CCl_3)_2(PPh_3)_2$ (^{31}P),[956] $M(O_2CR)_2(NO)(PPh_3)_2$ (M = Rh, Ir; ^{19}F, ^{31}P),[957] $[Co\{o\text{-}C_6H_4(PPh_2)_2\}_2X_2]^+$ (^{31}P),[958] $Co_6Se_8(PPh_3)_6$ (^{31}P),[959] $[Rh\{O{=}C(NMePF_2)_2\}]_2(C_2O_4)$ (^{31}P),[960] $Rh(acac)(PPh_3)_2Cl(O_2H)$ (^{31}P),[961] $[MeC(CH_2PPh_2)_3RhS_2CO]^+$ (^{31}P),[962] $[Rh(\mu\text{-}SC_6H_4F)(PPh_3)_2]_2$ (^{19}F, ^{31}P),[963] (2-6-η-2,6-Bu^t_2-4-$MeC_6H_2O)Rh(PPh_3)_2$ (^{31}P),[964] $MeC(CH_2PPh_2)_3RhCl(Se_2CO)$ (^{31}P),[965] and $RhCl_3(PEt_2Ph)_2(OH_2)$ (^{31}P).[966]

Complexes of Ni, Pd, and Pt.—The ^{195}Pt chemical shifts of 13 organoplatinum compounds in solution have been determined. The $\delta(^{195}Pt)$ values of the phosphine-Pt(II) and -Pt(O) compounds lie in separate ranges and allow the metal-diene systems to be characterised either as metallocyclopentenes or as η^2-bonded dienes. The chemical shift of $Pt(\eta-C_3H_5)_2$ lies in the Pt(O) region.[967]

The σ-donor strength of the hydridoaluminate anion and the hydride ion in some Ni(O) complexes has been compared with that of other σ-donors by ^{13}C n.m.r. spectroscopy.[968] The effect of $\underline{T}_1$ on metal hydride integrals has been investigated. In $[PtH(PEt_3)_3]^+$, the hydride proton relaxation is faster than the other ligand relaxation.[969] N.m.r. data have also been reported for $[Et_3Al\text{-}\mu\text{-}H\text{-}Ni(cdt)]^-$ (^{13}C),[970] $Pt_2H_2(PBu^t_3)_2(\mu\text{-}SPPh_2)_2$ (^{31}P),[971] $[Pt(\overline{COCH_2CH_2O})H(PPh_3)_2]^+$ (^{31}P),[972] $[Pt_2H_2(\mu\text{-}dppm)_3]^{2+}$ (^{31}P),[973] $[Pt_2(\mu\text{-}H)(\mu\text{-}CH\text{-}CH_2Ph)(dppe)_2]^+$ (^{13}C),[974] and $(C_9H_{14})PtCl(\mu\text{-}Cl)_2PtHCl$ (^{13}C, ^{195}Pt).[975]

The ^{31}P n.m.r. spectrum of $Pd_2ClMe(\mu\text{-}dppm)_2$ is $[AB]_2$ at low temperature.[976] It is possible to determine between dppm acting as a chelating or non-chelating ligand. In $PtR_2(dppm)$ the ^{195}Pt chemical shifts are 600 p.p.m. to high frequency of cis-$PtR_2(dppm)_2$, while the ^{13}C chemical shifts are δ 3.9 to 4.5 in the bridge and d 1.8 to 3.0 when terminal. The ^{31}P n.m.r. spectra were also reported.[977] ^{31}P, ^{119}Sn, and ^{195}Pt n.m.r. spectra have been recorded for $Pt(SnCl_3)R(PEt_3)_2$. It was found that $^1\underline{J}(^{195}Pt,^{119}Sn)$ lies between 2376 (R = COPh) and 11895 Hz (R = C_6Cl_5). For aryl complexes there is a linear correlation between $\delta(^{119}Sn)$, $^1\underline{J}(^{195}Pt,^{119}Sn)$, $^1\underline{J}(^{195}Pt,^{31}P)$, $^2\underline{J}(^{119}Sn,^{31}P)$, and Hammett substituent constants.[978] N.m.r. data have also been reported for $Ni(CO)_2\{B_2H_4(PMe_3)_2\}$ (^{11}B),[979] $[MeNi(C_2H_4)_3]^-$ (^{13}C),[980] $Ni(CH_2CH{=}CHCHMeCHMeC_3H_4\text{-}\eta^3)(PCy_3)$ (^{13}C, ^{31}P),[981] (Ph_3P)2-5-Ni-3,3,7,7-Me_4-trans-tricyclo[4.1.0.0.^{2,4}]heptane (^{31}P),[982] $R_3PCH_2Ni(CO)_3$ (^{13}C, ^{31}P),[983] $Ni(CH_2PMe_3)Ph(OCPh{=}CHPPh_2)$ (^{13}C, ^{31}P),[984] $Ni\{CH_2CH_2C(O)NR\}(dppe)$ (^{13}C, ^{31}P),[985] $Ni(CH_2C_6H_4Me)Cl(PMe_3)_2$ (^{13}C),[986] $Ni_2(CBu^t{=}CHBu^t)_2(CO)_2(\mu\text{-}Cl)_2$ (^{13}C),[987] $Ni(C_2Cl_3)(C_6H_4Me\text{-}2)(PMe_2Ph)_2$ (^{31}P),[988] $(MeOC_{10}H_6)Ni(PMe_2Ph)_2Cl$ (^{13}C, ^{31}P),[989] $[(C_5H_5)Ni(C_6F_5)_2]^-$ (^{13}C),[990] $(Et_3P)_2Ni(CONEt_2)I$ (^{31}P),[991] $\{Pd(CH_2CMe_2PBu^t_2)Cl\}_2(dppm)$ (^{31}P),[992] $Pd(CH_2X)Cl(CH_2{=}CHCMe_2CH_2NMe_2)$ (^{13}C),[993] $Pd_2(CH_2CH_2CMe{=}NOR)_2(\mu\text{-}Cl)_2$ (^{13}C),[994] $Me_2NCH_2CMe_2CH_2Pd(\mu\text{-}OAc)_2Pd(\mu\text{-}OAc)_2PdCH_2CMe_2CH_2NMe_2$ (^{13}C),[995] $\overline{Pd\{C(CO_2R)_2CH_2C_5H_3NC_5H_3NCH_2C}(CO_2R)_2\}$ (^{13}C),[996] $Pd\{C_{10}H_{12}(CO_2Me)_2\}(PR_3)$ (^{13}C, ^{31}P),[997] $Pd(CH{=}CHCO_2R)X(PPh_3)_2$ (^{31}P),[998] Pd{C-

(C_6F_5)=NMe}Cl(CNMe)$_2$ (^{19}F),[999] $Pd_2(Ph_2PC_6F_4C$=NMe)$_2Cl_2$ (^{19}F),[1000] $Pd(C_6H_4CH_2NMe_2)(R_2NCH_2CO_2)$ (^{13}C, ^{195}Pt),[1001] $Pd(C_6H_2R^1R^2CH_2NHR^3)$-$(PPh_3)$Br (^{13}C),[1002] $M(C_6H_4CH_2PPh_2)(C_6H_4CH_2NMe_2)$ (M = Pd, Pt; ^{31}P),[1003] PdClC(=NBu^t){C_5H_3N(6-Cl)}(dppe) (^{31}P),[1004] $[Pd(C_4H_3N_2)$-$(CNR)(PPh_3)_2$ (^{31}P),[1005] $[Pd(C_6F_5)(PR_3)_2(Ph_3PCHCOMe)]^+$ (^{31}P),[1006] (dppe)Pd($SnMe_2$Cl)(C_2R) (^{13}C, ^{31}P, ^{119}Sn),[1007] $Pd_2(\mu\text{-}C_2)(PMe_3)_4Cl_3$ (^{13}C, ^{31}P),[1008] $M\{Sn(NR_2)_2\}_3$ (M = Pd, Pt; ^{119}Sn, ^{195}Pt),[1009] $M_3\{\mu$-$M'(NR_2)_2\}(CO)_3$ (M = Pd, Pt; Ge, Sn; ^{195}Pt),[1010] Pd(CONHBu^t){$(Ph_2P$-$CH_2)_2NC_5H_3$}Cl (^{31}P),[1011] $[M(\overline{CNHCH_2CH_2O})_4]^{2+}$ (M = Pd, Pt; ^{13}C, ^{195}Pt),[1012] PtMe(CO_2Me)(dppe) (^{13}C, ^{31}P),[1013] $PtIMe_2Pr^i$(phen) (^{13}C),[1014] $[Me_3Pt(\mu\text{-}X)(\mu\text{-dmpm})_2PtMe]^+$ (^{31}P, ^{195}Pt),[1015] Me_2Pt-(CH_2X)X(phen) (^{13}C),[1016] $[Me_2Pt(\mu\text{-dppm})MMe]^+$ (M = Pd, Pt; ^{31}P),[1017] Me_3PtI(dppm) (^{31}P),[1018] $\{PtMe_2Cl(phen)\}_2C_2F_4$ (^{19}F),[1019] $PdX_2(\mu$-dpmp)Pd(dpmp)X] (^{31}P),[1020] $[(PR^1{}_3)PtX(\mu\text{-}SR^2)]_2$ (^{13}C, ^{31}P, ^{119}Sn),[1021] $(Et_3P)_2\overline{Pt(CH_2CMe_2CMe_2CH_2)}$ (^{31}P),[1022] $[Pt_2R(\mu\text{-dppm})_2$-(dppm-P)]$^+$ (^{31}P),[1023] $PtXR(PEt_3)_2$ (^{31}P),[1024] $Pt(CH_2C_5H_4N)Cl_2$(py) (^{13}C),[1025] $Pt(C_8H_{10}O)Cl_2(py)_2$ (^{13}C, ^{195}Pt),[1026] $Pt(C_{14}H_{10})(PPh_3)_2$ (^{13}C, ^{31}P),[1027] $PtPh(SC_6H_4F)(PPh_3)_2$ (^{19}F, ^{31}P),[1028] $Pt(C_6H_4X\text{-}4)_2$-$(PBu^n{}_3)_2$ (^{13}C, ^{31}P),[1029] $[Pt_2(\mu\text{-}P_2O_5H_2)_4ArX]^{4-}$ (^{31}P),[1030] $Pt_2(C_6H_3$-$ClCH{=}NCH_2)_2Cl_2(PEt_3)_4$ (^{13}C, ^{31}P),[1031] $Pt(C_6H_4X)_2(PBu^n{}_3)_2$ (^{13}C, ^{31}P, ^{195}Pt),[1032] $PtCl(C_6F_5)$(phen) (^{19}F),[1033] $Pt(C_2Ph)_2(PR_3)_2$ (^{13}C, ^{31}P),[1034] $Pt(C_2Me)_2(R_2PCH_2PR_2)$ (^{31}P),[1035] $Pt(C_2Ph)_2(PMePh_2)_2$ (^{31}P),[1038] $Pt_2R_2(\mu\text{-}MeO_2CC_2CO_2Me)(\mu\text{-dppm})_2$ (^{31}P),[1037] PtCl(COPh)-$(Ph_2PCH_2CH_2ER_n)$ (^{31}P),[1038] $Pt(CO_2H)(C_6H_9)$(dppe) (^{13}C, ^{31}P),[1039, 1040] $Pt(CO_2Me)_2$(dppe) (^{13}C, ^{31}P),[1041] $PtCl(PPh_3)_2CO_3Pt$-$(PPh_3)_2$ (^{31}P),[1042] $PtCl_2(PR_3)(=\overline{CNHC_6H_4C}{=}PR_3)$ (^{31}P),[1043] $Pt(SnCl_3)_2$-$Cl_2(PPh_3)_2$(MeOH) (^{31}P),[1044] Pt(II) complexes of $SnCl_3$ with nitrogen ligands (^{15}N, ^{31}P, ^{119}Sn, ^{195}Pt),[1045] (diop)Pt($SnCl_3$)Cl (^{31}P),[1046] $Pt(\mu\text{-Cl})(PEt_3)\{Sn(NR_2)_2Cl\}_2$ (^{31}P),[1047] and {(S,S)-chiraphos}PtCl-$(SnCl_3)$ (^{31}P).[1048]

Variable-temperature ^{1}H n.m.r. studies on cis-$PtCl_2$(p-toluidine)(olefin) have shown that the coordination of the olefin is stereoselective.[1049] A paramagnetic shift reagent analogue of tertiary phosphine ligands has been used to elucidate the structure of substituted allyl palladium complexes.[1050] The diastereomer ratio in [Pd(S,S-chiraphos)(allyl)]$^+$ has been determined by ^{31}P n.m.r. spectroscopy.[1051] ^{1}H n.O.e. measurements have been used to determine the structure of syn and anti isomers of $[\{1\text{-}(Me_3Si)\text{-}1\text{-}Me\text{-}C_3H_3\}PdCl]_2$.[1052] ^{13}C n.m.r. spectroscopy has been used to determine the stereochemistry of some $[(allyl)PdCl]_2$ complexes.[1053]

The ^{1}H chemical shift of the cyclopentadienyl group in $[(C_5H_5)Ni\{P(S)R_2\}_2]^-$ is very temperature dependent, due to the presence of a low-lying paramagnetic state.[1054] N.m.r. data have also been reported for $(dmpe)_{1+n}Ni_2(C_2H_4)_{4-n}$ (^{13}C, ^{31}P),[1055] $Ni_2(C_2H_4)_2(PPh_3)_2(C_2H_2)$ (^{13}C, ^{31}P),[1056] $Ni(\eta^2-C_6H_4)(Cy_2PCH_2CH_2PCy_2)$ (^{13}C, ^{31}P),[1057] $\{(C_5Me_5)Ni\}_2(C_8H_8)$ (^{13}C),[1058] $Ni(\eta-CH_2{=}C{=}O)(PPh_3)_2$ (^{13}C),[1059] $\{\eta^2-ClP{=}C(SiMe_3)_2\}_2Ni(CO)$ (^{13}C, ^{31}P),[1060] $Pd(OAc)_2(CH_2{=}CHCN)$ (^{13}C),[1061] $M\{\eta-O{=}C(CO_2Et)_2\}(PCy_3)_2$ (M = Pd, Pt; ^{13}C, ^{31}P),[1062] $Pt(\eta-CH_2{=}C{=}O)(PPh_3)_2$ (^{13}C),[1063] $Pt(\eta-C_6H_2Bu^t{}_2P{=}C{=}PC_6H_2Bu^t{}_3)(PPh_3)_2$ (^{31}P),[1064] $Pt(\eta^2-C_7H_6)(PPh_3)_2$ (^{13}C, ^{31}P),[1065] $[(C_3H_5)NiL_2]^+$ (^{13}C),[1066] $(\mu-C_3H_5)Ni_2(\mu-X)(PR^1R^2{}_2)_2$ (^{13}C, ^{31}P),[1067] $[(2-Me-C_3H_4)NiBr]_2$ (^{13}C),[1068] $[(allyl)PdCl]_2$ (^{13}C),[1069-1071] $Pd(\eta^3-1,1-Me_2-C_3H_3)_2$ (^{13}C),[1072] $[\{\eta^3-PhCHCHCHCH_2CH_2(CO_2Et)_2\}PdCl]_2$ (^{13}C),[1073] $[Pd(\eta^3-C_7H_{10}OMe)Cl]_2$ (^{13}C),[1074] $(\eta^3-2-MeC_3H_4)PdCl\{PPh(OCH_2CH_2)_2NR\}$ (^{31}P),[1075] $[(\eta^3-2-Me-C_3H_4)Pd(Ph_2PCH_2CH_2SR)]^+$ (^{31}P),[1076] $Pt(\eta^3-Ph-CHCPhC{=}CH_2)(PPh_3)_2$ (^{31}P),[1077] $(C_5H_5)_3ZnNi(C_5H_5)(PPh_3)$ (^{13}C),[1078] $[Ni(C_5H_5)(\mu-CNCF_3)]_2$ (^{13}C, ^{19}F),[1079] $(C_5H_5)_2Ni_2(Bu^tCP)$ (^{31}P),[1080] $(C_5H_5)Ni(C_3B_2Et_4Me)Pt(C_3B_2Et_4Me)$ (^{11}B, ^{13}C),[1081] $(C_5H_5)Pd_2(PR_3)_2-Br(\mu-SO_2)$ (^{13}C, ^{31}P),[1082] $Pd\{\eta^5-C_5H_4C(=NC_6H_4Me)Pr^i\}Cl(PREt_2)$ (^{13}C),[1083] $[Ni(CO)_2(PPh_3)(CNPr^i{}_2)]^+$ (^{13}C),[1084] $Ni_4(CNBu^t)_4(\mu_3-\eta^2-CNBu^t)\{\mu_3-\eta^2-(Me_3Si)_2NP{=}NSiMe_3\}_2$ (^{31}P),[1085] $Ni_2(\mu-P_3Pr^i{}_3)_2(CO)_4$ (^{13}C, ^{31}P),[1086] $Pd_4(CO)_5(PR_3)_4$ (^{13}C, ^{15}N, and ^{31}P),[1087] $[Pd_3(\mu-dppm)_3(\mu_3-CO)Cl]^+$ (^{31}P),[1088] $(dppe)M(\mu-E)Pt(PPh_3)(CS)$ (M = Pd, Pt; E = S, Se; ^{31}P),[1089] $PtCl_2(CO)(PR_3)$ (^{13}C),[1090] $Pt_3(\mu-CO)_3(PPr^i{}_3)_3$ (^{31}P, ^{195}Pt),[1091] $Pt_3(CO)_3(PBu^t{}_2Ph)_3$ (^{31}P),[1092] and $Pt_3(\mu-CO)(\mu-CNC_8H_9)_2(CNC_8H_9)(PCy_3)_2$ (^{31}P, ^{195}Pt).[1093]

The ^{1}H and ^{13}C n.m.r. spectra of Ni complexes of 1,8,9,20-tetrahydro-3,6-Me$_2$-1,8-diphenyldibenzo[*c*,*m*]dipyrazolo[3,4-*f*:4',3'-*j*][1,2,5,8,9,12]hexaazacyclotetradecine have indicated the presence of two isomers.[1094] ^{13}C n.m.r. spectroscopy has been used as a probe of the steric restriction to dioxygen binding for a series of polymethylene-bridged lacunar macrobicyclic complexes of Ni(II).[1095] In the ^{195}Pt n.m.r. spectrum of $Pt(2-Br-1,3,4-thiadiazole)_2Cl_2$ $^{1}J(^{195}Pt,^{14}N)$ is observed as 230 Hz. The ^{13}C n.m.r. spectrum was also recorded.[1096] ^{1}H n.m.r. spectroscopy has been used to show that $Pt(NH_3)_2(O_2C)_2C_4H_6$ forms a 1:1 adduct with α-cyclodextrin.[1097] The ^{1}H and ^{13}C n.m.r. spectra of platinum deuterioporphyrin IX dimethyl ester have been fully assigned.[1098] ^{13}C and ^{195}Pt T_1 has been determined for $[(py)PtCl_3]^-$. The correlation times were determined and calculated on the basis of a two-site random jump model.[1099] A method for determining enantiomer excess of an organic compound using ^{31}P n.m.r. spectroscopy has been described and applied to $Ni(Ph_2PCH_2CHMeS)_2$.[1100] The ^{31}P n.m.r. spectra of $Pd(PR_3)_2Cl_2$ have been determined. A correlation was found between $\delta(^{31}P)$ and steric parameters of tertiary phosphines. Consequently $\delta(^{31}P)$ is a useful measure of the steric bulkiness of phosphines.[1101] Atropism due to hindered rotation around the Pt-PPh_3 bond of $[Pt_2(\mu\text{-dppm})_2Cl(PPh_3)]^+$ has been observed by ^{31}P n.m.r. spectroscopy.[1102] $^2\underline{J}(^{195}Pt,^{195}Pt)$ has been determined as 7750 Hz in $Pt_2Cl_2(\mu\text{-}Ph_2PCH_2SMe)_2$ using ^{13}C and ^{31}P n.m.r. spectroscopy.[1103] The ^{195}Pt n.m.r. parameters of several Pt(O) complexes have been illustrated. In PtL_2 complexes the relaxation is dominated by chemical shift anisotropy.[1104] N.m.r. data have also been reported for Ni(II) and Zn(II) complexes of substituted pyridines (^{13}C),[1105] Zn(II), Cd(II), and Ni(II) complexes of 6-methylpyridine-2-phosphonic acid (^{13}C, ^{31}P),[1106] Ni(II), Zn(II), and Cd(II) complexes of tetraazaalkanes (^{13}C),[1107] Ni(II) complexes of 7-amino-4-methyl-5-aza-3-heptene-2-one (^{13}C),[1108] $Ni(OCMeCRCRNCH_2)_2$ (^{13}C),[1109] Pd(II) and Pt(II) complexes of 5-alkyl-2-thiopyrrolidinones (^{13}C),[1110] $\overline{M(O_2CCMe{=}N}CH_2COR)_2$ (M = Pd, Pt; ^{13}C, ^{195}Pt),[1111] Pd(II) complexes of glycyl-L-histidyl-L-lysine (^{13}C),[1112] $[Pd(PMe_2Ph)_3(2,9\text{-}R_2\text{-phen})]^{2+}$ (^{31}P),[1113] $PdCl(C_{10}H_7N_2)(PBu^n_3)$ (^{13}C),[1114] $Pd_2\{\mu\text{-}N_2C_4H_3N\text{-}(SO_2C_6H_4NH_2)\}_2Cl_2(PBu^n_3)_2$ (^{31}P),[1115] $PdX_2\{Ph_2P(CH_2)_2CN\}_2$ (^{13}C, ^{31}P),[1116] Pd(II)-thiaproline systems (^{13}C),[1117] Pd(2-pivaloylpyridine)$_2$ (^{13}C),[1118] $PtCl_2(NH_3)_2$ complexes with glycine (^{13}C),[1119, 1120] Pt(II)-L-histidine complexes (^{13}C),[1121] $PtCl_2(NH_3)_2$ complexes with DNA bases (^{13}C, ^{15}N, ^{195}Pt),[1122] (^{31}P),[1123-1125] (^{13}C),[1126] Pt(IV) complexes with amino acids (^{15}N, ^{195}Pt),[1127] $[\overline{PtCl_2(O_2CCH_2N}RCH_2CO_2H)]$ (^{13}C),[1128] Pt(II) complexes with (methylimino)diacetic acid (^{13}C, ^{15}N, ^{195}Pt),[1129] $Pt(C_3N_2\text{-}R^1R^2NO_2)_2X_2$ (^{195}Pt),[1130] Pt{cis-$(H_2N)_2C_6H_{10}$}(ascorbate) (^{13}C, ^{195}Pt),[1131] $Pt(S_2N_2)(PPh_3)_2$ (^{31}P),[1132] $(pyPh_2PO)PtBr_4$ (^{31}P),[1133] Ni(II) analogue of cobyrinic acid (^{13}C),[1134] $(pyPtCl_2)_2(\mu\text{-}SMe_2)$ (^{13}C),[1135] $PtCl_2$(8-R-quinoline)(PEt_3) (^{31}P),[1136] Pt(II) complexes of tripeptides (^{13}C, ^{195}Pt),[1137] $[Pt_2(P_2O_5H_2)_4(NO_2)_2]^{4-}$ (^{31}P),[1138] $Pt(NSNSiMe_3)_2(PPh_3)_2$ (^{31}P),[1139] $MCl_2(Ph_2PC_6H_4NHCH_2Ph)$, (M = Pd, Pt; ^{13}C, ^{31}P),[1140] (dppe)$Ni(\eta^2\text{-}RN{=}PNR_2)$ (^{31}P),[1141] Pt(II) complexes of octaethylporphyrin (^{13}C),[1142] Pt complexes of S-alkyl cysteine complexes (^{13}C),[1143] $\overline{Pt(NCOCH_2CH_2C}O)(OOH)(PPh_3)_2$ (^{31}P),[1144] Pt(II) complexes of guanosine (^{195}Pt),[1145] $\overline{M\{Ph_2PCH_2C(CF_3)_2O}\}_2$ (M = Ni, Pd, Pt; ^{19}F, ^{31}P),[1146] $[Ni\{S[(CH_2)_{\underline{n}}PPh(CH_2)_2PPh(CH_2)_2]_2S\}]^{2+}$

(^{31}P),[1147, 1148] $[Ni_3(\mu_3-S)_2(OH_2)(PPh_3)_5]^{2+}$ (^{31}P),[1149] $M(C_8Me_4H_6-P_2Ph_2)Cl_2$ (M = Ni, Pd, Pt; ^{13}C, ^{31}P),[1150] $Ni(\eta-Bu^tP=PBu^t)(P_4Bu^t_4)$ (^{31}P),[1151] $Ni(\eta^2-R^1P=PR^1)(R^2_2PCH_2CH_2PR^2_2)$ (^{31}P),[1152] $Pd_2Cl_2(\mu-SO)(\mu-dppm)_2$ (^{31}P),[1153] $Pd_2Cl_2(dmpm)_2$ (^{13}C, ^{31}P),[1154] $PdCl_2\{PhP(R_2P)-C_6H_7\}$ (^{13}C, ^{31}P),[1155] Pt(II) complexes with $C_6H_4(PPhCH_2CH_2CH_2E)_2-C_6H_4$ (E = PPh, AsMe, S, O, NMe; ^{31}P),[1156] $(Et_3P)ClPd(\mu-Cl)_2PtCl(PEt_3)$ (^{31}P, ^{195}Pt),[1157] $(Ph_3P)_2\overline{Pt\{O_2C(CH_2)_nCO_2}\}$ (^{13}C, ^{31}P, ^{195}Pt),[1158] $[Pt_3(\mu-SO_2)_3(PCy_3)_3]_2$ (^{31}P, ^{195}Pt),[1159] $(R_3P)_2Pt(S_2CH_2)$ (^{31}P),[1160] $PtCl_2(PEt_3)(PR_2Cl)$ (^{31}P, ^{195}Pt),[1161] $Pt(C_2O_4)(PR_3)_2$ (^{13}C, ^{19}F),[1162] $PtCl_2(PCy_3)\{(PhCH_2)_2SO\}$ (^{31}P),[1163] $Cl_2Pt(PBu^t_2)_2Te$ (^{31}P),[1164] $[PtPd_2Cl(PPh_2)_2(PPh_3)_3]^+$ (^{31}P, ^{195}Pt),[1165] $Pt(PR_3)_n$ (^{13}C, ^{31}P, ^{195}Pt),[1166] $Pt_2Cl_4(\mu-Bu^tPHCH_2PHBu^t)_2$ (^{31}P),[1167] $[Pt\{Ph-(C_6H_4NH_2-o)PCH_2\}_2CH_2]^{2+}$ (^{31}P),[1168] $[Pt[\{Ph(C_6H_4NH_2-o)PCH_2\}_2CH_2]Cl]^+$ (^{31}P),[1169] $Pt_2(\mu-R_2PCH_2PR_2)_3(PPh_3)_n$ (^{31}P),[1170] (dppe)Pt(μ-S)Pt-$(PPh_3)(CS)$ (^{31}P),[1171] $PtCl_2\{P(OPh)(OC_6H_4Cl)(OC_6H_4Me)\}_2$ (^{31}P),[1172] $MeC(CH_2PPh_2O)_3Ni(S_2CS)$ (^{31}P),[1173] $M(RSeCH_2CH_2SeR)X_2$ (M = Pd, Pt; ^{77}Se, ^{195}Pt),[1174] $[Pt\{MeC(CH_2SeMe)_3\}Cl]^+$ (^{77}Se, ^{195}Pt),[1175] $[Pt_2(P_2O_5H_2)_4]^{4-}$ (^{31}P),[1176] $[Pt_2(\mu-PO_4H)_4(OH_2)_2]^{2-}$ (^{31}P),[1177] and $[Pt_2(PO_4H)_4XY]^{4-}$ (^{13}C, ^{31}P, ^{195}Pt).[1178]

Complexes of Cu, Ag, and Au.—A review entitled 'What can we obtain from n.m.r. spectra of diamagnetic metallocomplexes in solution? Characterisation of reactive copper(I) complexes by ^1H and ^{13}C n.m.r. spectroscopy' has appeared.[1179] Experimental ^{109}Ag n.m.r. spectroscopy and theoretical M.O. methods have been used to investigate the electronic structure of silver complexes in aqueous solution.[1180] N.m.r. data have also been reported for [Cu(phen)-$(PPh_3)_2][HBO_2CH]$ (^{11}B, ^{13}C),[1181] and $[HCuP(NMe_2)_3]_n$ (^{31}P).[1182]

The ^{199}Hg n.m.r. spectrum of $[AgHg(\mu-dppm)_2]^+$ yields $J(^{199}Hg, ^{109}Ag)$ = 60 Hz. The ^{31}P n.m.r. spectrum was also recorded.[1183] N.m.r. data have also been reported for $MeCuPCy_3$ (^{13}C, ^{31}P),[1184] $MeCuPBu^t_3$ (^{13}C, ^{31}P),[1185] $AuMe_2ArL$ (^{13}C),[1186] $Me_2AuS_2CNR_2$ (^{13}C),[1187] $Au_2(\mu-CH_2PR_2CH_2)_2MeX$ (^2H, ^{31}P),[1188] $AuR(PPh_3)$ (^{13}C),[1189] $[Au\{CH(PPh_3)CO_2Et\}(PPh_3)]^+$ (^{31}P),[1190] $X(C_6F_5)_2Au\{Ph_2PCH(AuSC_4H_8)-PPh_2\}AuX$ (^{19}F, ^{31}P),[1191] $[M(C_2Ph)_2]^-$ (M = Ag, Au; ^{13}C),[1192] $Br_3Au=C(OEt)NMe_2$ (^{13}C),[1193] and $Cu\{C_5(CO_2Me)_5\}(PPh_3)_2$ (^{13}C).[1194]

In $[Ag\{(RC_4H_2SCH=N)_2C_6H_{10}\}_2]^+$, the ^{15}N n.m.r. spectrum indicates that all four nitrogen atoms are coordinated, but ^{109}Ag n.m.r. spectroscopy indicates that the sulphur atoms are not coordinated.[1195] ^{31}P n.m.r. data indicate that in $CuI(PH_2Ph)$ the PH_2Ph

remains coordinated but is exchanging rapidly.[1196] By the use of $^{109}Ag-\{^{31}P\}$ INEPT n.m.r. spectroscopy the ^{109}Ag n.m.r. spectra of some chelated tetrahedral bis(tertiary phosphine) silver(I) complexes have been recorded.[1197] The ^{31}P n.m.r. spectrum of $Ag_2(mnt)-(PPh_3)_4$ shows $^1\underline{J}(^{107,109}Ag, ^{31}P)$.[1198] ^{31}P n.m.r. spectra of a series of gold phosphine cluster compounds have been determined.[1199] ^2H and ^{13}C n.m.r. spectroscopy has been used to study the antiarthritic drug 'Ridaura'.[1200] N.m.r. data have also been reported for $[Ag_2\{\mu-(\underline{R})(\underline{S})-1,2-(thiophene-2-CH=N)_2cyclohexane\}_2]^{2+}$ (^{109}Ag),[1201] Ph_3PAu complexes with nucleosides (^{13}C),[1202] AuX_3-(alkylpyrazine) (^{13}C),[1203] $MeC(CH_2PPh_2)_3CuS_2CH$ (^{31}P),[1204] $CuBr\{P(mesityl)_3\}$ (^{13}C, ^{31}P),[1205] $MeC(CH_2PPh_2)_3Cu\{SC(O)H\}$ (^{13}C, ^{31}P),[1206] $Ag_2(NO_3)_2\{P(OPh)_3\}_4$ (^{31}P, ^{109}Ag),[1207] $Au(NCO)(PPh_3)$ (^{31}P),[1208] $Au_2Cl_2\{(Ph_2P)_2C=CH_2\}_2$ (^{13}C, ^{31}P),[1209] $[Cu_4\{\underline{o}-(SCH_2)_2-C_6H_4\}_3]^{2-}$ (^{13}C),[1210] $[(Bu^tO)_3SiSAg]_4$ (^{13}C, ^{29}Si),[1211] AuCl($\underline{N}$-R-substituted-1,3-imidazolidine-2-thione) (^{13}C),[1212, 1213] $AgOTeF_5$ (^{19}F),[1214] and $Au(OTeF_5)_3$ (^{19}F).[1215]

Complexes of Zn, Cd, and Hg.—^{13}C isotope effects on $\delta(^{199}Hg)$ have been measured for a wide variety of compounds. Shifts of up to 0.38 p.p.m. were observed.[1216]

The ^{13}C n.m.r. spectrum of Pr^i_2Zn shows a single environment for the methyl group, while $(MeEtCH)_2Zn$ is diastereomeric.[1217] The technique of bi-linear rotation pulse application in heteronuclear two-dimensional n.m.r. chemical shift correlation spectroscopy has been used to study Ph_4Pb and PhHgCl. The $\underline{J}(M,^{13}C)$ and $\underline{J}(M,^1H)$ were determined, including their relative signs.[1218] N.m.r. ^2H isotope shifts have been examined as a probe of electronic structure and applied to $[(CD_3)_2CC(CH_3)_2Hg]^{2+}$.[1219] ^{13}C and ^{199}Hg n.m.r. spectra indicate σ,π-conjugation of the C-Hg with the π-orbitals of the aromatic ring and intramolecular coordination of the mercury to the bromine.[1220] A comparison of the ^{19}F n.m.r. spectra of $RHgC_6H_4F$-$\underline{p}$ and $RCH_2C_6H_4F$-$\underline{p}$ shows that the bridging hydrogen is less effective as a transmitter of the electronic effects of the substituents than the CH_2 groups.[1221] A similar study has been performed on $ArHgE-C_6H_4F$ and $Ar_3SnEC_6H_4F$, E = CH_2, S, SO_2Ph, using ^{19}F n.m.r. spectroscopy.[1222] ^{19}F and ^{199}Hg n.m.r. spectroscopy has been used to examine the electron-withdrawing character of R^1 and R^2 in $R^1HgC_2-C_6H_4F$-4 and $4-FC_6H_4HgC_2R^2$.[1223] ^1H n.m.r. spectroscopy has been used to investigate the complexation of $[MeHg]^+$ by some dithiols.[1224] N.m.r. data have also been reported for $Zn(CF_3)_2$

(^{19}F),[1225] $MeHg\{S_2CN(CH_2CH_2OH)_2\}$, $M\{S_2CN(CH_2CH_2OH)_2\}_2$ (M = Zn, Cd, Hg; ^{13}C),[1226] $(Me_2PBH_2PMe_2CH_2)_2M$ (M = Zn, Cd; ^{13}C, ^{31}P),[1227] $(Me_2NCH_2CH_2CH_2)_2M$ (M = Cd, Hg; ^{13}C),[1228] CF_3CdX (^{19}F, ^{113}Cd),[1229] $Cd(SiF_3)_2$ (^{19}F),[1230] $\{CF_3SO_3CMe{=}C(OSO_2CF_3)\}_2Hg$ (^{19}F),[1231] Ar_2Hg (^{13}C, ^{199}Hg),[1232] $[MeHg(guanosine)]^+$ (^{13}C),[1233] $[MeHg]^+$ complexes with thymidine and guanosine (^{13}C),[1234] $(MeHg)_2$(xanthosine) (^{13}C),[1235] $RHg\{NC_4(SCF_3)_4\}$ (^{19}F),[1236] $RHgN(SO_2Ph)(C_6H_3FBr)$ (^{19}F),[1237] $(CF_3CO_2Hg)_2C_6H_4$ (^{199}Hg),[1238] $HgCl(NC_5H_4)$ (^{199}Hg),[1239] (1) (^{199}Hg),[1240] $MeCO_2(CH_2)_3HgBr$ (^{13}C),[1241] and $MeCH(HgBr)CH_2CH_2$-

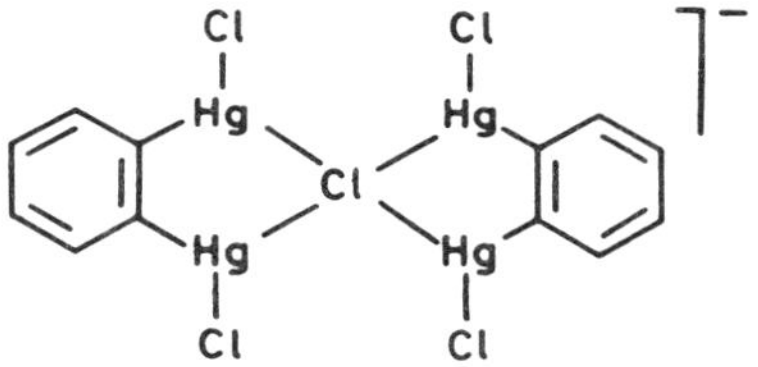

(1)

NMe_2 (^{13}C).[1242]

Molecular distortions in the solid- and solution-state ^{113}Cd n.m.r. spectra of the piperidine adduct of (tetraphenylporphyrin)Cd have been investigated. The ^{13}C n.m.r. spectrum was also determined.[1243] A phase-switching modification on the low-powered decoupler unit of a Bruker CXP-200 n.m.r. spectrometer has been described. Illustrative results on ^{113}Cd-^{1}H shift-correlated two-dimensional n.m.r. spectrum of $[Cd(edta)]^{2-}$ were presented.[1244] Long-range $\underline{J}(^{113}Cd,^{1}H)$ have been employed in the ^{1}H detected two-dimensional shift correlation for ^{1}H-^{113}Cd in $[Cd(edta)]^{2-}$, Cd_6-metallothionein,[1245-1247] and plastocyanin.[1248] Metallothionein has also been investigated using two-dimensional $\underline{J}$-resolved ^{113}Cd n.m.r. spectra.[1249] ^{31}P and ^{199}Hg n.m.r. spectroscopy has shown that $Hg\{N(SiMe_3)_2\}_2$ exists in solution both as cyclic dimers and as a species containing two equivalent ligands bound via phosphorus to mercury.[1250] For $HgX_2(PR_3)$, R = 2-furyl or 2-thienyl, comparison of $^{1}\underline{J}(^{199}Hg,^{31}P)$ indicates that both triheterophosphines are weaker σ-donors towards mercury than is PPh_3.[1251] In $[Hg_2(\mu\text{-dppm})_2(dmso)_4]^{4+}$ the ^{31}P and ^{199}Hg n.m.r. spectra yield $\underline{J}(^{199}Hg,^{199}Hg)$ = 601 to 1061 Hz.[1252] ^{31}P and ^{199}Hg n.m.r. spectroscopy has been used to investigate complexation of Hg^{2+} by $Ph_2PCHRPPh_2$ and 2:2, 1:2, and 1:3 complexes identified.[1253] ^{13}C and ^{14}N n.m.r. spectroscopy has been used to investigate $[M(SCN)_4]^{2-}$, M = Zn, Cd, Hg, in solution. It was found that the structures are $[Zn(NCS)_4]^{2-}$, [Cd-

$(NCS)_2(SCN)_2]^{2-}$, and $[Hg(SCN)_4]^{2-}$.[1254] The ^{113}Cd $\underline{T}_1$ of $Cd(acac)_2$ in DMSO and $\underline{d}_5$-pyridine has been measured as a function of temperature at three magnetic field strengths, and separated into the chemical shift anisotropy, dipole-dipole, and spin rotation terms.[1255] $^2\underline{J}(^{113}Cd,^{111,113}Cd)$ has been measured in the ^{113}Cd n.m.r. spectra of polycadmium thiolate/chalcogenide aggregates. ^{13}C n.m.r. spectra were also measured.[1256] ^{113}Cd n.m.r. spectroscopy has been used to show the formation of intermediates $[Cd_4(EPh)_{10-\underline{n}}X_{\underline{n}}]^{2-}$ when $[Cd_4(EPh)_{10}]^{2-}$ is mixed with $[Cd_4(EPh)_6X_4]^{2-}$.[1257] N.m.r. data have also been reported for $\{CH_2(CH_2NHCH_2CH_2NHCH_2)_2CH_2\}Zn(OCO_2R)(ClO_4)$ (^{13}C),[1258] $[M\{Me_2NCH_2CH_2N[(CH_2)_3NHCH_2)_3]_2NCH_2CH_2NMe_2\}]^{2+}$ (M = Zn, Cd; ^{13}C),[1259] $MX_2(C_{16}H_{28}N_6)$ (M = Zn, Cd; ^{13}C),[1260] $[Zn\{RN(CH_2CH_2NRCH_2CH_2CH_2)_2NCH_2CH_2NMe_2\}]^{2+}$ (^{13}C),[1261] Zn^{2+} complexes of cyclo($\underline{L}$-methionyl-$\underline{L}$-histidyl) (^{13}C),[1262] $[Zn\{C_5H_3N[CH_2N(CH_2Ph)CH_2CH_2CH_2]_2CH_2C_5H_4N\}]^{2+}$ (^{13}C),[1263] $Zn(OH_2)_4$(Me-5'-inosine monophosphate)$_2$ (^{31}P),[1264] [Zn(cyclo-$\underline{L}$-histidyl-$\underline{L}$-histidyl)$_2]^{2+}$ (^{13}C),[1265] (theophyllinato)$_2(H_2O)_4Cd$ (^{13}C),[1266] cadmium peptide complexes (^{13}C, ^{113}Cd),[1267] parvalbumin (^{13}C, ^{113}Cd),[1268] metallothionein (^{113}Cd),[1269] yeast enolase (^{113}Cd),[1270] $[Hg(O_2CCF_3)_2\{P(mesityl)_3\}]_2$ (^{31}P),[1271] $Hg\{P(O)(OEt)_2\}_2$ (^{31}P),[1272] Zn^{2+} and Hg^{2+} complexes of dithiophosphate (^{31}P),[1273] $Zn_2\{S_2P(OC_6H_4C_9H_{19})_2\}_3(OH)$ (^{31}P),[1274] $[ME_{\underline{n}}]^{\underline{Y}-}$ (M = Hg, Cd, Sn, Tl; E = Se, Te; ^{77}Se, ^{113}Cd, ^{119}Sn, ^{125}Te, ^{199}Hg, $^{203,205}Tl$),[1275] $Hg[\{R_2P(S)\}_2\{Ph_2P(S)\}C]X$ (^{31}P),[1276] and $\{(Bu^tO)_3SiS\}_2Hg$ (^{13}C, ^{29}Si).[1277]

3 Dynamic Systems

This section is in three main parts: (i) 'Fluxional Molecules', dealing with rate processes involving no molecular change, (ii) 'Equilibria', dealing with the use of n.m.r. spectroscopy to measure the position of equilibria and ligand-exchange reactions, including solvation, and (iii) 'Course of Reactions', dealing with the use of n.m.r. spectroscopy to monitor the course of reactions. Each section is ordered by the Periodic Table.

Five relevant reviews have appeared: 'Probing ligand substitution reactions in non-aqueous solvents by n.m.r. methods: stopped-flow and high-pressure flow Fourier transform n.m.r. and natural abundance ^{17}O n.m.r. line broadening',[1278] 'The structure of carbanions and organolithium reagents in solution',[1279] 'The structure and dynamic behaviour of organolithium compounds in solution', which includes ^{6}Li, ^{7}Li, and ^{13}C n.m.r. data,[1280] 'Fluxional organogold complexes',[1281] and 'Studies of ion-molecule interaction by n.m.r. spectroscopy'.[1282]

Fluxional Molecules.—A permutational analysis has been given of isomerism and rearrangements of cis-bis(chelate) complexes of the type $M(AB)_2XY$. Even in the absence of signal assignment, use of two-dimensional n.m.r. spectroscopy gives the number of modes of rearrangements and the type of exchange pattern responsible for observed isomerisations in the $M(AB)_2XY$ systems. The merits of both one-dimensional and two-dimensional n.m.r. spectroscopy have been discussed in the light of the dynamic stereochemistry of $M(AB)_2XY$ systems.[1283]

Beryllium. $\Delta G^{\ddagger}$ has been derived for the enantiomerisation of Be(substituted N-Pr^{i}-salicylaldiminate)$_2$.[1284]

Magnesium. ^{1}H and ^{13}C n.m.r. spectroscopy suggests that $Ar^{1}Ar^{2}$-$Mg(thf)_2$ are fluxional.[1285] Low-temperature ^{1}H and ^{13}C n.m.r. spectra of $(C_5H_5)Mg$(2-Me-allyl) show that the allyl group is σ-bonded. The activation energy for fluxionality was determined and the mechanism discussed. The ^{25}Mg n.m.r. spectrum was also measured.[1286]

Scandium. ^{1}H and ^{45}Sc n.m.r. studies of $(C_5H_5)_3Sc$ in thf have shown that it exists as one fluxional species with two fluxional processes. The high-energy process is exchange between σ- and π-bonded rings. The low-energy process is a [1,5]-shift. In toluene the exchange is complicated by dimer formation, and variable-temperature ^{45}Sc n.m.r. spectra show two species.[1287]

Thorium and Uranium. At low temperatures ^{13}C n.m.r. spectra show that, in $(C_5Me_5)ThXRu(C_5H_5)(CO)_2$, rotation about the Th-Ru bond can be frozen out.[1288] Variable-temperature ^{17}O n.m.r. spectroscopy has been used to show oxygen exchange in $[\{(C_5H_5)_2U\}_2(\mu\text{-O-TiW}_5\text{-}O_{19})_2]^{4-}$.[1289] ^{1}H n.m.r. spectroscopy has been used to determine the activation energy for ring inversion in $\{(Me_3Si)_2N\}_2M(\mu\text{-}CH_2)\text{-}(\mu\text{-}NSiMe_3)SiMe_2$, M = Th, U,[1290] and for C-N bond rotation in $[UO_2\{1,1\text{-}(MeNH)_2CO\}_5]^{2+}$.[1291]

Titanium, Zirconium, and Hafnium. ^{1}H, ^{11}B, and ^{31}P n.m.r. spectra have been used to show that rotation occurs at both ends of [Hf{N-$(SiMe_2CH_2PMe_2)_2$}]$_2$(μ-H)$_4$(BH_4)$_2$.[1292] ^1H n.O.e. difference n.m.r. spectroscopy has been used to determine the structure of (C_5H_5)Zr-(η-allyl)(η-C_4H_6). A $\Delta \underline{G}^‡$ of 19 kcal mol^{-1} was derived for the automerisation reaction of the allyl ligand.[1293] Zr($C_5H_4Bu^t$)$_2$-(C_4H_6) undergoes two fluxional processes in solution. Rotation of the two bulky $Bu^tC_5H_4$ rings is frozen out at low temperatures, with $\Delta \underline{G}^‡$ = 9.8 kcal mol^{-1}, while the zirconacyclopentene moiety undergoes ring inversion with $\Delta \underline{G}^‡$ = 13.5 kcal mol^{-1}. Rotamers of (Bu^t-C_5H_4)$_2$MCl_2, M = Zr, Hf, are not frozen out on the ^{13}C n.m.r. time scale down to -100 °C.[1294] [2,3-Bis(methylene)bicyclo[2.2.2]octane]zirconocene exhibits dynamic n.m.r. spectra due to diene ring flipping with $\Delta \underline{G}^‡$ = 7.1 kcal mol^{-1}.[1295] The ^1H n.m.r. spectra of M(OAr)$_3$Cl, M = Ti, Zr, Hf, show restricted rotation about the M-O-Ar bonds.[1296]

<u>Niobium</u>. In [(C_5H_5)$_2$Nb($S_2C_2R_2$)]$^+$ the cyclopentadienyl groups are inequivalent due to the NbS_2C_2 ring being non-planar. The activation energy for ring inversion was determined by ^1H and ^{13}C n.m.r. spectroscopy.[1297]

<u>Tantalum</u>. ^1H n.m.r. spectroscopy has been used to investigate TaS_2C_2 ring inversion in (C_5Me_5)Ta($S_2C_2H_2$)$_2$.[1298] According to ^{13}C and ^{31}P n.m.r. spectroscopy, TaX(CO)$_3$(PMe_3)$_3$ is fluxional at room temperature and frozen out on cooling.[1299]

<u>Chromium, Molybdenum, and Tungsten</u>. Cr{P(OMe)$_3$}$_5H_2$ is fluxional. The ^1H and ^{31}P n.m.r. lineshape changes were analysed completely. The most probable mechanism is closely related to the Berry process.[1300] The ^{13}C n.m.r. spectrum of (C_5H_5)MoH(CO)$_3$ shows one ^{13}CO signal at room temperature and two at -86 °C. $\Delta \underline{G}^‡$ was determined.[1301] The ^1H and ^{31}P n.m.r. spectrum of [WH_5Li(PMe_3)$_3$]$_4$ is temperature dependent.[1302] Cr(CO)$_2$(C_7H_{10})L has an agostic hydrogen interaction. The ^1H, ^{13}C, and ^{31}P n.m.r. spectra were recorded and $\Delta \underline{G}^‡$ determined for exchange across the ring and hydrogen transfer.[1303] The variable-temperature n.m.r. spectra of (2) shows hindered methyl group rotation and an independent 1,5-sigmatropic hydrogen shift. $\Delta \underline{G}^‡$ was determined.[1304]

(3) shows fluxional behaviour <u>via</u> a bridging carbene with $\Delta \underline{G}^‡$ = 18 kcal mol^{-1}.[1305] ^1H, ^{13}C, and ^{31}P n.m.r. spectra of (4) show fluxionality of the vinyl ligand.[1306] The ^{13}C n.m.r. spectrum of

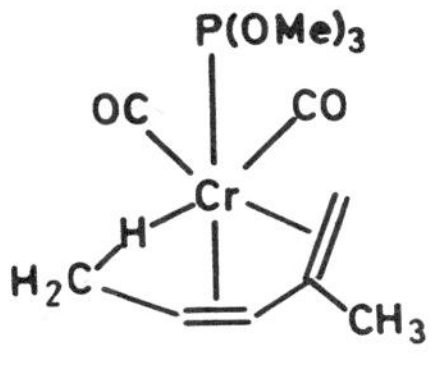

(2)

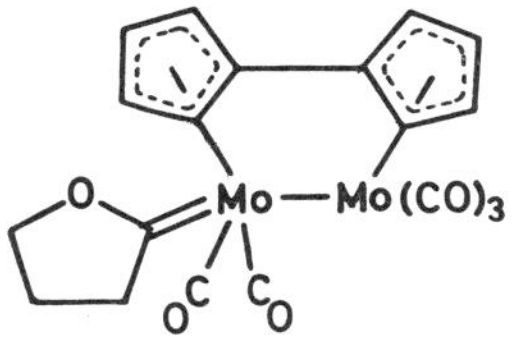

(3)

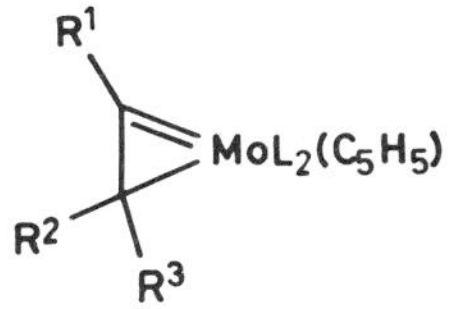

(4)

$W_2(CH_2SiMe_3)_4(\mu\text{-}CSiMe_3)_2(H^{13}C_2H)$ gives $^1\underline{J}(^{13}C,^{13}C)$ = 42 Hz, consistent with the acetylene acting as a 4-electron donor. Variable temperature n.m.r. spectra of $(Me_3SiCH_2)_2W(\mu\text{-}CSiMe_3)(\mu\text{-}CMeCMeCSiMe_3)W(CH_2SiMe_3)_2$ show that the μ-CMeCMeCSiMe$_3$ ligand is fluxional.[1307] Variable-temperature ^1H n.m.r. spectra show that for $W_4(CSiMe_3)_2(OEt)_{14}$ there is rapid bridge-terminal OEt site exchange.[1308] The ^{13}C n.m.r. spectrum of $(C_5Me_5)Ir(CO)_2W(CO)_5$ shows carbonyl exchange involving bridging carbonyls.[1309]

The ^{13}C n.m.r. spectrum of $[(C_5Me_5)M(CO)_2]_2(R^1C_2R^2)$, M = Cr, Mo, W, shows four CO and two C_5Me_5 environments. The barrier to racemisation depends on the bulk of substituents. The ^{119}Sn chemical shifts were also determined.[1310] ^{13}C n.m.r. spectroscopy has been used to investigate dynamic processes in $(C_5Ph_4R)W(CO)_2(\mu\text{-}EtC_2Et)\text{-}W(C_5H_5)(CO)_2$.[1311] ^1H and ^{31}P n.m.r. studies of $WBr_2(CO)L_2(R^1C_2R^2)$ show alkyne propeller rotation.[1312] ^1H and ^{13}C n.m.r. spectroscopy has been used to show restricted C=N rotation in $W(S)(PhC_2Ph)(S_2CNEt_2)_2$.[1313] The dynamic behaviour of the C_3 ring in $Mo(\eta\text{-}C_3Ph_3)(C_5H_5)(CO)_2$ has been examined by ^1H and ^{13}C n.m.r. spectroscopy.[1314] In cis-$[Mo(1\text{-}3\text{-}\eta\text{-}HCH_2CHCHCH_2)(\eta^4\text{-}C_4H_6)(PMe_3)_2]^+$ the presence of the Mo-H-C bond is shown by low-temperature n.m.r. spectroscopy. Variable-temperature n.m.r. spectroscopy shows that the agostic hydrogen can scramble between the four terminal carbon atoms of the two C_4 ligands. The activation parameters were determined and the ^{13}C and ^{31}P n.m.r. spectra recorded.[1315] Variable temperature ^1H n.m.r. spectroscopy of $pzB(\mu\text{-}pz)_2(\mu\text{-}NMe_2)Mo(CO)_2(\eta^3\text{-}C_3H_5)$ shows interconversion of the two types of pyrazolyl groups. The ^{11}B n.m.r. spectrum was also recorded.[1316] ^1H and ^{13}C n.m.r. spectroscopy has been used to investigate the fluxionality of the WC_3 ring in $(C_5H_5)W(CPhCBu^tCPh)Cl_2$.[1317] The ^1H, ^{13}C, and ^{31}P n.m.r. spectra of $Cr(CO)_3(dmpe)(diene)$ show temperature dependence due to diene rotation.[1318] For a series of $Cr(CO)_2(PMe_3)(\eta\text{-}diene)$ complexes the activation barriers for the ligand movement have been determined by dynamic ^{31}P n.m.r. spectroscopy. The mobility of the ligand spheres depends on the position of the alkyl groups of the diene ligands. A methyl group in the 1-*E*-position increases the barrier by 16 kJ mol^{-1}, while in the 2-position it lowers the barrier by 5 kJ mol^{-1}.[1319] ^1H and ^{13}C n.m.r. spectroscopy shows that $Mo(C_7H_2Bu^t{}_4)(O_2CMe)_2$ is fluxional.[1320] The ^{31}P n.m.r. spectrum of $(C_5H_5)Mo(NO)\{S_2P(OPr^i)_2\}_2$ shows two signals at low temperatures and one at high. Activation energies were determined and the ^{13}C n.m.r. spectrum recorded.[1321] $(C_5H_5)Mo(O)_2$-$(C_5H_5)NiFe(CO)_3(PhC_2CO_2Pr^i)$ has been shown by ^1H and ^{13}C n.m.r. spectroscopy to be dynamic.[1322] Barriers to rotation about the cycloheptatriene axis for $(7\text{-}R\text{-}C_7H_7)Cr(CO)_3$ have been determined using dynamic ^{13}C n.m.r. spectroscopy.[1323]

Two-dimensional ^{31}P n.m.r. spectroscopy has been used to determine the rearrangement mechanism in $Cr(CO)_{5-n}(CX)\{P(OR)_3\}_n$. The mechanism was shown to be a trigonol twist.[1324] Detailed variable temperature n.m.r. studies of $M(CO)_5(Me_2\overline{CCH_2EEC}H_2)$, M = Cr, Mo, W, E = S, Se, show pyramidal inversion of the coordinated sulphur and selenium ligand and intramolecular 1,2-metal shifts.[1325] Variable temperature ^1H n.m.r. spectra of $M(CO)_5(Me_3SiCH_2EECH_2SiMe_3)$, M = Cr, Mo, W, E = S, Se, have established the occurrence of two distinct internal dynamic phenomena. The coordinated sulphur or selenium atom undergoes facile pyramidal inversion. Above ambient, a novel 1,2-metal shift between adjacent pairs of chalcogen atoms of the ligands occurs.[1326] The fluxional processes in the 10-membered $M(\mu\text{-dppe})_2M'$ rings of $(OC)_4M(\mu\text{-dppe})_2M'(CO)_4$, M, M' = Cr, Mo, W, have been studied by variable-temperature ^{31}P n.m.r. spectroscopy. The spectrum changes from $[AX]_2$ at room temperature to AMPX at −90 °C.[1327] *Cis*- and *trans*-isomers of $Cr(CO)_4\{(MeS)_2CHCH(SMe)_2\}$ have been identified using NOESY. The activation energy for sulphur inversion was determined.[1328] The ^{31}P n.m.r. spectrum of $W(CO)_3$-$(PMe_3)(NO)(\mu\text{-}F)SbF_5$ shows $^2J(^{31}P,^{19}F)$ = 37.6 Hz at 194 K, proving the coordination of the SbF_6 group. On warming, exchange occurs, and activation parameters were determined.[1329] The ^1H and ^{31}P n.m.r. spectra of $WBr_2(CO)_2(PMe_2Ph)_3$ show two dynamic isomeric forms at room temperature which freeze out at −90 °C.[1330] W(CN)-$(CO)_2L(CNEt_2)$ is fluxional, and ^1H and ^{13}C n.m.r. spectra were recorded.[1331]

The dynamic behaviour of $[\alpha\text{-}Mo_8O_{26}]^{4-}$, $[PhAsMo_7O_{25}]^{4-}$, and $[(PhAs)_2Mo_6O_{24}]^{4-}$ has been studied using variable-temperature ^{17}O n.m.r. spectroscopy, including magnetisation transfer measurements.[1332] Rotation about the Mo=Mo bond in $1,2\text{-}Mo_2(OAr)_2(NMe_2)_4$

and related compounds has been investigated by variable-temperature ^1H n.m.r. spectroscopy, including magnetisation transfer and two dimensional techniques.[1333] ^1H n.m.r. spectroscopy has been used to investigate the fluxional behaviour of $W_2(OAr)_2(NMe_2)_4$ due to restricted rotation about the M-N bond.[1334]

Manganese. $Mn(\eta^3\text{-}MeCHCMeCHCH_3)(CO)_3$ undergoes two dynamic processes, observable by variable-temperature ^1H and ^{13}C n.m.r. spectroscopy. The lowest-energy process is exchange between the agostic hydrogen and the other two hydrogen atoms on the bridging methyl group with $\Delta G^\ddagger$ = 8.9 kcal mol^{-1}. At higher temperatures proton exchange between the two methyl groups occurs with $\Delta G^\ddagger$ = 17.3 kcal mol^{-1}.[1335] ^1H n.m.r. spectroscopy shows restricted rotation of the $Pt(PEt_3)_2$ moiety about the Mn-H bond in $MnPt(\mu\text{-}H)(\mu\text{-}CO)(CO)_4\text{-}(PEt_3)_2$.[1336] ^1H and ^{13}C n.m.r. spectroscopy has been used to determine the C-N bond rotation activation energy in $(C_5H_4Me)Mn(CO)_2\{CPh(NMe_2)\}$.[1337] ^1H and ^{13}C n.m.r. spectroscopy has shown that $Mn_2(CO)_6\{Bu^tN{=}CHCH(NBu^t)CH(NBu^t)CH{=}NBu^t\}$ is fluxional.[1338]

Rhenium. Variable-temperature ^1H n.m.r. spectroscopy of $ReH_3(C_5H_6)(PPh_3)_2$ shows at least two distinct fluxional processes. Magnetisation transfer measurements reveal a process in which the endo hydrogen exchanges with the hydride ligands.[1339] The fluxionality of $ReH_2(PMe_2Ph)_2(\eta^5\text{-}C_8H_{11})$ has been attributed to restricted rotation of the η^5-ring. The ^{13}C and ^{31}P n.m.r. spectra were recorded.[1340] ^{13}C n.m.r. measurements have been performed on $[Re_3(\mu\text{-}H)_3(CO)_{10}]^{2-}$ at various temperatures and field strengths. The spin-lattice relaxation is due to both scalar coupling and chemical shift anisotropy. Variable-temperature ^{13}C n.m.r. spectroscopy shows more than one mechanism of CO scrambling.[1341] The bridging alkenyl ligand in $(\mu\text{-}H)Re_2(CO)_6(\mu\text{-}dppm)(\mu\text{-}\eta^1,\eta^2\text{-}C_2R)$ is fluxional with $\Delta G^\ddagger$ = 10.5 kcal mol^{-1}. The ^{13}C n.m.r. spectrum was also recorded.[1342] Variable-temperature ^{13}C n.m.r. spectra of $(\mu\text{-}H)Re_2(CO)_4(dppm)_2(\mu\text{-}C_2H)$ show four carbonyl signals at -80 °C and one at 5 °C. $\Delta G^\ddagger$ for exchange is 10.3 kcal mol^{-1}.[1343] Similarly, in $(\mu\text{-}H)Re_2(CO)_6(dppe)(\mu\text{-}CH{=}CH_2)$ ^{13}C n.m.r. spectroscopy shows carbonyl scrambling occurs with $\Delta G^\ddagger$ = 9.8 kcal mol^{-1}.[1344]

$(\eta^1\text{-}C_9H_7)Re(CO)_3$ undergoes 1,3-rhenium migrations, which can be frozen out at low temperatures. ^{13}C and ^{31}P n.m.r. spectra were recorded.[1345] The ^{13}C n.m.r. spectrum of $(\eta^1\text{-}C_5H_5)ReMe(NO)(PMe_3)_3$ shows that the cyclopentadienyl group is fluxional, and an equilibrium constant for the formation of $(\eta^5-C_5H_5)ReMe(NO)(PMe_3)$ was determined.[1346] The variable-temperature ^1H, ^{13}C, and ^{31}P n.m.r. spectra of $(C_5H_5)Re(NO)(PPh_3)(PPhH)$ and $(C_5H_5)Re(NO)(PPh_3)\{P(C_6H_4-Me-4)_2\}$ have revealed phosphorus inversion barriers which are amongst the lowest known.[1347] Variable-temperature ^{13}C n.m.r. spectroscopy indicates restricted rotation of the acetylene in $(C_5H_5)Re(CO)_2(PhC_2Ph)$.[1348] The variable-temperature ^1H, ^{13}C, and ^{31}P n.m.r. spectra of $Re_2Cl_4(dppm)_2(CO)_n$ are consistent with an A-frame/non-A-frame fluxionality that keeps the CO ligand bound to a single rhenium atom.[1349]

Iron, Ruthenium, and Osmium. The fluxionality of trans-$[Fe(\eta^2-H_2)-H(dppe)_2]^+$ has been investigated by ^1H n.m.r. spectroscopy and $\Delta G^{\ddagger}$ determined for both η^2-H_2 rotation and exchange between terminal H and η^2-H_2. In the ruthenium analogue $^1J(^2H,^1H)$ has been determined for the η^2-H_2 as 32 Hz. The ^{31}P n.m.r. spectra were also recorded.[1350] A variable-temperature ^1H and ^{13}C n.m.r. spectroscopic study of $H_2Fe(CO)_4$ has been reported. The activation energy for axial-equatorial carbonyl exchange was determined as ca. 8.1 kcal mol^{-1}.[1351] The low-temperature ^1H and ^{31}P n.m.r. spectra of $FeH(BH_4)(Ph_2PCH_2)_3CMe$ are consistent with the crystal structure. On raising the temperature, scrambling of the three hydrogen ligands occurs, and the coordinating phosphorus atoms become magnetically equivalent.[1352] ^{13}C and ^{31}P n.m.r. spectra have been used to study carbonyl fluxionality in $Fe_3(CO)_9(\mu_2-PPr^i{}_2CH_2PPr^iH)(\mu-H)$.[1353] $[RuH_2Cl\{P(tol)_3\}_2]_2$ is dynamic, showing two ^{31}P n.m.r. signals at room temperature and four at low temperatures.[1354] A quantitative ^{13}C two-dimensional n.m.r. analysis of three-site exchange in $Ru_3(\mu-H)(CO)_9(MeCCHCMe)$ has been reported.[1355] The mechanism of localised site exchange of carbonyl groups in $HRu_3(CO)_8[PMe(CH_2Ph)Ph](\mu_3-C_2Bu^t)$ has been studied by ^{13}C n.m.r. spectroscopy. Axial-equatorial carbonyl exchange occurs via C_3 rotation.[1356] The dynamics of carbonyl exchange in $[HRu_3(CO)_{10}(SiEt_3)_2]^-$ has been investigated using ^{13}C n.m.r. spectroscopy.[1357] ^{13}C COSY and NOESY two-dimensional n.m.r. spectra of $Os_3H_2(CO)_{10}$ have been investigated to get coupling and exchange information.[1358] Variable-temperature ^{13}C n.m.r. spectra of $H_2Os_3(CO)_{10}(Bu^t{}_2E^1NSNE^2Bu^t)_2$, E = P, As, show cyclic ligand exchange involving both hydrides and carbonyls.[1359] The dynamics of scrambling on the $Os(CO)_4$ moiety in $(\mu-H)Os_3(CO)_{10}(\mu_3-CPh)$ have been studied using ^{13}C n.m.r. spectroscopy.[1360] Similar observations have been made on $(\mu-H)Os_3(CO)_{10}[\mu-NHSN(SiMe_3)_2]$

using ^{13}C n.m.r. spectroscopy.[1361] ^{13}C n.m.r. spectroscopy has been used to demonstrate carbonyl scrambling in $Os_3(\mu\text{-}H)(C_4H_3O)\text{-}(CO)_{10}$.[1362]

Low-temperature ^{1}H and ^{13}C n.m.r. spectra have shown that the $\eta^1\text{-}C_5H_5$ in $(C_5Me_5)Fe(CO)_2(\eta^1\text{-}C_5H_5)$ is more fluxional than in $(C_5H_5)Fe(CO)_2(\eta^1\text{-}C_5H_5)$.[1363] The variations in the ^{19}F n.m.r. spectrum of $[(C_5H_5)Fe(CO)(PPh_3)(CF_2)]^+$ as a function of temperature have been attributed to CF_2 rotation. The ^{13}C n.m.r. spectrum was also reported.[1364] ^{1}H n.m.r. spectroscopy has been used to demonstrate the fluxionality of $[(C_5H_5)_2(OC)_2Fe_2(\mu\text{-}CO)(\mu\text{-}\eta^1,\eta^2\text{-}(\underline{E})\text{-}CH{=}CH\text{-}Pr^n)]^+$ and $\Delta\underline{G}^{\ddagger}$ was determined. The ^{13}C n.m.r. spectrum was also reported.[1365] Variable-temperature ^{13}C n.m.r. spectra of $[Fe_4(\mu_4\text{-}\eta^3\text{-}CMeCHCH)(\mu\text{-}CO)_2(CO)_9]^-$ show carbonyl exchange.[1366] ^{1}H and ^{13}C n.m.r. studies of $(C_5H_5)Ru\{C[{=}C(CN)_2]CRC{=}C(CN)_2\}L^1L^2$ have shown the presence of rotational isomers.[1367] ^{1}H n.m.r. spectroscopy has been used to show CH_2 exchange in $Ru_3(CO)_{10}(\mu\text{-}CO)(\mu\text{-}CH_2)$. The ^{13}C n.m.r. spectrum was also recorded.[1368] ^{13}C n.m.r. spectroscopy has been used to demonstrate carbonyl fluxionality in $[Os_3\{\mu\text{-}O{=}C(L)\}\text{-}(CO)_{10}]^-$.[1369]

^{1}H and ^{13}C n.m.r. spectroscopy has shown that (5) is fluxional with $\underline{E}_a$ = 13.9 kcal mol^{-1}.[1370] Dynamic ^{1}H n.m.r. spectroscopy has been used to determine the rotation barrier for the alkyne in $RuCo_2(CO)_9(R^1C_2R^2)$.[1371] Exchange in $[Ru(\eta^3\text{-}C_7H_{11})\{P(OMe)_2Ph\}_3]^+$ has been investigated by ^{1}H n.m.r. spectroscopy. The ^{13}C and ^{31}P n.m.r. spectra were also reported.[1372] The fluxionality of $Fe(\eta^4\text{-}C_8H_8)(CO)_2L$ has been investigated by ^{13}C n.m.r. spectroscopy using the DANTE pulse sequence. It was shown that the mechanism of fluxionality is consistent with the Woodward-Hoffmann rules. The ^{31}P n.m.r. spectrum was also recorded.[1373] Variable-temperature ^{13}C n.m.r. studies of some ferrocene bis crown ethers show intramolecular dynamic processes due to restricted rotation about a N-CO bond.[1374] ^{1}H n.m.r. spectroscopy has been used to show that in (6) both the $Fe(CO)_3$ and CMe_2 groups migrate around the ring.[1375] According to ^{13}C n.m.r. data, both $RuRh_3(\mu_3\text{-}CO)_2(CO)_3(C_5Me_5)$ and $RuRh_2(\mu\text{-}CO)(\mu_3\text{-}CO)(CO)_2(C_5Me_5)(\eta^4\text{-bicyclo[4.2.0]octa-2,4-diene})$ undergo carbonyl exchange.[1376] ^{13}C magnetisation transfer measurements on $Os(\eta^6\text{-}C_8H_8)(\eta^4\text{-cod})$ show [1,5]-exchange is the dominant mechanism for the C_8H_8 ring, with the [1,3]-shift occurring more slowly. The two ligands also rotate with respect to each other, initially as a twist and then as a complete rotation.[1377] The ^{1}H, ^{13}C, and ^{31}P n.m.r. spectra of (7) show the rotation of the rings,

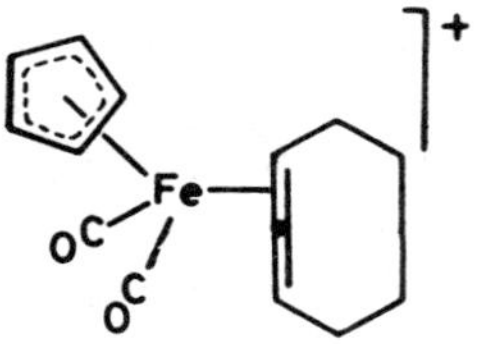

(5)

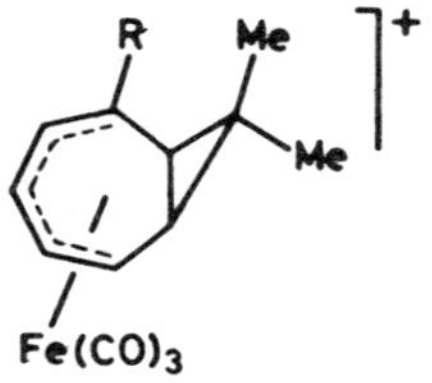

(6)

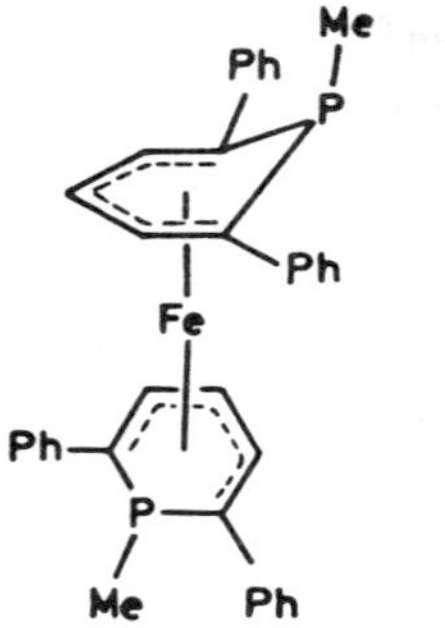

(7)

and the activation energy was determined.[1378] ^1H and ^{13}C n.m.r. spectroscopy shows dynamic selenium exchange in $[(C_5H_5)Fe(CO)_2(RSeSeR)]^+$, but the tellurium analogue could not be frozen out.[1379] The ^{13}C n.m.r. spectrum of $[Fe_2(CO)_6(\eta^5,\eta'^5-C_{16}H_{18})]^{2+}$ shows carbonyl exchange. ^{31}P n.m.r. data were also reported on the PPh_3 substituted analogue.[1380] The bridge reversal energy barrier has been determined for 1,3-dithia-2,2-dimethyl[3]ruthenocenophane.[1381]

Axial-equatorial exchange in $M(CO)_4L$, L = P, As, Sb, has been studied by ^{13}C n.m.r. spectroscopy.[1382] The ^{13}C n.m.r. spectrum of $FeRu(CO)_2(\mu-CO)_2(C_5H_5)_2$ shows cis-trans isomerism and bridge-terminal carbonyl exchange at room temperature, but is static at -80 °C.[1383] ^{13}C n.m.r. spectroscopy has been used to examine carbonyl exchange in $[RhFe_3(CO)_{12}C]^+$.[1384] Carbonyl exchange in $[Fe_4(CO)_{13}]^{2-}$ is fast even at -90 °C according to ^{13}C n.m.r. spectroscopy.[1385] The ^{31}P n.m.r. spectrum of $M_3(CO)_9(dppm)\{Ph_2PCH_2CH_2Si(OEt)_3\}$, M = Ru, Os, shows exchange of the ends of the dppm due to scrambling on the other metal.[1386] Evidence has been obtained from ^{13}C and ^{31}P n.m.r. spectroscopy of the trigonal twist mechanism in $Os_3(CO)_7\{P(OMe)_3\}_4$.[1387] The ^{31}P n.m.r. spectrum of $Os_3(CO)_{10}(PEt_3)_2$ shows two ^{31}P n.m.r. signals at -60 °C, which exchange on warming. The ^{13}C n.m.r. spectrum shows six carbonyl signals at low temperature, but there is rapid exchange on warming.[1388] Intramolecular tertiary phosphine exchange mechanisms in 1,2-$Os_3(CO)_{10}(PMe_2Ph)_2$ and 1,2,3-$Os_3(CO)_9\{P(OMe)_3\}L_2$ have been studied by ^{13}C and ^{31}P n.m.r. spectroscopy. The ^{13}C n.m.r. results rule out a mechanism involving bridging carbonyl forms in favour of one involving axial-equatorial carbonyl exchange at the same rate as phosphine exchange.[1389] Variable-temperature ^1H n.m.r. spectroscopy has been used to show that at 155 K four methyl groups are observed for $PtOs_3(CO)_9(PMe_2Ph)_2(\mu\text{-}S)_2$ but only two at 199 K. The ^{31}P n.m.r. spectrum was also reported.[1390] Variable-temperature ^{13}C n.m.r. spectroscopy has shown that $(Me_3P)(OC)_4OsOs_3(CO)_{11}$ is stereochemically nonrigid in solution and all carbonyl ligands are involved in the process.[1391]

The ^{1}H n.m.r. spectrum of $Fe_2(SMe)_2(NO)_4$ has been measured in 22 solvents, and the activation barrier for $\underline{C}_{2h}$-$\underline{C}_{2v}$ isomerisation was determined.[1392] The barrier for rotation about the coordinated N=C bond in $[Os(NH_3)_5(\text{benzoquinonediimine})]^{2+}$ is 52 kJ mol^{-1}. This weakening of the double bond is attributed to the strong π-back-bonding capacity of Os(II).[1393]

Cobalt, Rhodium, and Iridium. ^{59}Co chemical shifts of a variety of organo-cobalt(I) and cobalt(III) complexes with π-ligands have been determined in solution and range from 1100 to -1300 p.p.m. Molecular orbital calculations were used to explain the shifts. ^{1}H and ^{13}C n.m.r. spectroscopy was used to investigate the dynamic behaviour of $Co(allyl)_3$ and $(C_5H_5)Co(\eta^4\text{-}C_7H_8)$.[1394] ^{13}C n.m.r. spectroscopy has shown that in $(C_5Me_5)Co\{P(OMe)_3\}H(C_2H_4)$ the hydride is agostic with $^{1}\underline{J}(^{13}C,^{1}H)$ = 61 Hz. Dynamic n.m.r. spectroscopy has shown hydride exchange between the hydride and olefin protons.[1395] $[IrH_6(PCy_3)_2]^+$ has been shown by ^{1}H n.m.r. and $\underline{T}_1$ measurements at low temperature to give two signals due to the structure being $[Ir(H)_2(\eta\text{-}H_2)_2(PCy_3)_2]^+$. $\underline{T}_1$ of the coordinated H_2 is very short, 0.03 s.[1396] The dynamic behaviour of $[(Bu^t{}_2\overline{PCH_2CMe_2CH_2)Ir}H]_2(\mu\text{-}Cl)_2$ has been investigated by ^{13}C and ^{31}P n.m.r. spectroscopy.[1397] The ^{1}H n.m.r. spectrum of $IrH(CO)(Ph_2PCH_2)_3CMe$ shows a quartet for the hydride proton at room temperature and an AA'BX pattern at low temperature. The ^{31}P n.m.r. spectrum was also reported.[1398] The ^{13}C, ^{19}F, and ^{31}P n.m.r. spectra have been reported for $[\{(C_5Me_5)\text{-}Ir(PMe_3)H\}_2(\mu\text{-}H)][PF_6]$. Hydride exchange is observed in $[(C_5H_5)\text{-}Ir(PMe_3)H_3]^+$ and the activation energy was derived.[1399] Variable temperature ^{31}P n.m.r. data on $[IrAu_4H_2(PPh_3)_6]^+$ have shown that a process which interchanges axial and equatorial Au sites occurs.[1400]

$MeC(CH_2PPh_2)_3MR(CO)$, M = Rh, Ir, R = Me, CH_2Bu^t, CH_2SiMe_3, Ph, is fluxional due to exchange of the inequivalent phosphorus atoms. ^{31}P n.m.r. spectroscopy was used to investigate the process.[1401] ^{13}C n.m.r. studies of $[Rh(COMe)(CO)_2I_3]^-$ show restricted rotation about the Rh-COMe bond. Isotope labelling has shown that the formation of $[Rh_2(COMe)_2(CO)_2I_6]^{2-}$ from $[RhMe(CO)_2I_3]^-$ is irreversible.[1402] Variable-temperature ^{1}H and ^{19}F n.m.r. measurements have established that $(C_5Me_5)_2Rh_2(\mu\text{-}CO)\{\mu\text{-}\eta^2{:}\eta^2\text{-}C(O)C_2R_2\}$, R = CF_3, CO_2Me, is fluxional.[1403] ^{13}C and ^{31}P n.m.r. spectroscopy has shown syn-anti exchange in $[IrX(\mu\text{-}SBu^t)(CO)L]_2(\mu\text{-}CH_2)$.[1404]

^{13}C and ^{31}P n.m.r. studies have shown that $[Co(\eta^4\text{-}C_4H_6)(PMe_3)_3]^+$ is fluxional.[1405] According to ^{1}H and ^{13}C n.m.r. spectra, $(C_5H_5)Co(Pr^iN{=}CMeCMe{=}NPr^i)$ has conformers, which freeze out at low temperatures.[1406]

^{13}C and ^{59}Co n.m.r. spectra of $Co_4(CO)_{12}$ have been interpreted in terms of carbonyl exchange and ^{13}C relaxation of the second kind caused by $^{1}J(^{59}Co,^{13}C)$.[1407] The fluxionality of $[\{Rh[C_5H_3N(CH_2S\text{-}CH_2)_2CH_2](CO)_2\}_2]^{2+}$ in the ^{1}H and ^{13}C n.m.r. spectra have been attributed to ring flexing. Activation energies were determined.[1408] Variable-temperature ^{1}H and ^{31}P n.m.r. spectra of $Rh_2(CO)_3(dppm)_2$ have been recorded and discussed.[1409] ^{1}H, ^{13}C, and ^{31}P n.m.r. spectra have shown that $[Rh_3(CO)_2(\mu\text{-}CO)_3(dppm)(\mu\text{-}dppm)_2]^+$ is fluxional due to phosphorus exchange.[1410] Non-rigidity of $\{Me_2Ga(pz)_2\}Ir(cod)$ has been demonstrated by variable-temperature ^{1}H n.m.r. studies.[1411] The lowest-energy carbonyl scrambling process for $Ir_4(CO)_9(\mu_2\text{-}CO)_2(\mu_2\text{-}SO_2)_2$ has been determined by ^{13}C n.m.r. spectroscopy.[1412]

^{1}H and ^{13}C chemical shifts and T_1 have been measured for $[Co\{RC(NO)CMeNOH\}_2(py)_2]^+$ and the T_1 values interpreted in terms of pyridine rotation.[1413] The values of $\Delta G^{\ddagger}$ for inversion of $[Co(RSeCH_2CH_2NH_2)(tren)]^{3+}$, as determined using ^{1}H and ^{13}C n.m.r. spectroscopy, decrease in the order Me > Et > $PhCH_2$, whereas $\Delta H^{\ddagger}$ are almost the same.[1414] The fluxionality of $[(Bu^tNC)(OC)Ir(\mu\text{-}dppm)_2Au]^+$ has been investigated using ^{31}P n.m.r. spectroscopy.[1415]

Nickel, Palladium, and Platinum. Exchange of the inequivalent CH_2 protons in $[Pt_3(\mu_3\text{-}H)(\mu\text{-}dppm)_3]^+$ has been determined by ^{1}H n.m.r. spectroscopy to be 50 kJ mol^{-1}. The ^{13}C, ^{31}P, and ^{195}Pt n.m.r. spectra were also recorded.[1416] A low-temperature ^{1}H n.m.r. spectroscopic investigation of $[PdCl(C_4H_4N_2)(PMe_2Ph)_2]^+$ has shown restricted rotation of the Pd-C bond. The ^{13}C and ^{31}P n.m.r. spectra were also recorded.[1417] ^{1}H n.m.r. data for $(R_3P)_2Pd[\{CH(CO_2R)\}_2CO]$ have been interpreted in terms of inversion of the palladacyclobutanone ring via a planar transition state. ^{13}C and ^{31}P n.m.r. spectra were also recorded.[1418] Similar observations have been made on the platinum analogue.[1419] The ^{1}H and ^{13}C n.m.r. spectra of (8) show two NMe_2 groups at low temperatures, which exchange on heat-

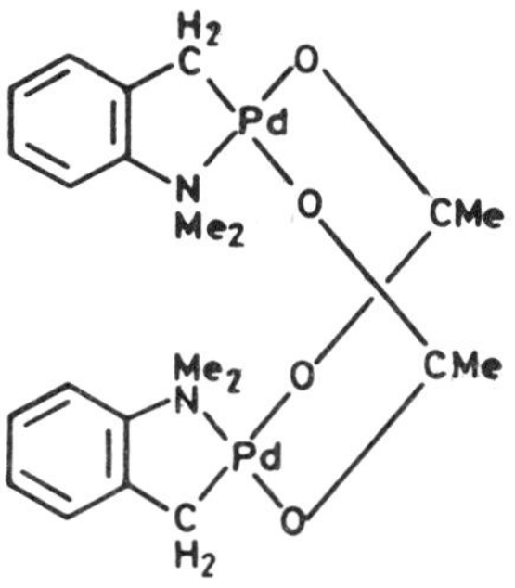

(8)

ing.[1420] The static structures of $PtMe_3X(MeECH=CHEMe)$, E = S, Se, have been characterised by ^1H, ^{13}C, ^{77}Se, and ^{195}Pt n.m.r. spectroscopy. The dynamic stereochemistry arising from low-temperature chalcogen inversion and high-temperature fluxional rearangements has been studied quantitatively by variable-temperature ^1H n.m.r. spectroscopy.[1421] $\{Me_2M(pz)_2\}PtMeL$, M = B, Ga, has been shown to be fluxional by variable-temperature ^1H n.m.r. spectroscopy.[1422] ^{31}P n.m.r. spectra show that $[(RC_2)Pt(\mu\text{-dppm})_2(\eta^1,\eta^2\text{-}C_2R)Rh(CO)]^+$ is fluxional, freezing out at low temperatures.[1423]

The olefin rotation in $M(PPh_3)_2(Ph_2PC_6H_4CH=CH_2)$, M = Ni, Pd, Pt, has been observed by ^{31}P n.m.r. spectroscopy.[1424] Olefin rotation in some palladium and platinum complexes of bicyclo[3.3.1]non-1-ene has been studied by ^{13}C n.m.r. spectroscopy.[1425] High-pressure ^1H n.m.r. spectroscopy has been used to investigate syn-anti exchange in $[Pd(2\text{-Me-}C_3H_4)Cl]_2$ and $Pd(2\text{-Me-}C_3H_4)Cl(PPh_3)$. $\Delta\underline{V}^{\ddagger}$ was found to be 0, and the mechanism was discussed.[1426] ^{31}P n.m.r. magnetisation transfer has been used to demonstrate phosphorus exchange in $[(diop)Pt(1\text{-Me-}C_3H_4)]^+$. The ^{13}C n.m.r. spectrum was also recorded.[1427] ^1H n.m.r. spectra of $Pt_2(\eta^1,\eta^2\text{-}C_3H_5)_2(acac)_2$ have shown pairwise allyl proton exchange with $\Delta\underline{G}^{\ddagger}$ of 15.8 kcal mol^{-1}. The slow formation of $Pt(\eta\text{-}C_3H_5)(acac)$ with $\Delta\underline{G}^{\ddagger}$ = 23 kcal mol^{-1} and $\Delta\underline{G}^{\circ}$ = 0.43 kcal mol^{-1} was also observed.[1428] The ^{31}P n.m.r. spectrum of $[Ni_2(CNMe)_4(dppm)_2]^{2+}$ shows phosphorus exchange with $\Delta\underline{G}^{\ddagger}$ = 10 kcal mol^{-1}.[1429]

The inversion barrier between the saddle conformers of Ni(II) octaethylpyrrocorphinate has been determined by ^1H n.m.r. spectroscopy.[1430] ^1H n.m.r. spectroscopy has been used to determine the barrier to racemisation of Ni(II) complexes of 4-aminomethylenepyrazole-4-thione(selene).[1431] The rotational barrier for Et_2N rotation in $Ni\{morpholinoC(S)=CPhNH\}_2$ has been determined using ^1H

n.m.r. spectroscopy.[1432] Rotation about the Pt-guanosine bond in (N,N'-dimethylethylenediamine)bis(guanosine)platinum(II) has been studied in detail by ^1H n.m.r. spectroscopy. The activation energy for exchange between the SR diastereomers is 65 kJ mol^{-1}, while for the RR or SS diastereomers it is about 55 kJ mol^{-1}.[1433] The mobility of the acetone molecule coordinated to platinum in trans-$PtCl_2$-$(NC_5H_4R)(OCMe_2)$ has been unambiguously established through ^1H and ^{13}C variable-temperature n.m.r. measurements.[1434] The barrier to inversion of cis-PtL_2(perfluoropinacolate) has been determined using ^1H, ^{19}F, and ^{31}P n.m.r. spectroscopy. The trans influence was determined from $^1J(^{195}Pt,^{31}P)$.[1435] ^1H, ^{13}C, ^{31}P, and ^{195}Pt n.m.r. spectroscopy has been used to determine $\Delta G^{\ddagger}$ in $Pt_3\{\mu\text{-}Bu^t(Me_3Si)N\text{-}P{=}NBu^t\}_3(CO)_3$ and $Pt_3(\mu\text{-}CNBu^t)(CNBu^t)_3\{\mu\text{-}Bu^t(Me_3Si)NP{=}NBu^t\}_2$.[1436]

Gold. ^1H n.m.r. spectroscopy has been used to show that $[AuMe_2\{B(pz)_3(Hpz)\}]^+$ is fluxional due to five-coordinate intermediates.[1437]

Zinc, Cadmium, and Mercury. ^{13}C n.m.r. spectroscopy has been used to demonstrate the order of fluxionality for the cyclopentadienyl ring is $Hg(C_5H_5)_2 > Hg(C_5H_5)Cl \sim Hg(C_5Cl_5)Ph > Hg(C_5Me_5)_2$.[1438]

The enantiomerisation of ML_2, M = Ni, Zn, Cd, Hg, Pb, HL = (9), has been studied by dynamic n.m.r. spectroscopy. The activation barrier of the process depends on M, E, and R^1, R^2, and R^3 in the order: Hg, Pb < Ni < Cd < Zn; O < Se $\leq$ S; CH_2Ph < Pr^i < cyclohexyl.[1439] The temperature dependence of the ^{13}C n.m.r. spectrum of $[Cd(O_2CNMe_2)\{CH_2(CH_2NMeCH_2CH_2NMeCH_2)_2CH_2\}]^+$ is due to methyl orientation.[1440] In $[Hg(\eta^1\text{-dppm})_2]^{2+}$ the ^{31}P n.m.r. spectrum is temperature dependent due to exchange between the coordinated and free ends of the dppm ligand.[1441]

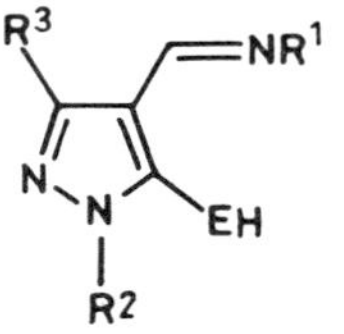

(9)

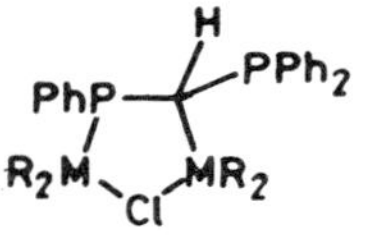

(10)

Boron. Hindered rotation of the Me_3Si group in $(Me_3Si)_2CHC(BR)_2H$ has been examined by ^1H, ^{11}B, and ^{13}C n.m.r. spectroscopy and $\Delta G^{\ddagger}$ determined.[1442] The L_3H groupings in 6,6,6,6-$(Me_2PhP)_3H$-nido-6-Re-

B_9H_{13} and related compounds show a dual pseudorotational fluxionality with $\Delta G^\ddagger$ of 45 to 60 and ca. 30 kJ mol^{-1}, respectively, according to ^1H, ^{11}B, and ^{31}P n.m.r. spectra.[1443] The ^{31}P n.m.r. spectrum of $[closo\text{-}2,2\text{-}(PPh_3)_2\text{-}2,1,7\text{-}RhC_2B_9H_{11}]^-$ shows two phosphorus signals at low temperature and one at high temperature. The activation energy was determined and the ^{11}B n.m.r. spectrum recorded.[1444] The ^1H n.m.r. spectrum of $(C_5Me_5)Rh(B_{10}H_{11}Cl)(PMe_2Ph)$ shows Rh-H-B and B-H-B exchange with $\Delta G^\ddagger$ = 33 kJ mol^{-1}. The ^{11}B n.m.r. spectrum was also reported.[1445]

^{11}B n.m.r. spectroscopy has been used to determine the isomerisation of $H_4C_4B_6Me_6$ and to determine its fluxionality. The ^{13}C n.m.r. spectrum was also recorded.[1446] $\Delta G^\ddagger$ for B-NMe_2 rotation in $PhB(NMe_2)(NHBu^t)$ has been determined using variable-temperature ^{13}C n.m.r. spectroscopy.[1447] Exchange in $R(HO)\overline{BC_6H_4CH_2N}Me_2$ has been studied and $\Delta G^\ddagger$ determined. The ^{11}B n.m.r. spectrum was also recorded.[1448] ^{13}C n.m.r. spectroscopy has been used to determine the barrier to rotation about the B-N bond in aminoboranes.[1449]

Aluminium and Gallium. Variable-temperature ^1H n.m.r. spectroscopy has been used to show that $Me_2Si(Bu^tN)_2(AlPh_2)_2$ is fluxional.[1450] Variable-temperature ^1H, ^{13}C, and ^{31}P n.m.r. spectroscopy has been used to demonstrate that (10), M = Al, Ga, is fluxional with the inequivalent phosphorus atoms exchanging roles.[1451, 1452]

Carbon. The rotational correlation times for $[NO_3]^-$ and $[CO_3]^{2-}$ anions in dilute aqueous solution have been determined by ^{14}N and ^{17}O n.m.r. spectroscopy. ^{23}Na n.m.r. spectroscopy provides evidence for Na^+ and $[CO_3]^{2-}$ association.[1453]

Silicon. Variable-temperature ^{13}C, ^{29}Si, and ^{35}Cl T_1 measurements of Cl_3SiH, Cl_2SiMeH, and $ClMe_2SiH$ have been used to investigate molecular dynamics.[1454] The fluxionality of (11) has been investi-

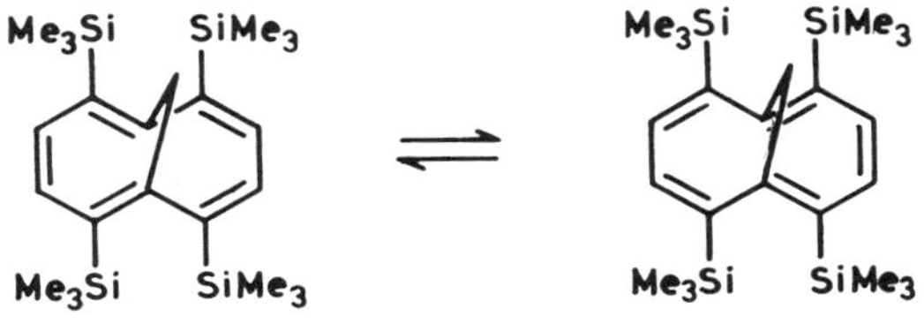

(11)

gated by ^1H and ^{13}C n.m.r. spectroscopy.[1455] ^1H, ^{13}C, and ^{29}Si

n.m.r. spectra of $XCH_2SiMe_2OCR^1=CHCOR^2$ have indicated the presence of mixtures of cis and trans isomers. The cis isomer undergoes intramolecular silylotropic rearrangement.[1456] The terminal-chain diffusional spectrum of fractionated, molten dimethyl siloxane has been characterised from relaxation of transverse magnetic components of ^1H or ^{13}C nuclei.[1457] A hindered twist-to-twist interconversion in $(PhO)(S)P(MeNNMe)_2SiPh_2$ has been detected at low temperatures by ^1H n.m.r. spectroscopy.[1458] The rates of ligand permutation in (12), M = Si, have been measured by ^{19}F n.m.r. spectroscopy for the CF_3 groups.[1459] A similar study of (12), M = SiO=CH-$C_6H_4NMe_2$, using ^1H, ^{19}F, and ^{29}Si n.m.r. spectroscopy, has shown that inversion of the silicon proceeds by a non-dissociative intramolecular pseudorotation with $\Delta G^‡$ = 10.2 kcal mol^{-1}.[1460] According to a ^{19}F n.m.r. study of (13), the energy barrier to intramolecular fluorine equilibration is 7.5 kcal mol^{-1}.[1461]

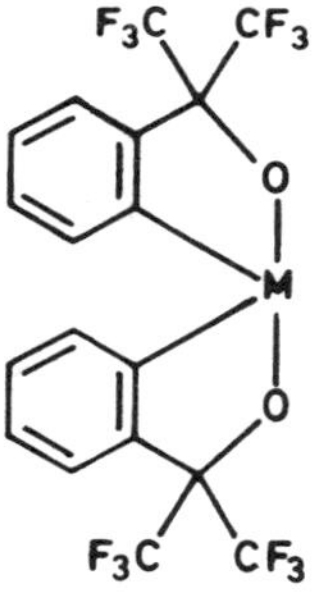

(12)

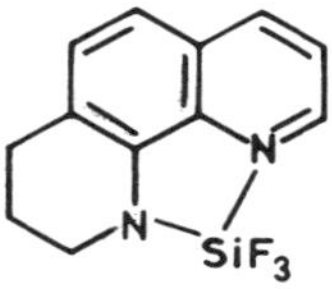

(13)

Tin. The behaviour of 2,2-dibutyl-1,3,2-dioxastannolane has been examined by ^1H, ^{13}C, and ^{119}Sn n.m.r. spectroscopy. There is a fast intramolecular shift with inversion at tin with $\Delta G^‡$ = 42 kJ mol^{-1}.[1462] ^{119}Sn two-dimensional NOESY has been used to demonstrate unambiguously that $CH_2\{PhSn(SCH_2CH_2)_2NMe\}_2$ isomerises at the tin centre in an uncorrelated way.[1463] ^{31}P and ^{119}Sn n.m.r. spectroscopy has been used to investigate $Bu^n{}_3PO$ adducts of Ph_2SnX_2. There is pseudorotation in $Ph_2SnClBr(OPBu^n{}_3)$.[1464] A two-dimensional n.m.r. study of the modes of rearrangement in PhSnCl(PhCOCHCO-$Me)_2$ has been reported. The results were explained in terms of the Bailar twist.[1465] The temperature-dependent ^1H n.m.r. spectra of $SnCl_2$(8-quinolinato)$_2$ have been analysed in terms of intramolecular exchange.[1466] The ^{31}P and ^{119}Sn n.m.r. spectra of (14) have been reported. A Berry pseudorotation is observed with $\Delta G^{\ddagger}$ = 12.7 kcal mol^{-1}.[1467] The ^{1}H, ^{13}C, ^{31}P, and ^{119}Sn n.m.r. spectra of $Sn_2(PR_2)_2(\mu\text{-}PR_2)_2$ and $Sn_2(SR)_2(\mu\text{-}SR)_2$ have been reported. In the case of $Sn_2(SR)_2(\mu\text{-}SR)_2$ SR exchange was observed.[1468] The mechanisms of Me_2S intermolecular exchange and trans-cis isomerism of $SnCl_4(SMe_2)_2$ have been studied by variable-pressure ^{1}H n.m.r. and variable-temperature ^{119}Sn magnetisation transfer n.m.r. spectroscopy. This work provides a rare example where both magnetisation transfer and line shape analysis have been performed at the same temperature.[1469]

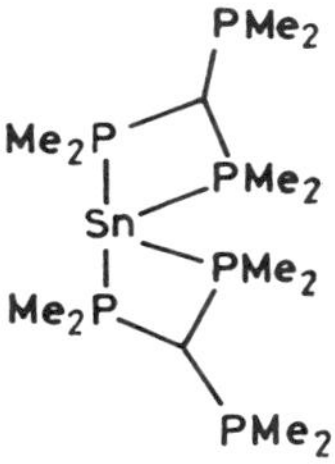

(14)

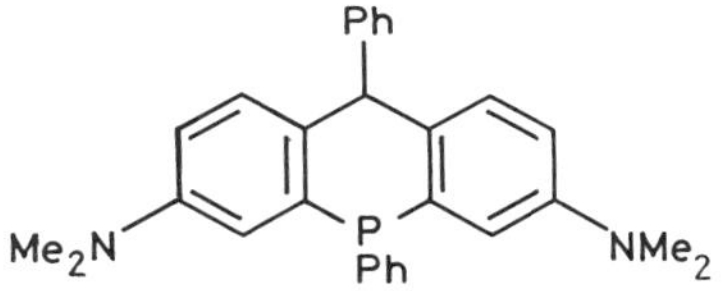

(15)

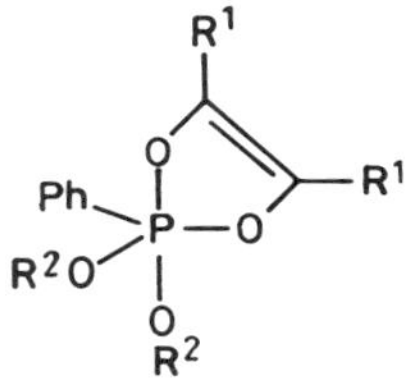

(16)

Phosphorus. ^{7}Li, ^{13}C, and ^{31}P chemical shifts and T_1 have been used to investigate $C_6H_4(PR_2)_2$ and the di-lithiated species. A phosphorus inversion barrier of 100 to 110 kJ mol^{-1} was determined.[1470] ^{1}H n.m.r. spectroscopy has been used to determine the activation energy for flipping in $RP(CH{=}CH)_2PR$. ^{1}H n.O.e. measurements were used to assign the structure and the ^{13}C and ^{31}P n.m.r. spectra measured.[1471] Similarly, the activation energy for ring inversion in (15) has been determined as 28.9 to 29.6 kcal mol^{-1}.[1472] ^{1}H, ^{13}C, and ^{31}P n.m.r. spectroscopy has been used to determine P-Bu^t rotation barriers in Bu^t_2PR and compared with molecular mechanics calculations.[1473]

Pseudorotation in (16) has been studied using ^{1}H and ^{31}P n.m.r. spectroscopy.[1474] Permutational isomerism in (17) has been studied by ^{1}H and ^{31}P n.m.r. spectroscopy. It proceeds via an irregular mechanism with dissociation of the P-N bond.[1475] ^{1}H and ^{13}C n.m.r. spectra of $(Me_2N)_2CHC({=}N_2)PR^1R^2(O)$ show restricted rotation about the C-N and C-C bond.[1476] The barrier to rotation of the P-N bond in $RP(X)NEt_2$ has been studied by ^{1}H, ^{13}C, ^{19}F, ^{31}P, and ^{77}Se n.m.r. spectroscopy. It is determined by steric factors.[1477]

^{1}H, ^{13}C, and ^{31}P n.m.r. spectroscopy has been used to study intramolecular ligand rearrangement in $\overline{OCAr^1{=}CAr^2OPN(CH_2CH_2CH_2N}$-

Me)$_2$.[1478] A similar study using ^1H, ^{13}C, ^{19}F, and ^{31}P n.m.r. spectroscopy has been performed on $\overline{OCAr^1{=}CAr^2OPN}(CH_2CH_2CH_2N$-Me)$_2$.[1479] The first example of an intramolecular sequential [1,3]-migration of a phosphoranyl group in an ambidentate N=CN=CN pentad has been demonstrated for (18) with an activation energy of

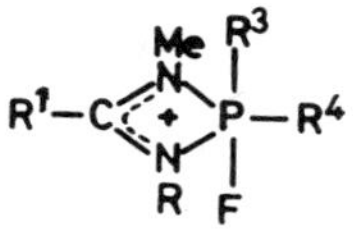

(17)

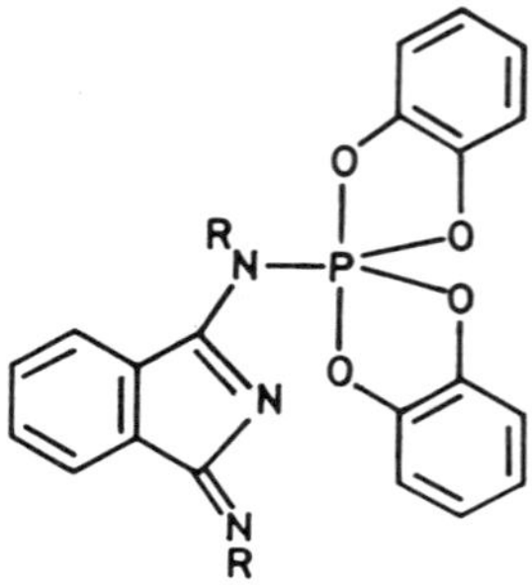

(18)

16.9 kcal mol^{-1}.[1480] $\Delta G^{\ddagger}$ has been determined for $CH_2(C_6H_4O)_2$-$P(OR)_3$ using ^1H, ^{13}C, ^{19}F, and ^{31}P n.m.r. spectroscopy.[1481] Different isotopic shifts are produced by equatorial and axial P-^{18}O single bonds in 5-coordinate, trigonal-bipyramidal oxyphosphoranes. This difference can be used to follow permutational isomerisation by variable-temperature ^{31}P n.m.r. spectroscopy.[1482] The rotational motions of $[PO_4]^{3-}$, $[SO_4]^{2-}$, and $[ClO_4]^-$ have been studied using ^{17}O spin-lattice relaxation times. The electric field gradients at the oxygen site were calculated by using the ab initio M.O. method, and hence the correlation time calculated. The activation energies were also derived.[1483]

Arsenic. The ^{13}C T_1 has been measured as a function of temperature for three 2-phenyl-1,3,2-dioxarsolanes. The data were interpreted in terms of isotropic overall molecular tumbling, rotation about the As-Ph bond, and internal methyl rotation.[1484]

Sulphur. ^{17}O n.m.r. spectroscopy has been used to study the rotational motion of the sulphate anion in aqueous solutions.[1485]

Selenium. The inversion barrier and the A value of the PhSe group in PhSe(cyclohexyl) have been determined by dynamic ^{77}Se n.m.r.

spectroscopy. The $\delta(^{77}Se)$ show large diamagnetic shifts in substituted analogues due to γ-gauche CH_2 interactions.[1486] The activation energy for $Pr^i{}_2N$ rotation in $Pr^i{}_2NC(Se)Ph$ has been determined.[1487] The barrier to rotation about the Se-Se bond in PhSe-Se-CH_2Ph has been determined as 6.3 kcal mol^{-1} as indicated by changes in the ^1H n.m.r. spectrum.[1488]

Tellurium. ^{19}F, ^{75}As, and ^{125}Te n.m.r. spectra have been observed for $Te(OTeF_5)_4$, $As(OTeF_5)_5$, and related species. In $As(OTeF_5)_5$ the ^{75}As signal is sharp enough to observe $^2\underline{J}(^{125}Te,^{75}As)$ = 430 Hz. The activation energy for intramolecular exchange in $Te(OTeF_5)_4$ was determined as 31 kJ mol^{-1}.[1489]

Equilibria.—Solvation Studies of Ions. The sources of disagreement of ion coordination numbers in aqueous solutions, determined by different methods, have been examined.[1490] The basic physical principles for determining the structure and dynamics of liquids from n.m.r. relaxation have been presented and applied to solvation structures around $^{19}F^-$, $^{23}Na^+$, and $^{127}I^-$.[1491] ^1H chemical shifts of aqueous electrolyte solutions have been measured as functions of salt concentration, ionic size, and ionic charge/radius ratio. The observed shifts were interpreted in terms of the breakdown of the hydrogen-bonded structure of water and the ability of the ions to polarize water molecules.[1492] Macroscopic counterion diffusion in solutions of cylindrical polyelectrolytes has been studied using ^7Li n.m.r. self-diffusion experiments.[1493] ^1H n.m.r. spectroscopy has been used to investigate methanol exchange on Mg(II), Co(II), and Zn(II) complexed by 15-crown-5.[1494] ^1H $\underline{T}_1$ measurements of ethylene glycol in the presence of Gd(II), Mn(II), Co(II), Ni(II), and Cu(II) have been reported and correlation times derived.[1495]

Group Ia. The pressure, temperature, and concentration dependence of ^2H $\underline{T}_1$ in supercooled LiCl-D_2O has been measured. Hydrostatic pressure suppresses anomalies and turns water into a normal viscous liquid.[1496] The water orientation in the hydration sphere of Li^+ has been determined by ^7Li relaxation measurements.[1497] The rates of ^1H magnetic relaxation have been determined at a variety of LiCl concentrations in aqueous solution containing 1.2 x 10^{-3} M $MnCl_2$. The temperature dependence of $\underline{T}_1$ was determined.[1498] Solvation effects of Li^+, Na^+, and Mg^{2+} in dmf/H_2O have been investigated by ^1H n.m.r. spectroscopy.[1499] The spin-lattice relaxation time of ^7Li of concentrated $LiNO_3$-NH_3 solutions has been measured and correlated with viscosity data.[1500] The ^{7}Li n.m.r. relaxation of molten (Li-Cs)NO_3 has been measured as a function of composition and temperature. The relaxation rates increase with increasing concentration of $CsNO_3$. This behaviour was explained in terms of the anion polarization effect due to the introduction of cations with different ionic radii.[1501]

The hydration of Na^+ and $[OH]^-$ in aqueous solutions of NaOH has been determined by ^{23}Na n.m.r. spectroscopy.[1502] Self-diffusion of water molecules in aqueous solutions of NaCl has been investigated.[1503] The ^{23}Na relaxation rates in aqueous $NaClO_4$, NaI, NaBr are proportional to (concentration)$^{1/2}$ at low concentration.[1504] The spin-lattice relaxation times of ^{2}H, ^{23}Na, ^{85}Rb, and ^{133}Cs in basic solution have been measured. The hydration numbers of Cs^+ and $[CO_3]^{2-}$ were estimated, and the hydration numbers of Br^-, Li^+, and Na^+, measured earlier, were confirmed.[1505] The coordination and mobility of water have been studied in NaSCN, $Ca(SCN)_2$, and $Ba(SCN)_2$ solutions by ^{2}H and ^{23}Na $\underline{T}_1$ measurements.[1506] The solvation of $^{23}Na^+$ in binary mixtures of solvents has been studied. The solvation abilities follow the order water > acetone $\geq$ acetonitrile $\geq$ EtOH and dmf > MeOH.[1507] The molecular reorientation and Na^+ solvation in $(Me_2N)_3PO$ have been studied using ^{2}H, ^{14}N, and ^{23}Na relaxation measurements.[1508] Solutions of the metals Na, K, Rb, and Cs in some amides have been investigated by ^{13}C, ^{23}Na, ^{39}K, ^{85}Rb, and ^{133}Cs n.m.r. spectroscopy. Although the ^{23}Na n.m.r. spectrum showed the presence of Na^-, solutions of the heavier alkali metals showed no evidence of genuine alkali ions.[1509]

Magnesium. Solvent exchange kinetics of $[Mg(15\text{-crown-}5)(dmf)_2]^{2+}$ have been studied in DMF-CD_3NO_2 by n.m.r. spectroscopy.[1510]

The Lanthanides. The solvation of La^{3+} in aqueous dmf has been studied as a function of dmf mole fraction and type of anion by ^{139}La n.m.r. spectroscopy.[1511] ^{1}H chemical shifts of aqueous solutions of rare-earth ions have been studied and only small paramagnetic shifts were found.[1512] The hydration of Lu^{3+} has been investigated by ^{1}H and ^{35}Cl n.m.r. spectroscopy in water-acetone mixtures.[1513]

Uranium. Low-temperature ^{15}N and ^{31}P n.m.r. spectroscopy has been used to investigate the species formed in the organic layer following the extraction of uranium from nitric acid solution with di-2-ethylhexyl phosphoric acid.[1514]

Vanadium. Water exchange with $[V(OH_2)_6]^{3+}$ has been studied as a function of temperature and pressure by ^{17}O n.m.r. spectroscopy.

The kinetics and activation parameters were determined and the mechanisms of exchange were discussed.[1515] ^{51}V n.m.r. spectroscopy has been used to investigate the interaction of vanadate with ethanol in water.[1516] DMSO exchange in $[VO(DMSO)_5]^{2+}$ has been investigated by ^1H n.m.r. spectroscopy and the kinetic parameters determined.[1517]

Ruthenium. The electron-exchange rate of the $[Ru(OH_2)_6]^{2+/3+}$ couple in acidic solutions has been measured directly by ^{17}O and ^{99}Ru n.m.r. spectroscopy at 252 to 366 K. The ^{17}O n.m.r. signals for the coordinated H_2O of $[Ru(OH_2)_6]^{2+}$ and $[Ru(OH_2)_6]^{3+}$ occur at -200 and 35 p.p.m. from H_2O, while the ^{99}Ru n.m.r. signal for $[Ru(OH_2)_6]^{2+}$ is at 16 050 p.p.m. from $[Ru(CN)_6]^{4-}$.[1518]

Cobalt. A recently reported n.m.r. method for measuring ^2H equilibrium isotope effects has been applied to the hydration of Co(II). An isotope effect of ca. 1.3% was measured.[1519] The catalytic effect of nitrate ion on water exchange on *cis*-$[Co(en)_2(OH_2)_2]^{3+}$ has been determined by ^{17}O n.m.r. spectroscopy.[1520] Activation energies have been determined from a ^1H n.m.r. study of $(Et_2N)_2NCHO$ exchange on $[Co\{2,2',2''-(Me_2N)_3triethylamine\}\{(Et_2N)_2NCHO\}]^{2+}$ and its Cu(II) analogue.[1521] ^1H n.m.r. spectroscopy has been used to show that the loss of solvent ligands at the two sites in $[Co\{N(CH_2CH_2NH_2)_3\}(OSMe_2)_2]^{3+}$ occurs at the same rates.[1522]

Rhodium. The activation energy for the interconversion of $[Rh(acetone)_2\{P(OPh)_3\}_2]^+$ and $[Rh(acetone)(OH_2)\{P(OPh)_3\}_2]^+$ has been determined by ^1H and ^{31}P n.m.r. spectroscopy.[1523]

Nickel. The ^1H T_1 of water in aqueous Ni^{2+} has been measured as a function of temperature, frequency, and pH.[1524]

Platinum. Water exchange on $[Pt(OH_2)_4]^{2+}$ has been studied by ^{17}O n.m.r. spectroscopy and activation parameters determined.[1525]

Copper. ^1H relaxation measurements of water in the presence of Cu^{2+} and ethylene glycol have been made as a function of temperature. Information on molecular motion was obtained.[1526]

Zinc. The hydration number of Zn^{2+} has been determined as 3.8 from ^1H n.m.r. measurements. The compounds formed in $(Bu^nO)_3PO-CCl_4$ extraction of $ZnCl_2$ have been determined using ^1H and ^{13}C n.m.r. spectroscopy.[1527]

Aluminium and Gallium. The ^1H T_1 of free and coordinated water signals in aqueous $AlClO_4$ solutions has been measured between 10 and -65 °C. It was concluded that the coordinated water molecules undergo anisotropic rotational motion as an axially symmetric ellipsoidal body and this motion is slowed at low aluminium concentrations.[1528] The transverse relaxation rate of H_2O in $[Al(OH_2)_6]^{3+}$ has been measured as a function of temperature and pressure using ^{17}O line broadening in the presence of Mn^{2+}. At high temperatures, the relaxation rate is governed by chemical exchange with bulk water, while at low temperatures quadrupole relaxation is prevailing. The water exchange parameters were obtained.[1529] The ^{27}Al n.m.r. chemical shift of $[Al(OH_2)_6]^{3+}$ is medium dependent due to possible outer-sphere effects.[1530] ^{27}Al n.m.r. spectroscopy has been used to investigate the preferential solvation of Al^{3+} in a DMF-DMSO mixed-solvent system. The effect of inert diluent on preferential solvation was also examined.[1531] The outer-sphere interactions of solvated Ga^{3+} and Al^{3+} with F^- and Cl^- have been studied in ethylene glycol by ^{19}F n.m.r. spectroscopy.[1532]

Oxygen. The ^2H and ^{17}O quadrupolar splittings of H_2O in a lyotropic mesophase have ratios widely different from all previous observations.[1533]

Sulphur. The ^{17}O $\underline{T}_1$ values for aqueous $NaHSO_3$ have been measured as a function of pH, temperature, and concentration. The rate law for oxygen exchange between $[HSO_3]^-$ and water was determined.[1534]

Fluorine. The hydration of F^- has been studied by ^{19}F n.m.r. spectroscopy. The dipole-dipole contributions of ^1H and ^{17}O to the ^{19}F $\underline{T}_1$ were calculated to test various models.[1535] The ^{19}F chemical shifts for solutions of various fluoride salts have been measured.[1536]

Ionic Equilibria. ^{31}P $\underline{T}_1$ and $\underline{T}_2$ have been used to investigate complex formation between $H_2N(CH_2)_2NH(CH_2)_2NH(CH_2)_2NH_2$, phosphate, and metal ions.[1537] The influence of micellization on complexing properties of amphiphilic ligands towards Mn^{2+}, Ni^{2+}, and $[VO]^{2+}$ has been studied using ^{31}P n.m.r. spectroscopy.[1538] ^{13}C n.m.r. spectroscopy has been used to investigate conformational changes of crown ethers on metal complexation.[1539] The complexes formed by Co(II), Mn(II), and Cu(II) with glycine, alanine, and proline have been studied by ^{14}N and ^{17}O n.m.r. spectroscopy.[1540]

Group Ia. Bu^nLi exists in THF as a tetramer in equilibrium with a dimer. The rates of exchange were measured by ^7Li n.m.r. spectroscopy at variable temperature.[1541] The influence of aggregation on the reactivity of Bu^nLi towards benzaldehyde and cyclopentadiene has been studied at -85 °C by rapid injection n.m.r. spectroscopy.[1542] The temperature and concentration dependences of ^{13}C n.m.r.

spectra of ^{6}Li and ^{13}C labelled organo-lithium compounds have been fitted and the thermodynamic parameters determined for $^{6}Li^{13}CHBr_2$, $^{6}Li^{13}CH_2SPh$, and $^{6}Li^{13}CH_2Pr^n$.[1543] Complex formation between diarylmethylenemalonaldehydes with Mg^{2+} and Li^+ has been studied by ^{1}H n.m.r. spectroscopy.[1544] ^{7}Li and ^{35}Cl n.m.r. spectroscopy has been used to investigate association of Li^+, [enolate]$^-$, and $[ClO_4]^-$.[1545] ^{13}C n.m.r. spectroscopy has been used to investigate equilibria in the PhM/$Mg(OCH_2CH_2OEt)_2$, M = Na, Li, K, system.[1546] ^{1}H n.m.r. spectroscopy has been used to determine the stoichiometry of the crown ether complexes $ML_{\underline{n}}$, M = Li, Na, K, Rb, Cs, Ca, Sr, and Ba, L = 12-crown-4, 15-crown-5, and 18-crown-6.[1547] ^{7}Li, ^{13}C, ^{17}O, ^{23}Na, and ^{39}K relaxation studies have been used to study the complexation of metal ions by crown ethers.[1548] The stability constants of LiSCN, NaSCN, KSCN, and NH_4SCN with poly(ethylene glycol) have been determined by ^{7}Li, ^{13}C, and ^{23}Na n.m.r. spectroscopy.[1549] The nuclear magnetic relaxation rates of ^{7}Li, ^{23}Na, ^{35}Cl, ^{81}Br, and ^{133}Cs have been measured for alkali-metal halide-poly(ethylene glycol)-water mixtures.[1550] ^{7}Li n.m.r. spectroscopy has been used to investigate Li^+ interaction with phosphatidylcholine-phosphatidylglycerol membranes.[1551] The interactions between ATP, monovalent cations, and divalent cations on rabbit muscle pyruvate kinase have been examined by ^{1}H, ^{7}Li, and ^{31}P n.m.r. spectroscopy.[1552]

An n.m.r. technique has been developed for the determination of formation constants of metal-ion complexes that uses competition of two metal cations for a given ligand and applied to ^{23}Na and ^{133}Cs.[1553] ^{1}H, ^{13}C, and ^{23}Na n.m.r. spectroscopy has been used to investigate the structure of $NaBu^n$ in thf.[1554] ^{23}Na $\underline{T}_1$ has been used to show that there is no interaction between Na^+ and $[MeSO_4]^-$.[1555] The kinetics of complexation of Na^+ with 18-crown-6 have been studied using ^{23}Na n.m.r. spectroscopy.[1556] ^{23}Na n.m.r. spectra of [Na(18-crown-6)]$^+Na^-$ solutions in $MeNH_2$ have been studied as functions of temperature and mole ratio of crown ether to Na. Na^+/Na^- exchange rates were determined and activation parameters determined.[1557] The influence of anions on the kinetics of complexation of Na^+ with 18-crown-6 has been investigated in thf by ^{23}Na n.m.r. spectroscopy and activation parameters determined.[1558] The effects of the salt concentration and counterion on the stability of alkali-ion-18-crown-6 complexes in aqueous and methanolic solution have been studied using ^{23}Na n.m.r. spectroscopy.[1559] The exchange kinetics for complexation-decomplexation of $NaBPh_4$ with dibenzo-24-crown-8 in $MeNO_2$ have been studied by ^{23}Na n.m.r. spectroscopy.[1560, 1561] ^{23}Na n.m.r. spectroscopy has been used to study sodium dibenzo-30-crown-10 aggregation in solution, and equilibrium constants have been determined.[1562] ^{23}Na T_1 and ^{1}H chemical shifts have been used to investigate the association of alkali-metal ions with polyene macrolides in methanol solution.[1563] Lamellar liquid crystalline phases containing sodium n-octanoate, water, and an alcohol have been studied by ^{2}H n.m.r. spectroscopy.[1564] The importance of the alcohol chain length and the nature of the hydrocarbon for the properties of ionic microemulsion systems have been studied by ^{1}H and ^{23}Na n.m.r. spectroscopy.[1565] ^{23}Na n.m.r. spectroscopy has been been used to investigate micelles.[1566] The effect of terminal unsaturation on the CMC values of sodium 11-dodecenoate and sodium 12-tridecenoate has been probed by ^{23}Na n.m.r. spectroscopy.[1567] Sodium diethyl-2-hexyl phosphate/H_2O/C_6H_6 inverted micelles have been investigated by ^{23}Na n.m.r. spectroscopy.[1568] The importance of the cosurfactant and the oil for the properties of microemulsions has been investigated by ^{23}Na n.m.r. spectroscopy.[1569] A ^{23}Na n.m.r. study of NaBr in methylamine solutions that contain macrocyclic polyethers has been reported.[1570] The donor number of some aprotic cyanamides has been determined by using ^{23}Na n.m.r. spectroscopy.[1571] The interaction of cysteamine with DNA has been studied using ^{23}Na n.m.r. spectroscopy.[1572] The monensin-mediated transport of Na^+ through phospholipid bilayers has been studied by ^{23}Na n.m.r. spectroscopy.[1573] The ^{23}Na n.m.r. spectrum of intact bovine lens and vitreous humor has been reported.[1574] The interaction of $^{23}Na^+$ with gramicidin channel has been studied by n.m.r. spectroscopy.[1575] Intra- and inter-cellular Na^+ and K^+ have been discriminated, in the ^{23}Na and ^{39}K n.m.r. spectrum, by using shift reagents.[1576] The interaction of group Ia cations with carrageenans has been studied by ^{23}Na, ^{39}K, ^{87}Rb, and ^{133}Cs n.m.r. spectroscopy.[1577] The state of intracellular Na^+ in human and dog erythrocytes has been characterised by ^{23}Na n.m.r. spectroscopy using dysprosium complexes as shift reagents. The extracellular volume was determined from ^{59}Co n.m.r. spectra of $[Co(CN)_6]^{3-}$.[1578] The effect of Ca^{2+} on vascular smooth muscle has been studied by ^{23}Na n.m.r. spectroscopy.[1578a] Intracellular sodium concentrations in erythrocytes[1579] and mammalian cardiac myocytes[1579a] have been determined by ^{23}Na n.m.r. spectroscopy.

The complexation of cyclohexanedioxydiacetamides[1580] and phenylenedioxydiacetamides[1581] by K^+ has been confirmed by ^{1}H and ^{13}C n.m.r. spectroscopy. A lyotropic ternary system containing potassium has been studied by ^{2}H n.m.r. spectroscopy.[1582] A ^{13}C n.m.r. study of K^{+} and Tl^{+} binding to gramicidin A transmembrane channel has been reported.[1583]

The kinetics of complexation of Cs^{+} with large crown ethers in acetone and methanol have been studied by ^{133}Cs n.m.r. spectroscopy.[1584] The interaction of Cs^{+} with gramicidin transmembrane channel has been investigated by ^{133}Cs relaxation [1585] and two dimensional ^{1}H n.m.r. spectroscopy.[1586]

Group IIa. ^{9}Be and ^{13}C n.m.r. spectroscopy has been used to investigate the Be^{2+}/malonic acid system and $Be\{(O_2C)_2CH_2\}$, $[Be\{(O_2C)_2CH_2\}_2]^{2-}$, and $[Be_3(OH)_3\{(O_2C)_2CH_2\}_3]^{3-}$ identified.[1587] The ^{19}F T_1 of molten $LiBeF_3$ has been measured. The relaxation is due to the dissociation of F^{-} from $[BeF_4]^{-}$.[1588]

Trans-1,2-diaminocyclohexane-*N*,*N*,*N*',*N*'-tetraacetate has been shown to be a superior ligand to edta for sequestering Mg^{2+} in ^{31}P n.m.r. experiments involving ATP.[1589] Charge-transfer interactions of chlorophyll *a*, pheophytin *a*, and their metal derivatives with 2,4,7-trinitrofluorenone have been studied by ^{1}H and ^{13}C n.m.r. spectroscopy.[1590] ^{13}C n.m.r. spectroscopy has been used to study the complexation of Mg^{2+} and Al^{3+} by cimetidine and ranitidine.[1591] A multicomponent self-diffusion n.m.r. study of aggregation of nucleotides, nucleosides, and nucleic acid bases with divalent metal ions has been reported.[1592] The exchange rate between free Mg^{2+} and Mg^{2+} complexed to ATP has been determined from ^{31}P n.m.r. spectra.[1593, 1594] A similar study has been made for Mg^{2+} with adenosine-5'-tetraphosphate using ^{31}P n.m.r. spectroscopy.[1595] Stability constants of Ca^{2+} or Pr^{3+} with aspartame have been determined by n.m.r. spectroscopy.[1596] ^{43}Ca, ^{25}Mg, ^{39}K, and ^{67}Zn n.m.r. spectroscopy has been used to study trifluoroperazine-calmodulin solutions.[1597]

^{1}H and ^{13}C n.m.r. spectroscopy has been used to study Ca^{2+} complexation by lasalocid-A.[1598] ^{2}H n.m.r. spectroscopy has been used to investigate the liquid crystallinity in barium surfactants.[1599]

The Lanthanides. A ^{1}H and ^{13}C n.m.r. study of La^{3+} and Lu^{3+} with triaza triacetic acid has been reported.[1600] ^{1}H and ^{13}C n.m.r. spectroscopy has been used to investigate the structure of $[R(C_8H_{17})_3N]^{+}$ ion pairs with some rare-earth-containing anions.[1601] ^{13}C n.m.r. spectroscopy has been used to investigate the binding of Gd^{3+} and Mn^{2+} to *D*-gluconamide.[1602] Mixed-ligand heteronuclear complexes of Gd^{3+} and Cu^{2+} with malic and mandelic acids have been studied by ^{1}H n.m.r. spectroscopy.[1603] A ^{1}H n.m.r. study of lanthanide complexes of aspartate and glutamate residues has been reported.[1604] A ^{6}Li, ^{7}Li, ^{17}O, ^{23}Na, and ^{31}P n.m.r. study of lanthanide cation binding to sodium triphosphate has been reported.[1605] The interaction of ATP with Sc, La, and Lu has been studied by ^{1}H, ^{17}O, and ^{31}P n.m.r. spectroscopy.[1606] The water relaxation properties of proteins labelled with Gd^{3+} and Mn^{2+} have been reported.[1607]

The Actinides. The kinetics of the ligand-exchange reactions in Th(2-thenoyltrifluoroacetonate)$_4$ and its DMSO complex have been determined using ^{1}H n.m.r. spectroscopy.[1608] The interaction of pyridoxal with $UO_2(OAc)_2$ has been studied by ^{1}H and ^{13}C n.m.r. measurements.[1609]

Vanadium, Niobium, and Tantalum. The system $VOSO_4$-histidine-glycol-H_2O has been studied by ^{13}C n.m.r. spectroscopy.[1610] ^{1}H and ^{13}C relaxation rates have been measured for the system $[VO]^{2+}$-glutathione in aqueous solution.[1611] In the catalysis of H_2O_2 disproportionation by NH_4VO_3, intermediate peroxo complexes of V^{5+} were detected by ^{51}V n.m.r. spectroscopy.[1612] A ^{31}P and ^{51}V n.m.r. study of the formation of high-vanadium vanadotungstophosphate heteropoly compounds has been reported.[1613] ^{31}P n.m.r. spectroscopy has been shown to be a useful analytical technique for determining V(V), Nb(V), and Ta(V) as the heteromolybdophosphate.[1614] The ^{17}O and ^{51}V n.m.r. spectra of $K_2S_2O_7.\underline{n}V_2O_5$ melts during the catalytic oxidation of SO_2 have been reported.[1615] ^{51}V n.m.r. spectroscopy has been used to study the binding of vanadium(V) oligoanions to sarcoplasmic reticulum[1616] and saccharomyces cerevisiae.[1617]

Chromium. Paramagnetic shifts and $\underline{T}_1$ have been determined for alcohols in CCl_4 solutions of $Cr(acac)_3$. Stability constants were calculated.[1618]

Molybdenum. ^{1}H n.m.r. spectroscopy has shown that $(C_5H_5)Mo(CO)_2$-$CH_2CH=CH_2$ and $[(C_5H_5)Mo(CO)_2(\eta\text{-}CH_2=CHMe)]^+$ are in equilibrium.[1619] Equilibria in the H^+-$[MoO_4]^{2-}$-$[HPO_4]^{2-}$ have been studied by ^{31}P n.m.r. spectroscopy.[1620]

Tungsten. Complexation of L-malic acid with W(IV) has been studied by ^{1}H n.m.r. spectroscopy.[1621] The temperature dependence of the bridging-hydrogen ^{1}H chemical shift of $Cl_3W(\mu\text{-}H)(\mu\text{-}Me_2S)_2WCl_2(SMe_2)$ has been attributed to an equilibrium between isomers. The ^{2}H n.m.r. spectrum was also recorded.[1622]

Manganese. The $[Mn(CNR)_6]^{+/2+}$ electron self-exchange in acetonitrile has been studied by ^{55}Mn n.m.r. spectroscopy.[1623] ^{13}C and ^{31}P n.m.r. relaxation has been used to study the stability and dynamics of complexes in the glycine/Mn(II)/ATP system.[1624] ^{13}C n.m.r.

spectroscopy has been used to determine the mode of interaction of Mn^{2+} and Cu^{2+} with cis-inositol.[1625] The conformation of the hydrocarbon chains of micellar amphiphilic molecules has been studied by n.m.r. relaxation of the surfactant nuclei, induced by Mn^{2+}.[1626] The volume of activation for electron transfer between $[MnO_4]^-$ and $[MnO_4]^{2-}$ ions has been determined by ^{55}Mn n.m.r. spectroscopy.[1627]

Iron. ^1H, ^{13}C, ^{31}P, and ^{199}Hg n.m.r. spectroscopy has been used to demonstrate the interconversion of isomers of $[Fe_4(HgMe)(CO)_{13}]^-$ and $Fe_4(AuPEt_3)(CO)_{12}(COMe)$.[1628] Electron self-exchange rate constants in a series of dicyanoiron porphyrins[1629] and bis(imidazole)-iron porphyrins[1630] have been measured by ^1H n.m.r. spectroscopy. ^1H and ^{13}C n.m.r. spectroscopy has been used to investigate carbon monoxide dissociation kinetics from haemoglobin.[1631] The pH-dependent axial ligation changes of monomer and dimer forms of iron(III) uroporphyrin I have been investigated by ^1H n.m.r. spectroscopy.[1632] The formation of Fe(III) mixed malic-mandelic acid complexes has been studied by ^1H n.m.r. spectroscopy as a function of pH.[1633] ^1H n.m.r. spectroscopy has been used to study Fe^{3+}-aspartic acid complexes.[1634] ^1H and ^{19}F n.m.r. spectroscopy has been used to investigate the equilibrium[1635]

$$[Fe_4(SPh)_{10}]^{2-} + 6[PhS]^- \rightleftharpoons 4[Fe(SPh)_4]^{2-}$$

Ruthenium. The rate constants for electron exchange between $[Ru(bipy)_2(acac)]^{+/2+}$ and $[Ru(bipy)_2(hfac)]^{+/2+}$ have been determined from ^1H and ^{19}F n.m.r. line broadening measurements.[1636]

Cobalt. The equilibrium constant for

$$2[Co(CNR)_4L]^+ \rightleftharpoons [Co(CNR)_3L_2]^+ + [Co(CNR)_5]^+$$

has been determined by ^1H n.m.r. spectroscopy.[1637] ^1H n.m.r. spectroscopy has been used to measure the hydrogen exchange rate of trans-$[Co(en)_2XY]^{n+}$.[1638] The absence of intramolecular proton transfer in the conjugate bases of trans-$[Co(en)_2XY]^{n+}$ has been established by ^1H n.m.r. spectroscopy.[1639] ^1H T_1 and chemical shifts have been used to characterise the outer-sphere complex formed between $[Co(MeOH)_6]^{2+}$ and pyridine in d_4-methanol.[1640] ^1H n.m.r. studies of the complex formation of Co(II) with 12-crown-4 has been investigated by ^2H n.m.r. spectroscopy.[1641] ^1H n.m.r. spectroscopy has been used to study the composition and stability constants of chloro complexes of Cu, Ni, and Co in HCl, H_2O-MeCN, H_2O-acetone, and aqueous dioxan.[1642]

Rhodium and Iridium. ^{31}P n.m.r. spectroscopy has been used to demonstrate the equilibrium[1643]

$$[Rh(dppe)_2]^- + [Rh(dppe)_2]^+ \rightleftharpoons 2Rh(dppe)_2$$

The PMe_2Ph ligands in mer-$[MCl_2(OH_2)(PMe_2Ph)_3]^+$ rapidly exchange on the n.m.r. time-scale, giving coalescence in the ^1H and ^{31}P n.m.r. spectra.[1644]

Nickel. The Evans' method has been used to investigate the planar-octahedral equilibrium in aqueous solutions of $[Ni(Me_4\text{-cyclam})]^{2+}$.[1645] Steric effects on the rates of redox reactions involving nickel(II)/(III) macrocycles have been studied by ^{13}C n.m.r. spectroscopy.[1646] The kinetics and mechanism of chelation of Ni(II) by α-(8-quinolylazo)-α-acetoacetonitrile have been investigated by ^1H and ^{13}C n.m.r. spectroscopy.[1647] The ^{13}C n.m.r. spectra of $[Ni(edta)]^{2-}$ show a pentadentate-hexadentate equilibrium.[1648] ^1H n.m.r. spectroscopy has been used to study the complexation of Zn(II), Cd(II), and Ni(II) by 6-methyl-pyridine-2-phosphonic acid.[1649] A ^1H and ^{17}O n.m.r. study of acetic acid exchange processes of nickel chloride, nitrate, and acetate have been reported.[1650] The binding of Cu^{2+} and Ni^{2+} to serum albumin has been studied by ^{35}Cl n.m.r. spectroscopy.[1651]

Platinum. ^{31}P n.m.r. spectroscopy has been used to examine the equilibrium between $PtPh(OH)(Ph_2PCH{=}CHPPh_2)$ and $[PtPh(OH_2)(Ph_2PCH{=}CHPPh_2)]^+$.[1652] ^1H n.m.r. spectroscopy has been used to demonstrate the simultaneous binding of Pt(II) to three different sites of a guanine nucleobase.[1653] ^1H n.m.r. spectroscopy has been used to determine the formation constant for $[(NH_3)_2Pt(C_5H_5N_2O_2)_2Zn(OH_2)_3]^{2+}$.[1654] ^1H n.m.r. spectroscopy has been used to investigate the unwinding of an undecamer DNA fragment on binding to cis-$PtCl_2(NH_3)_2$.[1655]

Copper. The equilibrium

$$2Me_2CuLi \rightleftharpoons Me_3Cu_2Li + MeLi$$

has been investigated by ^1H and ^7Li n.m.r. spectroscopy.[1656] ^{13}C n.m.r. spectroscopy has been used to investigate the complexation of Cu(II) by $H_2NCH_2CH_2OCH{=}CH_2$.[1657] ^1H n.m.r. spectroscopy has been used to investigate complexation between Cu(II)-histidine-L-ascorbic acid.[1658] The structure and dynamics of the Cu(II)-L-proline complex have been investigated by ^1H and ^{13}C relaxation measurements.[1659] ^1H n.m.r. broadening has been used to investigate complex formation between Cu(II) and $(HOCH_2CH_2)_3N$.[1660] ^{13}C relaxation has been used to investigate the binding of Cu(II) to $HO_2CC_5H_3CH_2N_2C_8H_{13}O$.[1661] The interaction of Cu(II)-histidine complexes with the vitamin C redox system has been investigated by ^1H n.m.r. spectroscopy.[1662] ^1H and ^{31}P T_1 and T_2 have been used to investigate Cu(II) binding to hydroxyethylidenediphosphonic acid.[1663]

N.m.r. spectroscopy has been used to investigate the binding of Cu(II) to hyaluronic acid.[1664]

Gold. A ^{31}P n.m.r. study of $Ph_2P(CH_2)_6PPh_2$ and $[AuBr_2]^-$ has shown the presence of 2-, 3-, and 4-coordinate gold, and their interchange above -50 °C.[1665] ^1H and ^{13}C n.m.r. spectroscopy has been used to investigate the interaction of $[CN]^-$ with aurothiomalate.[1666]

Zinc, Cadmium, and Mercury. ^{13}C n.m.r. spectroscopy has been used to determine the stability of complexes formed between trien and Zn(II),[1667] and between Zn(II) and 2,2'-ethylenediiminobis(ethylamine) and related ligands.[1668] The stability constants of mixed ligand complexes of the type $[Zn(phen)(amino\ acidate)]^+$ have been determined by ^1H n.m.r. spectroscopy.[1669] ^1H n.m.r. spectroscopy has been used to investigate the binding of N,N'-dipyridoxyethylenediamine-N,N'-diacetic acid to Zn^{2+}.[1670] ^1H and ^{13}C n.m.r. spectroscopy has been used to investigate the binding of Zn^{2+} to oxidised glutathione.[1671] The binding of Zn^{2+} to glycylhistidine and alanylhistidine has been investigated by ^1H n.m.r. spectroscopy.[1672] The complexation of zinc by glycylhistidyllysine[1673] and L-histidylglycylglycine[1674] has been studied by ^1H n.m.r. spectroscopy. The complexing of zinc with aspartic acid has been studied by ^1H n.m.r. spectroscopy.[1675] Ligand exchange kinetics of $(Me_2N)_3PS$ on Zn(II) and Cd(II) have been determined by ^{31}P and ^{35}Cl n.m.r. spectroscopy.[1676] ^1H, ^{13}C, ^{31}P and ^{113}Cd n.m.r. spectroscopy has been used to study the binding of Cd^{2+} to ATP.[1677] ^{13}C, ^{31}P, and ^{113}Cd n.m.r. spectroscopy has been used to study equilibria between Cd^{2+} and ethylenediamine-N,N,N',N'-tetraphosphonate.[1678] The binding of Cd^{2+} to osteocalcin has been studied by ^1H, ^{31}P, and ^{113}Cd n.m.r. spectroscopy.[1679] ^{113}Cd n.m.r. spectroscopy has been used to investigate the binding of the lanthanide ions to parvalbumins.[1680] ^{13}C and ^{113}Cd n.m.r. spectroscopy has been used to investigate the chelation of metal ions by parvalbumin.[1681] Metal-ion binding at the activating site of rabbit muscle phosphoglucomutase has been studied by ^7Li, ^{31}P, and ^{113}Cd n.m.r. spectroscopy.[1682] ^{113}Cd n.m.r. spectroscopy has been used to study exchange reactions between $Cd(S_2NR_2)_2$ and mercury.[1683] ^{113}Cd n.m.r. spectroscopy has been used to investigate cadmium iodide complexes in supercooled aqueous solutions.[1684] ^{199}Hg n.m.r. spectroscopy has been used to study the complexation of $[MeHg]^+$ with a series of pyridines.[1685] ^1H, ^{13}C, and ^{199}Hg n.m.r. spectroscopy has been used to investigate $[MeHg]^+$ derivatives of adenine with reference to disproportionation and syn-anti isomerism.[1686]

Boron. A ^1H n.m.r. study of the DMSO-H_3BO_3-KOH-H_2O system has shown the presence of $[B(OH)_4]^-$.[1687] ^{11}B n.m.r. spectroscopy has been used to study the equilibrium between borate and peroxide.[1688] ^{11}B and ^{13}C n.m.r. spectroscopy has been used to investigate the nature of ester formation between borate and D-mannitol, D-glucitol, D-fructose, and D-glucose.[1689] A ^{11}B and ^{27}Al n.m.r study of the H^+-Al^{3+}-$B(OH)_3$ system has indicated that no or very weak aluminium-borate complexes form in sea water.[1690]

Aluminium and Gallium. The ^{27}Al n.m.r. spectra of $Na[H_2Al(OCH_2CH_2OMe)_2]$ show that disproportionation has the solvent dependence C_6H_6 << Bu_2O $\leq$ Et_2O $\leq$ anisole < dioxan << THF < monoglyme $\leq$ diglyme.[1691] The system $NaAlEt_4$-benzene-HMPA has been investigated by ^1H and ^{27}Al n.m.r. spectroscopy.[1692] N.m.r. spectroscopy has been used to demonstrate the equilibrium between $[Al_2FEt_6]^-$, $[Al_3FEt_9]^-$, $AlEt_3$, and Al_2Et_6.[1693] Al(III) and Ga(III) complexes of 2-amino-3-phosphonopropionic acid have been studied by ^1H and ^{31}P n.m.r. spectroscopy.[1694] The various complexes formed in the system Al(III)-$N[CH_2P(O)(OH)_2]_3$ have been studied by ^1H and ^{31}P n.m.r. spectroscopy.[1695] ^{27}Al n.m.r. spectroscopy has provided evidence for the formation of $[Al(OH)]^{2+}$ from line broadening studies at high dilution.[1696] ^{27}Al n.m.r. spectroscopy has been used to study complex formation between Al(III) and the sulphate ion.[1697] The equilibrium between Al^{3+}, oxalic acid, and $[OH]^-$ has been studied by ^{27}Al n.m.r. spectroscopy.[1698] The nature of the ionic species formed when $AlCl_3$ or $Al(ClO_4)_3$ is dissolved in anhydrous alcohols has been studied by ^1H and ^{27}Al n.m.r. spectroscopy.[1699] The interaction of F^- with Al(III) and Ga(III) has been studied by ^{19}F n.m.r. spectroscopy.[1700] The formation of $[AlCl_4]^-$ and $[Al_2Cl_7]^-$ in chloroaluminate molten salts has been studied by ^{27}Al n.m.r. spectroscopy.[1701] The ^{27}Al n.m.r. spectrum of $[Bu^nNC_5H_5]Cl$ + $AlCl_3$ melts shows that the ions $[AlCl_4]^-$, $[Al_2Cl_7]^-$, $[Al_3Cl_{10}]^-$, and Al_2Cl_6 are present.[1702, 1703]

Indium. The synergic extraction of In(III) from chloride and perchlorate media with 1-Ph-3-Me-4-acylpyrazol-5-ones and $(C_8H_{17})_3PO$ mixtures has been studied by ^{31}P n.m.r. spectroscopy.[1704] ^{115}In n.m.r. spectroscopy has been used to investigate InCl, $InCl_2$, and $InCl_3$ in molten salts, and little support was found for $InCl_2$ being $In^I[In^{III}Cl_4]$.[1705]

Thallium. ^{205}Tl, ^{14}N, and ^1H n.m.r. spectroscopy has been used to investigate the association of Tl(I) and nitrate in liquid

$MeNH_2$.[1706] The kinetics of ligand exchange in aqueous thallium(III) halide solutions have been studied by ^{205}Tl n.m.r. spectroscopy.[1707]

Silicon. High-temperature high-resolution ^{23}Na, ^{27}Al, and ^{29}Si n.m.r. spectroscopy has been used to investigate molten silicates at 1300 °C.[1708] The formation constants for polysilicates have been determined by ^{29}Si n.m.r. spectroscopy.[1709]

Nitrogen. The relative motion of $[PF_6]^-/[ON(SO_3)_2]^{2-}$ and $F^-/[ON(SO_3)_2]^{2-}$ has been studied by ^{19}F n.m.r. spectroscopy.[1710]

Phosphorus. The extraction of phosphorus and fluorine from fluoride-phosphate-sulphate solutions by neutral oxygen-containing solvents has been investigated by ^{19}F and ^{31}P n.m.r. spectroscopy.[1711]

Selenium. ^{77}Se n.m.r. spectroscopy has been used to investigate the interaction α-chymotrypsin with $[PhSeCH_2CO_2]^-$.[1712]

Chlorine. ^{35}Cl n.m.r. spectroscopy has been used to study the association of $NaClO_4$ in aqueous solution.[1713]

Iodine. The binding of I^- to peroxidase has been studied using ^{127}I n.m.r. spectroscopy.[1714]

Equilibria Among Uncharged Species.—Titanium. ^1H n.m.r. studies have shown that $(C_5H_5)_2Ti(\mu\text{-}SH)_2Mo(CO)_4$ is fluxional, existing as a mixture of *syn* and *anti* isomers, which interconvert with $\Delta G^\ddagger = 72$ kJ mol^{-1}.[1715] ^1H n.m.r. spectroscopy has been used to determine the equilibrium[1716]

$$(C_5H_5)_2TiCl_2 \; + \; (C_5H_5)_2TiI_2 \; \rightleftharpoons \; 2\,(C_5H_5)_2TiClI$$

The ^1H, ^{13}C, and ^{31}P n.m.r. spectra of $[Ti(\eta\text{-}C_7H_7)(dmpe)(PMe_3)]^+$ show PMe_3 exchange, and $\Delta G^\ddagger$ was determined.[1717]

Chromium. ^1H and ^{13}C n.m.r. spectroscopy has been used to determine the equilibrium between axial and equatorial $C_6H_5Cr(CO)_3$ in $\{RCH(CH_2CH_2)_2CHC_6H_5\}Cr(CO)_3$.[1718] ^1H, ^{13}C, and ^{17}O n.m.r. spectroscopy has been used to study the coordination of cumene hydroperoxide to $Cr(acac)_3$.[1719]

Molybdenum. The ^1H and ^{13}C chemical shifts of cyclopentadienyl coordinated to $[(C_5H_5)_2Mo(\mu\text{-}SBu^t)_2Ni(C_5H_4R)]^+$ are temperature dependent due to an equilibrium between paramagnetic and diamagnetic isomers.[1720] ^1H and ^{13}C n.m.r. spectroscopy has been used to determine the temperature dependence of the equilibrium[1721]

$$(toluene)Mo(CO)_3 \; + \; 3THF \; \rightleftharpoons \; (THF)_3Mo(CO)_3 \; + \; toluene$$

The activation energy for *syn*-*anti* exchange at nitrogen in

$[Mo(N_2Ph)_2(acac)(OMe)]_2$ has been determined to be 57 kJ mol^{-1}.[1722]

Iron. Variable-temperature ^{13}C n.m.r. spectra of $Fe_4(MPR_3)(CO)_{13}$, M = Cu, Au, have been used to demonstrate an equilibrium between the tetrahedral and butterfly forms.[1723]

Ruthenium. ^{31}P n.m.r. spectroscopy has been used to investigate the interaction of $RuCl_2(PPh_3)_3$ with some Rh(III), Ir(III), and Pt(IV) tertiary phosphine complexes.[1724]

Cobalt. The ^{1}H n.m.r. spectrum of methylcobalamin shows concentration-dependent shifts which conform to a monomer-dimer equilibrium.[1725] Dynamic ^{1}H n.m.r. spectroscopic evidence for synergistic enhancement of *cis*- and *trans*-effects in vitamin B_{12} models has been presented.[1726] The equilibrium for base coordination to (dibenzo[*b*,*i*][1,4,8,11]tetraazacyclotetradecinato)cobalt(II) has been investigated by ^{13}C n.m.r. spectroscopy and $\Delta H°$ and $\Delta S°$ determined.[1727] Variable-temperature ^{1}H n.m.r. spectroscopy has been used to demonstrate that Co{1-hydroxy-2(^{1}H)-pyridone}$_3$ is conformationally labile.[1728]

Rhodium. ^{1}H n.m.r. spectroscopy has been used to determine the equilibrium between

$$Rh_2(OEP)_2 \quad + \quad PhCH{=}CH_2 \quad \rightleftharpoons \quad (OEP)RhCH_2CHPhRh(OEP)$$

where OEP = octaethylporphyrin.[1729] The relative labilities of the amine and ethylene ligands in *cis*-(amine)$Rh(CO)_2Cl$ and *trans*-(amine)$PtCl_2(C_2H_4)$ have been determined.[1730] It has been shown by ^{31}P n.m.r. spectroscopy that PPh_3 exchanges with that in Rh(anthranilate)(CO)(PPh_3) *via* an S_N2 mechanism.[1731] The redistribution of ligands in the $Rh(CO)(PPh_3)_2Cl$-$[Rh(CO)_2Cl]_2$ system has been studied by ^{13}C n.m.r. spectroscopy.[1732] Using ^{1}H, ^{13}C, ^{31}P, and ^{103}Rh variable-temperature n.m.r. spectroscopy, $P(OR)_3$ exchange has been found in $Rh_2(O_2CR)_4P(OR)_3$.[1733]

Iridium. $IrCl(CO)(PPh_3)_2(SO_2)$ displays a reversible exchange reaction in the presence of $IrCl(CO)(PPh_3)_2$. Thermodynamic parameters for the exchange were calculated from variable temperature ^{31}P n.m.r. data.[1734]

Nickel. The kinetics and mechanism of ligand substitution in four-coordinate nickel(II) chelate complexes have been studied by ^{1}H n.m.r. spectroscopy.[1735] The temperature dependence of the ^{13}C n.m.r. spectra of Ni(II) complexes of *o*-hydroxynaphthalidimine has been attributed to a tetrahedral-square-planar equilibrium.[1736]

Palladium and Platinum. ^{31}P n.m.r. spectroscopy has been used to investigate *cis*-*trans* isomerism in $PtH_2(PMe_3)_2(C_{10}H_8)$.[1737] The temperature dependence of the ^{1}H n.m.r. spectrum of $\overline{Pd\{CH_2C(=N}N$-

$Me_2)Bu^t\}Cl_2$ has been attributed to <u>cis</u>-<u>trans</u> isomerism.[1738] The interaction between <u>trans</u>-PtCl(CO-n-hexyl)$(PPh_3)_2$ and $SnCl_2$ has been investigated by ^{31}P n.m.r. spectroscopy.[1739] Solution studies using ^{31}P and ^{195}Pt n.m.r. spectra have revealed the equilibrium

$$2\ Cl_2PdL_2PtCl_2 \rightleftharpoons Cl_2PdL_2PdCl_2 + Cl_2PtL_2PtCl_2$$

where L = $Ph_2PCH{=}CHPPh_2$,[1740] and demonstrated the equilibrium[1741]

$$Pt(PPh_3)_2(PhNO) + CO_2 \rightleftharpoons (Ph_3P)_2\overline{PtONPhC(O)O}$$

The binding of $PtCl_2(NH_3)_2$ to guanosine monophosphate has been investigated by ^1H and ^{13}C n.m.r. spectroscopy.[1742]

<u>Copper</u>. ^1H and ^{13}C n.m.r. spectroscopy has been used to demonstrate an equilibrium between $Li[Me_2Cu]$ and $PhCH{=}CHCO_2R$.[1743] ^1H and ^{31}P n.m.r. spectroscopy has been used to investigate the interaction of {1-(α-pyridylmethylene)-2-(α-pyridyl)hydrazine}CuCl with DNA.[1744]

<u>Cadmium</u>. ^{31}P and ^{113}Cd n.m.r. spectroscopy has been used to investigate the interaction of $P(C_6H_{11})_3$ and $PBu^n{}_3$ with Cd(II) salts, and the formation of $CdX_2(PR_3)$ was demonstrated. $CdX_2(PCy_3)(PBu^n{}_3)$ was formed, and $^2\underline{J}(^{31}P,^{31}P)$ was measured as between 66 and 95 Hz.[1745]

<u>Boron</u>. ^1H and ^{11}B n.m.r. spectroscopy has been used to investigate the equilibrium[1746]

$$R^1OH + Ph_2BOR^2 \rightleftharpoons R^2OH + Ph_2BOR^1$$

<u>Aluminium</u>. ^1H n.m.r. spectroscopy has been used to investigate the equilibrium between $Bu^i{}_2AlH$ and an oligomer.[1747] The temperature dependence of the ^{13}C and ^{27}Al n.m.r. spectra of $[\overline{NC_5H_4CH_2OAl}R_{2-\underline{n}}X_{\underline{n}}]_2$ is due to a four-five-coordination equilibrium differing in nitrogen bonding.[1748]

<u>Gallium</u>. <u>Cis</u>-<u>trans</u> isomerism of $[Cl_2GaNMeSiMe_3]_2$ has been examined by ^1H, ^{13}C, and ^{29}Si n.m.r. spectroscopy.[1749]

<u>Silicon</u>. The activation energy for <u>syn</u>-<u>anti</u> isomerism in Me_2N-$N{=}CMeOSiMe_3$ and the equilibrium between $Me_3SiNMeN{=}CMeOSiMe_3$ and $Me_3SiNMeNAcSiMe_3$ have been investigated by ^1H, ^{13}C, ^{15}N, ^{17}O, and ^{29}Si n.m.r. spectroscopy.[1750] The activation energy for <u>cis</u>-<u>trans</u> isomerism in (mesityl)RSi=Si(mesityl)R has been determined, and the ^{29}Si n.m.r. spectrum recorded.[1751] The complexation of $Si(OH)_4$ by tropolone has been investigated by ^{29}Si n.m.r. spectroscopy.[1752]

<u>Tin</u>. The stability and self-association of $\{(neophyll)_3Sn\}_2O$, $(neophyll)_3SnOH$, and $\{(neophyll)_3Sn\}_2CO_3$ have been examined by ^{13}C and ^{119}Sn n.m.r. spectroscopy.[1753] ^1H n.m.r. spectroscopy has been used to examine the equilibrium between Ar_6Sn_3 and $Ar_2Sn{=}SnAr_2$. The ^{119}Sn n.m.r. spectrum was also recorded.[1754] ^1H, ^{13}C, and

^{119}Sn n.m.r. spectroscopy has been used to investigate the structure and equilibria of $R_2Sn(glygly)$.[1755] ^{77}Se, ^{119}Sn, and ^{125}Te n.m.r. spectroscopy has been used to determine the position of equilibrium in $Sn(SPh)_{\underline{x}}(SePh)_{\underline{y}}(TePh)_{4-\underline{x}-\underline{y}}$. The coupling constants and shifts were interpreted using pairwise additivity.[1756]

Phosphorus. The position of the chair-chair equilibrium in some 1,3,2-oxazaphosphorinanes has been determined by ^{1}H n.m.r. spectroscopy. The ^{31}P n.m.r. spectrum has also been recorded.[1757] $\Delta G^{\ddagger}$ for E-Z isomerism in $Ph_2P(S)C(SMe){=}NMe$ has been determined by ^{13}C and ^{31}P n.m.r. spectroscopy.[1758] The reversible isomerism of 1-methyl-bis-o-phenylenedioxyphosphoranylamino-3-methyliminoisoindolenine has been studied by dynamic ^{1}H and ^{31}P n.m.r. spectroscopy.[1759]

Antimony. ^{1}H n.m.r. spectroscopy has been used to show the formation of 1:1 compounds between $SbBr_3$ and alkylbenzenes.[1760]

Sulphur. The equilibrium

$$D_2S \quad + \quad H_2S \quad \rightleftharpoons \quad HDS$$

has been investigated by ^{1}H and ^{2}H n.m.r. spectroscopy.[1761]

Course of Reactions.—Alkali Metals. The cyclisation of 5-hexenyllithium to (cyclopentylmethyl)lithium has been determined using ^{1}H n.m.r. spectroscopy.[1762] ^{31}P n.m.r. spectroscopy has been used to investigate the reaction between P_4 and alkali-metal phosphides to give M_3P_7.[1763]

Uranium. An n.m.r. study of the reaction of alkylene oxides[1764] and alcohols[1765] with UCl_4 has been reported.

Titanium and Zirconium. Ethylene polymerisation by the system $(C_5H_5)_2TiMeCl/AlMeCl_2$ has been investigated by ^{13}C n.m.r. spectroscopy.[1766] The rate of hydrolysis of $(C_5H_5)_2TiCl_2$, $(C_5H_5)_2VCl_2$, and $(C_5H_5)_2ZrCl_2$ has been determined. The solid-state ^{13}C n.m.r. spectrum of '$Ti(C_5H_5)_{0.31}O_{0.30}(OH)$' has been recorded.[1767] ^{1}H and ^{13}C n.m.r. spectroscopy has been used to determine the rate of ligand redistribution reactions of $(C_5H_5)_2ZrR_2$, $(C_5H_5)_2ZrRX$, and $(C_5H_5)_2ZrX_2$.[1768] ^{1}H n.m.r. spectroscopy has been used to monitor the deuteriation of $\{R_2Si(C_5H_4)_2\}_2Zr_2H_2(\mu\text{-}H)_2$.[1769]

Niobium and Tantalum. The deuteriation of $(C_5H_5)_3NbH_3$ has been followed by ^{1}H n.m.r. spectroscopy. The ^{93}Nb n.m.r. spectrum was also recorded.[1770] The kinetics of hydride-olefin insertion in $(C_5Me_5)_2NbH(C_2H_4)$ have been investigated by ^{1}H and ^{13}C n.m.r. spectroscopy.[1771]

Molybdenum. ^{1}H n.m.r. spectroscopy has been used to investigate the reaction of $MoMe(CO)_3(C_5H_5)$ with $LiBEt_3H$ to give [MoMe(CHO)-

$(CO)(C_5H_5)]^-$, then $MoH(COMe)(CO)_2(C_5H_5)$, and finally $[Mo(\eta^2\text{-}MeCHO)\text{-}(CO)_2(C_5H_5)]^-$. The ^2H and ^{13}C n.m.r. spectra were also recorded.[1772] ^{31}P n.m.r. spectroscopy has been used to monitor the formation of ammonia during protonation of trans-$Mo(N_2)_2$(triphos)-(PR_3).[1773] The stepwise hydrolysis of MoF_6 in MeCN, containing H_2O, has been studied by ^{19}F n.m.r. spectroscopy.[1774]

Tungsten. ^2H and ^{13}C n.m.r. spectroscopy has been used to investigate the deuteriation of $(C_5H_5)_2WHMe$.[1775] ^1H and ^{13}C n.m.r. spectroscopy has been used to investigate the mechanism of CO_2 insertion into the W-R bond in threo-$[W(CO)_5CHDCHDPh]^-$.[1776] The reaction between $W(CHBu^t)(OCH_2Bu^t)_2Br_2$ and $GaBr_3$ and norbornenes to produce carbene complexes has been monitored by ^1H n.m.r. spectroscopy.[1777] ^{13}C and ^{31}P n.m.r. spectroscopy has been used to monitor the reaction of trans-$Pt(C_2H)_2(PMe_2Ph)_2$ and $W_2(OBu^t)_6$ to give trans-$Pt(C_2H)\{C_2W(OBu^t)_5\}(PMe_2Ph)_2$ and Bu^tOH. $^1J(^{13}C,^{13}C)$ was also determined.[1778] The thermal decomposition of 7-phenyl-7-phosphanorbornadiene-$W(CO)_5$ has been investigated by ^1H and ^{31}P n.m.r. spectroscopy.[1779] The reaction of WOF_4 with some amino alcohols has been investigated by ^{19}F n.m.r. spectroscopy.[1780] The effect of medium basicity on the products of WF_6 stepwise hydrolysis has been investigated by ^{19}F n.m.r. spectroscopy.[1781]

Rhenium. The reaction of $(C_5H_5)Re(NO)(PPh_3)H$ with Bu^nLi to give $(C_5H_4Li)Re(NO)(PPh_3)H$ has been monitored by ^{31}P n.m.r. spectroscopy.[1782] The reaction of $(C_5H_5)Re(NO)(PPh_3)(CH_2CN)$ with Bu^nLi has been investigated by ^{31}P n.m.r. spectroscopy.[1783] ^{31}P n.m.r. spectroscopy has been used to monitor the reaction between $Re_2Cl_4(dppm)_2$ with EtCN to give $[Re_2Cl_3(dppm)_2(NCEt)_2]Cl$.[1784]

Iron. The reaction of $FeMeI(CO)_2(PMe_3)_2$ with CO at 250 K has been monitored by ^{13}C n.m.r. spectroscopy.[1785] ^1H n.m.r. spectroscopy has been used to monitor the reaction of $Fe(tropone)(CO)_3$ with tcne, and activation parameters were determined. The ^{13}C n.m.r. data of the products were recorded.[1786] The conversion of $[(C_5Me_5)Fe(C_5Me_4CH_2)]^+$ to $[(C_5Me_5)Fe(C_5Me_4CH_2)]_2{}^{2+}$ has been monitored by ^1H and ^{13}C n.m.r. spectroscopy.[1787] ^1H n.m.r. spectroscopy has been used to investigate the oxidation of (tetramesitylporphyrin)FeO by halogen.[1788] The rate of conversion of Fe(III) tetra-p-tolylporphyrin hydroxide to the corresponding μ-oxo-bridged dimer has been studied by ^1H n.m.r. spectroscopy.[1789] The kinetics of the decomposition of H_2O_2 catalysed by $[Fe(edta)]^-$ have been determined by ^1H n.m.r. spectroscopy.[1790]

Ruthenium and Osmium. During the reaction of $MH(NO)(PPh_3)_3$, M =

Ru, Os, with RCO_2H, the intermediates $RuH(O_2CR)(NO)(PPh_3)_2$ and $[OsH_2(NO)(PPh_3)_3]^+$ were detected by ^1H and ^{31}P n.m.r. spectroscopy.[1791] ^{31}P n.m.r. spectroscopy has been used to investigate the reaction of $HOs_3(CO)_{10}(PPhH)$ with DBU.[1792] ^1H n.m.r. spectroscopy has been used to determine the rates and equilibrium constants for CD_3CN solvolysis of halide ion in $Ru(C_5H_5)XL_2$ complexes.[1793] ^{13}C n.m.r. spectroscopy has been used to follow the reduction of CO_2 to formate catalysed by $Ru(bipy)_3$.[1794]

Cobalt. The reaction of styrene with $HCo(CO)_4$ produces CIDNP.[1795] ^{13}C n.m.r. spectroscopy has been used to show that the transfer of a methyl group from base-off methylcobalamin to $[PtCl_6]^{2-}$ produces $[MePtCl_5]^{2-}$.[1796] A similar ^{13}C, ^{31}P, and ^1H n.m.r. spectroscopy study of methyl transfer from methylcobalamin to thiols has been reported.[1797] ^{13}C n.m.r. spectroscopy has been used to investigate the acid-induced decomposition of $HOCH_2CH_2Co(dmgH)_2(OH_2)$.[1798] The kinetics of the oxidation of Co(II) to Co(III) in ammoniacal solution have been studied by ^1H T_1 measurements.[1799] The activation energy for the intramolecular rearrangement of the secondary explosive $[2\text{-}(5\text{-cyanotetrazolato})Co(NH_3)_5][ClO_4]_2$ has been determined by variable-temperature ^1H n.m.r. spectroscopy.[1800] ^1H n.m.r. spectroscopy has been used to determine the rate of deuteriation of $[CoCl(NH_3)_5]^{2+}$.[1801] The acid dependence of the kinetics of ammoniation of trans-$[CoN_3(en)_2(DMSO)]^{2+}$ to give cis- and trans-$[Co(N_3)(en)_2(NH_3)]^{2+}$ has been determined.[1802] The rate of deuteriation of $[Co(en)_2XY]^+$ has been determined.[1803-1805] ^1H n.m.r. spectroscopy has been used to study the exchange of amine protons between syn,anti-cis-$[Co(1,4,7,10\text{-tetraazacyclodecane})Cl_2]^+$, and D_2O has been investigated.[1806]

Rhodium and Iridium. The reaction of $[IrH_2(PPh_3)_2(OCMe_2)_2]^+$ with $Au(PPh_3)(NO_3)$ has been monitored by ^1H and ^{31}P n.m.r. spectroscopy.[1807] The reaction between $[Rh(nbd)(Pr^i{}_2PCH_2CH_2PPr^i{}_2)]^+$ and H_2 has been monitored by ^1H n.m.r. spectroscopy.[1808] ^{31}P n.m.r. spectroscopy has been used to monitor the reaction of $Rh(acac)(CO)_2$ with $P(OPh)_3$.[1809] ^{31}P n.m.r. spectroscopy has been used to investigate the photocatalytic dehydrogenation of MeOH in the presence of cis-$Rh_2Cl_2(CO)_2(dppm)_2$ and $Pd_2Cl_2(dppm)_2$.[1810] ^1H n.m.r. spectroscopy has been used to monitor the reaction of IrH(CO)(dppe) with HBr. The ^{13}C and ^{31}P n.m.r. spectra of $IrHBr(dppe)\{Si(OEt)_3\}(CO)$ were reported.[1811] The role of O_2 in conversion of $M(NO)(PPh_3)_3$, M = Rh, Ir, to $M(O_2CCF_3)(NO)(PPh_3)_2$ by CF_3CO_2H has been investigated by ^1H and ^{31}P n.m.r. spectroscopy.[1812]

Nickel. The reaction of $\{(tol)_3P\}_2Ni(C_2H_4)$ with C_2H_4 and HCN at -40 °C to give $\{(tol)_3P\}Ni(CN)Et(C_2H_4)$ has been investigated by ^1H, ^{31}P, and ^{13}C n.m.r. spectroscopy.[1813] The reaction of $[Ni(\eta\text{-allyl})\{P(OR)_3\}_2]^+$ with butadiene has been monitored by ^{31}P n.m.r. spectroscopy. The ^{13}C n.m.r. spectrum was also recorded.[1814] The reactions of 1,1-dimethyl-allene with $Ni(PPh_3)_3$ have been studied by n.m.r. spectroscopy.[1815] ^{31}P n.m.r. spectroscopy has been used to investigate the role of $NiX_2(PEt_3)_2$ as a catalyst for the water gas shift reaction.[1816]

Palladium and Platinum. Cis-trans isomerism of cis-$Pd(CD_3)_2(PEt_2Ph)_2$ has been followed by ^1H and ^{31}P n.m.r. spectroscopy.[1817] ^1H n.m.r. spectroscopy has been used to show that $Me_4Pt_2(\mu\text{-}SMe_2)_2$ reacts with MeI in CD_3CN to give the intermediate fac-$[Me_3Pt(SMe_2)_2(NCCD_3)]I$.[1818] The reaction of PtMe(OMe)(dppm) with C_2F_4 has been investigated quantitatively by ^{19}F and ^{31}P n.m.r. spectroscopy.[1819] ^2H n.m.r. spectroscopy has been used to study the deuteriation of $[MePt(\mu\text{-dppm})_2(\mu\text{-}CH_2)PtL]^+$, and the ^{31}P n.m.r. spectrum was also recorded.[1820] ^1H and ^{31}P n.m.r. spectroscopy has been used to study the mechanism of decomposition of $PtEtCl(PEt_3)_2$.[1821] The thermolysis of (cod)$PtEt_2$, (cod)PtEtCl, and (cod)PtEtI has been studied by quantitative ^{13}C n.m.r. spectroscopy. $\Delta G^{\ddagger}$ and ^{13}C T_1 were determined.[1822] ^{31}P n.m.r. spectroscopy has been used to investigate the catalytic allylation of 1,1,3-triphenylprop-2-enyl acetate with sodium dimethyl malonate using a Pd(S,S-chiraphos) catalyst.[1823] ^{13}C and ^{31}P n.m.r. spectroscopy has been used to investigate the mechanism of dimerisation of 1,3-dienes catalysed by Pd(O),[1824] and the trimerisation of butadiene catalysed by $Pd(dba)_2$ in the presence of PR_3.[1825] ^{31}P n.m.r. spectroscopy has been used to investigate the reaction of $Pt(\eta\text{-}C_3H_5)_2$ with tcne.[1826] Photolysis of $[Pd_2(CNMe)_6]^{2+}$ and $[Pt_2(CNMe)_6]^{2+}$ gives $[PdPt(CNMe)_6]^{2+}$ as monitored by ^{195}Pt n.m.r. spectroscopy.[1827] Solvolysis of cis-$PtX_2(NH_3)_2$ in DMSO has been studied by ^{195}Pt n.m.r. spectroscopy.[1828] ^{13}C and ^{31}P n.m.r. spectroscopy has been used to investigate the reaction of $(Ph_3P)_2Pd(O_2CCF_3)_2$ to give $[(Ph_3P)_3PdH]^+$.[1829] The kinetics and mechanisms of Pt(II) promoted hydrolysis of inorganic polyphosphates have been determined by ^{31}P n.m.r. spectroscopy.[1830]

Gold. The reactions of $AuBr_3(PEt_3)$ with phthalimide have been followed by ^{31}P n.m.r. spectroscopy. The ^{15}N n.m.r. spectrum was also determined.[1831]

Boron. The pyrolysis reaction of $MBH_4\text{-}Et_4NX$ has been studied by

^{11}B n.m.r. spectroscopy.[1832] The reaction of BH_3.THF with As(N-Et_2)Me_2 has been monitored by ^{1}H, ^{11}B, and ^{13}C n.m.r. spectroscopy.[1833] The disproportionation of $[BH_3Cl]^-$ to $[BH_2Cl_2]^-$ and $[B_2H_7]^-$ has been observed by ^{1}H and ^{11}B n.m.r. spectroscopy.[1834] ^{2}H and ^{11}B n.m.r. spectroscopy has been used to investigate the deuteriation of B_2H_6 and B_5H_9.[1835] The reaction of (μ-Br)$Mn_2(CO)_6$-(B_3H_8) with $HSnBu_3$ has been monitored by ^{11}B and ^{119}Sn n.m.r. spectroscopy.[1835a] ^{11}B n.m.r. spectroscopy has been used to show that B_4Cl_4 is reduced by $[BH_4]^-$ to B_5H_9 and B_6H_{10}.[1836] ^{11}B n.m.r. spectroscopy has been used to monitor the thermal rearrangement of BB'Cl_2-2,4-$C_2B_5H_5$ [1837] and the reaction between $LiCHCl_2$ and boronic esters.[1838]

Silicon, Germanium, and Tin. ^{1}H, ^{13}C, ^{29}Si, and ^{119}Sn n.m.r. spectroscopy has been used to monitor the rate of cleavage of $Me_3SnCH_2MMe_3$, M = C, Si, Ge, Sn.[1839] ^{31}P n.m.r. spectroscopy has been used to investigate the reaction of $Bu^n{}_4PCl$ with $(Me_3Si)_3C$-SiPhBr.[1840] ^{1}H n.m.r. spectroscopy has been used to follow the protolyses of Me_3SnMMe_3, M = C, Si, Ge, Sn,[1841] and its reaction with $ClSnMe_3$ or Cl_2Me_2Sn.[1842] The thermal isomerism of 3-Me_3Si-1-pyrazoline to 1-Me_3Si-2-pyrazoline has been followed by ^{1}H n.m.r. spectroscopy. The ^{13}C n.m.r. spectra were also recorded.[1843] ^{31}P n.m.r. spectroscopy has been used to monitor the reaction of $(PhO)_2P(O)(OSiMe_3)$ with α-(naphthyl)PhMeSiCl.[1844] The presence of radicals in the reaction of $Et_3Ge^{\bullet}$ with RSO_2ER has been proven by ^{13}C CIDNP n.m.r. spectroscopy.[1845] The temperature dependence of the diastereoselectivity in the aldol-type reaction of CHR^1=CR^2OSn-Bu_3 with PhCHO has been monitored by ^{119}Sn n.m.r. spectroscopy.[1846] The kinetics of hydrolysis of some chlorodisiloxanes have been studied by ^{1}H n.m.r. spectroscopy,[1847] along with the acetolysis of $XSiMe_2Cl$.[1848] ^{13}C and ^{29}Si n.m.r. spectroscopy has been used to follow the hydrolysis and condensation of $R^1Si(OR^2)_3$.[1849] ^{29}Si n.m.r. spectroscopy has been used to monitor the polymerisation of supersaturated silica solutions [1850] and the hydrolysis of Si(O-$Me)_4$.[1851-1853] A ^{19}F n.m.r. study of the reaction of $[GeF_6]^{2-}$ in H_2O_2 solution has indicated the formation of $[GeF_{6-\underline{n}}(OOH)_{\underline{n}}]^{2-}$.[1854]

Phosphorus. ^{1}H, ^{13}C, and ^{31}P n.m.r. spectroscopy has been used to determine the rate of the Wittig reaction of nonstabilised phosphorus ylides.[1855] The kinetics of deuteriation of Bu^tSH with Ph_2PH have been investigated by ^{1}H n.m.r. spectroscopy.[1856] ^{31}P n.m.r. spectroscopy has been used to determine the rate of disproportionation of a series of phosphinic carboxylic mixed anhydrides.[1857]

The kinetics of hydrolysis of $RC_6H_4CH=C\{P(O)(OEt)_2\}_2$ have been determined by ^1H and ^{31}P n.m.r. spectroscopy.[1858] ^{16}O, ^{17}O, ^{18}O isotope effects in the ^{31}P n.m.r. spectrum have been used to investigate the mechanism of phosphoryl transfer.[1859] ^{18}O-induced shifts in the ^{31}P n.m.r. spectrum have been used to determine the mechanism of the reaction between 2-MeO-5,5-Me_2-1,3,2-dioxaphosphorinan-2-one and 2-chloro-5,5-Me_2-1,3,2-dioxaphosphorinan-2-one.[1860] ^{31}P n.m.r. spectroscopy has been used to investigate the mechanism of transfer of a phospho group from phenyl phosphate to Bu^tOH, using ^{16}O, ^{17}O, and ^{18}O labelling.[1861] ^{31}P n.m.r. spectroscopy has been used to investigate the reaction of cyclo-triphosphate with L-α- or L-β-alanine.[1862] ^1H and ^{13}C n.m.r. spectroscopy has been used to investigate the mechanism of alkaline hydrolysis of methyl ethylene phosphate.[1863] The preparation of phosphoric polyanion gels from metaphosphates has been studied by ^{31}P n.m.r. spectroscopy.[1864]

Chlorine. The kinetics of disproportionation of anhydrous $HClO_4$ have been studied by ^1H n.m.r. spectroscopy.[1865]

4 Paramagnetic Complexes

In this section compounds of *d*-block transition elements will be considered first and then those of the lanthanide and actinide elements. Papers concerning the use of paramagnetic complexes as 'shift reagents' are usually omitted.

Four reviews have appeared: 'Theory of nuclear spin relaxation in paramagnetic systems in solution',[1866] 'Recent developments in the theory of n.m.r. shifts in paramagnetic systems',[1867] 'Conformational studies using n.m.r. spectroscopy with lanthanide shift reagents',[1868] and 'N.m.r. of shift reagents'.[1869]

The calculation of the n.m.r. chemical shift for $3d^2$ and $4d^2$ systems in a strong crystal field of octahedral symmetry has been examined.[1870, 1871]

The Transition Metals.—Vanadium. ^1H n.m.r. spectra of $(C_5H_5)V(BH_4)(dmpe)$[1872] and $(C_5H_5)_2V(CMe{=}NC_6H_3Me_2)$[1873] have been reported. Solution ^1H and ^{31}P n.m.r. observations on $VCl_3(PMePh_2)_2$ are consistent with two unpaired electrons.[1874]

Chromium and Molybdenum. The stereochemistry of $[Cr(malonate)_2(nicotinate)_2]^-$ has been established by ^2H n.m.r. spectroscopy.[1875]

The ^{2}H n.m.r. spectra of $[Cr(edta)]^-$ have been assigned by using specific deuteriation.[1876] For $[Cr(oxalate)_2(py)_2]^-$ the computed ^{2}H chemical shift is in agreement with experiment.[1877] The $Cr(acac)_3$-induced increase in spin density on carbon atoms of halogenated alkanes has been attributed to electron donor-acceptor interaction.[1878] The ^{1}H n.m.r. spectrum of $[Mo_2Fe_6S_8(SEt)_3(SFc)_6]^{3-}$ has also been reported.[1879]

Manganese and Technetium. The ^{1}H n.m.r. spectrum of bis(2,6-lutidine N-oxide)(tetraphenylporphinato)Mn(III) has been used to establish the ground state to be d_4 high-spin.[1880] N.m.r. data have also been reported for $[Mn(tetraphenylporphyrin)(DMF)_2]^+$ (^{1}H)[1881] and $TcN(S_2PPh)_2$ (^{1}H, ^{31}P).[1882]

Iron, Ruthenium, and Osmium. High-, low-, and variable-spin-state five-coordinate aryl and perfluoroaryl iron(III) porphyrins have been characterised by ^{1}H n.m.r. spectroscopy.[1883] ^{1}H n.m.r. studies of iron porphyrins with a vinyl carbene inserted between iron and a pyrrole nitrogen have been reported.[1884] ^{1}H n.m.r. contact shifts have been measured for $[M(bipy)_3]^-$ and $[M(bipy)_3]^+$.[1885] ^{1}H and ^{2}H n.m.r. studies on Fe(II) complexes of N-substituted porphyrins have been reported.[1886] ^{1}H n.m.r. spectra have been measured for some high-spin Fe(III) porphyrin complexes.[1887] The aggregation of Fe(III)(mesoporphyrin)Cl has been studied using ^{1}H n.m.r. spectroscopy.[1888] ^{2}H n.m.r. studies of low-valent iron porphyrin species[1889] and Fe(octaethylporphyrin)SPh[1890] have been reported. The ^{1}H chemical shifts of paramagnetic mixed-ligand Fe(III) dithiocarbamates have been analysed and the temperature dependence of the isotropic shift determined.[1891] ^{19}F chemical shifts have been determined for paramagnetic Fe(III) porphyrins, substituted with CF_3 groups.[1892] The ^{1}H T_1 of methyl groups in a series of low-spin ferric porphyrin complexes has been analysed in terms of the paramagnetic influences.[1893] Some Fe(III) complexes of methylporphyrin have been characterised by ^{1}H and ^{2}H n.m.r. spectroscopy.[1894] Analysis of variable-temperature ^{1}H n.m.r. spectra has shown that (trans-7,8-dihydro-2,3,7,8,12,13,17,18-octaethylporphyrinato)iron(II) possesses rhombic magnetic anisotropy.[1895] ^{1}H n.m.r. studies have been carried out on some Fe(III) porphyrins. The isotropic ^{1}H chemical shifts show anomalous temperature dependence.[1896] ^{1}H and ^{2}H n.m.r. spectroscopy has been used to study axial amine coordination in ferryl porphyrin complexes.[1897] Evidence for dialkoxyiron(IV) porphyrins has been obtained using ^{1}H, ^{13}C, and ^{19}F n.m.r. spectroscopy.[1898] The susceptibility of ClFe(Et_8-porphyrin) has been determined using Evans' method.[1899] The hyperfine-shifted ^{2}H n.m.r. spectra of meso-$^{2}H_4$-haem have been studied.[1900] N.m.r. data have also been reported for some haems.[1901] The high-spin—low-spin transition in Fe^{3+} dithiocarbamates has been investigated and interpreted in terms of preferential solvation or second-coordination-sphere reorganisation effects.[1902]

Cobalt. Full assignment of the ^{1}H n.m.r. spectra of some Co(II) porphyrins has been achieved using selective deuteriation and chlorine substitution.[1903] The ^{1}H n.m.r. spectra of some Co(II) complexes of octaethyl porphyrin have been assigned.[1904] Variable temperature ^{1}H n.m.r. spectra of some Co(II) complexes of tetraphenyl porphyrin have been used to deduce the electronic structure of the Co(II) ion.[1905] The temperature dependence of the isotropic paramagnetic shifts of CoX_2(3,5-Et_2-4-Me-pyrazole)$_2$ has been reported.[1906] ^{1}H n.m.r. studies have been used to demonstrate complete electron delocalisation in $[Co_8S_6(SPh)_8]^{5-}$.[1907]

Nickel. The ^{1}H n.m.r. spectrum of $[Ni(Me_6[14]ane\text{-}N_4)X]_2^{2+}$ has been reported.[1908] Full assignment of the ^{1}H n.m.r. spectra of some Ni(II) porphyrins has been achieved using selective deuteriation and chlorine substitution.[1909] Contact shifts induced by 1,11-bis-(2'-oxyphenyl)-2,6,10-triazaundeca-1,10-dienato-Ni(II) have been analysed.[1910] The first application of 2D n.m.r. spectroscopy to a paramagnetic complex has been reported using bis[R-(+)-P,P-diethylthiophosphinic acid-N-α-methylbenzylamido]nickel(II).[1911]

Copper. ^{1}H n.m.r. spectroscopy has been used to study the hydrogen positions, molecular motions, and magnetic properties of $Cu(py)_4Cl_2$.[1912]

Compounds of the Lanthanides and Actinides.—Lanthanides. The ^{1}H n.m.r. spectra of $Me_2Si(C_5H_4)_2SmCH(SiMe_3)_2$[1913] and $(C_5H_5)_2Ln(NCR)_n$, M = Ce, Pr.[1914] ^{1}H n.m.r. data for ethylenediaminetetraphosphonic acid complexes with Yb^{3+}, Tm^{3+}, Pr^{3+}, and Ho^{3+} in D_2O indicate that the complexes have the same structure, independent of solution pH.[1915] The ^{31}P n.m.r. spectra of $LnCl_3\{R_2P(O)H\}_3$ have been reported.[1916] The use of $N(CH_2CO_2H)_3$ as a chelating agent for determining rare-earth metals and actinides by ^1H n.m.r. spectroscopy has been discussed.[1917] Paramagnetic induced shifts in $N(CH_2CH_2OCH_2CH_2OCH_2CH_2)(CH_2CH_2OCH_2CH_2)(CH_2CH_2OC_6H_4OCH_2CH_2)N$ by lanthanides have been observed.[1918] ^1H and ^{13}C chemical shifts have been reported for some lanthanide complexes of dicyclohexyl-18-crown-6.[1919] The structure of some Eu(III) crown ether complexes in solution has been investigated by ^1H n.m.r. spectroscopy.[1920]

A series of new achiral binuclear complexes formed from Yb(β-diketonate)$_3$ and Ag(β-diketonate) have been evaluated as olefin shift reagents.[1921] A method has been outlined to determine the dipolar contribution to the isotropic chemical shift for substrates attached to a lanthanide shift reagent.[1922] The origin of shifts induced by Ag-Ln shift reagents in the ^1H n.m.r. spectra of unsaturated hydrocarbons has been analysed.[1923] Binuclear lanthanide(III)-Ag(I) β-diketonate chelates have been shown to be effective ^1H, ^{13}C, and ^{15}N shift reagents for multiple-bonded nitrogen atoms.[1924] The shift of Eu(fod)$_3$(imidazole) has been reported.[1925] The ^1H, ^{13}C, and ^{17}O shifts have been determined for M(fod)$_3$, M = La, Pr, Eu, Gd, Dy, Yb, complexes of adamantone and separated into diamagnetic, contact, and pseudocontact shifts.[1926]

Actinides. N.m.r. data have also been reported for $U(BH_4)_3(PR_3)_2$ (^{11}B, ^{31}P),[1927] $U(C_5H_5)_3(H_3BR)$ (^{11}B),[1928] $UCl_2\{H_2B(pz)_2\}_2$.THF (^{11}B),[1929] $U(C_5H_5)_3(\eta^2$-CMe=NCy) (^1H),[1930] $U(C_5H_4Me)_3NSiMe_3$ (^{13}C),[1931] $(C_5H_5)_2U(\eta^2$-NRCNEt$_2$)(NEt$_2$) (^1H),[1932] and U(IV) complexes of 1,2-cyclohexane(dinitrilo)-N,N,N',N'-tetraacetate.[1933]

5 Solid-state N.M.R. Spectroscopy

This section consists of three main parts: 'Motion in Solids', 'Structure of Solids', and 'Molecules Sorbed Onto Solids'.

A number of reviews have appeared: 'Solid state n.m.r.',[1934] 'Solid-state high-resolution n.m.r.',[1935] 'Nuclear magnetic resonance for solids. Did you say for solids?',[1936] 'High resolution solid state n.m.r. - theory and applications',[1937] 'High-resolution nuclear magnetic resonance of inorganic solids',[1938] 'Nuclear magnetic resonance studies in solid electrolytes',[1939] 'N.m.r. and librational tunnelling in some ammonium compounds',[1940] 'N.m.r.-n.q.r. studies of structural transitions',[1941] 'N.m.r. diffusion studies in zeolites',[1942] 'High-resolution ^{27}Al n.m.r. spectroscopy and its use in the chemistry of inorganic solids',[1943] 'N.m.r. in exchange coupled systems',[1944] 'Study of metal hydrides by n.m.r.',[1945] 'Magnetic resonance in graphite intercalation compounds',[1946] 'High-resolution ^{29}Si and ^{31}P n.m.r. spectroscopy and its use in inorganic solid-state chemistry',[1947] 'N.m.r. of glass structure',[1948] 'Acid properties of a bi-dimensional zeolite',[1949] 'Application of solid high resolution n.m.r. to the study of zeolite',[1950] 'N.m.r. studies on the acidity of zeolites and related catalysts',[1951] 'N.m.r. of fluorides',[1952] 'N.m.r. on solid surfaces',[1953] 'Magnetic resonance in surface science',[1954] 'New possibilities of n.m.r. spectroscopy in the studies of adsorption and catalysis',[1955] 'N.m.r. surface area in porous substances and exchange processes',[1956] 'The role of intercalates in heterogeneous catalysis',[1957] 'N.m.r. self-diffusion studies in heterogeneous systems',[1958] 'N.m.r. in heterogeneous systems',[1959] 'New possibilities of n.m.r. spectroscopy in studies of adsorption and catalysis',[1960] 'Application of high resolution solid state n.m.r. spectroscopy to surface studies',[1961] 'High resolution n.m.r. on adsorbate-adsorbent systems',[1962] 'N.m.r. of catalyst surfaces',[1963] 'Use of n.m.r. for study of the distribution and acidity of constitution of solids',[1964] 'N.m.r. investigations of proton donator and electron acceptor properties in heterogeneous systems',[1965] '^{1}H and ^{2}H solid-state spectroscopy of physisorbed and chemisorbed species',[1966] 'Fast exchange approach in n.m.r. studies of specific adsorption in oxides',[1967] and '^{13}C n.m.r. studies of molecular conformations and interactions in the curved surfactant monolayers of Aerosol OT water-in-oil microemulsions'.[1968]

Motion in Solids.—The field-independent part of the spin-lattice relaxation of ^{7}Li in Li_3N is due to self-diffusion of Li.[1969, 1970] The characteristic features and analysis of the behaviour of magnetic relaxation, $\underline{T}_{1d}$, $\underline{T}_{1\rho}$, and $\underline{T}_2$ in disordered metal-hydrogen systems have been studied to determine hydrogen diffusion in metals.[1971] The enhancement of ionic conduction in $LiBr-H_2O-Al_2O_3$ composite solid electrolyte has been studied by means of ^{7}Li $\underline{T}_1$ measurements.[1972] ^{1}H and ^{7}Li n.m.r. spectroscopy has been used to investigate molecular and ionic mobility in $LiClO_4.3H_2O$.[1973] The ^{7}Li and ^{19}F diffusion coefficient of LiO_3SCF_3 has been measured.[1974] The activation energy for lithium diffusion in highly conductive $LiI-Li_2S-P_2S_5$ glasses with varying lithium concentrations has been measured by ^{7}Li line width techniques.[1975] The activation energy for methyl group rotation in NaO_2CMe has been determined.[1976] By the analysis of ^{7}Li T_1 measurements, lithium diffusion in Li_2O, Li_2SiO_3, and γ-$LiAlO_2$ has been investigated.[1977] The mobility of lithium ions in Li_4SiO_4 and Li_2SiO_3 has been investigated by ^{7}Li n.m.r. spectroscopy.[1978] Fast ^{23}Na and ^{7}Li exchange in channels of brush-border membranes of renal proximal tubule has been investigated.[1979] The rotational motion of dilute off-centre Ag^+ defect ions in RbX crystals has been studied using ^{109}Ag T_1 and $T_{1\rho}$ measurements.[1980] The correlation time of the rotational motion of dilute F^- off-centre defects in NaBr single crystals has been studied using ^{19}F relaxation measurements.[1981] The influence of cationic vacancies on the mobility of Li^+ in some cation deficient materials has been studied using ^{7}Li n.m.r. spectroscopy.[1982]

^{7}Li and ^{19}F T_1 measurements have been used to study motion and diffusion in Li_2BeF_4.[1983] The kinetics of hydrogen absorption in magnesium-rich hydride systems have been studied using ^{1}H n.m.r. spectroscopy.[1984]

Hydrogen vacancy diffusion parameters in ScH_{2-x} and ScD_{2-x} have been determined from ^{45}Sc T_1 measurements.[1985] The conduction electron density of states and ^{1}H T_1 in the dihydrides of scandium, yttrium, lanthanum, and lutetium have been determined.[1986] Hydrogen diffusion in Ce, Nd, Gd, Dy, and Er in YH_2 has been investigated using ^{1}H T_1 measurements.[1987] A ^{1}H n.m.r. study of the valence state of Yb in YbH_3 has been reported.[1988] The ^{17}O T_1 in CeO_2 doped with Y_2O_3 has been measured.[1989, 1990] A molecular dynamic simulation of Eu^{3+}-doped sodium borate glasses has been reported.[1991] The motion of oxygen atoms in $(NH_4)_2Ce(NO_3)_6$, NH_4ClO_4, $(NH_4)_2Cr_2O_7$, and $(NH_4)_2S_2O_8$ has been studied using ^{1}H $T_{1\rho}$ measurements.[1992] The diffusion of F^- in $Pb_{1-x}La_xF_{2+x}$ has been investigated using ^{19}F n.m.r. spectroscopy.[1993] The ionic conductivity of $NaYF_4$ and $NaBiF_4$ has been studied by ^{19}F and ^{23}Na n.m.r. spectroscopy.[1994] ^{19}F n.m.r. spectroscopy has also been used to study vacancy distribution and ionic motion in LiF_3.[1995]

The behaviour of Mn in TiH_x has been studied by ^{1}H n.m.r. spectroscopy in order to determine the effects on hydrogen distribution and diffusion.[1996] ^{1}H and ^{51}V n.m.r. spectroscopy has been used to study the dihydric phase of the Ti-V-H system.[1997] Hydrogen diffusion has been studied using ^{1}H T_1 measurements in $ZrH_{1.684}$[1998] and MV_2H_x, M = Zr, Hf.[1999] ^{1}H n.m.r. spectroscopy has been used to study Korringa relaxation in MV_2H_x, M = Zr, Hf.[2000] Spin relaxation associated with hydrogen diffusion has been observed in the ^{181}Ta n.m.r. spectrum of tantalum-doped $HfH_{1.64}$ above 100 °C.[2001]

A phase transition in $NiTiF_6.6D_2O$ has been detected by ^2H n.m.r. spectroscopy.[2002]

It has been shown from ^7Li n.m.r. measurements of the Li_3VO_4-$LiGeO_4$ solid solution that the activation energy is insensitive to change of composition over a wide range while it is highly sensitive to moisture.[2003] Lithium mobility in Li_3VO_4-Li_4GeO_4 and Li_3VO_4-Li_4SiO_4 solid-solution ionic conductors has been investigated using ^7Li n.m.r. spectroscopy.[2004] The ionic conductivity of Li_3NbSe_3 has been studied by ^7Li n.m.r. spectroscopy.[2005]

The structure and motion of hydrogen in H_xMoO_3 have been studied by ^1H n.m.r. spectroscopy.[2006] The proton mobility in polycrystalline $H_3PMo_{12}O_{40}.nH_2O$ has been studied by means of n.m.r. spectroscopy.[2007, 2008] The anisotropic diffusion coefficient of ^{109}Ag in $Ag_{26}I_{18}W_4O_{16}$ has been determined.[2009]

The molecular motions of water molecules in a $Na_2Fe(CN)_5(NO).2H_2O$ single crystal have been investigated by ^1H n.m.r. spectroscopy.[2010, 2011]

A ^1H n.m.r. study of the dynamic state of the benzene groups in $[PdAlCl_4(C_6H_6)]_2$ has been reported.[2012] Reorientation in crystalline $(guanidium)_2MCl_4$, M = Pd(II), Pt(II), Au(III), has been investigated by ^1H n.m.r. spectroscopy.[2013] ^1H T_1 values of tunnelling $[NH_4]^+$ and $[NH_3D]^+$ in $(NH_4)_2MCl_6$, M = Pd, Pt, Os, Sn, Pb, and Te, have been reported.[2014]

Methyl group dynamics in $Cd(O_2CMe)_2.2H_2O$ have been investigated using ^1H T_1 methods.[2015, 2016]

Dynamic properties of borate glasses have been investigated by ^{11}B T_1 and $T_{1\rho}$ measurements.[2017] Ammonium rotational tunnelling in ammonium β-alumina has been investigated using ^1H n.m.r. spectroscopy.[2018] ^{27}Al n.m.r. spectroscopy has been used to study water diffusion in edingtonite.[2019] The dynamical properties of $[Li(OH_2)_3]^+$ in interlamellar space of a swelling clay have been investigated using ^1H and ^7Li n.m.r. spectroscopy.[2020] The effects of iron impurities and Li^+ diffusion on T_1 in hydrated zeolite LiX have been studied by ^7Li n.m.r. spectroscopy.[2021] The potential energy of rare gases and of several alkanes with Na-X zeolites has been calculated and compared with self-diffusion measurements from n.m.r. spectroscopy.[2022] ^{23}Na T_1 and T_2 measurements on dehydrated NaX zeolites have been reported and related to the sodium jump rate.[2023] The state of water in perfluorinated membranes MF-4SK and MF-4KF has been studied by ^1H n.m.r. spectroscopy.[2024]

The molecular dynamics of the ternary graphite intercalation compounds with potassium, toluene and *o*-xylene have been investigated by ^{1}H T_1 measurements.[2025] Diffusion and phonon-like excitations in caesium-intercalated graphite have been investigated using ^{133}Cs n.m.r. spectroscopy.[2026] ^{1}H relaxation measurements have been used to investigate diffusion in $[Ph_2SiO]_4$.[2027] The interaction among tunnelling methyl groups in $GeMe_4$ has been investigated using n.m.r. polarization transfer experiments.[2028] An experimental method has been proposed for the investigation of Jahn-Teller effects using ^{7}Li n.m.r. spectroscopy and applied to lithium-doped SnTe.[2029] Rotational tunnelling and librations of the methyl group in $PbMe_4$, measured using inelastic neutron scattering, have been used to explain ^{1}H T_1 measurements.[2030] The structure of $PbO-B_2O_3-PbF_2-AlF_3$ glasses has been studied by ^{11}B and ^{19}F n.m.r. spectroscopy. The diffusion of anions is stimulated by the high polarizability of Pb^{2+}.[2031] ^{2}H and ^{14}N n.m.r. spectra of $MeNH_3PbX_3$ have revealed the existence of several phases.[2032]

^{15}N n.m.r. spectroscopy has been used to investigate phase transitions in NH_4NO_3.[2033] ^{1}H and ^{19}F temperature-dependent c.p.m.a.s.n.m.r. spectroscopy has been used to investigate pseudorotation in $[X_3PNR]_2$.[2034]

The mobility of water in xenon chlathrates in water has been studied by ^{1}H n.m.r. spectroscopy.[2035] Phase transitions in NH_4HSeO_4 and $RbHSeO_4$ have been studied using ^{2}H and ^{77}Se n.m.r. spectroscopy.[2036] Cation diffusion in NH_4HSeO_4, ND_4DSeO_4, and $RbHSeO_4$ has been investigated using ^{2}H and ^{77}Se n.m.r. spectroscopy.[2037] Methyl group tunnelling in tetramethyltetraselenafulvenium salts has been investigated by ^{1}H n.m.r. spectroscopy.[2038]

Structure of Solids.—Dilute ^{1}H magic-angle spinning n.m.r. spectroscopy has been used to obtain ^{1}H isotropic chemical shifts of solid inorganic species, including H_2O, $[OH]^-$, $[H_3O]^+$, and $[H_5O_2]^+$.[2039] The theory of nuclear spin relaxation due to paramagnetic ions in crystals has been treated rigorously.[2040] High-resolution ^{17}O solid-state n.m.r. spectra have been obtained for some metal carbonyls and chemical shift anisotropies of 600 to 700 p.p.m. found, while the ^{17}O quadrupole coupling constants are small, *ca*. 1 MHz.[2041]

The ^{13}C n.m.r. spectra of solid $CH_2(^{7}Li)_2$, $CH_2(^{6}Li)_2$, $CH_3{}^{7}Li$, and $CH_3{}^{6}Li$ have been obtained by the ^{1}H-^{13}C cross-polarization magic angle spinning technique. The ^{1}H and ^{6}Li decoupled ^{13}C n.m.r. spectrum of $CH_3{}^{6}Li$ was also obtained.[2042] The saturation-recovery pulse sequence has been applied to ^{1}H ^{23}Na rotating-frame imaging n.m.r. spectroscopy.[2043] The ^{1}H n.m.r. spectra of anhydrous LiOH, NaOH, KOH, and RbOH have been reported.[2044] The ^{23}Na and ^{2}H quadrupole couplings and ^{1}H dipolar coupling and motional narrowing in the orthorhombic and monoclinic phases of NaOH have been measured.[2045] The effect of dispersed γ-alumina second-phase particles on the ionic conduction of β-Li_2SO_4 has been studied.[2046] The ^{7}Li n.m.r. spectra of $LiKSO_4$[2047] and $LiCsSO_4$[2048] have been analysed. The asymmetry of the quadrupole satellite lines of ^{23}Na in $NaNO_2$ has been determined.[2049] An investigation of the interactions between electrolytes and a zwitterionic surfactant using ^{23}Na and ^{2}H n.m.r. measurements has appeared.[2050] ^{7}Li n.m.r. spectroscopy has been used to study LiI-Li_2S-As_2S_3 glasses.[2051] Subharmonic resonances are observed in the pulsed-Fourier transform n.m.r. spectra of ^{7}Li in LiCl, ^{19}F in KF, and ^{1}H in H_2O.[2052] The ^{23}Na line widths in NaCl have been investigated.[2053, 2054] The structural phase transition in $RbCaF_3$ has been investigated by ^{87}Rb n.m.r. spectroscopy.[2055]

^{17}O n.m.r. spectra have been determined on solid BeO, MgO, CaO, SrO, BaO, ZnO, CdO, and HgO, and range over 600 p.p.m. The chemical shift depends on (ionic radius of the metal ion)^3.[2056] ^{9}Be n.m.r. spectroscopy has been used to investigate beryllium silicates of the ZSM-5 type.[2057] ^{27}Al m.a.s.n.m.r. spectroscopy has been used to probe the coordination of aluminium in $MgAl_2O_4$ spinels, enabling the extent of Mg/Al order in the structure to be determined directly.[2058] ^{1}H and ^{27}Al n.m.r. spectroscopy has been used to study the hydration of $CaAl_2O_4$.[2059] The ^{29}Si chemical shift in a synthetic single crystal of Mg_2SiO_4 has been determined.[2060] ^{1}H and ^{29}Si n.m.r. spectroscopy has been used to show the presence of two intermediate phases in the dehydration of $Ca_3(SiO_3$-$OH)_2.2H_2O$.[2061] $BaO.Al_2O_3.7H_2O$ has been reconstituted as $Na_{\underline{n}}[Al_2$-$(OH)_8]_{\underline{n}}.3\underline{n}H_2O$ on the basis of ^{1}H and ^{27}Al n.m.r. spectroscopy.[2062] The cation distribution in $Zn_{3-\underline{x}}Mg_{\underline{x}}(PO_4)_2$ solid solutions has been investigated using ^{31}P n.m.r. spectroscopy.[2063] ^{31}P n.m.r. spectra have been determined from a variety of synthetic, solid calcium phosphate phases.[2064] ^{1}H n.m.r. spectroscopy has been used to study the setting of $CaSO_4.0.5H_2O$.[2065] ^{9}Be and ^{19}F n.m.r. techniques have been used to gain information about short-range order in a simple BeF_2 glass and a $NaF.2BeF_2$ glass.[2066] Nonadiabatic effects on ^{19}F rotating frame dipolar relaxation measurements have been studied for MgF_2 and BaF_2.[2067] ^{1}H and ^{19}F n.m.r. spectra in

H_2O and CaF_2 have been used to investigate relaxation in the doubly rotating frame.[2068] The cubic ^{19}F spin system in CaF_2:Tb crystals has been studied using the dynamic polarization of Tb. Transverse ordering was observed in the rate of polarisation of ^{43}Ca.[2069] The $\underline{T}_1$ of $^{173}Yb^{3+}$ in SrF_3 has been measured.[2070]

^2H $\underline{T}_1$ and $\underline{T}_2$ have been measured for $CeD_{\underline{x}}$.[2071] $\underline{T}_1$ of ^{169}Tm in $TmPO_4$ has been measured.[2072] ^1H and ^{19}F n.m.r. studies of $Ln(O_2CCF_3)_3.3H_2O$, Ln = Sm, Eu, Er, Yb, indicate that these compounds are isostructural with $Pr(O_2CCF_3)_3.3H_2O$.[2073] ^{53}Cr $\underline{T}_1$ has been measured for $YCrO_3$ and $EuCrO_3$ has been measured.[2074] The ^{57}Fe n.m.r. spectra of Al-, Ga-, and In-substituted $Y_3Fe_5O_{12}$ garnets[2075] and unsubstituted $Y_3Fe_5O_{12}$[2076, 2077] have been reported.

The ^{31}P n.m.r. spectrum in the ferromagnet U_3P_4 has been investigated.[2078] The magnetic shielding of ^{19}F in polycrystalline UF_6 has been measured,[2079] and analysed in terms of a spin-pair model.[2080]

^1H $\underline{T}_1$ measurements have been used to study the effects of hydrogen on TiCo.[2081] ^{51}V n.m.r. spectroscopy has been used to investigate the VTiH system and the effect of Ti on electric field gradients.[2082] The effects of thermal treatments on the lattice properties and electronic structure of $ZrH_{\underline{x}}$ have been investigated by ^1H $\underline{T}_1$ measurements.[2083] The n.m.r. powder spectrum of ^{51}V in the low-temperature tetragonal phase of HfV_2 has been studied.[2084] The effects of increasing H, (D), concentration in $HfV_2H_{\underline{x}}(D_{\underline{x}})$ and ZnV_2-$H_{\underline{x}}(D_{\underline{x}})$ on $\underline{T}_1$ of ^1H, ^2H, and ^{51}V have been studied.[2085] ^{13}C n.m.r. spectroscopy has been used to determine B_2O_3-SiO_2-TiO_2 glasses prepared by the sol-gel method.[2086] The ^{57}Fe n.m.r. spectra of the Li-Ti-Mg and Li-Ti-Zn ferrite systems have been studied.[2087] Oxygen-fluorine order in $PbZr_3F_6O_4$ has been investigated using ^{19}F n.m.r. spectroscopy.[2088] The effects of sodium intercalation on the electronic structure of a TiS_2 slab have been investigated by ^{23}Na n.m.r. spectroscopy.[2089] A ^2H n.m.r. study in a single crystal of $NiTiF_3.6D_2O$ has revealed a first-order structural phase transition.[2090] The characteristics of a ^{19}F n.m.r. spectrum of ZrF_4 have been reported.[2091]

^{51}V n.m.r. spectroscopy has shown the presence of tetrahedral V(V) in a vanadium-magnesium oxide butene oxidation catalyst.[2092] The thermal decomposition of 10-molybdo-2-vanadosilicic heteropolyacid has been studied using ^{29}Si and ^{51}V n.m.r. spectroscopy.[2093] ^{51}V n.m.r. studies of polycrystalline ferroelastic $BiVO_4$ have indicated that the ^{51}V electric field gradient asymmetry parameter is an order parameter in the ferroelastic transition.[2094, 2095] The ^{7}Li n.m.r. spectrum of Li_6VCl_8 has been recorded.[2096] A ^{51}V and ^{133}Cs n.m.r. study has been performed to obtain information on the spin state of a single crystal of $CsVCl_3$.[2097]

The temperature dependence of the ^{1}H Knight shift of (α,α')-$NbH_{\underline{x}}$ has been measured.[2098] The effects of chemical and quadrupole shifts of ^{93}Nb in the n.m.r. spectra of $LiNbO_3$ have been separated.[2099] No differences in the ^{7}Li n.m.r. parameters of $LiNbO_3$ are observed on Fe and Cu doping.[2100] ^{93}Nb n.m.r. spectra have been used to investigate niobium disorder in $LiNbO_3$.[2101] The central line in the ^{93}Nb n.m.r. spectrum has been investigated by two dimensional n.m.r. spectroscopy and applied to $LiNbO_3$ and NbO_2F.[2102] ^{93}Nb n.m.r. spectroscopy has been applied to single crystals of $(NbSe_4)_{3.33}I$.[2103, 2104]

A ^{7}Li n.m.r. investigation of nonstoichiometric ferroelectric phases derived from $LiTaO_3$ has been performed.[2105] A ^{7}Li n.m.r. study of members of the solid-solution system $Li_{1+5\underline{x}}Ta_{1-\underline{x}}O_3$ and $Li_{1+\underline{x}}Ta_{1-\underline{x}}Ti_{\underline{x}}O_3$ has indicated that the excess Li^{+} ions occupy interstitial tetrahedral sites in the former. In the latter system the Li^{+} ions appear to occupy interstitial holes for small values of $\underline{x}$, but mostly octahedral sites for $\underline{x} > 0.1$.[2106] The ^{13}C n.m.r. spectra of TaF_5-doped poly(phenylene sulphide) have been measured to investigate the conduction mechanism.[2107]

A well defined canted spin structure has been deduced from ^{53}Cr and ^{113}Cd n.m.r. data for $CdCr_2S_4$.[2108] ^{53}Cr $\underline{T}_1$ and $\underline{T}_2$ have been calculated for Rb_2CrCl_4 and compared with experiment.[2109] ^{7}Li and ^{133}Cs n.m.r. spectroscopy has been used to study the trigonal phase in $CsLiMoO_4$.[2110] A ^{87}Rb n.m.r. investigation of $Rb_{0.3}MoO_3$ has been reported.[2111, 2112] The presence of large Knight shifts in the ^{205}Tl n.m.r. spectrum of $TlMo_6Se_8$ has been interpreted in terms of the Tl s-band lying at the Fermi level.[2113] ^{95}Mo and ^{97}Mo n.m.r. measurements have been performed in MoF_6. Strong quadrupole effects were observed in the liquid as well as the plastic and orthorhombic phases.[2114] A ^{1}H n.m.r. study of tungsten trioxide electrochromic tin layers has been reported.[2115] $[MeAsW_7O_{27}H]^{7-}$ has been characterised in the solid state by ^{1}H and ^{13}C n.m.r. spectroscopy.[2116]

The measurement of paramagnetic relaxation time by the c.w. saturation method has been examined and applied to Mn^{2+} in sphalerite.[2117] The problems of the determination of local magnetic fields in paramagnetic hydrates have been examined for $MSO_4.H_2O$, M

= Mn, Fe, Ni, and $CoCl_2.6H_2O$.[2118] The theory of stochastic n.m.r. saturation in magnetically ordered crystals has been developed and applied to ^{55}Mn in $CsMnF_3$.[2119] ^{55}Mn n.m.r. signals have been observed for $RbMnCl_3$.[2120] ^{1}H n.m.r. measurements have been carried out to study the effects of the external field on the field-induced phase transition in a system with competing anisotropy, $CsMn_{1-\underline{x}}Co_{\underline{x}}Cl_3.2H_2O$.[2121]

$TbFe_2H_{\underline{x}}$ has been studied by zero-field ^{159}Tb n.m.r. spectroscopy.[2122] The ^{1}H, ^{13}C, and ^{15}N chemical shift tensors of $Na_2Fe(CN)_5$-(NO)$.2H_2O$ single crystal have been determined.[2123] A single-crystal ^{2}H n.m.r. study of $Fe_3O(O_2CCH_3)_6(4\text{-Me-py})_3.C_6D_6$ shows that the deuterated benzene solvate molecules are rapidly rotating.[2124] Acoustic ^{57}Fe n.m.r. spectroscopy has been used to study haematite.[2125] The structure of Fe_3O_4 in the low-temperature phase has been deduced from advanced analysis of the n.m.r. spectrum.[2126] The ^{57}Fe n.m.r. spectrum of $BaFe_{12}O_{19}$ has been recorded.[2127] High resolution c.p.m.a.s. ^{15}N n.m.r. spectroscopy has been used to investigate $[RuCl(NO)_2(PPh_3)_2][BF_4]$. There is a large chemical shift anisotropy for the bent as compared with the linear NO ligand.[2128]

^{59}Co n.m.r. spectroscopy has been used to study cobalt sulphides in cobalt-based hydrode-sulphuration catalysts. Four different cobalt sites were observed.[2129] The ^{2}H n.m.r. spectrum of $CoCl_2.2D_2O$ has been studied.[2130] Two ^{31}P n.m.r. signals are observed for $HRh(CO)(PPh_3)_3$ and attributed to α and ß modifications.[2131] The ^{31}P chemical shielding tensors have been derived from single crystals of $RhCl(PPh_3)_3$.[2132] N.m.r. studies of M_2IrCl_6, M = K, NH_4, Rb, Cs, have been reported.[2133]

The high-resolution solid-state ^{31}P n.m.r. spectrum of Pt(PR=C-$Ph_2)(PPh_3)_2$ shows the expected features for an η^1-coordinated phosphaalkene ligand and is completely different from that of the η^2-complex which exists in solution.[2134] The ^{133}Cs $\underline{T}_1$ of $CsNiF_3$ has been measured.[2135]

Anisotropic parameters for the valence p-orbitals of Cu, Ag, and Au have been calculated.[2136] $[Cu(PPh_3)_4][ClO_4]$, $[Cu(PPh_3)_3(NCMe)]$-$[ClO_4]$, and $[Cu(PPh_3)_2(NCMe)_2][ClO_4]$ have been characterised by solid-state ^{31}P n.m.r. spectroscopy.[2137] Copper precipitates in cuprous oxide have been detected by ^{63}Cu n.m.r. spectroscopy.[2138] Solid-state ^{31}P n.m.r. spectra of some novel $Cu_4X_4L_4$ cubane clusters have been reported.[2139] M.a.s.n.m.r. spectra of ^{63}Cu in CuCl, CuBr, and CuI have been measured.[2140] The ^{63}Cu n.m.r. spectrum of $CuCl_2.2H_2O$ has been reported.[2141] A $CuCl_2.2H_2O$ single crystal has been studied by ^{63}Cu and ^{65}Cu n.m.r. spectroscopy.[2142] The ^{1}H n.m.r. spectrum of $Rb_2CuCl_4.2H_2O$ and $Rb_2CuBr_4.2H_2O$ has been measured at room temperature.[2143] ^{109}Ag chemical shifts have been determined for solid AgX, $RbAg_4I_5$, KAg_4I_5, and $Ag_{26}I_{18}W_4O_{16}$.[2144] High-pressure ^{87}Rb T_1 measurements have been reported for $RbAg_4I_5$.[2145] Frequency-dependent ^{1}H and ^{19}F n.m.r. experiments have been performed on tetrathiafulvalene-$AuS_4C_4(CF_3)_4$.[2146]

^{1}H and ^{13}C n.m.r. spectral data have indicated that Et_2O forms structurally similar 2:1 complexes with $HAlCl_4$, $HTiCl_5$, $HSnCl_5$, and $HMCl_3$, M = Zn, Ge, Sn.[2147] The frequency distribution of the ^{87}Rb n.m.r. central transition has been measured as a function of temperature and crystal orientation.[2148] The possibility of discriminating true incommensurate phases from long-period commensurate phases *via* n.m.r. line shape and T_1 measurements has been applied to Rb_2ZnCl_4, Rb_2ZnBr_4, and $(Me_4N)_2ZnCl_4$.[2149]

The solid-state and solution ^{113}Cd n.m.r. spectra of some Cd(II) complexes have been reported.[2150] Solid-state ^{113}Cd c.p.m.a.s. n.m.r. spectra have been determined for parvalbumin and concanavalin A and the shielding tensors determined.[2151] The ^{113}Cd n.m.r. chemical shift tensor in Cd diethylphosphate has been determined.[2152] The ^{113}Cd solid-state n.m.r. spectrum of $CdPS_3$ is consistent with a single Cd environment.[2153] The hydrogen bonds in the structure of Hg(OH)F have been studied using ^{1}H and ^{19}F n.m.r. spectroscopy.[2154]

The chemical shift and quadrupole interaction parameter of ^{11}B, ^{23}Na, and ^{27}Al have been determined from the line shapes of powders.[2155] ^{1}H, ^{2}H, ^{7}Li, ^{13}C, and ^{27}Al n.m.r. spectra have been reported for a variety of polycrystalline materials at zero field.[2156] A selective 180° pulse in the *z*-direction refocusses the second-order broadening of the 1/2 to -1/2 central transition in the spin 5/2 ^{27}Al nucleus. Implications for high-resolution n.m.r. of weak quadrupolar nuclei in the solids were discussed.[2157]

The composition of some sodium borosilicate glasses has been investigated by n.m.r. spectroscopy.[2158] The structure of some sodium borate glasses has been investigated by ^{11}B n.m.r. spectroscopy.[2159] The effect of hydration and dehydration on the coordination of boron in H-Boralite has been investigated by ^{11}B m.a.s. n.m.r. spectroscopy.[2160] ^{7}Li T_1 measurements have been measured as a function of temperature in two mixed ($^{6,7}Li$) triborate glasses and one mixed-alkali (Li,Na) triborate.[2161] Modifications of the ^{11}B n.m.r. central line shape of four-fold coordinated borate groups have been considered probable if fluorine is incorporated into the glass network.[2162] Solid-state ^{11}B n.m.r. techniques have been used to study the structural nature of various phases obtained in the B-S, B-Se, B-S-Se, and B-Te systems.[2163] Glassy B_2S_3 and binary Li_2S-B_2S_3 glasses have been prepared and studied by using ^{11}B n.m.r. spectroscopy.[2164] ^1H and ^{19}F n.m.r. studies on conducting polymer polypyrrole-BF_4 have been reported.[2165]

The temperature dependence of ^{109}Ag and ^{205}Tl $\underline{T}_1$ for Ag-β-alumina and Tl-β-alumina has been reported.[2166] ^{27}Al m.a.s.n.m.r. spectroscopy has been used to investigate Al_2O_3 gels.[2167] The coordination of aluminium during the devitrification of aluminoborate glass has been examined using m.a.s.n.m.r. spectroscopy.[2168] The local environment of Al(III) ions in gibbsite, boehmite, and amorphous products has been studied by ^{27}Al n.m.r. spectroscopy.[2169] ^{27}Al and ^{31}P chemical shift data of the low quartz, low tridymite, and low cristobalite $AlPO_4$ phases have been obtained from m.a.s.n.m.r. spectroscopy.[2170] The ^{27}Al and ^{31}P n.m.r. spectra of the molecular sieve $AlPO_4$-5 have been reported.[2171] Several structural characteristics and the thermal decomposition of $MH(SO_4)_2.\underline{n}H_2O$, M = Al, Ga, In, Tl, have been studied by ^1H and ^{205}Tl n.m.r. spectroscopy.[2172] The correlation between the nuclear quadrupole interaction and the e.s.r. zero-field splitting parameter has been determined by n.m.r. monitoring of the quadrupole splitting of the ^{27}Al nucleus in single crystals of KAl alum.[2173] ^1H n.m.r. spectra have been studied for crystalline $Al(IO_3)_2NO_3.6H_2O$ and $Al(IO_3)_3.2HIO_3.6H_2O$.[2174]

Catalytic sites in ZSM-5 have been investigated by ^{11}B and ^{27}Al n.m.r. spectroscopy.[2175] The acidity of H-ZSM-5 zeolites has been investigated using m.a.s.n.m.r. spectroscopy.[2176] A series of synthetic mordenites have been investigated by solid-state ^{29}Si n.m.r. spectroscopy.[2177] An n.m.r. nutation experiment for quadrupolar nuclei has been applied to ^{27}Al spins in the H-ZSM-5 zeolite catalyst.[2178] The effect of temperature on the ^{29}Si m.a.s. n.m.r. spectrum of ZSM-5 has shown a phase transition.[2179] ^{29}Si m.a.s.n.m.r. spectroscopy has been used to examine the monoclinic/orthorhombic transition in silicalite.[2180] The generation of shape selective acidic sites in ZSM-5 and ZSM-11 zeolites by treatment with Al_2Cl_6 has been investigated by ^{27}Al m.a.s.n.m.r. spectroscopy.[2181] High-resolution ^{27}Al and ^{29}Si m.a.s.n.m.r. spectroscopy has been used to study the solid-state transformations occurring in ZSM-5 zeolites due to heating procedures.[2182] The reactivity of

Pr^iOH on potassium- and caesium-exchanged ZSM-5 and mordenite has been investigated by ^{13}C n.m.r. spectroscopy.[2183] ^{13}C, ^{27}Al, and ^{29}Si m.a.s.n.m.r. spectroscopy has been used to characterise intermediate phases progressively formed during two different crystallisation processes of ZSM-5 zeolite.[2184] ^{29}Si - enriched Na,$[Pr_4N]$-ZSM-5 zeolite has been studied using high - resolution ^{27}Al and ^{29}Si m.a.s.n.m.r. spectroscopy.[2185] The ^{27}Al m.a.s.n.m.r. spectrum of the zeolite $[Pr_4N]$-ZSM-5 has been investigated.[2186] ^{27}Al and ^{29}Si m.a.s.n.m.r. spectroscopy has been used to characterise intermediates formed during the crystallisation of $[Pr_4N]$-ZSM-5.[2187] The presence of $[Pr_4N]^+$ causes an increase in $\underline{T}_1$ for ^{29}Si in ZSM-5 zeolite.[2188] A high-resolution ^{13}C and ^{29}Si n.m.r. investigation of occluded templates in ZSM-5 has been reported.[2189] The mean Si-O-Si bond angles in dealuminated ZSM-5 zeolites have been determined from ^{29}Si m.a.s.n.m.r. spectroscopy.[2190] The uniqueness of highly siliceous ZSM-5 and -11, which have very similar lattice structures, has been shown by ^{29}Si m.a.s.n.m.r. spectroscopy.[2191] The ^{29}Si m.a.s.n.m.r. spectrum of ZSM-12 has been determined.[2192]

The influence of second neighbour aluminium atoms on the ^{29}Si chemical shift in a zeolite has been examined.[2193] Both ^{27}Al and ^{29}Si n.m.r. spectroscopy has been used to show that the properties of the aluminosilicate gels formed as intermediates from reactant mixtures in zeolite Na-A depend strongly on the chemical composition of the starting silicate and aluminate solutions.[2194] The silicon/aluminium ratio of single crystals of zeolite A has been determined from ^{29}Si m.a.s.n.m.r. spectroscopy.[2195] ^{27}Al m.a.s. n.m.r. spectroscopy has been used to study dealumination on heading zeolite Ca-A.[2196] Cation effects in Li,Na-A zeolites have been investigated by ^{7}Li and ^{29}Si m.a.s.n.m.r. spectroscopy.[2197] A ^{29}Si m.a.s.n.m.r. spectroscopic study of the crystalline-amorphous transition of zeolite A containing trapped krypton has been reported.[2198] For some hydrated faujasites, ^{23}Na $\underline{T}_1$, $\underline{T}_2$, and absolute intensity have been studied.[2199] A chemical structural analysis of Y-zeolites and mordenites has been made by solid-state ^{27}Al and ^{29}Si n.m.r. spectroscopy.[2200] A ^{13}C and ^{29}Si m.a.s.n.m.r. spectroscopic study has been made of the catalytic role of some zeolites for butane isomerisation.[2201] The quadrupolar corrections have been calculated for the position of the central transition in general and applied to ^{27}Al n.m.r. spectra of zeolites Y at different magnetic field strengths.[2202] The influence of the cation distribution in zeolites on the isotropic ^{29}Si chemical shift has been examined.[2203] The products of hydrothermal treatment of zeolite NH_4-Na-Y and of treatment of zeolite Na-Y with $SiCl_4$ vapour have been examined by ^{27}Al and ^{29}Si m.a.s.n.m.r. spectroscopy.[2204] The magnetic relaxation of ^{23}Na in completely hydrated NaX zeolite has been studied.[2205] The changes in ordering on heating of high-SiO_2 zeolites have been studied by ^{27}Al and ^{29}Si m.a.s.n.m.r. spectroscopy.[2206] ^{27}Al and ^{29}Si m.a.s.n.m.r. spectroscopy has been used to investigate the mechanism of formation of faujasite-type zeolites.[2207] The ^{27}Al and ^{29}Si m.a.s.n.m.r. spectra of the zeolites offretite and omega have been determined.[2208] High-resolution solid-state ^{13}C n.m.r. spectra have been measured for various zeolites containing $[Me_4N]^+$ ions.[2209]

The solid-state ^{13}C and ^{29}Si n.m.r. spectra of kaolinite intercalates have been interpreted.[2210] Outstanding problems in the kaolinite-mullite reaction sequence have been investigated by ^{27}Al and ^{29}Si m.a.s.n.m.r. spectroscopy.[2211, 2212] A ^{27}Al and ^{29}Si n.m.r. study of the mechanism of kaolinite transformation has been published.[2213] Thermal reactions of kaolinite have been studied by ^{27}Al and ^{29}Si m.a.s.n.m.r. spectroscopy.[2214] ^{29}Si n.m.r. spectra have been obtained from a variety of layer silicates and aluminosilicates.[2215] ^{27}Al and ^{29}Si m.a.s.n.m.r. spectra have been measured for some dealuminated faujasites.[2216] Several tobermorites have been studied by ^{27}Al and ^{29}Si m.a.s.n.m.r. spectroscopy.[2217] The coordination of aluminium in several synthetic aluminium substituted tobermorites has been shown to be tetrahedral by ^{27}Al m.a.s.n.m.r. spectroscopy.[2218] The ^{27}Al m.a.s.n.m.r. spectra have been recorded for albite glasses which had been prepared under pressure.[2219] The effect of water loss on the structure of pyrophyllite has been studied by ^{27}Al and ^{29}Si m.a.s.n.m.r. spectroscopy.[2220] The ^{27}Al and ^{29}Si m.a.s.n.m.r. spectra of protoimogolite have been reported.[2221] ^{29}Si m.a.s.n.m.r. experiments have been made to determine the SiO_2 distribution in albite and anorthite glasses.[2222] ^{27}Al m.a.s.n.m.r. spectroscopy has been used to determine tetrahedral aluminium in montmorillonite.[2223] The ^1H n.m.r. spectra of water in syngenite have been recorded.[2224] The thermal stability of hydroxyl groups in dealuminated mordenites has been investigated.[2225] The interpretation of ^{29}Si m.a.s.n.m.r. spectra of offretites has been assisted by computer.[2226] The frequency dependence of the ^{27}Al $\underline{T}_1$ value of a one-dimensional ionic conductor with the hollandite-type structure has been investigated.[2227] Aluminium-silicon ordering in corderite has been investigated using

^{29}Si m.a.s.n.m.r. spectroscopy.[2228, 2229] ^{27}Al and ^{29}Si m.a.s.n.m.r. spectroscopy has been used to investigate pillared clays.[2230, 2231]

A series of 50 glasses in the system $CaO-Al_2O_3-SiO_2$ has been investigated by ^{27}Al and ^{29}Si m.a.s.n.m.r. spectroscopy.[2232] Amorphous silica-alumina catalysts have been studied by ^{27}Al m.a.s.n.m.r. spectroscopy.[2233] A high-resolution ^{29}Si m.a.s.n.m.r. spectroscopic study of silicate and aluminosilicate glasses has been reported.[2234] The prediction of ^{29}Si chemical shifts using a group electronegativity approach has been examined and applied to silicate and aluminosilicate structures.[2235] A model calculation has been made of the tensor of the crystal field gradient at ^{27}Al in $Al_2(Si_4O_{10})(OH)_2$ and compared with n.m.r. data.[2236] High-resolution solid-state m.a.s.n.m.r. spectra of ^{29}Si in a $SiO_2-Al_2O_3$ glass has been used to infer neighbouring environments.[2237] The empirical correlation of ^{29}Si chemical shifts and the mean Si-O-M bond angles in silicates and aluminosilicates have been explained in terms of a simple quantum chemical model.[2238] The ^{27}Al m.a.s. n.m.r. spectra of some sodium aluminosilicate glasses have been recorded.[2239] ^{27}Al, ^{29}Si, and ^{31}P m.a.s.n.m.r. spectroscopy has been used to study the silicoaluminophosphates, SAPO-5, molecular sieve.[2240] Silicon and aluminium distribution in micas has been studied using ^{29}Si m.a.s.n.m.r. spectroscopy.[2241] Aluminium-silicon disorder and the effects of paramagnetic centres in feldspars has been studied by ^{29}Si m.a.s.n.m.r. spectroscopy.[2242] High-resolution solid-state ^{23}Na, ^{27}Al, and ^{29}Si m.a.s.n.m.r. spectra have been recorded for a variety of plagioclase and alkali feldspars and feldspar glasses.[2243] The ordering of aluminium in tetrahedral sites in mixed-layer 2:1 phyllosilicates has been studied by ^{27}Al and ^{29}Si m.a.s.n.m.r. spectroscopy.[2244] A solid-state ^{27}Al and ^{29}Si m.a.s.n.m.r. spectroscopic study of 1:1 dioctahedral layer silicates has been reported.[2245] High-resolution ^{27}Al and ^{29}Si m.a.s.n.m.r. spectra of layer silicates, including clay minerals, have been published.[2246] Tetrahedral cation ordering in layer silicates has been investigated by ^{29}Si m.a.s.n.m.r. spectroscopy.[2247] ^{11}B, ^{27}Al, and ^{29}Si m.a.s.n.m.r. spectroscopy has been used to investigate the exchange of Si^{4+} by M^{n+} in silicates.[2248] ^{29}Si m.a.s.n.m.r. spectroscopy has been used to study gallosilicates with zeolite structures.[2249]

The resonances in the ^{13}C m.a.s.n.m.r. spectrum of $Tl(O_2CCH_3)_3 \cdot 1.5H_2O$ have been assigned.[2250] Solid-state ^{205}Tl c.p.m.a.s.n.m.r.

spectra have been used to investigate Tl^+-lasolocid and Tl^+-gramicidin complexes.[2251] Glasses of the P-Se-Tl system have been investigated by ^{31}P, ^{203}Tl, and ^{205}Tl m.a.s.n.m.r. spectroscopy.[2252]

The concentration of sp^2 and sp^3 carbon atoms in amorphous carbon films has been determined directly from ^{13}C m.a.s.n.m.r. spectroscopy.[2253] The ^1H n.m.r. spectra of $KC_{24}(NH_3)_{4.3}$ intercalate indicate the preferred orientation of the C_3 axis of the NH_3 molecule with respect to the c-axis of graphite.[2254] $C_{26}BrF_{12.1}$ has been characterised by ^{19}F m.a.s.n.m.r. spectroscopy.[2255]

^1H and ^2H n.m.r. spectra have been measured in a series of plasma – deposited Si:H,D samples.[2256] Ammonium trimethylsilyl sulphite has been proposed as a standard for solid-state ^1H, ^{13}C, ^{15}N, and ^{29}Si c.p.m.a.s.n.m.r. spectroscopy.[2257] ^{13}C and ^{29}Si c.p.m.a.s.n.m.r. spectroscopy has been used to show that a plasma polymerised hexamethyldisiloxane is highly crosslinked.[2258] High resolution c.p.m.a.s.n.m.r. spectra have been obtained for $(Ph_3Si)_2$ and $(Ph_2Si)_n$.[2259] ^{13}C and ^{29}Si m.a.s.n.m.r. spectroscopy has been used to determine the number of non-equivalent sites in silicon carbide polymorphs.[2260] The coordination of silicon in silicon oxynitrides has been determined by ^{29}Si m.a.s.n.m.r. spectroscopy.[2261] The formation of a region of heterogeneous compensation in the process of preparing planar silicon p-n junctions has been studied by ^{29}Si m.a.s.n.m.r. spectroscopy.[2262]

^{29}Si m.a.s.n.m.r. spectra of $Li_2Si_2O_5$ glass has been investigated to study the effect of crystallinity.[2263] The ^{29}Si m.a.s.n.m.r. spectrum of $[H_2Si_2O_5]_x$ shows two phases.[2264] ^1H n.m.r. studies have been carried out on $[H_2Si_2O_5]_n$ by using combined multiple pulse techniques.[2265] ^{29}Si m.a.s.n.m.r. studies of calcium hydrogen silicate show that it contains end- and middle-group SiO_4 tetrahedra.[2266] The effect of the silica hydrosol surface on the structure of the dispersing medium at the boundary layer has been studied by ^1H n.m.r. spectroscopy.[2267] Silica-gel glass formation has been studied by ^1H n.m.r. spectroscopy.[2268] The effect of temperature, water content, and type of reagent used for silylation of fused-silica capillaries has been studied by ^{13}C and ^{29}Si c.p.m.a.s.n.m.r. spectroscopy.[2269] The phosphorus in silicate glass has been studied by ^{31}P n.m.r. spectroscopy.[2270] The structure of cristobalite has been studied by ^{29}Si n.m.r. spectroscopy.[2271] Atmospheric oxygen has been shown to be the dominant source of ^{29}Si relaxation in solid silicalite.[2272] ^{29}Si n.m.r. spectroscopy has been used to study the effect of surface groups on silica.[2273] A correlation between individual Si-O bond lengths and the principal values of the ^{29}Si chemical shift tensor in solid silicates has been found.[2274] ^{23}Na and ^{29}Si m.a.s.n.m.r. spectroscopy has been used to show that the introduction of Na_2O into SiO_2 causes the progressive replacement of SiO_4 tetrahedra containing four bridging oxygen atoms by SiO_4 tetrahedra with three bridging oxygen atoms.[2275] ^{29}Si m.a.s.n.m.r. spectroscopy has been used to investigate Si-O-Si bond angles in lithium silicate glasses.[2276] Ribbon magnesium silicates, sepiolite, and palygorskite have been investigated by ^{29}Si m.a.s.n.m.r. spectroscopy.[2277] ^{29}Si $\underline{T}_1$ values have been determined in several 2:1 phyllosilicate minerals.[2278]

High-resolution ^{29}Si m.a.s.n.m.r. spectroscopy has been used to study polymorphism of belites.[2279] The ^{29}Si m.a.s.n.m.r. spectrum of Ca_2SiO_4 has been measured.[2280] Fire-resistant concretes have been investigated by ^{29}Si m.a.s.n.m.r. spectroscopy.[2281] Hydration reactions of tricalcium silicate have been studied by ^{29}Si m.a.s. n.m.r. spectroscopy.[2282] The hydration of Portland cement has been followed by ^{29}Si m.a.s.n.m.r. spectroscopy.[2283] A ^{1}H $\underline{T}_1$ study of Portland cement hydration has been reported.[2284] ^{29}Si and ^{31}P m.a.s.n.m.r. spectra have been used to study $Si(HPO_4)_2.H_2O$ and SiP_2O_7.[2285]

$^{117,119}Sn$ satellites have been observed in ^{13}C c.p.m.a.s.n.m.r. spectra of $(Me_3Sn)_4C$ and a series of tetrastannacyclohexanes. The values found in solution and the solid state were the same.[2286] ^{119}Sn c.p.m.a.s.n.m.r. spectra of Me_3SnOH indicate two distinct chemical sites for the tin.[2287] Chemical shift and $^{1}\underline{J}(^{117,119}Sn,^{13}C)$ data from ^{13}C c.p.m.a.s.n.m.r. spectra have been given for $(Me_2SnS)_3$, $Me_3Sn(O_2CMe)$, $Me_2Sn(O_2CMe)_2$, $Me_2SmCl_2.2DMSO$, and amorphous $(Me_2SnO)_{\underline{n}}$. $^{1}\underline{J}(^{117,119}Sn,^{13}C)$ was compared with X-ray data.[2288] High-resolution ^{119}Sn c.p.m.a.s.n.m.r. spectroscopy of Bu_3SnF shows ^{119}Sn coupling to two fluorine atoms, which is consistent with a pentacoordinated tin with a chain structure.[2289] The magnitude of $^{1}\underline{J}(^{117,119}Sn,^{13}C)$ has been determined for 9 methyltin derivatives by ^{13}C c.p.m.a.s.n.m.r. spectroscopy and correlated with the C-Sn-C bond angle.[2290, 2291] The solid-state ^{13}C c.p.m.a.s.n.m.r. spectrum of $Me_2Sn\{S_2CN(CH_2)_4\}_2$ has been reported. The relationship between $^{1}\underline{J}(^{117,119}Sn,^{13}C)$ and the C-Sn-C bond angle gives an angle of 138.6° compared with the experimental angle of 137.3(3)°.[2292] The state of water in $SnO_2.\underline{n}H_2O$ has been investigated by ^{1}H n.m.r. spectroscopy.[2293] The distribution of the various forms of water in hydrated SnO_2 has been investigated by ^{1}H n.m.r. spectroscopy.[2294]

The ^{1}H T_1 in some polycrystalline ammonium salts has been measured.[2295] The ^{1}H T_{1d} and $T_{1\rho}$ has been measured in ammonium sulphate as a function of ^{17}O concentration.[2296] The effect of oxygen motion on ^{1}H T_1 of NH_4ClO_4 has been investigated.[2297] The high resolution ^{13}C c.p.m.a.s.n.m.r. spectra of $[Me_2Ph_2P]I$ and $[Me_3PhP]I$ have been recorded to determine the tensors.[2298] The solid-state ^{13}C and ^{31}P c.p.m.a.s.n.m.r. spectra of $Ph_2PCS(NMe_2)$ have been reported.[2299] A one-dimensional analogue of the two-dimensional exchange experiment has been applied to the measurement of the rate of intermolecular spin diffusion between ^{31}P in a single crystal of dipotassium α-D-glucose-1-phosphate dihydrate.[2300] ^{87}Rb n.m.r. spectroscopy has been used to investigate structural phase transitions in RbD_2PO_4.[2301] The anomalous ^{75}As n.m.r. spectrum of KH_2AsO_4 has been reinvestigated.[2302] The rigid lattice ^{1}H n.m.r. spectrum of $H(H_2O)_nSbO_3$ has been interpreted.[2303] Locally resolved ^{13}C Knight shifts in $(fluoranthenyl)_2SbF_6$ have been reported.[2304] T_1 and T_2 of ^{17}O in BiO_3 have been measured as a function of temperature.[2305]

The effect of intermolecular interactions on the ^{2}H and ^{17}O quadrupole coupling constants in ice and water has been determined.[2306] The pressure-induced magnetic state in tetramethylthiafulvalene hexafluorophosphate has been investigated using ^{1}H T_1 data,[2307, 2308] and the selenium analogue has been investigated using ^{77}Se n.m.r. spectroscopy.[2309-2311] Phase transitions in NH_4HSeO_4 and $RbHSeO_4$ have been investigated using ^{2}H and ^{77}Se n.m.r. spectroscopy.[2312] The ^{127}I n.m.r. spectra of some periodates have been measured.[2313] The temperature dependence of the ^{127}I quadrupole coupling constant in Me_4NIO_4 has been determined.[2314]

Molecules Sorbed Onto Solids.—This section is divided into two subsections: 'Water Sorbed Onto Solids' and 'Atoms and Other Molecules Sorbed Onto Solids'.

Water Sorbed Onto Solids.—^{1}H n.m.r. spectra have revealed specific rigidly bound water molecules on active centres of γ-Al_2O_3. These molecules do not interact with other adsorbed water molecules.[2315] The ^{1}H T_1 of water adsorbed on γ-Al_2O_3 is dominated by paramagnetic relaxation due to surface Fe^{3+} ions. If the alumina is impregnated with platinum, T_1 increases by an order of magnitude.[2316] ^{1}H

n.m.r. measurements of water intercalated into kaolinite have identified two types of interlayer water.[2317] ^{29}Si m.a.s.n.m.r. spectra of silica gel have been used to investigate surface OH groups and their trimethylsilylation.[2318] The deuteration of alumina has been determined by ^{1}H n.m.r. spectroscopy.[2319] ^{29}Si m.a.s.n.m.r. spectroscopy has been used to investigate the surface reactivity of hydroxyl groups to Me_3SiCl.[2320]

Atoms and Other Molecules Sorbed Onto Solids. The adsorption of hydrogen on reduced $RhCl_3/TiO_2$ catalysts produces a new signal in the ^{1}H n.m.r. spectrum.[2321] The temperature and nickel dependence of the Knight shift for hydrogen adsorbed on $Cu_{1-x}Ni_x$ has been investigated.[2322] The orientational behaviour of *ortho* hydrogen and *para* deuterium molecules adsorbed on graphite has been studied by ^{1}H and ^{2}H n.m.r. spectroscopy.[2323]

A high-resolution ^{13}C c.p.m.a.s.n.m.r. study of $(C_5Me_5)_2ThMe_2$ and $[(C_5Me_5)_2Th(\mu\text{-}H)H]_2$, adsorbed on alumina, has been reported.[2324] An n.m.r. study of the structure and molecular mobility of metal complexes immobilized on the surface of polymeric supports has appeared.[2325] A ^{13}C c.p.m.a.s.n.m.r. study of $Mo_2(C_3H_5)_4$ on alumina has shown that the complex is as fluxional as it is in solution.[2326] The $Mo(CO)_6$/alumina system has been studied by ^{13}C n.m.r. spectroscopy. Species such as $Mo(CO)_n$-alumina were identified.[2327] ^{13}C c.p.m.a.s.n.m.r. spectroscopy has been used to characterise substitution inert cobalt(III) complexes bonded to silica.[2328] The mechanism of H_2PtCl_6 fixation on γ-alumina has been studied by ^{195}Pt n.m.r. spectroscopy.[2329] ^{113}Cd and ^{27}Al n.m.r. spectroscopy has been used to investigate ions adsorbed onto silica gel and zeolite Na-Y.[2330]

The mobility of methane and ethane monolayers adsorbed onto graphite has been investigated by ^{1}H T_1 measurements.[2331] The adsorption of ethane, ethylene, and CO on zinc- and cadmium-exchanged A-zeolites has been investigated by ^{13}C n.m.r. spectroscopy.[2332] The self-diffusion of heptane and heptane/benzene in the intercrystalline space of NaX zeolite has been studied using the n.m.r. pulsed gradient technique.[2333] ^{1}H n.m.r. spectroscopy has been used to investigate the mobility of octdecyl chains on chemically modified silica gel.[2334] Solid-state ^{13}C and ^{29}Si n.m.r. spectroscopy has been used to characterise 3-chloropropyl and 3-aminopropyl modified silica gels.[2335] ^{13}C n.m.r. spectroscopy has been used to determine the structure of ethylene adsorbed on platinum clusters supported on alumina.[2336] ^{1}H n.m.r. spectroscopy has been used to study the chemisorption of ethylene on platinum powder.[2337] ^{2}H n.m.r. spectroscopy has been used to study the molecular mobility of C_6D_6 in the Na and Cs forms of zeolite X.[2338] ^{13}C n.m.r. spectra have been recorded for benzene molecules adsorbed on silicalite.[2339] The intracrystalline self-diffusion of benzene, toluene, and xylene isomers in zeolite Na-X have been studied by the n.m.r. pulsed field gradient technique.[2340] The adsorption and thermal decomposition of methanol on MgO have been studied by ^{13}C n.m.r. spectroscopy.[2341] The carbonaceous species on ruthenium catalysts have been characterised by ^{13}C n.m.r. spectroscopy.[2342]

Three types of adsorbed CO on ruthenium-exchanged zeolite Y have been identified by ^{13}C m.a.s.n.m.r. spectroscopy.[2343] The adsorption of ^{13}CO onto platinum particles on alumina has been investigated by ^{195}Pt and ^{13}C m.a.s.n.m.r. spectroscopy.[2344] ^{1}H, ^{13}C, and ^{195}Pt m.a.s.n.m.r. spectroscopy has been used to study the adsorption of CO and acetylene onto platinum surfaces.[2345] ^{13}C m.a.s. n.m.r. spectroscopy has been used to study the chemisorption bond of carbon monoxide on platinum.[2346]

^{13}C c.p.m.a.s.n.m.r. spectroscopy has been used to investigate the mobility of $ClMe_2Si(CH_2)_{11}Me$ on silica [2347] and $(MeO)_3Si(CH_2)_3$-SH and related compounds on silica.[2348] ^{13}C, ^{29}Si, and ^{31}P c.p. m.a.s.n.m.r. spectroscopy has been used to investigate $Ph_2PCH_2CH_2$-$SiH_3/C_{18}H_{38}SiH_3$ immobilised on silica.[2349] ^{1}H $\underline{T}_1$, $\underline{T}_{1\rho}$, and $\underline{T}_2$ have been measured for alkyl chains on SiO_2. The activation energy for chain motion was determined.[2350] Me_3SiH, immobilised on silica gel, has been shown to be an appropriate reference standard for ^{13}C c.p.m.a.s.n.m.r. spectroscopy.[2351] The acid sites in zeolite Y have been studied by ^{31}P c.p.m.a.s.n.m.r. spectroscopy, using PMe_3 as a probe molecule.[2352]

A quantum chemical study of the adsorption complex pyridine-hydroxyl group at the surface of silica has been compared with ^{13}C and ^{15}N n.m.r. data.[2353] The Mn^{2+} positions in MnCaNaY zeolites have been determined by ^{1}H broad line n.m.r. spectroscopy.[2354] ^{1}H n.m.r. spectroscopy has been used to investigate the surface properties of $NiO/Al_2O_3/SiO_2$ catalysts for pyridine and NH_3 adsorption.[2355]

The ^{13}C chemical shifts of graphite intercalation compounds have been calculated. For acceptor types, the shifts come mainly from the paramagnetic intraatomic terms.[2356, 2357] ^{19}F n.m.r. investigations have been performed on oriented first-stage $C_{\underline{x}}SbF_5$ graphite intercalation compounds.[2358]

^{3}He n.m.r. studies of ^{3}He adsorbed on Mylar film[2359] and grafoil[2360] have been reported. ^{129}Xe n.m.r. spectroscopy has been used to study xenon adsorbed on L, Z, and ZSM zeolites.[2361]

6 Group IIIB Compounds

Boron Hydrides and Carboranes.—The effect of methyl ring substituents on ^{11}B n.m.r. chemical shifts of pyrazineboranes have been reported.[2362, 2363] The ^{1}H, ^{11}B, ^{13}C, and ^{31}P n.m.r. chemical shifts and coupling constants have been reported for $XC_6H_4PMe_2BH_3$ and correlated with Hammett σ constants.[2364] The use of polarisation transfer for the observation of even ^{1}H n.m.r. resonances has been illustrated with experiments on polarisation transfer from ^{11}B to ^{1}H. Normal ^{1}H and ^{1}H{^{11}B} INEPT n.m.r. spectra of BH_3SMe_2 were reported.[2365] The ^{1}H, ^{11}B, and ^{13}C n.m.r. signals of $[Me_3NBH_2(\mu\text{-}pz)B(pz)(\mu\text{-}pz)_2B(pz)(\mu\text{-}pz)BH_2NMe_3]^{2+}$ have been assigned using two dimensional n.m.r. spectroscopy.[2366] N.m.r. data have also been reported for R_3NBH_3 (^{13}C),[2367] $(H_3B)_2PN_4(CH_2)_8$ (^{11}B, ^{13}C, ^{31}P),[2368] $(F_2HC)_3PBH_3$ (^{11}B, ^{13}C, ^{19}F, ^{31}P),[2369] $H_3BPPh_2(CH_2)_{\underline{n}}PPh_2BH_3$ (^{11}B),[2370] $ClPMe_2BH_3$ (^{11}B, ^{13}C, ^{31}P),[2371] $H_3BPH(NBu^t)_2SnMe_2$ (^{11}B, ^{14}N, ^{31}P, ^{119}Sn),[2372] $R_3NBH_2CO_2H$ (^{11}B, ^{13}C, ^{14}N),[2373] $[H_2B(N_2C_3H_2\text{-}Me)_2]^-$ (^{11}B),[2374] $R^1{}_2P(X)NR^2(CH_2)_3BH_2$ (^{11}B, ^{31}P),[2375] Me_3PBH_2Cl (^{13}C, ^{31}P),[2376] $H_2B(N_2C_3H_3)_2BH_2$ (^{11}B, ^{13}C),[2377] $(C_6H_4O_2)BH$ (^{11}B),[2378] and $HClBC_{11}H_{18}OMe$ (^{11}B, ^{13}C).[2379]

^{1}H and ^{11}B n.m.r. spectra of 97% ^{11}B-enriched B_2H_6 have been measured and analysed to give all the coupling constants, including $\underline{J}(^{11}B,^{11}B)$ = 3.8 Hz.[2380] The ^{1}H and ^{11}B n.m.r. spectra of $[B_3H_8]^-$ and its derivatives have been analysed.[2381] ^{10}B and ^{11}B n.m.r. spectral studies of base-catalysed intramolecular rearrangements in 2-methylpentaborane(9) have been reported.[2382] $(\eta^6\text{-}C_6H_6)Fe(Et_4C_4B_6H_6)$ has been characterised by ^{11}B two-dimensional n.m.r. spectroscopy.[2383] ^{11}B COSY and ^{1}H n.m.r. spectroscopy has been applied to $[B_9H_{12}]^-$.[2384] The ^{11}B and ^{1}H n.m.r. spectra of CB_8H_{14} and $[CB_8H_{13}]^-$ have been recorded and assigned unambiguously by decoupling and COSY two-dimensional n.m.r. spectroscopy.[2385] For μ-2,7-$(SCSNEt_2)$-7-(PMe_2Ph)-<u>nido</u>-7-$PtB_{10}H_{11}$ a linear relationship between δ(^{1}H) and δ(^{11}B) has been found.[2386] Two-dimensional COSY ^{11}B-^{11}B n.m.r. spectroscopy of $[TlMe_2]$[<u>nido</u>-$Me_2TlB_{10}H_{12}$] has correlated individual ^{11}B doublet components associated with particular ^{205}Tl spin states. This allowed the determination of the relative signs of $\underline{J}(^{205}Tl,^{11}B)$.[2387] The ^{11}B and ^{1}H n.m.r. spectra of $B_{12}H_{16}$ have been analysed with the aid of two-dimensional n.m.r. spectra.[2388] The structure of $(Et_4C_4B_8H_8)_2Co_2$ has been studied using two-dimensional ^{11}B n.m.r. spectroscopy.[2389] N.m.r. data have also been reported for $Me_3NB^2H_4PMe_3$ (^{11}B),[2390] 4,6-$(C_5H_5)_2Co_2$-3,5-$S_2B_2H_2$ (^{11}B),[2391] $\{Me_3SiHC{=}C(SiMe_3)\}_4B_2(\mu\text{-}H)_2$ (^{13}C, ^{29}Si),[2392] $[B_4H_7(PMe_3)_2]^+$ (^{11}B, ^{31}P),[2393] $[B_5H_9CN]^-$ (^{11}B),[2394] 2-$(Cl_2B)B_5H_8$ (^{11}B),[2395] $B_5H_{10}M(CO)_3$ (M = Mn, Re; ^{11}B),[2396] (dppe)Cu(B_5H_8) (^{11}B),[2397] $\{(C_5H_5)Co(C_3B_2R_5)\}_2Sn$ (^{11}B, ^{13}C),[2398] $Tl(C_3B_2HMe_4)Co(C_5H_5)$ (^{11}B),[2399] $[B_6X_6]^{2-}$ (X = H, Cl, Br, I; ^{11}B),[2400] $Pt_2(\mu,\eta^3\text{-}B_6H_9)_2(PMe_2Ph)_2$ (^{11}B, ^{31}P),[2401] 2-Me_3Si-3-R-2,3-$C_2B_4H_6$ (^{11}B, ^{13}C, ^{29}Si),[2402] 1-Br-1,5-$(Ph_3P)_2$-1,2,3-Ni$(Et_2C_2B_4H_3)$ (^{11}B),[2403] $(C_{10}H_8)Fe(Et_2C_2B_4H_4)$ (^{11}B),[2404] Fe(η^6-$PhCH_2CH_2CH_2C_2B_4H_5$-η^5) (^{11}B, ^{13}C),[2405] 5:1',2'-[1-(C_5H_5)Co-2,3-$(Me_3Si)_2C_2B_4H_3$][B_2H_5] (^{11}B),[2406] Co$(Et_2C_2B_4H_3)(Et_4C_4B_8H_6OC_4H_8)$ (^{11}B),[2407] _nido_-2,3,4,5-tetracarbaborane(6) (^{11}B, ^{13}C),[2408] X_2-2,4-$C_2B_5H_5$ (^{11}B),[2409] 2,4-$C_2B_5H_7$ (^{11}B),[2410] 5,8-$(C_5H_5)_2Co_2B_8H_{12}$ (^{11}B),[2411] 5,6-$C_2B_6H_{12}$ (^{11}B, ^{13}C),[2412] $Cu(PPh_3)_2(R_2C_2B_4H_5)$ (^{11}B),[2413] $Cu(PPh_3)_2(B_9H_{13}X)$ (^{11}B, ^{31}P),[2414] 2,5,6-$(C_6H_6)RuC_2B_7H_{11}$ (^{11}B),[2415] $(Ph_3P)HClRuB_9H_7(PPh_3)_2$ (^{11}B, ^{31}P),[2416] $(Ph_3P)_2HRuB_9H_{12}(PPh_3)$ (^{11}B, ^{31}P),[2417] 1,2'-(_nido_-$B_{10}H_{13})_2$ (^{11}B),[2418] $(B_{10}H_{13})_2O$ (^{11}B),[2419] $[B_{10}H_{13}L]^-$ (^{11}B),[2420] $MeB_{10}H_7(CO)_2$ (^{13}C),[2421] 7-(η-C_5Me_5)-8-Cl-11-(PMe_2Ph)-_nido_-7,12-RhO-$B_{10}H_9$ (^{11}B),[2422] $(H_{12}B_{10}Au)(\mu\text{-}AuPEt_3)_4(AuB_{10}H_{12})$ (^{11}B, ^{31}P),[2423] 1-X-6,9-$C_2B_8H_{13}$ (^{11}B),[2424] _arachno_-5-X-6,9-$C_2B_8H_{13}$ (^{11}B),[2425] 7-R_2-C=NH-7-$CB_{10}H_{12}$ (^{11}B),[2426] [_nido_-7-(_p_-C_6H_4NCS)-9-I-7,8-$C_2B_9H_{10}]^-$ (^{11}B),[2427] MoW(μ-CC_6H_4Me)$(CO)_3$(η-indenyl)(η-$C_2B_9H_9Me_2$) (^{11}B, ^{13}C, ^{31}P),[2428] [_closo_-3,3,3-$(OC)_3$-3,1,2-$ReC_2B_9H_{11}$-1-$CR^1R^2]^+$ (^{13}C),[2429] _closo_-1,1,1-$\{(MeO)_3P\}_3$-2,5-Me_2-1,2,4-$FeC_2B_9H_9$ (^{11}B, ^{13}C, ^{31}P),[2430] _closo_-3-(C_5H_4Et)-1,2-Me_2-3,1,2-$RhC_2B_9H_9$ (^{11}B),[2431] $[Rh(PR_3)(C_2B_9H_{11})]_2$ (^{11}B, ^{31}P),[2432] 3,1,2-$(C_5H_5)RhC_2B_9H_{11}$ (^{11}B),[2433] _closo_-2,2-$(Ph_3P)_2$-2-H-1-Me-7-Ph-2,1,7-$RhC_2B_9H_9$ (^{11}B, ^{31}P),[2434] [_closo_-3,3-(η^4-1,3-cod)-1,2-Me_2-3,1,2-$RhC_2B_9H_9]^-$ (^{11}B, ^{13}C),[2435] _closo_-3-(Ph_3P)-3,3-$(Ph_2PC_6H_4)$-3,1,2-$RhC_2B_9H_{11}$ (^{11}B, ^{13}C, ^{31}P),[2436] 3-(tmed)-3,1,2-$PdC_2B_9H_{11}$ (^{11}B),[2437] 9,9'-bis(_m_-carboranyl) (^{11}B),[2438] and $(BrAu)_2\{R_2P(C_2B_{10}H_{10})PR_2\}$ (^{11}B, ^{31}P).[2439]

Other Compounds of Boron.—The ^{11}B and ^{13}C n.m.r. spectra of $ArBMe_3Li$ show $^{1}\underline{J}(^{13}C,^{7}Li)$ and $^{1}\underline{J}(^{13}C,^{11}B)$.[2440] $^{1}\underline{J}(^{15}N,^{13}C)$, $^{1}\underline{J}(^{15}N,^{11}B)$, $^{1}\underline{J}(^{29}Si,^{15}N)$, $^{1}\underline{J}(^{29}Si,^{1}H)$, and $^{1}\underline{J}(^{29}Si,^{13}C)$ in ^{15}N labelled boratranes, silatranes, and germatranes have been measured and considered together with literature $^{1}\underline{J}(^{119}Sn,^{15}N)$ and

$^{1}J(^{119}Sn,^{13}C)$. The coupling constants correlate with structure.[2441] N.m.r. data have also been reported for 1,8-$(Me_2B)_2$-naphthalene (^{11}B, ^{13}C),[2442] $MeBC_8H_{16}$ (^{13}C),[2443] $MeC{=}CR^1BR^1CR^2R^3Sn$-Me_2 (^{11}B, ^{13}C),[2444] $B(CH_2CH_2CH_2)_3N$ (^{11}B, ^{13}C),[2445] 3,7,10-trimethylboratrane (^{13}C),[2446] $PhB(C_6H_4NMe_2)_2$ (^{11}B),[2447] and $R^1{}_2R^2{}_2C_2B_2$ (^{11}B, ^{13}C).[2448]

A ^{11}B, ^{13}C, ^{14}N, ^{15}N, ^{29}Si, ^{31}P, ^{119}Sn, ^{207}Pb n.m.r. study of $E(NC_4H_4)_n$, E = Me_2B, Et_2B, EtB, Me_3Si, Me_3Sn, Me_2Sn, Me_3Pb, has been carried out. The changes in chemical shifts and coupling constants were discussed in terms of the influence of various substituents. It was shown that $\delta(^{13}C)$ and $\delta(^{15}N)$ data act as indicators for mesomeric and/or inductive substituent effects.[2449] N.m.r. data have also been reported for $[C_{12}H_8BNC_{13}H_9][AlCl_4]$ (^{11}B, ^{27}Al),[2450] $R_3B_3N_3(BR_2)_3$ (^{11}B, ^{14}N),[2451] $C_5H_9(SiMe_3)BOMe$ (^{11}B, ^{13}C),[2452] R^1R^2BOEt (^{11}B),[2453] $(C_8H_{14}B)_2O$ (^{11}B),[2454-2456] $Bu^t{}_2(Pr^i$-$O)B$ (^{11}B),[2457] $C_{12}H_8BX$ (^{11}B, ^{13}C),[2458] C_5H_6BX (^{13}C),[2459] $C_{12}H_8$-$C{=}B{=}NC_5H_6Me_4$ (^{11}B, ^{13}C),[2460] $Me_2Si(NR^1)_2BR^2$ (^{11}B, ^{14}N, ^{31}P),[2461] $R^1{}_3N_3B_3R^2{}_3$ (^{11}B, ^{13}C),[2462] $MeB(NMeNMe)_2BMe$ (^{11}B),[2463] $MeB(OBMe)_2N$-$SiMe_3$ (^{11}B, ^{29}Si),[2464] $PhBOC_7H_{13}N$ (^{13}C),[2465] $SBPhN(C_6H_3Me_2)SN(SiMe_3)BPh$ (^{11}B, ^{13}C, ^{14}N, ^{29}Si),[2466] $PhNC_8H_4S_2BOMe$ (^{13}C),[2467] $Bu^tFBN(SiMe_3)_2$ (^{11}B, ^{13}C, ^{19}F, ^{29}Si),[2468] $C_6H_2Bu^t{}_3$-$BOSiMe_2CH_2CH_2SiMe_2O$ (^{11}B),[2469] $C_7H_{13}B(OCH_2)_2CH_2$ (^{11}B, ^{13}C),[2470] $Pr^iBOCHR^1CHR^2CHMeO$ (^{13}C),[2471] $C_3H_5BO_2C_6H_4$ (^{13}C),[2472] C_3H_5-$BOCH(CO_2R)CH(CO_2R)O$ (^{11}B, ^{13}C),[2473] $R^1B(OR^2)_2$ (^{13}C),[2474] Me_3Sn-$CHMeBOCMe_2CMe_2O$ (^{13}C),[2475] $(Me_2PhSi)_3CBF(OH)$ (^{11}B, ^{19}F),[2476] and $(Me_3Si)_3SiB{=}NSiMe_3$ (^{11}B, ^{13}C, ^{19}F, ^{29}Si).[2477]

^{11}B, ^{14}N, and ^{19}F n.m.r. spectroscopy has indicated that $N(BF_2)_3$ is planar.[2478] ^{11}B n.m.r. spectroscopy has been used to characterise a series of polyols and polyhydroxycarboxylates in aqueous alkaline solution.[2479] N.m.r. data have also been reported for BN_3-$(SCF_3)_3C_6H_{12}$ (^{11}B, ^{19}F),[2480] $C_5H_6Me_4NB(NBu^t)_2BNC_5H_6Me_4$ (^{11}B, ^{13}C),[2481] $B\{N(SCF_3)_2\}_3$ (^{11}B, ^{19}F),[2482] $C_5H_6Me_4N{=}B{=}NBu^tMX_3$ (M = Al, Ga; ^{11}B, ^{13}C),[2483] $[C_5H_6Me_4N{=}B{=}NBu^t(SiMe_3)]^+$ (^{11}B, ^{13}C),[2484] C_5H_6-$Me_4NB(OR^1)(NHR^2)$ (^{11}B, ^{13}C),[2485] $C_5H_6Me_4NB(NBu^t)_2C{=}E$ (E = O, S, Se; ^{13}C),[2486] $\{Bu^t(Me_3Si)N\}_2BF$ (^{11}B),[2487] $(F_3C)ClS{=}N(BCl_2)_2N{=}SCl(CF_3)$ (^{11}B, ^{13}C, ^{14}N, ^{19}F, ^{29}Si),[2488] $Br_2B(N_2C_3H_3)_2BBr_2$ (^{11}B, ^{13}C),[2489] $(Bu^tP)_2(BNMe_2)_2$ (^{11}B, ^{13}C, ^{31}P),[2490] (pyrazine)$(BX_3)_2$ (^{11}B),[2491] $C_4H_7NBF_3$ (^{11}B),[2492] $EtNH_2BF_3$ (^{11}B, ^{19}F),[2493] $P(OCMe_2CH_2)_2NBF_3$ (^{11}B, ^{19}F, ^{31}P),[2494] $B(OR)_3$ (^{11}B),[2495] $[BF_3(OOH)]^-$ (^{11}B, ^{19}F),[2496] $[BF_3OH]^-$ (^{11}B, ^{19}F),[2497] and $C_5H_5NHBF_4$ (^{19}F).[2498]

Complexes of Other Group IIIB Elements.—The temperature dependence of the ^{205}Tl chemical shift of Me_2TlNO_3 in aqueous solution has been proposed as a thermometer for ^{205}Tl n.m.r. spectroscopy.[2499] N.m.r. data have also been reported for (tmed)AlI$(Me_2PCH_2)_2AlMe_2$ (^{31}P),[2500] $[Et_2AlC_{14}H_8(SiMe_3)_2]^-$ (^{13}C),[2501] $Al_2Bu^i{}_4$ (^{27}Al),[2502] $Me_4Al_2(\mu-C_2Me)(\mu-OCMe_2C_2Me)$ (^{13}C),[2503] $[Me_2AlCH_2PMe_2]_2$ (^{13}C, ^{27}Al, ^{31}P),[2504] $Tl\{C_5H_2(SiMe_3)_3\}$ (^{13}C),[2505] $Bu^tNHSiMe_2NBu^tAlMe_2$ (^{13}C, ^{14}N, ^{15}N, ^{27}Al, ^{29}Si),[2506] and $[AlR^1(NR^2)]_{\underline{n}}$ (^{13}C).[2507]

^{27}Al n.m.r. spectroscopy has been used to investigate the nature of aluminium in highly basic and highly concentrated $AlCl_3$ solution.[2508] Aluminosilicates in solution have been characterised by ^{27}Al and ^{29}Si n.m.r. spectroscopy.[2509] A ^{205}Tl n.m.r. study of Tl(I) ions in molten binary mixtures of nitrates and chlorides has been reported.[2510] ^{13}C and ^{205}Tl n.m.r. spectroscopy has been used to study Tl(I) complexes of crown ethers.[2511] ^{19}F n.m.r. spectroscopy has been used to identify $[AlF_6]^{3-}$ in aqueous peroxide solutions, and $^1\underline{J}(^{27}Al,^{19}F)$ of 19 Hz was observed.[2512] N.m.r. data have also been reported for Ga(7-R-8-hydroxyquinolinate)$_3$ (^{13}C),[2513] $Al\{(PMe_2)_2CR\}_3$ (R = PMe_2, $SiMe_3$; ^{13}C, ^{27}Al, ^{31}P),[2514] $Ga[S_2P(OR)_2]_3$ (^{31}P),[2515] and $[(Bu^tO)_3SiSTl]_2$ (^{13}C, ^{29}Si).[2516]

7 Group IVB Elements

Several reviews have appeared: 'Polarisation transfer n.m.r. spectroscopy for ^{29}Si: the INEPT and DEPT techniques',[2517] 'Stannanedithiocarboylate: novel α-functional organotin ligands,' which includes ^{119}Sn n.m.r. spectra,[2517a] 'Utility of multi-nuclear n.m.r. in chemistry. Examples of ^{119}Sn n.m.r. spectroscopy'.[2518]

^{13}C and ^{29}Si $\underline{T}_1$ measurements in $Ph_{\underline{n}}SiH_{4-\underline{n}}$ have been used to obtain details of internal motion and tumbling.[2519] A good linear correlation exists between the ^{29}Si coupling constants of a silicon centred radical and $^1\underline{J}(^{29}Si,^1H)$ for the corresponding silane.[2520] The ^1H n.m.r. spectra of $HRSi(CH_2)_4O$ in the presence of Eu(fod)$_3$ show magnetic non-equivalence of the protons in the CH_2O moiety.[2521] The effect of stereochemistry of ^{29}Si chemical shifts and $\underline{T}_1$ have been determined for $Me_3SiO(MeHSiO)_{\underline{n}}SiMe_3$.[2522] Secondary $^{13}C/^{12}C$ isotope effects over one bond on $\delta(^{119}Sn)$ have been determined for 42 organotin compounds. They vary between +0.025 and -0.067 p.p.m. For ^{207}Pb the effects are more marked with -0.124 p.p.m. being observed.[2523] N.m.r. data have also been reported for $(Me_2SiH)_2C{=}C{=}O$ (^{13}C),[2524] $H_2C{=}CHCH{=}CHCH(SiMe_3)SiH_2Bu^t$ (^{13}C),[2525]

RPhMeSiH (^{13}C, ^{29}Si),[2526] C_5H_5SiH (^{13}C),[2527] Si_3Ph_7H (^{29}Si),[2528] $Me_3SiSiMeHX$ (^{19}F, ^{29}Si),[2529] Ph_2GeM_2 (M = H, K; ^{13}C),[2530] and $[SnMe_{3-n}H_n]^+$ (^{13}C, ^{119}Sn).[2531]

Ultra-high-resolution ^{13}C n.m.r. spectroscopy has been used to observe ^{13}C, ^{29}Si, and ^{30}Si isotope effects in the ^{13}C n.m.r. spectrum of Me_4Si.[2532] ^{17}O n.m.r. spectroscopy has been used to study (furyl)MMe_3, M = Si, Ge, and Sn.[2533] The ^{13}C n.m.r. signals of some Me_3Si derivatives have been assigned using INADEQUATE.[2534] Two-dimensional n.m.r. spectroscopy has been used to analyse ^1H and ^{13}C n.m.r. spectra of some Me_3Si derivatives.[2535] N.m.r. data have also been reported for $C(SnMe_3)_n(MMe_3)_{4-n}$ (M = Si, Pb; ^{13}C, ^{29}Si, ^{119}Sn, ^{207}Pb),[2536] $Me_3SiCO_2SiMe_3$ (^{13}C, ^{29}Si),[2537] Me_3SiCH_2X (^{13}C, ^{29}Si),[2538] $(Me_3Si)_3C_5H_2Li.L$ (^7Li, ^{13}C, ^{29}Si),[2539] $(Me_3Si)_2C{=}S$ (^{13}C),[2540] $Me_3SiCF_2CF_2OCF(CF_3)_2$ (^{19}F),[2541] Me_3MR (M = Si, Sn; ^{13}C),[2542, 2543] Me_3SiR (^{13}C, ^{29}Si),[2544, 2545] Me_3SiR (^{13}C),[2546-2561] $(Me_3Si)PhC{=}CSi(SiMe_3)_2C(SiMe_3){=}CPh$ (^{13}C, ^{29}Si),[2562] $Me_3SiC{=}C(SiMe_3)SiMe_2SiMe_2$ (^{13}C),[2563] $Me_3SiC{=}C(SiMe_3)SiMes_2SiMes_2$ (^{13}C),[2564] $PhSO_2CLi_2SiMe_3$ (^{13}C),[2565] $(Me_3Si)_2C(SiF_2Pr^i)_2$ (^{13}C, ^{19}F, ^{29}Si),[2566] $\{(Me_3Si)_2CH\}_2Sn{=}PC_6H_2Bu^t_3$ (^{31}P, ^{119}Sn),[2567] Me_3SiC_2-$PC(Se)(SiMe_3)_2$ (^{13}C, ^{31}P, ^{77}Se),[2568] $RP{=}CPh(SiMe_3)$ (^{13}C, ^{31}P),[2569] $RP{=}C(SiMe_3)_2$ (^{13}C, ^{31}P),[2570] $Me_3SiCR^1CH_2CMe{=}CMeCH_2PR^2$ (^{13}C, ^{31}P),[2571] $(Me_3Si)PhC{=}PNR^1R^2$ (^{13}C, ^{31}P),[2572] $[C_{12}H_8C{=}P\{{=}C$-$(SiMe_3)_2\}_2]^-$ (^{13}C, ^{31}P),[2573] $(Me_3Si)_2C{=}PPRSiMe_3$ (^{13}C, ^{31}P),[2574] $Ph_3PbCH_2SiMe_3$ (^{13}C),[2575] $(R_2N)_2PC(N_2)SiMe_3$ (^{31}P),[2576] $(Me_3Si)_2$-$CC(SiMe_3)ER$ (E = P, As; ^{13}C),[2577] $(Me_3Si)_3CTeTePBu^t_2$ (^{31}P, ^{125}Te),[2578] Me_2Bu^tSiR (^{13}C),[2579] $(Bu^tMe_2Si)_2CHSH$ (^{13}C),[2580] $(Me_3SiO)MeSi\{C(C_{10}H_{15})(SiMe_2Bu^t)\}_2SiMe(OSiMe_3)$ (^{13}C, ^{29}Si),[2581] $ClCH_2SiMe_2R$ (^{13}C),[2582] $Me_2SiCH_2CH_2CH_2$ (^{13}C),[2583] $Me_2SiCH{=}CHCH{=}CH$ (^{13}C, ^{29}Si),[2584] $Me_2SiC_2SiMe_2(CH2)_nSiMe_2C_2SiMe_2(CH_2)_n$ (^{13}C),[2585] polycarbosilane (^{29}Si),[2586] $(N{=}NCH)MeSiCPh{=}CPhCPh{=}CPh$ (^{13}C),[2587] $C_{10}H_8SiR^1R^2$ (^{13}C, ^{29}Si),[2588] $(Bu^t_2Si)_2(HC_2Ph)_2$ (^{13}C),[2589] $RSi(CH_2$-$CH_2CH_2)_3N$ (^{13}C, ^{29}Si),[2590] $(Me_3Si)_3SiLi$ (^{13}C, ^{29}Si),[2591] $(Me_3Si)_3$-$SiC(O)Ar$ (^{13}C, ^{29}Si),[2592] $(Me_3Si)_3SiSiMe(NBu^t)_2SiMeSi(SiMe_3)_3$ (^{19}F, ^{29}Si),[2593] $(Me_3SiO)_2SiMeSiMe_3$ (^{13}C),[2594] $R^1R^2SiSiR^1R^2SiR^3R^4SiR^3R^4$ (^{13}C, ^{29}Si),[2595] $Me_{10}Si_6S$ (^{29}Si),[2596] and $(Bu^n_2Si)_n$ (^{13}C).[2597]

^{13}C and ^{73}Ge n.m.r. spectra have been recorded for a variety of methyl-substituted germacyclohexanes. The chemical shifts are very sensitive to steric effects and can provide a useful means of conformational analysis of organogermanium compounds.[2598] For Et_3GeC_2H

and $Et_3GeC_2GeEt_3$, $\underline{J}(^{13}C,^{13}C)$ values have been measured.[2599] The preferred rotamers of distannylalkanes and $Me_3SnCH_2CH_2SiMe_3$ have been determined from ^{1}H, ^{13}C, ^{29}Si, and ^{119}Sn n.m.r. data.[2600] Deuterium isotope distribution in trimethylstannylated 1,4-dihalobicyclo[2.2.2]octanes has been determined by ^{1}H, ^{2}H, ^{13}C, and ^{119}Sn n.m.r. spectroscopy.[2601] ^{119}Sn n.m.r. spectroscopy has been used to determine the diastereomeric ratio of some optically active organotin compounds.[2602] ^{119}Sn n.m.r. spectroscopy has been used to investigate alkyl tin cations in strong acid solution. The chemical shift is very condition dependent due to the presence of several species.[2603] ^{13}C and ^{119}Sn n.m.r. spectra have been determined for chiral organotin enamines which show two sets of signals.[2604] The ^{119}Sn n.m.r. spectrum of $Sn_{\underline{n}}Bu^{n}{}_{2\underline{n}+2}$ shows long-range $J(^{119}Sn, {}^{117,119}Sn)$.[2605] N.m.r. data have also been reported for $Me_2\overline{GeCH_2CPh_2CPh_2C}H_2$ (^{13}C),[2606] Ph_3GeCHI_2 (^{13}C),[2607] $(C_5Me_5)_2M$ (M = Ge, Sn; ^{13}C, ^{119}Sn),[2608] Me_3SnR (^{13}C, ^{119}Sn),[2609] $RP{=}CX(SnMe_3)$ (^{31}P, ^{119}Sn),[2610] $R^1CH{=}C(CR^2R^3OH)SnMe_3$ (^{13}C),[2611] Me_3Sn(adamantyl) (^{119}Sn),[2612] $Me_3SnCR{=}CH_2$ (^{119}Sn),[2613] Sn_6Me_{12} (^{119}Sn),[2614] Bu^tMeR-$SnCH_2CH_2Bu^t$ (^{13}C),[2615] $Ph_2Sn(C_{10}H_{15})Me$ (^{119}Sn),[2616] $Bu^n{}_3SnR$ (^{13}C),[2617] $(CHCl_2)_2Sn.\underline{n}thf$ (^{13}C),[2618] $R_2Sn(allyl)_2$ (^{119}Sn),[2619] $(Bu^tC_5H_4)_2Sn$ (^{13}C),[2620] R_3Sn(norbornyl) (^{13}C, ^{119}Sn),[2621] tetraorganodistannoxanes (^{119}Sn),[2622] $I(Bu^t{}_2Sn)_4I$ (^{13}C, ^{119}Sn),[2623] $(Ph_3Pb)_2CHBr$ (^{13}C, ^{207}Pb),[2624] and $Pb_2Ph_3(C_6H_4Me)_3$ (^{13}C, ^{207}Pb).[2625]

^{13}C, ^{15}N, and ^{29}Si n.m.r. data for 60 organosilanes have given indications of $\underline{d}$-$\underline{p}$ bonding. $^{1}\underline{J}(^{29}Si,^{15}N)$ values were correlated with substituent electronegativity.[2626] $^{1}\underline{J}(^{29}Si,^{15}N)$ values have been determined in Me_3SiNR_2-type compounds, using INEPT, and depend on both the Si-N bond characteristics and the nitrogen hybridisation. A $^{14/15}N$ isotope shift in ^{29}Si chemical shifts has been observed.[2627] The ^{29}Si n.m.r. data for $Me_{\underline{n}}Si(NR_2)_{4-\underline{n}}$, $Me_{3-\underline{n}}(R^1O)_{\underline{n}}$-$SiNR^2{}_2$, and $(R^1O)_{\underline{n}}Si(NR^2{}_2)_{4-\underline{n}}$ have been discussed in terms of electronic and steric factors.[2628] $^{2,3}\underline{J}(^{29}Si,^{1}H)$ in silatranes and triethoxysilanes have been measured and discussed.[2629] ^{13}C and ^{29}Si assignments in Me_3SiO compounds have been made using selective ^{29}Si decoupling.[2630] A modification of the ^{29}Si INDOR technique has been made to selectively observe silicon nuclei on the basis of attached protons.[2631] A selective heteronuclear version of INADEQUATE has been described and applied to observing ^{13}C and ^{29}Si via $\underline{J}(^{29}Si,^{13}C)$ in per-trimethylsilylated monosaccharides.[2632] The two-dimensional ^{1}H-^{29}Si chemical shift correlated n.m.r. spectrum of methyl-2,3,4,6-tetra-O-trimethylsilyl-ß-$\underline{D}$-glucopyranoside[2633] and other sugars[2634] has been determined. ^{1}H n.m.r. spectroscopy has been used to determine the conformation of $R_2ClSi-SiClR_2$.[2635] N.m.r. data have also been reported for $RSi(NLiSiMe_3)_3$ (^{7}Li),[2636] $Me_3SiNMeR$ (^{13}C, ^{29}Si),[2637] Me_3SiNR_2 (^{13}C),[2638] $(Me_3Si)_2N\overline{P(S)NBu^tCH_2}$ (^{13}C, ^{29}Si, ^{31}P),[2639] $Me_3Si(NBu^t)_2SnCH_2PR_3$ (^{31}P),[2640] $Me_3SiNHAr$ (^{29}Si),[2641] $R(Me_3Si)PC(S)NPh(SiMe_3)$ (^{13}C, ^{29}Si, ^{31}P),[2642] $\{FSi(NMeSiMe_3)\}_2NH$ (^{19}F, ^{29}Si),[2643] Pr_3SiNR_2 (^{13}C),[2644] $Bu^t_2Si{=}NSiBu^t_3$ (^{29}Si),[2645] $(Me_3M)_2PCH_2CH{=}CH_2$ (M = Si, Ge; ^{13}C, ^{31}P),[2646] $C_6H_4\{P(SiMe_3)_2\}_2$ (^{13}C, ^{31}P),[2647] $R\overline{Si\{(CH_2)_3\}_2P}$ (^{31}P),[2648] Me_3SiOR (^{13}C),[2649-2652] (^{29}Si),[2653] $Me_3MOC(O)OMe$ (M = Si, Ge, Sn, Pb; ^{13}C),[2654] $(Me_3Si)_{14}Si_7O_{21}$ (^{29}Si),[2655] $RO_2CP(O-SiMe_3)_2$ (^{31}P),[2656] $X_2PC(CF_3)_2OSiMe_3$ (^{19}F, ^{31}P),[2657] $(Me_3SiO)_2P(O)SF_3$ (^{19}F, ^{31}P),[2658] $ArP(O)(OSiMe_3)_2$ (^{31}P),[2659] $Me_3SiOSO_2CF_3$ (^{13}C),[2660] trimethylsilylated sugars (^{13}C, ^{29}Si),[2661] Me_2Bu^tSiOR (^{13}C),[2662-2665] $C_5H_4(SiMe_2OMe)_2$ (^{13}C, ^{29}Si),[2666] MeSiPhR(O-norbornyl) (^{13}C),[2667] $R_2C(SiMe_2F)(SiPh_2OMe)$ (^{19}F),[2668] Pr^i_3SiOR (^{13}C),[2669] 1,5-$(Me_2N)C_{10}H_6SiPr^i_2OR$ (^{13}C),[2670] alkoxysilanes (^{13}C, ^{17}O, ^{29}Si),[2671] R_3SiSH (^{29}Si),[2672] $(mes)Bu^t\overline{SiSiBu^t(mes)S}$ (^{29}Si),[2673] $Me_3SiTeSeSeTeSiMe_3$ (^{77}Se, ^{125}Te),[2674] $C(SiMe_2F)_4$ (^{19}F),[2675] and $Cl_3SiCH_2CH_2SiMe_2Cl$ (^{13}C).[2676]

^{15}N chemical shifts have been used as a criterion for chelation in 8-quinolinato complexes of di- and tri-organotin compounds. The ^{13}C and ^{119}Sn n.m.r. spectra were also recorded.[2677] Both the Fermi contact term and electronegativity of X and the ligating atom contribute towards $^{3}\underline{J}\{^{119}$Sn,^{1}H(ortho)$\}$ in some phenyltin compounds.[2678, 2679] The ^{17}O n.m.r. spectra of $Bu^n_3SnO_2CR$ and $Bu^n_2Sn(O_2CR)_2$ show only one signal due to rapid exchange.[2680] The effect of R in $(R_3Sn)_2O$ on $\delta(^{119}Sn)$ and $^{2}\underline{J}(^{119}Sn,^{119}Sn)$ has been discussed in terms of the Sn-O-Sn bond angle.[2681] $\delta(^{13}C)$ and $^{1}\underline{J}(^{119}Sn,^{13}C)$ values for Me_3SnCl have been determined in 22 solvents. Linear correlation analysis of $^{1}\underline{J}(^{119}Sn,^{13}C)$ against solvochromic parameters gave only a fair correlation but a better correlation is obtained for $\delta(^{13}C)$ with these parameters.[2682] N.m.r. data have also been reported for $Ph_3GeS_2P(OMe)_2$ (^{13}C, ^{31}P),[2683] $E(GePh_2-GePh_2)_2E$ (E = O, S, Se; ^{13}C),[2684] (PhCOLeu.His.OMe)$SnMe_3$ (^{13}C),[2685] poly(Bu^n_3Sn methacrylate) (^{119}Sn),[2686] Bu^n_3Sn oxinate (^{13}C, ^{15}N, ^{119}Sn),[2687] $(R_3SnO)_2P(O)Cl$ (^{13}C, ^{31}P),[2688] $(C_6H_{11})_3SnX$ (^{119}Sn),[2689] $Ph_3MX\{O{=}C(NMeCH_2)_2CH_2\}$ (^{119}Sn, ^{207}Pb),[2690] $Me_3SnS_2P(OEt)_2$ (^{13}C, ^{31}P, ^{119}Sn),[2691] and Bu^n_3SnCl (^{119}Sn).[2692]

The ^{29}Si chemical shifts for a series of 2,2-diphenyl-1,3,2-dioxasilacycloalkanes reveal a clear dependence of shift on ring size.[2693] The molecular interaction of Me_2SnCl_2 with Lewis bases has been investigated using the concentration and temperature dependence of $^2\underline{J}(^{119}Sn,^1H)$.[2694] N.m.r. data have also been reported for $Me_2Si(NMeCH_2)_2CH_2$ (^{13}C, ^{29}Si),[2695] polydiorganosiloxane (^{29}Si),[2696] $[Me_2SiO]_{\underline{n}}$ (^{13}C),[2697] silicones (^{29}Si),[2698] $R^1R^2\overline{SiO(CH_2)_{\underline{n}}O}$ (^{13}C, ^{29}Si),[2699] poly(sulphone-b-siloxane) (^{13}C),[2700, 2701] $C_{10}H_7O\overline{SiMeO(SiMe_2O)_4}$ (^{13}C),[2702] $SiPh_2(acac)_2$ (^{13}C, ^{29}Si),[2703] $Bu^t{}_2SiP_2(C_6H_2Bu^t)_2$ (^{31}P),[2704] $PhGe(OGePh_2O)_3GePh$ (^{13}C),[2705] $(mes)_2GeX(PArH)$ (^{31}P),[2706] $M\{N(SiMe_3)_2\}_2$ (M = Sn, Pb; ^{119}Sn, ^{207}Pb),[2707] $R^1CH{=}CHNHCH_2CHR^2OSnBu_2NMe_2$ (^{119}Sn),[2708] R_2SnL (^{13}C, ^{119}Sn),[2709] $Sn_4Bu^n{}_8O_2Cl_4$ (^{119}Sn),[2710] $R^1{}_2\overline{SnPR^2(C_6H_4O)}_2$ (^{13}C, ^{31}P, ^{119}Sn),[2711] Bu_2SnL (^{13}C, ^{119}Sn),[2712] $R^1{}_2Sn(S_2COR^2)_2$ (^{119}Sn),[2713] $R_2SnCl_{\underline{n}}$(isooctyl thioglycollate)$_{2-\underline{n}}$ (^{119}Sn),[2714] and $E(CH_2CH_2S)_2PbPh_2$ (E = O, S; ^{13}C, ^{207}Pb).[2715]

The diamagnetic and paramagnetic terms of $\delta(^{29}Si)$ of $XSi(OCH_2CH_2)_3N$ and $XSi(OEt)_3$ have been estimated and discussed.[2716] The conformations of 3,7,10-methyl-substituted metallatranes have been studied by ^{13}C n.m.r. spectroscopy.[2717] The low-temperature ^{19}F n.m.r. spectra of $F_3\overline{SiCH_2CH_2CH_2SR{=}O}$ have shown that all the fluorine atoms are non-equivalent.[2718] The ^{13}C n.m.r. spectrum of $[Ge(CH_2Ph)(NR_2)(\mu\text{-}S)]_2$ shows a mixture of cis and trans isomers.[2719] N.m.r. data have also been reported for $R\overline{Si(OCH_2CH_2)_3N}$ (^{13}C),[2720] $MeC_6H_4Si(OCHMeCH_2)_3N$ (^{13}C, ^{15}N, ^{29}Si),[2721] $[RSi(O_2C_6H_4)_2]^-$ (^{13}C, ^{29}Si),[2722, 2723] $Ph\overline{SiF_2OCH_2CH_2N}(SiMe_3)_2$ (^{19}F),[2724] $C_6H_{11}CH_2SiCl_3$ (^{13}C),[2725] 2-substituted 1-trichlorogermylethanes (^{13}C),[2726] and $RSn\{S_2\overline{PO(CH_2)_2O}\}_{\underline{n}}Cl_{3-\underline{n}}$ (^{31}P, ^{119}Sn).[2727]

Evidence has been found for through-space $\underline{J}(^{13}C,^1H)$ coupling in dihydroxo-5,10,15,20-tetraphenylporphyrinatotin(IV).[2728] Linear relations between group electronegativity sums of ligands bonded to silicon and $\delta(^{29}Si)$ exist for silicates and aluminosilicates.[2729] The viscosity of phosphate ore melts has been determined by ^{29}Si n.m.r. spectroscopy.[2730] N.m.r. data have also been reported for $(R^1R^2N)_2Ge$ (^{13}C, ^{29}Si),[2731] $[Si_{1+\underline{n}}O_{2+3\underline{n}}(OH)_2]^{(2+2\underline{n})-}$ (^{29}Si),[2732] $(Si_8O_{12})(OMe)_8$ (^{13}C, ^{29}Si),[2733] $Sn(OBu^t)_2$ (^{119}Sn),[2734] a tin pyrophosphate complex (^{31}P, ^{99}Tc, ^{119}Sn),[2735] Sn(II) complexes of 1-hydroxy ethylidene-1,1-disodium phosphonate (^{31}P, ^{119}Sn),[2736] SnX_4 complexes of ethers (^{13}C, ^{119}Sn),[2737] $[Sn(EPh)_3]^-$ (E = S, Se; ^{119}Sn),[2738] SnX_4 complexes of $Ph_2P(CH_2)_{\underline{n}}PPh_2$ (^{31}P),[2739] SnX_4-$(PBu^n{}_3)_2$ (^{31}P),[2740] $[Si_2F_{11}]^{3-}$ (^{19}F),[2741, 2742] and $[SiF_5]^-$.[2743]

8 Compounds of Group VB Elements

A review entitled 'Organophosphorus chemistry. Physical methods', which contains n.m.r. data, has appeared.[2744]

The ^{15}N n.m.r. spectra of $^{15}N_2$ and $^{15}N^{14}N$ have been determined. There is an isotope shift of 0.06 p.p.m. and $^{1}J(^{15}N,^{14}N)$ = 1.8 Hz.[2745] It has been shown that ^{14}N and ^{17}O relaxation of N_2O in liquid CO_2 is much slower than in normal solvents.[2746]

The two-dimensional ^{31}P n.m.r. spectrum of $Li_2H_2P_{14}$ shows it to contain a $P_7(5)$ and a $P_9(3)$ group.[2747] ^{31}P n.m.r. spectroscopy has been used to determine the chirality of alcohols after reaction with PCl_3 to give $(RO)_2P(O)H$.[2748] ^{1}H, ^{19}F, and ^{31}P n.m.r. spectroscopy has been used to determine the structure of PF_2HS in liquid crystalline solvents.[2749] N.m.r. data have also been reported for $H_2\overline{P(OCHMeCH_2)_2N}$ (^{31}P),[2750] $[C_8H_{14}PHPC_8H_{14}]^+$ (^{31}P),[2751] $PhPH(O)OC_6H_3Bu^t{}_2Me$ (^{31}P),[2752] $R^1PH(O)NR^2{}_2$ (^{31}P),[2753] $HPh\overline{P(OCH_2CH_2)_2NC(O)O}$ (^{13}C, ^{31}P),[2754] $HPhP(OCH_2CH_2)_2N$ (^{13}C, ^{31}P),[2755] LiP_5H_4 (^{31}P),[2756] $H(O)\overline{PNR^1CHR^2C(O)N}R^3$ (^{31}P),[2757] $H(O)\overline{POCHRCHRO}$ (^{31}P),[2758] and $[HPF_5]^-$ (^{31}P).[2759]

A P.M.O. method has been used to study ^{31}P chemical shifts in phosphonium salts. The dependence of the paramagnetic component of the magnetic screening constant of the ^{31}P nucleus on the electronegativity of the substituents at phosphorus, the multiplicity of the bonds around phosphorus, and the resonance integral of the σ components of these bonds have been discussed.[2760] There appears to be a correlation between the ^{31}P chemical shift differences between A^+ and AX and between A^+ and $[AX_2]^-$ where A^+ is a phosphonium ion and X is an anionic ligand.[2761] Apical versus equatorial ^{18}O isotopic shifts of the ^{31}P chemical shift have been observed for $R^1R^2R^3\overline{PCMe_2C(CO_2Et)=CMeO}$.[2762] N.m.r. data have also been reported for $[P(C_{10}H_{21})_4]^+$ (^{31}P),[2763] $R_3PCHXCHXSO_3$ (^{13}C, ^{31}P),[2764] $Bu^n{}_3\overline{PCSeC(Se)SeC}Se$ (^{31}P),[2765] $[MePh_2PR]^+$ (^{13}C, ^{31}P),[2766] $R(NH_2)C{=}CHPPh_2{=}C(CO_2Me)C(CO_2Me){=}NPh$ (^{13}C),[2767] $ClPPh_2CH_2CH{=}C(SPh)CH_2PPh_3Cl$ (^{13}C, ^{31}P),[2768] $[Ph_3P(CH_2)_nR]^+$ (^{13}C, ^{31}P),[2769] $[PH_3PR]^+$ (^{13}C),[2770] $[Ph_3PCH_2C(PPh_3)CHCHPPh_3]^{2+}$ (^{31}P),[2771] and $Ph_{5-n}P(OCH_2CF_3)_n$ (^{31}P).[2772]

A ^{11}B, ^{13}C, ^{14}N, ^{15}N, ^{31}P, and ^{77}Se n.m.r. study of some N-azolylphosphorus compounds has been reported.[2773] The ^{31}P n.m.r. spectrum of $Ph_3P{=}NAr$ has been used to estimate Hammett σ-constants of Ar.[2774] In $Ph_3P{=}NC_6H_4R$ the ^{15}N and ^{31}P chemical shifts are not direct measures of $\underline{p}_{\pi}$-$\underline{d}_{\pi}$ bonding.[2775] $^{4}\underline{J}(^{31}P,^{31}P)$ has been determined for a series of phosphazenylcyclophosphazenes and related to conformation.[2776] The ^{17}O chemical shifts and $^{1}\underline{J}(^{31}P,^{17}O)$ have been determined for a series of 1-adamantylphosphoryl derivatives. It was found that a significant 2$\underline{p}$ electron transfer towards the phosphoryl oxygen atom in the phosphonic esters is clearly reflected by the ^{17}O n.m.r. parameters.[2777] Linear dual parameter equations relating the ^{31}P n.m.r. chemical shifts in a series of phosphoryl compounds with the sums of inductive and resonance constants of the substituents have been obtained.[2778] ($\underline{R}$)-(-)-$\underline{N}$-(3,5-dinitrobenzoyl)-α-phenylethylamine has been used as a chiral shift reagent for phosphine oxides.[2779] The ^{31}P chemical shifts of Et_3PO have been used as a probe for weak hydrogen-bond donor behaviour.[2780] ^{31}P n.m.r. spectroscopy has been used to characterise the products of the reaction between diphenylpolystyrylphosphine and Et_2O_2.[2781] A linear free-energy relationship has been found between the ^{31}P chemical shifts of $(ArO)_2PPh_3$ and the acid dissociation constants of the corresponding phenols.[2782] ^{77}Se n.m.r. studies of $(2\text{-furyl})_3P{=}Se$ have shown that selenium is shielded by 86 p.p.m. compared with $SePPh_3$, implying through-space interaction with oxygen atoms. The ^{31}P n.m.r. spectrum was also recorded.[2783] N.m.r. data have also been reported for (19) (^{13}C, ^{31}P),[2784] R_2PCN

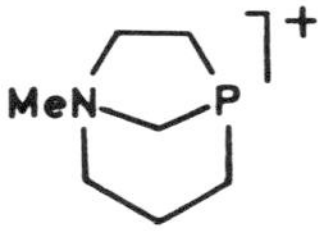

(19)

(^{13}C, ^{31}P),[2785] C_8H_8PR (^{13}C, ^{31}P),[2786] $Bu^{t}\overline{PCHMeCH_2}$ (^{13}C, ^{31}P),[2787] C_8H_6PMe (^{13}C, ^{31}P),[2788] $F_3CP(CF_2)_2PCF_3$ (^{19}F, ^{31}P),[2789] $C_5H_6CF_2PCF_3$ (^{19}F, ^{31}P),[2790] $\underline{o}$-$C_6F_4(PMe_2)_2$ (^{31}P),[2791] bicyclophosphorinanones (^{13}C, ^{31}P),[2792] $Ph_2PC_2PPh_2$ (^{31}P),[2793] $PhPR_2$ (^{31}P),[2794] (^{13}C, ^{31}P),[2795] benzo-7-phosphoranorbornane (^{13}C, ^{31}P),[2796] PAr_3 (^{31}P),[2797, 2798] $CH_2\{CH_2AsPh(CH_2)_3SCH_2\}_2CH_2$ (^{13}C),[2799] $PhAs\{(CH_2)_3$-$AsPh(CH_2)_3\}_2AsPh$ (^{13}C),[2800] $MePh_2As{=}CHAsPh_2$ (^{13}C),[2801] 1-Ph-arsole, 1-Ph-stibole (^{13}C),[2802] Ph_2AsCH_2I (^{13}C),[2803] $M_2(C_6F_4)_3$ (M = Sb, Bi; ^{13}C, ^{19}F),[2804] Ph_2BiR (^{13}C),[2805] $(Ph_2Bi)_2CH_2$ (^{13}C),[2806] $[R^{1}{}_3PNR^{2}{}_2]^+$ (^{13}C, ^{31}P),[2807] $PhCH_2PR^1R^2{=}NMe$ (^{31}P),[2808] $Ph_3P{=}NR$ (^{13}C),[2809] (^{31}P),[2810] $C_6H_4(CMe_2)_2C{=}NN{=}PPh_3$ (^{31}P),[2811] $N_3P_3(N{=}PPh_3)F_5$ (^{19}F, ^{31}P),[2812] $N_3P_3(N{=}PPh_3)(NC_2H_4)_5$ (^{13}C, ^{31}P),[2813] $Me_3SiNRPMe_2CS_2$ (^{13}C, ^{31}P),[2814] $TosN{=}PPh(CN)_2$ (^{31}P),[2815] $[R^1P(PR^2{}_3)_2]^{2+}$ (^{31}P),[2816] NCN-

$(Ph_2PPPPh_2)_2CCN$ (^{31}P),[2817] $[R_3PPPR_3]^+$ (^{31}P),[2818] (adamantyl)PR^1R^2O (^{13}C, ^{31}P),[2819] $R^1(O)PCH_2CR^2{=}CR^3CH_2$ (^{31}P),[2820] $C_5H_6Me_2OP(O)Ph$ (^{13}C),[2821] $Ph(E)PCH_2CH_2C(O)CHMeCH_2$ (E = O, S, Se; ^{13}C),[2822] ArC(O)-CHRP(E)Ph_2 (E = O, S; ^{13}C, ^{31}P),[2823] $Ph_3P(OEt)_2$ (^{13}C, ^{31}P),[2824] $Ph_3Sb(OMe)(acac)$ (^{13}C),[2825] $C_8H_4Me_2O_3P(Se)Ph$ (^{13}C, ^{31}P),[2826] 1-phenyl-1-seleno-2,5-dimethylphosphorinanone (^{13}C),[2827] Me_3PF_2 (^{19}F, ^{31}P),[2828] $CF_3P{=}CF_2$ (^{19}F, ^{31}P),[2829] Ph_2PNR_2 (^{13}C, ^{31}P),[2830] $(Ph_2P{-}N)_4(SN)_2$ (^{31}P),[2831] $(NPMe_2)_n$ (*n* = 3-12; ^{13}C, ^{31}P),[2832] $Ph_2P(E)NHR$ (^{13}C),[2833] $R^1R^2P(O)$(imidazole) (^{31}P),[2834] $Bu^t{}_2P(E)N{=}S{=}NP(E)Bu^t{}_2$ (^{13}C, ^{31}P),[2835] ArPhP(O)NHOH (^{31}P),[2836] $Sb\{C(CF_3){=}C(CF_3)C(COBu^t)\}_2N$ (^{13}C, ^{19}F),[2837] $C_6H_4(PPh)_2PR$ (^{31}P),[2838] $Ph_2P(O)(OR)$ (^{31}P),[2839] $Ph(EtO)P(O)CH_2C(O)NEt_2$ (^{13}C, ^{31}P),[2840] $OR^1C_6H_3P{=}CR^2$ (^{31}P),[2841] $Ph_2P(O)(OC_6H_4X)$ (^{31}P),[2842] $HO_2P(CH_2NH)_2C{=}S$ (^{13}C, ^{31}P),[2843] $[ArMeP{-}(OMe)_2]^+$ (^{31}P),[2844] $PhP(O)OCH_2CMe_2CH_2$ (^{13}C),[2845] $[MeBu^tP(S)]_2O$ (^{31}P),[2846] $Et_2P(S)(OR)$ (^{31}P),[2847] $CH_2CMe_2C_6H_4PS_2H$ (^{31}P),[2848] $[S_2P(SCHR)_2PS_2]^{2-}$ (^{31}P),[2849] (1-menthyl)PhP(S)Cl (^{13}C),[2850] and *p*-tolTePR_2 (^{13}C, ^{31}P, ^{125}Te).[2851]

The ^{31}P n.m.r. spectrum of RPPRPPRCl has been analysed using ^{31}P COSY. The ^{13}C n.m.r. spectrum was also recorded.[2852] The ^{1}H, ^{13}C, and ^{31}P n.m.r. spectra of $CH_2(CH_2S)_2CHP(O)(OMe)_2$ have been analysed using two-dimensional n.m.r. spectroscopy.[2853] N.m.r. data have also been reported for $(Me_2N)_3P{=}C(CO_2Me)CH(OMe)CO_2Me$ (^{31}P),[2854] $[\{O(CH_2CH_2)_2N\}_3PCH_2C(O)Ph]^+$ (^{13}C),[2855] $P{=}CBrCMe{=}NNR$ (^{13}C, ^{31}P),[2856] $P{=}CR^1SCR^2{=}N$ (^{13}C, ^{31}P),[2857] $RPN(CH_2Ph)(CH_2)_nNCH_2Ph$ (^{13}C, ^{31}P),[2858] FHXCP(S)NMeCHMeCHPhO (^{31}P),[2859] $F_3CP(NR^1)(OR^2)$, SeR_2 (^{13}C, ^{19}F, ^{31}P, ^{77}Se),[2860] $CF_3CH_2O\{SC(CF_3){=}C(CF_3)S\}P(OCH_2CH_2NMe)$ (^{13}C, ^{19}F, ^{31}P),[2861] MeN{P(O)Ph}2NMe (^{31}P),[2862] PhCH=CHP(O)OCHPhCHMeNMe (^{13}C, ^{31}P),[2863] $MeOC_6H_4P(S)SCMe{=}NNCO_2Et$ (^{13}C, ^{31}P),[2864] $MeP(O)(OR^1)N{=}CMe{-}OR^2$ (^{31}P),[2865] $OCMe_2CMe_2P(NMe_2)CH(COPh)CH{=}CPhO$ (^{13}C, ^{31}P),[2866] $MeC_6H_4P(S)(S_2Ph)NC_4H_8$ (^{31}P),[2867] $(MeNH)_2PF_2Bu^t$ (^{13}C, ^{19}F, ^{31}P),[2868] $N_3P_3F_5R$ (^{13}C, ^{31}P),[2869] $MePF_2(NMe_2)_2$ (^{13}C, ^{19}F, ^{31}P),[2870] Cl_2PCH_2Cl (^{13}C),[2871] $MePF(NMe_2)_2Cl$ (^{19}F, ^{31}P),[2872] $N_3P_3Cl_4R_2$ (^{13}C, ^{31}P),[2873] $(Bu^tP)_2PX$ (^{31}P),[2874] $P_7Bu^t{}_5$ (^{13}C, ^{31}P),[2875] $P_4(C_6H_2Bu^t{}_3)_2$ (^{31}P),[2876] P_8Et_6 (^{31}P),[2877] $P_9Bu^t{}_2X$ (^{13}C, ^{31}P),[2878] $As_{12}Bu^t{}_8$ (^{13}C),[2879] $RP(O)(OMe)_2$ (^{13}C),[2880, 2881] $RP(O)(OEt)_2$ (^{13}C, ^{31}P),[2882] (^{31}P),[2883] $(EtO)PR_2O$ (^{31}P),[2884] $(R^1O)_2P(O)R^2$ (^{13}C),[2885-2887] $(EtO)_2P(O)CH_2P(O)(OEt)_2$ (^{13}C, ^{31}P),[2888] $(R^1O)_2P(O)CH{=}CH(ER_2)CHClMe$ (^{13}C),[2889] H_2RPO_3 (^{31}P),[2890] $Ph(O)P(OCH_2)_2CMe_2$ (^{31}P),[2891] 1-phenyl-

1-thioxo-2,5-dimethylphosphorinan-4-ones (^{13}C, ^{31}P),[2892] R^1CO-$CH{=}CR^2P(O)(OR^3)_2$ (^{13}C),[2893] $H_2O_3PCR^1R^2PO_3H_2$ (^{31}P),[2894] vinylidene-diphosphonic acid (^{31}P),[2895] $R^1POCHR^2CR^3{}_2CH_2O$ (^{31}P),[2896] $(Pr^iO)_2$-$P(O)CH\{C(O)NEt_2\}\{CH_2C(O)NEt_2\}$ (^{13}C, ^{31}P),[2897] $(MeO)_3P{=}C(CO_2Me)CH$-$(OMe)CO_2Me$ (^{31}P),[2898] $(MeO)_2P\{C(CO_2Me){=}CHCO_2Me\}OCMe_2CMe_2O$ (^{31}P),[2899] $[(EtO)_2P(O)CHR]^-$ (^{13}C, ^{31}P),[2900] $(EtO)_2P(O)CH(SR^1)(OR^2)$ (^{31}P),[2901] 6-methylpyridine-2-phosphonic acid (^{13}C, ^{31}P),[2902] RPO_3Me_2 (^{13}C),[2903] $PhE\{S_2P(OPr^i)_2\}_2$ (E = As, Sb; ^{13}C, ^{31}P),[2904] $MeOC_6H_4P(SMe)(S)ON{=}CR^1R^2$ (^{13}C, ^{31}P),[2905] $R^1OP(O)ClR^2$ (^{31}P),[2906] $E(CH_2CH_2S)_2SbX$ (E = O, S; ^{13}C),[2907, 2908] $E(CH_2CH_2S)_2BiR$ (E = O, S; ^{13}C),[2908] $C_6F_5PCl_4$ (^{31}P),[2909] and $Cl_4SbCH_2CH_2C(O)OPr^i$ (^{13}C).[2910]

The relaxation of individual transitions in an AX spectrum has been examined in the ^{19}F and ^{31}P n.m.r. spectra of pyridoxyl fluorophosphate in the presence of glycogen phosphorylase.[2911] The ring conformation of six dialkylaminoarsolanes has been obtained from ^{1}H and ^{13}C n.m.r. spectroscopy.[2912] The structure of Li_2P_{16} has been determined using COSY ^{31}P n.m.r. spectroscopy.[2913] ^{31}P n.m.r. spectroscopy has been used to investigate the product of the reaction of $P(NMe_2)_3$ with P(V) halides.[2914] ^{31}P n.m.r. spectroscopy has been used to suggest that the $(CF_3CH_2O)_4P_3N_3X_3$ system is a strong electron acceptor.[2915] N.m.r. data have also been reported for $(Me_2N)_2PNR_2$ (^{31}P),[2916] $(EtO)_2P(O)NHR$ (^{31}P),[2917] C_6H_4-$O_2P(O)NR_2$ (^{31}P),[2918] $MeNCH_2CH_2NMePF$ (^{19}F, ^{31}P),[2919] $(F_3C)_2CHO$-$PX_2OC(CF_3)_2NR$ (^{19}F, ^{31}P),[2920] $[S_2PNHC_6H_4C(S)NR]^-$ (^{13}C, ^{15}N, ^{31}P),[2921] $(CF_3CH_2O)_3PNMeCH_2CH_2NMe$ (^{13}C, ^{19}F, ^{31}P),[2922] $ArOP(O)(SR)$-NH_2 (^{31}P),[2923] $Me_2C(CH_2O)_2P(SMe){=}NSO_2Ar$ (^{13}C, ^{31}P),[2924] $(R^1O)_2P(O)$-NHR_2 (^{31}P),[2925] $(R_2N)_2P(S)NMeP(O)(NMe)_2P(S)NR_2$ (^{13}C, ^{31}P),[2926] 5-phenylcyclophosphamides (^{31}P),[2927] phosphate esters (^{31}P),[2928] $R^1{}_2NPNR^2C(CF_3)_2O$ (^{13}C, ^{19}F, ^{31}P),[2929] $[(PhO)_3PNMe_2]^+$ (^{31}P),[2930] $C_2H_4NP(O)NRCH_2CH_2CH_2O$ (^{31}P),[2931] $(Me_2N)_2POR$ (^{31}P),[2932] CF_3CH_2O-$PCl_2{=}NP(O)Cl_2$ (^{13}C, ^{19}F, ^{31}P),[2933] $(EtO)_mCl_{2-m}(O)PNMeP(O)(OEt)_n$-$Cl_{2-n}$ (^{15}N, ^{31}P),[2934] $(MeO)_3POCPh{=}C(CH_2Ph)E$ (E = O, NPh; ^{13}C, ^{31}P),[2935] $(RO)(E)PR^1NCHR^2CR^3{}_2CR^4{}_2NR^1$ (^{13}C, ^{31}P),[2936] $(H_2NCONH)_2PS_2H$ (^{31}P),[2937] Et_2NPF_2 (^{19}F, ^{31}P),[2938] $[P(NEt_2)_3F]^+$ (^{31}P),[2939] MeN-$(PF_3)_2NMe$ (^{19}F, ^{31}P),[2940] $PF_4(NHCSMe)_2$ (^{19}F, ^{31}P),[2941] $Cl_4P(Pr^iN$-$CClNPr^i)$ (^{13}C, ^{31}P),[2942] $N{=}S{=}NSCH{=}CHS$ (^{13}C),[2943] $[P(NH_2)_4]^+$ (^{31}P),[2944] $(Pr^i{}_2N)_2PN$ (^{31}P),[2945] $(Pr^i{}_2N)_2P_2X_2$ (^{13}C),[2946] $[P_2C_{10}N_{10}]^{2-}$ (^{31}P),[2947] $[MeP\{MeNC(O)NMe\}_2PNEt_2]\{CF_3SO_3\}]_2$ (^{19}F, ^{31}P),[2948] spirocyclic fluorocyclotriphosphazenes (^{19}F, ^{31}P),[2949] dispirocyclotriphosphazenes (^{31}P),[2950] polyorganophosphazenes (^{31}P),[2951] $N_3P_3(NH_2)_2(OPr^n)_4$ (^{31}P),[2952] diaziridinocyclotriphosphazenes (^{31}P),[2953] $N_3P_3Cl_5HN(CH_2)_{\underline{n}}NHN_3P_3Cl_5$ (^{13}C, ^{31}P),[2954] $N_3P_3Cl_2\{HN(CH_2)_3NH\}_2$ (^{31}P),[2955] $OPCl_2N\bar{P}Cl_3$ (^{31}P),[2956] polyorganophosphazenes (^{31}P),[2957] $P_3N_3Cl_4(NHEt)_2$ (^{19}F),[2958] $N_3P_3(NPPh_3)(OMe)_4(OH)$ (^{31}P),[2959] $N_3P_3Cl_5(OR)$ (^{31}P),[2960] $[N_3P_3(OPh)_5]_2O$ (^{31}P),[2961] $N_3P_3Cl_5(NPPh_3)$ (^{31}P),[2962] $P_4N_4F_{8-\underline{n}}X_{\underline{n}}$ (^{19}F),[2963] $N_3P_3F_4(OCH{=}CH_2)_2$ (^{19}F, ^{31}P),[2964] $N_3P_3(OPh)_5NMeCH_2CH_2C\bar{N}$ (^{31}P),[2965] $P_4N_4(NMe_2)_2(NHEt)_4(NEt)$ (^{31}P),[2966] $N_5P_5Cl_{10-\underline{n}}(SR)_{\underline{n}}$ (^{31}P),[2967] $[Pr^i{}_2NPN{=}PClNPPr^i{}_2]^{2+}$ (^{31}P),[2968] and (vinyloxy)fluorocyclotriphosphazenes (^{19}F, ^{31}P).[2969]

The two-dimensional correlation of n.m.r. chemical shifts with nuclear spin relaxation rates has been described and applied to ^{31}P n.m.r. spectroscopy of NaH_2PO_4 and $Na_5P_3O_{10}$.[2970] The ^{34}S, ^{36}S, and ^{18}O isotope shifts on the ^{31}P chemical shift of $Me_2C(CH_2O)_2PXYP(O)(OCH_2)_2CMe_2$ have been reported.[2971] The changes in $\delta(^{17}O)$ and $\delta(^{31}P)$ on protonation of $POCl_3$ and HPO_2F_2 have been recorded.[2972] The ^{31}P n.m.r. spectrum of β-$P_4S_3I_2$ has been fully analysed.[2973] For NF_2Cl, $^1\underline{J}(^{19}F,^{14}N)$ has been observed in the ^{19}F n.m.r. spectrum.[2974] N.m.r. data have also been reported for $(RO)_3P{=}Se$ (^{31}P),[2975] 1,3,2-oxazaphosphorinanes (^{13}C),[2976] $\{(EtO)(EtS)P(S)\}_2O$ (^{13}C, ^{31}P),[2977] $\overline{OCMe_2CMe_2P(OMe)C(OMe)(CONMe_2)CH{=}C(OMe)O}$ (^{13}C, ^{31}P),[2978] $Bi(OCBu^tCHNCHCBu^tO)_3$ (^{13}C, ^{15}N, ^{17}O),[2979] $(MeO)_3\overline{POCPh{=}CPhO}$ (^{13}C, ^{31}P),[2980] $\{(Pr^iO)_2PO\}_2CRSCCl_{\underline{n}}F_{3-\underline{n}}$ (^{19}F, ^{31}P),[2981] $EtC(CH_2O)_3PO_3$ (^{31}P),[2982] $R^1C_6H_4O\overline{P(O)OCHR^2CH_2CH_2E}$ (E = Se; ^{13}C),[2983] (E = S; ^{13}C),[2984] cis-2-alkoxy-2-oxo-6-(iodomethyl)-1,2-oxaphosphorinanes (^{13}C, ^{31}P),[2985] 3,9-bis(alkylphenoxy)-2,4,8,10-tetraoxa-3,9-diphosphaspiro[5.5]undecanes (^{13}C),[2986] $[As(OTe_5)_6]^-$ (^{75}As),[2987] $\overline{OCH_2CH_2CHMeOP}(O)SCF_3$ (^{19}F, ^{31}P),[2988] $(EtO)_2P(O)F$ (^{31}P),[2989] $FP(OCH_2)_2CH_2$ (^{19}F, ^{31}P),[2990] organic phosphates (^{31}P),[2991, 2992] $(RO)_2P(O)Cl$ (^{31}P),[2993] $[As\{O_2C_6H_4CONH(CH_2)_3\}_2COC_6H_4(OH)_2]^-$ (^{13}C),[2994] aryl alkyl thiophosphates (^{13}C),[2995] 1,2-oxaphospholanes (^{13}C),[2996] $(EtO)_3PF_2$ (^{19}F, ^{31}P),[2997] $C_6H_4(O)(S)PER$ (^{13}C, ^{31}P),[2998] $ROPCl_2$ (^{31}P),[2999] di(alkylphenyl)dithiophosphoric acids (^{31}P),[3000] $ROP(O)(OEt)_2$ (^{31}P),[3001] polyorganophosphazenes (^{31}P),[3002] $[As(OCH_2CF_3)_6]^-$ (^{13}C, ^{75}As),[3003] 1,2-thiaphosphole 2-sulphides (^{13}C, ^{31}P),[3004] $P_{4-\underline{n}}As_{\underline{n}}Se_3$ (^{31}P),[3005] $[C_5H_5N][PF_6]$ (^{13}C, ^{31}P),[3006] $Te_2Se_2(Sb_3F_{14})(SbF_6)$ (^{77}Se, ^{125}Te),[3007] and $E(S_2COPr^i)_3$ (E = As, Sb; ^{13}C).[3008]

9 Compounds of Group VIB, Chlorine, Iodine, and Xenon

Two relevant reviews have appeared: '^{77}Se chemical shifts in organic compounds'[3009] and '^{125}Te N.M.R. and organotellurium compounds'.[3010]

A series of sulphur compounds have been examined by ^{33}S n.m.r. spectroscopy and the line widths discussed.[3011] A correlation between O-F bond energies and ^{19}F chemical shifts has been found for a variety of -OF compounds.[3012] ^{34}S isotope effects of ^{19}F chemical shifts have been correlated with bond lengths.[3013] N.m.r. data have also been reported for $(CF_3)_4Te$ (^{19}F),[3014] $CH_2(CH_2)_3TeRX$ (^{13}C, ^{125}Te),[3015] $Ph_3Te(R^1COCHR^2)$ (^{13}C),[3016] $Se(CH_2CH_2)_2CO$ (^{13}C, ^{77}Se),[3017] PhSeR (^{13}C),[3018] p-$RC_6H_4SeC(O)Ph$ (^{13}C, ^{77}Se),[3019] XC_6H_4-$CHBrCH_2SePh$ (^{13}C, ^{77}Se),[3020] $SeCR^1$=NCR^2=CH (^{13}C),[3021] ^{2}H-[1]-benzotellurin-2-one (^{13}C),[3022] RN=C=Se (^{13}C),[3023] HNC(Se)NHC(Se)CMe_2 (^{13}C),[3024] MeC(Se)NEt_2 (^{13}C),[3025] bis(thio)selenides (^{77}Se),[3026] selenones (^{77}Se),[3027] selenoimidates (^{13}C),[3028] selenocysteine ß-lyase (^{13}C),[3029] $(C_2F_5)_2TeF_4$ (^{19}F),[3030] $CF_3SF_4CHFCF_2Cl$ (^{19}F),[3031] F_5SC_2H (^{19}F),[3032] PhC(O)CMe_2SeSe(O)CMe_2C(O)Ph (^{13}C, ^{77}Se),[3033] CF_3-SeF_5 (^{13}C, ^{19}F, ^{77}Se),[3034] $C_8H_8N_2Se$ (^{13}C),[3035, 3036] 2-phenylselenoproprionic acid (^{77}Se),[3037] SeCMe=CMeN=N (^{13}C),[3038] $Cl_3TeC_6H_4NPh$=N (^{13}C),[3039] $X_3TeC_6H_3$(OMe)CH=NPh (^{13}C),[3040] $[Me_2NSF_4]^+$ (^{13}C, ^{19}F),[3041] $(F_5SN=)_2S$ (^{19}F),[3042] Cl_3SeN=PPh_3 (^{31}P, ^{77}Se),[3043] CF_3N=$SeCl_2$ (^{19}F, ^{77}Se),[3044] $F_5TeN(CF_3)$COMe (^{13}C, ^{19}F),[3045] F_5EOCN (E = Se, Te; ^{14}N, ^{15}N, ^{19}F, ^{33}S, ^{77}Se, ^{125}Te),[3046] and [NO][$OTeF_5$] (^{19}F).[3047]

^{35}Cl nuclear relaxation has been used to estimate anionic reactivity of Cl^- in organic solvents.[3048] Caesium-induced ^{129}Xe relaxation in N_2 and He has been measured.[3049, 3050] Stable oxygen bonded Xe(II) and Xe(IV) derivatives of O=IF_4O have been shown to exist in solution by ^{19}F and ^{129}Xe n.m.r. spectroscopy.[3051]

10 Appendix

This appendix contains a list of papers in which the use of nuclei other than ^{1}H has been described. The nuclei are ordered by increasing mass.

^{2}H 172, 524, 544, 735, 827, 1188, 1200, 1219, 1350, 1496, 1505, 1506, 1508, 1519, 1533, 1564, 1582, 1599, 1622, 1641, 1761,

1772, 1775, 1820, 1835, 1875-1877, 1886, 1889, 1894, 1897, 1900, 1966, 2002, 2032, 2036, 2037, 2045, 2050, 2071, 2085, 2090, 2124, 2130, 2156, 2256, 2306, 2312, 2323, 2338, and 2601.

^{3}He 2359 and 2360.

^{6}Li 22, 24, 26, 1280, 1543, 1605, 2042, and 2161.

^{7}Li 21, 23, 25, 26, 29, 38, 760, 1280, 1470, 1493, 1497, 1500, 1501, 1541, 1545, 1548-1552, 1605, 1656, 1682, 1969, 1972-1975, 1977-1979, 1982, 1983, 2003-2005, 2020, 2021, 2029, 2042, 2047, 2048, 2051, 2051, 2096, 2100, 2105, 2106, 2110, 2156, 2161, 2197, 2440, 2539, and 2636.

^{9}Be 1587.

^{10}B 2382.

^{11}B 70, 100, 120, 127, 202, 308, 623, 657, 691, 743, 744, 790, 871, 882, 979, 1081, 1181, 1292, 1316, 1442-1446, 1448, 1688-1690, 1746, 1832-1837, 1927-1929, 2017, 2031, 2155, 2159, 2160, 2162-2164, 2175, 2248, 2362-2366, 2368-2375, 2377-2391, 2393-2420, 2422-2428, 2430-2442, 2444, 2445, 2447-2458, 2460-2464, 2466, 2468-2470, 2473, 2476-2485, 2487-2497, and 2773.

^{13}C 6, 12, 14-17, 19-22, 14, 25, 27, 32-37, 40, 41, 45-51, 60-68, 70, 72, 73, 75, 76, 78-83, 85-103, 105-115, 119, 120, 125-127, 130,131, 133, 134, 137, 138, 140-143, 145, 146, 160, 167, 170, 171, 173, 175, 180, 181, 183, 184, 186, 187, 191-206, 208, 210-253, 256, 260, 262-291, 294-296, 298-300, 305, 306, 309, 311-319, 321-323, 326, 328, 330, 331, 333, 334, 336, 338, 339, 346-349, 351, 353-357, 360, 365, 367, 369, 374, 376, 378, 380-382, 384, 386, 391, 395, 399-401, 414, 422, 423, 425, 426, 431, 432, 435, 436, 438, 440, 441, 443-445, 450, 452-454, 456, 457, 459, 460, 462, 464-473, 475-481, 484-493, 499, 501-503, 505, 507, 509, 529, 527, 530-534, 537, 539, 542, 543, 547, 550-555, 557, 561, 564, 566-582, 585-589, 591-593, 596-598, 600, 602, 603, 605-607, 609-614, 617, 618, 620-628, 630, 632-638, 640-650, 653-656, 658, 660, 662-673, 675684, 689, 690, 693-697, 699, 700, 702, 704, 706-708, 711-714, 716, 721, 725, 727, 728, 730, 731, 735, 741, 745-749, 754, 760, 766, 767, 777, 786-788, 790, 792, 794, 795, 797, 799-812, 816, 817, 819-824, 827-829, 832, 837-839, 841-845, 847-850, 852-855, 857, 858, 861, 862, 864, 868-870, 872, 874-876, 881, 882, 887, 888, 893, 897, 902, 904, 910, 913, 914, 917, 918, 920-946, 948, 952, 954, 968,

970, 974, 975, 977, 980, 981, 983-987, 989, 990, 993-997, 1001, 1002, 1007, 1008, 1012-1014, 1016, 1021, 1025-1027, 1029, 1031-1034, 1039-1041, 1053, 1055-1063, 1065-1074, 1078, 1079, 1081-1084, 1086, 1087, 1090, 1094-1096, 1098, 1099, 1103, 1105-1112, 1114, 1116-1122, 1126, 1128, 1129, 1131, 1134, 1135, 1137, 1140, 1142, 1143, 1150, 1154, 1155, 1158, 1162, 1166, 1178, 1179, 1181, 1184-1187, 1189, 1192-1194, 1200, 1202, 1203, 1205, 1206, 1209-1213, 1217, 1218, 1220, 1226-1228, 1232-1235, 1241-1243, 1254, 1256, 1258-1263, 1265-1268, 1277, 1280, 1285, 1286, 1294, 1297, 1299, 1301, 1303, 1306, 1307, 1309-1311, 1313-1315, 1317, 1318, 1320-1323, 1331, 1335, 1337, 1338, 1340-1349, 1351, 1353, 1355-1370, 1372-1374, 1376-1380, 1382-1385, 1387, 1389, 1391, 1394, 1395, 1397, 1399, 1402, 1404-1408, 1410, 1412-1414, 1416-1418, 1420, 1421, 1425, 1427, 1434, 1436, 1438, 1440, 1442, 1446, 1447, 1449, 1451, 1452, 1454-1457, 1462, 1468, 1470, 1471, 1473, 1476-1479, 1481, 1484, 1509, 1527, 1539, 1543, 1546, 1548, 1549, 1554, 1582, 1583, 1587, 1590, 1591, 1598, 1600-1602, 1609-1611, 1624, 1625, 1628, 1631, 1646-1648, 1657, 1659, 1661, 1666, 1667, 1671, 1677, 1678, 1681, 1686, 1689, 1717-1721, 1723, 1727, 1732, 1733, 1736, 1742, 1743, 1748-1750, 1753, 1755, 1758, 1766-1768, 1771, 1772, 1775, 1776, 1778, 1785-1787, 1794, 1796-1798, 1811, 1813-1814, 1822, 1824, 1829, 1833, 1839, 1843, 1845, 1849, 1855, 1863, 1898, 1919, 1924, 1926, 1931, 1968, 2042, 2086, 2107, 2116, 2123, 2147, 2156, 2183, 2184, 2189, 2201, 2209, 2210, 2250, 2253, 2257, 2258, 2260, 2269, 2286, 2288, 2290-2292, 2298, 2299, 2304, 2324, 2326-2328, 2332, 2335, 2336, 2339, 2341-2347, 2349, 2351, 2353, 2356, 2357, 2364, 2366-2369, 2371, 2373, 2376, 2377, 2379, 2392, 2398, 2402, 2405, 2408, 2412, 2421, 2428-2430 2435, 2436, 2440-2446, 2448, 2449, 2452, 2458-2460, 2462, 2465-2468, 2470-2475, 2477, 2481, 2483-2486, 2488-2490, 2501, 2504-2507, 2511, 2513, 2514, 2516, 2519, 2523-2527, 2530-2532, 2534-2540, 2542-2566, 2568-2575, 2577, 2579-2585, 2587-2592, 2594, 2595, 2597-2601, 2604, 2606-2609, 2611, 2615, 2617, 2618, 2620, 2621, 2623-2626, 2630, 2632, 2637-2639, 2642, 2644, 2646, 2647, 2649-2652, 2654, 2660-2667, 2669-2671, 2676, 2677, 2682-2685, 2687, 2688, 2691, 2695, 2697, 2699-2703, 2705, 2709, 2711, 2712, 2715, 2717, 2719-2723, 2725, 2726, 2728, 2731, 2733, 2737, 2754, 2755, 2764, 2766-2770, 2773,

2784-2788, 2792, 2795, 2796, 2799-2807, 2809, 2813, 2814, 2819, 2821-2827, 2830, 2832, 2833, 2835, 2837, 2840, 2843, 2845, 2850-2853, 2855-2858, 2860, 2861, 2863, 2864, 2866, 2868-2871, 2873, 2875, 2878-2882, 2885-22889, 2892, 2893, 2897, 2900, 2902-2905, 2907, 2908, 2910, 2912, 2921, 2922, 2924, 2926, 2929, 2933, 2935, 2936, 2942, 2943, 2946, 2954, 2976-2980, 2983-2986, 2994-2996, 2998, 3003, 3004, 3006, 3008, 3015-3022, 3024, 3025, 3028, 3029, 3033-3036, 3038-3041, and 3045.

^{14}N 150, 323, 382, 383, 493, 515, 741, 750, 1096, 1254, 1453, 1508, 1540, 1706, 2032, 2372, 2373, 2449, 2451, 2461, 2466, 2478, 2488, 2507, 2745, 2746, 2773, 2974, and 3046.

^{15}N 12, 382, 385, 396, 443, 541, 700, 909, 911, 1045, 1087, 1122, 1127, 1129, 1195, 1514, 1750, 1831, 1924, 2033, 2123, 2128, 2257, 2353, 2440, 2449, 2506, 2626, 2627, 2677, 2687, 2721, 2745, 2773, 2775, 2921, 2934, 2979, and 3046.

^{17}O 54, 55, 69, 116, 153-155, 157, 164-166, 321, 326, 389, 411, 415, 416, 427, 428, 430, 453, 721, 1278, 1289, 1332, 1453, 1483, 1485, 1515, 1518, 1520, 1525, 1529, 1533-1535, 1540, 1548, 1605, 1606, 1615, 1650, 1719, 1750, 1859, 1861, 1926, 1989, 1990, 2041, 2056, 2296, 2305, 2306, 2533, 2671, 2680, 2746, 2777, 2972, and 2979.

^{19}F 51, 52, 57-59, 73, 139, 161, 163, 175, 204, 207, 282, 293, 304, 324, 361, 390, 397, 398, 404, 405, 420, 421, 434, 439, 587, 594, 625, 692, 761, 825, 827, 841, 842, 852, 856, 877, 920, 953, 954, 957, 963, 999, 1000, 1019, 1028, 1033, 1079, 1146, 1162, 1191, 1214, 1215, 1221-1223, 1225, 1229-1231, 1236, 1237, 1329, 1364, 1399, 1403, 1435, 1459-1461, 1477, 1479, 1481, 1489, 1491, 1532, 1535, 1536, 1588, 1635, 1636, 1700, 1710, 1711, 1774, 1780, 1781, 1819, 1854, 1892, 1898, 1974, 1981, 1983, 1993-1995, 2031, 2034, 2052, 2066-2069, 2073, 2079, 2088, 2091, 2146, 2154, 2165, 2255, 2358, 2369, 2468, 2476, 2477, 2478, 2480, 2482, 2488, 2493, 2494, 2496-2498, 2512, 2529, 2541, 2566, 2593, 2643, 2657, 2658, 2668, 2675, 2718, 2724, 2741-2743, 2749, 2789, 2790, 2804, 2812, 2828, 2829, 2837, 2860, 2861, 2868, 2870, 2872, 2911, 2919, 2920, 2922, 2929, 2933, 2938, 2940, 2941, 2948, 2949, 2958, 2963, 2964, 2969, 2974, 2981, 2988, 2990, 2997, 3012-3014, 3030-3032, 3034, 3041, 3042, 3044-3047, and 3051.

^{23}Na 30, 417, 1453, 1491, 1502, 1504-1509, 1548-1550, 1553-1563, 1565-1579a, 1605, 1709, 1979, 1994, 2023, 2043, 2045, 2049,

2050, 2053, 2054, 2089, 2155, 2199, 2205, 2243, and 2275.

^{25}Mg 45, 1286, and 1597.

^{27}Al 441, 1530, 1531, 1690-1692, 1696-1699, 1701-1703, 1708, 1748, 1943, 2019, 2058, 2059, 2062, 2155-2157, 2167, 2169-2171, 2173, 2175, 2178, 2181, 2182, 2184-2187, 2194, 2196, 2200, 2202, 2204, 2206-2208, 2211-2214, 2216-2221, 2223, 2227, 2230-2233, 2236, 2239, 2240, 2243-2246, 2248, 2330, 2450, 2502, 2504, 2506, 2508, 2509, 2512, and 2514.

^{31}P 12, 18, 29, 39, 71, 84, 95, 98, 104, 105, 110, 115, 117, 118, 121-125, 128, 135, 137, 154, 156, 162, 164, 170, 174, 176-183. 185, 188-190, 196, 206, 209, 212, 214, 221, 223, 225, 230-232, 249, 252, 253, 261, 262, 266, 271, 273, 280, 282, 292, 293, 297-304, 307, 308, 310, 312, 320, 323, 324, 326, 327, 329, 332-346, 349-355, 358, 359, 361-367, 370-378, 382, 387, 388, 395, 396, 402-410, 414, 416, 417, 426, 428, 429, 437, 441, 442, 446-449, 451, 452, 454, 455, 458, 460, 461, 463, 466, 467, 471, 474, 475, 477, 481-483, 495, 496, 498, 501-519, 521-523, 525, 527, 530, 533-5538, 540, 545-549, 553, 555, 556, 561-569, 576-578, 583, 584, 589, 590, 592-595, 598, 599 601, 604, 606, 608, 609, 626-629, 631, 639-641, 651, 652, 655, 657, 659, 661, 663, 671-674, 686-688, 694, 698, 699, 701, 709-717, 719, 720, 722-724, 726, 729, 730, 732-734, 736-739, 742, 750-757, 759, 761-766, 768, 769, 773-785, 787-789, 791-794, 796, 798, 805-807, 809, 813-816, 818-820, 826, 830-835, 846, 848, 849, 851, 854, 855, 857-861, 863, 865-867, 873, 875, 877-881, 883-908, 919, 947, 950, 951, 953-966, 971-973, 976-978, 981-985, 988, 989, 991, 992, 997, 998, 1003-1008, 1011, 1013, 1015, 1017, 1018, 1020-1024, 1027-1032, 1034-1048, 1051, 1055-1057, 1060, 1062, 1064, 1065, 1067, 1075-1077, 1080, 1082, 1085-1089, 1091-1093, 1100-1103, 1106, 1113, 1115, 1116, 1123-1125, 1132, 1133, 1136, 1138-1141, 1144, 1146-1161, 1163-1173, 1176-1178, 1182-1185, 1188, 1190, 1191, 1196-1199, 1204-1209, 1227, 1250-1253, 1264, 1271-1274, 1276, 1292, 1299, 1300, 1302, 1303, 1306, 1312, 1315, 1318, 1319, 1321, 1324, 1327, 1329, 1330, 1340, 1345, 1347, 1349, 1350, 1352-1354, 1372, 1373, 1378, 1380, 1386-1390, 1397-1401, 1404, 1405, 1409, 1410, 1415-1419, 1423, 1424, 1427, 1429, 1435, 1436, 1441, 1443, 1444, 1451, 1452, 1464, 1467, 1468, 1470-1475, 1477, 1478, 1479, 1481, 1482, 1514, 1523, 1537, 1538, 1552, 1589, 1593-1595, 1605, 1606, 1613, 1614, 1620, 1624, 1628, 1643,

1644, 1652, 1663, 1665, 1676-1679, 1682, 1694, 1695, 1704, 1711, 1717, 1724, 1731, 1733, 1734, 1737, 1739, 1740, 1744, 1745, 1757-1759, 1763, 1773, 1778, 1779, 1782-1784, 1791, 1792, 1797, 1807, 1809-1814, 1816, 1817, 1819-1821, 1823, 1824, 1826, 1829-1831, 1840, 1844, 1855, 1857-1862, 1864, 1874, 1882, 1916, 1927, 1947, 2063, 2064, 2078, 2131, 2132, 2134, 2137, 2139, 2170, 2171, 2240, 2252, 2270, 2285, 2299, 2300, 2349, 2352, 2364, 2368, 2369, 2371, 2372, 2375, 2376, 2393, 2401, 2414, 2416, 2417, 2423, 2428, 2430, 2432, 2434, 2436, 2439, 2449, 2461, 2490, 2494, 2500, 2504, 2514, 2515, 2567-2576, 2578, 2610, 2639, 2640, 2642, 2646-2648, 2656-2659, 2683, 2688, 2691, 2704, 2706, 2711, 2727, 2735, 2736, 2739, 2740, 2747-2766, 2768, 2769, 2771-2778, 2780-2798, 2807, 2808, 2810-2820, 2823, 2824, 2826, 2828-2832, 2834-2836, 2838-2844, 2846-2849, 2851-2854, 2856-2870, 28722878, 2882-2884, 2888, 2890-2892, 2894-2902, 2904-2906, 2909, 2911, 2913-2942, 2944, 2945, 2947-2957, 2959-2962, 2964-2973, 2975, 2977, 2978, 2980-2982, 2985, 2988-2993, 2997-3006, and 3043.

^{33}S 419, 3011, and 3047.

^{35}Cl 13, 1454, 1513, 1545, 1550, 1651, 1676, 1713, and 3048.

^{39}K 28, 31, 1509, 1548, 1576, 1577, and 1597.

^{43}Ca 43, 44, 45, 1597, and 2069.

^{45}Sc 1287 and 1985.

^{47}Ti 77.

^{49}Ti 77.

^{51}V 131, 132, 134-136, 144, 147-159, 163, 164, 1516, 1612, 1613, 1615-1617, 1997, 2082, 2084, 2085, 2092-2095, and 2097.

^{53}Cr 2074, 2108, and 2109.

^{55}Mn 136, 493, 494, 499, 500, 1623, 1627, 2119, and 2120.

^{57}Fe 619, 703-705, 2075-2077, 2087, 2125, and 2127.

^{59}Co 500, 521, 758, 836, 915, 916, 949, 1394, 1407, 1578, and 2129.

^{63}Cu 721, 2138, and 2140-2142.

^{65}Cu 2142.

^{67}Zn 1597.

^{73}Ge 2598.

^{75}As 1489, 2302, 2987, and 3003.

^{77}Se 700, 1174, 1175, 1275, 1420, 1477, 1486, 1712, 1756, 2036, 2037, 2309-2312, 2568, 2674, 2773, 2783, 2860, 3007, 3009, 3017, 3019, 3020, 3026, 3027, 3033, 3034, 3037, 3043, 3044, and 3046.

^{79}Br 55.

^{81}Br 1550.

^{85}Rb 1505, and 1509.

^{87}Rb 1577.

^{89}Y 53.

^{93}Nb 69, 157, 168, 1770, 2099, and 2101-2104.

^{95}Mo 157, 159, 169, 255, 321-323, 325, 329, 368, 379, 383, 386, 392-394, 411, 413, 418, 424, 426, 427, 430, 433, and 2114.

^{97}Mo 157, 169, 321, and 2114.

^{99}Tc 496 and 2735.

^{99}Ru 1518.

^{103}Rh 759, 767, 794, 838, and 1733.

^{109}Ag 1180, 1183, 1195, 1197, 1198, 1201, 1207, 1980, 2009, 2144, and 2166.

^{113}Cd 12, 43, 1229, 1243-1249, 1255-1257, 1267-1270, 1275, 1677-1684, 1745, 2108, 2150-2153, and 2330.

^{115}In 1705.

^{117}Sn 2286, 2288, and 2290-2292.

^{119}Sn 131, 202, 496, 770-773, 780, 978, 1007, 1009 1021, 1045, 1275, 1310, 1462-1464, 1467-1469, 1753-1756, 1835a, 1839, 1846, 2286-2292, 2372, 2441, 2449, 2517a, 2518, 2423, 2531, 2536, 2567, 2600-2605, 2608-2610, 2612-2614, 2616, 2619, 2621-2623, 2677-2679, 2681, 2682, 2686, 2687, 2689-2692, 2694, 2707-2714, 2727, and 2734-2738.

^{125}Te 306, 307, 334, 365, 1275, 1489, 1756, 2578, 2674, 2851, 3007, 3010, 3015, and 3046.

^{127}I 1491, 1714, 2313, and 2314.

^{129}Xe 2361 and 3049-3051.

^{133}Cs 1505, 1509, 1550, 1553, 1577, 1584, 1585, 2026, 2097, 2110, and 2135.

^{139}La 1511.

$^{???}$Yb 2070.

^{169}Tm 2072.

^{181}Ta 157 and 2001.

^{183}W 117, 129, 157, 308, 409, 414-416, 428, 429, and 438.

^{187}Re 382, 440, and 497.

^{187}Os 527 and 529.

^{195}Pt 556, 680, 718, 861, 967, 975, 977, 978, 1001, 1009, 1010, 1012, 1015, 1026, 1032, 1045, 1091, 1093, 1096, 1099, 1103, 1104, 1111, 1122, 1127, 1129-1131, 1137, 1145, 1157-1159,

1161, 1165, 1166, 1174, 1175, 1178, 1416, 1421, 1435, 1436, 1741, 1827, 1828, 2329, 2344, and 2345.

^{199}Hg 1183, 1216, 1220, 1223, 1232, 1238-1240, 1250-1253, 1275, 1628, 1685, and 1686.

^{203}Tl 1275 and 2252.

^{205}Tl 1275, 1706, 1707, 2113, 2166, 2172, 2252, 2387, 2499, 2510, and 2511.

^{207}Pb 202, 2449, 2523, 2536, 2624, 2625, 2690, 2707, and 2715.

References

1 Nuclear Magnetic Resonance', ed. R.J. Abraham (Specialist Periodical Reports), The Royal Society of Chemistry, London, 1985, Vol. 14; 1986, Vol. 15.

2 Y.C. Lin, Hua Hsueh, 1985, **43**, A70 (Chem. Abstr., 1985, **103**, 114 657).

3 J.J. Dechter, Prog. Inorg. Chem., 1985, **33**, 393.

4 V.P. Tarasov, Khim. Svyaz Str. Mol., 1984, 195. Ed. by V.I. Nefedov (Chem. Abstr., 1985, **103**, 46 891).

5 A. Schwenk, Prog. Nucl. Magn. Reson. Spectrosc., 1985, **17**, 69 (Chem. Abstr., 1985, **103**, 152 228).

6 L.A. Fedorov, Zh. Strukt. Khim., 1984, **25**, 35 (Chem. Abstr., 1985, **102**, 6570).

7 L.A. Fedorov, Zh. Anal. Khim., 1985, **40**, 5 (Chem. Abstr., 1985, **102**, 124 612).

8 J.J. Katz and C.E. Brown, Bull. Magn. Reson., 1983, **5**, 3 (Chem. Abstr., 1985, **102**, 159 143).

9 J. Krebs, Congr. AMPERE Magn. Reson. Relat., Phenom., Proc., 22nd, 1984, 492 (Chem. Abstr., 1985, **102**, 127 388).

10 D.C. Dalgarno and I.M. Armitage, Adv. Inorg. Biochem., 1984, **6**, 113 (Chem. Abstr., 1985, **103**, 100 487).

11 C.J. Jameson and H.-J. Osten, J. Am. Chem. Soc., 1985, **107**, 4158).

12 D.A. Laude, jun., R.W.K. Lee, and C.L. Wilkins, Anal. Chem., 1985, **57**, 1281.

13 J. Kowalewski and K.M. Larsson, Chem. Phys. Lett., 1985, **119**, 157 (Chem. Abstr., 1985, **103**, 152 423).

14 N.A. Ogorodnikova and D.N. Kravtsov, Polyhedron, 1985, **4**, 311.

15 X. You, R.F. Fenske, and D.G. Freier, Gaodeng Xuexiao Huaxue Xuebao, 1985, **6**, 146 (Chem. Abstr., 1985, **103**, 114 827).

16 K. Vrieze and G. van Koten, Inorg. Chim. Acta, 1985, **100**, 79.

17 R. Benn and A. Rufinska, Organometallics, 1985, **4**, 209.

18 J. Podlahova, J. Silha, and J. Podlaha, Collect. Czech. Chem. Commun., 1985, **50**, 445.

19 F. Teixidor, A. Llobet, L. Escriche, and J. Casabo, Polyhedron, 1985, **4**, 215.

20 K. Shinozuka, K. Wilkowski, B.L. Heyl, and L.G. Marzilli, Inorg. Chim. Acta, 1985, **100**, 141.

21 J. Bernard, C. Schnieders, and K. Muellen, J. Chem. Soc., Chem. Commun., 1985, 12.

22 A. Rajca and L.M. Tolbert, J. Am. Chem. Soc., 1985, **107**, 2969.

23 L. Assadourian, R. Faure, and G. Gau, J. Organomet. Chem., 1985, **280**, 153.

24 A.-D. Schlüter, H. Huber, and G. Szeimies, Angew. Chem., Int. Ed. Engl., 1985, **24**, 404.

25 E. Buncel, T.K. Venkatachalam, B. Eliasson, and U. Edlund, J. Am. Chem. Soc., 1985, **107**, 303.

26 U. Edlund, T. Lejon, T. Venkatachalam, and B.E. Krishnan, J. Am. Chem. Soc., 1985, **107**, 6408.

27 K. Müllen, W. Huber, G. Neumann, C. Schnieders, and H. Unterberg, J. Am. Chem. Soc., 1985, **107**, 801.

28 M.L. Tinkham and J.L. Dye, J. Am. Chem. Soc., 1985, **107**, 6129.

29 A. Zschunke, M. Riemer, F. Krech, and K. Issleib, Phosphorus Sulphur, 1985, **22**, 349.

30 H. Sterk, H. Schrunner, and A. Sax, J. Mol. Liq., 1985, **30**, 171 (Chem. Abstr., 1985, **103**, 67 723).

31 P.P. Edwards, A.S. Ellboudly, and D.M. Holton, Nature, 1985, **317**, 242.

32 S. Bywater, P. Lachance, and P. Black, J. Organomet. Chem., 1985, **280**, 159.

33 G.W. Klumpp, M. Vos, F.J.J. de Kanter, C. Slob, H. Krabbendam, and A.L. Spek, J. Am. Chem. Soc., 1985, **107**, 8292.

34 W. Neugebauer, G.A.P. Geiger, A.J. Kos, J.J. Stezowski, and P.v.R. Schleyer, Chem. Ber., 1985, **118**, 1504.

35 L.M. Engelhardt, W.-P. Leung, C.L. Raston, and A.H. White, J. Chem. Soc., Dalton Trans., 1985, 337.

36 J.T.B.H. Jastrzebski, G. van Koten, M.J.N. Christophersen, and C.H. Stam, J. Organomet. Chem., 1985, **292**, 319.

37 H. Schumann, I. Albrecht, and E. Hahn, Angew. Chem., Int. Ed. Engl., 1985, **24**, 985.

38 D. Barr, W. Clegg, R.E. Mulvey, D. Reed, and R. Snaith, Angew. Chem., Int. Ed. Engl., 1985, **24**, 328.

39 M.P. Teulade, P. Savignac, E.E. Aboujaoude, and N. Collignon, J. Organomet. Chem., 1985, **287**, 145.

40 V.C. Emery and M. Akhtar, J. Chem. Soc., Chem. Commun., 1985, 600.

41 T. Ohhama, H. Seto, and S. Miyachi, Arch. Biochem. Biophys., 1985, **237**, 72 (Chem. Abstr., 1985, **102**, 92 724).

42 R.J. Abraham, K.M. Smith, D.A. Goff, and F.W. Bobe, J. Am. Chem. Soc., 1985, **107**, 1085.

43 H.J. Vogel, T. Drakenberg, S. Forsén, J.D.J. O'Neil, and T. Hofmann, Biochemistry, 1985, **24**, 3870 (Chem. Abstr., 1985, **103**, 66 938).

44 H.J. Vogel and W.H. Braunlin, J. Magn. Reson., 1985, **62**, 42.

45 J. Kowalewski and A. Maliniak, J. Magn. Reson., 1985, **62**, 316.

46 R.A. Andersen, R. Blom, A. Haaland, B.E.R. Schilling, and H.V. Volden, Acta Chem. Scand., 1985, **A39**, 563.

47 A.W. Duff, P.B. Hitchcock, M.F. Lappert, R.G. Taylor, and J.A. Segal, J. Organomet. Chem., 1985, **293**, 271.

48 H. Lehmkuhl, A. Shakoor, K. Mehler, C. Krüger, K. Angermund, and Y.-H. Tsay, Chem. Ber., 1985, **118**, 4239.

49 B. Bogdanović, K. Janke, C. Krüger, R. Mynott, K. Schlichte, and U. Westeppe, Angew. Chem., Int. Ed. Engl., 1985, **24**, 960.

50 H.A. Tajmir-Riari and T. Theophanides, Can. J. Chem., 1985, **63**, 2065.

51 D.T. Haworth and M. Das, J. Fluorine Chem., 1984, 26, 331 (Chem. Abstr., 1985, **102**, 6589).

52 K.K. Bhasin, S. Jindal, S. Singh, and R.D. Verma, Synth. React. Inorg. Met-Org. Chem., 1985, **15**, 733.

53 W.J. Evans, J.H. Meadows, A.G. Kostka, and G.L. Closs, Organometallics, 1985, **4**, 324.

54 W.S. Jung, H. Tomiyasu, and H. Fukutomi, Bull. Chem. Soc. Jpn., 1985, **58**, 938.

55 V.A. Shcherbakov, Yu.K. Gusev, L.G. Mashirov, L.L. Shcherbakova, V.A. Vodovatov, and V.V. Gavrilov, Radiokhimiya, 1984, **26**, 700 (Chem. Abstr., 1985, **102**, 88 842).

56 V.A. Shcherbakov and L.G. Mashirov, Radiokhimiya, 1984, **26**, 703 (Chem. Abstr., 1985, **102**, 124 289).

57 I. Ursu, D.E. Demco, M. Bogdan, P. Fitori, and A. Darabont, Congr. AMPERE Magn. Reson. Relat. Phenom., Proc., 22nd, 1984, 307. Ed. by K.A. Mueller, R. Kind, and J. Roos (Chem. Abstr., 1985, **102**, 124 323).

58 I. Ursu, D.E. Demco, M. Bogdan, P. Fitori and A. Darabont, Congr. AMPERE Magn. Reson. Relat. Phenom., Proc., 22nd, 1984, 305. Ed. by K.A. Mueller, R. Kind, and J. Roos (Chem. Abstr., 1985, **102**, 139 348).

59 I. Ursu, D.E. Demco, M. Bogdan, P. Fitori, and A. Darabont, J. Phys. Lett., 1985, **46**, 493 (Chem. Abstr., 1985, **103**, 81 033).

60 I. Albrecht, E. Hahn, J. Pickardt, and H. Schumann, Inorg. Chim. Acta, 1985, **110**, 145.

61 A.L. Wayda and R.D. Rogers, Organometallics, 1985, **4**, 1440.

62 H. Mauermann, P.N. Swepston, and T.J. Marks, Organometallics, 1985, **4**, 200.
63 W.J. Evans, I. Bloom, W.E. Hunter, and J.L. Atwood, Organometallics, 1985, **4**, 112.
64 W.J. Evans, J.W. Grate, H.W. Choi, I. Bloom, W.E. Hunter, and J.L. Atwood, J. Am. Chem. Soc., 1985, **107**, 941.
65 H. Schumann, J. Organomet. Chem., 1985, **281**, 95.
66 H. Schumann, F.-W. Reier, and E. Palamidis, J. Organomet. Chem., 1985, **297**, C30.
67 H. Schumann, H. Lauke, E. Hahn, M.J. Heeg, and D. van der Helm, Organometallics, 1985, **4**, 321.
68 J.M. Boncella and R.A. Andersen, Organometallics, 1985, **4**, 205.
69 V.W. Day, W.G. Klemperer, and D.J. Maltbie, Organometallics, 1985, **4**, 104.
70 R. Shinomoto, J.G. Brennan, N.M. Edelstein, and A. Zalkin, Inorg. Chem., 1985, **24**, 2896.
71 J.M. Ritchey, A.J. Zozulin, D.A. Wrobleski, R.R. Ryan, H.J. Wasserman, D.C. Moody, and R.T. Paine, J. Am. Chem. Soc., 1985, **107**, 501.
72 L. De Cola, D.L. Smailes, and L.M. Vallarino, Inorg. Chim. Acta, 1985, **110**, L1.
73 D.T. Haworth and M. Das, Inorg. Chim. Acta, 1985, **110**, L3.
74 D.A. Lemenovskii, I.F. Urazowski, Yu. K. Grishin, and V.A. Roznyatovsky, J. Organomet. Chem., 1985, **290**, 301.
75 D.M. Roddick, M.D. Fryzuk, P.F. Seidler, G.L. Hillhouse, and J.E. Bercaw, Organometallics, 1985, **4**, 97.
76 B.J.J. Van de Heisteeg, G. Schat, O.S. Akkerman, and F. Bickelhaupt, Organometallics, 1985, **4**, 1141.
77 M.J. McGlinchey and D.G. Bickley, Polyhedron, 1985, **4**, 1147.
78 C.S. Bajgur, W. Tikkanen, and J.L. Petersen, Inorg. Chem., 1985, **24**, 2539.
79 C. McDade and J.E. Bercaw, J. Organomet. Chem., 1985, **279**, 281.
80 T.D. Tilley, Organometallics, 1985, **4**, 1452.
81 T. Ji, X. Han, G. Cheng, M. Bei, Y. Wu, W. Zhang, and H. Guo, Kexue Tongbao (Foreign Lang. Ed.), 1984, **29**, 1200 (Chem. Abstr., 1985, **102**, 113 645).
82 V. Dimitrov, K.-H. Thiele, and D. Schenke, Z. Anorg. Allg. Chem., 1985, **527**, 85.
83 J.W. Pattiasina, C.E. Hissink, J.L. de Boer, A. Meetsina, J.H. Teuben, and A.L. Spek, J. Am. Chem. Soc., 1985, **107**, 7758.
84 F. Senocq, M. Basso-Bert, R. Choukroun, and D. Gervais, J. Organomet. Chem., 1985, **297**, 155.
85 S.L. Latesky, A.K. McMullen, I.P. Rothwell, and J.C. Huffman, J. Am. Chem. Soc., 1985, **107**, 5981.
86 S.L. Latesky, A.K. McMullen, G.P. Niccolai, I.P. Rothwell, and J.C. Huffman, Organometallics, 1985, **4**, 902.
87 T.V. Lubben and P.T. Wolczanski, J. Am. Chem. Soc., 1985, **107**, 701.
88 A.K. McMullen, I.P. Rothwell, and J.C. Huffman, J. Am. Chem. Soc., 1985, **107**, 1072.
89 V. Dimitrov, J. Organomet. Chem., 1985, **282**, 321.
90 K. Mashima and H. Takaya, Organometallics, 1985, **4**, 1464.
91 S.A. Cohen and J.E. Bercaw, Organometallics, 1985, **4**, 1006.
92 J.A. Bandy, V.S.B. Mtetwa, K. Prout, J.C. Green, C.E. Davies, M.L.H. Green, N.J. Hazel, A. Izquierdo, and J.J. Martin-Polo, J. Chem. Soc., Dalton Trans., 1985, 2037.
93 H.G. Alt, H.E. Engelhardt, M.D. Rausch, and L.B. Kool, J. Am. Chem. Soc., 1985, **107**, 3717.
94 E.A. Mintz, A.S. Ward, and D.S. Tice, Organometallics, 1985, **4**, 1308.
95 G. Erker, P. Czisch, R. Mynott, Y.-H. Tsay, and C. Krüger, Organometallics, 1985, **4**, 1310.
96 S.B. Jones and J.L. Petersen, Organometallics, 1985, **4**, 966.
97 C.P. Casey, R.E. Palermo, R.F. Jordan, and A.L. Rheingold, J. Am. Chem. Soc., 1985, **107**, 4597.
98 G. Erker, P. Czisch, C. Krüger, and J.M. Wallis, Organometallics, 1985, **4**, 2059.
99 L. Clawson, J. Soto, S.L. Buchwald, M.L. Steigerwald, and R.H. Grubbs, J. Am. Chem. Soc., 1985, **107**, 3377.

100 P. Paetzold, K. Delpy, R.P. Hughes, and W.A. Herrmann, Chem. Ber., 1985, **118**, 1724.

101 L.B. Kool, M.D. Rausch, H.G. Alt, M. Herberhold, B. Wolf, and U. Thewalt, J. Organomet. Chem., 1985, **297**, 159.

102 F. Bottomley, G.O. Egharevba, I.J.B. Lin, and P.S. White, Organometallics, 1985, **4**, 550.

103 S.M. Clift and J. Schwartz, J. Organomet. Chem., 1985, **285**, C5.

104 S.J. Young, M.M. Olmstead, M.J. Knudsen, and N.E. Schore, Organometallics, 1985, **4**, 1432.

105 J.A. Marsella, J.C. Huffman, K. Folting, and K.G. Caulton, Inorg. Chim. Acta, 1985, **96**, 161.

106 T. Cuenca and P. Royo, J. Organomet. Chem., 1985, **295**, 159.

107 G. Erker, U. Dorf, R. Mynott, Y.-H. Tsay, and C. Krüger, Angew. Chem., Int. Ed. Engl., 1985, **24**, 584.

108 W. Frömberg and G. Erker, J. Organomet. Chem., 1985, **280**, 355.

109 T.D. Tilley, J. Am. Chem. Soc., 1985, **107**, 4084.

110 D.M. Roddick, B.D. Santarsiero, and J.E. Bercaw, J. Am. Chem. Soc., 1985, **107**, 4670.

111 C. Krüger, G. Müller, G. Erker, U. Dorf, and K. Engel, Organometallics, 1985, **4**, 215.

112 J. Blenkers, P. Bruin, and J.H. Teuben, J. Organomet. Chem., 1985, **297**, 61.

113 M.R. Awang, R.D. Barr, M. Green, J.A.K. Howard, T.B. Marder, and F.G.A. Stone, J. Chem. Soc., Dalton Trans., 1985, 2009.

114 C.T. Jekel-Vroegop and J.H. Teuben, J. Organomet. Chem., 1985, **286**, 309.

115 L.B. Kool, M.D. Rausch, H.G. Alt, M. Herberhold, U. Thewalt, and B. Wolf, Angew. Chem., Int. Ed. Engl., 1985, **24**, 394.

116 T.M. Che, V.W. Day, L.C. Francesconi, M.F. Fredrich, and W.G. Klemperer, Inorg. Chem., 1985, **24**, 4055.

117 J.F.W. Keana, M.C. Ogan, Y. Lü, M. Beer, and J. Varkey, J. Am. Chem. Soc., 1985, **107**, 6714.

118 G.S. White and D.W. Stephan, Inorg. Chem., 1985, **24**, 1499.

119 W. Frömberg and G. Erker, J. Organomet. Chem., 1985, **280**, 343.

120 D. Männing, H. Nöth, M. Schwaryz, S. Weber, and U. Wietelmann, Angew. Chem., Int. Ed. Engl., 1985, **24**, 998.

121 L. Gelmini, L.C. Matassa, and D.W. Stephan, Inorg. Chim. Acta, 1985, **24**, 2585.

122 T.S. Targos, R.P. Rosen, R.R. Whittle, and G.L. Geoffroy, Inorg. Chem., 1985, **24**, 1375.

123 R.T. Baker, T.H. Tulip, and S.S. Wreford, Inorg. Chem., 1985, **24**, 1379.

124 G.S. Ferguson and P.T. Wolczanski, Organometallics, 1985, **4**, 1601.

125 C.P. Casey and F. Nief, Organometallics, 1985, **4**, 1218.

126 T. Cuenca and P. Royo, J. Organomet. Chem., 1985, **293**, 61.

127 G. Schmid, D. Kampmann, W. Meyer, R. Boese, P. Paetzold, and K. Delpy, Chem. Ber., 1985, **118**, 2418.

128 M.D. Fryzuk, A. Carter, and A. Westerhaus, Inorg. Chem., 1985, **24**, 642.

129 F. Chauveau, J. Eberlé, and J. Lefebvre, Nouv. J. Chim., 1985, **9**, 315.

130 S.L. Latesky, A.K. McMullen, G.P. Niccolai, I.P. Rothwell, and J.C. Huffman, Organometallics, 1985, **4**, 1896.

131 M. Hoch and D. Rehder, J. Organomet. Chem., 1985, **288**, C25.

132 K. Ihmels and D. Rehder, Chem. Ber., 1985, **118**, 895.

133 M.D. Curtis and J. Real, Organometallics, 1985, **4**, 940.

134 K. Ihmels, D. Rehder, and V. Pank, Inorg. Chim. Acta, 1985, **96**, L69.

135 K. Ihmels and D. Rehder, Organometallics, 1985, **4**, 1334.

136 D. Rehder, K. Ihmels, D. Wenke, and P. Oltmanns, Inorg. Chim. Acta, 1985, **100**, L11.

137 M.L. Luetkens, jun., J.C. Huffman, and A.P. Sattelberger, J. Am. Chem. Soc., 1985, **107**, 3361.

138 S.I. Bailey, L.M. Engelhardt, W.-P. Leung, C.L. Raston, I.M. Ritchie, and A.H. White, J. Chem. Soc., Dalton Trans., 1985, 1747.

139 J. Amaudrut, J.-C. Leblanc, C. Moise, and J. Sala-Pala, J. Organomet. Chem., 1985, **295**, 167.

140 H. van der Heijden, A.W. Gal, P. Pasman, and A.G. Orpen, Organometallics,

1985, **4**, 1847.
141 R.E. La Pointe, P.T. Wolczanski, and G.D. Van Duyne, Organometallics, 1985, **4**, 1810.
142 J.-F. Reynoud, J.-C. Leblanc, and C. Moise, Organometallics, 1985, **4**, 1059.
143 H. Yasuda, K. Tatsumi, T. Okamoto, K. Mashima, K. Lee, A. Nakamura, Y. Kai, N. Kanehisa, and N. Kasai, J. Am. Chem. Soc., 1985, **107**, 2411.
144 N.J. Coville, G.W. Harris, and D. Rehder, J. Organomet. Chem., 1985, **293**, 365.
145 K. Seitz and U. Behrens, J. Organomet. Chem., 1985, **294**, C9.
146 J. Arnold and T.D. Tilley, J. Am. Chem. Soc., 1985, **107**, 6409.
147 F. Naeumann and D. Rehder, Z. Naturforsch., Teil B, 1984, **39**, 1654 (Chem. Abstr., 1985, **102**, 159 433).
148 K. Ihmels and D. Rehder, Organometallics, 1985, **4**, 1340.
149 J. Hanich, M. Krestel, U. Mueller, K. Dehnicke, and D. Rehder, Z. Naturforsch., Teil B, 1984, **39**, 1686 (Chem. Abstr., 1985, **102**, 196 729).
150 F. Preuss, W. Towae, V. Kruppa, and E. Fuchslocher, Z. Naturforsch., Teil B, 1984, **39**, 1510 (Chem. Abstr., 1985, **102**, 55 065).
151 A.S. Tracey and K. Radley, Can. J. Chem., 1985, **63**, 2181.
152 W. Priebsch and D. Rehder, Inorg. Chem., 1985, **24**, 3058.
153 A.T. Harrison and O.W. Howarth, J. Chem. Soc., Dalton Trans., 1985, 1173.
154 A.T. Harrison and O.W. Howarth, J. Chem. Soc., Dalton Trans., 1985, 1953.
155 R.I. Maksimovskaya, A.K. Il'yasova, D.U. Begalieva, D.F. Takezhanova, and A.K. Akhmetova, Izv. Akad. Nauk SSSR, Ser. Khim., 1984, 2169 (Chem. Abstr., 1985, **102**, 38 878).
156 A.K. Akhmetova and A.K. Il'yasova, Zh. Neorg. Khim., 1985, **30**, 901 (Chem. Abstr., 1985, **103**, 31 508).
157 C.J. Besecker, W.G. Klemperer, D.J. Maltbie, and D.A. Wright, Inorg. Chem., 1985, **24**, 1027.
158 F. Preuss, E. Fuchslocher, and W.S. Sheldrick, Z. Naturforsch., Teil B, 1985, **40**, 363 (Chem. Abstr., 1985, **102**, 196 839).
159 J. Anhaus, Z.A. Siddiqi, J. Schimkowiak, H.W. Roesky, and H. Lueken, Z. Naturforsch., Teil B, 1984, **39**, 1722 (Chem. Abstr., 1985, **102**, 196 730).
160 P.A. Bates, A.J. Nielson, and J.M. Waters, Polyhedron, 1985, **4**, 1391.
161 E.G. Il'in, M.E. Ignatov, Yu.A. Buslaev, T.Ya. Medved, N.P. Nesterova, and M.I. Kabachnik, Dokl. Akad. Nauk SSSR, 1985, **282**, 126 (Chem. Abstr., 1985, **103**, 114 993).
162 N. Hovnanian, L.G. Hubert-Pfalzgraf, and G. Le Borgne, Inorg. Chem., 1985, **24**, 4647.
163 R.C. Hibbert, J. Chem. Soc., Chem. Commun., 1985, 317.
164 R.I. Maksimovskaya and M.A. Fedotov, Zh. Neorg. Khim., 1985, **30**, 103 (Chem. Abstr., 1985, **102**, 230 725).
165 W.G. Klemperer and C. Schwartz, Inorg. Chem., 1985, **24**, 4461.
166 C.J. Besecker, V.W. Day, W.G. Klemperer, and M.R. Thompson, Inorg. Chem., 1985, **24**, 44.
167 L.N. Lewis and M.F. Garbauskas, Inorg. Chem., 1985, **24**, 363.
168 J. Sola, Y. Do, J.M. Berg, and R.H. Holm, Inorg. Chem., 1985, **24**, 1706.
169 R.T.C. Brownlee, M.J. O'Connor, B.P. Shehan, and A.G. Wedd, J. Magn. Reson., 1985, **61**, 516.
170 J.E. Salt, G.S. Girolami, G. Wilkinson, M. Motevalli, M. Thornton-Pett, and M.B. Hursthouse, J. Chem. Soc., Dalton Trans., 1985, 685.
171 D. Wormsbächer, K.M. Nicholas, and A.L. Rheingold, J. Chem. Soc., Chem. Commun., 1985, 721.
172 P.A. Tooley, L.W. Arndt, and M.Y. Darensbourg, J. Am. Chem. Soc., 1985, **107**, 2422.
173 M.J. Mockford and W.P. Griffith, J. Chem. Soc., Dalton Trans., 1985, 717.
174 C. Shortman and R.L. Richards, J. Organomet. Chem., 1985, **286**, C3.
175 L. Gomes de Lima, M. Cariou, H. Scordia, R. Kergoat, M.M. Kubicki, and J.E. Guerchais, J. Organomet. Chem., 1985, **290**, 321.
176 R.A. Jones, S.T. Schwab, A.L. Stuart, B.R. Whittlesey, and T.C. Wright, Polyhedron, 1985, **4**, 1689.
177 J. Powell, J.F. Sawyer, and S.J. Smith, J. Chem. Soc., Chem. Commun., 1985, 1312.

178 D. Lyons and G. Wilkinson, J. Chem. Soc., Dalton Trans., 1985, 587.
179 M.L.H. Green, D. O'Hare, and G. Parkin, J. Chem. Soc., Chem. Commun., 1985, 356.
180 M.L.H. Green and G. Parkin, J. Chem. Soc., Chem. Commun., 1985, 1383.
181 M.L.H. Green and G. Parkin, J. Chem. Soc., Chem. Commun., 1985, 771.
182 P.R. Sharp and K.G. Frank, Inorg. Chem., 1985, **24**, 1808.
183 V.C. Gibson, C.E. Graimann, P.M. Hare, M.L.H. Green, J.A. Bandy, P.D. Grebenik, and K. Prout, J. Chem. Soc., Dalton Trans., 1985, 2025.
184 F.G.N. Cloke, J.C. Green, M.L.H. Green, and C.P. Morley, J. Chem. Soc., Chem. Commun., 1985, 945.
185 P. Legzdins, J.T. Martin, and J.C. Oxley, Organometallics, 1985, **4**, 1263.
186 Y. Chi and J.R. Shapley, Organometallics, 1985, **4**, 1900.
187 M.H. Chisholm, B.W. Eichhorn, and J.C. Huffmann, J. Chem. Soc., Chem. Commun., 1985, 861.
188 L. Weber and D. Wewers, Chem. Ber., 1985, **118**, 3560.
189 L. Weber, D. Wewers, and R. Boese, Chem. Ber., 1985, **118**, 3570.
190 L. Weber and D. Wewers, Organometallics, 1985, **4**, 841.
191 H.G. Alt, H.E. Englehardt, U. Thewalt, and J.G. Riede, J. Organomet. Chem., 1985, **288**, 165.
192 F. Seitz, H. Fischer, and J. Riede, J. Organomet. Chem., 1985, **287**, 87.
193 H. Fischer, A. Motsch, R. Märkl, and K. Ackermann, Organometallics, 1985, **4**, 726.
194 K.H. Dötz and W. Kuhn, J. Organomet. Chem., 1985, **286**, C23.
195 K.H. Dötz, W. Kuhn, and U. Thewalt, Chem. Ber., 1985, **118**, 1126.
196 G. Zinner and W.P. Fehlhammer, Angew. Chem., Int. Ed. Engl., 1985, **24**, 978.
197 K.H. Dötz and M. Popall, Tetrahedron, 1985, **41**, 5797.
198 H. Fischer, F. Seitz, J. Riede, and J. Vogel, Angew. Chem., Int. Ed. Engl., 1985, **24**, 721.
199 H.G. Raubenheimer, G.J. Kruger, A. van A. Lombard, L. Linford, and J.C. Viljoen, Organometallics, 1985, **4**, 275.
200 J.C. Jeffery, D.B. Lewis, G.E. Lewis, and F.G.A. Stone, J. Chem. Soc., Dalton Trans., 1985, 2001.
201 P. Jutzi, B. Hampel, K. Stroppel, C. Krüger, K. Angermund, and P. Hofmann, Chem. Ber., 1985, **118**, 2789.
202 H. Schmid, F. Schmidt, and R. Boese, Chem. Ber., 1985, **118**, 1949.
203 M.J. Chetcuti, M.H. Chisholm, H.T. Chiu, and J.C. Huffman, Polyhedron, 1985, **4**, 1213.
204 M. Cariou, M.M. Kubicki, R. Kergoat, L.C. Gomes de Lima, H. Scordia, and J.E. Guerchais, Inorg. Chim. Acta, 1985, **104**, 185.
205 R. Poli, G. Wilkinson, M. Motevalli, and M.B. Hursthouse, J. Chem. Soc., Dalton Trans., 1985, 931.
206 S.E. Himmel, G.B. Young, D.C.M. Fung, and C. Hollingshead, Polyhedron, 1985, **4**, 349.
207 J.L. Davidson, W.F. Wilson, and K.W. Muir, J. Chem. Soc., Chem. Commun., 1985, 460.
208 J.S. Adams, M. Cunningham, and M.W. Whiteley, J. Organomet. Chem., 1985, **293**, C13.
209 K.J. Ahmed, M.H. Chisholm, and J.C. Huffman, Organometallics, 1985, **4**, 1168.
210 H. Fischer and R. Märkl, Chem. Ber., 1985, **118**, 3683.
211 H. Adams, N.A. Bailey, V.A. Osborn, and M.J. Winter, J. Organomet. Chem., 1985, **284**, C1.
212 H. Arzoumanian, A. Baldy, R. Lai, J. Metzger, M.-L.N. Peh, and M. Pierrot, J. Chem. Soc., Chem. Commun., 1985, 1151.
213 V.A. Osborn and M.J. Winter, J. Chem. Soc., Chem. Commun., 1985, 1744.
214 R.G. Beevor, M.J. Freeman, M. Green, C.E. Morton, and A.G. Orpen, J. Chem. Soc., Chem. Commun., 1985, 68.
215 O.A. Reutov, I.V. Barinov, V.A. Chertkov, and V.I. Sokolov, J. Organomet. Chem., 1985, **297**, C25.
216 E.N. Jacobsen and R.G. Bergman, J. Am. Chem. Soc., 1985, **107**, 2023.
217 D.J. Darensbourg, R.K. Hanckel, C.G. Bauch, M. Pala, D. Simmons, J.N. White, J. Am. Chem. Soc., 1985, **107**, 7463.
218 I. Feinstein-Jaffe, J.C. Dewan, and R.R. Schrock, Organometallics, 1985, **4**,

1189.
219 P. Legzdins, S.J. Rettig, and L. Sánchez, Organometallics, 1985, **4**, 1470.
220 P. Legzdins, S.J. Rettig, L. Sánchez, B.E. Bursten, and M.G. Gatter, J. Am. Chem. Soc., 1985, **107**, 1411.
221 K.J. Ahmed, M.H. Chisholm, K. Folting, and J.C. Huffman, J. Chem. Soc., Chem. Commun., 1985, 152.
222 G.A. Miller and N.J. Cooper, J. Am. Chem. Soc., 1985, **107**, 709.
223 N.H.T. Huy, E.O. Fischer, H.G. Alt, and K.H. Dötz, J. Organomet. Chem., 1985, **284**, C9.
224 H.G. Alt, J. Organomet. Chem., 1985, **288**, 149.
225 G.L. Hillhouse, J. Am. Chem. Soc., 1985, **107**, 7772.
226 M.I. Bruce, T.W. Hambley, M.R. Snow, and A.G. Swincer, Organometallics, 1985, **4**, 494.
227 K.R. Birdwhistell and J.L. Templeton, Organometallics, 1985, **4**, 2062.
228 M.H. Chisholm, K. Folting, J.A. Heppert, and W.E. Streib, J. Chem. Soc., Chem. Commun., 1985, 1755.
229 D. Hodgson, J.A.K. Howard, F.G.A. Stone, and M.J. Went, J. Chem. Soc., Dalton Trans., 1985, 1331.
230 J.C. Jeffery and J.G. Lawrence-Smith, J. Organomet. Chem., 1985, **280**, C34.
231 J.A.K. Howard, J.C.V. Laurie, O. Johnson, and F.G.A. Stone, J. Chem. Soc., Dalton Trans., 1985, 2017.
232 J. Levisalles, F. Rose-Munch, H. Rudler, J.C. Daran, and Y. Jeannin, J. Organomet. Chem., 1985, **279**, 413.
233 N.E. Kolobova, L.L. Ivanov, O.S. Zhvanko, A.S. Batsanov, and Ya.T. Struchkov, J. Organomet. Chem., 1985, **279**, 419.
234 E. Delgado, A.T. Emo, J.C. Jeffery, N.D. Simmons, and F.G.A. Stone, J. Chem. Soc., Dalton Trans., 1985, 1323.
235 D.J. Darensbourg, R. Kudaroski, and T. Delord, Organometallics, 1985, **4**, 1094.
236 J.A.K. Howard, J.C. Jeffery, J.C.V. Laurie, I. Moore, F.G.A. Stone, and A. Stringer, Inorg. Chim. Acta, 1985, **100**, 23.
237 C.P. Casey and A.J. Shusterman, Organometallics, 1985, **4**, 736.
238 N.H.T. Huy and E.O. Fischer, Nouv. J. Chim., 1985, **9**, 257.
239 R. Aumann, E. Kuckert, and H. Heinen, Angew. Chem., Int. Ed. Engl., 1985, **24**, 978.
240 I.R. Butler, W.R. Cullen, F.W.B. Einstein, and A.C. Willis, Organometallics, 1985, **4**, 603.
241 E.O. Fischer and D. Wittmann, J. Organomet. Chem., 1985, **292**, 245.
242 A. Aguero, J. Kress, and J.A. Osborn, J. Chem. Soc., Chem. Commun., 1985, 793.
243 J.H. Freudenberger and R.R. Schrock, Organometallics, 1985, **4**, 1937.
244 H.G. Alt and T. Frister, J. Organomet. Chem., 1985, **293**, C7.
245 M.R. Churchill, L.V. Biondi, J.R. Shapley, and C.H. McAteer, J. Organomet. Chem., 1985, **280**, C63.
246 A. Mayr, G.A. McDermott, and A.M. Dorries, Organometallics, 1985, **4**, 608.
247 A. Mayr, A.M. Dorries, G.A. McDermott, S.L. Geib, and A.L. Rheingold, J. Am. Chem. Soc., 1985, **107**, 7775.
248 M.L. Listemann and R.R. Schrock, Organometallics, 1985, **4**, 74.
249 K.R. Birdwhistell, T.L. Tonker, and J.L. Templeton, J. Am. Chem. Soc., 1985, **107**, 4474.
250 M.H. Chisholm, J.A. Heppert, J.C. Huffman, and W.E. Streib, J. Chem. Soc., Chem. Commun., 1985, 1771.
251 F.A. Cotton, W. Schwotzer, and E.S. Shamshoum, Organometallics, 1985, **4**, 461.
252 E.O. Fischer, A.C. Filippou, H.G. Alt, and U. Thewalt, Angew. Chem., Int. Ed. Engl., 1985, **24**, 203.
253 S. Özkär, A. Erdoǧan, and C.G. Kreiter, Inorg. Chim. Acta, 1985, **98**, 147.
254 J.W. Faller, R.H. Crabtree, and A. Habib, Organometallics, 1985, **4**, 929.
255 H. Brunner, P. Beier, E. Frauendorfer, M. Muschiol, D.K. Rastogi, and J. Wachter, Inorg. Chim. Acta, 1985, **96**, L5.
256 N. Mori, M. Takamori, and T. Takemura, J. Chem. Soc., Dalton Trans., 1985, 1065.

257 N.A. Ustynyuk, L.N. Novikova, V.K. Bel'skii, Yu.F. Oprunenko, S.G. Malyugina, O.I. Trifonova, and Yu.A. Ustynyuk, J. Organomet. Chem., 1985, **294**, 31.

258 A. Solladié-Cavallo and J. Suffert, Magn. Reson. Chem., 1985, **23**, 739.

259 J.-C. Boutonnet, J. Levisalles, F. Rose-Munch, E. Rose, G. Precigoux, and F. Leroy, J. Organomet. Chem., 1985, **290**, 153.

260 S. Top, G. Jaouen, A. Vessières, J.-P. Abjean, D. Davoust, C.A. Rodger, B.G. Sayer, and M.J. McGlinchey, Organometallics, 1985, **4**, 2143.

261 J.J. Frizzell, R.L. Luck, R.H. Morris, and S.H. Peng, J. Organomet. Chem., 1985, **284**, 243.

262 R. Alvarez, E. Carmona, D.J. Cole-Hamilton, A. Galindo, E. Gutiérrez-Puebla, A. Monge, M.L. Poveda, and C. Ruiz, J. Am. Chem. Soc., 1985, **107**, 5529.

263 R.P. Hughes, J.W. Reisch, and A.L. Rheingold, Organometallics, 1985, **4**, 241.

264 H.G. Alt and H.I. Hayen, Angew. Chem., Int. Ed. Engl., 1985, **24**, 497.

265 W.E. Vanarsdale, R.E.K. Winter, and J.K. Kochi, J. Organomet. Chem., 1985, **296**, 31.

266 R.T. DePue, D.B. Collum, J.W. Ziller, and M.R. Churchill, J. Am. Chem. Soc., 1985, **107**, 2131.

267 M. Kamata, N. Okura, K. Hirotsu, T. Higuchi, M. Kido, K. Tatsumi, T. Yoshida, and S. Otsuka, New Front. Organomet. Inorg. Chem., Proc. China-Jpn.-U.S.A. Trilateral Semin., 2nd., 1982 (Pub. 1984), 189 (Chem. Abstr., 1985, **102**, 204 072).

268 E. Roland, E.C. Walborsky, J.C. Dewan, and R.R. Schrock, J. Am. Chem. Soc., 1985, **107**, 5795.

269 K.J. Ahmed, M.H. Chisholm, and J.C. Huffman, Organometallics, 1985, **4**, 1312.

270 M.A. Bennett and I.W. Boyd, J. Organomet. Chem., 1985, **290**, 165.

271 E. Carmona, A. Galindo, M.L. Poveda, and R.D. Rogers, Inorg. Chem., 1985, **24**, 4033.

272 J.R. Morrow, T.L. Tonker, and J.L. Templeton, J. Am. Chem. Soc., 1985, **107**, 6956.

273 F.R. Kreissl, W.J. Sieber, P. Hofmann, J. Riede, and M. Wolfgruber, Organometallics, 1985, **4**, 788.

274 J.C. Jeffery, J.C.V. Laurie, and F.G.A. Stone, Polyhedron, 1985, **4**, 1135.

275 N.M. Doherty, C. Elschenbroich, H.-J. Kneuper, and S.A.R. Knox, J. Chem. Soc., Chem. Commun., 1985, 170.

276 G.E. Herberich and J. Okuda, Angew. Chem., Int. Ed. Engl., 1985, **24**, 402.

277 H. Fischer, S. Zeuner, and H.G. Alt, J. Organomet. Chem., 1985, **289**, C21.

278 A.D. Horton, M.J. Mays, and P.R. Raithby, J. Chem. Soc., Chem. Commun., 1985, 247.

279 M. Green, R.J. Mercer, C.E. Morton, and A.G. Orpen, Angew. Chem., Int. Ed. Engl., 1985, **24**, 422.

280 W.A. Schenke, D. Rüb, and C. Burschka, Angew. Chem., Int. Ed. Engl., 1985, **24**, 971.

281 R.P. Hughes, W. Kläui, J.W. Reisch, and A. Müller, Organometallics, 1985, **4**, 1761.

282 L.G. McCullough, R.R. Schrock, J.C. Dewan, and J.C. Murdzek, J. Am. Chem. Soc., 1985, **107**, 5987.

283 J.L. Templeton, R.S. Herrick, C.A. Rusik, C.E. McKenna, J.W. McDonald, and W.E. Newton, Inorg. Chem., 1985, **24**, 1383.

284 J.C. Jeffery, A.L. Ratermann, and F.G.A. Stone, J. Organomet. Chem., 1985, **289**, 367.

285 F.A. Cotton and E.S. Shamshoum, Polyhedron, 1985, **4**, 1727.

286 D.J. Wink, J.R. Fox, and N.J. Cooper, J. Am. Chem. Soc., 1985, **107**, 5012.

287 A.D. Hunter, P. Legzdins, C.R. Nurse, F.W.B. Einstein, and A.C. Willis, J. Am. Chem. Soc., 1985, **107**, 1791.

288 T.J. Chow, M.-Y. Wu, and L.-K. Liu, J. Organomet. Chem., 1985, **281**, C33.

289 M.A. Bennett, I.W. Boyd, G.B. Robertson, and W.A. Wickramasinghe, J. Organomet. Chem., 1985, **290**, 181.

290 J.R. Morrow, T.L. Tonker, and J.L. Templeton, J. Am. Chem. Soc., 1985, **107**, 5004.

291 A. Ceccon, A. Gambaro, and A. Venzo, J. Chem. Soc., Chem. Commun., 1985, 540.

292 J. Grobe and R. Haubold, Z. Anorg. Allg. Chem., 1985, **526**, 145.
293 J. Grobe and R. Haubold, Z. Anorg. Allg. Chem., 1985, **522**, 159.
294 L.Y. Goh, C. Wei, and E. Sinn, J. Chem. Soc., Chem. Commun., 1985, 462.
295 P. Legzdins and C.R. Nurse, Inorg. Chem., 1985, **24**, 327.
296 M. Herberhold, W. Kremnitz, A. Razavi, H. Schöllhorn, and U. Thewalt, Angew. Chem., Int. Ed. Engl., 1985, **24**, 601.
297 J. Febvay, F. Casabianca, and J.G. Riess, Inorg. Chem., 1985, **24**, 3235.
298 M. Tilset and K.P.C. Vollhardt, Organometallics, 1985, **4**, 2230.
299 D. Gudat, E. Niecke, W. Malisch, U. Hofmockel, S. Quashie, A.H. Cowley, A.M. Arif, B. Krebs, and M. Dartmann, J. Chem. Soc., Chem. Commun., 1985, 1687.
300 O.J. Scherer, H. Sitzmann, and G. Wolmershäuser, Angew. Chem., Int. Ed. Engl., 1985, **24**, 351.
301 K. Sünkel, G. Urban, and W. Beck, J. Organomet. Chem., 1985, **290**, 231.
302 K. Starker and M.D. Curtis, Inorg. Chem., 1985, **24**, 3006.
303 A.M. Arif, A.H. Cowley, and M. Pakulski, J. Chem. Soc., Chem. Commun., 1985, 1707.
304 M.F. Meidine, C.J. Meir, S. Morton, and J.F. Nixon, J. Organomet. Chem., 1985, **297**, 255.
305 M. Mlekuz, P. Bougeard, B.G. Sayer, R. Faggiani, C.J.L. Lock, M.J. McGlinchey, and G. Jaouen, Organometallics, 1985, **4**, 2046.
306 L.E. Bogan, jun., T.B. Rauchfuss, and A.L. Rheingold, J. Am. Chem. Soc., 1985, **107**, 3843.
307 L.E. Bogan, jun., T.E. Rauchfuss, and A.L. Rheingold, Inorg. Chem., 1985, **24**, 3720.
308 R. Maisch, E. Olt, W. Buchner, W. Malisch, I.J. Colquhoun, and W. McFarlane, J. Organomet. Chem., 1985, **286**, C31.
309 F. Götzfried, R. Grenz, G. Urban, and W. Beck, Chem. Ber., 1985, **118**, 4179.
310 R. Schemm and W. Malisch, J. Organomet. Chem., 1985, **288**, C9.
311 S.A. MacLaughlin, R.C. Murray, J.C. Dewan, and R.R. Schrock, Organometallics, 1985, **4**, 796.
312 J.C. Jeffery and J.G. Lawrence-Smith, J. Chem. Soc., Chem. Commun., 1985, 275.
313 A. Ceccon, A. Gambaro, and A. Venzo, J. Organomet. Chem., 1985, **281**, 221.
314 F. Effenberger and K. Schöllkopf, Chem. Ber., 1985, **118**, 4377.
315 R.C. Cambie, G.R. Clark, A.C. Gourdie, P.S. Rutledge, and P.D. Woodgate, J. Organomet. Chem., 1985, **297**, 177.
316 E.P. Kündig, C. Perret, S. Spichiger, and G. Bernardinelli, J. Organomet. Chem., 1985, **286**, 183.
317 E. Michels, W.S. Sheldrick, and C.G. Kreiter, Chem. Ber., 1985, **118**, 964.
318 S. Özkar and C.G. Kreiter, J. Organomet. Chem., 1985, **293**, 229.
319 R. Drews and U. Behrens, Chem. Ber., 1985, **118**, 888.
320 M.L.H. Green and R.B.A. Pardy, Polyhedron, 1985, **4**, 1035.
321 R.T.C. Brownlee, M.J. O'Connor, B.P. Shehan, and A.G. Wedd, J. Magn. Reson., 1985, **61**, 22.
322 R.T.C. Brownlee, M.J. O'Connor, B.P. Shehan, and A.G. Wedd, J. Magn. Reson., 1985, **64**, 142.
323 M. Minelli, J.H. Enemark, A. Bell, and R.A. Walton, J. Organomet. Chem., 1985, **284**, 25.
324 J. Grobe, D. Le Van, and J. Szameitat, J. Organomet. Chem., 1985, **289**, 341.
325 E.C. Alyea, A. Malek, and J. Malito, Inorg. Chim. Acta, 1985, **101**, 147.
326 G.M. Gray and K.A. Redmill, J. Organomet. Chem., 1985, **280**, 105.
327 J.A. Connor and C. Overton, J. Organomet. Chem., 1985, **282**, 349.
328 G.M. Gray and K.A. Redmill, Inorg. Chem., 1985, **24**, 1279.
329 J.G. Riess, J. Organomet. Chem., 1985, **281**, 1.
330 H. Fischer, R. Märkl, and S. Zeuner, J. Organomet. Chem., 1985, **286**, 17.
331 H. Fischer and S. Zeuner, J. Organomet. Chem., 1985, **286**, 201.
332 J. Borm, G. Huttner, O. Orama, and L. Zsolnai, J. Organomet. Chem., 1985, **282**, 53.
333 M. Yoshifuji, K. Toyota, and N. Inamoto, Tetrahedron Lett., 1985, **26**, 1727.
334 N. Kuhn, H. Schumann, and G. Wolmershäuser, J. Chem. Soc., Chem. Commun., 1985, 1595.
335 H. Lang, O. Orama, and G. Huttner, J. Organomet. Chem., 1985, **291**, 293.

336 A. Marinetti, J. Fischer, and F. Mathey, J. Am. Chem. Soc., 1985, **107**, 5001.
337 M. Yoshifuji, T. Hashida, N. Inamoto, K. Hirotsu, T. Horiuchi, T. Higuchi, K. Itoh, and S. Nagase, Angew. Chem., Int. Ed. Engl., 1985, **24**, 211.
338 N.D. Sadanani, Inorg. Chem., 1985, **24**, 3136.
339 B. Weinberger and W.P. Fehlhammer, Chem. Ber., 1985, **118**, 42.
340 P. Lemoine, M. Gross, P. Braunstein, F. Mathey, and B. Deschamps, J. Organomet. Chem., 1985, **295**, 189.
341 R.B. King, W.-K. Fu, and E.M. Holt, Inorg. Chem., 1985, **24**, 3094.
342 C. Zeiher, J. Mohyla, I.-P. Lorenz, and W. Hiller, J. Organomet. Chem., 1985, **286**, 159.
343 C. Zeiher, W. Hiller, and I.-P. Lorenz, Chem. Ber., 1985, **118**, 3127.
344 E. Ocando, S. Majid, J.P. Majoral, A. Baceiredo, and G. Bertrand, Polyhedron, 1985, **4**, 1667.
345 B. Sigwarth, U. Weber, L. Zsolnai, and G. Huttner, Chem. Ber., 1985, **118**, 3114.
346 C.C. Santini and F. Mathey, J. Org. Chem., 1985, **50**, 467.
347 J.M. Maher, R.P. Beatty, and N.J. Cooper, Organometallics, 1985, **4**, 1354.
348 D. Czarkie and Y. Shvo, J. Organomet. Chem., 1985, **280**, 123.
349 A. Marinetti and F. Mathey, J. Am. Chem. Soc., 1985, **107**, 4700.
350 S. Rosenberg, G.L. Geoffroy, and A.L. Rheingold, Organometallics, 1985, **4**, 1184.
351 B. Deschamps and F. Mathey, J. Chem. Soc., Chem. Commun., 1985, 1010.
352 F. Mercier and F. Mathey, Tetrahedron Lett., 1985, **26**, 1717.
353 B. Deschamps and F. Mathey, Tetrahedron Lett., 1985, **26**, 4595.
354 B. Deschamps and F. Mathey, Tetrahedron Lett., 1985, **26**, 3461.
355 W.C. Mercer, R.R. Whittle, E.W. Burkhardt, and G.L. Geoffroy, Organometallics, 1985, **4**, 68.
356 R. Aumann and H. Heinen, Chem. Ber., 1985, **118**, 952.
357 R. Aumann and H. Heinen, Chem. Ber., 1985, **118**, 4186.
358 J. Ellermann and W. Wend, J. Organomet. Chem., 1985, **281**, C29.
359 L. Weber and D. Wewers, Chem. Ber., 1985, **118**, 541.
360 G. Bremer, P. Klüfers, and T. Kruck, Chem. Ber., 1985, **118**, 4224.
361 J. Grobe, M. Köhne-Wächter, and D. Le Van, J. Organomet. Chem., 1985, **280**, 331.
362 E.C. Alyea, R.P. Shakya, and A.E. Vougioukas, Trans. Met. Chem., 1985, **10**, 435.
363 T.G. Hill, R.C. Haltiwanger, and A.D. Norman, Inorg. Chem., 1985, **24**, 3499.
364 M.L. Boyles, D.V. Brown, D.A. Drake, C.K. Hostetler, C.A. Kiesler, and J.A. Mosbo, Inorg. Chem., 1985, **24**, 3126.
365 R. Hensel, W.W. Du Mont, R. Boese, D. Wewers, and L. Weber, Chem. Ber., 1985, **118**, 1580.
366 J. Borm, G. Huttner, and L. Zsolnai, Angew. Chem., Int. Ed. Engl., 1985, **24**, 1069.
367 W.C. Mercer, G.L. Geoffroy, and A.L. Rheingold, Organometallics, 1985, **4**, 1418.
368 K. Wieghardt, M. Guttmann, P. Chaudhuri, W. Gebert, M. Minelli, C.G. Young, and J.H. Enemark, Inorg. Chem., 1985, **24**, 3151.
369 M.D. Curtis and K.B. Shiu, Inorg. Chem., 1985, **24**, 1213.
370 A. Blagg, S.W. Carr, G.R. Cooper, I.D. Dobson, J.B. Gill, B.L. Shaw, D.C. Goodall, N. Taylor, and T. Boddington, J. Chem. Soc., Dalton Trans., 1985, 1213.
371 J. Grobe, N. Krummen, and Le Van Duc, Z. Naturforsch., Teil B, 1984, **39**, 1711 (Chem. Abstr., 1985, **102**, 220 985).
372 G. Andolfatto, R. Granger, and L.K. Peterson, Inorg. Chim. Acta, 1985, **99**, L1.
373 K.D. Gallicano and N.L. Paddock, Can. J. Chem., 1985, **63**, 314.
374 E.P. Kyba, R.E. Davis, S.-T. Liu, K.A. Hassett, and S.B. Larson, Inorg. Chem., 1985, **24**, 4629.
375 F.S.M. Hassan, D.P. Markham, P.G. Pringle, and B.L. Shaw, J. Chem. Soc., Dalton Trans., 1985, 279.
376 B. Chaudret, F. Dahan, and S. Sabo, Organometallics, 1985, **4**, 1490.
377 A. Blagg, A.T. Hutton, B.L. Shaw, and M. Thornton-Pett, Inorg. Chim. Acta,

1985, **100**, L33.
378 P.H. Bird, A.A. Ismail, and I.S. Butler, Inorg. Chem., 1985, **24**, 2911.
379 S. Lincoln, S.-L. Soong, S.A. Koch, M. Sato, and J.H. Enemark, Inorg. Chem., 1985, **24**, 1355.
380 M.H. Chisholm, D.M. Hoffman, and J.C. Huffman, Organometallics, 1985, **4**, 986.
381 J.T. Guy, jun. and D.W. Bennett, Inorg. Chim. Acta, 1985, **96**, L83.
382 S. Donovan-Mtunzi, C.J. MacDonald, R.L. Richards, G.E. Hawkes, and J. Mason, J. Chem. Soc., Dalton Trans., 1985, 2473.
383 M. Minelli, C.G. Young, and J.H. Enemark, Inorg. Chem., 1985, **24**, 1111.
384 G.E. Bossard, T.A. George, R.K. Lester, R.C. Tisdale, and R.L. Turcotte, Inorg. Chem., 1985, **24**, 1129.
385 P.J. Blower, J.R. Dilworth, J. Hutchinson, T. Nicholson, and J.A. Zubieta, J. Chem. Soc., Dalton Trans., 1985, 2639.
386 S. Bristow, J.H. Enemark, C.D. Garner, M. Minelli, G.A. Morris, and R.B. Ortega, Inorg. Chem., 1985, **24**, 4070.
387 K.J. Ahmed, M.H. Chisholm, K. Folting, and J.C. Huffman, Inorg. Chem., 1985, **24**, 4039.
388 T.A. George and R.C. Tisdale, J. Am. Chem. Soc., 1985, **107**, 5157.
389 D.D. Devore and E.A. Maatta, Inorg. Chem., 1985, **24**, 2846.
390 P. Huppmann and K. Seppelt, Chem. Ber., 1985, **118**, 457.
391 M.D. Fitzroy, J.M. Frederiksen, K.S. Murray, and M.R. Snow, Inorg. Chem., 1985, **24**, 3265.
392 M.V. Caparelli, B. Piggott, S.D. Thorpe, and R.N. Sheppard, Inorg. Chim. Acta, 1985, **98**, L53.
393 M.V. Capparelli, B. Piggott, S.D. Thorpe, S.F. Wong, and R.N. Sheppard, Inorg. Chim. Acta, 1985, **106**, 19.
394 B. Piggott, S.D. Thorpe, and R.N. Sheppard, Inorg. Chim. Acta, 1985, **103**, L3.
395 S.M. Bowen, E.N. Duesler, D.J. McCabe, and R.T. Paine, Inorg. Chem., 1985, **24**, 1191.
396 H. Dadkhah, J.R. Dilworth, K. Fairman, C.T. Kan, R.L. Richards, and D.L. Hughes, J. Chem. Soc., Dalton Trans., 1985, 1523.
397 S.G. Sakharov, Yu.V. Kokunov, M.P. Gustyakova, and Yu.A. Buslaev, Dokl. Akad. Nauk SSSR, 1984, **278**, 637 (Chem. Abstr., 1985, **102**, 71 656).
398 N.K. Homsy, H.W. Roesky, M. Noltemeyer, and G.M. Sheldrick, J. Chem. Soc., Dalton Trans., 1985, 2205.
399 A.J. Nielson, J.M. Waters, and D.C. Bradley, Polyhedron, 1985, **4**, 285.
400 C.J. Donahue, D. Clark-Motia, and M.E. Harvey, Inorg. Chem., 1985, **24**, 801.
401 P.A. Bates, A.J. Nielson, and J.M. Waters, Polyhedron, 1985, **4**, 999.
402 S.D. Ittel, F.A. Van-Catledge, and C.A. Tolman, Inorg. Chem., 1985, **24**, 62.
403 T.W. Coffindaffer, G.P. Niccolai, D. Powell, and I.P. Rothwell, J. Am. Chem. Soc., 1985, **107**, 3572.
404 D.J. Santure and A.P. Sattelberger, Inorg. Chem., 1985, **24**, 3477.
405 M.H. Chisholm, D.L. Clark, and J.C. Huffman, Polyhedron, 1985, **4**, 1203.
406 P.J. Blower, J.R. Dilworth, G.J. Leigh, B.D. Neaves, F.B. Normanton, J. Hutchinson, and J.A. Zubieta, J. Chem. Soc., Dalton Trans., 1985, 2647.
407 P. Salagre and J.E. Sueiras, Transition Met. Chem., 1985, **10**, 191.
408 P. Salagre, J.-E. Sueiras, X. Solans, and G. Germain, J. Chem. Soc., Dalton Trans., 1985, 2263.
409 D.J. Santure, J.C. Huffman, and A.P. Sattelberger, Inorg. Chem., 1985, **24**, 371.
410 N. Agh-Atabay, F.M. Ashmawy, C.A. McAuliffe, and W.E. Hill, Inorg. Chim. Acta, 1985, **104**, 73.
411 R. Curci, G. Fusco, O. Sciacovelli, and L. Troisi, J. Mol. Catal., 1985, **32**, 251.
412 B.P. Shehan, M. Kony, R.T.C. Brownlee, M.J. O'Connor, and A.G. Wedd, J. Magn. Reson., 1985, **63**, 343.
413 C.G. Young and J.H. Enemark, Inorg. Chem., 1985, **24**, 4416.
414 W.H. Knoth, P.J. Domaille, and R.D. Farlee, Organometallics, 1985, **4**, 62.
415 R.I. Maksimovskaya and K.G. Burtseva, Polyhedron, 1985, **4**, 1559.
416 C.L. Hill and D.A. Bouchard, J. Am. Chem. Soc., 1985, **107**, 5148.

417 M.H. Alizadeh, S.P. Harmalker, Y. Jeannin, J. Martin-Frere, and M.T. Pope, J. Am. Chem. Soc., 1985, **107**, 2662.

418 S. Bristow, C.D. Garner, S.K. Hagyard, G.A. Morris, J.R. Nicholson, and C.F. Mills, J. Chem. Soc., Chem. Commun., 1985, 479.

419 P.S. Belton, I.J. Cox, and R.K. Harris, J. Chem. Soc., Faraday Trans. 2, 1985, **81**, 63.

420 I. Ursu, M. Bogdan, P. Fitori, A. Darabont, and D.E. Demco, Mol. Phys., 1985, **56**, 297.

421 E.V. Iorga, A.G. Kucheryaev, V.A. Lebedev, and I.N. Leont'eva, Zavod. Lab., 1984, **50**, 27 (Chem. Abstr., 1985, **102**, 105 208).

422 W.G. Klemperer, V.V. Mainz, R.-C. Wang, and W. Shum, Inorg. Chem., 1985, **24**, 1968.

423 J.O. Dzięgielewski, Polyhedron, 1985, **4**, 1901.

424 B. Piggott, S.F. Wong, and R.N. Sheppard, Inorg. Chim. Acta, 1985, **107**, 97.

425 V. Bilik, J. Alfoldi, and E. Sookyova, Chem. Zvesti, 1984, **38**, 499 (Chem. Abstr., 1985, **102**, 24 936).

426 H. Arzoumanian, A. Baldy, R. Lai, A. Odreman, J. Metzger, and M. Pierrot, J. Organomet. Chem., 1985, **295**, 343.

427 Y. Do, E.D. Simhon, and R.H. Holm, Inorg. Chem., 1985, **24**, 2827.

428 R.I. Maksimovskaya, M.A. Fedotov, and G.M. Maksimov, Zh. Neorg. Khim., 1985, **30**, 918 (Chem. Abstr., 1985, **103**, 15 807).

429 R. Contant and A. Tézé, Inorg. Chem., 1985, **24**, 4610.

430 Y. Do, E.D. Simhon, and R.H. Holm, Inorg. Chem., 1985, **24**, 1831.

431 N. Ueyama, K. Kamabuchi, and A. Nakamura, J. Chem. Soc., Dalton Trans., 1985, 635.

432 F.A. Colton, R. Llusar, D.O. Marler, W. Schwotzer, and Z. Dori, Inorg. Chim. Acta, 1985, **102**, L25.

433 E.C. Alyea, G. Ferguson, M. Parvez, and A. Somogyvari, Polyhedron, 1985, **4**, 783.

434 V.D. Butskii, M.E. Ignatov, and B.V. Golovanov, Zh. Neorg. Khim., 1985, **30**, 813 (Chem. Abstr., 1985, **102**, 178 031).

435 D. Sevdić and L. Fekete, Polyhedron, 1985, **4**, 1371.

436 S.J.N. Burgmayer and J.L. Templeton, Inorg. Chem., 1985, **24**, 3939.

437 E.M. Kinsch and D.W. Stephan, Inorg. Chim. Acta, 1985, **96**, L87.

438 S.A. Cohen and E.I. Stiefel, Inorg. Chem., 1985, **24**, 4657.

439 R. Bougon, W.W. Wilson, and K.O. Christe, Inorg. Chem., 1985, **24**, 2286.

440 T. Beringhelli, H. Molinari, and A. Pastore, J. Chem. Soc., Dalton Trans., 1985, 1899.

441 G.S. Girolami, C.G. Howard, G. Wilkinson, H.M. Dawes, M. Thornton-Pett, M. Motevalli, and M.B. Hursthouse, J. Chem. Soc., Dalton Trans., 1985, 921.

442 V. Riera, M.A. Ruiz, A. Tiripicchio, and M.T. Camellini, J. Chem. Soc., Chem. Commun., 1985, 1505.

443 A.J. Deeming and S. Donovan-Mtunzi, J. Chem. Soc., Dalton Trans., 1985, 1609.

444 W.A. Herrmann, E. Voss, E. Guggolz, and M.L. Ziegler, J. Organomet. Chem., 1985, **284**, 47.

445 J. Keijsper, P. Grimberg, G. van Koten, K. Vrieze, M. Christophersen, and C.H. Stam, Inorg. Chim. Acta, 1985, **102**, 29.

446 A.J. Deeming and S. Donovan-Mtunzi, Organometallics, 1985, **4**, 693.

447 D.G. DeWit, K. Folting, W.E. Streib, and K.G. Caulton, Organometallics, 1985, **4**, 1149.

448 H. Felkin and J. Zakrzewski, J. Am. Chem. Soc., 1985, **107**, 3374.

449 K.R. Grundy and K.N. Robertson, Inorg. Chem., 1985, **24**, 3898.

450 J. Keijsper, P. Grimberg, G. van Koten, K. Vrieze, B. Kojić-Prodić, and A.L. Spek, Organometallics, 1985, **4**, 438.

451 J. Powell, J.F. Sawyer, and M.V.R. Stainer, J. Chem. Soc., Chem. Commun., 1985, 1314.

452 C.S. Young, S.W. Lee, and C.P. Cheng, J. Organomet. Chem., 1985, **282**, 85.

453 T. Beringhelli, G. Ciani, G. D'Alfonso, H. Molinari, and A. Sironi, Inorg. Chem., 1985, **24**, 2666.

454 A.A. Koridze, O.A. Kizas, N.E. Kolobova, and P.V. Petrovskii, J. Organomet. Chem., 1985, **292**, Cl.

455 J.R. Bleeke and J.J. Kotyk, Organometallics, 1985, **4**, 194.
456 J.M. Mayer, D.L. Thorn, and T.H. Tulip, J. Am. Chem. Soc., 1985, **107**, 7454.
457 M.L.H. Green and D. O'Hare, J. Chem. Soc., Chem. Commun., 1985, 332.
458 E. Lindner and A. Broesamle, Chem. Ber., 1985, **118**, 2134.
459 E.J.M. de Boer, J. de With, N. Meijboom, and A.G. Orpen, Organometallics, 1985, **4**, 259.
460 G.A. Carriedo, J.B. Parra Soto, V. Riera, X. Solans, and C. Miravitlles, J. Organomet. Chem., 1985, **297**, 193.
461 D. Miguel and V. Riera, J. Organomet. Chem., 1985, **293**, 379.
462 S. Onaka, Y. Kondo, M. Yamashita, Y. Tatematsu, Y. Kato, M. Goto, and T. Ito, Inorg. Chem., 1985, **24**, 1070.
463 J. Grobe and R. Martin, Z. Anorg. Allg. Chem., 1985, **523**, 199.
464 W.A. Herrmann, H.-J. Kneuper, and E. Herdtweck, Angew. Chem., Int. Ed. Engl., 1985, **24**, 1062.
465 E.O. Fischer, J.K.R. Wanner, G. Müller, and J. Riede, Chem. Ber., 1985, **118**, 3311.
466 A.B. Antonova, S.V. Kovalenko, E.D. Korniets, P.V. Petrovskii, G.R. Gul'bis, and A.A. Ioganson, Inorg. Chim. Acta, 1985, **96**, 1.
467 A.B. Antonova, S.V. Kovalenko, E.D. Korniyets, P.V. Petrovsky, A.A. Johansson, and N.A. Deykhina, Inorg. Chim. Acta, 1985, **105**, 153.
468 R.H. Fong and W.H. Hersh, Organometallics, 1985, **4**, 1468.
469 M.M. Singh and R.J. Angelici, Inorg. Chim. Acta, 1985, **100**, 57.
470 P. Stavropoulos, P.G. Edwards, G. Wilkinson, M. Motevalli, K.M.A. Malik, and M.B. Hursthouse, J. Chem. Soc., Dalton Trans., 1985, 2167.
471 P.C. Heah and J.A. Gladysz, J. Mol. Cat., 1985, **31**, 207.
472 G.K. Yang and R.G. Bergman, Organometallics, 1985, **4**, 129.
473 W.P. Fehlhammer, P. Hirschmann, and A. Völkl, J. Organomet. Chem., 1985, **294**, 251.
474 D.E. Smith and J.A. Gladysz, Organometallics, 1985, **4**, 1480.
475 L. Weber and K. Reizig, Angew. Chem., Int. Ed. Engl., 1985, **24**, 53.
476 H. van der Heijden, A.G. Orpen, and P. Pasman, J. Chem. Soc., Chem. Commun., 1985, 1576.
477 C.P. Casey, J.M. O'Connor, and K.J. Haller, J. Am. Chem. Soc., 1985, **107**, 3172.
478 B. Buchmann and A. Salzer, J. Organomet. Chem., 1985, **295**, 63.
479 C.G. Kreiter and M. Leyendecker, J. Organomet. Chem., 1985, **280**, 225.
480 K. Raab, M. Appel, and W. Beck, J. Organomet. Chem., 1985, **291**, C28.
481 N.I. Pyshnograeva, A.S. Batsanov, Yu.T. Struchkov, A.G. Ginzburg, and V.N. Setkina, J. Organomet. Chem., 1985, **297**, 69.
482 H. Lang, G. Mohr, O. Schneidsteger, and G. Huttner, Chem. Ber., 1985, **118**, 574.
483 G. Kraft, C. Kalbas, and U. Schubert, J. Organomet. Chem., 1985, **289**, 247.
484 W.A. Herrmann, B. Koumbouris, A. Schäfer, T. Zahn, and M.L. Ziegler, Chem. Ber., 1985, **118**, 2472.
485 M. Herberhold, B. Schmidkonz, M.L. Ziegler, and T. Zahn, Angew. Chem., Int. Ed. Engl., 1985, **24**, 515.
486 W.E. Buhro, S. Georgiou, J.P. Hutchinson, and J.A. Gladysz, J. Am. Chem. Soc., 1985, **107**, 3346.
487 C.F. Barrientos-Penna, A.B. Gilchrist, A.H. Klahn-Oliva, A.J.L. Hanlan, and D. Sutton, Organometallics, 1985, **4**, 478.
488 R. Ferede, J.F. Hinton, W.A. Korfmacher, J.P. Freeman, and N.T. Allison, Organometallics, 1985, **4**, 614.
489 W.A. Herrmann, R. Serrano, U. Küsthardt, E. Guggolz, B. Nuber, and M.L. Ziegler, J. Organomet. Chem., 1985, **287**, 329.
490 H.Y. Parker, C.E. Klopfenstein, R.A. Wielesek, and T. Koenig, J. Am. Chem. Soc., 1985, **107**, 5276.
491 Y.K. Chung, D.A. Sweigart, N.G. Connelly, and J.B. Sheridan, J. Am. Chem. Soc., 1985, **107**, 2388.
492 N.G. Connelly, M.J. Freeman, A.G. Orpen, A.R. Sheehan, J.B. Sheridan, and D.A. Sweigart, J. Chem. Soc., Dalton Trans., 1985, 1019.
493 R.M. Nielson and S. Wherland, Inorg. Chem., 1985, **24**, 3458.
494 D. Rehder, V. Pank, U. Kunze, and A. Bruns, Inorg. Chim. Acta, 1985, **99**,

L35.
495 D. Rehder and A. Keçeci, Inorg. Chim. Acta, 1985, **103**, 173.
496 N. Alberola and R. Point, Int. J. Appl. Radiat. Isot., 1985, **36**, 152 (Chem. Abstr., 1985, **102**, 214 103).
497 Y. Do, E.D. Simhon, and R.H. Holm, Inorg. Chem., 1985, **24**, 4635.
498 F.J. Garcia-Alónso, V. Riera, and M.J. Misas, Transition Met. Chem., 1985, **10**, 19.
499 J. Nitschke, S.P. Schmidt, and W.C. Trogler, Inorg. Chem., 1985, **24**, 1972.
500 P. Oltmanns and D. Rehder, J. Organomet. Chem., 1985, **281**, 263.
501 K.E. Warner and J.R. Norton, Organometallics, 1985, **4**, 2150.
502 E. Lindner, F. Zinsser, W. Hiller, and R. Fawzi, J. Organomet. Chem., 1985, **288**, 317.
503 S.K. Chopra, S.S.C. Chu, P. de Meester, D.E. Geyer, M. Lattman, and S.A. Morse, J. Organomet. Chem., 1985, **294**, 347.
504 H. Schäfer, J. Zipfel, B. Gutekunst, and U. Lemmert, Z.Anorg. Allg. Chem., 1985, **529**, 157.
505 D.A. DuBois, E.N. Duesler, and R.T. Paine, Inorg. Chem., 1985, **24**, 3.
506 M.W. Kokkes, W.G.J. de Lange, D.J. Stufkens, and A. Oskam, J. Organomet. Chem., 1985, **294**, 59.
507 U. Kunze and A. Bruns, J. Organomet. Chem., 1985, **292**, 349.
508 H.C. Aspinall and A.J. Deeming, J. Chem. Soc., Dalton Trans., 1985, 743.
509 S. Sabo, B. Chaudret, and D. Gervais, J. Organomet. Chem., 1985, **292**, 411.
510 S.W. Carr, B.L. Shaw, and M. Thornton-Pett, J. Chem. Soc., Dalton Trans., 1985, 2131.
511 A.J.L. Pombeiro, Inorg. Chim. Acta, 1985, **103**, 95.
512 M.F.N.N. Carvalho, A.J.L. Pombeiro, U. Schubert, O. Orama, C.J. Pickett, and R.L. Richards, J. Chem. Soc., Dalton Trans., 1985, 2079.
513 T.J. Barder, D. Powell, and R.A. Walton, J. Chem. Soc., Chem. Commun., 1985, 550.
514 L.B. Anderson, T.J. Barder, and R.A. Walton, Inorg. Chem., 1985, **24**, 1421.
515 K. Raab and W. Beck, Chem. Ber., 1985, **118**, 3830.
516 N.J. Taylor, J. Chem. Soc., Chem. Commun., 1985, 476.
517 F.A. Cotton, K.R. Dunbar, L.R. Falvello, and R.A. Walton, Inorg. Chem., 1985, **24**, 4180.
518 J. Chatt, W. Hussain, G.J. Leigh, H. Mohd, C.J. Pickett, and D.A. Rankin, J. Chem. Soc., Dalton Trans., 1985, 1131.
519 A.R. Chakravarty, F.A. Cotton, A.R. Cutler, S.M. Tetrick, and R.A. Walton, J. Am. Chem. Soc., 1985, **107**, 4795.
520 R. Krentz and R.K. Pomeroy, Inorg. Chem., 1985, **24**, 2976.
521 T. Saito and S. Sawada, Bull. Chem. Soc. Jpn., 1985, **58**, 459.
522 W.A. Fordyce, R. Wilczynski, and J. Halpern, J. Organomet. Chem., 1985, **296** 115.
523 T.V. Ashworth, A.A. Chalmers, and E. Singleton, Inorg. Chem., 1985, **24** 2125.
524 A.M. Stolzenberg and E.L. Muetterties, Organometallics, 1985, **4**, 1739.
525 F. Morandini, G. Consiglio, and V. Lucchini, Organometallics, 1985, **4**, 1202
526 J.C. Bricker, C.C. Nagel, A.A. Bhattacharyya, and S.G. Shore, J. Am. Chem Soc., 1985, **107**, 377.
527 C. Conway, R.D.W. Kemmitt, A.W.G. Platt, D.R. Russell, and L.J.S. Sherry, J Organomet. Chem., 1985, **292**, 419.
528 J.A. Cabeza, B.E. Mann, C. Brevard, and P.M. Maitlis, J. Chem. Soc., Chem Commun., 1985, 65.
529 J.S. Holmgren, J.R. Shapley, and P.A. Belmonte, J. Organomet. Chem., 1985 **284**, C5.
530 D.J. Brauer, S. Hietkamp, H. Sommer, and O. Stelzer, J. Organomet. Chem., 1985, **281**, 187.
531 J.C. Vites, G. Jacobsen, T.K. Dutta, and T.P. Fehlner, J. Am. Chem. Soc., 1985, **107**, 5563.
532 S. Attali and R. Mathieu, J. Organomet. Chem., 1985, **291**, 205.
533 B. Chaudret and R. Poilblanc, Organometallics, 1985, **4**, 1722.
534 G.J. Baird, S.G. Davies, S.D. Moon, S.J. Simpson, and R.H. Jones, J. Chem Soc., Dalton Trans., 1985, 1479.

535 A.S.C. Chan and H.-S. Shieh, J. Chem. Soc., Chem. Commun., 1985, 1379.
536 C. White and E. Cesarotti, J. Organomet. Chem., 1985, **287**, 123.
537 B. Delavaux, B. Chaudret, F. Dahan, and R. Poilblanc, Organometallics, 1985, **4**, 935.
538 B. Chaudret, J. Devillers, and R. Poilblanc, Organometallics, 1985, **4**, 1727.
539 T.V. Ashworth, D.C. Liles, and E. Singleton, Inorg. Chim. Acta, 1985, **98**, L65.
540 B. Delavaux, B. Chaudret, N.J. Taylor, S. Arabi, and R. Poilblanc, J. Chem. Soc., Chem. Commun., 1985, 805.
541 J.A. Smieja, R.E. Stevens, D.E. Fjare, and W.L. Gladfelter, Inorg. Chem., 1985, **24**, 3206.
542 L.R. Beanan and J.B. Keister, Organometallics, 1985, **4**, 1713.
543 D.M. Dalton, D.J. Barnett, T.P. Duggan, J.B. Keister, P.T. Malik, S.P. Modi, M.R. Shaffer, and S.A. Smesko, Organometallics, 1985, **4**, 1854.
544 G. Süss-Fink and G. Herrmann, J. Chem. Soc., Chem. Commun., 1985, 735.
545 J.C. Huffman, M.A. Green, S.L. Kaiser, and K.G. Caulton, J. Am. Chem. Soc., 1985, **107**, 5111.
546 H. Werner and K. Roder, J. Organomet. Chem., 1985, **281**, C38.
547 R. Weinand and H. Werner, J. Chem. Soc., Chem. Commun., 1985, 1145.
548 J. Gotzig, R. Werner, and H. Werner, J. Organomet. Chem., 1985, **285**, 99.
549 J.A. Clucas, M.M. Harding, and A.K. Smith, J. Chem. Soc., Chem. Commun., 1985, 1280.
550 M.L.H. Green and D. O'Hare, J. Chem. Soc., Chem. Commun., 1985, 355.
551 J.R. Shapley, W.-Y. Yeh, M.R. Churchill, and Y.-j. Li, Organometallics, 1985, **4**, 1898.
552 G. Süss-Fink, K. Guldner, M. Herberhold, A. Gieren, and T. Huebner, J. Organomet. Chem., 1985, **279**, 447.
553 L.J. Farrugia, A.D. Miles, and F.G.A. Stone, J. Chem. Soc., Dalton Trans., 1985, 2437.
554 A. Eisenstadt, C.M. Giandomenico, M.F. Frederick, and R.M. Laine, Organometallics, 1985, **4**, 2033.
555 E.J. Ditzel, B.F.G. Johnson, J. Lewis, P.R. Raithby, and M.J. Taylor, J. Chem. Soc., Dalton Trans., 1985, 555.
556 L.J. Farrugia, M. Green, D.R. Hankey, M. Murray, A.G. Orpen, and F.G.A. Stone, J. Chem. Soc., Dalton Trans., 1985, 177.
557 E. Sappa, G. Predieri, A. Tiripicchio, and M.T. Camellini, J. Organomet. Chem., 1985, **297**, 103.
558 A.D. Cameron and M.C. Baird, J. Chem. Soc., Dalton Trans., 1985, 2691; J.I. Seeman and S.G. Davies, J. Chem. Soc., Dalton Trans., 1985, 2692.
559 B.K. Hunter and M.C. Baird, Organometallics, 1985, **4**, 1481.
560 P. Reveco, J.H. Medley, A.R. Garber, N.S. Bhacca, and J. Selbin, Inorg. Chem., 1985, **24**, 1096.
561 G.S. Girolami, G. Wilkinson, A.M.R. Galas, M. Thornton-Pett, and M.B. Hursthouse, J. Chem. Soc., Dalton Trans., 1985, 1339.
562 C.R. Jablonski and Y.P. Wang, Organometallics, 1985, **4**, 465.
563 M. Pankowski, W. Chodkiewicz, and M.P. Simonnin, Inorg. Chem., 1985, **24**, 533.
564 B. Weinberger, G. Tanguy, and H. des Abbayes, J. Organomet. Chem., 1985, **280**, C31.
565 W.K. Wong, K.W. Chiu, G. Wilkinson, A.J. Howes, M. Motevalli, and M.B. Hursthouse, Polyhedron, 1985, **4**, 603.
566 E. Lindner, E. Schauss, W. Hiller, and R. Fawzi, Chem. Ber., 1985, **118**, 3915.
567 H.W. Fruehaur, F. Seils, R.J. Goddard, and M.J. Romao, Organometallics, 1985, **4**, 948.
568 C.P. Casey, M.W. Meszaros, S.M. Neumann, I.G. Cesa, and K.J. Haller, Organometallics, 1985, **4**, 143.
569 E. Lindner, C.-P. Krieg, W. Hiller, and R. Fawzi, Chem. Ber., 1985, **118**, 1398.
570 R.S. Bly, G.S. Silverman, and R.K. Bly, Organometallics, 1985, **4**, 374.
571 N.E. Kolobova, T.V. Rozantseva, Yu.T. Struchkov, A.S. Batsanov, and V.I. Bakhmutov, J. Organomet. Chem., 1985, **292**, 247.

572 R.S. Bly, M.M. Hossain, and L. Lebioda, J. Am. Chem. Soc., 1985, **107**, 5549.
573 C.P. Casey, W.H. Miles, and H. Tukada, J. Am. Chem. Soc., 1985, **107**, 2924.
574 K.A.M. Kremer and P. Helquist, J. Organomet. Chem., 1985, **285**, 231.
575 M. Brookhart, W.B. Studabaker, and G.R. Husk, Organometallics, 1985, **4**, 943.
576 W. Malisch, H. Blau, K. Blank, and C. Krüger, J. Organomet. Chem., 1985, **296**, C32.
577 L. Weber, K. Reizig, and R. Boese, Chem. Ber., 1985, **118**, 1193.
578 V. Guerchais and D. Astruc, J. Chem. Soc., Chem. Commun., 1985, 835.
579 T.W. Bodnar and A.R. Cutler, Organometallics, 1985, **4**, 1558.
580 D.L. Reger and K.A. Belmore, Organometallics, 1985, **4**, 305.
581 S.G. Davies, I.M. Dordor-Hedgecock, P. Warner, R.H. Jones, and K. Prout, J. Organomet. Chem., 1985, **285**, 213.
582 J.R. Lisko and W.M. Jones, Organometallics, 1985, **4**, 944.
583 H. Lehmkuhl and G. Mehler, Chem. Ber., 1985, **118**, 2407.
584 S.G. Shyu and A. Wojcicki, Organometallics, 1985, **4**, 1457.
585 G. Sundararajan and J.S. Filippo, jun., Organometallics, 1985, **4**, 606.
586 C.P. Casey, M.S. Konings, R.E. Palermo, and R.E. Colborn, J. Am. Chem. Soc., 1985, **107**, 5296.
587 D. Lentz, I. Brüdgam, and H. Hartl, Angew. Chem., Int. Ed. Engl., 1985, **24**, 119.
588 D. Nuel, F. Dahan, and R. Mathieu, J. Am. Chem. Soc., 1985, **107**, 1658.
589 N.C. Schroeder, J.W. Richardson, jun., S.-L. Wang, R.A. Jacobson, and R.J. Angelici, Organometallics, 1985, **4**, 1226.
590 S.I. Hommeltoft, A.D. Cameron, T.A. Shackleton, M.E. Fraser, S. Fortier, and M.C. Baird, J. Organomet. Chem., 1985, **282**, C17.
591 H. Nagashima, K. Yamaguchi, K. Mukai, and K. Itoh, J. Organomet. Chem., 1985, **291**, C20.
592 H. Kletzin and H. Werner, J. Organomet. Chem., 1985, **291**, 213.
593 W. Faulkner, D.S. Barratt, D.C. Cupertino, and D.J. Cole-Hamilton, Polyhedron, 1985, **4**, 1993.
594 A. Dobson, D.S. Moore, S.D. Robinson, M.B. Hursthouse, and L. New, Polyhedron, 1985, **4**, 1119.
595 S. Gopinathan, I.R. Unny, and C. Gopinathan, Polyhedron, 1985, **4**, 1569.
596 M.A. Bennett, I.J. McMahon, S. Pelling, G.B. Robertson, and W.A. Wickramasinghe, Organometallics, 1985, **4**, 754.
597 K. Hiraki, N. Ochi, Y. Sasada, H. Hayashida, Y. Fuchita, and S. Yamanaka, J. Chem. Soc., Dalton Trans., 1985, 873.
598 L.N. Lewis, Inorg. Chem., 1985, **24**, 4433.
599 H. Werner and J. Gotzig, J. Organomet. Chem., 1985, **284**, 73.
600 M.I. Bruce, D.N. Duffy, M.G. Humphrey, and A.G. Swincer, J. Organomet. Chem., 1985, **282**, 383.
601 G.T. Fey, G.L. Rempel, B.R. James, and D.V. Plackett, Inorg. Chim. Acta, 1985, **96**, L65.
602 L. Weber, K. Reizig, and R. Boese, Organometallics, 1985, **4**, 1890.
603 L. Weber, K. Reizig, and R. Boese, Organometallics, 1985, **4**, 2097.
604 D.S. Barratt and D.J. Cole-Hamilton, J. Chem. Soc., Chem. Commun., 1985, 1559.
605 M.I. Bruce, C. Dean, D.N. Duffy, M.G. Humphrey, and G.A. Koutsantonis, J. Organomet. Chem., 1985, **295**, C40.
606 G. Consiglio, F. Bangerter, and F. Morandini, J. Organomet. Chem., 1985, **293**, C29.
607 J. Keijsper, L.H. Polm, G. van Koten, K. Vrieze, K. Goubitz, and C.H. Stam, Organometallics, 1985, **4**, 1876.
608 N. Lugan, J.-J. Bonnet, and J.A. Ibers, J. Am. Chem. Soc., 1985, **107**, 4484.
609 J.A. Cabeza and P.M. Maitlis, J. Chem. Soc., Dalton Trans., 1985, 573.
610 E.D. Morrison and G.L. Geoffroy, J. Am. Chem. Soc., 1985, **107**, 3541.
611 D.F.R. Gilson, J. Mol. Struct., 1985, **127**, 121.
612 A.G. Nagy, Hung. Acad. Sci., Cent. Res. Inst. Phys., KFKI, 1985, KFKI-1985-23, 12 pp. (Chem. Abstr., 1985, **103**, 123 662).
613 A.G. Nagy and S. Toma, J. Organomet. Chem., 1985, **282**, 267.
614 A.G. Nagy, J. Organomet. Chem., 1985, **291**, 335.
615 A. Harada and S. Takahashi, J. Inclusion Phenom., 1984, **2**, 791 (Chem.

Abstr., 1985, **103**, 105 114).
616 B. Fuchs, R. Fröhlich, and H. Musso, Chem. Ber., 1985, **118**, 1968.
617 M. Sato, S. Tanaka, S. Ebine, K. Morinaga, and S. Akabori, J. Organomet. Chem., 1985, **282**, 247.
618 E.-W. Koch, H.-U. Siehl, and M. Hanack, Tetrahedron Lett., 1985, **26**, 1493.
619 L. Baltzer and O.A. Gansow, Congr. AMPERE Magn. Reson. Relat. Phenom., Proc., 22nd, 1984, 343 (Chem. Abstr., 1985, **103**, 142 136).
620 Y.-R. Hu, T.W. Leung, S.-C.H. Su, A. Wojcicki, M. Calligaris, and G. Nardin, Organometallics, 1985, **4**, 1001.
621 J.R. Lisko and W.M. Jones, Organometallics, 1985, **4**, 612.
622 K. Broadley, N.G. Connelly, P.G. Graham, J.A.K. Howard, W. Risse, and M.W. Whiteley, J. Chem. Soc., Dalton Trans., 1985, 777.
623 J. Edwin, W. Siebert, and C. Krüger, J. Organomet. Chem., 1985, **282**, 297.
624 H. Umland, D. Wormsbächer, and U. Behrens, J. Organomet. Chem., 1985, **284**, 353.
625 J. Ros, G. Commenges, R. Mathieu, X. Solans, and M. Font-Altaba, J. Chem. Soc., Dalton Trans., 1985, 1087.
626 C.P. Casey and S.R. Marder, Organometallics, 1985, **4**, 411.
627 C. Weatherell, N.J. Taylor, A.J. Carty, E. Sappa, and A. Tiripicchio, J. Organomet. Chem., 1985, **291**, C9.
628 H. Werner and R. Werner, Chem. Ber., 1985, **118**, 4543.
629 H. Werner and K. Zenkert, J. Chem. Soc., Chem. Commun., 1985, 1607.
630 E.D. Morrison, G.L. Geoffroy, and A.L. Rheingold, J. Am. Chem. Soc., 1985, **107**, 254.
631 R. Bertani, A. Scrivanti, and G. Carturan, Inorg. Chim. Acta, 1985, **98**, L9.
632 C.P. Casey, W.H. Miles, P.J. Fagan, and K.J. Haller, Organometallics, 1985, **4**, 559.
633 D. Seyferth, G.B. Womack, and J.C. Dewan, Organometallics, 1985, **4**, 398.
634 C.P. Casey, S.R. Marder, and A.L. Rheingold, Organometallics, 1985, **4**, 762.
635 J.A. Hriljac and D.F. Shriver, Organometallics, 1985, **4**, 2225.
636 D. Nuel, F. Dahan, and R. Mathieu, Organometallics, 1985, **4**, 1436.
637 D.N. Cox, R. Roulet, and G. Chapuis, Organometallics, 1985, **4**, 2001.
638 M.D. Jones and R.D.W. Kemmitt, J. Chem. Soc., Chem. Commun., 1985, 811.
639 J. Lunniss, S.A. MacLaughlin, N.J. Taylor, A.J. Carty, and E. Sappa, Organometallics, 1985, **4**, 2066.
640 J. Gotzig, H. Otto, and H. Werner, J. Organomet. Chem., 1985, **287**, 247.
641 J.A.S. Howell and M.J. Thomas, Organometallics, 1985, **4**, 1054.
642 H. Curtis, B.F.G. Johnson, and G.R. Stephenson, J. Chem. Soc., Dalton Trans., 1985, 1723.
643 E. Tagliaferri, U. Hänisch, R. Roulet, P. Vogel, and K.J. Schenk, Helv. Chim. Acta, 1985, **68**, 1362.
644 J. Ioset and R. Roulet, Helv. Chim. Acta, 1985, **68**, 236.
645 R. Goddard, F.-W. Grevels, and R. Schrader, Angew. Chem., Int. Ed. Engl., 1985, **24**, 353.
646 Y. Butsugan, A. Yamashita, and S. Araki, J. Organomet. Chem., 1985, **287**, 103.
647 A. Hafner, W. von Philipsborn, and A. Salzer, Angew. Chem., Int. Ed. Engl., 1985, **24**, 126.
648 Z. Goldschmidt, S. Antebi, H.E. Gottlieb, D. Cohen, U. Shmueli, and Z. Stein, J. Organomet. Chem., 1985, **282**, 367.
649 G.M. Williams and D.E. Rudisill, J. Am. Chem. Soc., 1985, **107**, 3357.
650 A.R. Koray, C. Krieger, and H.A. Staab, Angew. Chem., Int. Ed. Engl., 1985, **24**, 521.
651 H. Lang, L. Zsolnai, and G. Huttner, Chem. Ber., 1985, **118**, 4426.
652 H.-W. Frühauf, D. Meyer, and J. Breuer, J. Organomet. Chem., 1985, **297**, 211.
653 J. Ros, X. Solans, C. Miravitiles, and R. Mathieu, J. Chem. Soc., Dalton Trans., 1985, 1981.
654 M. Crocker, M. Green, C.E. Morton, K.R. Nagle, and A.G. Orpen, J. Chem. Soc., Dalton Trans., 1985, 2145.
655 H. Schumann, J. Organomet. Chem., 1985, **293**, 75.
656 M. Morán and I. Cuadrado, J. Organomet. Chem., 1985, **295**, 353.

657 W. Angerer, W.S. Sheldrick, and W. Malisch, Chem. Ber., 1985, **118**, 1261.
658 R.J. Angelici and J.W. Dunker, Inorg. Chem., 1985, **24**, 2209.
659 J.D. Goodrich and J.P. Selegue, Organometallics, 1985, **4**, 798.
660 L. Weber, K. Reizig, R. Boese, and M. Polk, Angew. Chem., Int. Ed. Engl., 1985, **24**, 604.
661 W. Malish, W. Angerer, A.H. Cowley, and N.C. Norman, J. Chem. Soc., Chem. Commun., 1985, 1811.
662 B.N. Diel, J. Organomet. Chem., 1985, **284**, 257.
663 L. Weber and K. Reizig, Angew. Chem., Int. Ed. Engl., 1985, **24**, 865.
664 C.C. Tso and A.R. Cutler, Organometallics, 1985, **4**, 1242.
665 W.A. Herrmann, J. Rohrmann, and C. Hecht, J. Organomet. Chem., 1985, **290**, 53.
666 N. Kuhn and H. Schumann, J. Organomet. Chem., 1985, **288**, C51.
667 N.G. Connelly, M.J. Freeman, A.G. Orpen, J.B. Sheridan, A.N.D. Symonds, and M.W. Whiteley, J. Chem. Soc., Dalton Trans., 1985, 1027.
668 A.C. Bray, M. Green, D.R. Hankey, J.A.K. Howard, O. Johnson, and F.G.A. Stone, J. Organomet. Chem., 1985, **281**, C12.
669 E. Sappa, A.M.M. Lanfredi, G. Predieri, A. Tiripicchio, and A.J. Carty, J. Organomet. Chem., 1985, **288**, 365.
670 R.P. Hughes, A.L. Rheingold, W.A. Herrmann, and J.L. Hubbard, J. Organomet. Chem., 1985, **286**, 361.
671 E. Cesarotti, M. Angoletta, N.P.C. Walker, M.B. Hursthouse, R. Vefghi, P.A. Schofield, and C. White, J. Organomet. Chem., 1985, **286**, 343.
672 M. Draganjac, C.J. Ruffing, and T.B. Rauchfuss, Organometallics, 1985, **4**, 1909.
673 M.F. Joseph and M.C. Baird, Inorg. Chim. Acta, 1985, **96**, 229.
674 L.A. Oro, M.A. Ciriano, M. Campo, C. Foces-Foces, and F.H. Cano, J. Organomet. Chem., 1985, **289**, 117.
675 F. Bouachir, B. Chaudret, D. Neibecker, and I. Tkatchenko, Angew. Chem., Int. Ed. Engl., 1985, **24**, 347.
676 A.A. Chalmers, D.C. Liles, E. Meintjies, H.E. Oosthuizen, J.A. Pretorius, and E. Singleton, J. Chem. Soc., Chem. Commun., 1985, 1340.
677 M. Sato, S. Tanaka, S. Ebine, K. Morinaga, and S. Akabori, J. Organomet. Chem., 1985, **289**, 91.
678 S. Toma, E. Solčániová, and A.G. Nagy, J. Organomet. Chem., 1985, **288**, 331.
679 J.-H. Youn, R. Herrmann, and B. Floris, J. Organomet. Chem., 1985, **291**, 355.
680 B. McCulloch, D.L. Ward, J.D. Woolins, and C.H. Brubaker, jun., Organometallics, 1985, **4**, 1425.
681 M.-T. Lee, B.M. Foxman, and M. Rosenblum, Organometallics, 1985, **4**, 539.
682 M.-H. Desbois, J. Guillin, J.-P. Mariot, F. Varret, and D. Astruc, J. Chem. Soc., Chem. Commun., 1985, 447.
683 D.C. Liles, A. Shaver, E. Singleton, and M.B. Wiege, J. Organomet. Chem., 1985, **288**, C33.
684 A.G. Osborne, A.J. Blake, R.E. Hollands, R.F. Bryan, and S. Lockhart, J. Organomet. Chem., 1985, **287**, 39.
685 A.G. Osborne, R.E. Hollands, R.F. Bryan, and S. Lockhart, J. Organomet. Chem., 1985, **288**, 207.
686 I.R. Butler, W.R. Cullen, T.-J. Kim, S.J. Rettig, and J. Trotter, Organometallics, 1985, **4**, 972.
687 T.D. Appleton, W.R. Cullen, S.V. Evans, T.-J. Kim, and J. Trotter, J. Organomet. Chem., 1985, **279**, 5.
688 T. Wilczewski, J. Organomet. Chem., 1985, **297**, 331.
689 S. Akabori, H. Munegumi, Y. Habata, S. Sato, K. Kawazoe, C. Tamura, and M. Sato, Bull. Chem. Soc. Jpn., 1985, **58**, 2185.
690 U. Kölle and J. Grub, J. Organomet. Chem., 1985, **289**, 133.
691 G. Schmid, O. Boltsch, D. Blaeser, and R. Boese, Z. Naturforsch., Teil B, 1984, **39**, 1082 (Chem. Abstr., 1985, **102**, 24 787).
692 N.A. Vol'kenau, I.N. Bolesova, L.S. Shul'pina, A.N. Kitaigorodskii, and D.N. Kravtsov, J. Organomet. Chem., 1985, **288**, 341.
693 N. Mori and M. Takamori, J. Chem. Soc., Dalton Trans., 1985, 1661.
694 E. Roman, V. Castro, and M. Camus, J. Organomet. Chem., 1985, **293**, 93.

695 R.G. Sutherland, R.L. Chowdhury, A. Piórko, and C.C. Lee, J. Chem. Soc., Chem. Commun., 1985, 1298.
696 T.V. Ashworth, A.A. Chalmers, D.C. Liles, E. Meintjies, H.E. Oosthuizen, and E. Singleton, J. Organomet. Chem., 1985, **284**, C19.
697 D. Astruc, P. Michaud, A.M. Madonik, J.-Y. Saillard, and R. Hoffmann, Nouv. J. Chem., 1985, **9**, 41.
698 I.S. Thornburn, S.J. Rettig, and B.R. James, J. Organomet. Chem., 1985, **296**, 103.
699 Z. Dauter, R.J. Mawby, C.D. Reynolds, and D.R. Saunders, J. Chem. Soc., Dalton Trans., 1985, 1235.
700 A.R. Butler, C. Glidewell, A.R. Hyde, and J. McGinnis, Inorg. Chem., 1985, **24**, 2931.
701 H. Inoue, M. Sasagawa, and E. Fluck, Z. Naturforsch., Teil B, 1985, **40**, 22 (Chem. Abstr., 1985, **103**, 88 034).
702 E.J.S. Vichi, F.Y. Fujiwara, and E. Stein, Inorg. Chem., 1985, **24**, 286.
703 H.C. Lee, J.K. Gard, T.L. Brown, and E. Oldfield, J. Am. Chem. Soc., 1985, **107**, 4087.
704 G.N. La Mar, C.M. Dellinger, and S.S. Sankar, Biochem. Biophys. Res. Commun., 1985, **128**, 628 (Chem. Abstr., 1985, **102**, 217 066).
705 L. Baltzer, E.D. Becker, R.G. Tschudin, and O.A. Gansow, J. Chem. Soc., Chem. Commun., 1985, 1040.
706 M. Schaal, W. Weigand, U. Nagel, and W. Beck, Chem. Ber., 1985, **118**, 2186.
707 S.M. Tetrick and R.A. Walton, Inorg. Chem., 1985, **24**, 3363.
708 D.H. Champion and A.H. Cowley, Polyhedron, 1985, **4**, 1791.
709 H. Lang, L. Zsolnai, and G. Huttner, J. Organomet. Chem., 1985, **282**, 23.
710 W.J. Richter, Tetrahedron Lett., 1985, **26**, 1039.
711 E.P. Kyba, K.L. Hassett, B. Sheikh, J.S. McKennis, R.B. King, and R.E. Davis, Organometallics, 1985, **4**, 994.
712 R. Appel, C. Casser, and F. Knoch, J. Organomet. Chem., 1985, **293**, 213.
713 R.R. Ford, B.-L. Li, R.H. Neilson, and R.J. Thoma, Inorg. Chem., 1985, **24**, 1993.
714 R.B. King, F.-J. Wu, N.D. Sadanani, and E.M. Holt, Inorg. Chem., 1985, **24**, 4449.
715 H.R. Allcock, P.R. Suszko, L.J. Wagner, R.R. Whittle, and B. Boso, Organometallics, 1985, **4**, 446.
716 D.J. Brauer, S. Hietkamp, H. Sommer, O. Stelzer, G. Müller, M.J. Romão, and C. Krüger, J. Organomet. Chem., 1985, **296**, 411.
717 W.K. Wong, K.W. Chiu, G. Wilkinson, M. Motevalli, and M.B. Hursthouse, Polyhedron, 1985, **4**, 1231.
718 P. Braunstein, G. Predieri, F.J. Lahoz, and A. Tiripicchio, J. Organomet. Chem., 1985, **288**, C13.
719 V.D. Patel, A.A. Cherkas, D. Nucciarone, N.J. Taylor, and A.J. Carty, Organometallics, 1985, **4**, 1792.
720 C.M. Hay, B.F.G. Johnson, J. Lewis, R.C.S. McQueen, P.R. Raithby, R.M. Sorrell, and M.J. Taylor, Organometallics, 1985, **4**, 202.
721 G. Doyle, B.T. Heaton, and E. Occhiello, Organometallics, 1985, **4**, 1224.
722 K. Knoll, G. Hullner, L. Zsolnai, I. Jibril, and M. Wasiucionek, J. Organomet. Chem., 1985, **294**, 91.
723 A.M. Arif, A.H. Cowley, and M. Pakulski, J. Am. Chem. Soc., 1985, **107**, 2553.
724 A.M. Arif, A.H. Cowley, and S. Quashie, J. Chem. Soc., Chem. Commun., 1985, 428.
725 H. Umland and U. Behrens, J. Organomet. Chem., 1985, **287**, 109.
726 G.D. Williams, G.L. Geoffroy, and R.R. Whittle, J. Am. Chem. Soc., 1985, **107**, 729.
727 C.P. Horwitz, E.M. Holt, C.P. Brock, and D.F. Shriver, J. Am. Chem. Soc., 1985, **107**, 8136.
728 M.J. Sailor and D.F. Shriver, Organometallics, 1985, **4**, 1476.
729 P. Mura, B.G. Olby, and S.D. Robinson, Inorg. Chim. Acta, 1985, **97**, 45.
730 J.A. Clucas, M.M. Harding, B.S. Nicholls, and A.K. Smith, J. Chem. Soc., Dalton Trans., 1985, 1835.
731 U. Keppeler, W. Kobel, H.U. Siehl, and M. Hanack, Chem. Ber., 1985, **118**, 2095.

732 C.R. Jablonski, Y.-P. Wang, and N.J. Taylor, Inorg. Chim. Acta, 1985, **96**, L17.
733 P. Braunstein, J. Kervennal, and J.-L. Richert, Angew. Chem., Int. Ed. Engl., 1985, **24**, 768.
734 P. Mura, B.G. Olby, and S.D. Robinson, J. Chem. Soc., Dalton Trans., 1985, 2101.
735 G. Süss-Fink, G. Herrmann, P. Morys, J. Ellermann, and A. Veit, J. Organomet. Chem., 1985, **284**, 263.
736 J. Pursiainen, M. Ahlgren, and T.A. Pakkanen, J. Organomet. Chem., 1985, **297**, 391.
737 G. Smith, D.J. Cole-Hamilton, M. Thornton-Pett, and M.B. Hursthouse, J. Chem. Soc., Dalton Trans., 1985, 387.
738 J.A. Iggo, D.P. Markham, B.L. Shaw, and M. Thornton-Pett, J. Chem. Soc., Chem. Commun., 1985, 432.
739 S. Gopinathan, S.S. Deshpande, and C. Gopinathan, Z.Anorg. Allg. Chem., 1985, **527**, 203.
740 C.O. Dietrich-Buchecker, P.A. Marnot, J.P. Sauvage, J.P. Kintzinger, and P. Maltese, Nouv. J. Chim., 1984, **8**, 573.
741 N.N. Gerasimchuk, V.V. Skopenko, and S.I. Tyukhtenko, Ukr. Khim. Zh. (Russ. Ed.), 1985, **51**, 232 (Chem. Abstr., 1985, **103**, 47 218).
742 E. Tsuchida, H. Nishide, M. Yuasa, E. Hasegawa, Y.-i. Matsushita, and K. Eshima, J. Chem. Soc., Dalton Trans., 1985, 275.
743 J.J. Grzybowski, Inorg. Chem., 1985, **24**, 1125.
744 M.K. Robbins, D.W. Naser, J.L. Heiland, and J.J. Grzybowski, Inorg. Chem., 1985, **245**, 3381.
745 H.E. Toma, E. Giesbrecht, and R.L. Espinoza Rojas, J. Chem. Soc., Dalton Trans., 1985, 2469.
746 T. Akasheh and K. Barqawi, Inorg. Chim. Acta, 1985, **102**, L33.
747 P. Bernhard and A.M. Sargeson, J. Chem. Soc., Chem. Commun., 1985, 1516.
748 M.J. Root, B.P. Sullivan, T.J. Meyer, and E. Deutsch., Inorg. Chem., 1985, **24**, 2731.
749 Y. Yoshino, M. Kasahara, and M. Saito, Polyhedron, 1985, **4**, 1019.
750 M.M. Taqui Khan, H.C. Bajaj, Md.R.H. Siddiqui, B.T. Khan, M.S. Reddy, and K.V. Reddy, J. Chem. Soc., Dalton Trans., 1985, 2603.
751 R. Cini, A. Cinquantini, M. Sabat, and L.G. Marzilli, Inorg. Chem., 1985, **24**, 3903.
752 M.B. Hursthouse, N.P.C. Walker, C.P. Warrens, and J.D. Woollins, J. Chem. Soc., Dalton Trans., 1985, 1043.
753 T. Ikariya, Y. Ishii, H. Kawano, T. Arai, M. Saburi, S. Yoshikawa, and S. Akutagawa, J. Chem. Soc., Chem. Commun., 1985, 922.
754 O.H. Bailey and A. Ludi, Inorg. Chem., 1985, **24**, 2582.
755 A.V. Rivera, E.R. de Gil, and B. Fontal, Inorg. Chim. Acta, 1985, **98**, 153.
756 T. Easton, R.O. Gould, G.A. Heath, and T.A. Stephenson, J. Chem. Soc., Chem. Commun., 1985, 1741.
757 V.T. Coombe, G.A. Heath, T.A. Stephenson, J.D. Whitelock, and L.J. Yellowlees, J. Chem. Soc., Dalton Trans., 1985, 947.
758 R. Bramley, M. Brorson, A.M. Sargeson, and C.E. Schaeffer, J. Am. Chem. Soc., 1985, **107**, 2780.
759 H. Werner, J. Wolf, and A. Höhn, J. Organomet. Chem., 1985, **287**, 395.
760 T.M. Gilbert and R.G. Bergman, J. Am. Chem. Soc., 1985, **107**, 6391.
761 H. Bauer, U. Nagel, and W. Beck, J. Organomet. Chem., 1985, **290**, 219.
762 L.F. Rhodes, J.C. Huffman, and K.G. Caulton, J. Am. Chem. Soc., 1985, **107**, 1759.
763 D.G. Holah, A.N. Hughes, S. Maciaszek, V.R. Magnuson, and K.O. Parker, Inorg. Chem., 1985, **24**, 3956.
764 D.J. Elliot, D.G. Holah, A.N. Hughes, S.I. Khan, and S. Maciaszek, Inorg. Chim. Acta, 1985, **96**, L29.
765 D. Bawdry, P. Boydell, M. Ephritikhine, H. Felkin, J. Guilhem, C. Pascard, and E.T.H Dau, J. Chem. Soc., Chem. Commun., 1985, 670.
766 D.M. Haddleton and R.N. Perutz, J. Chem. Soc., Chem. Commun., 1985, 1372.
767 P.O. Bentz, J. Ruiz, B.E. Mann, C.M. Spencer, and P.M. Maitlis, J. Chem. Soc., Chem. Commun., 1985, 1374.

768 M.A. Bennett and T.R.B. Mitchell, J. Organomet. Chem., 1985, **295**, 223.
769 J. Ott, L.M. Venanzi, C.A. Ghilardi, S. Midollini, and A. Orlandini, J. Organomet. Chem., 1985, **291**, 89.
770 T. Yamakawa, S. Shinoda, Y. Saito, H. Moriyama, and P.S. Pregosin, Magn. Reson. Chem., 1985, **23**, 202.
771 D.P. Krut'ko, A.B. Permin, V.S. Petrosyan, and O.A. Reutov, Izv. Akad. Nauk SSSR, Ser. Khim., 1984, 2787 (Chem. Abstr., 1985, **102**, 105 068).
772 K.R. Koch and J.M. Wyrley-Birch, Inorg. Chim. Acta, 1985, **102**, L5.
773 D.P. Krut'ko, A.B. Permin, V.S. Petrosyan, and O.A. Reutov, Izv. Akad. Nauk SSSR, Ser. Khim., 1985, 909.
774 S.L. Schiavo, G. Bruno, F. Nicolò, P. Piraino, and F. Faraone, Organometallics, 1985, **4**, 2091.
775 S.P. Deraniyagala and K.R. Grundy, Inorg. Chim. Acta, 1985, **101**, 103.
776 A. Albinati, H. Lehner, and L.M. Venanzi, Inorg. Chem., 1985, **24**, 1483.
777 A.L. Balch, J.C. Linehan, and M.M. Olmstead, Inorg. Chem., 1985, **24**, 3975.
778 J. Ott and L.M. Venanzi, J. Am. Chem. Soc., 1985, **107**, 1760.
779 L.F. Rhodes and K.G. Caulton, J. Am. Chem. Soc., 1985, **107**, 259.
780 M. Kretschmer, P.S. Pregosin, A. Albinati, and A. Togni, J. Organomet. Chem., 1985, **281**, 365.
781 M. Drouin and J.F. Harrod, Can. J. Chem., 1985, **63**, 353.
782 S. Zecchin, G. Zotti, and G. Pilloni, J. Organomet. Chem., 1985, **294**, 379.
783 E. Arpac and L. Dahlenburg, Chem. Ber., 1985, **118**, 3188.
784 L. Dahlenburg and N. Höck, Inorg. Chim. Acta, 1985, **104**, L29.
785 E.A.V. Ebsworth and R. Mayo, Angew. Chem., Int. Ed. Engl., 1985, **24**, 68.
786 P.O. Stoutland and R.G. Bergman, J. Am. Chem. Soc., 1985, **107**, 4581.
787 L.J. Newman and R.G. Bergman, J. Am. Chem. Soc., 1985, **107**, 5314.
788 K. Tani, E. Tanigawa, Y. Tatsuno, and S. Otsuka, J. Organomet. Chem., 1985, **279**, 87.
789 B.R. Sutherland and M. Cowie, Organometallics, 1985, **4**, 1801.
790 T.M. Gilbert, F.J. Hollander, and R.G. Bergman, J. Am. Chem. Soc., 1985, **107**, 3508.
791 V.G. Albano, D. Braga, R. Ros, and A. Scrivanti, J. Chem. Soc., Chem. Commun., 1985, 866.
792 N. Bresciani-Pahor, L. Randaccio, E. Zangrando, M.F. Summers, J.H. Ramsden, jun., P.A. Marzilli, and L.G. Marzilli, Organometallics, 1985, **4**, 2086.
793 M. Rossi, J.P. Glusker, L. Randaccio, M.F. Summers, P.J. Toscano, and L.G. Marzilli, J. Am. Chem. Soc., 1985, **107**, 1729.
794 B.E. Mann, N.J. Meanwell, C.M. Spencer, B.F. Taylor, and P.M. Maitlis, J. Chem. Soc., Dalton Trans., 1985, 1555.
795 I.Ya. Levitin, A.N. Kitaigorodskii, A.T. Nikitaev, V.I. Bakhmutov, A.L. Sigan, and M.E. Vol'pin, Inorg. Chim. Acta, 1985, **100**, 65.
796 L. Randaccio, N. Bresciani-Pahor, J.D. Orbell, M. Calligaris, M.F. Summers, B. Snyder, P.J. Toscano, and L.G. Marzilli, Organometallics, 1985, **4**, 469.
797 M.R. Ashcroft, P. Bougeard, A. Bury, C.J. Cooksey, and M.D. Johnson, J. Organomet. Chem., 1985, **289**, 403.
798 H. Werner and R. Zolk, Organometallics, 1985, **4**, 601.
799 B. Heiser, A. Kühn, and H. Werner, Chem. Ber., 1985, **118**, 1531.
800 H. Werner and B. Heiser, Chem. Ber., 1985, **118**, 2282.
801 U. Plaia, H. Stolzenberg, and W.P. Fehlhammer, J. Am. Chem. Soc., 1985, **107**, 2171.
802 D.W. Macomber and R.D. Rogers, Organometallics, 1985, **4**, 1485.
803 G. Balavoine, J. Collin, J.J. Bonnet, and G. Lavigne, J. Organomet. Chem., 1985, **280**, 429.
804 D.L. Thorn, Organometallics, 1985, **4**, 192.
805 H. Werner, R. Feser, and L. Hofmann, J. Organomet. Chem., 1985, **292**, 361.
806 W. Paul and H. Werner, Chem. Ber., 1985, **118**, 3032.
807 H. Werner, W. Paul, R. Feser, R. Zolk, and P. Thometzek, Chem. Ber., 1985, **118**, 261.
808 M. Gómez, J.M. Kisenyi, G.J. Sunley and P.M. Maitlis, J. Organomet. Chem., 1985, **296**, 197.
809 H. Werner, L. Hofmann, R. Feser, and W. Paul, J. Organomet. Chem., 1985, **281**, 317.

810 T. Yoshida, T. Okana, S. Otsuka, I. Miura, T. Kubota, K. Kafuku, and K. Nakatsu, Inorg. Chim. Acta, 1985, **100**, 7.
811 J.M. Kisenyi, J.A. Cabeza, A.J. Smith, H. Adams, G.J. Sunley, N.J.S. Salt, and P.M. Maitlis, J. Chem. Soc., Chem. Commun., 1985, 770.
812 W.A. Herrmann, C. Weber, M.L. Ziegler, and O. Serhadli, J. Organomet. Chem., 1985, **297**, 245.
813 S.P. Deranlyagala and K.R. Grundy, Organometallics, 1985, **4**, 424.
814 D.H. Berry and R. Eisenberg, J. Am. Chem. Soc., 1985, **107**, 7181.
815 T.G. Schenck, C.R.C. Milne, J.F. Sawyer, and B. Bosnick, Inorg. Chem., 1985, **24**, 2338.
816 J.W. Suggs, M.J. Wovkulich, and S.D. Cox, Organometallics, 1985, **4**, 1101.
817 S.L. Van Voorhees and B.B. Wayland, Organometallics, 1985, **4**, 1887.
818 M.J. Auburn and S.R. Stobart, Inorg. Chem., 1985, **24**, 318.
819 A.R. Sanger, Inorg. Chim. Acta, 1985, **99**, 95.
820 A.W. Coleman, P.B. Hitchcock, M.F. Lappert, R.K. Maskell, and J.H. Müller, J. Organomet. Chem., 1985, **296**, 173.
821 F.J.G. Alonso, A. Höhn, J. Wolf, H. Otto, and H. Werner, Angew. Chem., Int. Ed. Engl., 1985, **24**, 406.
822 W.A. Herrmann and C. Weber, J. Organomet. Chem., 1985, **282**, C31.
823 H. Werner, F.J.G. Alonso, H. Otto, K. Peters, and H.G. von Schnering, J. Organomet. Chem., 1985, **289**, C5.
824 J.C. Jeffery, C. Marsden, and F.G.A. Stone, J. Chem. Soc., Dalton Trans., 1985, 1315.
825 R.S. Dickson, G.D. Fallon, R.J. Nesbit, and G.N. Pain, Organometallics, 1985, **4**, 355.
826 F. Barceló, P. Lahuerta, M.A. Ubeda, C. Foces-Foces, F.H. Cano, and M. Martinez-Ripoli, J. Chem. Soc., Chem. Commun., 1985, 43.
827 M. Green, A.G. Orpen, C.J. Schaverien, and I.D. Williams, J. Chem. Soc., Dalton Trans., 1985, 2483.
828 M. Gómez, P.I.W. Yarrow, D.J. Robinson, and P.M. Maitlis, J. Organomet. Chem., 1985, **279**, 115.
829 K.A. King, P.J. Spellane, and R.J. Watts, J. Am. Chem. Soc., 1985, **107**, 1431.
830 M.A. Lilga and J.A. Ibers, Organometallics, 1985, **4**, 590.
831 K. Bowman, A.J. Deeming, and G.P. Proud, J. Chem. Soc., Dalton Trans., 1985, 857.
832 M.D. Fryzuk, P.A. MacNeil, and S.J. Rettig, J. Am. Chem. Soc., 1985, **107**, 6708.
833 M.J. Auburn, S.L. Grundy, S.R. Stobart, and M.J. Zaworotko, J. Am. Chem. Soc., 1985, **107**, 266.
834 A.L. Casalnuovo, T. Laska, P.V. Nilsson, J. Olofson, and L.H. Pignolet, Inorg. Chem., 1985, **24**, 233.
835 C. Bianchini, D. Masi, C. Mealli, A. Meli, and M. Sabat, Organometallics, 1985, **4**, 1014.
836 H. Bönnemann, Angew. Chem., Int. Ed. Engl., 1985, **24**, 248.
837 E. Dunach, R.L. Halterman, and K.P.C. Vollhardt, J. Am. Chem. Soc., 1985, **107**, 1664.
838 S.M. Socol and D.W. Meek, Inorg. Chim. Acta, 1985, **101**, L45.
839 H. Brunner, P. Beier, G. Riepl, I. Bernal, G.M. Reisner, R. Benn, and A. Rufińska, Organometallics, 1985, **4**, 1732.
840 H. Brunner, A. Knott, R. Benn, and A. Rufińska, J. Organomet. Chem., 1985, **295**, 211.
841 F.Y. Pétillon, J.L. le Quéré, F. le Floch-Pérennou, J.E. Guerchais, and P. L'Haridon, J. Organomet. Chem., 1985, **281**, 305.
842 R.S. Dickson, G.S. Evans, G.D. Fallon, and G.N. Pain, J. Organomet. Chem., 1985, **295**, 109.
843 M.L. Deem, Inorg. Chim. Acta, 1985, **103**, L19.
844 A. Salzer and C. Täschler, J. Organomet. Chem., 1985, **294**, 261.
845 W.A. Herrmann, C.E. Barnes, T. Zahn, and M.L. Ziegler, Organometallics, 1985, **4**, 172.
846 J.M. Brown, R.A. John, and A.R. Lucy, J. Organomet. Chem., 1985, **279**, 245.
847 G.W. Bushnell, M.J. Decker, D.T. Eadie, S.R. Stobart, R. Vefghi, J.L.

Atwood, and M.J. Zaworotko, Organometallics, 1985, **4**, 2106.
848 R. Appel, W. Schuhn, and F. Knock, Angew. Chem., Int. Ed. Engl., 1985, **24**, 420.
849 M.D. Fryzuk, W.E. Piers, and S.J. Rettig, J. Am. Chem. Soc., 1985, **107**, 8259.
850 I.T. Chizhevsky, N.V. Rastova, N.E. Kolobova, P.V. Petrovskii, L.E. Vinogradova, and L.A. Leites, Tetrahedron Lett., 1985, **26**, 3605.
851 S.D. Chappell, D.J. Cole-Hamilton, A.M.R. Galas, M.B. Hursthouse, and N.P.C. Walker, Polyhedron, 1985, **4**, 121.
852 R.S. Dickson, R.J. Nesbit, H. Pateras, W. Baimbridge, J.M. Patrick, and A.H. White, Organometallics, 1985, **4**, 2128.
853 C.F. Jewell, jun., L.S. Liebeskind, and M. Williamson, J. Am. Chem. Soc., 1985, **107**, 6715.
854 J. Bakos, I. Tóth, B. Heil, and L. Markó, J. Organomet. Chem., 1985, **279**, 23.
855 E.P. Kyba and S.-T. Liu, Inorg. Chem., 1985, **24**, 1613.
856 N.M. Doherty, B.E. Ewels, R.P. Hughes, D.E. Samkoff, W.D. Saunders, R.E. Davis, and B.B. Laird, Organometallics, 1985, **4**, 1606.
857 D.H.M.W. Thewissen, K. Timmer, J.G. Noltes, J.W. Marsman, and R.M. Laine, Inorg. Chim. Acta, 1985, **97**, 143.
858 K. Timmer and D.H.M.W. Thewissen, Inorg. Chim. Acta, 1985, **100**, 235.
859 A. Christofides and I. Labrinou, Inorg. Chim. Acta, 1985, **99**, 195.
860 A.A. Bahsoun and R. Ziessel, Nouv. J. Chim., 1985, **9**, 225.
861 C.E. Briant, D.I. Gilmour, M.A. Luke, and D.M.P. Mingos, J. Chem. Soc., Dalton Trans., 1985, 851.
862 G.W. Bushnell, D.O.K. Fjeldsted, S.R. Stobart, M.J. Zaworotko, S.A.R. Knox, and K.A. Macpherson, Organometallics, 1985, **4**, 1107.
863 Q.B. Bao, S.J. Landon, A.L. Rheingold, T.M. Haller, and T.B. Brill, Inorg. Chem., 1985, **24**, 900.
864 S.J. Carter and L.S. Stuhl, Organometallics, 1985, **4**, 197.
865 R. Zolk and H. Werner, Angew. Chem., Int. Ed. Engl., 1985, **24**, 576.
866 H. Werner, W. Hofmann, R. Zolk, L.F. Dahl, J. Kocal, and A. Kühn, J. Organomet. Chem., 1985, **289**, 173.
867 L. Chen, D.J. Kountz, and D.W. Meek, Organometallics, 1985, **4**, 598.
868 U. Schubert, B. Heiser, L. Hee, and H. Werner, Chem. Ber., 1985, **118**, 3151.
869 P. Braunstein, H. Lehner, D. Matt, A. Tiripicchio, and M.T. Camellini, Nouv. J. Chim., 1985, **9**, 597.
870 E.J. Miller, S.J. Landon, and T.B. Brill, Organometallics, 1985, **4**, 533.
871 G.E. Herberich and H. Ohst, Chem. Ber., 1985, **118**, 4303.
872 M. Kajitani, T. Suetsugu, R. Wakabayashi, A. Igarashi, T. Akiyama, and A. Sugimori, J. Organomet. Chem., 1985, **293**, C15.
873 G. Bruno, S.L. Schiavo, P. Piraino, and F. Faraone, Organometallics, 1985, **4**, 1098.
874 Z.L. Lutsenko, G.G. Aleksandrov, P.V. Petrovskii, E.S. Shubina, V.G. Andrianov, Yu.T. Struchkov, and A.Z. Rubezhov, J. Organomet. Chem., 1985, **281**, 349.
875 C.J. Besecker, V.W. Day, and W.G. Klemperer, Organometallics, 1985, **4**, 564.
876 R.H. Crabtree and C.P. Parnell, Organometallics, 1985, **4**, 519.
877 A.M. Trzeciak and J.J. Ziolkowski, Inorg. Chim. Acta, 1985, **96**, 15.
878 C. Woods and C. Daffron, Inorg. Chim. Acta, 1985, **101**, 13.
879 R.V. Parish and S.M. Razzoki, Inorg. Chim. Acta, 1985, **96**, 49.
880 S. Jans-Buerli and P.S. Pregosin, Magn. Reson. Chem., 1985, **23**, 198.
881 M. Lattman, S.A. Morse, A.H. Cowley, J.G. Lasch, and N.C. Norman, Inorg. Chem., 1985, **24**, 1364.
882 P. Paetzold and K. Delpy, Chem. Ber., 1985, **118**, 2552.
883 G.L. Geoffroy, W.C. Mercer, R.R. Whittle, L. Markó, and S. Vastag, Inorg. Chem., 1985, **24**, 3771.
884 F.R. Askham, G.G. Stanley, and E.C. Marques, J. Am. Chem. Soc., 1985, **107**, 7423.
885 N.J. Bailey, G. De Leeuw, J.S. Field, R.J. Haines, I.C.D. Stuckenberg, and R.B. English, S. Afr. J. Chem., 1985, **38**, 139 (Chem. Abstr., 1985, **103**, 188 412).

886 M.G. Richmond, J.D. Korp, and J.K. Kochi, J. Chem. Soc., Chem. Commun., 1985, 1102.
887 D.J. Darensbourg and D.J. Zalewski, Organometallics, 1985, **4**, 92.
888 A.R. Sanger, Can. J. Chem., 1985, **63**, 571.
889 J. Ellermann and G. Szucsanyi, Z.Anorg. Allg. Chem., 1985, **520**, 113.
890 A.L. Balch, R.R. Guimerans, J. Lineham, and F.E. Wood, Inorg. Chem., 1985, **24**, 2021.
891 A.L. Balch, R.R. Guimerans, J. Linehan, M.M. Olmstead, and D.E. Oram, Organometallics, 1985, **4**, 1445.
892 F. Bonati, L.A. Oro, and M.T. Pinillos, Polyhedron, 1985, **4**, 357.
893 W. Willis and K.M. Nicholas, Report, 1984, TR-3; Order No. AD-A147784/3/GAR, 9 pp. Avail. NTIS. From Gov. Rep. Announce. Index (U.S.), 1985, **85**, 81 (Chem. Abstr., 1985, **102**, 196 869).
894 E.E. Nifantyev, A.T. Teleshev, G.M. Grishina, and A.A. Borisenko, Phosphorus Sulphur, 1985, **24**, 333.
895 A.L. Balch, H. Hope, and F.E. Wood, J. Am. Chem. Soc., 1985, **107**, 6936.
896 J. Ellermann, G. Szucsányi, and E. Wilhelm, Chem. Ber., 1985, **118**, 1588.
897 J.G. Gaudiello, T.C. Wright, R.A. Jones, and A.J. Bard, J. Am. Chem. Soc., 1985, **107**, 888.
898 L.A. Oro, M.T. Pinillos, A. Tiripicchio, and M. Tiripicchio-Camellini, Inorg. Chim. Acta, 1985, **99**, L13.
899 C. Woodcock and R. Eisenberg, Organometallics, 1985, **4**, 4.
900 S.P. Deraniyagala and K.R. Grundy, Inorg. Chem., 1985, **24**, 50.
901 T.G. Schenck, J.M. Downes, C.R.C. Milne, P.B. MacKenzie, H. Boucher, J. Whelan, and B. Bosnich, Inorg. Chem., 1985, **24**, 2334.
902 A.L. Balch, L.A. Fossett, M.M. Olmstead, D.E. Oram, and P.E. Reedy, jun., J. Am. Chem. Soc., 1985, **107**, 5272.
903 J. Ellermann and G. Szucsanyi, Chem. Ber., 1985, **118**, 1868.
904 A.L. Balch, L.A. Fossett, R.R. Guimerans, and M.M. Olmstead, Organometallics, 1985, **4**, 781.
905 L.A. Oro, D. Carmona, P.L. Pérez, M. Esteban, A. Tiripicchio, and M. Tiripicchio-Camellini, J. Chem. Soc., Dalton Trans., 1985, 973.
906 R. Mutin, W. Abboud, J.M. Basset, and D. Sinou, J. Mol. Catal., 1985, **33**, 47.
907 C.E. Johnson and R. Eisenberg, J. Am. Chem. Soc., 1985, **107**, 3148.
908 B.R. Sutherland and M. Cowie, Organometallics, 1985, **4**, 1637.
909 R.E. Stevens, P.C.C. Liu, and W.L. Gladfelter, J. Organomet. Chem., 1985, **287**, 133.
910 C.B. Storm, A.H. Turner, and N.S. Rowan, Inorg. Chem., 1985, **24**, 1269.
911 Y. Odaka, A. Yamasaki, and M. Watabe, Bull. Chem. Soc. Japan., 1985, **58**, 3407.
912 B.D. Gupta and S. Roy, Tetrahedron Lett., 1985, **26**, 3609.
913 R. Benn and R. Mynott, Angew. Chem., Int. Ed. Engl., 1985, **24**, 333.
914 N.W. Alcock, J.M. Brown, A.E. Derome, and A.R. Lucy, J. Chem. Soc., Chem. Commun., 1985, 575.
915 A.M. Bond, R. Colton, J.E. Moir, and D.R. Page, Inorg. Chem., 1985, **24**, 1298.
916 A.M. Bond, R. Colton, Y. Ho, J.E. Moir, D.R. Page, and R. Stott, Inorg. Chem., 1985, **24**, 4402.
917 J. Shaw and G.W. Everett, Inorg. Chem., 1985, **24**, 1917.
918 S. Bagger, I Kristjansson, I Soetofte, and A. Thorlacius, Acta Chem. Scand., Ser. A, 1985, **A39**, 125.
919 G.P. Haight, jun., T.W. Hambley, P. Hendry, G.A. Lawrance, and A.M. Sargeson, J. Chem. Soc., Chem. Commun., 1985, 488.
920 S. Aygen, Y. Kitamura, K. Kuroda, R. Kume, H. Kelm, and R. Van Eldik, Inorg. Chem., 1985, **24**, 423.
921 L.R. Solujik, R. Herak, B. Prelesnik, and M.B. Celap, Inorg. Chem., 1985, **24**, 32.
922 A.R. Murdock, T. Tyree, W. Otterbein, L. Kinney, M. Carreras, J.N. Cooper, and R.C. Elder, Inorg. Chem., 1985, **24**, 3674.
923 L.R. Gahan, J.M. Harrowfield, A.J. Herlt, L.F. Lindoy, P.O. Whimp, and A.M. Sargeson, J. Am. Chem. Soc., 1985, **107**, 6231.

924 V. McKee, M. Harnett, and D.A. House, Inorg. Chim. Acta, 1985, **102**, 83.
925 M. Harnett, D.A. House, and W.T. Robinson, Inorg. Chim. Acta, 1985, **102**, 87.
926 D.A. House, Helv. Chim. Acta, 1985, **68**, 1872.
927 G.R. Brubaker, M.G. Henk, and D.W. Johnson, Inorg. Chim. Acta, 1985, **100**, 201.
928 M. Noji, H. Toida, and Y. Kidani, Bull. Chem. Soc. Jpn., 1985, **58**, 1279.
929 M. Sato, Y. Yokoo, M. Yashiro, S. Yano, S. Yoshikawa, K. Kobayashi, and T. Sakurai, Bull. Chem. Soc. Jpn., 1985, **58**, 1392.
930 Y. Yamamoto, E. Toyota, and S. Tsukuda, Bull. Chem. Soc. Jpn., 1985, **58**, 1595.
931 S. Utsuno, Y. Yoshikawa, and Y. Tahata, Inorg. Chem., 1985, **24**, 2724.
932 K. Kanamori, Y. Nunomura, M. Takashima, K. Nagata, K. Kawai, and J. Hidaka, Bull. Chem. Soc. Jpn., 1985, **58**, 92.
933 M. Saburi, S. Obata, M. Yamaguchi, and S. Yoshikawa, Bull. Chem. Soc. Jpn., 1985, **58**, 763.
934 R.J. Geue, M.G. McCarthy, A.M. Sargeson, P. Jørgensen, R.G. Hazell, and F.K. Larsen, Inorg. Chem., 1985, **24**, 2559.
935 M. Atoh, K. Kashiwabara, and J. Fujita, Bull. Chem. Soc. Jpn., 1985, **58**, 2793.
936 M. Yashiro, S. Yano, M. Ajioka, and S. Yoshikawa, Inorg. Chem., 1985, **24**, 3607.
937 M. Saburi, Y. Ishii, S. Obata, M. Yamaguchi, and S. Yoshikawa, Inorg. Chem., 1985, **24**, 527.
938 R.J. Geue, M.G. McCarthy, A.M. Sargeson, B.W. Skelton, and A.H. White, Inorg. Chem., 1985, **24**, 1607.
939 Y. Yamamoto and S. Tsukuda, Bull. Chem. Soc. Jpn., 1985, **58**, 1509.
940 J.K. Das and K.C. Dash, Indian J. Chem., Sect. A, 1985, **24A**, 387 (Chem. Abstr., 1985, **103**, 133 894).
941 J.K. Das and K.C. Dash, Polyhedron, 1985, **4**, 1109.
942 J.K. Das and K.C. Dash, Indian J. Chem., Sect. A, 1985, **24A**, 696 (Chem. Abstr., 1985, **103**, 204 663).
943 B. Kräutler and R. Stepánek, Angew. Chem., Int. Ed. Engl., 1985, **24**, 62.
944 A. Ruhe, L. Walder, and R. Scheffold, Helv. Chim. Acta, 1985, **68**, 1301.
945 J.K. Das and K.C. Dash, Z. Naturforsch., Teil B, 1985, **40**, 470 (Chem. Abstr., 1985, **103**, 63 816).
946 T. Konno, S. Aizawa, K. Okamoto, and J. Hidaka, Chem. Lett., 1985, 1017.
947 G. Brewer and C.M. Grisham, Inorg. Chim. Acta, 1985, **106**, 37.
948 K. Okamoto, M. Suzuki, H. Einaga, and J. Hidaka, Bull. Chem. Soc. Jpn., 1985, **58**, 1807.
949 V.P. Tarasov, T.Sh. Kapanadze, G.V. Tsintsadze, I.B. Baranovskii, and Yu.A. Buslaev, Koord. Khim., 1985, **11**, 655 (Chem. Abstr., 1985, **103**, 97 913).
950 K. Tani, Pure Appl. Chem., 1985, **57**, 1845.
951 R.A. Henderson, J. Chem. Soc., Dalton Trans., 1985, 2067.
952 A.Y.A. Mohamed, R.D. Gillard, and F.L. Wimmer, Polyhedron, 1985, **4**, 1239.
953 P.B. Hitchcock, S. Morton, and J.F. Nixon, J. Chem. Soc., Dalton Trans., 1985, 1295.
954 G. Pimblett, C.D. Garner, and W. Clegg, J. Chem. Soc., Dalton Trans., 1985, 1977.
955 R.J. McNair, P.V. Nilsson, and L.H. Pignolet, Inorg. Chem., 1985, **24**, 1935.
956 E.B. Boyar and S.D. Robinson, J. Chem. Soc., Dalton Trans., 1985, 2113.
957 A. Dobson, D.S. Moore, S.D. Robinson, A.M.R. Galas, and M.B. Hursthouse, J. Chem. Soc., Dalton Trans., 1985, 611.
958 A.L. Hale, H.C. Jewiss, and W. Levason, Inorg. Chim. Acta, 1985, **97**, 85.
959 D. Fenske, J. Ohmer, and J. Hachgenei, Angew. Chem., Int. Ed. Engl., 1985, **24**, 993.
960 S. Morton and J.F. Nixon, J. Organomet. Chem., 1985, **282**, 123.
961 H. Suzuki, S. Matsuura, Y. Moro-Oka, and T. Ikawa, J. Organomet. Chem., 1985, **286**, 247.
962 C. Bianchini, C. Mealli, A. Meli, and M. Sabat, J. Chem. Soc., Chem. Commun., 1985, 1024.
963 K. Osakada, K. Matsumoto, T. Yamamoto, and A. Yamamoto, Organometallics, 1985, **4**, 857.

964 L. Dahlenburg and N. Höck, J. Organomet. Chem., 1985, **284**, 129.
965 C. Bianchini, C. Mealli, A. Meli, and M. Sabat, J. Am. Chem. Soc., 1985, **107**, 5317.
966 E.B. McAslan and T.A. Stephenson, Inorg. Chim. Acta, 1985, **96**, L49.
967 R. Benn, R.D. Reinhardt, and A. Rufinska, J. Organomet. Chem., 1985, **282**, 291.
968 W. Kleimann, K.R. Pörschke, and G. Wilke, Chem. Ber., 1985, **118**, 323.
969 R.H. Crabtree, B.E. Segmuller, and R.J. Uriarte, Inorg. Chem., 1985, **24**, 1949.
970 K.-R. Pörschke and G. Wilke, Chem. Ber., 1985, **118**, 313.
971 A.F.M.M. Rahman, C. Ceccarelli, J.P. Oliver, B. Messbauer, H. Meyer, and B. Walther, Inorg. Chem., 1985, **24**, 2355.
972 R.A. Michelin, G. Facchin, and R. Ros, J. Organomet. Chem., 1985, **279**, C25.
973 L. Manojlovic-Muir, I.R. Jobe, S.S.M. Ling, A.J. McLennan, and R.J. Puddephatt, J. Chem. Soc., Dalton Trans., 1985, 1725.
974 G. Minghetti, A. Albinati, A.L. Bandini, and G. Banditelli, Angew. Chem., Int. Ed. Engl., 1985, **24**, 120.
975 E.J. Parsons, R.D. Larsen, and P.W. Jennings, J. Am. Chem. Soc., 1985, **107**, 1793.
976 B. Kellenberger, S.J. Young, and J.K. Stille, J. Am. Chem. Soc., 1985, **107**, 6105.
977 F.S.M. Hassan, D.M. McEwan, P.G. Pringle, and B.L. Shaw, J. Chem. Soc., Dalton Trans., 1985, 1501.
978 A. Albinati, U. von Gunten, P.S. Pregosin, and H.J. Ruegg, J. Organomet. Chem., 1985, **295**, 239.
979 S.A. Snow and G. Kodama, Inorg. Chem., 1985, **24**, 795.
980 K.-R. Pörschke, K. Jonas, G. Wilke, R. Benn, R. Mynott, R. Goddard, and C. Krüger, Chem. Ber., 1985, **118**, 275.
981 R. Benn, B. Büssemeier, S. Holle, P.W. Jolly, R. Mynott, I. Tkatchenko, and G. Wilke, J. Organomet. Chem., 1985, **279**, 63.
982 T.A. Peganova, P.V. Petrovskii, L.S. Isaeva, D.N. Kravtsov, D.B. Furman, A.V. Kudryashev, A.O. Ivanov, S.V. Zotova, and O.V. Bragin, J. Organomet. Chem., 1985, **282**, 283.
983 K.-R. Pörschke, G. Wilke, and R. Mynott, Chem. Ber., 1985, **118**, 298.
984 K.A.O. Starzewski and J. Witte, Angew. Chem., Int. Ed. Engl., 1985, **24**, 599.
985 H. Hoberg, K. Sümmermann, and A. Milchereit, J. Organomet. Chem., 1985, **288**, 237.
986 E. Carmona, J.M. Marín, P. Palma, M. Paneque, and M.L. Poveda, Organometallics, 1985, **4**, 2053.
987 B. Fell, H.-U. Hög, and C. Krüger, J. Mol. Cat., 1985, **30**, 57.
988 J.M. Coromas, G. Muller, M. Rocamora, C. Miravitlles, and X. Solans, J. Chem. Soc., Dalton Trans., 1985, 2333.
989 C. Arlen, F. Maassarani, M. Pfeffer, and J. Fischer, Nouv. J. Chim., 1985, **9**, 249.
990 M.M. Brezinski, K.J. Klabunde, S.K. Janikowski, and L.J. Radonovich, Inorg. Chem., 1985, **24**, 3305.
991 H. Hoberg, F.J. Fañanás, K. Angermund, C. Krüger, and M.J. Romão, J. Organomet. Chem., 1985, **281**, 379.
992 A.B. Goel and S. Goel, Inorg. Chim. Acta, 1985, **98**, 67.
993 R. McCrindle and D.W. Sneddon, J. Organomet. Chem., 1985, **282**, 413.
994 H. Nishiyama, M. Matsumoto, T. Matsukura, R. Miura, and K. Itoh, Organometallics, 1985, **4**, 1911.
995 Y. Fuchita, K. Hiraki, and Y. Matsumoto, J. Organomet. Chem., 1985, **280**, C51.
996 G.R. Newkome, W.E. Puckett, G.E. Kiefer, V.K. Gupta, F.R. Fronczek, D.C. Pantaleo, G.L. McClure, J.B. Simpson, and W.A. Deutsch, Inorg. Chem., 1985, **24**, 811.
997 R. Benn, G. Gabor, P.W. Jolly, R. Mynott, and B. Raspel, J. Organomet. Chem., 1985, **296**, 443.
998 L.V. Rybin, E.A. Petrovskaya, M.I. Rubinskaya, L.G. Kuz'mina, Yu.T. Struchkov, V.V. Kaverin, and N.Yu. Koneva, J. Organomet. Chem., 1985, **288**, 119.

999 R. Usón, J. Forniés, P. Espinet, E. Lalinde, P.G. Jones, and G.M. Sheldrick, J. Organomet. Chem., 1985, **288**, 249.
1000 R. Usón, J. Fornies, P. Espinet, A. Garcia, M. Tomas, C. Foces-Foces, and F.H. Cano, J. Organomet. Chem., 1985, **282**, C35.
1001 E. Ambach and W. Beck, Chem. Ber., 1985, **118**, 2722.
1002 P.W. Clark and S.F. Dyke, J. Organomet. Chem., 1985, **281**, 389.
1003 H.-P. Abicht and K. Issleib, J. Organomet. Chem., 1985, **289**, 201.
1004 A. Campagnaro, A. Mantovani, and P. Uguagliati, Inorg. Chim. Acta, 1985, **99**, L15.
1005 B. Crociani, F. Dibianca, R. Bertani, and C.B. Castellani, Inorg. Chim. Acta, 1985, **101**, 161.
1006 R. Usón, J. Forniés, R. Navarro, P. Espinet, and C. Mendívil, J. Organomet. Chem., 1985, **290**, 125.
1007 C. Stader and B. Wrackmeyer, J. Organomet. Chem., 1985, **295**, C11.
1008 H. Ogawa, T. Joh, S. Takahashi, and K. Sonogashira, J. Chem. Soc., Chem. Commun., 1985, 1220.
1009 P.B. Hitchcock, M.F. Lappert, and M.C. Misra, J. Chem. Soc., Chem. Commun., 1985, 863.
1010 G.K. Campbell, P.B. Hitchcock, M.F. Lappert, and M.C. Misra, J. Organomet. Chem., 1985, **289**, C1.
1011 G. Vasapollo, C.F. Nobile, and A. Sacco, J. Organomet. Chem., 1985, **296**, 435.
1012 W.P. Fehlhammer, K. Bartel, U. Plaia, A. Völkl, and A.T. Liu, Chem. Ber., 1985, **118**, 2235.
1013 H.E. Bryndza, Organometallics, 1985, **4**, 1686.
1014 G. Ferguson, P.K. Monaghan, M. Parvez, and R.J. Puddephatt, Organometallics, 1985, **4**, 1669.
1015 S.S.M. Ling, N.C. Payne, and R.J. Puddephatt, Organometallics, 1985, **4**, 1546.
1016 P.K. Monaghan and R.J. Puddephatt, Organometallics, 1985, **4**, 1406.
1017 A.T. Hutton, P.G. Pringle, and B.L. Shaw, J. Chem. Soc., Dalton Trans., 1985, 1677.
1018 S.S.M. Ling, I.R. Jobe, L. Manojlović-Muir, K.W. Muir, and R.J. Puddephatt, Organometallics, 1985, **4**, 1198.
1019 J.E. Hux and R.J. Puddephatt, Inorg. Chim. Acta, 1985, **100**, 1.
1020 A.L. Balch, R.R. Guimerans, and J. Linehan, Inorg. Chem., 1985, **24**, 290.
1021 H.C. Clark, V.K. Jain, and G.S. Rao, J. Organomet. Chem., 1985, **279**, 181.
1022 G.M. Whitesides, M. Hackett, R.L. Brainard, J.-P.P.M. Lavalleye, A.F. Sowinski, A.N. Izumi, S.S. Moore, D.W. Brown, and E.M. Staudt, Organometallics, 1985, **4**, 1819.
1023 R.P. Brown, A. Yavari, R.H. Hill, and R.J. Puddephatt, J. Chem. Soc., Dalton Trans., 1985, 2421.
1024 G. Alibrandi, D. Minniti, R. Romeo, P. Uguagliati, L. Calligaro, U. Belluco, and B. Crociani, Inorg. Chim. Acta, 1985, **100**, 107.
1025 P.W. Jennings, R.E. Ekeland, M.D. Waddington, and T.W. Hanks, J. Organomet. Chem., 1985, **285**, 429.
1026 R.A. Ekeland and P.W. Jennings, J. Organomet. Chem., 1985, **281**, 397.
1027 W.R. Winchester, M. Gawron, G.J. Palenik, and W.M. Jones, Organometallics, 1985, **4**, 1894.
1028 S.I. Pombrik, A.A. Bezrukova, L.S. Golovchenko, A.S. Peregudov, A.Z. Rubezhov, and D.N. Kravtsov, J. Organomet. Chem., 1985, **293**, C1.
1029 W.-D. Müller and H.A. Brune, Chem. Ber., 1985, **118**, 4347.
1030 D.M. Roundhill, J. Am. Chem. Soc., 1985, **107**, 4354.
1031 R.M. Ceder and J. Sales, J. Organomet. Chem., 1985, **294**, 389.
1032 W.-D. Müller, G. Schmidtberg, and H.-A. Brune, Chem. Ber., 1985, **118**, 4653.
1033 G.B. Deacon and I.L. Grayson, J. Organomet. Chem., 1985, **292**, 1.
1034 G.K. Anderson and G.J. Lumetta, J. Organomet. Chem., 1985, **295**, 257.
1035 A.J. McLennan and R.J. Puddephatt, Organometallics, 1985, **4**, 485.
1036 R.J. Cross and M.F. Davidson, Inorg. Chim. Acta, 1985, **97**, L35.
1037 C.R. Langrick, P.G. Pringle, and B.L. Shaw, J. Chem. Soc., Dalton Trans., 1985, 1015.
1038 G.K. Anderson and G.J. Lumetta, Organometallics, 1985, **4**, 1542.

1039 M.A. Bennett and A. Rokicki, Organometallics, 1985, **4**, 180.
1040 M.A. Bennett and A. Rokicki, Aust. J. Chem., 1985, **38**, 1307.
1041 H.E. Bryndza, S.A. Kretchmar, and T.H. Tulip, J. Chem. Soc., Chem. Commun., 1985, 977.
1042 M. Kubota, F.S. Rosenberg, and M.J. Sailor, J. Am. Chem. Soc., 1985, **107**, 4558.
1043 G. Facchin, R. Campostrini, and R.A. Michelin, J. Organomet. Chem., 1985, **294**, C21.
1044 N.S. Akl and H.A. Tayim, J. Organomet. Chem., 1985, **297**, 371.
1045 A. Albinati, H. Moriyama, H. Rüegger, P.S. Pregosin, and A. Togni, Inorg. Chem., 1985, **24**, 4430.
1046 P. Haelg, G. Consiglio, and P. Pino, J. Organomet. Chem., 1985, **296**, 281.
1047 T.A.K. Al-Allaf, C. Eaborn, P.B. Hitchcock, M.F. Lappert, and A. Pidcock, J. Chem. Soc., Chem. Commun., 1985, 548.
1048 G. Consiglio, F. Morandini, M. Scalone, and P. Pino, J. Organomet. Chem., 1985, **279**, 193.
1049 R. Lazzaroni, G. Uccello-Barretta, S. Bertozzi, and P. Salvadori, J. Organomet. Chem., 1985, **297**, 117.
1050 J.W. Faller, C. Blankenship, B. Whitmore, and S. Sena, Inorg. Chem., 1985, **24**, 4483.
1051 P.R. Auburn, P.B. Mackenzie, and B. Bosnick, J. Am. Chem. Soc., 1985, **107**, 2033.
1052 T. Ohta, T. Hosokawa, S.-I. Murahashi, K. Miki, and N. Kasai, Organometallics, 1985, **4**, 2080.
1053 S. Imaizumi, T. Matsuhisa, and Y. Senda, J. Organomet. Chem., 1985, **280**, 441.
1054 W. Kläui, K. Schmidt, A. Bockmann, P. Hofmann, H.R. Schmidt, and P. Stauffert, J. Organomet. Chem., 1985, **286**, 407.
1055 K.R. Pörschke and R. Mynott, Z. Naturforsch., Teil B, 1984, **39**, 1565 (Chem. Abstr., 1985, **102**, 166 922).
1056 K.R. Pörschke, Y.-H. Tsay, and C. Krüger, Angew. Chem., Int. Ed. Engl., 1985, **24**, 323.
1057 M.A. Bennett, T.W. Hambley, N.K. Roberts, and G.B. Robertson, Organometallics, 1985, **4**, 1992.
1058 U. Kölle, B. Fuss, F. Khouzami, and J. Gersdorf, J. Organomet. Chem., 1985, **290**, 77.
1059 A. Miyashita, H. Shitara, and H. Nohira, J. Chem. Soc., Chem. Commun., 1985, 850.
1060 R. Appel, C. Cassar, and F. Knock, J. Organomet. Chem., 1985, **297**, 21.
1061 M. Lenarda, G. Nardin, G. Pellizer, E. Braye, and M. Graziani, J. Chem. Soc., Chem. Commun., 1985, 1536.
1062 A. Grassi, P. Longo, A. Musco, W. Porzio, and A. Scrivanti, J. Organomet. Chem., 1985, **289**, 439.
1063 A. Miyashita, H. Shitara, and H. Nohira, Organometallics, 1985, **4**, 1463.
1064 C.A. Akpan, M.F. Meidine, J.F. Nixon, M. Yoshifuji, K. Toyota, and N. Inamoto, J. Chem. Soc., Chem. Commun., 1985, 946.
1065 W.R. Winchester and W.M. Jones, Organometallics, 1985, **4**, 2228.
1066 R. Taube, J.-P. Gehrke, and U. Schmidt, J. Organomet. Chem., 1985, **292**, 287.
1067 R. Hanko, Angew. Chem., Int. Ed. Engl., 1985, **24**, 704.
1068 L.S. Hegedus and D.H.P. Thompson, J. Am. Chem. Soc., 1985, **107**, 5663.
1069 D. Wilhelm, J.-E. Bäckvall, R.E. Nordberg, and T. Norin, Organometallics, 1985, **4**, 1296.
1070 B. Akermark and A. Vitagliano, Organometallics, 1985, **4**, 1275.
1071 G.R. Wiger, S.S. Tomita, M.F. Rettig, and R.M. Wing, Organometallics, 1985, **4**, 1157.
1072 P.W. Jolly, Angew. Chem., Int. Ed. Engl., 1985, **24**, 283.
1073 W. Fischetti and R.F. Heck, J. Organomet. Chem., 1985, **293**, 391.
1074 J.-E. Bäckvall, R.E. Nordberg, and D. Wilhelm, J. Am. Chem. Soc., 1985, **107**, 6892.
1075 S. Agbossou, M.C. Bonnet, and I. Tkatchenko, Nouv. J. Chim., 1985, **9**, 311.
1076 M. Bressan, L.D. Mas, F. Morandini, and B. Prozzo, J. Organomet. Chem., 1985, **287**, 275.

1077 R.P. Hughes, J.M.J. Lambert, and A.L. Rheingold, Organometallics, 1985, **4**, 2055.
1078 P.H.M. Budzelaar, J. Boersma, G.J.M. van der Kerk, A.L. Spek, and A.J.M. Duisenberg, J. Organomet. Chem., 1985, **287**, C13.
1079 D. Lentz, Chem. Ber., 1985, **118**, 560.
1080 R. Bartsch, J.F. Nixon, and N. Sarjudeen, J. Organomet. Chem., 1985, **294**, 267.
1081 H. Wadepohl, H. Pritzkow, and W. Siebert, Chem. Ber., 1985, **118**, 729.
1082 P. Thometzek, K. Zenkert, and H. Werner, Angew. Chem., Int. Ed. Engl., 1985, **24**, 516.
1083 M. Onishi, K. Hiraki, Y. Ohama, and A. Kurosaki, J. Organomet. Chem., 1985, **284**, 403.
1084 E.O. Fischer and J.R. Schneider, J. Organomet. Chem., 1985, **295**, C29.
1085 O.J. Scherer, R. Walter, and W.S. Sheldrick, Angew. Chem., Int. Ed. Engl., 1985, **24**, 115.
1086 M. Baudler, F. Salzer, J. Hahn, and E. Daerr, Angew. Chem., Int. Ed. Engl., 1985, **24**, 415.
1087 R.D. Feltham, G. Elbaze, R. Ortega, C. Eck, and J. Dubrawski, Inorg. Chem., 1985, **24**, 1503.
1088 L. Manojlović-Muir, K.W. Muir, B.R. Lloyd, and R.J. Puddephatt, J. Chem. Soc., Chem. Commun., 1985, 536.
1089 M. Ebner, H. Otto, and H. Werner, Angew. Chem., Int. Ed. Engl., 1985, **24**, 518.
1090 R.C. Srivastava, J. Chem. Res. (S), 1985, 330.
1091 A. Albinati, K.-H. Dahmann, A. Togni, and L.M. Venanzi, Angew. Chem., Int. Ed. Engl., 1985, **24**, 766.
1092 C.S. Browning, D.H. Farrar, R.R. Gukathasan, and S.A. Morris, Organometallics, 1985, **4**, 1750.
1093 C.E. Briant, D.I. Gilmour, D.M.P. Mingos, and R.W.M. Wardle, J. Chem. Soc., Dalton Trans., 1985, 1693.
1094 Yu.S. Ryabokobylko, V.M. Dziomko, L.V. Shmelev, V.K. Vigran, G.M. Adamova, and R.V. Poponova, Koord. Khim., 1985, **11**, 661 (Chem. Abstr., 1985, **103**, 63 850).
1095 K.A. Goldsby, T.J. Meade, M. Kojima, and D.H. Busch, Inorg. Chem., 1985, **24**, 2588.
1096 A. Benedetti, A.C. Fabretti, and G.C. Franchini, Polyhedron, 1985, **4**, 2059.
1097 D.R. Alston, T.H. Lilley, and J.F. Stoddart, J. Chem. Soc., Chem. Commun., 1985, 1600.
1098 L.R. Milgrom, Polyhedron, 1985, **4**, 1661.
1099 K. Lystbaek and H. Bildsoe, Magn. Reson. Chem., 1985, **23**, 263.
1100 M.L. Pasquier and W. Marty, Angew. Chem., Int. Ed. Engl., 1985, **24**, 315.
1101 T. Bartik and T. Himmler, J. Organomet. Chem., 1985, **293**, 343.
1102 R.J. Blau, Report, 1985, IS-T-1165; Order No. DE85009224, 289 pp. Avail. NTIS. From Energy Res. Abstr., 1985, **10**, Abstr. No. 21297 (Chem. Abstr., 1985, **103**, 129 835).
1103 G.K. Anderson and R. Kumar, Inorg. Chim. Acta, 1985, **97**, L21.
1104 R. Benn, H.M. Buech, and R.D. Reinhardt, Magn. Reson. Chem., 1985, **23**, 559.
1105 C.E. Holloway and C.D. Hubbard, Inorg. Chim. Acta, 1985, **98**, 131.
1106 E. Matczak, B. Boduszek, and W. Wojciechowski, Bull. Pol. Acad. Sci., Chem., 1984, **32**, 277 (Chem. Abstr., 1985, **102**, 105 084).
1107 H. Ito and T. Ito, Chem. Lett., 1985, 1251 (Chem. Abstr., 1985, **103**, 152 615).
1108 J.P. Costes, Transition Met. Chem. (Weinheim, Ger.), 1985, **10**, 185.
1109 J.-P. Costes, G. Cros, and J.-P. Laurent, Inorg. Chim. Acta, 1985, **97**, 211.
1110 A.D. Shebaldova, V.I. Bystrenina, G.V. Bespalova, and V.I. Shebaldova, Koord. Khim., 1984, **10**, 1559 (Chem. Abstr., 1985, **102**, 38 619).
1111 E. Ambach and W. Beck, Chem. Ber., 1985, **118**, 2722.
1112 J.P. Laussac, R. Haran, and N. Hadjiliadis, C.R. Acad. Sci., Ser. 2, 1985, **300**, 137 (Chem. Abstr., 1985, **103**, 105 288).
1113 S. Yamazaki, Polyhedron, 1985, **4**, 1915.
1114 E.C. Constable, J. Chem. Soc., Dalton Trans., 1985, 1719.
1115 W. Beck, E. Ambach, and U. Nagel, Chem. Ber., 1985, **118**, 444.

1116 M. Habib, H. Trujillo, C.A. Alexander, and B.N. Storhoff, Inorg. Chem., 1985, **24**, 2344.
1117 B. Radomska, T. Tatarowski, J.P. Morawiec, and H. Kozlowski, Inorg. Chim. Acta, 1985, **106**, L29.
1118 K. Hiraki, Y. Fuchita, and M. Nakashima, Inorg. Chim. Acta, 1985, **97**, L15.
1119 V. Saudek, H. Pivcova, D. Noskova, and J. Drobnik, J. Inorg. Biochem., 1985, **24**, 13.
1120 H. Pivcova, V. Saudek, D. Noskova, and J. Drobnik, J. Inorg. Biochem., 1985, **23**, 43.
1121 V. Saudek, H. Pivcova, D. Noskova, and J. Drobnik, J. Inorg. Biochem., 1985, **23**, 55.
1122 S.G. Ward and R.C. Taylor, NATO ASI Ser., Ser. C, 1984, **124**, 767 (Chem. Abstr., 1985, **102**, 217).
1123 J.H.J. Den Hartog, C. Altona, J.H. Van Boom, and J. Reedijk, FEBS Lett., 1984, **176**, 393 (Chem. Abstr., 1985, **102**, 72 479).
1124 M.D. Reily and L.G. Marzilli, J. Am. Chem. Soc., 1985, **107**, 4916.
1125 L.G. Marzilli, M.D. Reily, B.L. Heyl, C.T. McMurray, and W.D. Wilson, FEBS Lett., 1984, **176**, 389 (Chem. Abstr., 1985, **102**, 17 264).
1126 M.D. Reily, K. Wilkowski, K. Shinozuka, and L.G. Marzilli, Inorg. Chem., 1985, **24**, 37.
1127 H. Bissinger and W. Beck, Z. Naturforsch., Teil B, 1985, **40**, 507 (Chem. Abstr., 1985, **103**, 47 192).
1128 T.G. Appleton, R.D. Berry, and J.R. Hall, Inorg. Chem., 1985, **24**, 666.
1129 T.G. Appleton, J.R. Hall,and S.F. Ralph, Inorg. Chem., 1985, **24**, 673.
1130 J.R. Bales, M.A. Mazid, P.J. Sadler, A. Aggarwal, R. Kuroda, S. Neidle, D.W. Gilmour, B.J. Peart, and C.A. Ramsden, J. Chem. Soc., Dalton Trans., 1985, 795.
1131 L.S. Hollis, A.R. Amundsen, and E.W. Stern, J. Am. Chem. Soc., 1985, **107**, 274.
1132 R. Jones, P.F. Kelly, D.J. Williams, and J.D. Woolins, Polyhedron, 1985, **4**, 1947.
1133 F.E. Wood, M.M. Olmstead, J.P. Farr, and A.L. Balch, Inorg. Chim. Acta, 1985, **97**, 77.
1134 G. Holze and H.H. Inhoffen, Angew. Chem., Int. Ed. Engl., 1985, **24**, 867.
1135 P. Courtot, R. Pichon, R. Rumin, and J.Y. Salaün, Inorg. Chim. Acta, 1985, **103**, L7.
1136 C.G. Anklin and P.S. Pregosin, Magn. Reson. Chem., 1985, **23**, 671.
1137 G.E. Kirvan and D.W. Margerum, Inorg. Chem., 1985, **24**, 3017.
1138 D. Hedden, D.M. Roundhill, and M.D. Walkinshaw, Inorg. Chem., 1985, **24**, 3146.
1139 N.P.C. Walker, M.B. Hursthouse, C.P. Warrens, and J.D. Woolins, J. Chem. Soc., Chem. Commun., 1985, 227.
1140 D. Hedden and D.M. Roundhill, Inorg. Chem., 1985, **24**, 4152.
1141 O.J. Scherer, R. Walter, and W.S. Sheldrick, Angew. Chem., Int. Ed. Engl., 1985, **24**, 525.
1142 L.R. Milgrom, Polyhedron, 1985, **4**, 1279.
1143 V. Theodorou and N. Hadjiliadis, Polyhedron, 1985, **4**, 1283.
1144 S. Cenini, F. Porta, and M. Pizzotti, J. Organomet. Chem., 1985, **296**, 291.
1145 S.K. Miller and L.G. Marzilli, Inorg. Chem., 1985, **24**, 2421.
1146 R.T. Boeré, C.D. Montgomery, N.C. Payne, and C.J. Willis, Inorg. Chem., 1985, **24**, 3680.
1147 M. Ciampolini, N. Nardi, P.L. Orioli, S. Mangani, and F. Zanobini, J. Chem. Soc., Dalton Trans., 1985, 1425.
1148 M. Ciampolini, N. Nardi, P.L. Orioli, S. Mangani, and F. Zanobini, J. Chem. Soc., Dalton Trans., 1985, 1179.
1149 C.A. Ghilardi, P. Innocenti, S. Midollini, and A. Orlandini, J. Chem. Soc., Dalton Trans., 1985, 2209.
1150 F. Mercier, F. Mathey, J. Fischer, and J.H. Nelson, Inorg. Chem., 1985, **24**, 4141.
1151 R.A. Jones, M.H. Seeberger, and B.R. Whittlesey, J. Am. Chem. Soc., 1985, **107**, 6424.
1152 H. Schäfer, D. Binder, and D. Fenske, Angew. Chem., Int. Ed. Engl., 1985,

24, 522.
1153 C.-L. Lee, G. Besenyei, B.R. James, D.A. Nelson, and M.A. Lilga, J. Chem. Soc., Chem. Commun., 1985, 1175.
1154 M.L. Kullberg, F.R. Lemke, D.R. Powell, and C.P. Kubiak, Inorg. Chem., 1985, **24**, 3589.
1155 M.S. Holt, J.H. Nelson, P. Savignac, and N.W. Alcock, J. Am. Chem. Soc., 1985, **107**, 6396.
1156 E.P. Kyba, C.N. Clubb, S.B. Larson, V.J. Schueler, and R.E. Davis, J. Am. Chem. Soc., 1985, **107**, 2141.
1157 H.C. Clark, G. Ferguson, V.K. Jain, and M. Parvez, Inorg. Chem., 1985, **24**, 1477.
1158 O.J. Scherer, K. Hussong, and G. Wolmershäuser, J. Organomet. Chem., 1985, **289**, 215.
1159 M.F. Hallam, N.D. Nowells, M.P. Mingos, and R.W.M. Wardle, J. Chem. Soc., Dalton Trans., 1985, 845.
1160 A. Shaver, R.D. Lai, P.H. Bird, and W. Wickramasinghe, Can. J. Chem., 1985, **63**, 2555.
1161 D.E. Berry, K.A. Beveridge, G.W. Bushnell, and K.R. Dixon, Can. J. Chem., 1985, **63**, 2949.
1162 R.S. Paonessa, A.L. Prignano, and W.C. Troger, Organometallics, 1985, **4**, 647.
1163 J.A. Davies and A. Sood, Inorg. Chem., 1985, **24**, 4213.
1164 R. Hensel and W.-W. du Mont, Chem. Ber., 1985, **118**, 5016.
1165 D.E. Berry, G.W. Bushnell, K.R. Dixon, P.M. Moroney, and C. Wan, Inorg. Chem., 1985, **24**, 2625.
1166 D.H. Goldsworthy, F.R. Hartley, G.L. Marshall, and S.G. Murray, Inorg. Chem., 1985, **24**, 2849.
1167 K.A. Azam, G. Ferguson, S.S.M. Ling, M. Parvez, R.J. Puddephatt, and D. Srokowski, Inorg. Chem., 1985, **24**, 2799.
1168 C.W.G. Ansell, M.K. Cooper, K.P. Dancey, P.A. Duckworth, K. Henrick, M. McPartlin, and P.A. Tasker, J. Chem. Soc., Chem. Commun., 1985, 439.
1169 C.W.G. Ansell, M.K. Cooper, K.P. Dancey, P.A. Duckworth, K. Henrick, M. McPartlin, G. Organ, and P.A. Tasker, J. Chem. Soc., Chem. Commun., 1985, 437.
1170 S.S.M. Ling, I.R. Jobe, A.J. McLennan, L. Manojlović-Muir, K.W. Muir, and R.J. Puddephatt, J. Chem. Soc., Chem. Commun., 1985, 566.
1171 E. Ma, G. Semelhago, A. Walker, D.H. Farrar, and R.R. Gukathasan, J. Chem. Soc., Dalton Trans., 1985, 2595.
1172 H.-P. Abicht, J.T. Spencer, and J.G. Verkade, Inorg. Chem., 1985, **24**, 2132.
1173 C. Bianchini, C.A. Ghilardi, A. Meli, and A. Orlandini, J. Organomet. Chem., 1985, **286**, 259.
1174 D.J. Gulliver, E.G. Hope, W. Levason, S.G. Murray, and G.L. Marshall, J. Chem. Soc., Dalton Trans., 1985, 1265.
1175 E.G. Hope, W. Levason, S.G. Murray, and G.L. Marshall, J. Chem. Soc., Dalton Trans., 1985, 2185.
1176 C.-M. Che, L.G. Butler, P.J. Grunthaner, and H.B. Gray, Inorg. Chem., 1985, **24**, 4662.
1177 R. El-Mehdawi, S.A. Bryan, and D.M. Roundhill, J. Am. Chem. Soc., 1985, **107**, 6285.
1178 T.G. Appleton, J.R. Hall, and D.W. Neale, Inorg. Chim. Acta, 1985, **104**, 19.
1179 S. Kitagawa, M. Munakata, and M. Yonezawa, JEOL News, [Ser.] Anal. Instrum., 1983, **19A**, 17 (Chem. Abstr., 1985, **102**, 88 916).
1180 K. Endo, K. Yamamoto, K. Matsushita, K. Deguchi, K. Kanda, and H. Nakatsuji, J. Magn. Reson., 1985, **65**, 268.
1181 G. La Monica, G.A. Ardizzoia, F. Cariati, S. Cenini, and M. Pizzotti, Inorg. Chem., 1985, **24**, 3920.
1182 T.H. Lemmon, K. Folting, J.C. Huffman, and K.G. Caulton, J. Am. Chem. Soc., 1985, **107**, 7774.
1183 M. Lusser and P. Peringer, Polyhedron, 1985, **4**, 1997.
1184 S. Pasynkiewicz and J. Poplawska, J. Organomet. Chem., 1985, **282**, 427.
1185 S. Pasynkiewicz, S. Pikul, and J. Popławska, J. Organomet. Chem., 1985, **293**, 125.

1186 S. Komiya and A. Shibue, Organometallics, 1985, **4**, 684.
1187 H.W. Chen, C. Paparizos, and J.P. Fackler, jun., Inorg. Chim. Acta, 1985, **96**, 137.
1188 J.D. Basil, H.H. Murray, J.P. Fackler, jun., J. Tocher, A.M. Mazany, B. Trzcinska-Bancroft, H. Knackel, D. Dudis, T.J. Delord, and D.O. Marler, J. Am. Chem. Soc., 1985, **107**, 6908.
1189 E.G. Perevalova, I.G. Bolesov, Yu.T. Struchkov, I.F. Leschova, Ye.S. Kalyuzhnaya, T.I. Voyevodskaya, Yu.L. Slovokhotov, and K.I. Grandberg, J. Organomet. Chem., 1985, **286**, 129.
1190 J. Vicente, M.T. Chicote, J.A. Cayuelas, J. Fernandez-Baeza, P.G. Jones, G.M. Sheldrick, and P. Espinet, J. Chem. Soc., Dalton Trans., 1985, 1163.
1191 R. Usón, A. Laguna, M. Laguna, B.R. Manzano, P.G. Jones, and G.M. Sheldrick, J. Chem. Soc., Dalton Trans., 1985, 2417.
1192 O.M. Abu-Salah, A.-R.A. Al-Ohaly, S.S. Al-Showiman, and I.M. Al-Najjar, Transition Met. Chem., 1985, **10**, 207.
1193 E.O. Fischer and M. Böck, J. Organomet. Chem., 1985, **287**, 279.
1194 M.I. Bruce, P.A. Humphrey, J.M. Patrick, B.W. Skelton, A.H. White, and M.L. Williams, Aust. J. Chem., 1985, **38**, 1441.
1195 G.C. Van Stein, G. Van Koten, and K. Vrieze, Inorg. Chem., 1985, **24**, 1367.
1196 R.J. Batchelor, T. Birchall, and R. Faggiani, Can. J. Chem., 1985, **63**, 928.
1197 S.J.B. Price, C. Brevard, A. Pagelot, and P.J. Sadler, Inorg. Chem., 1985, **24**, 4278.
1198 F. Estevan, P. Lahuerta, and L. Solana, Polyhedron, 1985, **4**, 1965.
1199 N.J. Clayden, C.M. Dobson, K.P. Hall, D.M.P. Mingos, and D.J. Smith, J. Chem. Soc., Dalton Trans., 1985, 1811.
1200 I.C.P. Smith, A. Joyce, H. Jarrell, B.M. Sutton, and D.T. Hill, Bioinorg. Chem. Gold Coord. Compd., Proc. Symp., 1981 (Pub. 1983), 47 (Chem. Abstr., 1985, **102**, 32 023).
1201 G.C. Van Stein, G. Van Koten, F. Blank, L.C. Taylor, K. Vrieze, A.L. Spek, A.J.M. Duisenberg, A.M.M. Schreurs, B. Kojic-Prodic, and C. Brevard, Inorg. Chim. Acta, 1985, **98**, 107.
1202 Y. Rosopulos, U. Nagel, and W. Beck, Chem. Ber., 1985, **118**, 931.
1203 A. Shamim, A. Adeyamo, A. Turner, A. Carty, and A.B. Hussain, J. Chem. Soc. Pak., 1984, **6**, 173 (Chem. Abstr., 1985, **103**, 31 522).
1204 C. Bianchini, C.A. Ghilardi, A. Meli, S. Midollini, and A. Orlandini, Inorg. Chem., 1985, **24**, 932.
1205 E.C. Alyea, G. Ferguson, J. Malito, and B. Ruhl, Inorg. Chem., 1985, **24**, 3719.
1206 C. Bianchini, C.A. Ghilardi, A. Meli, S. Midollini, and A. Orlandini, Inorg. Chem., 1985, **24**, 924.
1207 M. Ahlgrén, T. Pakkanen, and I. Tahvanainen, Acta Chem. Scand., 1985, **A39**, 651.
1208 W. Bos, J.J. Bour, J.J. Steggerda, and L.H. Pignolet, Inorg. Chem., 1985, **24**, 4298.
1209 H. Schmidbaur, R. Herr, G. Müller, and J. Riede, Organometallics, 1985, **4**, 1208.
1210 J.R. Nicholson, I.L. Abrahams, W. Clegg, and C.D. Garner, Inorg. Chem., 1985, **24**, 1092.
1211 W. Wojnowski, M. Wojnowski, K. Peters, E.-M. Peters, and H.G. von Schnering, Z. Anorg. Allg. Chem., 1985, **530**, 79.
1212 M.S. Hussain and A.A. Isab, J. Coord. Chem., 1985, **14**, 17 (Chem. Abstr., 1985, **103**, 170 972).
1213 A.A. Isab and M.S. Hussain, Polyhedron, 1985, **4**, 1683.
1214 S.H. Strauss, M.D. Noirot, and O.P. Anderson, Inorg. Chem., 1985, **24**, 4307.
1215 P. Huppmann, H. Hartl, and K. Seppelt, Z. Anorg. Allg. Chem., 1985, **524**, 26.
1216 A. Sebald and B. Wrackmeyer, J. Magn. Reson., 1985, **63**, 397.
1217 R. Mynott, B. Gabor, H. Lehmkuhl, and I. Doering, Angew. Chem., Int. Ed. Engl., 1985, **24**, 335.
1218 T.C. Wong, J. Magn. Reson., 1985, **63**, 179.
1219 K.L. Servis and R.L. Domenick, J. Am. Chem. Soc., 1985, **107**, 7186.
1220 Yu.K. Grishin, O.K. Sokolikova, Yu.A. Ustynyuk, A.A. Ivkina, KP. Butin, and O.A. Reutov, Izv. Akad. Nauk SSSR, Ser. Khim., 1985, 793 (Chem. Abstr.,

1985, **103**, 123 626).
1221 L.S. Golovchenko, S.I. Pombrik, A.S. Peregudov, and D.N. Kravtsov, Izv. Akad. Nauk SSSR, Ser. Khim., 1985, 668 (Chem. Abstr., 1985, **103**, 88 014).
1222 S.I. Pombrik, L.S. Golovchenko, E.V. Polunkin, A.S. Peregudov, and D.N. Kravtsov, J. Organomet. Chem., 1985, **292**, 81.
1223 A.S. Peregudov, V.F. Ivanov, E.I. Fedin, and D.N. Kravtsov, Izv. Akad. Nauk SSSR, Ser. Khim., 1985, 1524 (Chem. Abstr., 1985, **103**, 215 441).
1224 A.P. Arnold, A.J. Canty, R.S. Reid, and D.L. Rabenstein, Can. J. Chem., 1985, **63**, 2430.
1225 H. Lange and D. Naumann, J. Fluorine Chem., 1984, **26**, 435 (Chem. Abstr., 1985, **102**, 95 749).
1226 A. Pages, J.S. Casas, A. Sanchez, J. Sordo, J. Bravo, and M. Gayoso, J. Inorg. Biochem., 1985, **25**, 35.
1227 H. Schmidbaur, E. Weiss, and W. Graf, Organometallics, 1985, **4**, 1233.
1228 E. Langguth and K.-H. Thiele, Z. Anorg. Allg. Chem., 1985, **530**, 69.
1229 D.J. Burton and D.M. Wiemers, J. Am. Chem. Soc., 1985, **107**, 5014.
1230 M.A. Guerra, T.R. Bierschenk, and R.J. Lagow, J. Chem. Soc., Chem. Commun., 1985, 1550.
1231 G. Maas, R. Brückmann, and W. Lorenz, J. Organomet. Chem., 1985, **289**, 9.
1232 K.E. Rowland and R.D. Thomas, Magn. Res. Chem., 1985, **23**, 916.
1233 A. Passini and E. Bersanetti, Inorg. Chim. Acta, 1985, **107**, 259.
1234 E. Buncel, C. Boone, H. Joly, R. Kumar, and A.R. Norris, J. Inorg. Biochem., 1985, **25**, 61.
1235 F. Allaire, M. Simard, and A.L. Beauchamp, Can. J. Chem., 1985, **63**, 2054.
1236 A. Haas and T. Maciej, Z. Anorg. Allg. Chem., 1985, **524**, 33.
1237 S.I. Pombrik, E.V. Polunkin, A.S. Peregudov, E.I. Fedin, and D.N. Kravtsov, Izv. Akad. Nauk SSSR, Ser. Khim., 1985, 900 (Chem. Abstr., 1985, **103**, 141 325).
1238 J.D. Wuest and B. Zacharie, J. Am. Chem. Soc., 1985, **107**, 6121.
1239 G.B. Deacon and G.N. Stretton, Aust. J. Chem., 1985, **38**, 419.
1240 J.D. Wuest and B. Zacharie, Organometallics, 1985, **4**, 410.
1241 A.J. Bloodworth and C.J. Cooksey, J. Organomet. Chem., 1985, **295**, 131.
1242 V.V. Bashilov, E.V. Maskaeva, P.V. Petrovskii, and V.I. Sokolov, J. Organomet. Chem., 1985, **292**, 89.
1243 P.F. Rodesiler, E.A.H. Griffith, N.G. Charles, L. Lebioda, and E.L. Amma, Inorg. Chem., 1985, **24**, 4595.
1244 I.M. Armitage, D.C. Dalgarno, and F.E. Davis, J. Magn. Reson., 1985, **61**, 145.
1245 D. Live, I.M. Armitage, D.C. Dalgarno, and D. Cowburn, J. Am. Chem. Soc., 1985, **107**, 1775.
1246 M.H. Frey, G. Wagner, M. Vasák, O.W. Sørensen, D. Neuhaus, E. Wörgötter, J.H.R. Kägi, R.R. Ernst, and K. Würthrich, J. Am. Chem. Soc., 1985, **107**, 6847.
1247 J.D. Otvos, H.R. Engeseth, and S. Wehrli, J. Magn. Reson., 1985, **61**, 579.
1248 D.H. Live, C.L. Kojiro, D. Cowburn, and J.L. Markley, J. Am. Chem. Soc., 1985, **107**, 3043.
1249 M. Vasak, G.E. Hawkes, J.K. Nicholson, and P.J. Sadler, Biochemistry, 1985, **24**, 740 (Chem. Abstr., 1985, **102**, 91 588).
1250 P. Peringer and N. Baumann, J. Chem. Soc., Dalton Trans., 1985, 223.
1251 D.W. Allen, N.A. Bell, S.T. Fong, L.A. March, and I.W. Nowell, Inorg. Chim. Acta, 1985, **99**, 157.
1252 M. Lusser and P. Peringer, Chem. Ber., 1985, **118**, 2140.
1253 P.A.W. Dean and R.S. Srivastava, Can. J. Chem., 1985, **63**, 2829.
1254 T. Yamaguchi, K. Yamamoto, and H. Ohtaki, Bull. Chem. Soc. Japan, 1985, **58**, 3235.
1255 J.J. Dechter, O.R. Polanco, and J. Kowalewski, J. Magn. Reson., 1985, **65**, 395.
1256 I.G. Dance and J.K. Saunders, Inorg. Chim. Acta, 1985, **96**, L71.
1257 P.A.W. Dean and J.J. Vittal, Inorg. Chem., 1985, **24**, 3722.
1258 M. Kato and T. Ito, Inorg. Chem., 1985, **24**, 504.
1259 N.W. Alcock, E.H. Curzon, P. Moore, H.A.A. Omar, and C. Pierpoint, J. Chem. Soc., Dalton Trans., 1985, 1361.

1260 F. Paap, W.L. Driessen, J. Reedijk, B. Kojić-Prodić, and A.L. Spek, Inorg. Chim. Acta, 1985, **104**, 55.
1261 N.W. Alcock, H.A.A. Omar, P. Moore, and C. Pierpoint, J. Chem. Soc., Dalton Trans., 1985, 219.
1262 Y. Kojima, K. Hirotsu, T. Yamashita, and T. Miwa, Bull. Chem. Soc. Jpn., 1985, **58**, 1894.
1263 N.W. Alcock, K.P. Balakrishnan, and P. Moore, J. Chem. Soc., Chem. Commun., 1985, 1731.
1264 S.K. Miller, D.G. Van Derveer, and L.G. Marzilli, J. Am. Chem. Soc., 1985, **107**, 1048.
1265 Y. Kojima, T. Yamashita, S. Nishide, K. Hirotsu, and T. Higuchi, Bull. Chem. Soc. Jpn., 1985, **58**, 409.
1266 E. Buncel, R. Kumar, A.R. Norris, and A.L. Beauchamp, Can. J. Chem., 1985, **63**, 2575.
1267 S.M. Wang and R.K. Gilpin, Talanta, 1985, **32**, 329 (Chem. Abstr., 1985, **103**, 67 708).
1268 M.E. Bjornson, D.C. Corson, and B.D. Sykes, J. Inorg. Biochem., 1985, **25**, 141.
1269 D.C. Dalgarno, R.W. Olafson, and I.M. Armitage, Inorg. Chem., 1985, **24**, 3439.
1270 S.G. Spencer, J.M. Brewer, and P.D. Ellis, J. Inorg. Biochem., 1985, **24**, 47.
1271 E.C. Alyea, S.A. Dias, G. Ferguson, and P.Y. Siew, Croat. Chem. Acta, 1984, **57**, 457 (Chem. Abstr., 1985, **102**, 124 521).
1272 P. Peringer, P.-P. Winkler, G. Huttner, and L. Zsolnai, J. Chem. Soc., Dalton Trans., 1985, 1061.
1273 J.A. McCleverty, Stud. Inorg. Chem., 1984, **5**, 311 (Chem. Abstr., 1985, **102**, 26 076).
1274 Z. Wittmann and E. Pudmer, ASLE Trans., 1985, **28**, 426 (Chem. Abstr., 1985, **103**, 198 209).
1275 R.C. Burns, L.A. Devereux, P. Granger, and G.J. Schrobilgen, Inorg. Chem., 1985, **24**, 2615.
1276 S.O. Grim, P.H. Smith, S. Nittolo, H.L. Ammon, L.C. Satek, S.A. Sangokoya, R.K. Khanna, I.J. Colquhoun, W. McFarlane, and J.R. Holdon, Inorg. Chem., 1985, **24**, 2889.
1277 W. Wojnowski, M. Wojnowski, H.G. von Schnering, and M. Noltemeyer, Z. Anorg. Allg. Chem., 1985, **531**, 153.
1278 P. Moore, Pure Appl. Chem., 1985, **57**, 347.
1279 J.B. Grutzner, Prepr. - Am. Chem. Soc., Div. Pet. Chem., 1985, **30**, 604 (Chem. Abstr., 1985, **103**, 160 551).
1280 G. Fraenkel, H. Hsu, and B.M. Su, Lithium: Curr. Appl. Sci., Med. Technol., 1985, 273 (Chem. Abstr., 1985, **103**, 105 020).
1281 K.C. Dash, Indian J. Chem., Sect. A, 1985, **24A**, 265 (Chem. Abstr., 1985, **103**, 142 044).
1282 C. Detellier and P. Laszlo, Proc. - Indian Acad. Sci., Chem. Sci., 1985, **94**, 291 (Chem. Abstr., 1985, **103**, 129 189).
1283 R. Willem, M. Gielen, H. Pepermans, J. Brocas, D. Fastenakel, and P. Finocchiaro, J. Am. Chem. Soc., 1985, **107**, 1146).
1284 F.A. Boltino and P. Finocchiaro, Polyhedron, 1985, **4**, 1507.
1285 C.J. Screttas and M. Micha-Screttas, J. Organomet. Chem., 1985, **292**, 325.
1286 R. Benn, H. Lehmkuhl, K. Mehler, and A. Rufińska, J. Organomet. Chem., 1985, **293**, 1.
1287 P. Bougeard, M. Mancini, B.G. Sayer, and M.J. McGlinchey, Inorg. Chem., 1985, **24**, 93.
1288 R.S. Sternal, C.P. Brock, and T.J. Marks, J. Am. Chem. Soc., 1985, **107**, 8270.
1289 V.W. Day, C.W. Earley, W.G. Klemperer, and D.J. Maltbie, J. Am. Chem. Soc., 1985, **107**, 8261.
1290 A. Dormond, A. El Bouadili, A. Aaliti, and C. Moise, J. Organomet. Chem., 1985, **288**, C1.
1291 S.F. Lincoln and A. White, Inorg. Chim. Acta, 1985, **110**, 107.
1292 M.D. Fryzuk, S.J. Rettig, A. Westerhaus, and H.D. Williams, Inorg. Chem., 1985, **24**, 4316.

1293 G. Erker, K. Berg, R. Benn, and G. Schroth, Chem. Ber., 1985, **118**, 1383.
1294 G. Erker, T. Muehlenbernd, R. Benn, A. Rufinska, Y.H. Tsay, and C. Krueger, Angew. Chem., Int. Ed. Engl., 1985, **24**, 321.
1295 G. Erker, K. Engel, C. Krueger, Y.H. Tsay, E. Samuel, and P. Vogel, Z. Naturforsch., Teil B, 1985, **40**, 150 (Chem. Abstr., 1985, **103**, 142 113).
1296 S. Lateskey, J. Keddington, A.E. McMullen, I.P. Rothwell, and J.C. Huffman, Inorg. Chem., 1985, **24**, 995.
1297 B. Viard, J. Amaudrut, J. Sala-Pala, A. Fakhr, Y. Mugnier, and C. Moise, J. Organomet. Chem., 1985, **292**, 403.
1298 K. Tatsumi, J. Takeda, Y. Sekiguchi, M. Kohsaka, and A. Nakamura, Angew. Chem., Int. Ed. Engl., 1985, **24**, 332.
1299 M.L. Luetkens, jun., D.J. Santure, J.C. Huffman, and A.P. Sattelberger, J. Chem. Soc., Chem. Commun., 1985, 552.
1300 F.A. Van-Catledge, S.D. Ittel, and J.P. Jesson, Organometallics, 1985, **4**, 18.
1301 B. Longato, B.D. Martin, J.R. Norton, and O.P. Anderson, Inorg. Chem., 1985, **24**, 1389.
1302 J.A. Bandy, A. Berry, M.L.H. Green, and K. Prout, J. Chem. Soc., Chem. Commun., 1985, 1462.
1303 G. Michael, J. Kaub, and C.G. Kreiter, Chem. Ber., 1985, **118**, 3944.
1304 G. Michael, J. Kaub, and C.G. Kreiter, Angew. Chem., Int. Ed. Engl., 1985, **24**, 502.
1305 J.S. Drago and K.P.C. Vollhardt, Organometallics, 1985, **4**, 191.
1306 S.R. Allen, R.G. Beevor, M. Green, N.C. Norman, A.G. Orpen, and I.D. Williams, J. Chem. Soc., Dalton Trans., 1985, 435.
1307 M.H. Chisholm, J.C. Huffman, and J.A. Heppert, J. Am. Chem. Soc., 1985, **107**, 5116.
1308 M.H. Chisholm, K. Folting, J.A. Heppert, D.M. Hoffman, and J.C. Huffman, J. Am. Chem. Soc., 1985, **107**, 1234.
1309 F.W.B. Einstein, R.K. Pomeroy, P. Rushman, and A.C. Willis, Organometallics, 1985, **4**, 250.
1310 P. Bougeard, S. Peng, M. Mlekuz, and M.J. McGlinchey, J. Organomet. Chem., 1985, **296**, 383.
1311 G.A. Carriedo, J.A.K. Howard, D.B. Lewis, G.E. Lewis, and F.G.A. Stone, J. Chem. Soc., Dalton Trans., 1985, 905.
1312 J.L. Davidson and G. Vasapollo, J. Chem. Soc., Dalton Trans., 1985, 2239.
1313 J.R. Morrow, T.L. Tonker, and J.L. Templeton, Organometallics, 1985, **4**, 745.
1314 R.P. Hughes, J.W. Reisch, and A.L. Rheingold, Organometallics, 1985, **4**, 1754.
1315 M. Brookhart, K. Cox, F.G.N. Cloke, J.C. Green, M.L.H. Green, P.M. Hare, J. Bashkin, A.E. Derome, and P.D. Grebenik, J. Chem. Soc., Dalton Trans., 1985, 423.
1316 K. Niedenzu and S. Trofimenko, Inorg. Chem., 1985, **24**, 4222.
1317 R.R. Schrock, J.H. Freudenberger, M.L. Listemann, and L.G. McCullough, J. Mol. Catal., 1985, **28**, 1.
1318 C.G. Kreiter and M. Kotzian, J. Organomet. Chem., 1985, **289**, 295.
1319 C.G. Kreiter, M. Kotzian, U. Schubert, R. Bau, and M.A. Bruck, Z. Naturforsch., Teil B, 1984, **39**, 1553 (Chem. Abstr., 1985, **102**, 185 217).
1320 H. Strutz, J.C. Dewan, and R.R. Schrock, J. Am. Chem. Soc., 1985, **107**, 5999.
1321 W. de Oliveira, J.-L. Migot, M.B.G. de Lima, J. Sala-Pala, J.-E. Guerchais, and J.-Y. Le Gall, J. Organomet. Chem., 1985, **284**, 313.
1322 M. Mlekuz, P. Bourgeard, B.G. Sayer, S. Peng, M.J. McGlinchey, A. Marinetti, J.Y. Saillard, J. Ben Naceur, B. Mentzen, and G. Jaouen, Organometallics, 1985, **4**, 1123.
1323 S.D. Reynolds and T.A. Albright, Organometallics, 1985, **4**, 980.
1324 A.A. Ismail, F. Sauriol, J. Sedman, and I.S. Butler, Organometallics, 1985, **4**, 1914.
1325 E.W. Abel, P.K. Mittal, K.G. Orrell, and V. Šik, J. Chem. Soc., Dalton Trans., 1985, 1569.
1326 E.W. Abel, S.K. Bhargava, P.K. Mittal, K.G. Orrell, and V. Šik, J. Chem. Soc., Dalton Trans., 1985, 1561.
1327 J.A. Iggo and B.L. Shaw, J. Chem. Soc., Dalton Trans., 1985, 1009.

1328 E.W. Abel, K.M. Higgins, K.G. Orrell, V. Šik, E.H. Curzon, and O.W. Howarth, J. Chem. Soc., Dalton Trans., 1985, 2195.
1329 W.H. Hersh, J. Am. Chem. Soc., 1985, **107**, 4599.
1330 J.L. Davidson and G. Vasapollo, J. Chem. Soc., Dalton Trans., 1985, 2231.
1331 E.O. Fischer, A.C. Filippou, and H.G. Alt, J. Organomet. Chem., 1985, **296**, 69.
1332 W.G. Klemperer, C. Schwartz, and D.A. Wright, J. Am. Chem. Soc., 1985, **107**, 6941.
1333 T.W. Coffindaffer, W.M. Westler, and I.P. Rothwell, Inorg. Chem., 1985, **24**, 4565.
1334 T.W. Coffindaffer, I.P. Rothwell, and J.C. Huffmann, Inorg. Chem., 1985, **24**, 1643.
1335 F. Timmers and M. Brookhart, Organometallics, 1985, **4**, 1365.
1336 P. Braunstein, G.L. Geoffroy, and B. Metz., Nouv. J. Chim., 1985, **9**, 221.
1337 M.J. McGeary, T.L. Tonker, and J.L. Templeton, Organometallics, 1985, **4**, 2102.
1338 J. Keijsper, G. Van Koten, K. Vrieze, M. Zoutberg, and C.H. Stam, Organometallics, 1985, **4**, 1306.
1339 W.D. Jones and J.A. Maguire, Organometallics, 1985, **4**, 951.
1340 M.C.L. Trimarchi, M.A. Green, J.C. Huffman, and K.G. Caulton, Organometallics, 1985 4, 514.
1341 T. Beringhelli, G. D'Alfonso, and H. Molinari, J. Organomet. Chem., 1985, **295**, C35.
1342 K.-W. Lee, W.T. Pennington, A.W. Cordes, and T.L. Brown, J. Am. Chem. Soc., 1985, **107**, 631.
1343 K.-W. Lee and T.L. Brown, Organometallics, 1985, **4**, 1025.
1344 K.-W. Lee and T.L. Brown, Organometallics, 1985, **4**, 1030.
1345 C.P. Casey and J.M. O'Connor, Organometallics, 1985, **4**, 384.
1346 C.P. Casey, J.M. O'Connor, and K.J. Haller, J. Am. Chem. Soc., 1985, **107**, 1241.
1347 W.E. Buhro and J.A. Gladysz, Inorg. Chem., 1985, **24**, 3505.
1348 F.W.B. Einstein, K.G. Tyers, and D. Sutton, Organometallics, 1985, **4**, 489.
1349 F.A. Cotton, L.M. Daniels, K.R. Dunbar, L.R. Falvello, S.M. Tetrick, and R.A. Walton, J. Am. Chem. Soc., 1985, **107**, 3524.
1350 R.H. Morris, J.F. Sawyer, M. Shiralian, and J. Zubkowski, J. Am. Chem. Soc., 1985, **107**, 5581.
1351 K.H. Whitmire and T.R. Lee, J. Organomet. Chem., 1985, **282**, 95.
1352 C.A. Ghilardi, P. Innocenti, S. Midollini, and A. Orlandini, J. Chem. Soc., Dalton Trans., 1985, 605.
1353 D.J. Brauer, S. Hietkamp, H. Sommer, O. Stelzer, G. Müller, and C. Krüger, J. Organomet. Chem., 1985, **288**, 35.
1354 T.W. Dekleva, I.S. Thorburn, and B.R. James, Inorg. Chim. Acta, 1985, **100**, 49.
1355 G.E. Hawkes, L.Y. Lian, E.W. Randall, K.D. Sales, and S. Aime, J. Magn. Reson., 1985, **65**, 173.
1356 E. Rosenberg, C.B. Thorsen, L. Milone, and S. Aime, Inorg. Chem., 1985, **24**, 231.
1357 H.-P. Klein, U. Thewalt, G. Herrmann, G. Süss-Fink, and C. Moinet, J. Organomet. Chem., 1985, **286**, 225.
1358 G.E. Hawkes, L.Y. Lian, E.W. Randall, and K.D. Sales, J. Chem. Soc., Dalton Trans., 1985, 225.
1359 W. Ehrenreich, M. Herberhold, G. Herrmann, G. Süss-Fink, A. Gieren, and T. Hübner, J. Organomet. Chem., 1985, **294**, 183.
1360 W.Y. Yeh, J.R. Shapley, Y.J. Li, and M.R. Churchill, Organometallics, 1985, **4**, 767.
1361 G. Süss-Fink, W. Buehlmeyer, M. Herberhold, A. Gieren, and T. Hübner, J. Organomet. Chem., 1985, **280**, 129.
1362 D. Himmelreich and G. Müller, J. Organomet. Chem., 1985, **297**, 341.
1363 M.E. Wright, G.O. Nelson, and R.S. Glass, Organometallics, 1985, **4**, 245.
1364 A.M. Crespi and D.F. Shriver, Organometallics, 1985, **4**, 1830.
1365 C.P. Casey, S.R. Marder, and B.R. Adams, J. Am. Chem. Soc., 1985, **107**, 7700.
1366 M.K. Alami, F. Dahan, and R. Mathieu, Organometallics, 1985, **4**, 2122.

1367 M.I. Bruce, T.W. Hambley, M.R. Snow, and A.G. Swincer, Organometallics, 1985, **4**, 501.

1368 J.S. Holmgren and J.R. Shapley, Organometallics, 1985, **4**, 793.

1369 H.D. Kaesz, A. Mayr, Y.C. Lin, N.M. Boag, and C. Kampe, New Front. Organomet. Inorg. Chem., Proc. China-Jpn.-U.S.A. Trilateral Semin., 2nd., 1982 (Pub. 1984), 301 (Chem. Abstr., 1985, **102**, 185 247).

1370 F.J. Manganiello, S.M. Oon, M.D. Radcliffe, and W.M. Jones, Organometallics, 1985, **4**, 1069.

1371 E. Roland, W. Bernhardt, and H. Vahrenkamp, Chem. Ber., 1985, **118**, 2858.

1372 T.V. Ashworth, A.A. Chalmers, E. Meintjies, H. Oosthuizen, and E. Singleton, J. Organomet. Chem., 1985, **286**, 237.

1373 M.J. Hails, B.E. Mann, and C.M. Spencer, J. Chem. Soc., Dalton Trans., 1985, 693.

1374 P.D. Beer, J. Organomet. Chem., 1985, **297**, 313.

1375 R.F. Childs and A. Varadarajan, Can. J. Chem., 1985, **63**, 418.

1376 L.J. Farrugia, J.C. Jeffery, C. Marsden, and F.G.A. Stone, J. Chem. Soc., Dalton Trans., 1985, 645.

1377 M. Grassi, B.E. Mann, and C.M. Spencer, J. Chem. Soc., Chem. Commun., 1985, 1169.

1378 T. Dave, S. Berger, E. Bilger, H. Kaletsch, J. Pebler, J. Knecht, and K. Dimroth, Organometallics, 1985, **4**, 1565.

1379 N. Kuhn and H. Schumann, J. Organomet. Chem., 1985, **287**, 345.

1380 N.G. Connelly, A.R. Lucy, R.M. Mills, J.B. Sheridan, and P. Woodward, J. Chem. Soc., Dalton Trans., 1985, 699.

1381 S. Akabori, T. Kumagai, and M. Sato, J. Organomet. Chem., 1985, **286**, 69.

1382 L.R. Martin, F.W.B. Einstein, and R.K. Pomeroy, Inorg. Chem., 1985, **24**, 2777.

1383 B.P. Gracey, S.A.R. Knox, K.A. Macpherson, A.G. Orpen, and S.R. Stobart, J. Chem. Soc., Dalton Trans., 1985, 1935.

1384 J.A. Hriljac, P.N. Swepston, and D.F. Shriver, Organometallics, 1985, **4**, 158.

1385 G. Van Buskirk, C.B. Knobler, and H.D. Kaesz, Organometallics, 1985, **4**, 149.

1386 D.F. Foster, J. Harrison, B.S. Nicholls, and A.K. Smith, J. Organomet. Chem., 1985, **295**, 99.

1387 R.F. Alex and R.K. Pomeroy, J. Organomet. Chem., 1985, **284**, 379.

1388 A.J. Deeming, S. Donovan-Mtunzi, S.E. Kabir, and P.J. Manning, J. Chem. Soc., Dalton Trans., 1985, 1037.

1389 A.J. Deeming, S. Donovan-Mtunzi, and S.E. Kabir, J. Organomet. Chem., 1985, **281**, C43.

1390 R.D. Adams and S. Wang, Inorg. Chem., 1985, **24**, 4447.

1391 F.W.B. Einstein, L.R. Martin, R.K. Pomeroy, and P. Rushman, J. Chem. Soc., Chem. Commun., 1985, 345.

1392 C. Glidewell and A.R. Hyde, Polyhedron, 1985, **4**, 1155 (Chem. Abstr., 1985, **103**, 166 878).

1393 S. Joss, P. Bigler, and A. Ludi, Inorg. Chem., 1985, **24**, 3487.

1394 R. Benn, K. Cibura, P. Hofmann, K. Jonas, and A. Rufińska, Organometallics, 1985, **4**, 2214.

1395 G.F. Schmidt and M. Brookhart, J. Am. Chem. Soc., 1985, **107**, 1443.

1396 R.H. Crabtree and M. Lavin, J. Chem. Soc., Chem. Commun., 1985, 1661.

1397 L. Dahlenburg and A. Yardimcioglu, J. Organomet. Chem., 1985, **291**, 371.

1398 P. Janser, L.M. Venanzi, and F. Bachechi, J. Organomet. Chem., 1985, **296**, 229.

1399 T.M. Gilbert and R.G. Bergman, J. Am. Chem. Soc., 1985, **107**, 3502.

1400 A.L. Casalnuovo, J.A. Casalnuovo, P.V. Nilsson, and L.H. Pignolet, Inorg. Chem., 1985, **24**, 2554.

1401 L. Dahlenburg and F. Mirzaei, Inorg. Chim. Acta, 1985, **97**, L1.

1402 A.G. Kent, B.E. Mann, and C.P. Manuel, J. Chem. Soc., Chem. Commun., 1985, 728.

1403 R.S. Dickson, G.S. Evans, and G.D. Fallon, Aust. J. Chem., 1985, **38**, 273.

1404 M. El Amane, A. Maisonnat, F. Dahan, R. Pince, and R. Poilblanc, Organometallics, 1985, **4**, 773.

1405 L.C.A. De Carvahlo, Y. Peres, M. Dartiguenave, Y. Dartiguenave, and A.L.

Beauchamp, Organometallics, 1985, **4**, 2021.
1406 H. tom Dieck, J. Organomet. Chem., 1985, **291**, 71.
1407 S. Aime, G. Gobetto, D. Osella, L. Milone, G.E. Hawkes, and E.W. Randall, J. Magn. Reson., 1985, **65**, 308.
1408 D. Parker, J.-M. Lehn, and J. Rimmer, J. Chem. Soc., Dalton Trans., 1985, 1517.
1409 C. Woodcock and R. Eisenberg, Inorg. Chem., 1985, **24**, 1285.
1410 S.P. Deraniyagala and K.R. Grundy, J. Chem. Soc., Dalton Trans., 1985, 1577.
1411 S. Nussbaum, S.J. Rettig, A. Storr, and J. Trotter, Can. J. Chem., 1985, **63**, 692.
1412 D. Braga, R. Ros, and R. Roulet, J. Organomet. Chem., 1985, **286**, C8.
1413 S. Kinoshita and I. Masuda, Polyhedron, 1985, **4**, 1245.
1414 K. Nakajima, K. Tozaki, M. Kojima, and J. Fujita, Bull. Chem. Soc. Jpn., 1985, **58**, 1130.
1415 C.R. Langrick and B.L. Shaw, J. Chem. Soc., Dalton Trans., 1985, 511.
1416 B.R. Lloyd and R.J. Puddephatt, J. Am. Chem. Soc., 1985, **107**, 7785.
1417 B. Crociani, F. Dibianca, A. Giovenco, and A. Scrivanti, J. Organomet. Chem., 1985, **291**, 259.
1418 R.D.W. Kemmitt, P. McKenna, D.R. Russell, and L.J.S. Sherry, J. Chem. Soc., Dalton Trans., 1985, 259.
1419 A. Imran, R.D.W. Raymond, A.J.W. Markwick, P. McKenna, D.R. Russell , and L.J.S. Sherry, J. Chem. Soc., Dalton Trans., 1985, 549.
1420 M. Pfeffer, E. Wehman, and G. van Koten, J. Organomet. Chem., 1985, **282**, 127.
1421 E.W. Abel, S.K. Bhargava, K.G. Orrell, A.W.G. Platt, V. Šik, and T.S. Cameron, J. Chem. Soc., Dalton Trans., 1985, 345.
1422 S. Nussbaum and A. Storr, Can. J. Chem., 1985, **63**, 2550.
1423 A.T. Hutton, C.R. Langrick, D.M. McEwan, P.G. Pringle, and B.L. Shaw, J. Chem. Soc., Dalton Trans., 1985, 2121.
1424 M.A. Bennett and C. Chiraratvatana, J. Organomet. Chem., 1985, **296**, 255.
1425 S.A. Godleski, K.B. Gundlach, and R.S. Valpey, Organometallics, 1985, **4**, 296.
1426 R.L. Batstone-Cunningham, H.W. Dodgen, J.P. Hunt, and D.M. Roundhill, J. Organomet. Chem., 1985, **289**, 431.
1427 J.M. Brown and J.E. MacIntyre, J. Chem. Soc., Perkin Trans. 2, 1985, 961.
1428 B.P. Gracey, B.E. Mann, and C.M. Spencer, J. Organomet. Chem., 1985, **297**, 375.
1429 D.L. DeLaet, D.R. Powell, and C.P. Kublak, Organometallics, 1985, **4**, 954.
1430 R. Waditschatka, C. Kratky, B. Jaun, J. Heinzer, and A. Eschenmoser, J. Chem. Soc., Chem. Commun., 1985, 1604.
1431 L.E. Nivorozhkin, A.L. Nivorozhkin, M.S. Korobov, L.E. Konstantinovsky, and V.I. Minken, Polyhedron, 1985, **4**, 1701.
1432 J. Hartung, G. Weber, L. Beyer, and R. Szargan, Z. Anorg. Allg. Chem., 1985, **523**, 153.
1433 R.E. Cramer and P.L. Dahlstrom, Inorg. Chem., 1985, **24**, 3420.
1434 P. Courtot, R. Pichon, and J.Y. Salaun, J. Organomet. Chem., 1985, **286**, C17.
1435 R.T. Boere and C.J. Willis, Inorg. Chem., 1985, **24**, 1059.
1436 O.J. Scherer, R. Konrad, E. Guggolz, and M.L. Ziegler, Chem. Ber., 1985, **118**, 1.
1437 P.K. Byers, A.J. Canty, N.J. Minchin, J.M. Patrick, B.W. Skelton, and A.H. White, J. Chem. Soc., Dalton Trans., 1985, 1183.
1438 A.G. Davies, J.P. Goddard, M.B. Hursthouse, and N.P.C. Walker, J. Chem. Soc., Dalton Trans., 1985, 471.
1439 L.E. Nivorozhkin, M.S. Korobov, A.L. Nivorozhkin, L.E. Konstantinovskii, and V.I. Minkin, Zh. Obshch. Khim., 1985, **55**, 1125 (Chem. Abstr., 1985, **103**, 133 846).
1440 H. Ito and T. Ito, Bull. Chem. Soc. Jpn., 1985, **58**, 1755.
1441 P. Peringer and M. Lusser, Inorg. Chem., 1985, **24**, 109.
1442 R. Wehrmann, H. Meyer, and A. Berndt, Angew. Chem., Int. Ed. Engl., 1985, **24**, 788.
1443 M.A. Beckett, N.N. Greenwood, J.D. Kennedy, and M. Thornton-Pett, J. Chem. Soc., Dalton Trans., 1985, 1119.

1444 J.A. Walker, C.B. Knobler, and M.F. Hawthorne, Inorg. Chem., 1985, **24**, 2688.
1445 X.L.R. Fontaine, H. Fowkes, N.N. Greenwood, J.D. Kennedy, and M. Thornton-Pett, J. Chem. Soc., Chem. Commun., 1985, 1165.
1446 R. Köster, G. Seidel, and B. Wrackmeyer, Angew. Chem., Int. Ed. Engl., 1985, **24**, 326.
1447 R.H. Cragg, T.J. Miller, and D.O. Smith, J. Organomet. Chem., 1985, **291**, 273.
1448 M. Lauer, H. Böhnke, R. Grotstollen, M. Salehnia, and G. Wulff, Chem. Ber., 1985, **118**, 246.
1449 C. Brown, R.H. Cragg, T.J. Miller, and D. O'N. Smith, J. Organomet. Chem., 1985, **296**, C17.
1450 M. Veith, H. Lange, O. Recktenwald, and W. Frank, J. Organomet. Chem., 1985, **294**, 273.
1451 H. Schmidbaur, S. Lauteschläger, and G. Müller, J. Organomet. Chem., 1985, **281**, 33.
1452 H. Schmidbaur, S. Lauteschläger, and G. Müller, J. Organomet. Chem., 1985, **281**, 25.
1453 A.M. de P. Nicholas and R.E. Wasylishen, J. Phys. Chem., 1985, **89**, 5446.
1454 K.M. Larsson, J. Kowalewski, and U. Henriksson, J. Magn. Reson., 1985, **62**, 260.
1455 R. Neidlein, W. Wirth, A. Gieren, N. Lamm, and T. Hübner, Angew. Chem., Int. Ed. Engl., 1985, **24**, 587.
1456 I.D. Kalikhman, O.B. Bannikova, L.I. Belousova, B.A. Gostevskii, O.A. Vyazankina, and N.S. Vyazankin, Zh. Obshch. Khim., 1984, **54**, 2609 (Chem. Abstr., 1985, **102**, 149 340).
1457 J.P. Cohen-Addad and R. Dupeyre, Macromolecules, 1985, **18**, 1101 (Chem. Abstr., 1985, **103**, 6919).
1458 U. Engelhardt, T. Buenger, and H. Viertel, J. Crystallogr. Spectrosc. Res., 1984, **14**, 603 (Chem. Abstr., 1985, **102**, 113 725).
1459 W.H. Stevenson, III, S. Wilson, J.C. Martin, and W.B. Farnham, J. Am. Chem. Soc., 1985, **107**, 6340.
1460 W.H. Stevenson, III and J.C. Martin, J. Am. Chem. Soc., 1985, **107**, 6352.
1461 G. Klebe and K. Hensen, J. Chem. Soc., Dalton Trans., 1985, 5.
1462 S. Roelens and M. Taddei, J. Chem. Soc., Perkin Trans. 2, 1985, 799.
1463 C. Wynants, G. Van Binst, C. Mügge, K. Jurkschat, A. Tzschach, H. Pepermans, M. Gielen, and R. Willem, Organometallics, 1985, **4**, 1906.
1464 R. Colton and D. Dakternieks, Inorg. Chim. Acta, 1985, **102**, L17.
1465 R. Willem, M. Gielen, H. Pepermans, K. Hallenga, A. Recca, and P. Finocchiaro, J. Am. Chem. Soc., 1985, **107**, 1153.
1466 J.L. Nieto, F. Galindo, and A.M. Gutiérrez, Polyhedron, 1985, **4**, 1611.
1467 H.H. Karsch, A. Appelt, and G. Müller, Angew. Chem., Int. Ed. Engl., 1985, **24**, 403.
1468 W.W. Du Mont and M. Grenz, Chem. Ber., 1985, **118**, 1045.
1469 C.T.G. Knight and A.E. Merbach, Inorg. Chem., 1985, **24**, 576.
1470 A. Zschunke, M. Riemer, E. Leissring, and K. Issleib, Z. Anorg. Allg. Chem., 1985, **525**, 35.
1471 G. Märkl, W. Weber, and W. Weiss, Chem. Ber., 1985, **118**, 2365.
1472 V.V. Negrebetskii, N.N. Bychkov, T.I. Koroleva, I.E. Konareva, and B.I. Stepanov, Zh. Obshch. Khim., 1984, **54**, 2649 (Chem. Abstr., 1985, **102**, 149 372).
1473 C.D. Rithner and C.H. Bushweller, J. Am. Chem. Soc., 1985, **107**, 7823.
1474 L.H. Koole, W.J.M. Van der Hofstad, and H.M. Buck, J. Org. Chem., 1985, **50**, 4381.
1475 V.V. Negrebetskii, V.I. Kal'chenko, R.B. Rudyi, and L.N. Markovskii, Zh. Obshch. Khim., 1985, **55**, 271.
1476 G. Maas, H. Gümbel, G. Weise, and M. Regitz, Chem. Ber., 1985, **118**, 2105.
1477 D. Dakternieks and R. di Giacomo, Phosphorus Sulphur, 1985, **24**, 217.
1478 D.B. Denney, D.Z. Denney, and S.D. Pastor, Phosphorus Sulphur, 1985, **22**, 183.
1479 D.B. Denney, D.Z. Denney, and S.D. Pastor, Phosphorus Sulphur, 1985, **22**, 191.
1480 V.V. Negrebetskii, V.I. Kal'chinko, O.V. Balitskaya, M.Yu. Kornilov, and

L.N. Markovskii, Zh. Obshch. Khim., 1984, **54**, 1764 (Chem. Abstr., 1985, **102**, 6677).
1481 W.M. Abdou, D.B. Denney, D.Z. Denney, and S.D. Pastor, Phosphorus Sulphur, 1985, **22**, 99.
1482 P.B. Kay and S. Trippett, J. Chem. Soc., Chem. Commun., 1985, 135.
1483 Y. Masuda, M. Sano, and H. Yamatera, J. Phys. Chem., 1985, **84**, 3086.
1484 D.W. Aksnes and K. Ramstad, Magn. Reson. Chem., 1985, **23**, 253.
1485 Y. Masuda, M. Sano, and H. Yamatera, J. Chem. Soc., Faraday Trans. 1, 1985, **81**, 127.
1486 H. Duddeck, P. Wagner, and S. Gegner, Tetrahedron Lett., 1985, **26**, 1205.
1487 U. Berg and I. Pettersson, Magn. Reson. Chem., 1985, **23**, 536.
1488 J.E. Anderson and L. Henriksen, J. Chem. Soc., Chem. Commun., 1985, 1397.
1489 M.J. Collins and G.J. Schrobilgen, Inorg. Chem., 1985, **24**, 2608.
1490 E.I. Akhumov, Zh. Fiz. Khim., 1985, **59**, 2375.
1491 H.G. Hertz, J. Chim. Phys. Phys.-Chim. Biol., 1985, **82**, 557.
1492 K. Seomoon and M.D. Lee, Kongdae Yon'gu Pogo (Soul Taehakkyo), 1985, **17**, 135 (Chem. Abstr., 1985, **103**, 93 825).
1493 L.G. Nilsson, L. Nordenskioeld, P. Stilbs, and W.H. Braunlin, J. Phys. Chem., 1985, **89**, 3385.
1494 F.L. Dickert, W. Gumbrecht, S.W. Hellmann, and M.F. Waidkas, Ber. Bunsen. Phys. Chem., 1985, **89**, 875.
1495 L. Banci, I. Bertini, and C. Luchinat, Inorg. Chim. Acta, 1985, **100**, 173.
1496 E.W. Lang and H.D. Luedemann, Ber. Bunsen-Ges. Phys. Chem., 1985, **89**, 508.
1497 R.K. Mazitov, K.J. Mueller, and H.G. Hertz, Z. Phys. Chem. (Munich), 1984, **140**, 55 (Chem. Abstr., 1985, **102**, 51 785).
1498 M.K. Khripun, A.Yu. Efimov, L.S. Lilich, and A.G. Shmatko, Zh. Strukt. Khim., 1985, **26**, 88 (Chem. Abstr., 1985, **103**, 130 070).
1499 P. Kyselka and I. Sláma, Collect. Czech. Chem. Commun., 1985, **50**, 2325.
1500 K. Uchibayashi, M. Niibe, and Y. Nakamura, J. Chem. Eng. Data, 1985, **30**, 214 (Chem. Abstr., 1985, **102**, 155 787).
1501 Y. Nakamura and S. Shimokawa, Ber. Bunsen-Ges. Phys. Chem., 1985, **89**, 371.
1502 J. Kunze, A. Ebert, H. Lang, and B. Philipp, Z. Phys. Chem. (Leipzig), 1985, **266**, 49 (Chem. Abstr., 1985, **102**, 138 586).
1503 V.P. Yagodarov and Ya.P. Glazkova, Zh. Fiz. Khim., 1985, **59**, 199 (Chem. Abstr., 1985, **102**, 120 141).
1504 A. Sacco, M. Holz, and H.G. Hertz, J. Magn. Reson., 1985, **65**, 82.
1505 V.I. Chizhik, V.I. Mikhailov, and S.C. Pak, Vestn. Leningr. Univ., Fiz., Khim., 1985, 78 (Chem. Abstr., 1985, **102**, 196 646).
1506 V.I. Baranovskii, T.T. Ermolaeva, V.I. Chizhik, O.V. Sizova, and N.V. Ivanova, Khim. Fiz., 1985, **4**, 528.
1507 Y. Huang, Z. Li, and R. Liu, Wuli Huaxue Xuebao, 1985, **1**, 109 (Chem. Abstr., 1985, **103**, 28 160).
1508 M. Holz and A. Sacco, Mol. Phys., 1985, **54**, 149.
1509 D.M. Holton, P.P. Edwards, D.C. Johnson, C.J. Page, W. McFarlane, and B. Wood, J. Am. Chem. Soc., 1985, **107**, 6499.
1510 F.L. Dickert and M.F. Waidhas, Angew. Chem., Int. Ed. Engl., 1985, **24**, 575.
1511 V.P. Tarasov, G.A. Kirakosyan, Yu.A. Buslaev, S.V. Trots, and V.T. Panyushkin, Koord. Khim., 1985, **11**, 913 (Chem. Abstr., 1985, **103**, 110 877).
1512 X. Mao, L. Shen, Y. Gao, and J. Ni, Huaxue Xuebao, 1985, **43**, 566 (Chem. Abstr., 1985, **103**, 114 834).
1513 E. Brucher, J. Glaser, I. Grenthe, and I. Puigdomenech, Inorg. Chim. Acta, 1985, **109**, 111.
1514 A.E. Lemire, A.F. Janzen, and K. Marat, Inorg. Chim. Acta, 1985, **110**, 237.
1515 A.D. Hugi, L. Helm, and A.E. Merbach, Helv. Chim. Acta, 1985, **68**, 508.
1516 M.J. Gresser and A.S. Tracey, J. Am. Chem. Soc., 1985, **107**, 4215.
1517 K. Kaya, Y. Ikeda, and H. Fukutomi, Bull. Chem. Soc. Jpn., 1985, **58**, 2648.
1518 P. Bernhard, L. Helm, A. Ludi, and A.E. Merbach, J. Am. Chem. Soc., 1985, **107**, 312.
1519 R.F. Evilia and M. Saunders, Spectrosc. Lett., 1985, **18**, 105 (Chem. Abstr., 1985, **102**, 159 345).
1520 S. Aygen, H. Hanssum, and R. Van Eldik, Inorg. Chem., 1985, **24**, 2853.
1521 S.F. Lincoln, A.M. Hounslow, B.G. Doddridge, J.H. Coates, A.E. Merbach, and

D. Zbinden, Inorg. Chim. Acta, 1985, **100**, 207.
1522 N.J. Curtis and G.A. Lawrance, J. Chem. Soc., Dalton Trans., 1985, 1923.
1523 E. Bittersmann, K. Hildenbrand, A. Cervilla, and P. Lahuerta, J. Organomet. Chem., 1985, **287**, 255.
1524 H.G. Hertz and M. Holz, J. Magn. Reson., 1985, **63**, 64.
1525 L. Helm, L.I. Elding, and A.E. Merbach, Inorg. Chem., 1985, **24**, 1719.
1526 L. Banci, I. Bertini, and C. Luchinat, Chem. Phys. Lett., 1985, **118**, 345 (Chem. Abstr., 1985, **103**, 133 739).
1527 B. Yu, Z. Chen, T. Teng, and Q. Shen, Jinshu Xuebao, 1985, **21**, B51 (Chem. Abstr., 1985, **103**, 148 114).
1528 K. Miura, K. Hashimoto, H. Fukui, E. Yamada, and S. Shimokawa, J. Phys. Chem., 1985, **89**, 5098.
1529 D. Hugi-Cleary, L. Helm, and A.E. Merbach, Helv. Chim. Acta, 1985, **68**, 545.
1530 J.W. Akitt and J.M. Elders, J. Magn. Reson., 1985, **63**, 587.
1531 J.B. Sloan, S.A. Cannon, E.C. Delionback, and J.J. Dechter, Inorg. Chem., 1985, **24**, 883.
1532 S.P. Petrosyants, E.R. Tsabel, and Yu.A. Buslaev, Koord. Khim., 1985, **11**, 330 (Chem. Abstr., 1985, **102**, 192 231).
1533 T.C. Wong and M.R. Hakala, Chem. Phys. Lett., 1985, **119**, 85 (Chem. Abstr., 1985, **103**, 147 658).
1534 D.A. Horner, Report, 1984, LBL-18297; Order No. DE85001866, 232 pp. Avail. NTIS. From Energy Res. Abstr., 1984, **9**, Abstr. No. 50302 (Chem. Abstr., 1985, **102**, 120 729).
1535 R. Ahlrichs, H.J. Boehm, H.G. Hertz, and K.J. Müller, Z. Phys. Chem. (Munich), 1984, **142**, 67 (Chem. Abstr., 1985, **103**, 184 684).
1536 M. Hudlicky, J. Fluorine Chem., 1985, **28**, 461 (Chem. Abstr., 1985, **103**, 114 830).
1537 G.C. Mei and C.D. Gutsche, J. Am. Chem. Soc., 1985, **107**, 7959.
1538 Y. Chevalier and C. Chachaty, J. Am. Chem. Soc., 1985, **107**, 1102.
1539 R. Tabeta and H. Saitô, Bull. Chem. Soc. Japan, 1985, **58**, 3215.
1540 E.D. Gotsis and D. Fiat, Congr. AMPERE Magn. Reson. Relat. Phenom., Proc., 22nd, 1984, 502. Ed. by K.A. Mueller, R. Kind, and J. Roos (Chem. Abstr., 1985, **102**, 138 578).
1541 J.F. McGarrity and C.A. Ogle, J. Am. Chem. Soc., 1985, **107**, 1805.
1542 J.F. McGarrity, C.A. Ogle, Z. Brich, and H.-R. Loosli, J. Am. Chem. Soc., 1985, **107**, 1810.
1543 J. Heinzer, J.F.M. Oth, and D. Seebach, Helv. Chim. Acta, 1985, **68**, 1848.
1544 R. Laatikainen and V. Král, J. Chem. Soc., Perkin Trans. 2, 1985, 1091.
1545 L.M. Jackman and T.S. Dunne, J. Am. Chem. Soc., 1985, **107**, 2805.
1546 C.G. Screttas and M. Micha-Screttas, J. Organomet. Chem., 1985, **290**, 1.
1547 T. Iwachido, K. Shibuya, and K. Toei, Chem. Lett., 1985, 307.
1548 B. Eliasson, K.M. Larsson, and J. Kowalewski, J. Phys. Chem., 1985, **89**, 258.
1549 E.A. Bekturov, S.E. Kudaibergenov, V.Zh. Ushanov, and S.S. Saltybaeva, Makromol. Chem., Rapid Commun., 1985, **6**, 515 (Chem. Abstr., 1985, **103**, 88 320).
1550 E. Florin, Macromolecules, 1985, **18**, 360 (Chem. Abstr., 1985, **102**, 114 325).
1551 E.T. Fossel, M.M. Sarasua, and K.A. Koehler, J. Magn. Reson., 1985, **64**, 536.
1552 J.M. Van Divender and C.M. Grisham, J. Biol. Chem., 1985, **260**, 14060 (Chem. Abstr., 1985, **103**, 192 297).
1553 R.D. Boss and A.I. Popov, Inorg. Chem., 1985, **24**, 3660.
1554 C. Schade, W. Bauer, and P. v. R. Schleyer, J. Organomet. Chem., 1985, **295**, C25.
1555 S. Koda and H. Nomura, J. Solution Chem., 1985, **14**, 355.
1556 B.O. Strasser and A.I. Popov, J. Am. Chem. Soc., 1985, **107**, 7921.
1557 R.C. Phillips, S. Khazaeli, and J.L. Dye, J. Phys. Chem., 1985, **89**, 606.
1558 B.O. Strasser, K. Hallenga, and A.I. Popov, J. Am. Chem. Soc., 1985, **107**, 789.
1559 J. Zavada, V. Pechanec, J. Zajicek, I. Stibor, and A. Vitek, Collect. Czech. Chem. Commun., 1985, **50**, 1184.
1560 A. Delville, H.D.H. Stoever, and C. Detellier, J. Am. Chem. Soc., 1985, **107**, 4172.
1561 H.D.H. Stoever, A. Delville, and C. Detellier, J. Am. Chem. Soc., 1985, **107**,

4167.
1562 H.D.H. Stover, L.J. Maurice, A. Delville, and C. Detellier, Polyhedron, 1985, **4**, 1091.
1563 J.M. Brown, A.E. Derome, and S.J. Kimber, Tetrahedron Lett., 1985, **26**, 253.
1564 T. Waernheim, U. Henriksson, and T. Klason, Mol. Cryst. Liq. Cryst., 1985, **126**, 247 (Chem. Abstr., 1985, **103**, 28 086).
1565 E. Sjoeblom and U. Henriksson, Surfactants Solution, [Proc. Int. Symp.], 4th 1982 (Pub. 1984), **3**, 1867 (Chem. Abstr., 1985, **102**, 138 254).
1566 K. Takagi, K. Aoshima, Y. Sawaki, and H. Iwamura, J. Am. Chem. Soc., 1985, **107**, 47.
1567 B. Durairaj and F.D. Blum, J. Colloid Interface Sci., 1985, **106**, 561 (Chem. Abstr., 1985, **103**, 147 650).
1568 A. Faure, A.M. Tistchenko, T. Zemb, and C. Chachaty, J. Phys. Chem., 1985, **89**, 3373 (Chem. Abstr., 1985, **103**, 110 387).
1569 E. Sjoeblom, T. Waernheim, U. Henriksson, and P. Stenius, Tenside Deterg., 1984, **21**, 303 (Chem. Abstr., 1985, **102**, 32 916).
1570 R.C. Phillips, S. Khazaeli, and J.L. Dye, J. Phys. Chem., 1985, **89**, 600.
1571 H.D. Schaedler, M. Riemer, and W. Schroth, Z. Chem., 1984, **24**, 407 (Chem. Abstr., 1985, **102**, 131 455).
1572 D. Vasilescu and G. Mallet, Biopolymers, 1985, **24**, 1845 (Chem. Abstr., 1985, **103**, 156 450).
1573 F.G. Riddell and M.K. Hayer, Biochim. Biophys. Acta, 1985, **817**, 313 (Chem. Abstr., 1985, **103**, 66 896).
1574 J.W. Pettegrew, T. Glonek, N.J. Minshew, and D.E. Woessner, J. Magn. Reson., 1985, **63**, 439.
1575 H. Monoi, Biophys. J., 1985, **48**, 643 (Chem. Abstr., 1985, **103**, 191 741).
1576 M.M. Pike, J.C. Frazer, D.F. Dedrick, J.S. Ingwall, P.D. Allen, C.S. Springer, jun., and T.W. Smith, Biophys. J., 1985, **48**, 159 (Chem. Abstr., 1985, **103**, 101 252).
1577 P.S. Belton, V.J. Morris, and S.F. Tanner, Int. J. Biol. Macromol., 1985, **7**, 53 (Chem. Abstr., 1985, **102**, 147 654).
1578 H. Shinar and G. Navon, Biophys. Chem., 1984, **20**, 275 (Chem. Abstr., 1985, **102**, 21 970).
1578a G. Siegel, A. Adler, E. Mildenberger, and A. Walter, Inst. Natl. Sante Rech. Med., [Colloq.], 1985, **124**, 101 (Chem. Abstr., 1985, **103**, 51 898).
1579 Y. Boulanger, P. Vinay, and M. Desroches, Biophys. J., 1985, **47**, 553 (Chem. Abstr., 1985, **102**, 217 742).
1579a B.A. Wittenberg and R.K. Gupta, J. Biol. Chem., 1985, **260**, 2031 (Chem. Abstr., 1985, **102**, 129 577).
1580 W.O. Lin and H.G. Alt, Z. Naturforsch., Teil B, 1985, **40**, 441 (Chem. Abstr., 1985, **102**, 196 841).
1581 W.O. Lin, M.C.B.V. De Souza, and H.G. Alt, Z. Naturforsch., Teil B, 1984, **39**, 1375 (Chem. Abstr., 1985, **102**, 16 506).
1582 G. Chidichimo, A. Golemme, J.W. Doane, and P.W. Westerman, J. Chem. Phys., 1985, **82**, 536.
1583 D.W. Urry, T.L. Trapane, C.M. Venkatachalam, and K.U. Prasad, Can. J. Chem., 1985, **63**, 1976.
1584 B.O. Strasser, M. Shamsipur, and A.I. Alexander, J. Phys. Chem., 1985, **89**, 4822.
1585 D.W. Urry, T.L. Trapane, R.A. Brown, C.M. Venkatachalam, and K.U. Prasad, J. Magn. Reson., 1985, **65**, 43.
1586 A.S. Arseniev, I.L. Barsukov, and V.F. Bystrov, FEBS Lett., 1985, **180**, 33 (Chem. Abstr., 1985, **102**, 127 509).
1587 M. Jaber and G. Thomas-David, Bull. Soc. Chim. Fr., 1985, 644.
1588 T. Matsuo and H. Ohno, Nippon Kagaku Kaishi, 1985, 1851 (Chem. Abstr., 1985, **103**, 226 118).
1589 M.B. Bass and H.J. Fromm, Anal. Biochem., 1985, **145**, 292 (Chem. Abstr., 1985, **102**, 181 777).
1590 P.R. Droupadi and V. Krishnan, J. Phys. Chem., 1985, **89**, 909.
1591 F. Teixeira, C.F.G.C. Geraldes, V.M.S. Gil, M. Helena, and S.F. Teixeira, Gastroenterol. Clin. Biol., 1984, **8**, 879 (Chem. Abstr., 1985, **102**, 105 743).
1592 I. Stokkeland and P. Stilbs, Biophys. Chem., 1985, **22**, 65 (Chem. Abstr.,

1985, **103**, 118 388).

1593 H. Terada, M. Ikuno, Y. Shinohara, and K. Yoshikawa, Biochim. Biophys. Acta, 1984, **767**, 648 (Chem. Abstr., 1985, **102**, 57 953).

1594 R.V. Prigodich and P. Haake, Inorg. Chem., 1985, **24**, 89.

1595 W. Klaus, P. Roesch, and R.S. Goody, Res. Exp. Med., 1985, **185**, 145 (Chem. Abstr., 1985, **103**, 50 684).

1596 L. Asso and M. Asso, Thermochim. Acta, 1985, **87**, 373 (Chem. Abstr., 1985, **103**, 60 226).

1597 T. Shimizu and M. Hatano, Inorg. Chem., 1985, **24**, 2003.

1598 C.K. Vishwanath and K.R.K. Easwaran, J. Chem. Soc., Perkin Trans. 2, 1985, 65.

1599 A. Khan, K. Fontell, and B. Lindman, Prog. Colloid Polym. Sci., 1985, **70**, 30 (Chem. Abstr., 1985, **103**, 166 600).

1600 C.F.G.C. Geraldes, M.C. Alpoim, M.P.M. Marques, A.D. Sherry, and M. Singh, Inorg. Chem., 1985, **24**, 3876.

1601 V.V. Bagreev and S.O. Popov, Polyhedron, 1985, **4**, 929.

1602 K. Dill, M.E. Daman, E. Decoster, J.M. Lacombe, and A.A. Pavia, Inorg. Chim. Acta, 1985, **106**, 203.

1603 Yu.I. Sal'nikov and V.V. Ustyak, Koord. Khim., 1984, **10**, 1636 (Chem. Abstr., 1985, **102**, 51 748).

1604 N. Jamin, D. Baron, and N. Lumbroso-Bader, J. Chem. Soc., Perkin Trans. 2, 1985, 1.

1605 M.S. Nieuwenhuizen, J.A.Peters, A. Sinnema, A.P.G. Kieboom, and H. Van Bekkum, J. Am. Chem. Soc., 1985, **107**, 12.

1606 Y.J. Shyy, T.C. Tsai, and M.D. Tsai, J. Am. Chem. Soc., 1985, **107**, 3478.

1607 R.B. Lauffer and T.J. Brady, Magn. Reson. Imaging, 1985, **3**, 11 (Chem. Abstr., 1985, **103**, 101 218).

1608 N. Fujiwara, H. Tomiyasu, and H. Fukutomi, Bull. Chem. Soc. Jpn., 1985, **58**, 1386.

1609 A. Marzotto and H. Kozlowski, Inorg. Chim. Acta, 1985, **110**, 31.

1610 X. Li and Y. Xu, Kexue Tongbao (Foreign Lang. Ed.), 1985, **30**, 340 (Chem. Abstr., 1985, **103**, 188 377).

1611 M. Delfini, E. Gaggelli, A. Lepri, and G. Valensin, Inorg. Chim. Acta, 1985, **107**, 87 (Chem. Abstr., 1985, **103**, 100 587).

1612 A.P. Makarov, A.E. Gekhman, V.M. Nekipelov, E.P. Talzi, O.Ya. Polotnyuk, K.I. Zamaraev, and I.I. Moiseev, Izv. Akad. Nauk SSSR, Ser. Khim., 1985, 1914 (Chem. Abstr., 1985, **103**, 166 930).

1613 A.K. Il'yasova and A.K. Akhmetova, Zh. Neorg. Khim., 1985, **30**, 653 (Chem. Abstr., 1985, **103**, 97 764).

1614 T. Fukumoto, J. Iyoda, K. Murata, and S. Ikeda, Chem. Lett., 1985, 1311 (Chem. Abstr., 1985, **103**, 226 594).

1615 V.M. Mastikhin, O.B. Lapina, and L.G. Simonova, React. Kinet. Catal. Lett., 1984, **26**, 431 (Chem. Abstr., 1985, **102**, 120 671).

1616 P. Csermely, A. Martonosi, G.C. Levy, and A.J. Ejchart, Biochem. J., 1985, **230**, 807 (Chem. Abstr., 1985, **103**, 192 178).

1617 G.R. Willsky, J.O. Leung, P.V. Offermann, jun., E.K. Plotnick, and S.F. Dosch, J. Bacteriol., 1985, **164**, 611 (Chem. Abstr., 1985, **103**, 211 019).

1618 A.G. Stepanov, V.M. Nekipelov, and K.I. Zamaraev, Dokl. Akad. Nauk SSSR, 1985, **281**, 400 (Chem. Abstr., 1985, **103**, 93 782).

1619 J. Markham, K. Menard, and A. Cutler, Inorg. Chem., 1985, **24**, 1581.

1620 L. Pettersson, I. Andersson, and L.O. Oehman, Acta Chem. Scand., Ser. A, 1985, **A39**, 53.

1621 A. Cervilla, J.A. Ramirez, and E. Llopis, Can. J. Chem., 1985, **63**, 1041.

1622 P.M. Boorman, K.J. Moynihan, V.D. Patel, and J.F. Richardson, Inorg. Chem., 1985, **24**, 2989.

1623 R.M. Nielson and S. Wherland, J. Am. Chem. Soc., 1985, **107**, 1505.

1624 J.J. Led, J. Am. Chem. Soc., 1985, **107**, 6755.

1625 R.D. Carter and K. Dill, Inorg. Chim. Acta, 1985, **108**, 83.

1626 Y. Chevalier and C. Chachaty, J. Phys. Chem., 1985, **89**, 875.

1627 L. Spiccia and T.W. Swaddle, J. Chem. Soc., Chem. Commun., 1985, 67.

1628 C.P. Horwitz and D.F. Shriver, J. Am. Chem. Soc., 1985, **107**, 8147.

1629 D.W. Dixon, M. Barbush, and A. Shirazi, Inorg. Chem., 1985, **24**, 1081.

1630 A. Shirazi, M. Barbush, S. Ghosh, and D.W. Dixon, Inorg. Chem., 1985, **24**, 2495.

1631 K. Simolo, G. Stucky, S. Chen, M. Bailey, C. Scholes, and G. McLendon, J. Am. Chem. Soc., 1985, **107**, 2865.

1632 J.D. Satterlee and J.A. Shelnutt, Inorg. Chim. Acta, 1985, **106**, 165.

1633 Yu.I. Sal'nikov and V.V. Ustyak, Koord. Khim., 1985, **11**, 1225 (Chem. Abstr., 1985, **103**, 167 107).

1634 N.N. Tananaeva, M.Ya. Gorokhovatskaya, and N.A. Kostromina, Ukr. Khim. Zh. (Russ. Ed.), 1984, **50**, 1014 (Chem. Abstr., 1985, **102**, 57 992).

1635 W.C. Stevens and D.M. Kurtz, jun., Inorg. Chem., 1985, **24**, 3444.

1636 M.S. Chan and A.C. Wahl, J. Phys. Chem., 1985, **89**, 5829.

1637 C.A.L. Becker, R. Rezaaiyan, and F. Moattar, Inorg. Chim. Acta, 1985, **100**, 227.

1638 S. Balt, H.J. Gamelkoorn, and K.R. Lammers, Transition Met. Chem. (Weinheim, Ger.), 1985, **10**, 164.

1639 S. Balt and H.J. Gamelkoorn, Inorg. Chim. Acta, 1985, **98**, L57.

1640 J.R. Vriesenga and G.H. Fairchild, Inorg. Chim. Acta, 1985, **98**, 195.

1641 F.L. Dickert and W. Gumbrecht, Z. Phys. Chem. (Munich), 1984, **141**, 169.

1642 Z.A. Saprykova and N.D. Chichirova, Zh. Neorg. Khim., 1985, **30**, 939 (Chem. Abstr., 1985, **102**, 226 797).

1643 A.J. Kunin, E.J. Nanni, and R. Eisenberg, Inorg. Chem., 1985, **24**, 1852.

1644 A.J. Deeming and G.P. Proud, Inorg. Chim. Acta, 1985, **100**, 223.

1645 J.K. Beattie, M.T. Kelso, W.E. Moody, and P.A. Tregloan, Inorg. Chem., 1985, **24**, 415.

1646 M.G. Fairbank, P.R. Norman, and A. McAuley, Inorg. Chem., 1985, **24**, 2639.

1647 G.A. Meyers, F.M. Michaels, R.L. Reeves, and P.J. Trotter, Inorg. Chem., 1985, **24**, 731.

1648 R.F. Evilia, Inorg. Chem., 1985, **24**, 2076.

1649 E. Matczak-Jon, B. Boduszek, T. Tatarowski, and W. Wojciechowski, Ser. Fiz. (Uniw. im. Adama Mickiewicza Poznaniu), 1985, **53**, 145 (Chem. Abstr., 1985, **103**, 148 128).

1650 A. Hioki, S. Funahashi, and M. Tanaka, J. Phys. Chem., 1985, **89**, 5057.

1651 P. Mohanakrishnan, C.F. Chignell, and R.H. Cox, J. Pharm. Sci., 1985, **74**, 61 (Chem. Abstr., 1985, **102**, 91 055).

1652 G. Strukul and R.A. Michelin, J. Am. Chem. Soc., 1985, **107**, 7563.

1653 G. Raudaschl-Sieber, H. Schoelhorn, U. Thewalt, and B. Lippert, J. Am. Chem. Soc., 1985, **107**, 3591.

1654 H. Schöllhorn, U. Thewalt, and B. Lippert, Inorg. Chim. Acta, 1985, **108**, 77.

1655 J.H.J. den Hartog, C. Altona, H. van den Elst, G.A. van der Marel, and J. Redijk, Inorg. Chem., 1985, **24**, 986.

1656 B.H. Lipshutz, J.A. Kozlowski, and C.M. Breneman, J. Am. Chem. Soc., 1985, **107**, 3197.

1657 S.E. Kudaibergenov, E.M. Shaikhutdinov, G.M. Zhaimina, and E.A. Bekturov, Koord. Khim., 1985, **11**, 1329 (Chem. Abstr., 1985, **103**, 226 336).

1658 H. Sapper, H.H. Paul, K. Beinhauer, and W. Lohmann, Inorg. Chim. Acta, 1985, **106**, 25.

1659 B. Henry, J.C. Boubel, and J.J. Delpuech, Polyhedron, 1985, **4**, 1069.

1660 R. Tauler, E. Casassas, M.J.A. Rainer, and B.M. Rode, Inorg. Chim. Acta, 1985, **105**, 165.

1661 J.T. Groves and J.R. Olson, Inorg. Chem., 1985, **24**, 2717.

1662 H. Sapper, H.H. Paul, K. Beinhauer, and W. Lohmann, Congr. AMPERE Magn. Reson. Relat. Phenom., Proc., 22nd, 1984, 471. Ed. by K.A. Mueller, R. Kind, and J. Roos (Chem. Abstr., 1985, **102**, 127 518).

1663 E.L. Gogolashvili, R.R. Amirov, Z.A. Saprykova, and A.V. Zakharov, Zh. Obshch. Khim., 1985, **55**, 730 (Chem. Abstr., 1985, **102**, 226 811).

1664 H. Sterk, M. Braun, and H. Schmut, Congr. AMPERE Magn. Reson. Relat. Phenom., Proc., 22nd, 1984, 347 (Chem. Abstr., 1985, **102**, 104 901).

1665 S. Al-Baker, W.E. Hill, and C.A. McAuliffe, J. Chem. Soc., Dalton Trans., 1985, 2655.

1666 G.G. Graham, J. Inorg. Biochem., 1985, **25**, 163.

1667 S.P. Dagnall, D.N. Hague, M.E. McAdam, and A.D. Moreton, J. Chem. Soc., Dalton Trans., 1985, 2381.

1668 S.P. Dagnall, D.N. Hague, M.E. McAdam, and A.D. Moreton, J. Chem. Soc., Faraday Trans. 1, 1985, **81**, 1483.
1669 H. Sigel, R. Tribolet, and K.H. Scheller, Inorg. Chim. Acta, 1985, **100**, 151.
1670 C.H. Taliaferro and A.E. Martell, Inorg. Chem., 1985, **24**, 2408 (Chem. Abstr., 1985, **103**, 43 589).
1671 W.S. Postal, E.J. Vogel, C.M. Young, and F.T. Greenaway, J. Inorg. Biochem., 1985, **25**, 25.
1672 D.L. Rabenstein, S.A. Daignault, A.A. Isab, A.P. Arnold, and M.M. Shoukry, J. Am. Chem. Soc., 1985, **107**, 6435.
1673 S.A. Daignault, A.P. Arnold, A.A. Isab, and D.L. Rabenstein, Inorg. Chem., 1985, **24**, 3984.
1674 J. Ueda, A. Hanaki, N. Yoshida, and T. Nakajima, Chem. Pharm. Bull., 1985, **33**, 3096.
1675 N.N. Tanaeva, M.Ya. Gorokhovatskaya, and N.A. Kostromina, Khimiya Poverkh.-aktiv. Veshchestv i Kompleksonov, Kalinin, 1984, 10. From Ref. Zh. Khim., 1985, Abstr. No. 5V73 (Chem. Abstr., 1985, **102**, 210 200).
1676 K. Alouani, M.R. Khaddar, L. Rodehueser, P.R. Rubini, and J.J. Delpuech, Polyhedron, 1985, **4**, 643.
1677 S. Biagini, M. Casu, A. Lai, R. Caminiti, and G. Crisponi, Chem. Phys., 1985, **93**, 461 (Chem. Abstr., 1985, **102**, 162 374).
1678 A.D. Keller, T. Drakenberg, R.W. Briggs, and I.M. Armitage, Inorg. Chem., 1985, **24**, 1170.
1679 R.V. Prigodich, T. O'Connor, and J.E. Coleman, Biochemistry, 1985, **24**, 6291.
1680 T. Drakenberg, M. Swaerd, A. Cave, and J. Parello, Biochem. J., 1985, **227**, 711 (Chem. Abstr., 1985, **103**, 2389).
1681 M.E. Bjornson, D.C. Corson, and B.D. Sykes, J. Inorg. Biochem., 1985, **25**, 141.
1682 G.I. Rhyu, W.J. Ray, jun, and J.L. Markley, Biochemistry, 1985, **24**, 2536 (Chem. Abstr., 1985, **102**, 200 241).
1683 A.M. Bond, R. Colton, M.L. Dillon, A.F. Hollenkamp, and J.E. Moir, Inorg. Chem., 1985, **24**, 1591.
1684 M.J.B. Ackerman and J.J.H. Ackerman, J. Am. Chem. Soc., 1985, **107**, 6413.
1685 R.D. Bach, H.B. Vardhan, A.F.M.M. Rahman, and J.P. Oliver, Organometallics, 1985, **4**, 846.
1686 J.-P. Charland, M.T.P. Viet, M. St-Jacques, and A.L. Beauchamp, J. Am. Chem. Soc., 1985, **107**, 8202.
1687 A. Kamars and E. Schwartz, Latv. PSR Zinat. Akad. Vestis, Kim. Ser., 1985, 629 (Chem. Abstr., 1985, **103**, 221 954).
1688 B.N. Chernyshov, G.P. Shchetinina, O.V. Brovkina, and E.G. Ippolitov, Koord. Khim., 1985, **11**, 31 (Chem. Abstr., 1985, **102**, 105 127).
1689 M. Makkee, A.P.G. Kieboom, and H. van Bekkum, Recl. Trav. Chim. Pays-Bas, 1985, **104**, 230.
1690 L.O. Oehman and S. Sjoeberg, Mar. Chem., 1985, **17**, 91 (Chem. Abstr., 1985, **103**, 128 650).
1691 S. Hermanek, O. Kriz, J. Fusek, B. Casensky, and Z. Cerny, Z. Naturforsch., Teil B, 1985, **40**, 582 (Chem. Abstr., 1985, **103**, 43 620).
1692 J.H. Medley, N. Ahmad, and M.C. Day, Can. J. Chem., 1985, **63**, 2906.
1693 M.S. Su, J.S. Gentry, C.B. Bass, C.G. Moreland, and K.J. Bachmann, J. Electrochem. Soc., 1985, **132**, 802 (Chem. Abstr., 1985, **102**, 211 565).
1694 B. Szpoganicz and A.E. Martell, Inorg. Chem., 1985, **24**, 2414.
1695 V.E. Larchenko, K.I. Popov, A.I. Grigor'ev, and N.M. Dyatkova, Koord. Khim., 1984, **10**, 1187 (Chem. Abstr., 1985, **102**, 24 732).
1696 J.W. Akitt and J.M. Elders, J. Chem. Soc., Faraday Trans. 1, 1985, **81**, 1923.
1697 J.W. Akitt and J.A. Farnsworth, J. Chem. Soc., Faraday Trans. 1, 1985, **81**, 193.
1698 S. Sjöberg and L.-O. Öhman, J. Chem. Soc., Dalton Trans., 1985, 2665.
1699 J.W. Akitt and J. Lelievre, J. Chem. Soc., Dalton Trans., 1985, 591.
1700 S.P. Petrosyants and E.R. Buslaeva, Koord. Khim., 1985, **11**, 755 (Chem. Abstr., 1985, **103**, 60 206).
1701 L.P. Davis, C.J. Dymek, jun., J.J.P. Stewart, H.P. Clark, and W.J. Lauderdale, J. Am. Chem. Soc., 1985, **107**, 5041.
1702 I. Ichikawa and T. Matsumoto, J. Magn. Reson., 1985, **63**, 445.

1703 T. Matsumoto and K. Ichikawa, Proc. Int. Symp. Molten Salt Chem. Technol., 1st, 1983, 383 (Chem. Abstr., 1985, **102**, 138 616).
1704 J.P. Brunette, M. Taheri, G. Goetz-Grandmont, and M.J.F. Leroy, Solvent Extr. Ion Exch., 1985, **3**, 309 (Chem. Abstr., 1985, **103**, 130 001).
1705 C. Margheritis, Z. Naturforsch., Tiel A, 1984, **39**, 1112 (Chem. Abstr., 1985, **102**, 38 369).
1706 K.R. Metz, J.F. Hinton, and T.L. Gosforth, J. Phys. Chem., 1985, **89**, 2097.
1707 U. Henriksson and J. Glaser, Acta Chem. Scand., Ser. A, 1985, **A39**, 355.
1708 J.F. Stebbins, J.B. Murdoch, E. Schneider, I.S.E. Carmichael, and A. Pines, Nature (London), 1985, **314**, 250.
1709 S. Sjoeberg, L.O. Oehman, and N. Ingri, Acta Chem. Scand., Ser. A, 1985, **A39**, 93.
1710 P.H. Fries, J. Rendell, E.E. Burnell, and G.N. Patey, J. Chem. Phys., 1985, **83**, 307.
1711 A.I. Nikolaev, N.I. Kasikova, M.E. Ignatov, A.G. Babkin, and E.G. Il'in, Zh. Neorg. Khim., 1985, **30**, 176 (Chem. Abstr., 1985, **102**, 101 369).
1712 G.P. Mullen, R.B. Dunlap, and J.D. Odom, J. Am. Chem. Soc., 1985, **107**, 7187.
1713 A.G. Miller, J.A. Franz, and J.W. Macklin, J. Phys. Chem., 1985, **89**, 1190.
1714 J. Sakurada, T. Hosoya, T. Shimizu, and M.Hatano, Chem. Lett., 1985, 211 (Chem. Abstr., 1985, **102**, 108 890).
1715 C.J. Ruffing and T.B. Rauchfuss, Organometallics, 1985, **4**, 524.
1716 M.R.M. Bruce and D.R. Tyler, Organometallics, 1985, **4**, 528.
1717 C.E. Davies, I.M. Gardiner, J.C. Green, M.L.H. Green, N.J. Hazel, V.S.B. Mtetwa, and K. Prout, J. Chem. Soc., Dalton Trans., 1985, 669.
1718 E.L. Eliel, M. Manoharan, S.G. Levine, and A. Ng, J. Org. Chem., 1985, **50**, 4978.
1719 Y.S. Zimin, A.G. Stepanov, and V.M. Nekipelov, React. Kinet. Catal. Lett., 1985, **28**, 145 (Chem. Abstr., 1985, **103**, 188 455).
1720 H. Werner, B. Ulrich, U. Schubert, P. Hofmann, and B. Zimmer-Gasser, J. Organomet. Chem., 1985, **297**, 27.
1721 C.D. Hoff, J. Organomet. Chem., 1985, **282**, 201.
1722 D. Carrillo, P. Gouzerh, and Y. Jeannin, Nouv. J. Chim., 1985, **9**, 749.
1723 C.P. Horwitz, E.M. Holt, and D.F. Shriver, J. Am. Chem. Soc., 1985, **107**, 281.
1724 A.-R. Al-Ohaly and J.F. Nixon, Inorg. Chim. Acta, 1985, **103**, 83.
1725 J.J. Pignatello and Y.-T. Fanchiang, J. Chem. Soc., Dalton Trans., 1985, 1381.
1726 L.G. Marzilli, M.F. Summers, N. Bresciani-Pahor, E. Zangrando, J.-P. Charland, and L. Randaccio, J. Am. Chem. Soc., 1985, **107**, 6880.
1727 K. Sakata, M. Hashimoto, and H. Yoshino, Inorg. Chim. Acta, 1985, **99**, 231
1728 R.C. Scarrow, P.E. Riley, K. Abu-Dari, D.L. White, and K.N. Raymond, Inorg. Chem., 1985, **24**, 954.
1729 R.S. Paonessa, N.C. Thomas, and J. Halpern, J. Am. Chem. Soc., 1985, **107**, 4333.
1730 R.A. Newmark and A.R. Siedle, Magn. Reson. Chem., 1985, **23**, 67.
1731 E. Kwaskowska-Chec, A.M. Trzeciak, and J.J. Ziolkowski, React. Kinet. Catal. Lett., 1984, **26**, 21 (Chem. Abstr., 1985, **102**, 155 571).
1732 Yu.S. Varshavskii, T.G. Cherkasova, N.V. Kiseleva, L.S. Bresler, N.A. Buzina, and A.S. Khachaturov, Zh. Neorg. Khim., 1985, **30**, 152 (Chem. Abstr., 1985, **102**, 124 582).
1733 E.B. Boyar and S.D. Robinson, J. Chem. Soc., Dalton Trans., 1985, 629.
1734 R.A. Vanderpool and H.B. Abrahamson, Inorg. Chem., 1985, **24**, 2985.
1735 M. Schumann and H. Elias, Inorg. Chem., 1985, **24**, 3187.
1736 F. Cariati, M.L. Ganadu, M.A. Zoroddu, R. Mansani, and R. Quidacciolu, Inorg. Chem., 1985, **24**, 4030.
1737 D.L. Packett, C.M. Jensen, R.L. Cowan, C.E. Strouse, and W.C. Trogler, Inorg. Chem., 1985, **24**, 3578.
1738 B. Galli, F. Gasparrini, B.E. Mann, L. Maresca, G. Natile, A.M. Manotti-Lanfredi, and A. Tiripicchio, J. Chem. Soc., Dalton Trans., 1985, 1155.
1739 A. Scrivanti, G. Cavinato, L. Toniolo, and C. Botteghi, J. Organomet. Chem., 1985, **286**, 115.

1740 H.C. Clark, G. Ferguson, P.N. Kapoor, and M. Parvez, Inorg. Chem., 1985, **24**, 3924.
1741 S. Cenini, F. Porta, M. Pizzotti, and C. Crotti, J. Chem. Soc., Dalton Trans., 1985, 165.
1742 M. Polissiou, M.T.P. Viet, M. St-Jacques, and T. Theophanides, Inorg. Chim. Acta, 1985, **107**, 203.
1743 G. Hallnemo, T. Olsson, and C. Ullenius, J. Organomet. Chem., 1985, **282**, 133.
1744 D.L. Banville, W.D. Wilson, and L.G. Marzilli, Inorg. Chem., 1985, **24**, 2479.
1745 D. Dakternieks and C.L. Rolls, Inorg. Chim. Acta, 1985, **105**, 213.
1746 P.J. Domaille, J.D. Druliner, L.W. Gosser, J.M. Read, jun., E.R. Schmelzer, and W.R. Stevens, J. Org. Chem., 1985, **50**, 189.
1747 R. Vestin, U. Vestin, and J. Kowalewski, Acta Chem. Scand., 1985, **A39**, 767.
1748 M.R.P. van Vliet, P. Buysingh, G. van Koten, and K. Vrieze, Organometallics, 1985, **4**, 1701.
1749 W.R. Nutt, J.A. Anderson, J.D. Odom, M.M. Williamson, and B.H. Rubin, Inorg. Chem., 1985, **24**, 159.
1750 I.D. Kalikhman, O.B. Bannikova, A.V. Kalinin, B.N. Khasapov, S.L. Ioffe, V.A. Tartakovskii, and M.G. Voronkov, Izv. Akad. Nauk SSSR, Ser. Khim., 1985, 464.
1751 M.J. Michaelczyk, R. West, and J. Michl, Organometallics, 1985, **4**, 826.
1752 S. Sjöberg, N. Ingri, A.-M. Nenner, and L.-O. Öhman, J. Inorg. Biochem., 1985, **24**, 267.
1753 T.P. Lockhart, J. Organomet. Chem., 1985, **287**, 179.
1754 S. Masamune and L.R. Sita, J. Am. Chem. Soc., 1985, **107**, 6390.
1755 G. Ruisi, A. Silvestri, M.T. Lo Giudice, R. Barbieri, G. Atassi, F. Huber, K. Grätz, and L. Lamartini, J. Inorg. Biochem., 1985, **25**, 229.
1756 P.A.W. Dean and R.S. Srivastava, Inorg. Chim. Acta, 1985, **105**, 1.
1757 W.N. Setzer, A.E. Sopchik, and W.G. Wesley, J. Am. Chem. Soc., 1985, **107**, 2083.
1758 U. Kunze, A. Bruns, W. Hiller, and J. Mohyla, Chem. Ber., 1985, **118**, 227.
1759 V.V. Negrebetskii, V.I. Kal'chenko, O.V. Balitskaya, M.Yu. Kornilov, and L.N. Markovskii, Zh. Obshch. Khim., 1984, **54**, 2217 (Chem. Abstr., 1985, **102**, 149 356).
1760 V.G. Torgov, G.V. Verevkhin, V.V. Denisov, and S.V. Tkachev, Zh. Neorg. Khim., 1985, **30**, 1012 (Chem. Abstr., 1985, **102**, 226 785).
1761 J.D. Halliday and P.E. Bindner, Can. J. Chem., 1985, **63**, 2821.
1762 W.F. Bailey, J.J. Patricia, V.C. DelGobbo, R.M. Jarret, and P.J. Okarma, J. Org. Chem., 1985, **50**, 1999.
1763 A. Schmidpeter and G. Burget, Phosphorus Sulphur, 1985, **22**, 323.
1764 G.M. Khvostik, V.N. Sokolov, V.Ya. Minevich, A.N. Shchupik, and G.P. Kondratenkov, Katalit. i Initsiiruyushch. Sistemy dlya Sinteza i Modifik. Polimerov, L., 1984, 13. From Ref. Zh., Khim., 1985, Abstr. No. 9S260 (Chem. Abstr., 1985, **103**, 215 086).
1765 S.S. Lalayan, A.N. Kitaigorodskii, V.Ya. Minevich, Kh.T. Tursunov, E.A. Fushman, A.N. Shupik, and G.M. Khvostik, Z. Fiz. Khim., 1985, **59**, 2516 (Chem. Abstr., 1985, **103**, 221 980).
1766 G. Fink, W. Fenzl, and R. Mynott, Z. Naturforsch., Teil B, 1985, **40**, 158 (Chem. Abstr., 1985, **102**, 167 210).
1767 J.H. Toney and T.J. Marks, J. Am. Chem. Soc., 1985, **107**, 947.
1768 R.F. Jordan, J. Organomet. Chem., 1985, **294**, 321.
1769 C.S. Bajgur, S.B. Jones, and J.L. Petersen, Organometallics, 1985, **4**, 1929.
1770 M.D. Curtis, L.G. Bell, and W.M. Butler, Organometallics, 1985, **4**, 701.
1771 N.M. Doherty and J.E. Bercaw, J. Am. Chem. Soc., 1985, **107**, 2670.
1772 J.T. Gauntlett, B.F. Taylor, and M.J. Winter, J. Chem. Soc., Dalton Trans., 1985, 1815.
1773 J.A. Baumann, G.E. Bossard, T.A. George, D.B. Howell, L.M. Koczon, R.K. Lester, and C.M. Noddings, Inorg. Chem., 1985, **24**, 3568.
1774 E.G. Il'in, B.V. Golovanov, M.E. Ignatov, V.D. Butskii, and Yu.A. Buslaev, Dokl. Akad. Nauk SSSR, 1984, **278**, 1376 (Chem. Abstr., 1985, **102**, 124 573).
1775 R.H. Bullock, C.E.L. Headford, S.E. Kegley, and J.R. Norton, J. Am. Chem. Soc., 1985, **107**, 727.

1776 D.J. Darensbourg and G. Grötsch, J. Am. Chem. Soc., 1985, **107**, 7473.
1777 J. Kress, J.A. Osborn, R.M.E. Greene, K.J. Ivin, and J.J. Rooney, J. Chem. Soc., Chem. Commun., 1985, 874.
1778 R.J. Blau, M.H. Chisholm, K. Folting, and R.J. Wang, J. Chem. Soc., Chem. Commun., 1985, 1582.
1779 A. Marinetti, C. Charrier, F. Mathey, and J. Fischer, Organometallics, 1985, **4**, 2134.
1780 Yu.V. Kokunov, V.A. Bochkareva, S.G. Sakharov, and Yu.A. Buslaev, Koord. Khim., 1985, **11**, 36 (Chem. Abstr., 1985, **102**, 105 085).
1781 E.G. Golovanov, B.V. Ignatov, V.D. Butskii, and I.A. Buslaev, Dokl. Akad. Nauk SSSR, 1985, **282**, 1147 (Chem. Abstr., 1985, **103**, 226 350).
1782 G.L. Crocco and J.A. Gladysz, J. Chem. Soc., Chem. Commun., 1985, 283.
1783 G.L. Crocco and J.A. Gladysz, J. Am. Chem. Soc., 1985, **107**, 4103.
1784 T.J. Barder, F.A. Cotton, L.R. Falvello, and R.A. Walton, Inorg. Chem., 1985, **24**, 1258.
1785 S.C. Wright and M.C. Baird, J. Am. Chem. Soc., 1985, **107**, 6899.
1786 Z. Goldschmidt, H.E. Gottlieb, and D. Cohen, J. Organomet. Chem., 1985, **294**, 219.
1787 E.I. Fedin, A.L. Blumenfeld, P.V. Petrovskii, A.Z. Kreindlin, S.S. Fadeeva, and M.I. Rybinskaya, J. Organomet. Chem., 1985, **292**, 257.
1788 A.L. Balch, L. Latos-Grazynski, and M.W. Renner, J. Am. Chem. Soc., 1985, **107**, 2983.
1789 L. Fielding, G.R. Eaton, and S.S. Eaton, Inorg. Chem., 1985, **24**, 2309.
1790 K.C. Francis, D. Cummins, and J. Oakes, J. Chem. Soc., Dalton Trans., 1985, 493.
1791 E.B. Boyar, A. Dobson, S.D. Robinson, B.L. Haymore, and J.C. Huffman, J. Chem. Soc., Dalton Trans., 1985, 621.
1792 S.B. Colbran, B.F.G. Johnson, J. Lewis, and R.M. Sorrell, J. Organomet. Chem., 1985, **296**, Cl.
1793 P.M. Treichel and P.J. Vincenti, Inorg. Chem., 1985, **24**, 228.
1794 J. Hawecker, J.-M. Lehn, and R. Ziessel, J. Chem. Soc., Chem. Commun., 1985, 56.
1795 T.M. Bockman, J.F. Garst, R.B. King, L. Marko, and F. Ungvary, J. Organomet. Chem., 1985, **279**, 165.
1796 Y.-T. Fanchiang, Inorg. Chim. Acta, 1985, **105**, 209.
1797 H.P.C. Hogenkamp, G.T. Bratt, and S.Z. Sun, Biochemistry, 1985, **24**, 6428 (Chem. Abstr., 1985, **103**, 192 082).
1798 K.L. Brown, S. Ramamurthy, and D.S. Marynick, J. Organomet. Chem., 1985, **287**, 377.
1799 C. Cabal, L. Fiffe, and A. Guzman, Rev. Cubana Fis., 1984, **4**, 9 (Chem. Abstr., 1985, **103**, 166 932).
1800 E.M. Pitre and A. Attalla, Report, 1985, MLM-3226; Order No. DE85008915, 11 pp. Avail. NTIS. From Energy Res. Abstr., 1985, **10**, Abstr. No. 21299 (Chem. Abstr., 1985, **103**, 166 877).
1801 S. Balt and H.J. Gamelkoorn, J. Am. Chem. Soc., 1985, **107**, 3596.
1802 S. Balt and H.J. Gamelkoorn, J. Chem. Soc., Dalton Trans., 1985, 1495.
1803 S. Balt, H.J. Gamelkoorn, and S. Oosterink, Inorg. Chem., 1985, **24**, 2384.
1804 S. Balt, H.J. Gamelkoorn, and K.R. Lammers, Transition Met. Chem., 1985, **10**, 164.
1805 S. Balt and H.J. Gamelkoorn, Inorg. Chem., 1985, **24**, 3347.
1806 M.E. Sosa and M.L. Tobe, J. Chem. Soc., Dalton Trans., 1985, 475.
1807 A.L. Casalnuovo, T. Laska, P.V. Nilsson, J. Olofson, and L.H. Pignolet, Inorg. Chem., 1985, **24**, 182.
1808 I.R. Butler, W.R. Cullen, B.E. Mann, and C.R. Nurse, J. Organomet. Chem., 1985, **280**, C47.
1809 R. van Eldik, S. Aygen, H. Kelm, A.M. Trzeciak, and J.J. Ziółkowski, Transition Met. Chem., 1985, **10**, 167.
1810 H. Yamamoto, S. Shinoda, and Y. Saito, J. Mol. Catal., 1985, **30**, 259.
1811 C.E. Johnson and R. Eisenberg, J. Am. Chem. Soc., 1985, **107**, 6531.
1812 E.B. Boyar, D.S. Moore, S.D. Robinson, B.R. James, M. Preece, and I. Thorburn, J. Chem. Soc., Dalton Trans., 1985, 617.
1813 R.J. McKinney and D.C. Roe, J. Am. Chem. Soc., 1985, **107**, 261.

1814 R. Taube, J.-P. Gehrke, and R. Radeglia, J. Organomet. Chem., 1985, **291**, 101.
1815 D.J. Pasto and N.Z. Huang, Organometallics, 1985, **4**, 1386.
1816 P. Giannoccaro, E. Pannacciulli, and G. Vasapollo, Inorg. Chim. Acta, 1985, **96**, 179.
1817 F. Ozawa, K. Kurihara, T. Yamamoto, and A. Yamamoto, J. Organomet. Chem., 1985, **279**, 233.
1818 R.J. Puddephatt and J.D. Scott, Organometallics, 1985, **4**, 1221.
1819 H.E. Bryndza, Organometallics, 1985, **4**, 406.
1820 K.A. Azam, A.A. Frew, B.R. Lloyd, L. Manojlovic-Muir, K.W. Muir, and R.J. Puddephatt, Organometallics, 1985, **4**, 1400.
1821 R.L. Brainard and G.M. Whitesides, Organometallics, 1985, **4**, 1550.
1822 H.E. Bryndza, J. Chem. Soc., Chem. Commun., 1985, 1696.
1823 P.B. Mackenzie, J. Whelan, and B. Bosnich, J. Am. Chem. Soc., 1985, **107**, 2046.
1824 R. Benn, P.W. Jolly, R. Mynott, B. Raspel, G. Schenker, K.-P. Schick, and G. Schroth, Organometallics, 1985, **4**, 1945.
1825 R. Benn, P.W. Jolly, R. Mynott, and G. Schenker, Organometallics, 1985, **4**, 1136.
1826 R. Bertani, G. Carturan, A. Scrivanti, and A. Wojcicki, Organometallics, 1985, **4**, 2139.
1827 M.K. Reinking, M.L. Kullberg, A.R. Cutler, and C.P. Kubiak, J. Am. Chem. Soc., 1985, **107**, 3517.
1828 S.J.S. Kerrison and P.J. Sadler, Inorg. Chim. Acta, 1985, **104**, 197.
1829 V.N. Zudin, V.D. Chinakov, V.M. Nekipelov, V.A. Likholobov, and Yu.I. Yermakov, J. Organomet. Chem., 1985, **289**, 425.
1830 R.N. Bose, R.D. Cornelius, and R.E. Viola, Inorg. Chem., 1985, **24**, 3989.
1831 S.J.B. Price, M.J. DiMartino, D.T. Hill, R. Kuroda, M.A. Mazid, and P.J. Sadler, Inorg. Chem., 1985, **24**, 3425.
1832 A. Ouassas and B. Frange, Bull. Soc. Chim. Fr., 1984, 336.
1833 R.K. Kanjolia, L.K. Krannich, and C.L. Watkins, Inorg. Chem., 1985, **24**, 445.
1834 S.H. Lawrence, S.G. Shore, T.F. Koetzle, J.C. Huffman, C.Y. Wei, and R. Bau, Inorg. Chem., 1985, **24**, 3171.
1835 D.F. Gaines, J.A. Heppert, and J.C. Kunz, Inorg. Chem., 1985, **24**, 621.
1835a D.F. Gaines, J.C. Kunz, and M.J. Kulzick, Inorg. Chem., 1985, **24**, 3336.
1836 S.L. Emery and J.A. Morrison, Inorg. Chem., 1985, **24**, 1612.
1837 Z.J. Abdou, M. Soltis, B. Oh, G. Siwap, T. Banuelos, W. Nam, and T. Onak, Inorg. Chem., 1985, **24**, 2363.
1838 H.C. Brown, R.G. Naik, B. Singaram, and C. Pyun, Organometallics, 1985, **4**, 1925.
1839 D.W. Hawker and P.R. Wells, Organometallics, 1985, **4**, 821.
1840 R. Damrauer, P. Wheelen, H. Razavi, M. Whalen, and L. Aran, Inorg. Chim. Acta, 1985, **99**, 69.
1841 M.J. Cuthbertson, D.W. Hawker, and P.R. Wells, J. Organomet. Chem., 1985, **287**, 7.
1842 M.J. Cuthbertson and P.R. Wells, J. Organomet. Chem., 1985, **287**, 25.
1843 R.T. Conlin and Y.-W. Kwak, J. Organomet. Chem., 1985, **293**, 177.
1844 J. Chojnowski, M. Cypryk, J. Michalski, and L. Woźniak, J. Organomet. Chem., 1985, **288**, 275.
1845 M. Kobayashi, M. Kobayashi, and M. Yoshida, Bull. Chem. Soc. Jpn., 1985, **58**, 473.
1846 M. Kawanisi, S. Kozima, K. Kobayashi, and T. Hitomi, Kenkyu Hokoku - Asahi Garasu Kogyo Gijutsu Shoreikai, 1984, **44**, 109 (Chem. Abstr., 1985, **103**, 22 675).
1847 K. Rühlmann, J. Brumme, U. Scheim, and H. Grosse-Ruyken, J. Organomet. Chem., 1985, **291**, 165.
1848 U. Scheim, H. Grosse-Ruyken, K. Rühlmann, and A. Porzel, J. Organomet. Chem., 1985, **293**, 29.
1849 E.R. Pohl and F.D. Osterholtz, Polym. Sci. Technol. (Plenum), 1985, **27**, 157 (Chem. Abstr., 1985, **103**, 22 959).
1850 R.H. Meinhold, H.P. Rothbaum, and R.H. Newman, J. Colloid Interface Sci., 1985, **108**, 234 (Chem. Abstr., 1985, **103**, 184 669).

1851 I. Artaki, S. Sinha, A.D. Irwin, and J. Jonas, J. Non-Cryst. Solids, 1985, **72**, 391 (Chem. Abstr., 1985, **103**, 91 861).
1852 I. Artaki, T.W. Zerda, and J. Jonas, Mater. Lett., 1985, **3**, 493 (Chem. Abstr., 1985, **103**, 201 468).
1853 I. Artaki, M. Bradley, T.W. Zerda, and J. Jonas, J. Phys. Chem., 1985, **89**, 4399.
1854 B.N. Chernyshov, N.A. Didenko, and E.G. Ippolitov, Koord. Khim., 1984, **10**, 1640 (Chem. Abstr., 1985, **102**, 88 981).
1855 B.E. Maryanoff, A.B. Reitz, M.S. Mutter, R.R. Inners, and H.R. Almond, jun., J. Am. Chem. Soc., 1985, **107**, 1068.
1856 A. Wawer and I. Wawer, React. Kinet. Catal. Lett., 1984, **26**, 167 (Chem. Abstr., 1985, **103**, 22 105).
1857 R. Ramage, B. Atrash, D. Hopton, and M.J. Parrott, J. Chem. Soc., Perkin Trans. 2, 1985, 1617.
1858 A.V. Moskvin, M.V. Korsakov, N.A. Smorygo, and B.A. Ivin, Zh. Obshch. Khim., 1984, **54**, 2223 (Chem. Abstr., 1985, **102**, 149 357).
1859 P.M. Cullis and A.J. Rous, J. Am. Chem. Soc., 1985, **107**, 6721.
1860 P.M. Cullis, P.B. Kay, and S. Trippett, J. Chem. Soc., Chem. Commun., 1985, 1329.
1861 J.M. Friedman and J.R. Knowles, J. Am. Chem. Soc., 1985, **107**, 6121.
1862 M. Tsuhako, A. Nakajima, T. Miyajima, S. Ohashi, H. Nariai, and I. Motooka, Bull. Chem. Soc. Jpn, 1985, **58**, 3092.
1863 R. Kluger and G.R.J. Thatcher, J. Am. Chem. Soc., 1985, **107**, 6006.
1864 M. Draoui, P. Vast, and G. Palavit, Rev. Chim. Miner., 1985, **22**, 256 (Chem. Abstr., 1985, **103**, 147 805).
1865 G.V. Lagodzinskaya, G.B. Manelis, Z.K. Nikitina, V.I. Shestov, and V.Ya. Rosolovskii, Izv. Akad. Nauk SSSR, Khim., 1985, 781 (Chem. Abstr., 1985, **102**, 226 686).
1866 J. Kowalewski, L. Nordenskioeld, N. Benetis, and P.O. Westlund, Prog. Nucl. Magn. Reson. Spectrosc., 1985, **17**, 141 (Chem. Abstr., 1985, **103**, 152 229).
1867 R.M. Golding and R.O. Pascual, Bull. Magn. Reson., 1983, **5**, 126 (Chem. Abstr., 1985, **102**, 177 926).
1868 E. Butkus and P. Kadzeuskas, Zh. Strukt. Khim., 1985, **26**, 145 (Chem. Abstr., 1985, **103**, 141 050).
1869 M.C.M. Gribnau, C.P. Keijzers, and E. De Boer, Magn. Reson. Rev., 1985, **10**, 161.
1870 S. Ahn, D.H. Kim, and E. Park, Bull. Korean Chem. Soc., 1985, **6**, 63 (Chem. Abstr., 1985, **103**, 31 420).
1871 S. Ahn, S. Oh, and J.H. Yang, Bull. Korean Chem. Soc., 1985, **6**, 255 (Chem. Abstr., 1985, **103**, 226 132).
1872 B. Hessen, J.H. Teuben, T.H. Lemmen, J.C. Huffman, and K.G. Caulton, Organometallics, 1985, **4**, 946.
1873 J. Nieman and J.H. Teuben, J. Organomet. Chem., 1985, **287**, 207.
1874 R.L. Bansemer, J.C. Huffman, and K.G. Caulton, Inorg. Chem., 1985, **24**, 3003.
1875 W.E. Broderick and J.I. Legg, Inorg. Chem., 1985, **24**, 3724.
1876 W.D. Wheeler and J.I. Legg, Inorg. Chem., 1985, **24**, 1292.
1877 W.D. Wheeler and R.D. Poshusta, Inorg. Chem., 1985, **24**, 3100.
1878 A.N. Kitaigorodskii, Izv. Akad. Nauk SSSR, Ser. Khim., 1985, 1518 (Chem. Abstr., 1985, **103**, 214 672).
1879 K. Tanaka, M. Nakamoto, Y. Tashiro, and T. Tanaka, Bull. Chem. Soc. Jpn., 1985, **58**, 316.
1880 C.L. Hill and M.W. Williamson, Inorg. Chem., 1985, **24**, 3024.
1881 C.L. Hill and M.M. Williamson, Inorg. Chem., 1985, **24**, 2836.
1882 U. Abram, H. Spies, W. Görner, R. Kirmse, and J. Stack, Inorg. Chim. Acta, 1985, **109**, L9.
1883 R. Guilard, B. Boisselier-Cocolios, A. Tabard, P. Cocolios, B. Simonet, and K.M. Kadish, Inorg. Chem., 1985, **24**, 2509.
1884 A.L. Balch, R.J. Cheng, G.N. La Mar, and L. Latos-Grazynski, Inorg. Chem., 1985, **24**, 2651.
1885 Y. Ohsawa, M.K. DeArmond, K.W. Hanck, and C.G. Moreland, J. Am. Chem. Soc., 1985, **107**, 5383.
1886 A.L. Balch, Y.W. Chan, G.N. La Mar, L. Latos-Grazynski, and M.W. Renner,

Inorg. Chem., 1985, **24**, 1437.
1887 G.M. Godziela, L.A. Ridnour, and H.M. Goff, Inorg. Chem., 1985, **24**, 1609.
1888 L. Banci, Inorg. Chim. Acta, 1985, **98**, L23.
1889 D.L. Hickman, A. Shirazi, and H.M. Goff, Inorg. Chem., 1985, **24**, 563.
1890 J.H. Osborne and W.C. Trogler, Inorg. Chem., 1985, **24**, 3095.
1891 K. Drabent and L. Latos-Grazynski, Polyhedron, 1985, **4**, 1637.
1892 H. Toi, M. Homma, A. Suzuki, and H. Ogoshi, J. Chem. Soc., Chem. Commun., 1985, 1791.
1893 S.W. Unger, T. Jue, and G.N. La Mar, J. Magn. Reson., 1985, **61**, 448.
1894 A.L. Balch, G.N. La Mar, L. Latos-Grazynski, and M.W. Renner, Inorg. Chem., 1985, **24**, 2432.
1895 S.H. Strauss, M.E. Silver, K.M. Long, R.G. Thompson, R.A. Hudgens, K. Spartalian, and J.A. Ibers, J. Am. Chem. Soc., 1985, **107**, 4207.
1896 L.B. Dugad, V.R. Marathe, and S. Mitra, Proc. - Indian Acad. Sci., Chem. Sci., 1985, **95**, 189 (Chem. Abstr., 1985, **103**, 129 317).
1897 A.L. Balch, G.N. La Mar, L. Latos-Grazynski, M.W. Renner, and V. Thanabal, J. Am. Chem. Soc., 1985, **107**, 3003.
1898 J.T. Groves, R. Quinn, T.J. McMurry, M. Nakamura, G. Lang, and B. Boso, J. Am. Chem. Soc., 1985, **107**, 354.
1899 C.E. Castro and S.E. Anderson, Inorg. Chem., 1985, **24**, 1113.
1900 I. Morishima, Y. Shiro, and T. Wakino, J. Am. Chem. Soc., 1985, **107**, 1063.
1901 R. Young and C.K. Chang, J. Am. Chem. Soc., 1985, **107**, 898.
1902 P. Ganguli, Z. Naturforsch., Teil A, 1985, **40**, 79 (Chem. Abstr., 1985, **102**, 124 303).
1903 L. Latos-Grazynski, Inorg. Chem., 1985, **24**, 1104.
1904 A. Botulinski, J.W. Buchler, B. Tonn, and M. Wicholas, Inorg. Chem., 1985, **24**, 3239.
1905 L.B. Dugad, V.R. Marathe, and S. Mitra, Chem. Phys. Lett., 1985, **120**, 239.
1906 A.V. Kessenikh and G.N. Boiko, Khim. Fiz., 1985, **4**, 919 (Chem. Abstr., 1985, **103**, 114 833).
1907 G. Christou, K.S. Hagen, J.K. Bashkin, and R.H. Holm, Inorg. Chem., 1985, **24**, 1010.
1908 A. Caneschi and A. Dei, Inorg. Chim. Acta, 1985, **107**, L23.
1909 L. Latos-Grazynski, Inorg. Chem., 1985, **24**, 1681.
1910 E. Kwiatkowski and T. Ossowski, J. Coord. Chem., 1985, **14**, 9 (Chem. Abstr., 1985, **103**, 130 033).
1911 W. Peters, M. Fuchs, H. Sicius, and W. Kuchen, Angew. Chem., Int. Ed. Engl., 1985, **24**, 231.
1912 K.V.G. Reddy, K.V.R. Chary, B.A. Sastry, G. Ponticelli, and M. Massacessi, Indian J. Phys., A, 1985, **59A**, 68 (Chem. Abstr., 1985, **102**, 124 279).
1913 G. Jeske, L.E. Schock, P.N. Swepston, H. Schumann, and T.J. Marks, J. Am. Chem. Soc., 1985, **107**, 8103.
1914 L. Xing-Fu, S. Eggers, J. Kopf, W. Jahn, R.D. Fischer, C. Apostolidis, B. Kanellakopulos, F. Benetollo, A. Polo, and G. Bombieri, Inorg. Chim. Acta, 1985, **100**, 183.
1915 T.A. Babushkina, L.B. Lazukova, V.F. Zolin, and L.G. Koreneva, Koord. Khim., 1985, **11**, 610 (Chem. Abstr., 1985, **103**, 47 275).
1916 P.N. Kapoor, R. Saraswati, and I.J. McMahon, Inorg. Chim. Acta, 1985, **110**, 63.
1917 L.A. Fedorov, P.V. Petrovskii, L.V. Ruzaikina, A.N. Ermakov, and B.F. Myasoedov, Dokl. Akad. Nauk SSSR, 1985, **280**, 649 (Chem. Abstr., 1985, **102**, 230 975).
1918 O.A. Gansow and A.R. Kausar, Inorg. Chim. Acta, 1985, **109**, 1.
1919 Z.H. Xu, J.Y. Guo, J. Huang, J.G. Wu, and G.X. Xu, Huaxue Xuebao, 1985, **43**, 733 (Chem. Abstr., 1985, **103**, 152 649).
1920 J.D. Simon, W.R. Moomaw, and T.M. Ceckler, J. Phys. Chem., 1985, **89**, 5659.
1921 T.J. Wenzel, A.C. Ruggles, and D.R. Lalonde, jun., Magn. Reson. Chem., 1985, **23**, 778.
1922 P.C.M. van Zijl, R.P. van Wezel, C. MacLean, and A.A. Bothner-By, J. Phys. Chem., 1985, **89**, 204.
1923 A. Dambska and A. Janowski, Ser. Fiz. (Uniw. im. Adama Mickiewicza Poznaniu), 1985, **53**, 93 (Chem. Abstr., 1985, **103**, 114 849).

1924 M. Audit, D. Davoust, N. Goasdoue, and N. Platzer, Magn Reson. Chem., 1985, **23**, 33.
1925 K. Iftikhar and N. Ahmad, Polyhedron, 1985, **4**, 333.
1926 J.A. Peters, M.S. Nieuwenhuizen, and D.J. Raber, J. Magn. Reson., 1985, **65**, 417.
1927 B. Ban, G. Folcher, H. Marquet-Ellis, and P. Rigny, Nouv. J. Chim., 1985, **9**, 51 (Chem. Abstr., 1985, **103**, 105 093).
1928 G. Rossetto, M. Porchia, F. Ossola, P. Zanella, and R.D. Fischer, J. Chem. Soc., Chem. Commun., 1985, 1460.
1929 I. Santos, N. Marques, and A.P. de Matos, Inorg. Chim. Acta, 1985, **110**, 149.
1930 P. Zanella, G. Paolucci, G. Rossetto, F. Benetollo, A. Polo, R.D. Fischer, and G. Bombieri, J. Chem. Soc., Chem. Commun., 1985, 96.
1931 J.G. Brennan and R.A. Andersen, J. Am. Chem. Soc., 1985, **107**, 514.
1932 A. Dormond, A. Aaliti, and C. Moise, J. Chem. Soc., Chem. Commun., 1985, 1231.
1933 L. Latos-Grazynski, Polyhedron, 1985, **4**, 731.
1934 P.S. Belton, S.F. Tanner, and K.M. Wright, Nucl. Magn. Reson., 1985, **14**, 137 (Chem. Abstr., 1985, **103**, 225 743).
1935 A.R. Grimmer, Tagungsber., Wiss. Tag. - Tech. Hochsch. Karl-Marx-Stadt, 1984, (12, Festkoerperanal., Band 2), 415 (Chem. Abstr., 1985, **102**, 55 128).
1936 A.P. Legrand, Actual. Chim., 1985, 27 (Chem. Abstr., 1985, **102**, 177 751).
1937 A.D.H. Clague, Catalysis (London), 1985, **7**, 75 (Chem. Abstr., 1985, **103**, 225 741).
1938 E. Oldfield and R.J. Kirkpatrick, Science (Washington, D.C., 1883-), 1985, **227**, 1537 (Chem. Abstr., 1985, **102**, 177 741).
1939 D. Brinkmann, Helv. Phys. Acta, 1985, **58**, 167 (Chem. Abstr., 1985, **103**, 56 671).
1940 M. Punkkinen, E.E. Ylinen, and L.P. Ingman, Bull. Magn. Reson., 1983, **5**, 133 (Chem. Abstr., 1985, **102**, 177 737).
1941 A. Rigamonti, Adv. Phys., 1984, **33**, 115 (Chem. Abstr., 1985, **102**, 15 293).
1942 J. Kaerger, H. Pfeifer, and W. Heink, Proc. Int. Zeolite Conf., 6th 1983 (Pub. 1984), 184 (Chem. Abstr., 1985, **103**, 43 088).
1943 D. Mueller, W. Gessner, and G. Scheler, Mitteilungsbl. - Chem. Ges. D.D.R., 1984, **31**, 107 (Chem. Abstr., 1985, **102**, 71 312).
1944 J.P. Renard, NATO ASI Ser., Ser. C, 1985, **140**, 207 (Chem. Abstr., 1985, **103**, 31 290).
1945 S. Hayashi, Kagaku Kogyo Shiryo (Tsukuba-gun, Jpn.), 1985, **19**, 166 (Chem. Abstr., 1985, **102**, 230 481).
1946 H. Estrade-Szwarckopf, Helv. Phys. Acta, 1985, **58**, 139.
1947 A.R. Grimmer, Mitteilungsbl. - Chem. Ges. D.D.R., 1984, **31**, 200 (Chem. Abstr., 1985, **102**, 71 316).
1948 P.J. Bray, J. Non-Cryst. Solids, 1985, **73**, 19 (Chem. Abstr., 1985, **103**, 129 200).
1949 D. Plee, A. Schutz, G. Poncelet, and J.J. Fripiat, Stud. Surf. Sci. Catal., 1985, **20**, 343 (Chem. Abstr., 1985, **102**, 226 554).
1950 T. Sheng, Shiyou Huagong, 1985, **14**, 52 (Chem. Abstr., 1985, **103**, 31 285).
1951 H. Pfeifer, D. Freude, and M. Hunger, Zeolites, 1985, **5**, 274 (Chem. Abstr., 1985, **103**, 201 308).
1952 G. Villeneuve, Inorg. Solid Fluorides, 1985, 493 (Chem. Abstr., 1985, **103**, 114 665).
1953 H. Pfeifer, Congr. AMPERE Magn. Reson. Relat. Phenom., Proc., 22nd, 1984, 269. Ed. by K.A. Mueller, R. Kind, and J. Roos (Chem. Abstr., 1985, **102**, 104 660).
1954 R.F. Howe, Springer Ser. Chem. Phys., 1984, **35**, 39 (Chem. Abstr., 1985, **102**, 32 979).
1955 K.I. Zamaraev and V.M. Mastikhin, Proc. Indian Natl. Sci. Acad., Part A, 1985, **51**, 180 (Chem. Abstr., 1985, **102**, 191 987).
1956 L. Barbic, Congr. AMPERE Magn. Reson. Relat. Phenom., Proc., 22nd, 1984, 281. Ed. by K.A. Mueller, R. Kind, and J. Roos (Chem. Abstr., 1985, **102**, 104 661).
1957 W. Jones, J.M. Thomas, T.B. Tennakoon, R. Schlogl, and P. Diddams, ACS Symp. Ser., 1985, **288**, 472 (Chem. Abstr., 1985, **103**, 201 492).

1958 J. Kaerger, Adv. Colloid Interface Sci., 1985, **23**, 129 (Chem. Abstr., 1985, **103**, 147 629).
1959 T. Cosgrove, Nucl. Magn. Reson., 1984, **13**, 318 (Chem. Abstr., 1985, **102**, 38 170).
1960 K.I. Zamaraev and V.M. Mastikhin, Colloids Surf., 1984, **12**, 401 (Chem. Abstr., 1985, **102**, 88 706).
1961 A.M. Zaper and J.L. Koenig, Adv. Colloid Interface Sci., 1985, **22**, 113 (Chem. Abstr., 1985, **103**, 147 615).
1962 J.B. Nagy, G. Engelhardt, and D. Michel, Adv. Colloid Interface Sci., 1985, **23**, 67 (Chem. Abstr., 1985, **103**, 147 628).
1963 S. Shinoda and Y. Saito, Hyomen, 1984, **22**, 627 (Chem. Abstr., 1985, **102**, 71 308).
1964 C. Doremieux-Morin and J. Fraissard, Stud. Surf. Sci. Catal., 1985, **20**, 299 (Chem. Abstr., 1985, **102**, 226 381).
1965 D. Freude, Adv. Colloid Interface Sci., 1985, **23**, 21 (Chem. Abstr., 1985, **103**, 147 627).
1966 B. Boddenberg, R. Grosse, W. Horstmann, and G. Neue, Colloids Surf., 1984, **11**, 265 (Chem. Abstr., 1985, **102**, 32 841).
1967 V.B. Kazanskii and V.Yu. Borovkov, Colloids Surf., 1984, **12**, 429 (Chem. Abstr., 1985, **102**, 84 784).
1968 L.J. Magid and C.A. Martin, Reverse Micelles, [Proc. Eur. Sci. Found. Workshop], 4th, 1982 (Pub. 1984), 181 (Chem. Abstr., 1985, **102**, 138 206).
1969 I.A. Nesterov, V.I. Volkov, V.I. Muromtsev, K.K. Pukhov, and S.F. Timashev, Sovrem. Metody YaMR EPR Khim. Tverd. Tela, [Mater. Vses. Koord. Soveshch.], 4th, 1985, 107 (Chem. Abstr., 1985, **103**, 221 519).
1970 H. Birli, L. Schimmele, and R. Messer, Congr. AMPERE Magn. Reson. Relat. Phenom., Proc., 22nd, 1984, 171. Ed. by K.A. Mueller, R. Kind, and J. Roos (Chem. Abstr., 1985, **102**, 124 317).
1971 J. Shinar, J. Less-Common Met., 1984, **104**, 87 (Chem. Abstr., 1985, **102**, 101 030).
1972 T. Asai and S. Kawai, Zairyo, 1984, **33**, 1395 (Chem. Abstr., 1985, **102**, 37 639).
1973 K. Larsson and J. Tegenfeldt, J. Magn. Reson., 1985, **64**, 451.
1974 M. Mali, J. Roos, and D. Brinkmann, Congr. AMPERE Magn. Reson. Relat. Phenom., Proc., 22nd, 1984, 177. Ed. by K.A. Mueller, R. Kind, and J. Roos (Chem. Abstr., 1985, **102**, 191 581).
1975 S.J. Visco, P. Spellane, and J.H. Kennedy, J. Electrochem. Soc., 1985, **132**, 751 (Chem. Abstr., 1985, **102**, 169 724).
1976 A.S. Montjoie and W. Mueller-Warmuth, Z. Naturforsch., Teil A, 1985, **40**, 596 (Chem. Abstr., 1985, **103**, 63 752).
1977 H. Ohno, S. Konishi, T. Nagasaki, T. Kurasawa, H. Katsuta, and H. Watanabe, J. Nucl. Mater., 1985, **133**, 181 (Chem. Abstr., 1985, **103**, 148 975).
1978 I.S. Pronin, A.A. Vashman, and A.S. Nikiforov, At. Energ., 1985, **59**, 227 (Chem. Abstr., 1985, **103**, 222 649).
1979 G.A. Elgavish and A. Elgavish, Biochem. Biophys. Res. Commun., 1985, **128**, 746 (Chem. Abstr., 1985, **102**, 216 977).
1980 O. Kanert, R. Kuechler, and M. Olthoff, Congr. AMPERE Magn. Reson. Relat. Phenom., Proc., 22nd, 1984, 200. Ed. by K.A. Mueller, R. Kind, and J. Roos (Chem. Abstr., 1985, **102**, 123 357).
1981 O. Kanert, R. Kuechler, and W. Kuechler, Solid State Commun., 1985, **54**, 999 (Chem. Abstr., 1985, **103**, 79 730).
1982 L. Latie, G. Villeneuve, C. Cros, and P. Hagenmuller, Phys. Status Solidi B, 1985, **128**, 475 (Chem. Abstr., 1985, **102**, 195 942).
1983 T. Matsuo and H. Ohno, J. Chem. Phys., 1985, **82**, 3968.
1984 M. Stioui, A. Grayevsky, A. Moran, S. Kreitzman, N. Kaplan, and D. Shaltiel, J. Less-Common Met., 1984, **104**, 119 (Chem. Abstr., 1985, **102**, 101 255).
1985 M. Jerosch-Herold, L.T. Lu, D.R. Torgeson, D.T. Peterson, R.G. Barnes, and P.M. Richards, Z. Naturforsch., Teil A, 1985, **40**, 222 (Chem. Abstr., 1985, **102**, 191 609).
1986 D.R. Torgeson, L.T. Lu, T.T. Phua, R.G. Barnes, D.T. Peterson, and E.F.W. Seymour, J. Less-Common Met., 1984, **104**, 79 (Chem. Abstr., 1985, **102**, 88 860).

1987 T.T. Phua, D.R. Torgeson, R.G. Barnes, R.J. Schoenberger, D.T. Peterson, M. Belhoul, G.A. Styles, and E.F.W. Seymour, J. Less-Common Met., 1984, **104**, 105 (Chem. Abstr., 1985, **102**, 88 861).
1988 K. Fuda, K. Kishio, S. Yamauchi, K. Fueki, and Y. Onoda, J. Phys. Chem. Solids, 1984, **45**, 1253 (Chem. Abstr., 1985, **102**, 159 308).
1989 K. Fuda, K. Kishio, S. Yamauchi, and K. Fueki, J. Phys. Chem. Solids, 1985, **46**, 1141 (Chem. Abstr., 1985, **103**, 203 934).
1990 O.J. Zogal, K. Hoffmann, W. Petrynski, H. Drulis, and B. Stalinski, J. Less-Common Met., 1984, **101**, 259 (Chem. Abstr., 1985, **102**, 104 876).
1991 K. Hirao and N. Soga, J. Am. Ceram. Soc., 1985, **68**, 515 (Chem. Abstr., 1985, **103**, 201 287).
1992 K. Morimoto and K. Shimomura, J. Phys. C, 1985, **18**, 403 (Chem. Abstr., 1985, **102**, 88 863).
1993 V.A. Vopilov, E.A. Vopilov, V.M. Buznik, I.V. Murin, and S.V. Chernov, Sovrem. Metody YaMR EPR Khim. Tverd. Tela, [Mater. Vses. Koord. Soveshch.], 4th, 1985, 73 (Chem. Abstr., 1985, **103**, 201 271).
1994 V.A. Vopilov, A.I. Livshits, V.M. Vuznik, P.P. Fedorov, and B.P. Sobolev, Yad. Magn. Rezon. Strukt. Krist., 1984, 158 (Chem. Abstr., 1985, **103**, 225 230).
1995 A.F. Aalders, A.F.M. Arts, and H.W. De Wijn, Phys. Rev. B: Condens. Matter, 1985, **32**, 5412 (Chem. Abstr., 1985, **103**, 188 299).
1996 M. Belhoul, G.A. Styles, E.F.W. Seymour, T.T. Phua, R.G. Barnes, D.R. Torgeson, R.J. Schoenberger, and D.T. Peterson, J. Phys. F, 1985, **15**, 1045 (Chem. Abstr., 1985, **102**, 191 659).
1997 B. Nowak and M. Minier, J. Less-Common Met., 1984, **101**, 245 (Chem. Abstr., 1985, **102**, 104 875).
1998 C. Korn and S. Goren, J. Less-Common Met., 1984, **104**, 113 (Chem. Abstr., 1985, **102**, 88 862).
1999 J. Shinar, D. Davidov, and D. Shaltiel, Phys. Rev. B: Condens. Matter, 1984, **30**, 6331 (Chem. Abstr., 1985, **102**, 38 363).
2000 J. Shinar, J. Less-Common Met., 1984, **104**, 125 (Chem. Abstr., 1985, **102**, 104 848).
2001 O. de O. Damasceno, A.F. De Oliveira, J. De Oliveira, A. Baudry, and P. Boyer, Solid State Commun., 1985, **53**, 363 (Chem. Abstr., 1985, **102**, 155 149).
2002 M. Bose, K. Roy, and A. Ghoshray, Proc. Nucl. Phys. Solid State Phys. Symp. 1982 (Pub. 1984), **25C**, 219 (Chem. Abstr., 1985, **102**, 104 869).
2003 M. Bose, A. Basu, D. Mazumdar, and D.N. Bose, Solid State Ionics, 1985, **15**, 101 (Chem. Abstr., 1985, **103**, 8977).
2004 Z. Li, G. Wang, L. Chen, G. Che, and L. Wang, Wuli Xuebao, 1985, **34**, 263 (Chem. Abstr., 1985, **102**, 158 747).
2005 Yu.G. Kriger, A.V. Mishchenko, S.P. Gabuda, and V.E. Fedorov, Sovrem Metody YaMR EPR Khim. Tverd. Tela, [Mater. Vses. Koord. Soveshch.], 4th, 1985, 89 (Chem. Abstr., 1985, **103**, 225 224).
2006 Cl. Ritter, W. Müller-Warmuth, and R. Schöllhorn, J. Chem. Phys., 1985, **83**, 6130.
2007 L.O. Avtovmyan, L.N. Erofeev, A.I. Korosteleva, L.S. Leonova, V.P. Tarasov, E.A. Ukshe, and V.G. Shteinberg, Khim. Fiz., 1984, **3**, 1518 (Chem. Abstr., 1985, **102**, 15 750).
2008 A. Hardwick, P.G. Dickens, and R.C.T. Slade, Solid State Ionics, 1984, **13**, 345 (Chem. Abstr., 1985, **102**, 15 648).
2009 A. Dommann, M. Mali, J. Roos, and D. Brinkmann, Congr. AMPERE Magn. Reson. Relat. Phenom., Proc., 22nd, 1984, 173. Ed. by K.A. Mueller, R. Kind, and J. Roos (Chem. Abstr., 1985, **102**, 141 592).
2010 J. Tritt-Goc, N. Pislewski, and U. Haeberlen, Ser. Fiz. (Uniw. im. Adama Mickiewicza Poznaniu), 1985, **53**, 107 (Chem. Abstr., 1985, **103**, 152 405).
2011 J. Tritt-Goc and N. Pislewski, Phys. Status Solidi A, 1985, **87**, K41 (Chem. Abstr., 1985, **102**, 155 173).
2012 E.R. Mognaschi and A. Chierico, J. Phys. C, 1985, **18**, 3815 (Chem. Abstr., 1985, **103**, 80 742).
2013 Y. Furukawa, S.-e. Gima, and D. Nakamura, Ber. Bunsen Phys. Chem., 1985, **89**, 863.

2014 I. Svare, A.M. Raaen, and B.O. Fimland, Congr. AMPERE Magn. Reson. Relat. Phenom., Proc., 22nd, 1984, 359. Ed. by K.A. Mueller, R. Kind, and J. Roos (Chem. Abstr., 1985, **102**, 104 902).
2015 A.S. Mahgoub and S. Clough, Congr. AMPERE Magn. Reson. Relat. Phenom., Proc., 22nd, 1984, 414. Ed. by K.A. Mueller, R. Kind, and J. Roos (Chem. Abstr., 1985, **102**, 124 326).
2016 A.S. Mahgoub and S. Clough, J. Phys. C, 1985, **18**, 647 (Chem. Abstr., 1985, **102**, 88 879).
2017 G. Balzer, H. Jain, O. Kanert, and D. Wolf, Congr. AMPERE Magn. Reson. Relat. Phenom., Proc., 22nd, 1984, 95. Ed. by K.A. Mueller, R. Kind, and J. Roos (Chem. Abstr., 1985, **102**, 142 067).
2018 H. Arribart, J. Phys. (Les Ulis, Fr.), 1985, **46**, 83 (Chem. Abstr., 1985, **102**, 71 461).
2019 N.K. Moroz and S.G. Kozlova, Sovrem. Metody YaMR EPR Khim. Tverd. Tela, [Mater. Vses. Koord. Soveshch.], 4th, 1985, 124 (Chem. Abstr., 1985, **103**, 224 950).
2020 J. Conard, H. Hestrade-Szwarkopf, C. Poinsignon, and A.J. Dianoux, J. Phys., Colloq., 1984, 169 (Chem. Abstr., 1985, **102**, 28 599).
2021 W.D. Basler, J. Chem. Phys., 1985, **82**, 5297.
2022 A.G. Bezus, M. Kocirik, and A.A. Lopatkin, Zeolites, 1984, **4**, 346 (Chem. Abstr., 1985, **102**, 32 955).
2023 M. Hunger, Z. Phys. Chem. (Leipzig), 1985, **266**, 135 (Chem. Abstr., 1985, **102**, 124 370).
2024 V.G. Khutsishvili, Yu.S. Bogachev, V.I. Volkov, B.V. Tarasova, N.A. Dreiman, N.N. Shapet'ko, and S.F. Timashev, Zh. Fiz. Khim., 1984, **58**, 2633 (Chem. Abstr., 1985, **102**, 33 064).
2025 A.L. Blumfeld, Yu.V. Isaev, and Yu.N. Novikov, Synth. Met., 1985, **10**, 193 (Chem. Abstr., 1985, **102**, 172 938).
2026 F. Borsa, M. Corti, A. Rigamonti, and S. Torre, Phys. Rev. Lett., 1984, **53**, 2102 (Chem. Abstr., 1985, **102**, 15 441).
2027 T.C. Moore, W.D. McCormick, and C.G. Wade, J. Phys. Chem., 1985, **89**, 3936.
2028 K.R. Sridharan, W.T. Sobol, and M.M. Pintar, J. Chem. Phys., 1985, **82**, 4886.
2029 S.D. Letyuchenko, E.I. Slin'ko, K.D. Tovstyuk, A.G. Khandozhko, and N.K. Tovstyuk, Phys. Status Solidi B, 1985, **128**, K87 (Chem. Abstr., 1985, **102**, 157 081).
2030 M. Prager and M. Mueller-Warmuth, Z. Naturforsch., Teil A, 1984, **39**, 1187 (Chem. Abstr., 1985, **102**, 204 059).
2031 V.A. Vopilov, V.M. Buznik, A.N. Matsulev, V.L. Bogdanov, A.Kh. Karapetyan, and V.D. Khalilev, Fiz. Khim. Stekla, 1985, **11**, 162 (Chem. Abstr., 1985, **102**, 224 817).
2032 R.E. Wasylishen, O. Knop, and J.B. Macdonald, Solid State Commun., 1985, **56**, 581 (Chem. Abstr., 1985, **103**, 224 808).
2033 R.A. Marino and S. Bulusu, J. Energ. Mater., 1985, **3**, 57 (Chem. Abstr., 1985, **103**, 89 867).
2034 G.E. Kibrik, I.A. Kjuntsel, E.S. Kozlov, V.A. Mokeeva, and G.B. Soifer, Magn. Reson. Chem., 1985, **23**, 903.
2035 V.V. Rodin, A.N. Arkhangel'skii, F.Sh. Isangalin, and V.Ya. Volkov, Kriobiol. Kriomed., 1984, **14**, 8 (Chem. Abstr., 1985, **102**, 71 533).
2036 Yu.N. Moskvich, O.V. Rozanov, A.A. Sukhovskii, and I.P. Aleksandrova, Sovrem. Metody YaMR EPR Khim. Tverd. Tela, [Mater. Vses. Koord. Soveshch.], 4th, 1985, 97 (Chem. Abstr., 1985, **103**, 225 559).
2037 Yu.N. Moskvich, O.V. Rozanov, A.A. Sukhovskii, and I.P. Aleksandrova, Ferroelectrics, 1985, **63**, 83 (Chem. Abstr., 1985, **103**, 97 256).
2038 Z. Toffano, K. Bechgaard, and A. Moradpour, Mol. Cryst. Liq. Cryst., 1985, **119**, 151 (Chem. Abstr., 1985, **102**, 159 330).
2039 C.I. Ratcliffe, J.A. Ripmeester, and J.S. Tse, Chem. Phys. Lett., 1985, **120**, 427 (Chem. Abstr., 1985, **103**, 226 076).
2040 E.F.W. Seymour and C.A. Sholl, J. Phys. C, 1985, **18**, 4521 (Chem. Abstr., 1985, **103**, 114 852).
2041 E. Oldfield, M.A. Keniry, S. Shinoda, S. Schramm, T.L. Brown, and H.S. Gutowsky, J. Chem. Soc., Chem. Commun., 1985, 791.
2042 J.A. Gurak, J.W. Chinn, jun., R.J. Lagow, R.D. Kendrick, and C.S. Yannoni,

Inorg. Chim. Acta, 1985, **96**, L75.
2043 J.P. Boehmer, K.R. Metz, and R.W. Briggs, J. Magn. Reson., 1985, **62**, 322.
2044 D.T. Amm, S.L. Segel, R.D. Heyding, and B.K. Hunter, J. Chem. Phys., 1985, **82**, 2529.
2045 D.T. Amm and S.L. Segel, Congr. AMPERE Magn. Reson. Relat. Phenom., Proc., 22nd, 1984, 52. Ed. by K.A. Müller, R. Kind, and J. Roos (Chem. Abstr., 1985, **102**, 104 897).
2046 L. Chen, Z. Zhao, C. Wang, and Z. Li, Wuli Xuebao, 1985, **34**, 1027 (Chem. Abstr., 1985, **103**, 151 585).
2047 F. Holuj, Ferroelectrics, 1985, **65**, 63 (Chem. Abstr., 1985, **103**, 80 754).
2048 F. Holuj, Ferroelectrics, 1985, **65**, 55 (Chem. Abstr., 1985, **103**, 80 753).
2049 T. Kanashiro, T. Ohno, and M. Satoh, J. Phys. Soc. Jpn., 1985, **54**, 2720 (Chem. Abstr., 1985, **103**, 97 698).
2050 G.J.T. Tiddy, K. Rendall, and M.A. Trevethan, Comun. Jorn. Com. Esp. Deterg., 1984, **15**, 51 (Chem. Abstr., 1985, **103**, 11 861).
2051 S.J. Visco, P.J. Spellane, and J.H. Kennedy, J. Electrochem. Soc., 1985, **132**, 1766 (Chem. Abstr., 1985, **103**, 73 815).
2052 C. Ye, L. Li, J. Qiu, J. Yang, and X. Wu, Kexue Tongbao (Foreign Lang. Ed.), 1985, **30**, 598 (Chem. Abstr., 1985, **103**, 152 411).
2053 M. Alla and I. Heinmaa, Congr. AMPERE Magn. Reson. Relat. Phenom., Proc., 22nd, 1984, 560. Ed. by K.A. Müller, R. Kind, and J. Roos (Chem. Abstr., 1985, **102**, 124 332).
2054 Q. Meng, Kexue Tongbao (Foreign Lang. Ed.), 1984, **29**, 1174 (Chem. Abstr., 1985, **102**, 38 637).
2055 A. Trokiner, P.P. Man, H. Theveneau, and P. Papon, Solid State Commun., 1985, **55**, 929 (Chem. Abstr., 1985, **103**, 132 751).
2056 G.L. Turner, S.E. Chung, and E. Oldfield, J. Magn. Reson., 1985, **64**, 316.
2057 V.N. Romannikov, L.S. Chumachenko, V.M. Mastikhin, and K.G. Ione, J. Catal., 1985, **94**, 508 (Chem. Abstr., 1985, **103**, 43 383).
2058 G.C. Gobbi, R. Christoffersen, M.T. Otten, B. Miner, P.R. Buseck, G.J. Kennedy, and C.A. Fyfe, Chem. Lett., 1985, 771.
2059 A. Rettel, W. Gessner, D. Müller, and G. Scheler, Br. Ceram. Trans. J., 1985, **84**, 25 (Chem. Abstr., 1985, **103**, 10 488).
2060 N. Weiden and H. Rager, Z. Naturforsch., Teil A, 1985, **40**, 126 (Chem. Abstr., 1985, **102**, 159 347).
2061 R. Stodolski, D. Heidemann, W. Wieker, and W. Pilz, Z. Anorg. Allg. Chem., 1985, **527**, 150.
2062 W. Gessner, D. Müller, and H. Schubert, Z. Anorg. Allg. Chem., 1985, **524**, 59.
2063 R.J.B. Jakeman, A.K. Cheetham, N.J. Clayden, and C.M. Dobson, J. Am. Chem. Soc., 1985, **107**, 6249.
2064 W.P. Aue, A.H. Roufosse, M.J. Glimcher, and R.G. Griffin, Biochemistry, 1984, **23**, 6110 (Chem. Abstr., 1985, **102**, 21 982).
2065 A. Lasys, TIZ, 1985, **109**, 44 (Chem. Abstr., 1985, **102**, 153 857).
2066 W.J. Dell, R.V. Mulkern, P.J. Bray, M.J. Weber, and S.A. Brawer, Phys. Rev. B: Condens. Matter, 1985, **31**, 2624 (Chem. Abstr., 1985, **102**, 142 092).
2067 B. Gunther and D.C. Ailion, Congr. AMPERE Magn. Reson. Relat. Phenom., Proc., 22nd, 1984, 588. Ed. by K.A. Müller, R. Kind, and J. Roos (Chem. Abstr., 1985, **102**, 124 339).
2068 V.A. Atsarkin and A.E. Mefed, Congr. AMPERE Magn. Reson. Relat., Phenom., Proc., 22nd, 1984, 558. Ed. by K.A. Müller, R. Kind, and J. Roos (Chem. Abstr., 1985, **102**, 124 331).
2069 M. Goldman, Bull. Magn. Reson., 1983, **5**, 110 (Chem. Abstr., 1985, **102**, 177 924).
2070 M. Velter-Stefanescu and S.V. Nistor, Congr. AMPERE Magn. Reson. Relat. Phenom., Proc., 22nd, 1984, 225. Ed. by K.A. Müller, R. Kind, and J. Roos (Chem. Abstr., 1985, **102**, 142 036).
2071 A. Raizman, D. Zamir, and R.M. Cotts, Phys. Rev. B: Condens. Matter, 1985, **31**, 3384 (Chem. Abstr., 1985, **102**, 159 340).
2072 Y. Roinel, V. Bouffard, J.F. Jacquinot, C. Fermon, and G. Fournier, J. Phys. (Les Ulis, Fr.), 1985, **46**, 1699 (Chem. Abstr., 1985, **103**, 226 084).
2073 V.Ya. Kavun, V.I. Sergienko, V.I. Kostin, B.N. Chernyshov, and N.I.

Steblevskaya, Zh. Neorg. Khim., 1985, **30**, 341 (Chem. Abstr., 1985, **102**, 214 101).
2074 A.S. Karnachev, N.M. Kovtun, M.I. Kurkin, E.E. Solov'ev, A.A. Tkachenko, and M.M. Lukina, Zh. Eksp. Teor. Fiz., 1985, **88**, 988 (Chem. Abstr., 1985, **102**, 196 650).
2075 J. Englich, M. Brabenec, P. Novak, and H. Luetgemeier, J. Magn. Magn. Mater., 1985, **50**, 74 (Chem. Abstr., 1985, **103**, 80 686).
2076 O.E. Koptsov, V.N. Lukashin, I.V. Starostenko, and M.N. Uspenskii, Yader. Izluch. v Nauke i Tekhn., M., 1984, 33. From Ref. Zh., Fiz. (A-D), 1985, Abstr. No. 4N1109 (Chem. Abstr., 1985, **103**, 63 633).
2077 V.A. Borodin, V.D. Doroshev, and T.N. Tarasenko, Fiz. Tverd. Tela (Leningrad), 1985, **27**, 583 (Chem. Abstr., 1985, **102**, 177 855).
2078 S. Takagi, N. Niitsuma, T. Suzuki, and T. Kasuya, J. Magn. Magn. Mater., 1985, **52**, 301 (Chem. Abstr., 1985, **103**, 226 040).
2079 E.P. Zeer, O.V. Falaleev, and V.E. Zobov, Sovrem. Metody YaMR EPR Khim. Tverd. Tela, [Mater. Vses. Koord. Soveshch.], 4th, 1985, 85 (Chem. Abstr., 1985, **103**, 226 093).
2080 V.E. Zobov, Yad. Mag. Rezon. Strukt. Krist., 1984, 3. Ed. by E.P. Zeer (Chem. Abstr., 1985, **103**, 226 105).
2081 R.C. Bowman, jun., B.D. Craft, W.E. Tadlock, E.L. Venturini, and J.S. Cantrell, J. Appl. Phys., 1985, **57**, 3036 (Chem. Abstr., 1985, **102**, 230 548).
2082 A.F. Shevakin and E.Yu. Kulefeeva, Metallofizika (Akad. Nauk Ukr. SSR, Otd. Fiz.), 1984, **6**, 91 (Chem. Abstr., 1985, **102**, 38 356).
2083 R.C. Bowman, jun., B.D. Craft, J.S. Cantrell, and E.L. Venturini, Phys. Rev. B: Condens. Matter, 1985, **31**, 5604 (Chem. Abstr., 1985, **102**, 230 738).
2084 D.T. Ding, A.P. Sedee, T.O. Klaassen, and N.J. Poulis, Physica B+C (Amsterdam), 1985, **132**, 170 (Chem. Abstr., 1985, **103**, 152 424).
2085 M.Yu. Belyaev and A.P. Stepanov, Sovrem. Metody YaMR EPR Khim. Tverd. Tela, [Mater. Vses. Koord. Soveshch.], 4th, 1985, 65 (Chem. Abstr., 1985, **103**, 226 092).
2086 S. Dave and R.K. Maccrone, J. Non-Cryst. Solids, 1985, **71**, 303 (Chem. Abstr., 1985, **103**, 31 438).
2087 G. Li, H. Wang, Y. Qu, D. Jiao, R. Qiao, and Z. Li, Bopuxue Zazhi, 1985, **2**, 1 (Chem. Abstr., 1985, **103**, 225 975).
2088 A. Mikou, J.P. Laval, B. Frit, and J. Senegas, Rev. Chim. Miner., 1985, **22**, 115 (Chem. Abstr., 1985, **103**, 132 793).
2089 M.H. Whangbo, J. Rouxel, and L. Trichet, Inorg. Chem., 1985, **24**, 1824.
2090 M. Bose, K. Roy, and A. Ghoshray, J. Phys. C, 1984, **17**, 5277 (Chem. Abstr., 1985, **102**, 15 430).
2091 V.N. Kravchenko, I.V. Shtambur, and A.Ya. Yakunin, Materialy i Pribory Radioelektron., Dnepropetrovsk, 1984, 83. From Ref. Zh., Khim., 1985, Abstr. No. 15B2231 (Chem. Abstr., 1985, **103**, 226 133).
2092 A.V. Simakov and S.A. Veniaminov, React. Kinet. Catal. Lett., 1985, **28**, 67 (Chem. Abstr., 1985, **103**, 201 504).
2093 R.I. Maksimovskaya, N.N. Chumachenko, and D.V. Tarasova, React. Kinet. Catal. Lett., 1985, **28**, 111 (Chem. Abstr., 1985, **103**, 221 603).
2094 S. Segel, Solid State Commun., 1985, **54**, 403 (Chem. Abstr., 1985, **102**, 230 742).
2095 S.L. Segel, Congr. AMPERE Magn. Reson. Relat. Phenom., Proc., 22nd, 1984, 50. Ed. by K.A. Müller, R. Kind, and J. Roos (Chem. Abstr., 1985, **102**, 123 438).
2096 G. Villeneuve, L. Latie, C. Cros, and P. Hagenmuller, Mater. Res. Bull., 1984, **19**, 1515 (Chem. Abstr., 1985, **102**, 54 927).
2097 T. Tsuda, M. Itoh, and H. Yasuoka, J. Phys. Soc. Jpn., 1985, **54**, 1599 (Chem. Abstr., 1985, **103**, 31 436).
2098 W. Baden, P.C. Schmidt, and A. Weiss, J. Less-Common Met., 1984, **104**, 99 (Chem. Abstr., 1985, **102**, 88 742).
2099 A.V. Yatsenko, N.A. Sergeev, and O.V. Falaleev, Yad. Magn. Rezon. Strukt. Krist., 1984, 75 (Chem. Abstr., 1985, **103**, 226 111).
2100 A.V. Yatsenko and N.A. Sergeev, Ukr. Fiz. Zh. (Russ. Ed.), 1985, **30**, 118 (Chem. Abstr., 1985, **102**, 88 891).
2101 A.V. Yatsenko and N.A. Sergeev, Fiz. Tverd. Tela (Leningrad), 1985, **27**, 1239

(Chem. Abstr., 1985, **103**, 15 368).
2102 P.P. Man, H. Theveneau, and P. Papon, J. Magn. Reson., 1985, **64**, 271.
2103 P. Butaud, P. Segransan, C. Berthier, and A. Meerschaut, Lect. Notes Phys., 1985, **217**, 121 (Chem. Abstr., 1985, **102**, 159 339).
2104 P. Butaud, P. Segransan, C. Berthier, and A. Meerschaut, Congr. AMPERE Magn. Reson. Relat. Phenom., Proc., 22nd, 1984, 40. Ed. by K.A. Müller, R. Kind, and J. Roos (Chem. Abstr., 1985, **102**, 124 309).
2105 J.Senegas and M. Zriouil, J. Solid State Chem., 1985, **58**, 137 (Chem. Abstr., 1985, **103**, 80 394).
2106 M. Zriouil, J. Senegas, B. Elouadi, and J.B. Goodenough, Mater. Res. Bull., 1985, **20**, 679 (Chem. Abstr., 1985, **103**, 96 756).
2107 J. Tsukamoto and K. Matsumura, Jpn. J. Appl. Phys., Part 1, 1985, **24**, 974 (Chem. Abstr., 1985, **103**, 133 036).
2108 M.C. Mery, P. Veillet, and K. Le Dang, Phys. Rev. B: Condens. Matter, 1985, **31**, 2656 (Chem. Abstr., 1985, **102**, 142 093).
2109 K. Tsuru, N. Uryu, K. Le Dang, and P. Veillet, J. Phys. Soc. Jpn., 1985, **54**, 3111 (Chem. Abstr., 1985, **103**, 152 426).
2110 Yu.N. Ivanov, V.N. Voronov, O.V. Falaleev, P.V. Klevtsov, E.P. Zeer, and K.S. Aleksandrov, Fiz. Tverd. Tela (Leningrad), 1985, **27**, 13 (Chem. Abstr., 1985, **102**, 104 886).
2111 P. Butaud, P. Segransan, C. Berthier, J. Dumas, and C. Schlenker, Phys. Rev. Lett., 1985, **55**, 253 (Chem. Abstr., 1985, **103**, 80 088).
2112 D.C. Douglass, L.F. Schneemeyer, and S.E. Spengler, Phys. Rev. B: Condens. Matter, 1985, **32**, 1813 (Chem. Abstr., 1985, **103**, 114 032).
2113 R.F. Bell, F.J. DiSalvo, R.G. Dunn, and J.M. Tarascon, Phys. Rev. B: Condens. Matter, 1985, **32**, 1461 (Chem. Abstr., 1985, **103**, 110 184).
2114 F. Brunet, H. Le Bail, and J. Virlet, Ann. Chim. (Paris), 1984, **9**, 771 (Chem. Abstr., 1985, **102**, 159 336).
2115 P. Molinie and S. Paoli, C.R. Acad. Sci., Ser. 2, 1984, **299**, 1243 (Chem. Abstr., 1985, **102**, 140 227).
2116 G.B. Jameson, M.T. Pope, and S.H. Wasfi, J. Am. Chem. Soc., 1985, **107**, 4911.
2117 D. Yang, D. Xia, T. Yao, and X. Chen, Xiamen Daxue Xuebao, Ziran Kexueban, 1985, **24**, 182 (Chem. Abstr., 1985, **103**, 188 213).
2118 J. Kassovicova, J. Murin, and M. Rakos, Czech. J. Phys., 1984, **B34**, 1015 (Chem. Abstr., 1985, **102**, 104 893).
2119 V.I. Kesaev and A.I. Ugulava, Fiz. Tverd. Tela (Leningrad), 1985, **27**, 1259 (Chem. Abstr., 1985, **102**, 230 745).
2120 T. Tsuda and H. Yasuoka, J. Phys. Soc. Jpn., 1984, **53**, 3994 (Chem. Abstr., 1985, **102**, 38 371).
2121 H. Kubo, N. Uryu, H. Nishihara, and H. Yasuoka, J. Phys. Soc. Jpn., 1984, **53**, 3747 (Chem. Abstr., 1985, **102**, 38 370).
2122 Y. Berthier, T. De Saxce, D. Fruchart, and P. Vulliet, Physica B+C (Amsterdam), 1985, **130**, 520 (Chem. Abstr., 1985, **103**, 46 956).
2123 J. Tritt-Goc, N. Pislewski, and U. Haeberlen, Ser. Fiz. (Uniw. im. Adama Mickiewicza Poznaniu), 1985, **53**, 111 (Chem. Abstr., 1985, **103**, 133 743).
2124 S.M. Oh, T. Kambara, D.N. Hendrickson, M. Sorai, K. Kaji, S.E. Woehler, and R.J. Wittebort, J. Am. Chem. Soc., 1985, **107**, 5540.
2125 R.A. Bagautdinov, Kh.G. Bogdanova, V.A. Golenishchev-Kutuzov, and G.R. Enikeeva, Zh. Eksp. Teor. Fiz., 1984, **87**, 2008 (Chem. Abstr., 1985, **102**, 54 944).
2126 M. Mizoguchi, J. Phys. Soc. Jpn., 1985, **54**, 2168 (Chem. Abstr., 1985, **103**, 63 760).
2127 S.P. Kuntsevich, A.A. Bezlepkin, and Yu.A. Popkov, Zh. Eksp. Teor. Fiz., 1985, **88**, 1820 (Chem. Abstr., 1985, **103**, 47 042).
2128 J. Mason, D.M.P. Mingos, J. Schaefer, D. Sherman, and E.O. Stejskal, J. Chem. Soc., Chem. Commun., 1985, 444.
2129 M.J. Ledoux, O. Michaux, G. Agostini, and P. Panissod, J. Catal., 1985, **96**, 189 (Chem. Abstr., 1985, **103**, 184 320).
2130 K. Okada, S. Kajiwara, I. Inaba, and K. Kawano, Phys. Rep. Kumamoto Univ., 1984, **6**, 103 (Chem. Abstr., 1985, **102**, 159 349).
2131 H.L. Pelt, R.P.J. Verburg, and J.J.F. Scholten, J. Mol. Catal., 1985, **32**, 77.

2132 A. Naito, D.L. Sastry, and C.A. McDowell, Chem. Phys. Lett., 1985, **115**, 19.
2133 I. Svare and A.M. Raaen, Ser. Fiz. (Uniw. im. Adama Mickiewicza Poznaniu), 1985, **53**, 63 (Chem. Abstr., 1985, **103**, 133 742).
2134 H.W. Kroto, S.I. Klein, M.F. Meidine, J.F. Nixon, R.K. Harris, K.J. Packer, and P. Reams, J. Organomet. Chem., 1985, **280**, 281.
2135 H. Seitz and H. Benner, Congr. AMPERE Magn. Reson. Relat. Phenom., Proc., 22nd, 1984, 143. Ed. by K.A. Müller, R. Kind, and J. Roos (Chem. Abstr., 1985, **102**, 142 069).
2136 D.M. Lindsay and P.H. Kasai, J. Magn. Reson., 1985, **64**, 278.
2137 P.F. Barron, J.C. Dyason, L.M. Engelhardt, P.C. Healy, and A.H. White, Aust. J. Chem., 1985, **38**, 261.
2138 D.G. Hughes, S. Mohanty, L. Pandey, and F.L. Weichman, Can. J. Phys., 1985, **63**, 397 (Chem. Abstr., 1985, **102**, 177 947).
2139 J.C. Dyason, P.C. Healy, L.M. Engelhardt, C. Pakawatchai, V.A. Patrick, C.L. Raston, and A.H. White, J. Chem. Soc., Dalton Trans., 1985, 831.
2140 K. Deguchi, K. Matsushita, T. Fujito, K. Endo, and K. Yamamoto, JEOL News, [Ser.] Anal. Instrum., 1984, **20A**, 15 (Chem. Abstr., 1985, **102**, 88 847).
2141 T. Kubo and H. Yamahaku, Phys. Rev. B: Condens. Matter, 1985, **32**, 1831 (Chem. Abstr., 1985, **103**, 97 697).
2142 T. Kubo, H. Yamahaku, and Y. Tabuchi, Nara Kyoiku Daigaku Kiyo, Shizen Kaguku, 1984, **33**, 1 (Chem. Abstr., 1985, **103**, 226 117).
2143 N.D. Moon, K. Lee, and S.H. Choh, Can. J. Phys., 1985, **63**, 946 (Chem. Abstr., 1985, **103**, 226 060).
2144 H. Looser and D. Brinkmann, J. Magn. Reson., 1985, **64**, 76.
2145 H. Huber, M. Mali, J. Roos, and D. Brinkmann, Congr. AMPERE Magn. Reson. Relat. Phenom., Proc., 22nd, 1984, 175. Ed. by K.A. Müller, R. Kind, and J. Roos (Chem. Abstr., 1985, **102**, 124 318).
2146 T.W. Hijmans, H.B. Brom, and L.J. De Jongh, Phys. Rev. Lett., 1985, **54**, 1714 (Chem. Abstr., 1985, **102**, 177 840).
2147 S.P. Kolesnikov, I.V. Lyudkovskaya, M.Yu. Antipin, Yu.T. Struchkov, and O.M. Nefedov, Izv. Akad. Nauk SSSR, Ser. Khim., 1985, 79 (Chem. Abstr., 1985, **102**, 220 400).
2148 J. Petersson and E. Schneider, Z. Phys. B: Condens. Matter, 1985, **61**, 33 (Chem. Abstr., 1985, **103**, 151 939).
2149 R. Blinc, S. Zumer, D.C. Ailion, and J. Nicponski, J. Phys. (Les Ulis, Fr.), 1985, **46**, 1205 (Chem. Abstr., 1985, **103**, 62 868).
2150 P.F. Rodesiler, E.A.H. Griffith, N.G. Charles, and E.L. Amma, Acta Crystallogr., Sect. C: Cryst. Struct. Commun., 1985, **C41**, 673 (Chem. Abstr., 1985, **103**, 46 209).
2151 P.S. Marchetti, P.D. Ellis, and R.G. Bryant, J. Am. Chem. Soc., 1985, **107**, 8191.
2152 V.W. Miner and J.H. Prestegard, J. Am. Chem. Soc., 1985, **107**, 2177.
2153 J. Covino, P. Dragovich, C.K. Lowe-Ma, R.F. Kubin, and R.W. Schwartz, Mater. Res. Bull., 1985, **20**, 1099 (Chem. Abstr., 1985, **103**, 226 211).
2154 V.I. Trofimov and G.L. Trofimov, Yad. Magn. Rezon. Strukt. Krist., 1984, 122 (Chem. Abstr., 1985, **103**, 224 927).
2155 E. Kundla, Eesti NSV Tead. Akad. Toim., Fuus., Mat., 1985, **34**, 68 (Chem. Abstr., 1985, **103**, 31 419).
2156 D.B. Zax, A. Bielecki, K.W. Zilm, A. Pines, and D.P. Weitekamp, J. Chem. Phys., 1985, **83**, 4877.
2157 R.D. Walker and B.C. Gerstein, Phys. Rev. B: Condens. Matter, 1985, **31**, 3167 (Chem. Abstr., 1985, **102**, 142 097).
2158 H. Wakabayashi and R. Terai, Yogyo Kyokaishi, 1985, **93**, 13 (Chem. Abstr., 1985, **102**, 136 477).
2159 S. Simon, V. Simon, and A. Nicula, Congr. AMPERE Magn. Reson. Relat. Phenom., Proc., 22nd, 1984, 97. Ed. by K.A. Müller, R. Kind, and J. Roos (Chem. Abstr., 1985, **102**, 124 249).
2160 K.F.M.G.J. Scholle and W.S. Veeman, Zeolites, 1985, **5**, 118 (Chem. Abstr., 1985, **102**, 226 663).
2161 H. Jain, G. Balzer-Joellenbeck, and O. Kanert, J. Am. Ceram. Soc., 1985, **68**, C24 (Chem. Abstr., 1985, **102**, 171 223).
2162 C. Jaeger and U. Haubenreisser, Phys. Chem. Glasses, 1985, **26**, 152 (Chem.

Abstr., 1985, **103**, 226 066).
2163 H.U. Huerter, B. Krebs, H. Eckert, and W. Müller-Warmuth, Inorg. Chem., 1985, **24**, 1288.
2164 D.E. Hintenlang and P.J. Bray, J. Non-Cryst. Solids, 1985, **69**, 243 (Chem. Abstr., 1985, **102**, 124 366).
2165 H. Lecavelier, F. Devreux, M. Nechtschein, and G. Bidan, Mol. Cryst. Liq. Cryst., 1985, **118**, 183 (Chem. Abstr., 1985, **102**, 141 602).
2166 T. Suemoto and M. Ishigame, Phys. Rev. B: Condens. Matter, 1985, **32**, 4126 (Chem. Abstr., 1985, **103**, 149 836).
2167 S. Komarneni, R. Roy, C.A. Fyfe, and G.J. Kennedy, J. Am. Ceram. Soc., 1985, **68**, C-243 (Chem. Abstr., 1985, **103**, 165 024).
2168 R. Dupree, D. Holland, and D.S. Williams, Phys. Chem. Glasses, 1985, **26**, 50 (Chem. Abstr., 1985, **102**, 224 799).
2169 O.P. Krivoruchko, V.M. Mastikhin, B.P. Zolotovskii, S.M. Paramzin, D.P. Klevtsov, and R.A. Buyanov, Kinet. Katal., 1985, **26**, 763 (Chem. Abstr., 1985, **103**, 97 757).
2170 D. Mueller, E. Jahn, G. Ladwig, and U. Haubenreisser, Chem. Phys. Lett., 1984, **109**, 332 (Chem. Abstr., 1985, **102**, 104 854).
2171 D. Mueller, E. Jahn, B. Fahlke, G. Ladwig, and U. Haubenreisser, Zeolites, 1985, **5**, 53 (Chem. Abstr., 1985, **102**, 84 874).
2172 Z.N. Prozorovskaya, V.F. Chuvaev, and A.B. Yaroslavtsev, Zh. Neorg. Khim., 1985, **30**, 1143 (Chem. Abstr., 1985, **103**, 80 909).
2173 S. Sinha and R. Srinivasan, Proc. Nucl. Phys. Solid State Phys. Symp., 1982 (Pub. 1984), **25C**, 223 (Chem. Abstr., 1985, **102**, 104 871).
2174 T.G. Balicheva, G.A. Petrova, and V.S. Kasperovich, Zh. Neorg. Khim., 1984, **29**, 2510 (Chem. Abstr., 1985, **102**, 35 628).
2175 E.G. Derouane, L. Baltusis, R.M. Dessau, and K.D. Schmitt, Stud. Surf. Sci. Catal., 1985, **20**, 135 (Chem. Abstr., 1985, **102**, 226 581).
2176 A.G. Ashton, S. Batmanian, D.M. Clark, J. Dwyer, F.R. Fitch, A. Hinchliffe, and F.J. Machado, Stud. Surf. Sci. Catal., 1985, **20**, 101 (Chem. Abstr., 1985, **102**, 226 580).
2177 G.R. Hays, W.A. Van Erp, N.C.M. Alma, P.A. Couperus, R. Huis, and A.E. Wilson, Zeolites, 1984, **4**, 377 (Chem. Abstr., 1985, **102**, 32 956).
2178 F.M.M. Geurts, A.P.M. Kentgens, and W.S. Veeman, Chem. Phys. Lett., 1985, **120**, 206 (Chem. Abstr., 1985, **103**, 226 061).
2179 C.A. Fyfe, G.J. Kennedy, G.T. Kokotailo, J.R. Lyerla, and W.W. Fleming, J. Chem. Soc., Chem. Commun., 1985, 740.
2180 D.G. Hay, H. Jaeger, and G.W. West, J. Phys. Chem., 1985, **89**, 1070.
2181 R.M. Dessau and G.T. Kerr, Zeolites, 1984, **4**, 315 (Chem. Abstr., 1985, **102**, 33 245).
2182 P.A. Jacobs, M. Tielen, J.B. Nagy, G. Debras, E.G. Derouane, and Z. Gabelica, Proc. Int. Zeolite Conf., 6th, 1983 (Pub. 1984), 783 (Chem. Abstr., 1985, **103**, 93 518).
2183 J.B. Nagy, J.P. Lange, A. Gourgue, P. Bodart, and Z. Gabelica, Stud. Surf. Sci. Catal., 1985, **20**, 127 (Chem. Abstr., 1985, **103**, 141 131).
2184 Z. Gabelica, J.B. Nagy, G. Debras, and E.G. Derouane, Proc. Int. Zeolite Conf., 6th, 1983 (Pub. 1984), 914.
2185 G. Engelhardt, B. Fahlke, M. Maegi, and E. Lippmaa, Z. Phys. Chem. (Leipzig), 1985, **266**, 239 (Chem. Abstr., 1985, **102**, 192 030).
2186 K.F.M.G.J. Scholle and W.S. Veeman, J. Phys. Chem., 1985, **89**, 1850.
2187 K.F.M.G.J. Scholle, W.S. Veeman, P. Frenken, and G.P.M. Van der Velden, Appl. Catal., 1985, **17**, 233 (Chem. Abstr., 1985, **103**, 184 290).
2188 L.J.M. Van de Ven, J.G. Post, J.H.C. Van Hooff, and J.W. De Haan, J. Chem. Soc., Chem. Commun., 1985, 214.
2189 G. Boxhoorn, R.A. Van Santen, W.A. Van Erp, G.R. Hays, N.C.M. Alma, R. Huis, and A.D.H. Clague, Proc. Int. Zeolite Conf., 6th, 1983 (Pub. 1984), 694 (Chem. Abstr., 1985, **103**, 47 071).
2190 G. Engelhard, R. Radeglia, U. Lohse, A. Samoson, and E. Lippmaa, Z. Chem., 1985, **25**, 252 (Chem. Abstr., 1985, **103**, 187 217).
2191 C.A. Fyfe, G.T. Kokotailo, G.J. Kennedy, and C. De Schutter, J. Chem. Soc., Chem. Commun., 1985, 306.
2192 J.C. Trewella, J.L. Schlenker, D.E. Woessner, and J.B. Higgins, Zeolites,

1985, **5**, 130 (Chem. Abstr., 1985, **103**, 43 182).
2193 J.M. Newsam, J. Phys. Chem., 1985, **89**, 2002.
2194 G. Engelhardt, B. Fahlke, M. Magi, and E. Lippmaa, Zeolites, 1985, **5**, 49 (Chem. Abstr., 1985, **102**, 84 927).
2195 C.S. Blackwell, J.J. Pluth, and J.V. Smith, J. Phys. Chem., 1985, **89**, 4420.
2196 D. Freude, J. Haase, H. Pfeifer, D. Prager, and G. Scheler, Chem. Phys. Lett., 1985, **114**, 143 (Chem. Abstr., 1985, **102**, 138 367).
2197 M.T. Melchior, D.E.W. Vaughan, A.J. Jacobson, and C.F. Pictroski, Proc. Int. Zeolite Conf., 6th, 1983 (Pub. 1984), 684 (Chem. Abstr., 1985, **103**, 43 216).
2198 J. Klinowski, X. Liu, R.D. Penzhorn, P. Schuster, N.J. Clayden, and C.M. Dobson, J. Chem. Soc., Faraday Trans. 1, 1985, **81**, 1435.
2199 W.D. Basler, Colloids Surf., 1984, **12**, 59 (Chem. Abstr., 1985, **102**, 67 901).
2200 S. Nakata, S. Asaoka, and H. Takahashi, Nippon Kagaku Kaishi, 1985, 1372 (Chem. Abstr., 1985, **103**, 110 579).
2201 J.H. Raeder, Finn. Chem. Lett., 1984, 159 (Chem. Abstr., 1985, **102**, 120 603).
2202 D. Freude, J. Haase, J. Klinowski, T.A. Carpenter, and G. Ronikier, Chem. Phys. Lett., 1985, **119**, 365 (Chem. Abstr., 1985, **103**, 152 428).
2203 P.J. Grobet, W.J. Mortier, and K. Van Genechten, Chem. Phys. Lett., 1985, **119**, 361 (Chem. Abstr., 1985, **103**, 152 427).
2204 J. Klinowski, C.A. Fyfe, and G.C. Gobbi, J. Chem. Soc., Faraday Trans. 1, 1985, **81**, 3003.
2205 M. Hunger, Actas Simp. Iberoam. Catal., 9th, 1984, **2**, 1665 (Chem. Abstr., 1985, **102**, 101 107).
2206 Ch. Gullyev, B.L. Khusid, G.D. Chukin, B.K. Nefedov, and E.D. Radchenko, Sovrem. Metody YaMR EPR Khim. Tverd. Tela, [Mater. Vses. Koord. Soveshch.], 4th, 1985, 76 (Chem. Abstr., 1985, **103**, 224 841).
2207 F. Pechar, Acta Mont., 1984, **68**, 231 (Chem. Abstr., 1985, **103**, 198 682).
2208 C.A. Fyfe, G.C. Gobbi, G.J. Kennedy, J.D. Graham, R.S. Ozubko, W.J. Murphy, A. Bothner-By, J. Dadok, and A.S. Chesnick, Zeolites, 1985, **5**, 179 (Chem. Abstr., 1985, **103**, 43 184).
2209 S. Hayashi, K. Suzuki, S. Shin, K. Hayamizu, and O. Yamamoto, Chem. Phys. Lett., 1985, **113**, 368 (Chem. Abstr., 1985, **102**, 104 891).
2210 J.G. Thompson, Clays Clay Miner., 1985, **33**, 173 (Chem. Abstr., 1985, **103**, 39 990).
2211 K.J.D. MacKenzie, I.W.M. Brown, R.H. Meinhold, and M.E. Bowden, J. Am. Ceram. Soc., 1985, **68**, 293.
2212 I.W.M. Brown, K.J.D. Mackenzie, M.E. Bowden, and R.H. Meinhold, J. Am. Ceram. Soc., 1985, **68**, 298.
2213 D.P. Klevtsov, O.P. Krivoruchko,, V.M. Mastikhin, B.P. Zolotvskii, and R.A. Buyanov, Izv. Sib. Otd. Akad. Nauk SSSR, Ser. Khim. Nauk, 1985, 75 (Chem. Abstr., 1985, **102**, 176 901).
2214 R.H. Meinhold, K.J.D. MacKenzie, and I.W.M. Brown, J. Mater. Sci. Lett., 1985, **4**, 163 (Chem. Abstr., 1985, **102**, 142 276).
2215 V.Ya. Proshko, A.S. Litovchenko, I.V. Matyash, and A.S. Povarennykh, Dokl. Akad. Nauk SSSR, 1985, **280**, 201 (Chem. Abstr., 1985, **102**, 169 914).
2216 P.J. Grobet and P.A. Jacobs, Congr. AMPERE Magn. Reson. Relat. Phenom., Proc., 22nd, 1984, 379. Ed. by K.A. Müller, R. Kind, and J. Roos (Chem. Abstr., 1985, **102**, 120 622).
2217 S. Komarneni, R. Roy, D.M. Roy, C.A. Fyfe, G.J. Kennedy, A.A. Bothner-By, J. Dadok, and A.S. Chesnick, J. Mater. Sci., 1985, **20**, 4209 (Chem. Abstr., 1985, **103**, 226 103).
2218 S. Komarneni, R. Roy, D.M. Roy, C.A. Fyfe, and G.J. Kennedy, Cem. Concr. Res., 1985, **15**, 723 (Chem. Abstr., 1985, **103**, 75 372).
2219 E. Ohtani, F. Taulelle, and C.A. Angell, Nature (London), 1985, **314**, 78 (Chem. Abstr., 1985, **102**, 152 482).
2220 K.J.D. MacKenzie, I.W.M. Brown, R.H. Meinhold, and M.E. Bowden, J. Am. Ceram. Soc., 1985, **68**, 266 (Chem. Abstr., 1985, **102**, 224 983).
2221 M.A. Wilson and S.A. McCarthy, Anal. Chem., 1985, **57**, 2733 (Chem. Abstr., 1985, **103**, 152 520).
2222 B.H.W.S. De Jong, C.M. Schramm, and V.E. Parziale, Geochim. Cosmochim. Acta, 1984, **48**, 2619 (Chem. Abstr., 1985, **102**, 65 061).

2223 B.A. Goodman and J.W. Stucki, Clay Miner., 1984, **19**, 663 (Chem. Abstr., 1985, **102**, 71 979).
2224 Yu.G. Elizar'ev, S.P. Prozorov, Yu.N. Moskvich, A.M. Vakhrameev, and I.A. Belitskii, Zh. Strukt. Khim., 1984, **25**, 149 (Chem. Abstr., 1985, **102**, 88 841).
2225 P. Fejes, I. Hannus, I. Kiricsi, H. Pfeifer, D. Freude, and W. Oehme, Zeolites, 1985, **5**, 45 (Chem. Abstr., 1985, **102**, 84 873).
2226 J.H. Raeder, Zeolites, 1984, **4**, 311 (Chem. Abstr., 1985, **102**, 33 078).
2227 Y. Onoda, Y. Fujiki, M. Takigawa, H. Yasuoka, S. Yoshikado, T. Ohachi, and I. Taniguchi, Solid State Ionics, 1985, **17**, 127 (Chem. Abstr., 1985, **103**, 226 088).
2228 A. Putnis and R.J. Angel, Phys. Chem. Miner., 1985, **12**, 217.
2229 A. Putnis, C.A. Fyfe, and G.C. Gobbi, Phys. Chem. Miner., 1985, **12**, 211 (Chem. Abstr., 1985, **103**, 188 265).
2230 D. Plee, F. Borg, L. Gatineau, and J.J. Fripiat, J. Am. Chem. Soc., 1985, **107**, 2362.
2231 T.J. Pinnavaia, S.D. Landau, M.-S. Tzou, and I.D. Johnson, J. Am. Chem. Soc., 1985, **107**, 7222.
2232 G. Engelhardt, M. Nofz, K. Forkel, F.G. Wihsmann, M. Magi, A. Samoson, and E. Lippmaa, Phys. Chem. Glasses, 1985, **26**, 157 (Chem. Abstr., 1985, **103**, 184 005).
2233 L.B. Welsh, J.P. Gilson, and M.J. Gattuso, Appl. Catal., 1985, **15**, 327 (Chem. Abstr., 1985, **102**, 173 274).
2234 J.B. Murdoch, J.F. Stebbins, and I.S.E. Carmichael, Am. Mineral., 1985, **70**, 332 (Chem. Abstr., 1985, **102**, 170 067).
2235 N. Janes and E. Oldfield, J. Am. Chem. Soc., 1985, **107**, 6769.
2236 A.S. Litovchenko, O.D. Ishutina, I.V. Matyash, and A.A. Mel'nikov, Sovrem. Metody YaMR EPR Khim. Tverd. Tela, [Mater. Vses. Koord. Soveshch.], 4th, 1985, 122 (Chem. Abstr., 1985, **103**, 226 094).
2237 B.C. Gerstein and A.T. Nicol, J. Non-Cryst. Solids, 1985, **75**, 423 (Chem. Abstr., 1985, **103**, 219 675).
2238 R. Radeglia and G. Engelhardt, Chem. Phys. Lett., 1985, **114**, 28 (Chem. Abstr., 1985, **102**, 124 369).
2239 E. Hallas and M. Haehnert, Cryst. Res. Technol., 1985, **20**, K25 (Chem. Abstr., 1985, **102**, 196 642).
2240 I.P. Appleyard, R.K. Harris, and F.R. Fitch, Chem. Lett., 1985, 1747 (Chem. Abstr., 1985, **103**, 226 130).
2241 C.P. Herrero, J. Sanz, and J.M. Serratosa, J. Phys. C, 1985, **18**, 13 (Chem. Abstr., 1985, **102**, 135 138).
2242 B. Sherriff and J.S. Hartman, Can. Mineral., 1985, **23**, 205 (Chem. Abstr., 1985, **103**, 90 572).
2243 R.J. Kirkpatrick, R.A. Kinsey, K.A. Smith, D.M. Henderson, and E. Oldfield, Am. Mineral., 1985, **70**, 106 (Chem. Abstr., 1985, **102**, 65 078).
2244 P.F. Barron, P. Slade, and R.L. Frost, J. Phys. Chem., 1985, **89**, 3880.
2245 S. Komarneni, C.A. Fyfe, and G.J. Kennedy, Clay Miner., 1985, **20**, 327 (Chem. Abstr., 1985, **103**, 198 731).
2246 R.A. Kinsey, R.J. Kirkpatrick, J. Hower, K.A. Smith, and E. Oldfield, Am. Mineral., 1985, **70**, 537 (Chem. Abstr., 1985, **103**, 25 111).
2247 C.P. Herrero, J. Sanz, and J.M. Serratosa, Solid State Commun., 1985, **53**, 151 (Chem. Abstr., 1985, **102**, 104 894).
2248 K.G. Ione, L.A. Vostrikova, and V.M. Mastikhin, J. Mol. Catal., 1985, **31**, 355.
2249 S. Hayashi, K. Suzuki, S. Shin, K. Hayamizu, and O. Yamamoto, Bull. Chem. Soc. Jpn., 1985, **58**, 52.
2250 P.F. Barron, R.K. Harris, P. Jonsen, and K.J. Packer, J. Magn. Reson., 1985, **61**, 371.
2251 J.F. Hinton, K.R. Metz, G.L. Turner, D.L. Bennett, and S.F. Millett, J. Magn. Reson., 1985, **64**, 120.
2252 H. Eckert and W. Müller-Warmuth, J. Non-Cryst. Solids, 1985, **70**, 199 (Chem. Abstr., 1985, **102**, 191 671).
2253 R.H. Jarman and G.J. Ray, J. Chem. Soc., Chem. Commun., 1985, 1153.
2254 H.A. Resing, B.R. York, S.A. Solin, and R.M. Fronko, Ext. Abstr. Program -

Bienn. Conf. Carbon, 1985, **17th**, 192 (Chem. Abstr., 1985, **103**, 80 806).
2255 V.F. Sukhoverkhov, Yu.I. Nikonorov, and E.L. Zhuzhgov, Zh. Neorg. Khim., 1985, **30**, 1391 (Chem. Abstr., 1985, **103**, 63 899).
2256 D.J. Leopold, P.A. Fedders, R.E. Norberg, J.B. Boyce, and J.C. Knights, Phys. Rev. B: Condens. Matter, 1985, **31**, 5642 (Chem. Abstr., 1985, **102**, 230 739).
2257 L.D. Spicer, D.W. Bennett, and J.F. Davis, U.S. Pat. Appl. US 492,926 14 Sep 1984, Appl. 09 May 1983; 20 pp. Avail. NTIS Order No. PAT-APPL-6-492 926 (Chem. Abstr., 1985, **102**, 100 136).
2258 I. Tajima and M. Yamamoto, J. Polym. Sci., Polym. Chem. Ed., 1985, **23**, 615 (Chem. Abstr., 1985, **102**, 167 434).
2259 R.K. Harris, K. Metcalfe, and E. Hengge, Polyhedron, 1985, **4**, 1319.
2260 G.R. Finlay, J.S. Hartman, M.F. Richardson, and B.L. Williams, J. Chem. Soc., Chem. Commun., 1985, 159.
2261 R. Dupree, M.H. Lewis, G. Leng-Ward, and D.S. Williams, J. Mater. Sci. Lett., 1985, **4**, 393.
2262 N.T. Bagraev, V.A. Mashkov, R.P. Seisyan, V.L. Sukhanov, and N.M. Shmidt, Zh. Tekh. Fiz., 1985, **55**, 2064 (Chem. Abstr., 1985, **103**, 225 411).
2263 C.N.R. Rao, J.M. Thomas, J. Klinowski, U. Selvaraj, K.J. Rao, and G.R. Millward, Angew. Chem., Int. Ed. Engl., 1985, **24**, 61.
2264 D. Heidemann, A.R. Grimmer, C. Huebert, P. Starke, and Mägi, Z. Anorg. Allg. Chem., 1985, **528**, 22.
2265 D. Heidemann, A.R. Grimmer, P. Starke, C. Huebert, and G. Scheler, Z. Anorg. Allg. Chem., 1985, **528**, 37.
2266 H. Stade, A.-R. Grimmer, G. Engelhardt, M. Mägi, and E. Lippmaa, Z. Anorg. Allg. Chem., 1985, **528**, 147.
2267 A.A. Valyukhov, V.V. Nazarov, V.I. Ermakov, and Yu.G. Frolov, Kolloidn. Zh., 1985, **47**, 660 (Chem. Abstr., 1985, **103**, 147 820).
2268 H. Rosenberger, G. Scheler, H. Buerger, and M. Jakob, Colloids Surf., 1984, **12**, 53 (Chem. Abstr., 1985, **102**, 33 143).
2269 G. Rutten, A. Van de Ven, J. De Haan, L. Van de Ven, and J. Rijks, HRC CC, J. High Resolut. Chromatogr. Chromatogr. Commun., 1984, **7**, 607 (Chem. Abstr., 1985, **102**, 55 371).
2270 D.C. Douglass, T.M. Duncan, K.L. Walker, and R. Csencsits, J. Appl. Phys., 1985, **58**, 197 (Chem. Abstr., 1985, **103**, 63 749).
2271 J.J. Pluth, J.V. Smith, and J. Faber, jun., J. Appl. Phys., 1985, **57**, 1045 (Chem. Abstr., 1985, **102**, 104 020).
2272 D.J. Cookson and B.E. Smith, J. Magn. Reson., 1985, **63**, 217.
2273 M.L. Miller, R.W. Linton, G.E. Maciel, and B.L. Hawkins, J. Chromatogr., 1985, **319**, 9 (Chem. Abstr., 1985, **102**, 120 443).
2274 A.R. Grimmer, Chem. Phys. Lett., 1985, **119**, 416 (Chem. Abstr., 1985, **103**, 188 258).
2275 R. Dupree, D. Holland, P.W. McMillan, and R.F. Pettifer, J. Non-Cryst. Solids, 1984, **68**, 399 (Chem. Abstr., 1985, **102**, 66 197).
2276 U. Selvaray, K.J. Rao, C.N.R. Rao, J. Klinowski, and J.M. Thomas, Chem. Phys. Lett., 1985, **114**, 24 (Chem. Abstr., 1985, **102**, 124 368).
2277 P.F. Barron and R.L. Frost, Am. Mineral., 1985, **70**, 758 (Chem. Abstr., 1985, **103**, 90 592).
2278 P.F. Barron, P. Slade, and R.L. Frost, J. Phys. Chem., 1985, **89**, 3305.
2279 F. Von Lampe, A.R. Grimmer, M. Mägi, and E. Lippmaa, Silikattechnik, 1984, **35**, 273 (Chem. Abstr., 1985, **102**, 99 847).
2280 A.R. Grimmer, F. Von Lampe, M. Mägi, and E. Lippmaa, Cem. Concr. Res., 1985, **15**, 467 (Chem. Abstr., 1985, **103**, 41 594).
2281 M. Mägi, E. Lippmaa, E.D. Podloznyi, and I.I. Karpunin, Zh. Prikl. Spektrosk., 1985, **43**, 315 (Chem. Abstr., 1985, **103**, 146 174).
2282 N.J. Clayden, C.M. Dobson, C.J. Hayes, and S.A. Rodger, Congr. AMPERE Magn. Reson. Relat. Phenom., Proc., 22nd, 1984, 370. Ed. by K.A. Müller, R. Kind, and J. Roos (Chem. Abstr., 1985, **102**, 171 583).
2283 J.R. Barnes, A.D.H. Clague, N.J. Clayden, C.M. Dobson, C.J. Hayes, G.W. Groves, and S.A. Rodger, J. Mater. Sci. Lett., 1985, **4**, 1293 (Chem. Abstr., 1985, **103**, 200 055).
2284 L.J. Schreiner, J.C. Mactavish, L. Miljkovic, M.M. Pintar, R. Blinc, G.

Lahajnar, D. Lasic, and L.W. Reeves, J. Am. Ceram. Soc., 1985, **68**, 10 (Chem. Abstr., 1985, **102**, 136 696).
2285 I.L. Mudrakovskii, V.M. Mastikhin, V.P. Shmachkova, and N.S. Kotsarenko, Chem. Phys. Lett., 1985, **120**, 424 (Chem. Abstr., 1985, **103**, 226 075).
2286 R.K. Harris, T.N. Mitchell, and G.J. Nesbitt, Magn. Reson. Chem., 1985, **23**, 1080).
2287 R.K. Harris, K.J. Packer, and P. Reams, J. Magn. Reson., 1985, **61**, 564.
2288 W.F. Manders and T.P. Lockhart, J. Organomet. Chem., 1985, **297**, 143.
2289 R.K. Harris, K.J. Packer, and P. Reams, Chem. Phys. Lett., 1985, **115**, 16.
2290 T.P. Lockhart, W.F. Manders, and J.J. Zuckerman, J. Am. Chem. Soc., 1985, **107**, 4546.
2291 T.P. Lockhart and W.F. Manders, J. Am. Chem. Soc., 1985, **107**, 5863.
2292 T.P. Lockhart, W.F. Manders, and E.O. Schlemper, J. Am. Chem. Soc., 1985, **107**, 7451.
2293 V.V. Parshutkin, A.B. Yaroslavtsev, Z.N. Prozorovskaya, and A.I. Mun, Zh. Neorg. Khim., 1985, **30**, 56 (Chem. Abstr., 1985, **102**, 124 426).
2294 L.M. Sharygin, T.A. Denisova, S.M. Vovk, T.N. Perekhozheva, R.N. Pletnev, and V.F. Gonchar, Zh. Neorg. Khim., 1985, **30**, 1968 (Chem. Abstr., 1985, **103**, 226 171).
2295 F. Koksal, J. Phys. Chem. Solids, 1984, **45**, 1201 (Chem. Abstr., 1985, **102**, 142 089).
2296 T. Chiba and S. Miyajima, J. Chem. Phys., 1985, **83**, 6385.
2297 K. Morimoto and K. Shimomura, J. Phys. Soc. Jpn., 1985, **54**, 3244 (Chem. Abstr., 1985, **103**, 188 296).
2298 R.K. Harris, K.J. Packer, and A.M. Thayer, J. Magn. Reson., 1985, **62**, 284.
2299 K. Albert, A. Bruns, H. Förster, W. Hiller, and U. Kunze, Angew. Chem., Int. Ed. Engl., 1985, **24**, 715).
2300 C. Connor, A. Naito, K. Takegoshi, and C.A. McDowell, Chem. Phys. Lett., 1985, **113**, 123.
2301 B. Topic and R. Blinc, Congr. AMPERE Magn. Reson. Relat. Phenom., Proc., 22nd, 1984, 58. Ed. by K.A. Müller, R. Kind, and J. Roos (Chem. Abstr., 1985, **102**, 123 439).
2302 M. Mali, J. Roos, E. Courtens, and K.A. Müller, Congr. AMPERE Magn. Reson. Relat. Phenom., Proc., 22nd, 1984, 32. Ed. by K.A. Müller, R. Kind, and J. Roos (Chem. Abstr., 1985, **102**, 104 896).
2303 C. Doremieux-Morin, J.P. Fraissard, J.P. Besse, and R. Chevalier, Solid State Ionics, 1985, **17**, 93 (Chem. Abstr., 1985, **103**, 226 087).
2304 M. Mehring and J. Spengler, Phys. Rev. Lett., 1984, **53**, 2441 (Chem. Abstr., 1985, **103**, 15 803).
2305 K. Fuda, K. Kishio, S. Yamauchi, and K. Fueki, Solid State Commun., 1985, **53**, 83 (Chem. Abstr., 1985, **102**, 88 853).
2306 P.L. Cummins, G.B. Bacskay, N.S. Hush, B. Halle, and S. Engstroem, J. Chem. Phys., 1985, **82**, 2002.
2307 F. Creuzet, D. Jerome, and A. Moradpour, Mol. Cryst. Liq. Cryst., 1985, **119**, 297 (Chem. Abstr., 1985, **102**, 159 331).
2308 P.C. Stein, A. Moradpour, and D. Jerome, J. Phys. Lett., 1985, **46**, 241 (Chem. Abstr., 1985, **102**, 159 343).
2309 F. Creuzet, D. Jerome, C. Bourbonnais, and A. Moradpour, J. Phys. C: Solid State Phys., 1985, **18**, L821 (Chem. Abstr., 1985, **103**, 204 667).
2310 T. Takahashi, Mol. Cryst. Liq. Cryst., 1985, **119**, 113 (Chem. Abstr., 1985, **102**, 159 329).
2311 F. Creuzet, C. Bourbonnais, D. Jerome, K. Bechgaard, and A. Moradpour, Mol. Cryst. Liq. Cryst., 1985, **119**, 45 (Chem. Abstr., 1985, **102**, 159 328).
2312 Yu.N. Moskvich, O.V. Rozanov, A.A. Sukhovskii, and I.P. Aleksandrova, Sovrem. Metody YaMR EPR Khim. Tverd. Tela, [Mater. Vses. Koord. Soveshch.], 4th, 1985, 97 (Chem. Abstr., 1985, **103**, 225 559).
2313 P.K. Burkert and D.G. Klobasa, Z. Naturforsch., Teil A, 1985, **40**, 274 (Chem. Abstr., 1985, **102**, 196 644).
2314 D.G. Klobasa and P.K. Burkert, Z. Naturforsch., Teil A, 1984, **39**, 1222 (Chem. Abstr., 1985, **102**, 88 868).
2315 P.P. Olodovskii and S.A. Kumashov, Inzh.-Fiz. Zh., 1985, **48**, 467 (Chem. Abstr., 1985, **102**, 191 794).

2316 H. Winkler and H. Spindler, Z. Phys. Chem. (Leipzig), 1985, **266**, 911 (Chem. Abstr., 1985, **103**, 221 615).
2317 M. Lipsicas, C. Straley, P.M. Costanzo, and R.F. Giese, jun., J. Colloid Interface Sci., 1985, **107**, 221 (Chem. Abstr., 1985, **103**, 166 755).
2318 C.A. Fyfe, G.C. Gobbi, and G.J. Kennedy, J. Phys. Chem., 1985, **89**, 277.
2319 K. Gaetano, R.M. Pagni, G.W. Kabalka, P. Bridwell, E. Walsh, J. True, and M. Underwood, J. Org. Chem., 1985, **50**, 499.
2320 R.W. Linton, M.L. Miller, G.E. Maciel, and B.L. Hawkins, SIA, Surf. Interface Anal., 1985, **7**, 196 (Chem. Abstr., 1985, **103**, 110 459).
2321 J. Sanz and J.M. Rojo, J. Phys. Chem., 1985, **89**, 4974.
2322 S. Modak and B.C. Khanra, J. Phys. C: Solid State Phys., 1985, **18**, L897 (Chem. Abstr., 1985, **103**, 201 380).
2323 P.R. Kubik, W.N. Hardy, and H. Glattli, Can. J. Phys., 1985, **63**, 605 (Chem. Abstr., 1985, **103**, 110 410).
2324 P.J. Toscano and T.J. Marks, J. Am. Chem. Soc., 1985, **107**, 653.
2325 I.I. Nazarova, N.D. Golubeva, N.M. Bravaya, A.D. Pomogailo, and S.M. Baturin, Sovrem. Metody YaMR EPR Khim. Tverd. Tela, [Mater. Vses. Koord. Soveshch.], 4th, 1985, 102 (Chem. Abstr., 1985, **103**, 201 532).
2326 W.P. McKenna and E.M. Eyring, J. Mol. Catal., 1985, **29**, 363.
2327 W.M. Shirley, B.R. McGarvey, B. Maiti, A. Brenner, and A. Cichowlas, J. Mol. Catal., 1985, **29**, 259.
2328 C.A. Chang, C.S. Huang, and P.D. Murphy, Inorg. Chem., 1985, **24**, 4216.
2329 V.K. Duplyakin, V.P. Doronin, T.V. Tsymbal, L.J. Al't, and A.S. Belyi, Dokl. Akad. Nauk SSSR, 1985, **281**, 89 (Chem. Abstr., 1985, **102**, 226 640).
2330 M. Monduzzi, A. Lai, and L. Burlamacchi, Chem. Phys., 1985, **96**, 419 (Chem. Abstr., 1985, **103**, 59 877).
2331 J.Z. Larese and R.J. Rollefson, Phys. Rev. B: Condens. Matter, 1985, **31**, 3048 (Chem. Abstr., 1985, **102**, 142 095).
2332 I.T.A. Emesh and I.D. Gay, J. Chem. Technol. Biotechnol., Chem. Technol., 1985, **35A**, 115 (Chem. Abstr., 1985, **102**, 209 870).
2333 P. Lorenz, M. Buelow, and J. Kaerger, Colloids Surf., 1984, **11**, 353 (Chem. Abstr., 1985, **102**, 32 978).
2334 K. Albert, B. Evers, and E. Bayer, J. Magn. Reson., 1985, **62**, 428.
2335 E.J.R. Sudhoelter, R. Huis, G.R. Hays, and N.C.M. Alma, J. Colloid Interface Sci., 1985, **103**, 554 (Chem. Abstr., 1985, **102**, 138 296).
2336 P.K. Wang, C.P. Slichter, and J.H. Sinfelt, J. Phys. Chem., 1985, **89**, 3606.
2337 T. Shibanuma and T. Matsui, Surf. Sci., 1985, **154**, L215 (Chem. Abstr., 1985, **102**, 226 469).
2338 D.L. Hasha, V.W. Miner, J.M. Garces, and S.C. Rocke, ACS Symp. Ser., 1985, **288**, 485 (Chem. Abstr., 1985, **103**, 201 403).
2339 B. Zibrowius, M. Buelow, and H. Pfeifer, Chem. Phys. Lett., 1985, **120**, 420 (Chem. Abstr., 1985, **103**, 184 250).
2340 A. Germanus, J. Kaerger, H. Pfeifer, N.N. Samulevic, and S.P. Zdanov, Zeolites, 1985, **5**, 91 (Chem. Abstr., 1985, **102**, 209 757).
2341 S.H. Liang and I.D. Gay, Langmuir, 1985, **1**, 593 (Chem. Abstr., 1985, **103**, 122 822).
2342 T.M. Duncan, P. Winslow, and A.T. Bell, J. Catal., 1985, **93**, 1 (Chem. Abstr., 1985, **102**, 173 289).
2343 R.K. Shoemaker and T.M. Apple, J. Phys. Chem., 1985, **89**, 3185.
2344 C.D. Makowka and C.P. Slichter, Phys. Rev. B: Condens. Matter, 1985, **31**, 5663 (Chem. Abstr., 1985, **102**, 226 577).
2345 C.P. Slichter, Congr. AMPERE Magn. Reson. Relat. Phenom., Proc., 22nd, 1984, 16. Ed. by K.A. Müller, R. Kind, and J. Roos (Chem. Abstr., 1985, **102**, 138 279).
2346 S.L. Rudaz, J.P. Ansermet, P.K. Wang, C.P. Slichter, and J.H. Sinfelt, Phys. Rev. Lett., 1985, **54**, 71 (Chem. Abstr., 1985, **102**, 51 385).
2347 R.K. Gilpin and M.E. Gangoda, J. Magn. Reson., 1985, **64**, 408.
2348 A.M. Zaper, A. Cholli, and J.L. Koenig, Polym. Sci. Technol. (Plenum), 1985, **27**, 299 (Chem. Abstr., 1985, **102**, 150 244).
2349 W.E. Rudzinski, T.L. Montgomery, J.S. Frye, B.L. Hawkins, and G.E. Maciel, J. Chromatogr., 1985, **323**, 281 (Chem. Abstr., 1985, **102**, 209 917).
2350 J.M. Chezeau, A.V. Chapman, J.H. Strange, and D.R. Kumar-Rao, Bull. Soc.

Chim. Fr., 1985, 1155.
2351 M.A. Ditzler, P.T. Inglefield, and R. Villani, Anal. Lett., 1984, **17**, 1783 (Chem. Abstr., 1985, **102**, 88 843).
2352 J.H. Lunsford, W.P. Rothwell, and W. Shen, J. Am. Chem. Soc., 1985, **107**, 1540.
2353 T. Bernstein and H. Pfeifer, Z. Phys. Chem. (Leipzig), 1985, **266**, 94 (Chem. Abstr., 1985, **102**, 120 417).
2354 B. Staudte, H. Pfeifer, K.H. Steinberg, and F. Roessner, Z. Phys. Chem. (Leipzig), 1985, **266**, 151.
2355 G. Wendt, J. Gottschling, B. Staudte, D. Hentschel, and R. Schöllner, Z. Anorg. Allg. Chem., 1985, **528**, 175.
2356 T. Tsang and H.A. Resing, Solid State Commun., 1985, **53**, 39 (Chem. Abstr., 1985, **102**, 124 293).
2357 T. Tsang and H.A. Resing, NASA Contract. Rep., 1985, NASA-CR-174261, NAS1.26:174261, 15 pp. Avail. NTIS. From Sci. Tech. Aerosp. Rep., 1985, **23**, Abstr. No. N85-16601 (Chem. Abstr., 1985, **103**, 15 813).
2358 I. Stang, G. Roth, K. Lueders, and H.J. Guentherodt, Congr. AMPERE Magn. Reson., Relat. Phenom., Proc., 22nd, 1984, 56. Ed. by K.A. Müller, R. Kind, and J. Roos (Chem. Abstr., 1985, **102**, 104 898).
2359 B.P. Cowan and A.J. Kent, J. Low Temp. Phys., 1984, **57**, 589 (Chem. Abstr., 1985, **102**, 38 302).
2360 B.P. Cowan and A.J. Kent, Phys. Lett. A, 1984, **106A**, 54 (Chem. Abstr., 1985, **102**, 12 711).
2361 T. Ito, L.C. De Menorval, E. Guerrier, and J.P. Fraissard, Chem. Phys. Lett., 1984, **111**, 271 (Chem. Abstr., 1985, **102**, 32 953).
2362 C.M. Merkel and D.R. Martin, Inorg. Chim. Acta, 1985, **96**, L59; **99**, 248.
2363 D.R. Martin, C.M. Merkel, J.U. Mondal, and C.R. Rushing, jun., Inorg. Chim. Acta, 1985, **99**, 81.
2364 J.A. Albanese, D.G. Kreider, C.D. Schaeffer, jun., C.H. Yoder, and M.S. Samples, J. Org. Chem., 1985, **50**, 2059.
2365 P.L. Rinaldi and N.J. Baldwin, J. Magn. Reson., 1985, **61**, 165.
2366 C.M. Clarke, K. Niedenzu, P.M. Niedenzu, and S. Trofimenko, Inorg. Chem., 1985, **24**, 2648.
2367 M.A. Paz-Sandoval, F. Santiesteban, and R. Contreras, Mag. Reson. Chem., 1985, **23**, 428.
2368 J.-M. Dupart, G. Le Borgne, S. Pace, and J.G. Riess, J. Am. Chem. Soc., 1985, **107**, 7202.
2369 A.B. Burg, Inorg. Chem., 1985, **24**, 3342.
2370 D.R. Martin, C.M. Merkel, J.P. Ruiz, and J.U. Mondal, Inorg. Chim. Acta, 1985, **100**, 293.
2371 H. Schmidbaur, E. Weiss, and G. Müller, Synth. React. Inorg. Met.-Org. Chem., 1985, **15**, 415.
2372 M. Bürklin, E. Hanecker, H. Nöth, and W. Storch, Angew. Chem., Int. Ed. Engl., 1985, **24**, 999.
2373 M.K. Das and P. Mukherjee, J. Chem. Res. (S), 1985, 66.
2374 K. Niedenzu, P.M. Niedenzu, and K.R. Warner, Inorg. Chem., 1985, **24**, 1604.
2375 Z. Benmaarouf-Khallaayoun, M. Baboulene, V. Speziale, and A. Lattes, J. Organomet. Chem., 1985, **289**, 309.
2376 H. Schmidbaur, E. Weiss, and G. Müller, Synth. React. Inorg. Met.-Org. Chem., 1985, **15**, 401.
2377 W.J. Layton, K. Niedenzu, P.M. Niedenzu, and S. Trofimenko, Inorg. Chem., 1985, **24**, 1454.
2378 D. Männig and H. Nöth, J. Chem. Soc., Dalton Trans., 1985, 1689.
2379 C.S. Shiner, C.M. Garner, and R.C. Haltwanger, J. Am. Chem. Soc., 1985, **107**, 7167.
2380 T.C. Farrar and G.R. Quinting, Inorg. Chem., 1985, **24**, 1941.
2381 M. Arunchaiya and J.H. Morris, Inorg. Chim. Acta, 1985, **103**, 31.
2382 D.F. Gaines and D.E. Coons, J. Am. Chem. Soc., 1985, **107**, 3266.
2383 R.G. Swisher, E. Sinn, R.J. Butcher, and R.N. Grimes, Organometallics, 1985, **4**, 882.
2384 G.B. Jacobsen, D.G. Meina, J.H. Morris, C. Thomson, S.J. Andrews, D. Reed, A.J. Welch, and D.F. Gaines, J. Chem. Soc., Dalton Trans., 1985, 1645.

2385 O.W. Howarth, M.J. Jasztal, J.G. Taylor, and M.G.H. Wallbridge, Polyhedron, 1985, **4**, 1461.

2386 M.A. Beckett, N.N. Greenwood, J.D. Kennedy, and M. Thornton-Pett, Polyhedron, 1985, **4**, 505.

2387 M.A. Beckett, J.D. Kennedy, and O.W. Howarth, J. Chem. Soc., Chem. Commun., 1985, 855.

2388 C.T. Brewer, R.G. Swisher, E. Sinn, and R.N. Grimes, J. Am. Chem. Soc., 1985, **107**, 3558.

2389 Z. Wang, E. Sinn, and R.N. Grimes, Inorg. Chem., 1985, **24**, 826.

2390 R. DePoy and G. Kodama, Inorg. Chem., 1985, **24**, 2871.

2391 R.P. Micciche, P.J. Carroll, and L.G. Sneddon, Organometallics, 1985, **4**, 1619.

2392 N.S. Hosmane, M.N. Mollenhauer, A.H. Cowley, and N.C. Norman, Organometallics, 1985, **4**, 1194.

2393 M. Kameda and G. Kodama, Inorg. Chem., 1985, **24**, 2712.

2394 J.G. Taylor and M.G.H. Wallbridge, Polyhedron, 1985, **4**, 321.

2395 S.A. Snow and G. Kodama, Inorg. Chem., 1985, **24**, 3339.

2396 D.E. Coons and D.F. Gaines, Inorg. Chem., 1985, **24**, 3774.

2397 P.K. Rush and L. Barton, Polyhedron, 1985, **4**, 1741.

2398 G. Schmid, D. Zaika, and R. Boese, Angew. Chem., Int. Ed. Engl., 1985, **24**, 602.

2399 K. Stumpf, H. Pritzkow, and W. Siebert, Angew. Chem., Int. Ed. Engl., 1985, **24**, 71.

2400 W. Preetz and J. Fritze, Z. Naturforsch., Teil B, 1984, **39**, 1472 (Chem. Abstr., 1985, **102**, 105 103).

2401 N.N. Greenwood, M.J. Hails, J.D. Kennedy, and W.S. McDonald, J. Chem. Soc., Dalton Trans., 1985, 953.

2402 N.S. Hosmane, N.N. Sirmokadam, and M.N. Mollenhauer, J. Organomet. Chem., 1985, **279**, 359.

2403 H.A. Boyter, jun., R.G. Swisher, E. Sinn, and R.N. Grimes, Inorg. Chem., 1985, **24**, 3810.

2404 R.G. Swisher, E. Sinn, and R.N. Grimes, Organometallics, 1985, **4**, 896.

2405 R.G. Swisher, E. Sinn, and R.N. Grimes, Organometallics, 1985, **4**, 890.

2406 J.J. Briguglio and L.G. Sneddon, Organometallics, 1985, **4**, 721.

2407 Z.-T. Wang, E. Sinn, and R.N. Grimes, Inorg. Chem., 1985, **24**, 834.

2408 R. Schlögl and B. Wrackmeyer, Polyhedron, 1985, **4**, 885.

2409 B. Ng, T. Onak, T. Banuelos, F. Gomez, and E.W. DiStefano, Inorg. Chem., 1985, **24**, 4091.

2410 B. Ng, T. Onak, and K. Fuller, Inorg. Chem., 1985, **24**, 4371.

2411 C.T. Brewer and R.N. Grimes, J. Am. Chem. Soc., 1985, **107**, 3552.

2412 E.W. Corcoran, jun. and L.G. Sneddon, J. Am. Chem. Soc., 1985, **107**, 7446.

2413 L. Barton and P.K. Rush, Inorg. Chem., 1985, **24**, 3413.

2414 D.G. Meina and J.H. Morris, J. Chem. Soc., Dalton Trans., 1985, 1903.

2415 T.P. Hanusa, J.C. Huffman, T.L. Curtis, and L.J. Todd, Inorg. Chem., 1985, **24**, 787.

2416 J.E. Crook, M. Elrington, N.N. Greenwood, J.D. Kennedy, M. Thornton-Pett, and J.D. Woolins, J. Chem. Soc., Dalton Trans., 1985, 2407.

2417 N.N. Greenwood, J.D. Kennedy, M. Thornton-Pett, and J.D. Woolins, J. Chem. Soc., Dalton Trans., 1985, 2397.

2418 S.A. Barrett, N.N. Greenwood, J.D. Kennedy, and M. Thornton-Pett, Polyhedron, 1985, **4**, 1981.

2419 B. Bonnetot, A. Tangi, M. Colombier, and H. Mongeot, Inorg. Chim. Acta, 1985, **105**, L15.

2420 M. Arunchaiya, P.C. Choi, and J.H. Morris, J. Chem. Res. (S), 1985, 216.

2421 W.H. Knoth, Organometallics, 1985, **4**, 207.

2422 X.L.R. Fontaine, H. Fowkes, N.N. Greenwood, J.D. Kennedy, and M. Thornton-Pett, J. Chem. Soc., Chem. Commun., 1985, 1722.

2423 A.J. Wynd, S.E. Robins, D.A. Welch, and A.J. Welch, J. Chem. Soc., Chem. Commun., 1985, 819.

2424 Z. Janoušek, J. Plešek, S. Heřmánek, and B. Štíbr, Polyhedron, 1985, **4**, 1797.

2425 B. Štíbr, Z. Janoušek, J. Plešek, T. Jelinek, and S. Heřmánek, J. Chem.

Soc., Chem. Commun., 1985, 1365.
2426 T. Jelínek, J. Plešek, S. Heřmánek, and B. Štíbr, Collect. Czech. Chem. Commun., 1985, **50**, 1376.
2427 E.A. Mizusawa, M.R. Thompson, and M.F. Hawthorne, Inorg. Chem., 1985, **24**, 1911.
2428 M. Green, J.A.K. Howard, A.P. Jones, A.N. de M. Jelfs, C.M. Nunn, and F.G.A. Stone, J. Chem. Soc., Chem. Commun., 1985, 1778.
2429 L.I. Zakharkin, V.V. Kobak, I.V. Pisareva, V.A. Antonovich, V.A. Ol'shevskaya, A.I. Yanovsky, and Yu.T. Struchkov, J. Organomet. Chem., 1985, **297**, 77.
2430 M.P. Garcia, M. Green, F.G.A. Stone, R.G. Somerville, A.J. Welch, C.E. Briant, D.N. Cox, and D.M.P. Mingos, J. Chem. Soc., Dalton Trans., 1985, 2343.
2431 D.M. Speckman, C.B. Knobler, and M.F. Hawthorne, Organometallics, 1985, **4**, 1692.
2432 P.E. Behnken, T.B. Marder, R.T. Baker, C.B. Knobler, M.R. Thompson, and M.F. Hawthorne, J. Am. Chem. Soc., 1985, **107**, 932.
2433 T.P. Hanusa and L.J. Todd, Polyhedron, 1985, **4**, 2063.
2434 J.D. Hewes, M. Thompson, and M.F. Hawthorne, Organometallics, 1985, **4**, 13.
2435 D.M. Speckman, C.B. Knobler, and M.F. Hawthorne, Organometallics, 1985, **4**, 426.
2436 R.E. King, tert., D.C. Busby, and M.F. Hawthorne, J. Organomet. Chem., 1985, **279**, 103.
2437 H.M. Colquhoun, T.J. Greenhough, and M.G.H. Wallbridge, J. Chem. Soc., Dalton Trans., 1985, 761.
2438 L.I. Zakharkin, N.Yu. Krainova, I.V. Pisareva, and V.A. Antonovich, Synth. React. Inorg. Met.-Org. Chem., 1985, **15**, 655.
2439 S. Al-Baker, W.E. Hill, and C.A. McAuliffe, J. Chem. Soc., Dalton Trans., 1985, 1387.
2440 E. Kalbarczyk and S. Pasynkiewicz, J. Organomet. Chem., 1985, **292**, 119.
2441 E. Kupce, E. Liepins, A. Lapsina, I. Urtane, G. Zelcans, and E. Lukevics, J. Organomet. Chem., 1985, **279**, 343.
2442 H.E. Katz, J. Am. Chem. Soc., 1985, **107**, 1420.
2443 Yu.N. Bubnov, A.I. Grandberg, M.Sh. Grigorian, V.G. Kiselev, M.I. Struchkova, and B.M. Mikhailov, J. Organomet. Chem., 1985, **292**, 93.
2444 S. Kerschl and B. Wrackmeyer, J. Chem. Soc., Chem. Commun., 1985, 1199.
2445 K.-J. Lee, P.D. Livant, M.L. McKee, and S.D. Worley, J. Am. Chem. Soc., 1985, **107**, 5901.
2446 M. Grassi, G. Di Silvestro, and M. Farina, Tetrahedron, 1985, **41**, 177.
2447 L. Horner, U. Kaps, and G. Simons, J. Organomet. Chem., 1985, **287**, 1.
2448 M. Hildenbrand, H. Pritzkow, and W. Siebert, Angew. Chem., Int. Ed. Engl., 1985, **24**, 759.
2449 B. Wrackmeyer, J. Organomet. Chem., 1985, **297**, 265.
2450 C.K. Narula and H. Nöth, Inorg. Chem., 1985, **24**, 2532.
2451 H. Nöth, P. Otto, and W. Storch, Chem. Ber., 1985, **118**, 3020.
2452 R. Boese, N. Finke, J. Henkelmann, G. Maier, P. Paetzold, H.P. Reisenauer, and G. Schmid, Chem. Ber., 1985, **118**, 1644.
2453 H.C. Brown, B. Singaram, and T.E. Cole, J. Am. Chem. Soc., 1985, **107**, 460.
2454 R. Köster and M. Yalpani, Angew. Chem., Int. Ed. Engl., 1985, **24**, 572.
2455 H.C. Brown, G.A. Molander, S.M. Singh, and U.S. Racherla, J. Org. Chem., 1985, **50**, 1577.
2456 H.C. Brown, J.S. Cha, B. Nazeer, and C.A. Brown, J. Org. Chem., 1985, **50**, 549.
2457 H.C. Brown, T.E. Cole, and M. Srebnik, Organometallics, 1985, **4**, 1788.
2458 C.K. Narula and H. Nöth, J. Organomet. Chem., 1985, **281**, 131.
2459 G. Maier, J. Henkelmann, and H.P. Reisenauer, Angew. Chem., Int. Ed. Engl., 1985, **24**, 1065.
2460 B. Glaser and H. Nöth, Angew. Chem., Int. Ed. Engl., 1985, **24**, 416.
2461 W. Jacksties, H. Nöth, and W. Storch, Chem. Ber., 1985, **118**, 2030.
2462 S. Allaoud and B. Frange, Inorg. Chem., 1985, **24**, 2520.
2463 F. Kumpfmüller, D. Nölle, H. Nöth, H. Pommerening, and R. Staudigl, Chem. Ber., 1985, **118**, 483.

2464 R. Oesterle, W. Maringgele, and A. Meller, J. Organomet. Chem., 1985, **284**, 281.
2465 R.H. Cragg and T.J. Miller, J. Organomet. Chem., 1985, **294**, 1.
2466 C. Habben, A. Meller, M. Noltemeyer, and G.M. Sheldrick, J. Organomet. Chem., 1985, **288**, 1.
2467 P. Zanirato, J. Organomet. Chem., 1985, **293**, 285.
2468 H.-A. Steuer, A. Meller, and G. Elter, J. Organomet. Chem., 1985, **295**, 1.
2469 B. Pachaly and R. West, J. Am. Chem. Soc., 1985, **107**, 2987.
2470 H.C. Brown, T. Imai, P.T. Perumal, and B. Singaram, J. Org. Chem., 1985, **50**, 4032.
2471 V.V. Kuznetsov and A.I. Gren, Zh. Obshch. Khim., 1984, **54**, 2263 (Chem. Abstr., 1985, **102**, 166 802).
2472 R. Danheiser and A.C. Savoca, J. Org. Chem., 1985, **50**, 2401.
2473 W.R. Roush, A.E. Walts, and L.K. Hoong, J. Am. Chem. Soc., 1985, **107**, 8186.
2474 R.W. Hoffmann, B. Kemper, R. Metternich, and T. Lehmeier, Ann., 1985, 2246.
2475 D.S. Matteson and J.W. Wilson, Organometallics, 1985, **4**, 1690.
2476 J.L. Atwood, S.G. Bott, C. Eaborn, M.N.A. El-Kheli, and J.D. Smith, J. Organomet. Chem., 1985, **294**, 23.
2477 M. Haase and U. Klingebiel, Angew. Chem., Int. Ed. Engl., 1985, **24**, 324.
2478 R. Lang, H. Nöth, P. Otto, and W. Storch, Chem. Ber., 1985, **118**, 86.
2479 M. Van Duin, J.A. Peters, A.P.G. Kieboom, and H. Van Bekkum, Tetrahedron, 1985, **41**, 3411.
2480 O.D. Gupta, W.A. Kamil, and J.M. Shreeve, Inorg. Chem., 1985, **24**, 2126.
2481 H. Nöth, M. Schwartz, and S. Weber, Chem. Ber., 1985, **118**, 4716.
2482 A. Haas and M. Willert-Porada, Chem. Ber., 1985, **118**, 1463.
2483 H. Nöth and S. Weber, Chem. Ber., 1985, **118**, 2554.
2484 H. Nöth and S. Weber, Chem. Ber., 1985, **118**, 2144.
2485 A. Brandl and H. Nöth, Chem. Ber., 1985, **118**, 3759.
2486 D. Männig, C.K. Narula, H. Nöth, and U. Wietelmann, Chem. Ber., 1985, **118**, 3748.
2487 P. Paetzold, E. Schröder, G. Schmid, and R. Boese, Chem. Ber., 1985, **118**, 3205.
2488 A. Haas and R. Walz, Chem. Ber., 1985, **118**, 3248.
2489 E. Hanecker, T.G. Hodgkins, K. Niedenzu, and H. Nöth, Inorg. Chem., 1985, **24**, 459.
2490 M. Baudler and M. Hintze, Z. Anorg. Allg. Chem., 1985, **522**, 184.
2491 D.R. Martin, C.M. Merkel, C.B. Drake, J.U. Mondal, and J.B. Iwamoto, Inorg. Chim. Acta, 1985, **97**, 189.
2492 H.C. Brown, J.V.N.V. Prasad, and S.-H. Zee, J. Org. Chem., 1985, **50**, 1582.
2493 J.A. Happe, R.J. Morgan, and C.M. Walkup, Polymer, 1985, **26**, 827 (Chem. Abstr., 1985, **103**, 124 385).
2494 D. Grec, L.G. Hubert-Pfalzgraf, A. Grand, and J.G. Riess, Inorg. Chem., 1985, **24**, 4642.
2495 D. Männig and H. Nöth, Angew. Chem., Int. Ed. Engl., 1985, **24**, 878.
2496 G.P. Shchetinina, O.V. Brovkina, B.N. Chernyshov, and E.G. Ippolitov, Zh. Neorg. Khim., 1985, **30**, 819 (Chem. Abstr., 1985, **102**, 196 836).
2497 G.P. Shchetinina, O.V. Brovkina, and B.N. Chernyshov, Zh. Neorg. Khim., 1985, **30**, 2161 (Chem. Abstr., 1985, **103**, 188 477).
2498 R.G. Kalbandkeri, K. Syed Mohamed, D.K. Padma, and A.R.V. Murthy, Polyhedron, 1985, **4**, 787.
2499 M.J. Forster, D.G. Gillies, and R.W. Matthews, J. Magn. Reson., 1985, **65**, 497.
2500 H.H. Karsch, A. Appelt, and G. Müller, Organometallics, 1985, **4**, 1624.
2501 H. Lehmkuhl, K. Mehler, A. Shakoor, C. Krüger, Y.-H. Tsay, R. Benn, A. Rufińska, and G. Schroth, Chem. Ber., 1985, **118**, 4248.
2502 M.A. Miller and E.P. Schram, Organometallics, 1985, **4**, 1362.
2503 K.B. Starowieyski, A. Bęcalska, and A. Okniński, J. Organomet. Chem., 1985, **293**, 7.
2504 H.H. Karsch, A. Appelt, F.H. Koehler, and G. Mueller, Organometallics, 1985, **4**, 231.
2505 P. Jutzi and W. Leffers, J. Chem. Soc., Chem. Commun., 1985, 1735.
2506 M. Veith, H. Lange, A. Belo, and O. Recktenwald, Chem. Ber., 1985, **118**,

1600.
2507 A.-A.I. Al-Wassil, P.B. Hitchcock, S. Sarisaban, J.D. Smith, and C.L. Wilson, J. Chem. Soc., Dalton Trans., 1985, 1929.
2508 R. Bertram, W. Gessner, D. Müller, H. Görz, and S. Schönherr, Z. Anorg. Allg. Chem., 1985, **525**, 14.
2509 G.L.S. Dent and G. Harvey, Proc. Int. Zeolite Conf., 6th, 1983 (Pub. 1984), 925.
2510 Y. Nakamura, S. Shimokawa, Y. Kitazawa, and M. Shimoji, Proc. Int. Symp. Molten Salt Chem. Technol., 1st, 1983, 387 (Chem. Abstr., 1985, **102**, 138 617).
2511 Y.C. Lee, J. Allison, and A.I. Popov, Polyhedron, 1985, **4**, 441.
2512 V.V. Kon'shin, B.N. Chernyshov, and E.G. Ippolitov, Dokl. Akad. Nauk SSSR, 1984, **278**, 370 (Chem. Abstr., 1985, **102**, 38 365).
2513 L. Bokobza and G. Cote, Polyhedron, 1985, **4**, 1499.
2514 H.H. Karsch and A. Appelt, J. Chem. Soc., Chem. Commun., 1985, 1083.
2515 R. Ahmad, G. Srivastava, and R.C. Mehrotra, Inorg. Chim. Acta, 1985, **97**, 159.
2516 W. Wojnowski, K. Peters, E.-M. Peters, and H.G. v. Schnering, Z. Anorg. Allg. Chem., 1985, **531**, 147.
2517 T.A. Blinka, B.J. Helmer, and R. West, Adv. Organomet. Chem., 1984, **23**, 193.
2517a U.R. Kunze, J. Organomet. Chem., 1985, **281**, 79.
2518 J. Otera, Kagaku (Kyoto), 1984, **39**, 738 (Chem. Abstr., 1985, **103**, 114 659).
2519 R.K. Harris and R.H. Newman, Mol. Phys., 1985, **54**, 1021.
2520 A. Hudson, R.A. Jackson, C.J. Rhodes, and A.L. del Vecchio, J. Organomet. Chem., 1985, **280**, 173.
2521 T.M. Kim, V.V. Yastrebov, G.A. Grigor'ev, N.A. Yashtulov, I.A. Al'tshul, and L.M. Volkova, Zh. Obshch. Khim., 1984, **54**, 2725 (Chem. Abstr., 1985, **102**, 185 143).
2522 Y.-M. Pai, W.P. Weber, and K.L. Servis, J. Organomet. Chem., 1985, **288**, 269.
2523 S. Kerschl, A. Sebald, and B. Wrackmeyer, Magn. Reson. Chem., 1985, **23**, 514.
2524 T.J. Barron and B.L. Groh, J. Am. Chem. Soc., 1985, **107**, 7221.
2525 P. Jutzi, G. Berg, C. Otto, and T. Wippermann, Z. Naturforsch., Teil B, 1984, **39**, 1368 (Chem. Abstr., 1985, **102**, 95 705).
2526 G.L. Larson and E. Torres, J. Organomet. Chem., 1985, **293**, 19.
2527 G. Maier, K. Schöttler, and H.P. Reisenauer, Tetrahedron Lett., 1985, **26**, 4079.
2528 E. Hengge and F.K. Mitter, Z. Anorg. Allg. Chem., 1985, **529**, 22.
2529 C. Eaborn and D.E. Reed, J. Chem. Soc., Perkin Trans. 2, 1985, 1687.
2530 K. Mochida, N. Matsushige, and M. Hamashima, Bull. Chem. Soc. Jpn., 1985, **58**, 1443.
2531 T. Birchall and V. Manivannan, J. Chem. Soc., Dalton Trans., 1985, 2671.
2532 A. Allerhand and M. Dohrenwend, J. Am. Chem. Soc., 1985, **107**, 6684.
2533 J. Popelis, E. Liepins, N.P. Erchak, and E. Lukevics, Latv. PSR Zinat. Akad. Vestis, Kim. Ser., 1985, 111 (Chem. Abstr., 1985, **103**, 88 004).
2534 T. Bartik, P. Heimbach, T. Himmler, and R. Mynott, Angew. Chem., Int. Ed. Engl., 1985, **24**, 313.
2535 R.L. Halterman, N.H. Nguyen, and K.P.C. Vollhardt, J. Am. Chem. Soc., 1985, **107**, 1379.
2536 T.N. Mitchell and R. Wickenkamp, J. Organomet. Chem., 1985, **291**, 179.
2537 A. Ricci, M. Fiorenza, A. Degl'Innocenti, G. Seconi, P. Dembech, K. Witzgall, and H.J. Bestmann, Angew. Chem., Int. Ed. Engl., 1985, **24**, 1068.
2538 G.A. Olah, H. Doggweiler, J.D. Felberg, and S. Frohlich, J. Org. Chem., 1985, **50**, 4847.
2539 P. Jutzi, E. Schlüter, S. Pohl, and W. Saak, Chem. Ber., 1985, **118**, 1959.
2540 A. Ricci, A. Degl'Innocenti, M. Fiorenza, P. Dembech, N. Ramadan, G. Seconi, and D.R.M. Walton, Tetrahedron Lett., 1985, **26**, 1091.
2541 G.J. Chen and C. Tamborski, J. Organomet. Chem., 1985, **293**, 313.
2542 C.A. Parnell and K.P.C. Vollhardt, Tetrahedron, 1985, **41**, 5791.
2543 B.A. Pearlman, S.R. Putt, and J.A. Fleming, J. Org. Chem., 1985, **50**, 3622.
2544 K. Mach, H. Antropiusová, L. Petrusová, F. Tureček, V. Hanuš, P. Sedmera, and J. Schraml, J. Organomet. Chem., 1985, **289**, 331.
2545 E. Wenschuh, W. Radeck, A. Porzel, A. Kolbe, and S. Edelmann, Z. Anorg.

Allg. Chem., 1985, **528**, 138.
2546 M. Grignon-Dubois and M. Ahra, J. Organomet. Chem., 1985, **280**, 313.
2547 A.H.-B. Cheng, P.R. Jones, M.E. Lee, and P. Roussi, Organometallics, 1985, **4**, 581.
2548 S.S. Nikam and K.K. Wang, J. Org. Chem., 1985, **50**, 2193.
2549 J.-M. Wulff and H.M.R. Hoffmann, Angew. Chem., Int. Ed. Engl., 1985, **24**, 605.
2550 K.H. Dötz, J. Mühlemeier, and B. Trenkle, J. Organomet. Chem., 1985, **289**, 257.
2551 G.J. Kang and T.H. Chan, Can. J. Chem., 1985, **63**, 3102.
2552 P.R. Jones, J.M. Rozell, jun., and B.M. Campbell, Organometallics, 1985, **4**, 1321.
2553 T.J. Barton and B.L. Groh, J. Org. Chem., 1985, **50**, 158.
2554 A. Padwa, Y.-Y. Chen, W. Dent, and H. Nimmesgern, J. Org. Chem., 1985, **50**, 4006.
2555 K.K. Wang and C. Liu, J. Org. Chem., 1985, **50**, 2578.
2556 Pham Chiem Van, R.S. Macomber, H.B. Mark, jun., and H. Zimmer, J. Org. Chem., 1984, **49**, 5250.
2557 H. Yasuda, T. Nishi, S. Miyanaga, and A. Nakamura, Organometallics, 1985, **4**, 359.
2558 T.J. Barton and B.L. Groh, Organometallics, 1985, **4**, 575.
2559 N. Chatani and T. Hanafusa, J. Chem. Soc., Chem. Commun., 1985, 838.
2560 E.-i. Negishi, S.J. Holmes, J.M. Tour, and J.A. Miller, J. Am. Chem. Soc., 1985, **107**, 2568.
2561 B.C. Berris, G.H. Hovakeemian, Y.-H. Lai, H. Mestdagh, and K.P.C. Vollhardt, J. Am. Chem. Soc., 1985, **107**, 5670.
2562 M. Ishikawa, S. Matsuzuwa, T. Higuchi, S. Kamitori, and K. Hirotsu, Organometallics, 1985, **4**, 2040.
2563 M. Ishikawa and S. Matsuzawa, J. Chem. Soc., Chem. Commun., 1985, 588.
2564 M. Ishikawa, S. Matsuzawa, H. Sugisawa, F. Yano, S. Kamitori, and T. Higuchi, J. Am. Chem. Soc., 1985, **107**, 7706.
2565 J. Volhardt, H.J. Gais, and K.L. Lukas, Angew. Chem., Int. Ed. Engl., 1985, **24**, 696.
2566 U. Klingebiel, S. Pohlmann, and L. Skoda, J. Organomet. Chem., 1985, **291**, 277.
2567 C. Couret, J. Escudie, J. Satge, A. Raharinirina, and J.D. Andriamizaka, J. Am. Chem. Soc., 1985, **107**, 8280.
2568 R. Appel and C. Casser, Chem. Ber., 1985, **118**, 3419.
2569 R. Appel, J. Menzel, and F. Knoch, Chem. Ber., 1985, **118**, 4068.
2570 Z.-M. Xie, P. Wislan-Neilson, and R.H. Neilson, Organometallics, 1985, **4**, 339.
2571 R. Appel, F. Knoch, and R. Zimmermann, Chem. Ber., 1985, **118**, 814.
2572 R. Appel, U. Kündgen, and F. Knoch, Chem. Ber., 1985, **118**, 1352.
2573 R. Appel, E. Gaitzsch, and F. Knoch, Angew. Chem., Int. Ed. Engl., 1985, **24**, 589.
2574 V.D. Romanenko, A.V. Ruban, S.V. Iksanova, L.K. Polyachenko, and L.N. Markovski, Phosphorus Sulphur, 1985, **22**, 365.
2575 T. Kauffmann, R. Kriegesmann, A. Rensing, R. König, and F. Steinseifer, Chem. Ber., 1985, **118**, 370.
2576 A. Baceiredo, G. Bertrand, and G. Sicard, J. Am. Chem. Soc., 1985, **107**, 4780.
2577 R. Appel, T. Gaitzsch, and F. Knoch, Angew. Chem., Int. Ed. Engl., 1985, **24**, 419.
2578 F. Sladky, B. Bildstein, and D. Obendorf, J. Organomet. Chem., 1985, **295**, C1.
2579 R.L. Danheiser, C.A. Kwasigroch, and Y.-M. Tsai, J. Am. Chem. Soc., 1985, **107**, 7233.
2580 E. Block and M. Aslam, J. Am. Chem. Soc., 1985, **107**, 6729.
2581 A.G. Brook, K.D. Safa, P.D. Lickiss, and K.M. Baines, J. Am. Chem. Soc., 1985, **107**, 4338.
2582 L.R. Robinson, G.T. Burns, and T.J. Barton, J. Am. Chem. Soc., 1985, **107**, 3935.

2583 R.T. Conlin, H.B. Huffaker, and Y.-W. Kwak, J. Am. Chem. Soc., 1985, **107**, 731.

2584 J. Dubac, A. Laporterie, and H. Iloughmane, J. Organomet. Chem., 1985, **293**, 295.

2585 E. Kloster-Jensen and G.A. Eliassen, Angew. Chem., Int. Ed. Engl., 1985, **24**, 565.

2586 Y. Hasegawa and K. Okamura, Nippon Kagaku Kaishi, 1985, 564 (Chem. Abstr., 1985, **103**, 161 043).

2587 A. Sekiguchi, H. Tanikawa, and W. Ando, Organometallics, 1985, **4**, 584.

2588 H. Appler, L.W. Gross, B. Mayer, and W.P. Neumann, J. Organomet. Chem., 1985, **291**, 9.

2589 A. Schäfer, M. Weidenbruch, and S. Pohl, J. Organomet. Chem., 1985, **282**, 305.

2590 K. Jurkschat, C. Mugge, J. Schmidt, and A. Tzschach, J. Organomet. Chem., 1985, **287**, C1.

2591 G. Becker, H.-M. Hartmann, A. Münch, and H. Riffel, Z. Anorg. Allg. Chem., 1985, **530**, 29.

2592 A.G. Brook and H.-J. Wessely, Organometallics, 1985, **4**, 1487.

2593 M. Haase and U. Klingebiel, Z. Anorg. Allg. Chem., 1985, **524**, 106.

2594 T.J. Barton and G.P. Hussmann, J. Am. Chem. Soc., 1985, **107**, 7581.

2595 H. Matsumoto, K. Takatsuna, M. Minemura, Y. Nagai, and M. Goto, J. Chem. Soc., Chem. Commun., 1985, 1366.

2596 W. Wojnowski, B. Dręczewski, A. Herman, K. Peters, E.-M. Peters, and H.G. von Schnering, Angew. Chem., Int. Ed. Engl., 1985, **24**, 992.

2597 P. Trefonas, tert., R. West, and R.D. Miller, J. Am. Chem. Soc., 1985, **107**, 2737.

2598 Y. Takeuchi, M. Shimoda, and S. Tomoda, Magn. Reson. Chem., 1985, **23**, 580.

2599 K. Kamienska-Trela, H. Ilcewicz, H. Baranska, and A. Labudzinska, Bull. Pol. Acad. Sci., Chem., 1984, **32**, 143 (Chem. Abstr., 1985, **102**, 204 058).

2600 T.N. Mitchell, W. Reimann, and C. Nettelbeck, Organometallics, 1985, **4**, 1044.

2601 W. Adcock, V.S. Iyer, W. Kitching, and D. Young, J. Org. Chem., 1985, **50**, 3706.

2602 J. Otera and T. Yano, Bull. Chem. Soc. Jpn., 1985, **58**, 387.

2603 T. Birchall and V. Manivannan, Can. J. Chem., 1985, **63**, 2211.

2604 C. Stetin, D. De Jeso, and J.-C. Pommier, J. Org. Chem., 1985, **50**, 3863.

2605 B. Jousseaume, E. Chanson, M. Bevilacqua, A. Saux, M. Pereyre, B. Barbe, and M. Petraud, J. Organomet. Chem., 1985, **294**, C41.

2606 J. Köcher and W.P. Neumann, Organometallics, 1985, **4**, 400.

2607 T. Kauffmann, G. Ilchmann, R. König, and M. Wensing, Chem. Ber., 1985, **118**, 391.

2608 P. Jutzi and B. Hielscher, J. Organomet. Chem., 1985, **291**, C25.

2609 D. Young and W. Kitching, J. Org. Chem., 1985, **50**, 4098.

2610 R. Appel, C. Casser, and M. Immenkeppel, Tetrahedron Lett., 1985, **26**, 3551.

2611 T.N. Mitchell and W. Reimann, J. Organomet. Chem., 1985, **281**, 163.

2612 C.S. Frampton, R.M.C. Roberts, J. Silver, J.F. Warmsley, and B. Yavari, J. Chem. Soc., Dalton Trans., 1985, 169.

2613 S. Matsubara, J.-I. Hibino, Y. Morizawa, K. Oshima, and H. Nozaki, J. Organomet. Chem., 1985, **285**, 163.

2614 B. Watta, W.P. Neumann, and J. Sauer, Organometallics, 1985, **4**, 1954.

2615 D. Farah, T.J. Karol, and H.G. Kuivila, Organometallics, 1985, **4**, 662.

2616 C.S. Frampton, R.M.G. Roberts, and J. Silver, J. Organomet. Chem., 1985, **297**, 273.

2617 B.A. Pearlman, S.R. Putt, and J.A. Fleming, J. Org. Chem., 1985, **50**, 3625.

2618 R. Hani and R.A. Geanangel, Inorg. Chim. Acta, 1985, **96**, 225.

2619 J. Otera, Y. Yoshinaga, T. Yamaji, T. Yoshioka, and Y. Kawasaki, Organometallics, 1985, **4**, 1213.

2620 R. Hani and R.A. Geanangel, J. Organomet. Chem., 1985, **293**, 197.

2621 A. Rahm, J. Grimeau, M. Petraud, and B. Barbe, J. Organomet. Chem., 1985, **286**, 297.

2622 T. Yano, K. Nakashima, J. Otera, and R. Okawara, Organometallics, 1985, **4**, 1501.

2623 S. Adams and M. Dräger, J. Organomet. Chem., 1985, **288**, 295.
2624 T. Kauffmann and A. Rensing, Chem. Ber., 1985, **118**, 380.
2625 N. Kleiner and M. Dräger, J. Organomet. Chem., 1985, **293**, 323.
2626 P.J. Kanyha and W.S. Brey, Congr. AMPERE Magn. Reson. Relat. Phenom., Proc., 22nd, 1984, 341 (Chem. Abstr., 1985, **103**, 123 567).
2627 E. Kupče, E. Liepiņš, and E. Lukevics, Angew. Chem., Int. Ed. Engl., 1985, **24**, 568.
2628 J. Pikies and W. Wojnowski, Z. Anorg. Allg. Chem., 1985, **521**, 173.
2629 E. Liepiņš, I. Birģele, P. Tomsons, and E. Lukevics, Magn. Reson. Chem., 1985, **23**, 485.
2630 J. Past, J. Puskar, J. Schraml, and E. Lippmaa, Collect. Czech. Chem. Commun., 1985, **50**, 2060.
2631 J. Schraml, M.F. Larin, and V.A. Pestunovich, Collect. Czech. Chem. Commun., 1985, **50**, 343.
2632 J. Past, J. Puskar, M. Alla, E. Lippmaa, and J. Schraml, Magn. Reson. Chem., 1985, **23**, 1076.
2633 R. Nardin and M. Vincendon, J. Magn. Reson., 1985, **61**, 338.
2634 E. Liepinsh, I. Sekacis, and E. Lukevics, Magn. Reson. Chem., 1985, **23**, 10.
2635 M. Weidenbruch, K. Kramer, A. Schäfer, and J.K. Blum, Chem. Ber., 1985, **118**, 107.
2636 D.J. Brauer, H. Bürger, G.R. Liewald, and J. Wilke, J. Organomet. Chem., 1985, **287**, 305.
2637 W. Walter and M. Akram, Phosphorus Sulphur, 1985, **21**, 291.
2638 C. Wentrup, S. Fischer, A. Maquestiau, and R. Flammang, Angew. Chem., Int. Ed. Engl., 1985, **24**, 56.
2639 E. Niecke, J. Böske, B. Krebs, and M. Dartmann, Chem. Ber., 1985, **118**, 3227.
2640 M. Veith and V. Huch, J. Organomet. Chem., 1985, **293**, 161.
2641 O.A. Vyazankina, B.A. Gostevskii, and N.S. Vyazankin, J. Organomet. Chem., 1985, **292**, 145.
2642 G. Becker, J. Härer, G. Uhl, and H.-J. Wessely, Z. Anorg. Allg. Chem., 1985, **520**, 120.
2643 U. Kliebisch, U. Klingebiel, and N. Vater, Chem. Ber., 1985, **118**, 4561.
2644 L. Birkofer and W. Quittmann, Chem. Ber., 1985, **118**, 2874.
2645 N. Wiberg, K. Schurz, and G. Fischer, Angew. Chem., Int. Ed. Engl., 1985, **24**, 1053.
2646 D.M. Schubert and A.D. Norman, Inorg. Chem., 1985, **24**, 1107.
2647 K. Issleib, H. Schmidt, and P. Bergmann, Z. Anorg. Allg. Chem., 1985, **529**, 216.
2648 K. Issleib, U. Kühne, and F. Krech, Phosphorus Sulphur, 1985, **21**, 367.
2649 L.M. Baigrie, H.R. Seiklay, and T.T. Tidwell, J. Am. Chem. Soc., 1985, **107**, 5391.
2650 R.F. Cunico and L. Bedell, J. Organomet. Chem., 1985, **281**, 135.
2651 M. Yalpani and G. Wilke, Chem. Ber., 1985, **118**, 661.
2652 H. Kleijn and P. Vermeer, J. Organomet. Chem., 1985, **292**, 437.
2653 T.H. Chan and M.A. Brook, Tetrahedron Lett., 1985, **26**, 2943.
2654 H. Yildirimyan and G. Gattow, Z. Anorg. Allg. Chem., 1985, **521**, 135.
2655 D. Hoebbel, K. Ujszászy, T. Reiher, G. Engelhardt, and A. Winkler, Z. Anorg. Allg. Chem., 1985, **524**, 51.
2656 K. Issleib, W. Mögelin, and A. Balszuweit, Z. Anorg. Allg. Chem., 1985, **530**, 16.
2657 H. Kischkel and G.-V. Röschenthaler, Chem. Ber., 1985, **118**, 4842.
2658 G.M. Blackburn and T.W. Maciej, J. Chem. Soc., Perkin Trans. 1, 1985, 1419.
2659 K. Issleib, A. Balszuweit, J. Kötz, St. Richter, and R. Leutloff, Z. Anorg. Allg. Chem., 1985, **529**, 151.
2660 R. Tacke, M. Link, and H. Zilch, Chem. Ber., 1985, **118**, 4637.
2661 J. Schraml, S. Kučár, J. Zelený, and V. Chvalovský, Collect. Czech. Chem. Commun., 1985, **50**, 1176.
2662 B.M. Trost and J.Y.L. Chung, J. Am. Chem. Soc., 1985, **107**, 4586.
2663 K.J. Stone and R.D. Little, J. Am. Chem. Soc., 1985, **107**, 2495.
2664 A. Goliaszewski and J. Schwartz, Tetrahedron, 1985, **41**, 5779.
2665 T. Takahashi, A. Ootake, and J. Tsuji, Tetrahedron, 1985, **41**, 5747.
2666 J.M. Rozell, jun. and P.R. Jones, Organometallics, 1985, **4**, 2206.

2667 W.J. Richter, J. Organomet. Chem., 1985, **286**, 1.

2668 C. Eaborn, P.D. Lickiss, S.T. Najim, and M.N. Romanelli, J. Chem. Soc., Chem. Commun., 1985, 1754.

2669 L. Horner and J. Mathias, J. Organomet. Chem., 1985, **282**, 155.

2670 L. Horner and J. Mathias, J. Organomet. Chem., 1985, **282**, 175.

2671 E. Liepins, I. Zicmane, and E. Lukevics, Kremniiorg. Soedin. Mater. Ikh Osn., Tr. Soveshch. Khim. Prakt. Primen. Kremniiorg. Soedin., 5th, 1981 (Pub. 1984), 16 (Chem. Abstr., 1985, **102**, 24 699).

2672 H.G. Horn and M. Hemeke, Chem.-Ztg., 1985, **109**, 1 (Chem. Abstr., 1985, **102**, 142 251).

2673 R. West, D.J. De Young, and K.J. Haller, J. Am. Chem. Soc., 1985, **107**, 4942.

2674 F. Sladky, B. Bildstein, C. Rieker, A. Gieren, H. Betz, and T. Hübner, J. Chem. Soc., Chem. Commun., 1985, 1800.

2675 C. Eaborn and P.D. Lickiss, J. Organomet. Chem., 1985, **294**, 305.

2676 B. Marciniec, E. Maćkowska, J. Gulińska, and W. Urbaniak, Z. Anorg. Allg. Chem., 1985, **529**, 222.

2677 V.K. Jain, J. Mason, M.S. Saraswat, and R.C. Mehrotra, Polyhedron, 1985, **4**, 2089.

2678 P.C. Srivastava and S.K. Srivastava, Spectrochim. Acta, Part A, 1985, **41A**, 687.

2679 P.C. Srivastava and S.K. Srivastava, Spectrochim. Acta, Part A, 1984, **40A**, 607.

2680 A. Lyčka and J. Holeček, J. Organomet. Chem., 1985, **294**, 179.

2681 T.P. Lockhart, W.F. Manders, and F.E. Brinckman, J. Organomet. Chem., 1985, **286**, 153.

2682 R.M. Davidson, H.G. Grant, and D. O'Smith, Spectrochim. Acta, Part A, 1985, **41A**, 581 (Chem. Abstr., 1985, **103**, 88 009).

2683 R.K. Chadha, J.E. Drake, and A.B. Sarkar, Inorg. Chem., 1985, **24**, 3156.

2684 M. Draeger and K. Haeberle, J. Organomet. Chem., 1985, **280**, 183.

2685 P.G. Harrison and N.W. Sharpe, Inorg. Chim. Acta, 1985, **108**, 7.

2686 M. Zeldin and J.J. Lin, J. Polym. Sci., Polym. Chem. Ed., 1985, **23**, 2333 (Chem. Abstr., 1985, **103**, 160 936).

2687 A. Lycka, J. Holecek, M. Nadvornik, and K. Handlir, J. Organomet. Chem., 1985, **280**, 323.

2688 S.P. Narula, S. Vashisht, and S. Lata, Chem. Ind. (London), 1985, 760.

2689 S.J. Blunden and R. Hill, Inorg. Chim. Acta, 1985, **98**, L7.

2690 C.T. Aitken and M. Onyszchuk, J. Organomet. Chem., 1985, **295**, 149.

2691 H.C. Clark, V.K. Jain, R.C. Mehrotra, B.P. Singh, G. Srivastava, and T. Birchall, J. Organomet. Chem., 1985, **279**, 385.

2692 D.W. Allen, S. Bailey, J.S. Brooks, and B.F. Taylor, Chem. Ind. (London), 1985, 826.

2693 R.H. Cragg and R.D. Lane, J. Organomet. Chem., 1985, **291**, 153.

2694 H. Fujiwara, F. Sakai, Y. Mikawa, and Y. Sasaki, Bull. Chem. Soc. Jpn., 1985, **58**, 1741.

2695 R.H. Cragg and R.D. Lane, J. Organomet. Chem., 1985, **294**, 7.

2696 J.H. Wengrovius, Ger. Offen. DE 3,428,840 (Cl. C08G77/06), 21 Feb 1985, US Appl. 520,978, 08 Aug 1983; 12 pp (Chem. Abstr., 1985, **102**, 186 000).

2697 J.P. Cohen-Addad and R. Dupeyre, Macromolecules, 1985, **18**, 1612 (Chem. Abstr., 1985, **103**, 71 896).

2698 D.A. Laude, jun., R.W.-K. Lee, and C.L. Wilkins, Anal. Chem., 1985, **57**, 1286.

2699 R.H. Cragg and R.D. Lane, J. Organomet. Chem., 1985, **289**, 23.

2700 D. Gagnebien, P.J. Madec, and E. Marechal, Eur. Polym. J., 1985, **21**, 301 (Chem. Abstr., 1985, **103**, 71 781).

2701 D. Gagnebien, P.J. Madec, and E. Marechal, Eur. Polym. J., 1985, **21**, 289 (Chem. Abstr., 1985, **103**, 71 780).

2702 C.G. Francis, P.D. Morand, and N.J. Spare, Organometallics, 1985, **4**, 1958.

2703 K.M. Taba and W.V. Dahlhoff, J. Organomet. Chem., 1985, **280**, 27.

2704 M. Weidenbruch, M. Herrndorf, A. Schäfer, K. Peters, and H.G. von Schnering, J. Organomet. Chem., 1985, **295**, 7.

2705 K. Haeberle and M. Draeger, Z. Naturforsch., Teil B, 1984, **39**, 1541 (Chem. Abstr., 1985, **102**, 185 186).

2706 J. Escudie, C. Couret, J. Satge, M. Andrianarison, and J.-D. Andriamizaka, J. Am. Chem. Soc., 1985, **107**, 3378.
2707 B. Wrackmeyer, J. Magn. Reson., 1985, **61**, 536.
2708 B. Nebout, B. de Jeso, and J.-C. Pommier, J. Chem. Soc., Chem. Commun., 1985, 504.
2709 W.F. Howard, jun., R.W. Crecely, and W.H. Nelson, Inorg. Chem., 1985, **24**, 2204.
2710 J. Otera, T. Yano, K. Nakashima, and R. Okawara, Chem. Lett., 1984, 2109 (Chem. Abstr., 1985, **102**, 185 263).
2711 A. Tzschach, E. Nietzschmann, and C. Mügge, Z. Anorg. Allg. Chem., 1985, **523**, 21.
2712 A. Saxena, J.P. Tandon, and A.J. Crowe, Polyhedron, 1985, **4**, 1085.
2713 D. Dakternieks, B.F. Hoskins, P.A. Jackson, E.R.T. Tiekink, and G. Winter, Inorg. Chim. Acta, 1985, **101**, 203.
2714 S.J. Blunden, L.A. Hobbs, P.J. Smith, A.G. Davies, and S.B. Teo, J. Organomet. Chem., 1985, **282**, 9.
2715 M. Dräger and N. Kleiner, Z. Anorg. Allg. Chem., 1985, **522**, 48.
2716 V.F. Sidorkin, V.A. Pestunovich, and M.G. Voronkov, Magn. Reson. Chem., 1985, **23**, 491.
2717 S.N. Tandura, M.G. Voronkov, A.V. Kisin, N.V. Alekseev, E.E. Shestakov, Z.A. Ovchinnikova, and V.P. Baryshok, Zh. Obshch. Khim., 1984, **54**, 2012 (Chem. Abstr., 1985, **102**, 62 368).
2718 V.A. Pestunovich, M.F. Larin, M.S. Sorokin, A.I. Albanov, and M.G. Voronkov, J. Organomet. Chem., 1985, **280**, C17.
2719 P.B. Hitchcock, H.A. Jasim, R.E. Kelly, and M.F. Lappert, J. Chem. Soc., Chem. Commun., 1985, 1776.
2720 J.C. Zhu, H.J. Wu, L.C. Chiang, G.J. Martin, P.C. Chen, G.L. Wu, and Z.G. Lai, J. Chim. Phys.-Chim. Biol., 1984, **81**, 407.
2721 L. Párkányi, P. Hencsel, L. Bihátsi, I. Kovács, and A. Szöllösy, Polyhedron, 1985, **4**, 242.
2722 R.R. Holmes, R.O. Day, V. Chandrasekhar, J.J. Harland, and J.M. Holmes, Inorg. Chem., 1985, **24**, 2016.
2723 R.R. Holmes, R.O. Day, V. Chandrasekhar, and J.M. Holmes, Inorg. Chem., 1985, **24**, 2009.
2724 R. Krebs, D. Schomburg, and R. Schmutzler, Z. Naturforsch., Teil B, 1985, **40**, 282 (Chem. Abstr., 1985, **103**, 123 568).
2725 A.R. Siedle, K.R. Mann, D.A. Bohling, G. Filipovich, P.E. Toren, F.J. Palensky, R.A. Newmark, R.W. Duerst, W.L. Stebbings, H.E. Mishmash, and K. Melancon, Inorg. Chem., 1985, **24**, 2216.
2726 N. Kakimoto, Magn. Reson. Chem., 1985, **23**, 140.
2727 R.J. Rao, G. Srivastava, and R.C. Mehrotra, Phosphorus Sulphur, 1985, **25**, 183.
2728 L.R. Milgrom and R.N. Sheppard, J. Chem. Soc., Chem. Commun., 1985, 350.
2729 N. Janes and E. Oldfield, J. Am. Chem. Soc., 1985, **107**, 6769.
2730 J. Mu, W.C. Park, B. DeJong, R.A. Hard, and F. Leder, Appl. Mineral., Proc. Int. Congr. Appl. Mineral. Miner. Ind., 2nd, 1984 (Pub. 1985), 699 (Chem. Abstr., 1985, **103**, 89 787).
2731 A. Meller and C.-P. Gräbe, Chem. Ber., 1985, **118**, 2020.
2732 J. Felsche, B. Ketterer, and R.L. Schmid, Thermochim. Acta, 1985, **88**, 443 (Chem. Abstr., 1985, **103**, 97 925).
2733 V.W. Day, W.G. Klemperer, V.V. Mainz, and D.M. Millar, J. Am. Chem. Soc., 1985, **107**, 8262.
2734 T. Fjeldberg, P.B. Hitchcock, M.F. Lappert, S.J. Smith, and A.J. Thorne, J. Chem. Soc., Chem. Commun., 1985, 939.
2735 L. Mathieu, P. Thivolle, G. Galy, and M. Berger, J. Biophys. Biomec., 1985, **9**, 95 (Chem. Abstr., 1985, **103**, 209 313).
2736 N. Alberola, Polyhedron, 1985, **4**, 1853.
2737 S.J. Blunden, D. Searle, and P.J. Smith, Inorg. Chim. Acta, 1985, **98**, 185.
2738 P.A.W. Dean, J.J. Vittal, and N.C. Payne, Can. J. Chem., 1985, **63**, 394.
2739 N.S. Yashina, E.I. Gefel, V.S. Petrosyan, and O.A. Reutov, Vestn. Mosk. Univ., Ser. 2: Khim., 1985, **26**, 90 (Chem. Abstr., 1985, **102**, 159 482).
2740 E.I. Gefel, Deposited Doc., 1983, VINITI 7085-83, Pt. 2, 275 (Chem. Abstr.,

1985, **102**, 78 995).
2741 V.O. Gel'mbol'dt, L.A. Gavrilova, S.P. Petrosyants, and A.A. Ennan, Koord. Khim., 1985, **11**, 1077 (Chem. Abstr., 1985, **103**, 226 238).
2742 V.O. Gel'mbol'dt, L.A. Gavrilova, A.A. Zinan, and S.P. Petrosyants, Zh. Neorg. Khim., 1985, **30**, 1173.
2743 V.O. Gel'mbol'dt, L.A. Gavrilova, P.N. Yurkevich, and A.A. Ennan, Zh. Neorg. Khim., 1985, **30**, 1597 (Chem. Abstr., 1985, **103**, 97 768).
2744 J.C. Tebby, Organophosphorus Chem., 1985, **15**, 286 (Chem. Abstr., 1985, **103**, 215 333).
2745 J.O. Friedrich and R.E. Wasylishen, J. Chem. Phys., 1985, **83**, 3707.
2746 J.M. Robert and R.F. Evilia, J. Am. Chem. Soc., 1985, **107**, 3733.
2747 M. Baudler, R. Heumüller, J. Germeshausen, and J. Hahn, Z. Anorg. Allg. Chem., 1985, **526**, 7.
2748 B.L. Feringa, A. Smaardijk, and H. Wynberg, J. Am. Chem. Soc., 1985, **107**, 4798.
2749 P.D. Blair, S. Cradock, and D.W.H. Rankin, J. Chem. Soc., Dalton Trans., 1985, 755.
2750 D. Houalla, H. Ramdane, and R. Wolf, Nouv. J. Chim., 1985, **9**, 189.
2751 T.V. Harris and W.R. Pretzer, Inorg. Chem., 1985, **24**, 4437.
2752 S.D. Pastor, J.D. Spivack, and L.P. Steinhuebel, Phosphorus Sulphur, 1985, **22**, 169.
2753 P.C. Thorstenson and H.S. Aaron, Phosphorus Sulphur, 1985, **22**, 145.
2754 M. Aresta, D. Ballivet-Tkatchenko, and M. Bonnet, Nouv. J. Chim., 1985, **9**, 321.
2755 M. Aresta, D. Ballivet-Tkatchenko, M.C. Bonnet, R. Faure, and H. Loiseleur, J. Am. Chem. Soc., 1985, **107**, 2994.
2756 M. Baudler and R. Heumueller, Z. Naturforsch., Teil B, 1984, **39**, 1306 (Chem. Abstr., 1985, **102**, 113 592).
2757 M. Mulliez, Bull. Soc. Chim. Fr., 1985, 1211.
2758 I. Hoppe, U. Schöllkopf, M. Nieger, and E. Egert, Angew. Chem., Int. Ed. Engl., 1985, **24**, 1067.
2759 L. Riesel and M. Kant, Z. Chem., 1984, **24**, 382 (Chem. Abstr., 1985, **102**, 71 707).
2760 A.S. Tarasevich, Teor. Eksp. Khim., 1985, **21**, 165 (Chem. Abstr., 1985, **103**, 71 412).
2761 K.B. Dillon and R.N. Reeve, Polyhedron, 1985, **4**, 1533.
2762 P.B. Kay and S. Trippett, J. Chem. Res. (S), 1985, 156.
2763 A.V. Kopytin, P. Gabor-klatsmanyi, V.P. Izvekov, E. Pungor, and E.G. Il'in, Anal. Chim. Acta, 1984, **162**, 133 (Chem. Abstr., 1985, **102**, 89 095).
2764 D.V. Griffiths and J.C. Tebby, Phosphorus Sulphur, 1985, **25**, 177.
2765 H.P. Fritz, G. Müller, G. Reber, and M. Weis, Angew. Chem., Int. Ed. Engl., 1985, **24**, 1058.
2766 H. Schmidbaur, R. Herr, T. Pollok, A. Schier, G. Müller, and J. Riede, Chem. Ber., 1985, **118**, 3105.
2767 J. Barluenga, F. Lopez, and F. Palacios, J. Chem. Soc., Chem. Commun., 1985, 1681.
2768 R.J. Pariza and P.L. Fuchs, J. Org. Chem., 1985, **50**, 4252.
2769 B.E. Maryanoff, A.B. Reitz, and B.A. Duhl-Emswiler, J. Am. Chem. Soc., 1985, **107**, 217.
2770 H.J. Bestmann, G. Schade, G. Schmid, and T. Mönius, Chem. Ber., 1985, **118**, 2635.
2771 H.J. Christau, F. Plénat, G. Duc, and A. Bennamara, Tetrahedron, 1985, **41**, 2717.
2772 N. Lowther and C.D. Hall, J. Chem. Soc., Chem. Commun., 1985, 1303.
2773 B. Wrackmeyer, Spectrochim. Acta, Part A, 1984, **40A**, 963.
2774 J. Bödeker, A. Köckritz, P. Köckritz, and R. Radeglia, J. Prakt. Chem. 1985, **327**, 723.
2775 M. Pomerantz, B.T. Ziemnicka, Z.M. Merchant, W.N. Chou, W.B. Perkins, and S. Bittner, J. Org. Chem., 1985, **50**, 1757.
2776 M. Biddlestone, R. Keat, H.G. Parkes, H. Rose, D.S. Rycroft, and R.A. Shaw, Phosphorus Sulphur, 1985, **25**, 25.
2777 H. Duddeck and A.G. Hanna, Magn. Reson. Chem., 1985, **23**, 533.

2778 A.I. Rezvukhin, G.N. Dolenko, and S.A. Krupoder, Magn. Reson. Chem., 1985, **23**, 221.
2779 E. Duñach and H.B. Kagan, Tetrahedron Lett., 1985, **26**, 2649.
2780 O.W. Kolling, Trans. Kans. Acad. Sci., 1984, **87**, 115 (Chem. Abstr., 1985, **102**, 6700).
2781 J.W. Kelly, P.L. Robinson, and S.A. Evans, jun., J. Org. Chem., 1985, **50**, 5009).
2782 M. Von Itzstein and I.D. Jenkins, Aust. J. Chem., 1984, **37**, 2447.
2783 D.W. Allen, I.W. Nowell, and B.F. Taylor, J. Chem. Soc., Dalton Trans., 1985, 2505.
2784 K. Issleib, U. Kühne, and F. Krech, Z. Anorg. Allg. Chem., 1985, **523**, 7.
2785 A. Schmidpeter, G. Burget, F. Zwaschka, and W.S. Sheldrick, Z. Anorg. Allg. Chem., 1985, **527**, 17.
2786 W.J. Richter, Chem. Ber., 1985, **118**, 97.
2787 M. Baudler and J. Germeshausen, Chem. Ber., 1985, **118**, 4285.
2788 L.D. Quin and K.C. Caster, Phosphorus Sulphur, 1985, **25**, 117.
2789 A.B. Burg, Inorg. Chem., 1985, **24**, 148.
2790 J. Grobe and D. Le Van, Tetrahedron Lett., 1985, **26**, 3681.
2791 S.J. Higgins and W. Levason, Inorg. Chem., 1985, **24**, 1105.
2792 G.M. Isaeva, B.M. Butin, and K.B. Erzhanov, Izv. Akad. Nauk Kaz. SSR, Ser. Khim., 1984, 67 (Chem. Abstr., 1985, **102**, 149 368).
2793 B. Demerseman, P. Le Coupanec, and P.H. Dixneuf, J. Organomet. Chem., 1985, **287**, C35.
2794 C. Mealli, M. Sabat, F. Zanobini, M. Ciampolini, and N. Nardi, J. Chem. Soc., Dalton Trans., 1985, 479.
2795 H. Brunner and H. Weber, Chem. Ber., 1985, **118**, 3380.
2796 L.D. Quin and F.C. Bernhardt, Magn. Reson. Chem., 1985, **23**, 929.
2797 G.P. Schiemenz and P. Nielsen, Phosphorus Sulphur, 1985, **21**, 267.
2798 G.P. Schiemenz and P. Nielsen, Phosphorus Sulphur, 1985, **21**, 259.
2799 T. Kauffmann and J. Ennan, Chem. Ber., 1985, **118**, 2703.
2800 T. Kauffmann and J. Ennan, Chem. Ber., 1985, **118**, 2692.
2801 H. Schmidbaur and P. Nusstein, Organometallics, 1985, **4**, 344.
2802 A.J. Ashe, tert., and F.J. Drone, Organometallics, 1985, **4**, 1478.
2803 T. Kauffmann, B. Altpeter, N. Klas, and R. Kriegesmann, Chem. Ber., 1985, **118**, 2353.
2804 N.A.A. Al-Jabar and A.G. Massey, J. Organomet. Chem., 1985, **287**, 57.
2805 T. Kauffmann and F. Steinseifer, Chem. Ber., 1985, **118**, 1031.
2806 T. Kauffmann, F. Steinseifer, and N. Klas, Chem. Ber., 1985, **118**, 1039.
2807 L.K. Krannich, R.K. Kanjolia, and C.L. Watkins, Inorg. Chim. Acta, 1985, **103**, 1.
2808 R. Münstedt and U. Wannagat, Monatshefte, 1985, **116**, 7.
2809 H. Wamhoff and G. Hendrikx, Chem. Ber., 1985, **118**, 863.
2810 S. Bittner, Y. Assaf, P. Krief, M. Pomerantz, B.T. Ziemnicka, and C.G. Smith, J. Org. Chem., 1985, **50**, 1712.
2811 T. Akasaka, R. Sato, and W. Ando, J. Am. Chem. Soc., 1985, **107**, 5539.
2812 K.C.K. Swamy, P. Ramabrahman, and S.S. Krishnamurthy, Synth. React. Inorg. Met.-Org. Chem., 1985, **15**, 1023.
2813 K.C.K. Swamy, M.D. Poojary, S.S. Krishnamurthy, and H. Manohar, J. Chem. Soc., Dalton Trans., 1985, 1881.
2814 D.W. Morton and R.H. Neilson, Phosphorus Sulphur, 1985, **25**, 315.
2815 A. Baceiredo, G. Bertrand, J.-P. Majoral, and K.B. Dillon, J. Chem. Soc., Chem. Commun., 1985, 562.
2816 A. Schmidpeter, S. Lockschmidt, K. Karaghiosoff, and W.S. Sheldrick, J. Chem. Soc., Chem. Commun., 1985, 1447.
2817 A. Schmidpeter and G. Burget, Angew. Chem., Int. Ed. Engl., 1985, **24**, 580.
2818 A. Schmidpeter, S. Lochschmidt, and W.S. Sheldrick, Angew. Chem., Int. Ed. Engl., 1985, **24**, 226.
2819 H. Duddeck and A.G. Hanna, Magn. Reson. Chem., 1985, **23**, 41.
2820 L.D. Quin and J.C. Kisalus, Phosphorus Sulphur, 1985, **22**, 35.
2821 A. Rudi, I. Goldberg, and Y. Kashman, Tetrahedron, 1985, **41**, 5267.
2822 L.P. Krasnomolova, S.K. Shishkin, O.V. Agashkin, and Yu.G. Bosyakov, Izv. Akad. Nauk Kaz. SSR, Ser. Khim., 1984, 60 (Chem. Abstr., 1985, **102**, 95 730).

2823 J. Barluenga, F. Lopez, and F. Palacios, J. Chem. Res. (S), 1985, 211.
2824 P.L. Robinson, C.N. Barry, J.W. Kelly, and S.A. Evans, jun., J. Am. Chem. Soc., 1985, **107**, 5210.
2825 N.K. Jha and D.M. Joshi, Polyhedron, 1985, **4**, 2083.
2826 R. Hussong, H. Heydt, and M. Regitz, Phosphorus Sulphur, 1985, **25**, 201.
2827 V.B. Rozhnov, O.V. Agashkin, L.P. Krasnomolova, E.I. Khokhlova, A.P. Logunov, and Yu.G. Bosyakov, Zh. Fiz. Khim., 1985, **59**, 1699 (Chem. Abstr., 1985, **103**, 132 799).
2828 A.M. Forster and A.J. Downs, Polyhedron, 1985, **4**, 1624.
2829 H.E. Hosseini, H.W. Kroto, J.F. Nixon, and O. Ohashi, J. Organomet. Chem., 1985, **296**, 351.
2830 H. Brunner, L. Grötzinger, and H. Weber, Inorg. Chim. Acta, 1985, **103**, 19.
2831 T. Chivers, M.N.S. Rao, and J.F. Richardson, Inorg. Chem., 1985, **24**, 2237.
2832 R.T. Oakley, S.J. Rettig, N.L. Paddock, and J. Trotter, J. Am. Chem. Soc., 1985, **107**, 6923.
2833 J.M.A. Al-Rawi and N. Ayed, Magn. Reson. Chem., 1985, **23**, 285.
2834 W. Dabkowski, J. Michalski, and Z. Skrzypczynski, Chem. Ber., 1985, **118**, 1809.
2835 M. Herberhold, W. Ehrenreich, A. Gieren, H. Betz, and T. Huebner, Chem. Ber., 1985, **118**, 1476.
2836 M.J.P. Harger and A. Smith, J. Chem. Soc., Perkin Trans. 1, 1985, 1787.
2837 C.A. Stewart, R.L. Harlow, and A.J. Arduengo, tert., J. Am. Chem. Soc., 1985, **107**, 5543.
2838 A. Schmidpeter, G. Burget, and W.S. Sheldrick, Chem. Ber., 1985, **118**, 3849.
2839 J.R. Lloyd, N. Lowther, and C.D. Hall, J. Chem. Soc., Perkin Trans. 2, 1985, 245.
2840 L.J. Caudle, E.N. Duesler, and R.T. Paine, Inorg. Chim. Acta, 1985, **110**, 91.
2841 J. Heinicke and A. Tzschach, Phosphorus Sulphur, 1985, **25**, 345.
2842 S. Hoz, E.J. Dunn, E. Buncel, R.A.B. Bannard, and J.G. Purdon, Phosphorus Sulphur, 1985, **24**, 321.
2843 A.W. Frank, Phosphorus Sulphur, 1985, **22**, 265.
2844 R.A. McClelland, G.H. McGall, and G. Patel, J. Am. Chem. Soc., 1985, **107**, 5204.
2845 G. Singh, G.S. Reddy, and J.C. Calabrese, J. Org. Chem., 1984, **49**, 5132
2846 G. Hägele, W. Kückelhaus, G. Tossing, D. Mootz, and H.-G. Wussow, Phosphorus Sulphur, 1985, **22**, 241.
2847 H. Keck, W. Kuchen, and S. Kühlborn, Phosphorus Sulphur, 1985, **24**, 343.
2848 J. Navech, M. Revel, and R. Kraemer, Tetrahedron Lett., 1985, **26**, 207.
2849 P.J. Retuert, E. Fluck, H. Riffel, and H. Hess, Z. Anorg. Allg. Chem., 1985, **521**, 153.
2850 R. Boese, G. Hägele, W. Kückelhaus, and G. Tossing, Phosphorus Sulphur, 1985, **25**, 103.
2851 W.W. DuMont, S. Kubiniok, and T. Severengiz, Z. Anorg. Allg. Chem., 1985, **531**, 21.
2852 P. Jutzi and T. Wippermann, J. Organomet. Chem., 1985, **287**, C5.
2853 M. Engelhardt, G. Hägele, M. Mikolajczyk, P. Balczewski, and D. Wendisch, Magn. Reson. Chem., 1985, **23**, 18.
2854 R. Burgada, Y.O. El Khoshnieh, and Y. Leroux, Tetrahedron, 1985, **41**, 1207.
2855 G.O. Neustad, K. Maartmann-Moe, C. Rømming, and J. Songstad, Acta. Chem. Scand., A, 1985, **39**, 523.
2856 J. Högel and A. Schmidpeter, Chem. Ber., 1985, **118**, 1621.
2857 A. Schmidpeter, K. Karaghiosoff, C. Cleve, and D. Schomburg, Angew. Chem., Int. Ed. Engl., 1985, **24**, 123.
2858 E.E. Nifant'ev, S.F. Sorokina, L.A. Vorob'eva, and A.R. Bekker, Zh. Obshch. Khim., 1985, **55**, 308 (Chem. Abstr., 1985, **103**, 105 058).
2859 C.R. Hall, T.D. Inch, and N.E. Williams, J. Chem. Soc., Perkin Trans. 1, 1985, 233.
2860 H.W. Roesky, J. Lucas, K.-L. Weber, H. Djarrah, E. Egert, M. Noltemeyer, and G.M. Sheldrick, Chem. Ber., 1985, **118**, 2396.
2861 L. Liu, Huaxue Xuebao, 1984, **42**, 1204 (Chem. Abstr., 1985, **102**, 113 612).
2862 S. Kleemann and E. Fluck, Z. Anorg. Allg. Chem., 1985, **526**, 141.
2863 K.C. Calvo, J. Am. Chem. Soc., 1985, **107**, 3690.

2864 P.B. Rasmussen, U. Pedersen, I. Thomsen, B. Yde, and S.O. Lawesson, Bull. Soc. Chim. Fr., 1985, 62.
2865 A.J. Muller and H.S. Aaron, Phosphorus Sulphur, 1985, **25**, 339.
2866 D. El Manouni, Y. Leroux, and R. Burgada, Phosphorus Sulphur, 1985, **25**, 319.
2867 R. Shabana, N.M. Yousif, and S.-O. Lawesson, Phosphorus Sulphur, 1985, **24**, 327.
2868 G. Heckmann, E. Fluck, and J. Svara, Z. Naturforsch., Teil B, 1984, **39B**, 1355 (Chem. Abstr., 1985, **102**, 95 731).
2869 C.W. Allen, S. Bedell, W.T. Pennington, and A.W. Cordes, Inorg. Chem., 1985, **24**, 1653.
2870 J. Svara and E. Fluck, Phosphorus Sulphur, 1985, **25**, 129.
2871 A.A. Prishchenko, N.S. Yashina, E.I. Gefel, and A.V. Gromov, Zh. Obsch. Khim., 1985, **55**, 340 (Chem. Abstr., 1985, **103**, 6 444).
2872 J. Svara and E. Fluck, Z. Anorg. Allg. Chem., 1985, **529**, 137.
2873 H.R. Allcock, J.L. Desorcie, and L.J. Wagner, Inorg. Chem., 1985, **24**, 333.
2874 M. Baudler and B. Makowka, Z. Anorg. Allg. Chem., 1985, **528**, 7.
2875 M. Baudler, M. Michels, J. Hahn, and M. Pieroth, Angew. Chem., Int. Ed. Engl., 1985, **24**, 504.
2876 R. Riedel, H.-D. Hausen, and E. Fluck, Angew. Chem., Int. Ed. Engl., 1985, **24**, 1056.
2877 M. Baudler, E. Daerr, G. Binseh, and D.S. Stephenson, Z. Naturforsch., Teil B, 1984, **39**, 1671 (Chem. Abstr., 1985, **102**, 204 046).
2878 M. Baudler and W. Göldner, Chem. Ber., 1985, **118**, 3268.
2879 M. Baudler and S. Wietfeldt-Haltenhoff, Angew. Chem., Int. Ed. Engl., 1985, **24**, 991.
2880 W. Theis and M. Regitz, Tetrahedron, 1985, **41**, 2625.
2881 W. Danikiewicz and M. Makosza, J. Chem. Soc., Chem. Commun., 1985, 1792.
2882 F. Effenberger and H. Kottmann, Tetrahedron, 1985, **41**, 4171.
2883 E.E. Aboujaoude, N. Collignon, M.-P. Teulade, and P. Savignac, Phosphorus Sulphur, 1985, **25**, 57.
2884 L.D. Quin and B.G. Marsi, J. Am. Chem. Soc., 1985, **107**, 3389.
2885 E. Öhler and E. Zbiral, Chem. Ber., 1985, **118**, 2917.
2886 G. Penz and E. Zbiral, Chem. Ber., 1985, **118**, 4131.
2887 G. Salbeck, Phosphorus Sulphur, 1985, **22**, 353.
2888 T. Bottin-Strzalko, J. Corset, F. Froment, M.J. Pouet, J. Seyden-Penne, and M.P. Simonnin, Phosphorus Sulphur, 1985, **22**, 217.
2889 S.L. Spassov, L. Markova, D.M. Mondeshka, Ch.N. Tancheva, and Ch.M. Angelov, Magn. Reson. Chem., 1985, **23**, 578.
2890 B. Cook and J.G. Dingwall, Phosphorus Sulphur, 1985, **22**, 211.
2891 G.H. McGall and R.A. McClelland, J. Am. Chem. Soc., 1985, **107**, 5198.
2892 V.B. Rozhnov and B.P. Nosov, Zh. Obshch. Khim., 1984, **54**, 1993 (Chem. Abstr., 1985, **102**, 78 978).
2893 E. Öhler, M. El-Badawi, and E. Zbiral, Monashefte, 1985, **116**, 77.
2894 D.W. Hutchinson and G. Semple, J. Organomet. Chem., 1985, **291**, 145.
2895 E.A. Berdnikov, Ya.A. Levin, L.Kh. Gazizova, and P.P. Chernov, Zh. Obshch. Khim., 1985, **55**, 579 (Chem. Abstr., 1985, **103**, 87 993).
2896 E.E. Nifant'ev, A.T. Teleshev, Yu.I. Blokhin, M.Yu. Antipin, and Yu.T. Struchkov, Zh. Obshch. Khim., 1985, **55**, 1265 (Chem. Abstr., 1985, **103**, 97 886).
2897 D.J. McCabe, E.N. Duesler, and R.T. Paine, Inorg. Chem., 1985, **24**, 4626.
2898 R. Burgada, Y.O. El Khoshnieh, and Y. Leroux, Phosphorus Sulphur, 1985, **22**, 225.
2899 R. Burgada, Y.O. El Khoshnieh, and Y. Leroux, Tetrahedron, 1985, **41**, 1223.
2900 G.A. Webb, M.P. Simonnin, J. Seyden-Penne, and T. Bottin-Strzalko, Magn. Reson. Chem., 1985, **23**, 48.
2901 T.H. Kim and D.Y. Oh, Tetrahedron Lett., 1985, **26**, 3479.
2902 E. Matczak-Jon and W. Wojciechowski, Pr. Nauk. Akad. Ekon. im. Oskara Langego Wroclawiu, 1984, **267**, 361 (Chem. Abstr., 1985, **102**, 124 498).
2903 W. Theis and M. Regitz, Chem. Ber., 1985, **118**, 3396.
2904 R.K. Gupta, A.K. Rai, R.C. Mehrotra, V.K. Jain, B.F. Hoskins, and E.R.T. Tiekink, Inorg. Chem., 1985, **24**, 3280.
2905 A.A. El-Barbary, R. Shabana, and S.-O. Lawesson, Phosphorus Sulphur, 1985,

21, 375.
2906 N. Ayed, B. Baccar, F. Mathis, and R. Mathis, Phosphorus Sulphur, 1985, **21**, 335.
2907 H.M. Hoffmann and M. Dräger, J. Organomet. Chem., 1985, **295**, 33.
2908 M. Dräger and B.M. Schmidt, J. Organomet. Chem., 1985, **290**, 133.
2909 K.B. Dillon and J. Lincoln, Polyhedron, 1985, **4**, 1333.
2910 E. Nakamura, J.-i. Shimada, and I. Kuwajima, Organometallics, 1985, **4**, 641.
2911 S.G. Withers, N.B. Madsen, and B.D. Sykes, J. Magn. Reson., 1985, **61**, 545.
2912 L.K. Krannich, C.L. Watkins, and R.K. Kanjolia, Polyhedron, 1985, **4**, 1023.
2913 M. Baudler, R. Heumüller, and J. Hahn, Z. Anorg. Allg. Chem., 1985, **529**, 7.
2914 J. Gloede, Z. Anorg. Allg. Chem., 1985, **531**, 17.
2915 R.C. Haddon, S.L. Mayo, S.V. Chichester, and J.H. Marshall, J. Am. Chem. Soc., 1985, **107**, 7585.
2916 M. Cypryk, J. Chojnowski, and J. Michalski, Tetrahedron, 1985, **41**, 2471.
2917 T. Gajda and A. Zwierzak, Tetrahedron, 1985, **41**, 4953.
2918 Yu.G. Gololobov, N.I. Gusar', and M.P. Chaus, Tetrahedron, 1985, **41**, 793.
2919 S.S. Snow, D.-X. Jiang, and R.W. Parry, Inorg. Chem., 1985, **24**, 1460.
2920 R. Francke, D. Dakternieks, R.W. Gable, B.F. Hoskins, and G.-V. Röschenthaler, Chem. Ber., 1985, **118**, 922.
2921 K. Nagarajan, S.J. Shenoy, H. Fritz, O. Hosang, and W.J. Richter, Helv. Chim. Acta, 1985, **68**, 900.
2922 D.B. Denney, D.Z. Denney, and L.T. Liu, Phosphorus Sulphur, 1985, **22**, 71.
2923 X. Lu, L. Li, L. Shen, S. Hu, C. Liu, and J. Zhang, Fenzi Kexue Yu Huaxue Yanjiu, 1985, **5**, 107 (Chem. Abstr., 1985, **103**, 66 751).
2924 I. Fenesan, V. Muresan, A. Barabas, and M. Bogdan, Rev. Roum. Chim., 1984, **29**, 761 (Chem. Abstr., 1985, **102**, 113 609).
2925 J. Steinbach, E. Herrmann, and L. Riesel, Z. Anorg. Allg. Chem., 1985, **523**, 180.
2926 S. Kleeman and E. Fluck, Z. Anorg. Allg. Chem., 1985, **526**, 133.
2927 Y.E. Shih, J.S. Wang, and L.K. Liu, Heterocycles, 1984, **22**, 2799 (Chem. Abstr., 1985, **102**, 166 713).
2928 D.F. Bushey, B.F. Johnson, and J. Huang, J. Org. Chem., 1985, **50**, 2091.
2929 E. Niecke, D. Gudat, W.W. Schoeller, and P. Rademacher, J. Chem. Soc., Chem. Commun., 1985, 1050.
2930 J. Gloede and S. Schramm, Phosphorus Sulphur, 1985, **22**, 369.
2931 K. Misiura, J.V. Silverton, and W.J. Stec, J. Org. Chem., 1985, **50**, 1815.
2932 H.R. Hudson, L. Powroznyk, and A.R. Qureshi, Phosphorus Sulphur, 1985, **25**, 289.
2933 A. Ouassini, R. de Jaeger, and J. Heubel, Z. Anorg. Allg. Chem., 1985, **531**, 188.
2934 B. Thomas, A. John, E. Pfützner, G. Grossmann, and E. Herrmann, Z. Anorg. Allg. Chem., 1985, **525**, 7.
2935 A.V. Aganov, N.A. Polezhaeva, A.I. Khayarov, and B.A. Arbuzov, Izv. Akad. Nauk SSSR, Ser. Khim., 1984, 2219 (Chem. Abstr., 1985, **102**, 132 135).
2936 J.M.A. Al-Rawi, G.Q. Behnam, N. Ayed, and R. Kraemer, Magn. Reson. Chem., 1985, **23**, 728.
2937 J. Neels, M. Meisel, and R. Moll, Z. Chem., 1984, **24**, 389 (Chem. Abstr., 1985, **102**, 71 708).
2938 W. Albers, W. Krueger, W. Storzer, and R. Schmutzler, Synth. React. Inorg. Met.-Org. Chem., 1985, **15**, 187).
2939 L. Riesel and M. Kant, Z. Anorg. Allg. Chem., 1985, **530**, 207.
2940 H. Hahn, W. Meindl, and K. Utvary, Monashefte, 1985, **116**, 279.
2941 U. Calov, L. Kolditz, E.G. Il'in, and Yu.A. Buslaev, Dokl. Akad. Nauk SSSR, 1984, **276**, 1405 (Chem. Abstr., 1985, **102**, 24 720).
2942 W. Schwarz, G. Rajca, and J. Weidlein, Z. Anorg. Allg. Chem., 1985, **525**, 143.
2943 J.L. Morris, C.W. Rees, and D.J. Rigg, J. Chem. Soc., Chem. Commun., 1985, 396.
2944 H. Schadow, B. Thomas, H.A. Lehmann, and L. Meznik, Z. Chem., 1984, **24**, 445 (Chem. Abstr., 1985, **102**, 214 178).
2945 A. Baceiredo, G. Bertrand, J.-P. Majoral, F. El Anba, and G. Manuel, J. Am. Chem. Soc., 1985, **107**, 3945.

2946 R.B. King and N.D. Sadanani, J. Org. Chem., 1985, **50**, 1719.
2947 A. Schmidpeter, F. Zwaschka, and W.S. Sheldrick, Chem. Ber., 1985, **118**, 1078.
2948 D. Schomburg, G. Bettermann, L. Ernst, and R. Schmutzler, Angew. Chem., Int. Ed. Engl., 1985, **24**, 975.
2949 K.C.K. Swamy and S.S. Krishnamurthy, Indian J. Chem., Sect. A, 1984, **23A**, 717 (Chem. Abstr., 1985, **102**, 24 824).
2950 K.V. Katti and S.S. Krishnamurthy, J. Polym. Sci., Polym. Chem. Ed., 1984, **22**, 3115 (Chem. Abstr., 1985, **102**, 25 149).
2951 M. Willson, L. Lafaille, G. Commenges, and J.-F. Labarre, Phosphorus Sulphur, 1985, **25**, 273.
2952 J.K. Fincham, M.B. Hursthouse, H.G. Parkes, L.S. Shaw, and R.A. Shaw, J. Chem. Soc., Chem. Commun., 1985, 252.
2953 J.-L. Sassus, M. Graffeuil, P. Castera, and J.-F. Labarre, Inorg. Chim. Acta, 1985, **108**, 23.
2954 P. Castera, J.-P. Faucher, G. Guerch, R. Lahana, A. Mahmoun, F. Sournies, and J.-F. Labarre, Inorg. Chim. Acta, 1985, **108**, 29.
2955 F. Sournies, R. Lahana, and J.-F. Labarre, Inorg. Chim. Acta, 1985, **101**, 31.
2956 H.R. Allcock, N.M. Tollefson, R.A. Arcus, and R.R. Whittle, J. Am. Chem. Soc., 1985, **107**, 5166.
2957 H.R. Allcock, P.E. Austin, and T.X. Neenan, U.S. US 4,495,174 (Cl. 424-78; A61K9/58, 22 Jan 1985, Appl. 390,345, 21 Jun 1982; 11 pp. (Chem. Abstr., 1985, **102**, 172 651).
2958 T.T. Bamgboye and O.A. Bamgboye, Spectrochim. Acta, Part A, 1985, **41A**, 981.
2959 K.C.K. Swamy and S.S. Krishnamurthy, J. Chem. Soc., Dalton Trans., 1985, 1431.
2960 V.V. Kireev, M. Poetzsch, and G.I. Mitropol'skaya, Deposited Doc., 1983, VINITI 5089-83, 10 pp. Avail. VINITI (Chem. Abstr., 1985, **102**, 79 398).
2961 K.V. Katt, S.S. Krishnamurthy, and M. Woods, Phosphorus Sulphur, 1985, **25**, 167.
2962 S.S. Krishnamurthy, P. Ramabrahmam, A.R.V. Murthy, R.A. Shaw, and M. Woods, Z. Anorg. Allg. Chem., 1985, **522**, 226.
2963 T.T. Bamgboye, Polyhedron, 1985, **4**, 657.
2964 C.W. Allen and R.P. Bright, Inorg. Chim. Acta, 1985, **99**, 107.
2965 H.R. Allcock, T.X. Neeman, and B. Boso, Inorg. Chem., 1985, **24**, 2656.
2966 P.Y. Narayanaswamy, K.S. Dhathathreyan, and S.S. Krishnamurthy, Inorg. Chem., 1985, **24**, 640.
2967 B. Thomas and G. Grossmann, Z. Anorg. Allg. Chem., 1985, **523**, 112.
2968 M.R. Marre-Mazières, M. Sanchez, R. Wolf, and J. Bellan, Nouv. J. Chim., 1985, **9**, 605.
2969 C.W. Allen and R.P. Bright, Report, 1984, TR15; Order No. AD-A147448/5/GAR, 16 pp. Avail. NTIS. From Gov. Rep. Announce. Index (U.S.), 1985, **85**, 52 (Chem. Abstr., 1985, **103**, 88 064).
2970 K.R. Metz and R.W. Briggs, J. Magn. Reson., 1985, **63**, 193.
2971 C. Roeske, P. Paneth, M.H. O'Leary, and W. Reimschuessel, J. Am. Chem. Soc., 1985, **107**, 1409.
2972 R.C. Hibbert and N. Logan, J. Chem. Soc., Dalton Trans., 1985, 865.
2973 B.W. Tattershall, J. Chem. Soc., Dalton Trans., 1985, 1707.
2974 K.O. Christe, W.W. Wilson, C.J. Schack, and R.D. Wilson, Inorg. Chem., 1985, **24**, 303.
2975 R.S. Edmundson, O. Johnson, D.W. Jones, and T.J. King, J. Chem. Soc., Perkin Trans. 2, 1985, 69.
2976 J.M.A. Al-Rawi, N. Ayed, R. Kraemer, and J. Navech, Magn. Reson. Chem., 1985, **23**, 925.
2977 R. Shabana, H.J. Meyer, and S.-O. Lawesson, Phosphorus Sulphur, 1985, **25**, 297.
2978 L. Labaudinière and R. Burgada, Phosphorus Sulphur, 1985, **24**, 235.
2979 C.A. Stewart, J.C. Calabrese, and A.J. Arduengo, tert., J. Am. Chem. Soc., 1985, **107**, 3397.
2980 A.V. Aganov, N.A. Polezhaeva, A.I. Khayarov, and B.A. Arbuzov, Phosphorus Sulphur, 1985, **22**, 303.
2981 G.M. Blackburn and T.W. Maciej, J. Chem. Soc., Perkin Trans. 1, 1985, 1935.

2982 G.D. Mendenhall, R.F. Kessick, and M. Dobrzelewski, J. Photochem., 1984, **25**, 227 (Chem. Abstr., 1985, **102**, 6640).
2983 R.P. Arshinova, O.V. Ovodova, O.I. Danilova, O.N. Nuretdinova, Yu.Yu. Samitov, and B.A. Arbuzov, Izv. Akad. Nauk SSSR, Ser. Khim., 1985, 575 (Chem. Abstr., 1985, **103**, 54 160).
2984 B.A. Arbuzov, R.P. Arshinova, O.V. Ovodova, O.I. Danilova, O.N. Nuretdinova, and Yu.Yu. Samitov, Izv. Akad. Nauk SSSR, Ser. Khim., 1985, 570 (Chem. Abstr., 1985, **103**, 54 159).
2985 Y. Zhao, S. Yan, and C. Zhai, J. Org. Chem., 1985, **50**, 2136.
2986 V. Patoprsty, L. Malik, I. Goljer, M. Goghova, M. Karvas, and J. Durmis, Magn. Reson. Chem., 1985, **23**, 122.
2987 M.J. Collins, U.R.K. Rao, and G.J. Schrobilgen, J. Magn. Reson., 1985, **61**, 137.
2988 A. Lopusiński and A. Haas, Chem. Ber., 1985, **118**, 4623.
2989 L. Riesel and M. Kant, Z. Anorg. Allg. Chem., 1985, **531**, 73.
2990 H. Hacklin, W. Offermann, and G.-V. Röschenthaler, Phosphorus Sulphur, 1985, **25**, 79.
2991 Z. Wittmann and Z. Kovacs, Talanta, 1985, **32**, 581 (Chem. Abstr., 1985, **103**, 204 658).
2992 P.A. Manninen, Ann. Acad. Sci. Fenn., Ser. A2, 1985, **203**, 48 pp. (Chem. Abstr., 1985, **103**, 142 042).
2993 D. Bouchu, F. Tardy, M. Moreau, J. Dreux, A. Skowronska, and J. Michalski, Tetrahedron Lett., 1985, **26**, 443.
2994 R.H. Fish and R.S. Tannous, Inorg. Chim. Acta, 1985, **104**, 137.
2995 S. Yang, Z. Wang, Y. Yang, and Y. Bi, Kexue Tongbao (Foreign Lang. Ed.), 1984, **29**, 1490 (Chem. Abstr., 1985, **103**, 71 401).
2996 A.E. Wroblewski, Z. Naturforsch., Teil B, 1985, **40B**, 407 (Chem. Abstr., 1985, **103**, 215 659).
2997 T. Mahmood and J.M. Shreeve, Inorg. Chem., 1985, **24**, 1395.
2998 D. Stollberg, J. Gloede, and K. Andrä, Z. Anorg. Allg. Chem., 1985, **527**, 180.
2999 C.A.A. Claesen, R.P.A.M. Segers, and G.I. Tesser, Recl. Trav. Chim. Pays-Bas, 1985, **104**, 119.
3000 Zs. Wittmann, Z. Decsy, and E. Pudmer, Talanta, 1985, **32**, 86 (Chem. Abstr., 1985, **102**, 151 205).
3001 A. Pomfret, K.D. Bartle, S. Barrett, N. Taylor, and J.W. Stadelhofer, Erdoel Kohle, Erdgas, Petrochem., 1984, **37**, 515 (Chem. Abstr., 1985, **102**, 113 607).
3002 K.V. Katti and S.S. Krishnamurthy, J. Polym. Chem. Ed., 1984, **22**, 3115 (Chem. Abstr., 1985, **102**, 25 149).
3003 D.B. Denney, D.Z. Denney, and K.-S. Tseng, Phosphorus Sulphur, 1985, **22**, 33.
3004 H. Tanaka, S. Kametani, T. Saito, and S. Motoki, Bull. Chem. Soc. Jpn., 1985, **58**, 667.
3005 R. Blachnik, P. Schroeter, and U. Wickel, Z. Anorg. Allg. Chem., 1985, **525**, 150.
3006 K.S. Mohamed and D.K. Padma, Spectrochim. Acta, Part A, 1985, **41A**, 725 (Chem. Abstr., 1985, **103**, 45 048).
3007 P. Boldrini, I.D. Brown, M.J. Collins, R.J. Gillespie, E. Maharajh, D.R. Slim, and J.F. Sawyer, Inorg. Chem., 1985, **24**, 4302.
3008 B.F. Hoskins, E.R.T. Tiekink, and G. Winter, Inorg. Chim. Acta, 1985, **99**, 177.
3009 M. Baiwir, Proc. Int. Conf. Org. Chem. Slenium Tellurium, 4th, 1983, 406. Ed. by F.J. Berry and W.R. McWhinnie (Chem. Abstr., 1985, **102**, 196 500).
3010 D.H. O'Brien, K.J. Irgolic, and C.K. Huang, Proc. Int. Conf. Org. Chem. Selenium Tellurium, 4th, 1983, 468. Ed. by F.J. Berry and W.R. McWhinnie (Chem. Abstr., 1985, **102**, 196 501).
3011 P.P. Mahendroo and A.D. Sherry, Magn. Reson. Chem., 1985, **23**, 503.
3012 E. Ghibaudi, A.J. Colussi, and K.O. Christe, Inorg. Chem., 1985, **24**, 2869.
3013 W. Gombler, Z. Naturforsch., Teil B, 1985, **40**, 782 (Chem. Abstr., 1985, **103**, 63 750).
3014 D. Naumann and B. Wilkes, J. Fluorine Chem., 1985, **27**, 115 (Chem. Abstr., 1985, **102**, 148 691).
3015 A.Z. Al-Rubaie, H.A.Y. Alshirayda, P. Granger, and S. Chapelle, J.

Organomet. Chem., 1985, **287**, 321.
3016 B.L. Khandelwal, A.K. Singh, H.B. Singh, K.M. Prasad, N.S. Bhandari, and W.R. McWhinnie, J. Organomet. Chem., 1985, **291**, 185.
3017 M.D. Thompson, E.M. Holt, K.D. Berlin, and B.J. Scherlag, J. Org. Chem., 1985, **50**, 2580.
3018 A. Toshimitsu, T. Aoai, H. Owada, S. Uemura, and M. Okano, Tetrahedron, 1985, **41**, 5301.
3019 G.P. Mullen, N.P. Luthra, R.B. Dunlap, and J.D. Odom, J. Org. Chem., 1985, **50**, 811.
3020 T.Y. Luh, W.H. So, K.S. Cheung, and S.W. Tam, J. Org. Chem., 1985, **50**, 3051.
3021 W.J. Hennen, B.C. Hinshaw, T.A. Riley, S.G. Wood, and R.K. Robins, J. Org. Chem., 1985, **50**, 1741.
3022 L. Christiaens, J.L. Piette, A. Luxen, and M. Renson, J. Heterocycl. Chem., 1984, **21**, 1281 (Chem. Abstr., 1985, **102**, 113 201).
3023 Z.J. Witczak, Tetrahedron, 1985, **41**, 4781.
3024 G. Cerioni, A. Diaz, F. Isaia, P. Mascia, and G. Verani, Phosphorus Sulphur, 1985, **25**, 53.
3025 N. Harrit, S. Rosenkilde, B.D. Larsen, and A. Holm, J. Chem. Soc., Perkin Trans. 1, 1985, 907.
3026 H. Sakurai, H. Tanaka, M. Nakayama, and M. Chikuma, Ganryu Aminosan, 1984, **7**, 215 (Chem. Abstr., 1985, **103**, 81 092).
3027 A. Krief, W. Dumont, J.N. Denis, G. Evrard, and B. Norberg, J. Chem. Soc., Chem. Commun., 1985, 569.
3028 C.O. Meese and W. Walter, Magn. Reson. Chem., 1985, **23**, 327.
3029 N. Esaki, N. Karai, T. Nakamura, H. Tanaka, and K. Soda, Arch. Biochem. Biophys., 1985, **238**, 418 (Chem. Abstr., 1985, **102**, 200 217).
3030 C. Lau, J. Passmore, E.K. Richardson, T.K. Whidden, and P.S. White, Can. J. Chem., 1985, **63**, 2273.
3031 K.D. Gupta and J.M. Shreeve, Inorg. Chem., 1985, **24**, 1457.
3032 J.A.M. Canich, M.M. Ludwig, W.W. Paudler, G.L. Gard, and J.M. Shreeve, Inorg. Chem., 1985, **24**, 3668.
3033 H.J. Reich, C.A. Hoeger, and W.W. Willis, jun., Tetrahedron, 1985, **41**, 4777.
3034 A. Haas and H.-U. Weiler, Chem. Ber., 1985, **118**, 943.
3035 N. Hanold and H. Meier, Chem. Ber., 1985, **118**, 198.
3036 T. Echter and H. Meier, Chem. Ber., 1985, **118**, 182.
3037 P. Michelsen, U. Annby, and S. Gronowitz, Chem. Scr., 1984, **24**, 251 (Chem. Abstr., 1985, **103**, 104 643).
3038 R. Arad-Yellin and F. Wudl, J. Organomet. Chem., 1985, **280**, 197.
3039 M.A.K. Ahmed, W.R. McWhinnie, and T.A. Hamor, J. Organomet. Chem., 1985, **281**, 205.
3040 H.B. Singh and W.R. McWhinnie, J. Chem. Soc., Dalton Trans., 1985, 821.
3041 T. Meier and R. Mews, Angew. Chem., Int. Ed. Engl., 1985, **24**, 344.
3042 R. Hoppenheit and R. Mews, Chem. Ber., 1985, **118**, 4276.
3043 H.W. Roesky, K.L. Weber, U. Seseke, W. Pinkert, M. Noltemeyer, W. Clegg, and G.M. Sheldrick, J. Chem. Soc., Dalton Trans., 1985, 565.
3044 J.S. Thrasher, C.W. Bauknight, jun., and D.D. DesMarteau, Inorg. Chem., 1985, **24**, 1598.
3045 J.S. Thrasher and K. Seppelt, Inorg. Chem., 1985, **24**, 4171.
3046 K. Seppelt and H. Oberhammer, Inorg. Chem., 1985, **24**, 1227.
3047 J.S. Thrasher and K. Seppelt, Z. Anorg. Allg. Chem., 1985, **529**, 85.
3048 G. Remaud, G.J. Martin, and M.L. Martin, J. Phys. Chem., 1985, **89**, 2082.
3049 J. Hsu, Z. Wu, and W. Happer, Phys. Lett. A, 1985, **112A**, 141 (Chem. Abstr., 1985, **103**, 226 071).
3050 Z.B. Iskandarov, L.S. Kornienko, A.L. Kotkin, and R.M. Umarkhodzhaev, Opt. Spektrosk., 1984, **57**, 773 (Chem. Abstr., 1985, **102**, 38 357).
3051 R.G. Syvret and G.J. Schrobilgen, J. Chem. Soc., Chem. Commun., 1985, 1529.

2
Nuclear Quadrupole Resonance Spectroscopy

BY K. B. DILLON

1 Introduction

This chapter reports on the pure nuclear quadrupole resonance (n.q.r.) spectra of quadrupolar ($\underline{I} > \frac{1}{2}$) nuclei in inorganic or organometallic solids. In an important paper this year on the basic theory and practice of obtaining high-resolution n.m.r. and n.q.r. spectra at zero magnetic field,[1] the examples given included the ^{7}Li n.q.r. spectrum of polycrystalline $Li_2SO_4.H_2O$ and the ^{27}Al n.q.r. spectrum of polycrystalline $KAl(SO_4)_2.12H_2O$. The techniques described were compared with previous frequency domain and high-field methods. The effects of molecular motion and sample rotation were also described, and various extensions of the method, including demagnetized initial conditions and the correlation by two-dimensional Fourier transformation of zero-field spectra with themselves or with high-field spectra, were discussed. An extensive review of n.q.r., including theory, instrumentation, magnetic interactions, chemical applications, and nuclear quadrupole double resonance, has appeared.[2] N.q.r. studies on inorganic solids have been reviewed, including instrumental developments, new frequency observations, the analysis of bonding and structure, and theoretical calculations.[3] The general principles of n.q.r. spectroscopy have been briefly reviewed,[4] while more specialised reviews have appeared on n.q.r. of hydrogen-bonded crystals under high pressure, particularly KD_2PO_4-type ferroelectrics,[5] on the n.m.r., n.q.r. and dielectric properties of clathrates,[6] and on n.m.r., e.s.r. and n.q.r. methods in the qualitative and quantitative analysis of explosives.[7] A review of ^{14}N n.q.r. and orientational ordering in smectic liquid crystals of mainly organic compounds included ^{14}N data for $(C_{10}H_{21}NH_3)_2CdCl_4$.[8]

The usual format is adopted in the more detailed sections, i.e. results for main-group elements (^{2}H, Groups I, III, V, VI and VII), followed by those for transition elements and lanthanides.

2 Main-group Elements

2.1 Deuterium.- The theory has been given for the dipolar interaction between two quadrupolar nuclei of $\underline{I} = 1$ and its effects on n.q.r. spectra.[9] The results were compared with experimental data for ^{2}H at 77K in various deuterated compounds, mainly organic but including phenylphosphonic acid - d_2. The 'forbidden' lines observed provided useful information about the relative orientations of the field gradient tensors at the coupled nuclei. Double-resonance methods have been used to obtain various n.q.r. data, including ^{2}H results for a series of inorganic and organic crystal hydrates, ^{2}H and ^{14}N spectra for a series of urea adducts, as a probe of charge transfer in the peptide linkage on H-bond formation, and ^{14}N and ^{17}O spectra for some substituted aromatic amine oxides.[10] The ^{2}H n.q.r. spectra for water of hydration appeared to fall into three categories, one of which was characterised by a six-line multiplet ν_+ and ν_- pattern, and two which gave rise to two separate ν_+ and ν_- peaks.

2.2 Group I (Lithium-7, Sodium-23 and Caesium-133).- The ^{7}Li result has been mentioned in the introduction.[1] Several studies have been described of ^{23}Na n.q.r. in NaOH or its deutero analogue.[11-13] In an effort to resolve a discrepancy in literature values for the asymmetry parameter η of ^{23}Na in NaOH, a powder Zeeman investigation has been carried out, using a CW spectrometer.[11] The resonance frequency was recorded at 77, 90, 195 and 291.5K in zero magnetic field. Zeeman studies gave lineshapes consistent with a very low or zero value of η. A low-temperature phase transition has been detected near 150K in NaOD by various physical techniques, including ^{23}Na n.q.r., calorimetry, ^{23}Na n.m.r. and ^{2}H n.m.r.[12] The n.q.r. results gave the transition temperature as 160K. This phase transition did not occur in NaOH down to 10K. It was concluded from the resonance experiments that the phase transition was continuous and that in the lower temperature phase there was one Na and one D crystallographic site per unit cell. X-ray powder diffraction suggested a monoclinic structure, and the probable cell parameters were given. The temperature (T)-dependence of the ^{23}Na frequency of NaOH from 77 - 550K showed a change in slope and a discontinuous change in frequency of 6 ± 2 kHz at the orthorhombic to monoclinic phase transition (505K).[13] The linewidth also increased, and this was deduced to arise from two slightly overlapping resonances, one for each phase. The data were consistent with a first-order transition. ^{2}H n.m.r. spectra were also obtained at various temperatures for NaOD. The T-dependence of the deuterium quadrupole coupling was fitted to a librating model between 293 and 570K and confirmed that the frequency of the OD^- libration decreased in the high-temperature (monoclinic) phase. Double-resonance methods (either double resonance by level-crossing d.r.l.c. or double resonance due to the solid-effect d.r.s.e.) have been used to obtain n.q.r. spectra at 120K from ^{23}Na nuclei in $NaHCO_3$, ^{10}B, ^{11}B and ^{23}Na nuclei in $Na_2B_4O_7.10H_2O$, ^{27}Al nuclei in $Al(OH)_3$ (all by d.r.l.c.) and ^{23}Na nuclei in $Na_2SiO_3.9H_2O$ (d.r.s.e.).[14] Magnetic satellites were observed from $Na_2SiO_3.9H_2O$ in the presence of a weak magnetic field. The advantages and limitations of the methods used were discussed. The quadrupolar coupling constant and asymmetry parameter were evaluated for ^{23}Na nuclei in $NaHCO_3$. A circuit design has been described for an autodyne spin detector based on field-effect transistors for high-sensitivity n.m.r., n.q.r. and e.s.r. spectrometers.[15] Successful tests were made on ^{133}Cs nuclei.

2.3 Group III (Boron-10 and -11 and Aluminium-27).- The ^{10}B and ^{11}B n.q.r. spectra of decaborane [14] have been obtained at 77 and 200K by a field cycling double-resonance method.[16] Four inequivalent boron sites were distinguished, and their e^2Qq/h and η parameters were evaluated. Analysis of the effect of ^{11}B-^{1}H dipolar coupling on the ^{11}B lineshape, carried out for the first time, showed that all the ^{11}B quadrupole coupling constants were negative. This work differed from an earlier report on the same compound in that some transitions were reassigned. The intensities of the lines in ^{10}B n.q.r. spectra obtained from d.r.l.c. experiments have been calculated for all possible asymmetry parameters.[17] Best agreement with experiment, as exemplified by H_3BO_3, was found when allowance was made for distortion of the n.q.r. levels by the magnetic field in the crossing of the n.m.r. and n.q.r. levels, incomplete saturation of the n.q.r. transitions by the radiofrequency field, and partial thermal mixing. The ^{10}B and ^{11}B n.q.r. frequencies in $Na_2B_4O_7.10H_2O$ at 120K have been measured.[18] From the calculated intensities of the double-resonance signals the ^{10}B spectrum was constructed, enabling η to be evaluated. Boron n.q.r. data obtained by double-resonance methods have been similarly reported for $Na_2B_4O_7.10H_2O$ and $K_2B_4O_7.4H_2O$ at 110K.[19] The quadrupole coupling constants were given. Boron atoms were found to be located at both trigonal and tetragonal sites in the two compounds. Other results for ^{10}B, ^{11}B and ^{27}Al nuclei have been described in the sub-section on Group I.[14]

^{81}Br and ^{27}Al n.q.r. data at 77K have been obtained for a series of MAl_2Br_7 compounds (M = NH_4, NMe_4 or NEt_4), the ^{27}Al results by a spin-echo double-resonance (SEDOR) method.[20] The mean frequency for the terminal bromines increased slightly with increasing cation radius. The bridging bromine frequency increased from 76.500 MHz in the NEt_4 salt to 86.330 MHz in the NH_4 compound; this change was considered too large to arise from crystal field effects only. The cause was suggested to be polarization of the anion induced by neighbouring cations, with the negative charges of the anions more localised at terminal than bridging positions as the polarizing power of the cation increased. The quadrupole coupling constants of the ^{27}Al nuclei were found to increase with the cation radius, supporting the anion polarization model. Other results for ^{27}Al have been mentioned in the introduction.[1]

2.4 Group V (Nitrogen-14, Arsenic-75, Antimony-121 and -123, and Bismuth-209).- Zero-field n.m.r. and n.q.r. spectra have been obtained by application of d.c. magnetic field pulses to a demagnetized sample.[21] These fields allowed for selective excitation of isotopic species and provided a means for coherent manipulation of the spin system in zero field. The theory of the method was described. By using these selective pulses and level-crossing techniques, indirect detection of a quadrupolar nucleus could be accomplished via protons, without obtaining the proton background signal in the n.q.r. spectrum. Its applications were shown by various examples, mainly organic but including ^{14}N n.q.r. from $(NH_4)_2SO_4$, where all the expected resonance frequencies for the two inequivalent N sites were clearly apparent. Values for e^2Qq/h and η were obtained in excellent agreement with those previously reported. A d.r.l.c. method has been used to obtain the ^{14}N n.q.r. spectrum of hydrazine sulphate at 120K.[22] It was shown that two-frequency saturation of the multi-level quadrupole system of the ^{14}N nuclei could be used to assign the lines to two non-equivalent positions of these nuclei in the crystal lattice. The quadrupole coupling constants and asymmetry parameters were evaluated for each of these.

The influence of the WAHUHA-4 multipulse sequence on a system of dipolar coupled spin-1 nuclei, each subject to a local electric field gradient (e.f.g.) in zero external magnetic field, has been studied theoretically and compared with experimental results for ^{14}N nuclei in a single crystal of $NaNO_2$ at 77K.[23] To interpret the experimental results the two equivalent spin-1 model was used, leading to reproduction of the experimentally observed quasi-steady state of magnetization. Good agreement was found between theory and experiment, showing the possibility of utilisation of the WAHUHA sequence for relaxation measurements in n.q.r. ^{14}N n.q.r. frequencies of 77K have been measured for the amine group of 1:1 salts of HNO_3 with methylurea, 1,3-dimethylurea and 1,1-dimethylurea and for the H_3PO_4 salt of 1,3-dimethylurea, enabling η values to be calculated.[24] The data were analysed on the basis of Townes-Dailey theory to obtain information on the changes in charge distribution caused by protonation. It was shown that on O protonation the N lost π-electron population, which was partially compensated by an increase in the population of the N-C σ-bonds. This loss of π-population was enhanced when the H atoms of the amine group were substituted by electron-donating groups such as Me. A double-exponential expression has been obtained for the relaxation of the longitudinal magnetization in the case of spin locking.[25] This was confirmed by experimental results for ^{14}N nuclei and made it possible to increase the accuracy and decrease the time of relaxation measurements in ^{14}N n.q.r. Other ^{14}N results have been mentioned in the sub-section on ^2H.[10]

A T-dependence study of the ^{75}As n.q.r. from yellow arsenic in the range 77 - 180K has revealed the existence of four new crystalline modifications.[26] The reorientation barrier height was estimated for the As_4 molecules in phase I crystals. Splitting of the spectral lines was observed, attributed to As-As spin-spin interactions. Assignment of lines in the n.q.r. spectrum of phase I was achieved by means of the interrelation between n.q.r. frequencies and inverse spin-lattice relaxation times. In further studies on yellow arsenic, the n.q.r. spectrum has been shown to be changed by external effects, such as annealing of the specimen or the application of static or oscillating magnetic fields.[27] As well as the normal spectrum, an additional two-component spectrum was observed in freshly prepared specimens which could not arise from impurities or from other known modifications of As studied previously. It was deduced from the annealing experiments that there could be a 'supercooled' high-symmetry phase of molecular arsenic.

T-dependence studies in the range 77 - 400K have been carried out for ^{121}Sb, ^{123}Sb nuclei in M_2SbF_5 (M = Na, Rb, Cs or CN_3H_6) and $(CN_3H_6)_2SbF_3SO_4$.[28] There was no evidence of phase transitions in the alkali-metal salts. For $(CN_3H_6)_2SbF_5$ the Sb atoms were equivalent at 77K. There was no change in the number of lines up to 270K, where the signals disappeared, but the T-dependence plot of the quadrupole coupling constant showed changes of slope at 183, 235 and 248K, characteristic of second-order phase transitions. A change in the T-dependence of η, including a sign change, was also apparent. A single type of Sb site was also found in $(CN_3H_6)_2SbF_3SO_4$ from 77 - 344K, where the signals disappeared, and no anomalies were detected in the T-dependence of e^2Qq/h. Anomalies found previously in dielectric permeability and electrical conductivity at 175 and 263K, and in ultrasound absorption at 263K, were explained by changes in the mechanism making the most important contribution to relaxation. Phase transitions in $SbCl_3.Ph_2NH_2^+Cl^-$(I) and $2SbCl_3$.PhOMe(II) have been investigated by means of the T-dependence of the ^{121}Sb, ^{123}Sb and ^{35}Cl n.q.r. signals in the range 77 - 300K.[29] In (I) there was a phase transition at 108K, and the ^{35}Cl signals disappeared above ca. 275K. The Sb lines could not be observed below 108K. Possible mechanisms for the phase transition and signal disappearance were suggested, and the structure of the compound in both phases was discussed with reference to the known crystal structure of the high-temperature phase. The transition point in II was determined as 141.3 ± 0.8K. Zeeman effect experiments on single crystals of II showed that both the low- and high-temperature phases were monoclinic. It was suggested from the results that the high-temperature phase has two types of symmetry elements, a two-fold screw axis and a centre of symmetry relating two $SbCl_3$ molecules to each other, and that the centre of symmetry is lost in the low-temperature phase. Multipulse spin-locking has been observed in n.q.r. measurements on ^{121}Sb and ^{123}Sb nuclei in $SbCl_3$ and on ^{35}Cl nuclei in C_2Cl_6 and CCl_3COOH.[30]

^{35}Cl and ^{121}Sb, ^{123}Sb (where observable) quadrupole resonance spectra, including T-dependence studies of the ^{35}Cl resonances from 77 - 300K, have been recorded for some five-membered ring chelate derivatives of $SbCl_5$, of general formula $SbCl_4$.L (L derived from tropolone, N-methyloxamide or N,N'-dimethyloxadiamide) and for the doubly chelated derivatives of oxalic acid and N,N'-dimethyloxadiamide.[31] The ^{35}Cl frequencies showed a normal temperature dependence. Two distinct groups of frequencies for ^{35}Cl were observed, associated with axial and equatorial chlorines (lower and higher frequency respectively). An approximately linear correlation was found for the first four compounds, where the crystal structures were known, between the Sb-Cl bond length and the ^{35}Cl n.q.r.

frequency. The ionic characters of the Sb-Cl bonds were estimated from the results; there was a ca. 10% difference between the axial and equatorial chlorines in the tropolone complex. Antimony resonances were obtained only for the doubly chelated complexes, and the T-dependence was studied for the oxalic acid derivative. The large value of η for ^{121}Sb in the oxalate complex was satisfactorily explained by the observed small angular deformations of the geometry from regular octahedral. Nuclear spin-lattice relaxation and lattice dynamics in SbSI have been studied by means of (presumably) the T-dependence of ^{121}Sb n.q.r. in the ferroelectric phase.[32] Soft-mode dynamics were observed, and a lattice potential was constructed which reproduced previous studies of high-pressure effects. The effect of an external electric field on a quadrupolar system has been considered theoretically.[33,34] The existence of a maximum shift direction was deduced, and has been demonstrated experimentally via ^{121}Sb, ^{123}Sb resonances from $SbCl_3$ [33,34] and ^{209}Bi resonances from $BiBr_3$, $BiCl_3$ and $Bi(Ph)_3$.[33] The theory also described the shape of the spin-echo envelope in pulsed investigations of electric field effects, and a convenient approximating formula that described the experiments quite accurately was derived. The new experimental procedure also showed several advantages over the previous technique.[33]

2.5 Group VI (Oxygen-17).- N.q.r. results for this nucleus have been mentioned in the sub-section on ^{2}H.[10]

2.6 Group VII (Chlorine-35 and -37, Bromine-79 and -81, and Iodine-127).- ^{35}Cl n.q.r. data, including the T-dependence of the resonance frequencies from 77 to 300K, have been reported for p-dichlorobenzene clathrates with $M(\gamma\text{-picoline})_4(SCN)_2$ (M = Ni, Co or Fe) and $Ni(\gamma\text{-picoline})_4(SeCN)_2$ and for 1,2-bis(diphenylphosphinoselenoyl)ethane/p-dichlorobenzene.[35] The results showed that useful information could be deduced by means of this technique regarding the nature of the guest-host interactions. ^{35}Cl n.q.r. frequencies at 77K have been measured for the perchloropolysilanes Si_2Cl_6, Si_3Cl_8, Si_4Cl_{10}, Si_5Cl_{12}, $Si(SiCl_3)_4$, Si_5Cl_{10} and a mixture of n-Si_4Cl_{10} and i-Si_4Cl_{10}.[36] The number of lines observed was usually smaller for the silicon compounds than for their carbon analogues, ascribed to a weaker field interaction of their non-bridged atoms compared with those in the carbon compounds. ^{35}Cl n.q.r. measurements on $C_6F_5PCl_4$ at 77 and 195K have confirmed that the compound has a trigonal-bipyramidal structure, with the C_6F_5 group axial.[37] Three signals attributable to equatorial chlorines were seen at each temperature, in contrast with a literature report where only one signal was detected in this region at 77K. The resonance frequency of the axial chlorine at 77K was lower (25.300 MHz) than previously reported (27.1 MHz); the latter value was shown to be as expected for the ^{37}Cl signal from the equatorial chlorines. The electron density of the Cl bonds in FCl, Cl_2, CH_2ClPCl_2, $CH_2ClSiMe_3$, CH_2ClSMe, $SiCl_3(OMe)$, $SiCl_4$, $SiCl_3Me$, $SiCl_2(OMe)_2$, $SiHCl_2Me$, $SiCl(OMe)_3$, $SiMe_3Cl$ and various organic compounds has been investigated by ^{35}Cl n.q.r. at 77K.[38] The Cl p-orbital populations were estimated by a CNDO/2 method and correlated with the resonance frequencies. The bonding to chlorine in the compounds was discussed on the basis of the results. The inverse linear relationship between the ^{35}Cl n.q.r. frequency and the p-σ-electron density of the chlorine atom was confirmed, indicating that the Townes-Dailey approximation was valid for these compounds.

The pressure dependence of the four ^{35}Cl signals at room temperature from $N_3P_3Cl_6$ has been shown to be linear up to a pressure of ca. 3500 kg cm^{-2}, when a significant change in slope for all four lines was observed, ascribed to a possible phase transition.[39] In contrast, the two lines from the K-form (meta-stable) and the four lines from the T-form of $N_4P_4Cl_8$ showed a non-linear pressure dependence at this temperature. The results were discussed on the basis of competing dynamic and static effects. Various thermodynamic quantities were derived from the results, and the frequencies of the torsional modes in $N_4P_4Cl_8$ (T-form) were estimated. The T-dependence of the ^{35}Cl n.q.r. frequencies of Cs_2HgCl_4 from 77 - 300K, together with the T-dependence of the line intensities, has indicated the existence of three stable phases.[40] At 115K there was a transition from a ferroelectric to an incommensurate phase, and at 219K the latter underwent transition to a paraelectric phase. The results were compared with ^{81}Br data from Cs_2HgBr_4, suggesting that these compounds have the same structure, and with those obtained by other methods for the present compound. Theoretical calculations of dipolar contributions to the linewidth of the ^{35}Cl resonance from a single crystal of $NaClO_3$ at room temperature have been carried out for various interactions with and without a Zeeman field, and compared with experimental results.[41] It was concluded that dipolar broadening was the predominant source of the linewidth in $NaClO_3$. A d.c. superconducting quantum interference device (SQUID) has been used as a tuned r.f. amplifier at 4.2K to detect pulsed ^{35}Cl n.q.r. from a powdered sample of $NaClO_3$.[42] Very high sensitivity with good tolerance for large r.f. pulses was achieved. Details of the system were described and its advantages were discussed.

The T-dependence of the ^{35}Cl and ^{37}Cl resonances from three samples of K_2OsCl_6, powder, powder recrystallised from D_2O, and single crystal, has been studied in the range 6 - 300K.[43-6] A detailed study of the lineshapes from 70 - 300K showed that they reflect the strain fields produced by lattice dislocations and point defects.[43,44] All of the observed spectra could be represented by one or two symmetric components with a T-dependent mixture of Lorentzian and Gaussian character. The temperature evolution of these strain fields gave rise to a satellite line, previously attributed to H^+ ion impurities. A comprehensive study of the T-dependence in the vicinity of the phase transition revealed a drop in intensity and the progressive appearance of an extra broad resonance component.[43,45] Both effects were associated with the existence of precursor dynamic clusters at temperatures higher than $\underline{T}_c$. Qualitatively similar but quantitatively different behaviour was observed for the powder and single-crystal samples. The results suggested that the detailed dynamics associated with the precursor clusters were sensitive to whether point defects or dislocations were dominant as the primary cause of line broadening. A comprehensive investigation of the lineshape below $\underline{T}_c$ (ca. 43K) in the tetragonal phase, and detailed analysis of the data, showed that the results were consistent with the occurrence of a commensurate structural phase transition at $\underline{T}_c$ and that additional strains were introduced into the crystal as a result of the structural alteration.[43,46] The phase shift effect previously observed in the tetragonal phase of K_2ReCl_6 was again apparent, and was explained as an experimental artefact introduced by truncation of the FID because of the spectrometer dead time.

The T-dependence of the ^{93}Nb and ^{35}Cl n.q.r. signals from the low-temperature (monoclinic) modification of niobium pentachloride has been followed from 4.2 to 440K, where the signals disappeared.[47] The niobium results suggested a phase transition around 20K, although this change was barely apparent in the chlorine lines. The data were as expected for bridged dimers (Nb_2Cl_{10}) and were compared with those from the high-temperature modification. The bond character in both forms was discussed. ^{35}Cl n.q.r. at 77K has been recorded for some 2:2 salts of trichloroacetic acid with organic bases and compared with literature data for the 2:1 salts $(CCl_3COO)_2HM$ (M = Tl, Rb, NH_4, K or Na).[48] The results showed that the crystal field makes a small contribution to the frequency shifts produced by salt formation. Analysis of the spectra of the 2:1 salts, including a plot of the average frequency shift against the ionic radius of the counterion, indicated that the Sternheimer factor was quenched. The inductive effect was found to determine the frequency shift in the salts, whereas solid-state effects were dominant in the addition complexes of trichloroacetic acid. In a further extension of previous work on substituted benzenesulphonyl chlorides, the T-dependence of the ^{35}Cl n.q.r. frequency and spin-lattice relaxation time in the range 77 - ca. 300K has been determined for the γ-phase of benzenesulphonyl chloride [49] and for p-bromobenzenesulphonyl chloride.[50] Mechanisms of spin-lattice relaxation in the compounds were discussed. T-dependence data for ^{35}Cl and ^{37}Cl n.q.r. from solid chlorine in the intervals 20 - 170K and 20 - 120K respectively have been re-analysed by a method which takes advantage simultaneously of data for both isotopes.[51] A fourth lattice librational frequency was calculated which should be Raman active, although it had not been detected so far, possibly because it was deduced to be strongly temperature dependent. A lattice contribution to the electric field gradient (e.f.g.) at the chlorine site of about 27% was computed. A sudden change in slope of the T-dependence plot for the ^{35}Cl n.q.r. frequency was detected at ca. 125K, possibly due to a decrease in the lattice contribution to the e.f.g.

^{35}Cl n.q.r. data at 77K for some substituted diaza- and triaza-phosphorins have been used to calculate the p-orbital populations of the chlorine atoms, via the Townes-Dailey approximation.[52] The spectral parameters were also calculated by a CNDO/2 method; the results were in satisfactory agreement for the P-Cl bonds, but not for the C-Cl bonds. Comparison of the electronic structures with those of model compounds showed that the λ^5-diaza- and -triaza-phosphorins were ylides, with a rather small contribution from the ylene structure. ^{35}Cl n.q.r. frequencies have been measured for a series of pentafluorophenyl compounds C_6F_5R (R = PCl_2, CON=PCl_3, N=PCl_3, SO_2NPCl_3, COCl, CH_2Cl, SO_2NCl_2, $CONCl_2$, SCl or SO_2Cl), for the cyclic compound $(C_6F_5NPCl_3)_2$, and for two tetrafluoro-4-pyridyl derivatives.[53] Comparison with the data for the corresponding phenyl compounds showed that the n.q.r. frequencies increased on replacement of Ph by C_6F_5, and a linear relation was found between the frequencies in the two series. The electronegativity of the element to which Cl was attached was also linearly related to the ionic character of the bond between Cl and the element. ^{35}Cl frequencies at 77K have been determined for the compounds PCl_3, $RPCl_2$ (R = Me or Et), R_2PCl (R = Et or Ph), $ROPCl_2$ (R = Me, Et, n-Pr or n-Bu), $\overline{CH_2CHCHCH_2P}OCl$, $(MeO)_2PCl$, Me_2C=$CHPCl_2$, PhCH=$CHPCl_2$ and R_2NPCl_2 (R = Me or Et).[54] The average n.q.r. frequency was shown to be linearly correlated with the overall Taft constant. Compounds where the inductive effect was predominant lay on one straight line, and those in which conjugation played a significant role lay on another straight line.

The T-dependence of the appropriate halogen (^{35}Cl, ^{81}Br or ^{127}I) n.q.r. from 77K to room temperature has been determined for α-Me_2TeX_2 (X = Cl, Br or I) and Me_2TeI_4.[55] The bromo compound gave no signals at 77K, but two signals were detected from 110K upwards. The characteristic T-dependence of the ^{127}I n.q.r. from α-Me_2TeI_2 was explained in terms of a 3c-4e bond of a linear I-Te-I group. The lowest ^{127}I frequency from Me_2TeI_4 showed an anomalous (positive) T-coefficient. The results were correlated with the crystal structures, and the electron populations were calculated from a Townes-Dailey treatment. ^{35}Cl and ^{127}I frequencies at 77K have been recorded for a series of compounds $XC_6H_4ICl_2$ (X = various substituents).[56] Data for $CsICl_4$ and $(ICl_3)_2$ were included for comparison. Values of η for ^{127}I were used to calculate the interorbital angle Θ at the iodine atom. The structures of the compounds were discussed on the basis of the results. The complexing of halotrinitromethanes $XC(NO_2)_3$ (X = Cl, Br or I) with various donors, mainly organic but including HMPA and Ph_3PO, has been studied by means of halogen n.q.r. at 77K.[57] Some bases formed crystalline adducts with the iodo- and bromo-compounds, but none was crystalline under normal conditions with the chloro-analogue. Some relaxation time measurements were also reported, together with ^1H n.m.r. data. Soft bases were found to cause comparatively small changes in the n.q.r. parameters, showing that the electronic and geometric structure was not greatly affected by complexation, whereas harder bases caused considerable changes in these parameters. The results made it possible to predict the reactivity of a halotrinitromethane in a coordinating solvent. Further data have since been obtained on ^{127}I in the $IC(NO_2)_3.L$ complexes, including the higher transition frequency, enabling $\underline{e}^2Qq/h$ and η to be evaluated.[58] The asymmetry parameters were found to be low, and it was deduced from the quadrupole coupling constants that the iodine carried a positive charge in the complexes. Other results for ^{35}Cl have been described in the sub-section on Group V.[29-31]

An ^{81}Br n.q.r. investigation of the 1:1 and 1:2 complexes of MeCOBr with $AlBr_3$ at 77K has shown that the compounds have ionic structures $MeCO^+AlBr_4^-$ and $MeCO^+Al_2Br_7^-$ respectively.[59] The unit cell of the 1:1 complex was deduced from the multiplicity and relative line intensities to contain two crystallographically independent $AlBr_4^-$ anions. On raising the temperature, a second-order phase transition was observed at 178K, with an increase in the symmetry of the anion. No spectra were observed above this temperature, attributed to molecular motion in the crystal. The 1:1 complex in excess CH_2Br_2 was shown by n.q.r. to retain the acylium salt structure, but with additional interaction of the Al atom with the CH_2Br_2 molecule. $AlBr_3$ did not interact with excess CH_2Br_2 at 77K, although the results showed that the dimer (Al_2Br_6) consisted of a mixture of two crystalline modifications. Bromine n.q.r. measurements on a series of hexabromostannates have shown that the anion in $[NHMe_3]_2SnBr_6$ has a regular octahedral structure from 77 - 300K and that $[SMe_3]_2SnBr_6$ may have a similar structure at room temperature, but that $[NMe_4]_2SnBr_6$ undergoes a phase transition to a less regular structure between 178 and 77K.[60] A variable-temperature study of $[pyH]_2SnBr_6$ from 77 - 335K showed three types of bromine, with one at a much lower frequency than the other two, probably because of hydrogen bonding. The only evidence for a phase transition was a cusp at 296 ± 1K in the T-dependence plot for one of the ^{81}Br resonances. The crystal structure of this compound at room temperature showed three significantly different Sn-Br distances in the anion, in agreement with the n.q.r. results where the three signals at 300K were in a 3:2:1 intensity ratio. The ^{79}Br frequencies for these and other hexabromostannates at 300K were correlated with the cation sizes, and the relationship between ^{79}Br n.q.r. frequencies and Sn-Br bond lengths in hexabromostannates at 300K was discussed.

^{79}Br and ^{127}I signals (as appropriate) have been observed at various temperatures from 77K upwards for α-Ga_2Br_4, β-Ga_2Br_4 and Ga_2I_4.[61] The signals disappeared at 353, 335 and 448K respectively. (The compounds were known to have ionic structures $Ga^+GaX_4^-$.) The results for α-Ga_2Br_4 were in good agreement with literature data. Two ^{79}Br resonances were found for β-Ga_2Br_4, more widely separated than from the α-form, suggesting that the $GaBr_4^-$ ion had a distorted tetrahedral shape. Eight widely separated ^{127}I lines (four for each transition) were found for Ga_2I_4. The separations could not be explained by distortion of the tetrahedron, as shown by the crystal structure, and were attributed mainly to interaction of the Ga^+ ion with the iodines. The thermal isomerisation of cis-$PtI_2(BuNH_2)_2$ has been followed by recording ^{127}I n.q.r. spectra (at 77K).[62] The results showed that isomerisation to the trans form began prior to heating, since signals from both isomers were observed, and was complete at 383K. The complexes cis-$PtBr_2L_2$ (L = NH_3, $EtNH_2$ or $MeNH_2$) and their trans analogues were prepared by standard methods and characterised by their ^{79}Br and ^{81}Br n.q.r. frequencies. Measurements were made at 77 and 300K, apart from cis-$PtBr_2(MeNH_2)_2$ where data were given at 77K only. During heating of cis-$PtBr_2(MeNH_2)_2$ the trans isomer began to form at 363K, and at 413K only the trans isomer was present. This process was also followed by bromine n.q.r. T-dependence studies on the halogen n.q.r. ($^{79,81}Br$ or ^{127}I) from 77 to ca. 400K have been carried out on the hydrobromides of glycine, ℓ-alanine, ℓ-leucine, ℓ-proline and semicarbazide and on the hydroiodides of ℓ-leucine and semicarbazide, together with the compound diglycine sodium iodide monohydrate.[63] Data were also presented for some deuterated analogues of the bromo compounds. The range of observation for $(glycine)_2.HBr$ and $(glycine)_2.DBr$ was limited to T > 175K, since the signals disappeared at lower temperatures; the reason for this behaviour was not clear. A reversible phase transition was observed for ℓ-leucine hydrobromide at 331.5K. The bromine n.q.r. frequencies were found to be strongly influenced by H-bonding. The ^{79}Br frequencies could be grouped into two bands, one between 15 and 19 MHz and the other between 24 and 26 MHz. The higher-frequency group was characterised by both N^+-H---Br^- and O-H---Br^- H-bonds, whereas the low-frequency group had a bond scheme determined by N^+-H---Br^- contacts only. A similar grouping was found from more limited evidence in the hydroiodides. Other ^{81}Br results have been described in the sub-section on Group III.[20]

^{127}I n.q.r. data, mainly for the $\frac{1}{2} \leftrightarrow \frac{3}{2}$ transition but including T-dependence studies from 77 - 300K, have been obtained for 1,4-dithiane.$2I_2$, thiane.I_2, 1,4-thioxane.I_2 and 1,3,5-trithiane.I_2.[64] Two distinct types of iodine atoms were apparent in the first three complexes but only one type in the last compound. Charge densities on the donors and the iodine atoms were obtained via the Townes-Dailey approximation. From the results the order of donor strength towards the iodine molecule was deduced as thiane > 1,4-thioxane > 1,4-dithiane >> 1,3,5-trithiane. These S donors were found to be almost as strong as the N donors in amine.iodine complexes, although the no-bond ionic structure of the ground state was less important in the thioether derivatives than in the amine complexes. The terminal iodine atoms in the complexes with 1,4-dithiane, thiane and 1,4-thioxane showed anomalous (positive) T-coefficients, ascribed to the increasing shift of the dative structure to the no-bond covalent structure with rise in temperature. The Zeeman effect on the $\frac{1}{2} \leftrightarrow \frac{3}{2}$ n.q.r. transition for nuclei of I = $\frac{5}{2}$ in polycrystalline samples has been considered theoretically, and has been investigated experimentally for ^{127}I nuclei in H_5IO_6 at 294K.[65] The experimental Zeeman frequency splittings were compared with the theoretical values in order to evaluate η. The value of 0.19 agreed with earlier values obtained from measurements of both transition frequencies and from a single-crystal Zeeman study on the $\frac{1}{2} \leftrightarrow \frac{3}{2}$ transition.

^{127}I n.q.r. frequencies, together with $\underline{e}^2Qq/h$ and η values, have been reported for various polycoordinated iodine compounds, including several carborane derivatives.[66] Data for other compounds were included for comparison. The changes in the quadrupole coupling constant followed the sequence $ICl_4^- > ICl_3 > PhICl_2 > Ph_2I^+ > PhCBI^+$ (CB = carboranyl), while the η values increased along this series. It was concluded that the distortion of the coordination environment was greater in the carborane derivatives than in the corresponding phenyl derivatives. ^{127}I spectral parameters at 77K have been obtained for a series of diphenyliodonium compounds $Ph_2I^+X^-$ with different counterions X^-.[67] An increase in the coordination interaction was found to cause a systematic decrease in the quadrupole coupling constants, asymmetry parameters and field constants $\partial\nu/\partial\underline{E}_Z$. ^{127}I data (both transition frequencies) have similarly been recorded for a series of iodonium compounds $ArAr'I^+X^-$.[68] $\underline{e}^2Qq/h$, η and the field constant varied regularly with the nature of the cation and the character of substitution in the ring. Strengthening of the coordination interactions led to a sharp jump in the field constant and a significant change in η. A linear correlation was established in both series between the values of $\underline{e}^2Qq/h$ and η.[67,68]

The electronic structures of IO_3^- ions in crystals of α- and ß-$LiIO_3$ have been calculated by a quantum mechanical (LCAO MO) method.[69] The n.q.r. parameters for these compounds and for α-HIO_3 and α-DIO_3 were also evaluated and were compared with experimental results at 77K and room temperature (all but α-DIO_3). The geometrical distortion of the IO_3^- ion was estimated. The significant asymmetry (η 0.435) of the e.f.g. tensor at the ^{127}I nuclei in α-HIO_3 could be explained only by taking into account the effect of a proton on the electron density distribution. The structure and dynamics of NH_4IO_3 in the temperature interval 295 - 400K, and particularly near the ferroelectric phase transition at 367K, have been studied by ^{127}I n.q.r. (both transitions) and vibrational spectroscopy.[70] It was shown that heating of the crystal within the range 320 - 335K induced prephase orientation fusion, followed by rearrangement of the H-bond system, rotation of the NH_4^+ ion (with a decrease in the reorientation barrier from 18 to 4.3 kcal mol^{-1}) and by an increase of NH_4^+ and IO_3^- ion symmetry to $\underline{T}_d$ and $\underline{C}_{3v}$ respectively. The T-dependence of η for ^{127}I nuclei in NH_4IO_3 between 300 and 400K has also been investigated.[71] An allowance for the pretransitional thermal expansion of the unit cell of the NH_4IO_3 crystal made it possible to explain the observed results, to predict the nature of the dependence of the reorientation barrier $\underline{U}(\underline{T})$ for the IO_3^- ions, and to estimate the height of this barrier $\underline{U}$ as ca. 20 kcal mol^{-1}.

3 Transition Metals and Lanthanides

3.1 Cobalt-59.- ^{59}Co n.q.r. data at 298K from a series of complexes having *fac*-a_3b_2c, *fac*-a_3b_3 or *cis*-a_3b_2 geometry, where a_3 = cyclopentadienyl (Cp) or pentamethylcyclopentadienyl ($CpMe_5$), have shown that $CpMe_5$ is a better inductive donor of electron density than Cp but that the increase felt by the metal is small (< 10%).[72,73] The electronic effects in $CpCoI_2c$ complexes (c = unidentate ligand) having *fac*-a_3b_2c geometry were linear when $CpMe_5$ replaced Cp and could be readily explained. The n.q.r. spectra of $CpCo(b,b')C_6H_4$ (b,b' = O, S, N or Se) complexes having *cis*-a_3b_2 geometry were controlled mainly by the electronegativity of b,b'. Replacement of Cp by $CpMe_5$ caused a non-linear variation in the field gradient, implying differences in the electronic structures of complexes with b = O and S compared with those with b = N. T-dependence studies on $CpMe_5Co(CO)_2$ from 77 - 300K revealed a phase transition at 177K but did not indicate a change in the 'allyl-ene' distortion of the Cp ring. The field gradient at Co in $CpCob_3$ complexes (b = unidentate ligand) was more sensitive to b than to replacement of Cp by $CpMe_5$. This effect was used to demonstrate that $P(OMe)_3$ is a slightly better electron donor than $P(O)(OMe)_2^-$. Other ^{59}Co results are mentioned in the sub-section on Cu.[74]

3.2 Copper-63 and -65.- The T-dependence at low temperatures of the nuclear spin-lattice relaxation rate $1/T_1$ for ^{63}Cu nuclei in the superconducting state of $CeCu_2Si_2$ has been measured down to 50mK by n.q.r. and n.m.r. techniques, for both a polycrystalline (T_c 0.72K) and a single-crystal (T_c 0.65K) sample.[74] For comparison, data were also given for the T-dependence of $1/T_1$ for ^{59}Co nuclei in $CeCo_2$ (T_c 0.95K), obtained by n.q.r. Although some differences in relaxation behaviour for the two samples of $CeCu_2Si_2$ were observed and discussed, the most striking feature was a rapid decrease of $1/T_1$, with no enhancement just below T_c. This behaviour was quite anomalous compared with that of the cobalt compond, which is a normal superconductor. The results showed that the superconductivity in $CeCu_2Si_2$ was consistent with an anisotropic energy gap model. ^{63}Cu and ^{65}Cu n.q.r. spectra have been obtained from two different samples of $CeCu_6$ at 4.2K.[75] The resonance frequencies and linewidths were found to depend strongly on the sample quality. ^{63}Cu nuclear spin-lattice relaxation rates were also measured as a function of T down to 60mK. $1/T_1$ was almost temperature-independent above 6K, reflecting the Curie-Weiss behaviour of the magnetic susceptibility. Below 6K, $1/T_1$ became T-dependent with decreasing temperature. Below ca. 0.2K, $1/T_1$ followed the relationship T_1T = constant, showing that the Fermi liquid state had set in. Relaxation behaviour at these low temperatures was discussed. The T-dependence of the spin-lattice relaxation times of ^{63}Cu nuclei between 77 and 500K has been studied for samples of cuprous oxide prepared at high and low temperatures.[76] Differences in the relaxation rates for the samples were correlated with their semiconducting properties. It was deduced that the relaxation processes for the 'low-temperature' form were predominantly influenced by an electron hopping mechanism, in which the mobility of the charge carriers was an exponential function of temperature.

3.3 Niobium-93.- Changes in the local e.f.g. in 2H-$NbSe_2$ in the temperature range 7 - 77K related to the charge-density-wave transition have been studied by means of ^{93}Nb n.q.r.[77] ^{93}Nb spectra have been recorded at 10K for $M_{0.5}NbS_2$, formed by intercalation of NbS_2 with Group III elements (M = B, Al, Ga or In).[78] Intercalation of elements other than boron reduced the quadrupole coupling constant of the parent compound 2H-NbS_2, but B caused line broadening only. Charge transfer in the compounds was discussed on the basis of the results. Other ^{93}Nb data have been given in the sub-section on Group VII.[47]

3.4 Praseodymium-141.- Optically detected n.q.r. spectra have been obtained for Pr^{3+} ions at 1.6K (from a single crystal of YPO_4 containing nominally 1% Pr^{3+}) in laser holeburning experiments.[79] The two resonance frequencies were very nearly in a 2:1 ratio in the ground state, giving $\eta < 0.01$. Results were also obtained for the lowest component of the excited state, with $\eta < 0.002$. Magnetic splitting factors for both states were measured and compared with values calculated on the assumption of pure $4\underline{f}^2$ electron states. The results agreed well in the ground state but rather poorly in the excited state.

3.5 Tantalum-181.- Evidence for hexadecapole interactions (HDI) has been obtained from ^{181}Ta n.q.r. spectra at 77K for TaF_5, and the values of the HDI constants have been calculated.[80] Principles for the selection of specimens, and optimum measurement conditions, for the study of HDI were also suggested. ^{181}Ta n.q.r. spectra have been recorded at 4.2K from 1T-TaS_2.[81] N.m.r. data for this nucleus were also obtained, and the results helped in the interpretation of the n.q.r., enabling groups of lines to be assigned to α-sites and ß-or γ-sites. Semi-conduction in 1T-TaS_2 was discussed in terms of the stacking of charge density waves. N.q.r. data at 4.2K have been similarly reported for ^{181}Ta nuclei in 1T-$TaSe_2$, from both single-crystal and powder samples, together with ^{181}Ta n.m.r. results.[82] Six fairly strong and sharp n.q.r. lines were observed between 90 and 115 MHz and another six lines between 160 and 175 MHz, corresponding to the $\frac{1}{2} \leftrightarrow \frac{3}{2}$ and $\frac{3}{2} \leftrightarrow \frac{5}{2}$ transitions respectively, together with additional weak resonances at 63.6 and 113.7 MHz. The $\underline{e}^2\underline{Qq}/\underline{h}$ and η values were evaluated for each site. The results were discussed on the basis of the star-shaped cluster model, with emphasis on the fine structure due to the three-dimensional ordering of clusters.

References

1 D.B. Zax, A. Bielecki, K.W. Zilm, A. Pines and D.P. Weitekamp, J. Chem. Phys., 1985, 83, 4877.

2 A.L. Porte, Annu. Rep. Progr. Chem., Sect. C, 1984, 80(C), 149.

3 K.V.S. Rama Rao and S. Ramaprabhu, Phys. Status Solidi A, 1986, 93, 17.
4 K.H. Hausser, Bull. Magn. Reson., 1984, 6, 107.
5 J. Stankowski, Ser. Fiz. (Uniw. im. Adama Mickiewicza Poznaniu), 1985, 53, 5.
6 D.W. Davidson and J.A. Ripmeester, Inclusion Compd., 1984, 3, 69.
7 W.L. Rollwitz and J.D. King, Proc. Int. Symp. Anal. Detect. Explos., 1983, 371; Chem. Abst., 1985, 102, 90835.
8 R. Blinc, M. Vilfan and J. Seliger, Bull. Magn. Reson., 1983, 5, 51.
9 N. Hadipour and J.L. Ragle, Z. Naturforsch. A, 1985, 40A, 355.
10 P.M. Woyciesjes, Diss. Abstr., Int. B, 1985, 45, 2153.
11 T.J. Bastow, J. Magn. Reson., 1985, 63, 590.
12 D.T. Amm, T.J. Bastow, K. Jeffrey, R.D. Heyding and S.L. Segel, Thermochim. Acta, 1985, 95, 447.
13 D.T. Amm, S.L. Segel, T.J. Bastow and K.R. Jeffrey, J. Chem. Phys., 1986, 84, 1.
14 V.S. Grechishkin and V.M. Mikhal'kov, Russian J. Phys. Chem., 1985, 59, 524.
15 V.V. Lisin, Yad. Magn. Rezon. Strukt. Krist., 1984, 164; Chem. Abstr., 1985, 103, 226116.
16 Y. Hiyama, L.G. Butler and T.L. Brown, J. Magn. Reson., 1985, 65, 472.
17 G.S. Beloglazov and V.S. Grechishkin, Russian J. Phys. Chem., 1984, 58, 1370.
18 G.S. Beloglazov, V.S. Grechishkin and V.M. Mikhal'kov, Russian J. Phys. Chem., 1984, 58, 1737.
19 V.P. Anferov, G.S. Beloglazov and V.S. Grechishkin, Sovrem. Metody YaMR EPR Khim. Tverd. Tela, [Mater. Vses. Koord. Soveshch.], 4th, 1985, 135; Chem. Abstr., 1985, 103, 226096.
20 K. Yamada and T. Okuda, J. Phys. Chem., 1985, 89, 4269.
21 J.M. Millar, A.M. Thayer, A. Bielecki, D.B. Zax and A. Pines, J. Chem. Phys., 1985, 83, 934.
22 V.S. Grechishkin and V.M. Mikhal'kov, Russian J. Phys. Chem., 1985, 59, 386.
23 V.L. Ermakov and D.Ya. Osokin, Mol. Phys., 1984, 53, 1335.
24 J. Murgich, M. Santana R., and J. Abanero, Magn. Reson. Chem., 1985, 23, 145.
25 D.Ya. Osokin, Sovrem. Metody YaMR EPR Khim. Tverd. Tela, [Mater. Vses. Koord. Soveshch.], 4th, 1985, 145; Chem. Abstr., 1985, 103, 226100.
26 G.K. Semin, G.V. Chikvaidze, A.V. Parygin and Ya.A. Eidus, Latv. PSR Zinat. Akad. Vestis, Fiz. Teh. Zinat. Ser., 1985, (3), 51.
27 A.V. Parygin, G.K. Semin, G.V. Chikvaidze and Ya.A. Eidus, Chem. Phys. Lett., 1985, 113, 314.
28 L.A. Zemnukhova, R.L. Davidovich and S.I. Kuznetsov, Bull. Acad. Sci. USSR, Div. Chem. Sci., 1985, 34, 512.
29 H. Ishihara, T. Okuda, K. Yamada and H. Negita, Bull. Chem. Soc. Jpn., 1985, 58, 2614.
30 N.E. Ainbinder, G.E. Kibrik, A.Yu. Polyakov, V.P. Tarasov and G.B. Furman, Sovrem. Metody YaMR EPR Khim. Tverd. Tela, [Mater. Vses. Koord. Soveshch.], 4th, 1985, 137; Chem. Abstr., 1985, 103, 226097.
31 C. Gerard-Dion and E.A.C. Lucken, J. Chem. Soc., Perkin Trans. II, 1985, 545.
32 M.A. Popov, Yad. Magn. Rezon. Strukt. Krist., 1984, 40; Chem. Abst., 1985, 103, 226108.
33 I.Yu. Amiantov, A.A. Boguslavskii and G.K. Semin, Bull. Acad. Sci. USSR, Phys. Series, 1984, 48(2), 182.
34 S.N. Aristov, I.Yu. Amiantov and A.A. Boguslavskii, Russian J. Phys. Chem., 1985, 59, 135.
35 E.A.C. Lucken and D. Rupp, J. Inclusion Phenom., 1985, 3, 157.
36 M.G. Voronkov, V.P. Feshin and P.A. Nikitin, J. Mol. Struct., 1985, 127, 181.
37 K.B. Dillon and J. Lincoln, Polyhedron, 1985, 4, 1333.
38 V.P. Feshin, P.A. Nikitin and M.G. Voronkov, Dokl. Akad. Nauk SSSR (Engl. transl.), 1985, 281, 275.
39 V. Krishnan, K.R. Sridharan and J. Ramakrishna, Polyhedron, 1985, 4, 739.
40 A.A. Boguslavskii, R.Sh. Lotfullin, M.V. Simonov, V.V. Kirikenko, V.I. Pakhomov and A.Ya. Mikhailova, Sov. Phys. Solid State, 1985, 27, 321.
41 G. Krishna Kishore, D. Krishna Rao and G. Satyanandam, Indian J. Pure Appl. Phys., 1985, 23, 38.
42 C. Hilbert, J. Clarke, T. Sleator and E.L. Hahn, Appl. Phys. Lett., 1985, 47, 637.

43 M.E. Ramia, Diss. Abstr., Int. B, 1985, 45, 2594.
44 M.E. Ramia and R.L. Armstrong, Can. J. Phys., 1985, 63, 350.
45 R.L. Armstrong and M.E. Ramia, J. Phys. C, 1985, 18, 2977.
46 M.E. Ramia and R.L. Armstrong, J. Phys. C, 1985, 18, 5367.
47 N. Okubo and Y. Abe, Phys. Lett. A, 1984, 106A, 435.
48 J. Murgich and B. Nogaj, J. Mol. Struct., 1985, 131, 291.
49 A.H. Brunetti and D.J. Pusiol, J. Phys. C, 1985, 18, 413.
50 D.J. Pusiol and A.H. Brunetti, J. Phys. C, 1985, 18, 3819.
51 C.A. Martín, J. Magn. Reson., 1985, 64, 1.
52 E.A. Romanenko, Yu. E. Sapozhnikov and Ya.B. Yasman, Teor. Éksp. Khim. (Engl. transl.), 1984, 20, 507.
53 O.Kh. Poleshchuk, G.G. Furin and G.G. Yakobson, Izv. Sib. Otd. Akad. Nauk SSSR, Ser. Khim. Nauk, 1985, (1), 116; Chem. Abstr., 1985, 103, 5832.
54 I.P. Biryukov and Zh. Zhuzhi, Russian J. Phys. Chem., 1985, 59, 1055.
55 T. Okuda, H. Kando, H. Ishihara, K. Yamada and H. Negita, J. Mol. Struct., 1985, 130, 309.
56 S.I. Gushchin, G.K. Semin, S.A. Petukhov, V.V. Saatsazov, T.L. Khotsyanova, N.N. Lisichkina and T.P. Tolstaya, Bull. Acad. Sci. USSR, Phys. Series, 1984, 48(2), 188.
57 V.D. Surkov, S.A. Petukhov, M.I. Vakhrin and G.K. Semin, Bull. Acad. Sci. USSR, Phys. Series, 1984, 48(2), 177.
58 G.K. Semin, S.I. Gushchin, S.A. Petukhov and V.D. Surkov, Bull. Acad. Sci. USSR, Div. Chem. Sci., 1984, 33, 2556.
59 I.Yu. Amiantov, T.L. Khotsyanova, I.S. Akhrem, A.V. Orlinkov and M.E. Vol'pin, Russ. J. Struct. Chem., 1984, 25, 706.
60 K.B. Dillon, J. Halfpenny and A. Marshall, J. Chem. Soc., Dalton Trans., 1985, 1399.
61 T. Okuda, H. Hamamoto, H. Ishihara and H. Negita, Bull. Chem. Soc. Jpn., 1985, 58, 2731.
62 E.A. Kravchenko, V.G. Morgunov, M.I. Gel'fman and T.V. Tsukshverdt, Koord. Khim., 1984, 10, 114.
63 S. Fleck and A. Weiss, Ber. Bunsenges. Phys. Chem., 1984, 88, 956.
64 H. Terao, H. Sakai and H. Negita, Bull. Chem. Soc. Jpn., 1985, 58, 2318.
65 S. Ramaprabhu and K.V.S. Rama Rao, Z. Naturforsch. A, 1985, 40A, 112.
66 G.K. Semin, S.I. Gushchin, S.A. Petukhov, V.V. Grushin, T.P. Tolstaya and I.N. Lisichkina, Bull. Acad. Sci. USSR, Div. Chem. Sci., 1984, 33, 2401.
67 G.K. Semin, T.L. Khotsyanova, S.I. Gushin and S.A. Petukhov, Chem. Phys. Lett., 1985, 114, 147.
68 G.K. Semin, S.I. Gushchin, S.A. Petukhov, V.V. Saatsazov, T.L. Khotsyanova, I.N. Lisichkina, L.D. Egorova and T.P. Tolstaya, Bull. Acad. Sci. USSR, Div. Chem. Sci., 1985, 34, 913.
69 D.F. Baisa, A.I. Barabash, I.G. Vertegel, B.G. Vekhter, V.P. Zenchenko and S.S. Stavrov, Khim. Fiz., 1984, 3, 18.
70 D.F. Baisa, A.I. Barabash, G.A. Puchkovskaya and O.V. Kozachek, Ukr. Fiz. Zh. (Russ. Ed.), 1985, 30, 1239.
71 D.F. Baisa, A.I. Barabash and E.A. Shadchin, Sov. Phys. Solid State, 1985, 27, 1414.
72 E.J. Miller, Diss. Abstr., Int. B, 1985, 45, 3501.
73 E.J. Miller, S.J. Landon and T.B. Brill, Organometallics, 1985, 4, 533.
74 Y. Kitaoka, K. Ueda, T. Kohara, K. Asayama, Y. Ōnuki and T. Komatsubara, J. Magn. Magn. Mater., 1985, 52, 341.
75 Y. Kitaoka, K. Fujiwara, Y. Kohori, K. Asayama, Y. Ōnuki and T. Komatsubara, J. Phys. Soc. Jpn., 1985, 54, 3686.
76 J. Lus, J. Kasprzak and J. Pietrzak, Ser. Fiz. (Uniw. im. Adama Mickiewicza Poznaniu), 1985, 53, 175.
77 V.L. Matukhin and I.A. Safin, Sovrem. Metody YaMR EPR Khim. Tverd. Tela, [Mater. Vses. Koord. Soveshch.], 4th, 1985, 140; Chem. Abstr., 1985, 103, 226098.
78 V.L. Matukhin, I.A. Safin, V.F. Shamraĭ and G.M. Leĭtus, Sov. Phys. Solid State, 1985, 27, 117.
79 R.M. Shelby, D.P. Burum and R.M. Macfarlane, Opt. Commun., 1984, 52, 283.
80 G.K. Semin, A.M. Raevsky and S.I. Gushin, Chem. Phys. Lett., 1985, 121, 214.

81 M. Naito and S. Tanaka, Proc. Int. Conf. Phys. Semicond. 17th, 1984, Springer, New York, 1985, p.1469.
82 M. Naito, H. Nishihara and S. Tanaka, J. Phys. Soc. Jpn., 1985, 54, 3946.

3
Rotational Spectroscopy

BY S. CRADOCK

1 Introduction

This year's Chapter on Rotational Spectroscopy and the derivation of information relating to rotation constants from other spectroscopic techniques is organised on the same lines as those in recent years, with inter-molecular complexes considered first, followed by sections on species with successively greater numbers of atoms, beginning with a Table dealing with diatomics. The number of reports found this year is very similar to that found last year, and among the smaller species (with 2, 3 or 4 atoms) about 20% concern studies of ions, almost all positively charged. This is a somewhat higher proportion than was the case a few years ago, reflecting the recent realisation that the electric charge on an ion makes it feasible to selectively record signals from ions in small concentrations in complex flowing systems by velocity-modulation techniques using fluctuating electric fields.

2 van der Waals and Hydrogen-bonded Complexes

The Ar.HCl complex has been studied in more detail, using molecular beam electric resonance spectroscopy (mbers) and laser Stark spectroscopy in the far-infrared[1] and infrared absorption spectroscopy with a difference frequency laser in the mid-infrared[2]. These studies both allow more information about the bending mode potential to be deduced. Microwave and radio-frequency mbers spectra of the Ar.H_2S complex[3] show that the Ar atom lies well out of the plane of the H_2S molecule. Complexes of HF with Ar_2[4] and Ar_3[5] clusters have been identified using pulsed-nozzle Fourier-transform microwave spectra; the HF molecule appears to be attached to the centre of more-or-less unperturbed Ar_2, or to the centre of a planar triangular Ar_3. For the Ar_3 complex, with three $I = 0$ atoms off-axis in a C_{3v} structure, only states with K=0 or a multiple of 3 exist, giving rather simple spectra. The OCS ν_3 band of complexes with Ne, Ar and Kr have been studied using supersonically cooled molecular beam samples and a diode laser spectrometer[6], effective structures in all cases having the rare-gas atom beside the C=O bond. Laser-induced fluorescence (l.i.f.) spectra of Hg atoms complexed to He and Xe have been identified, giving rotation constants and "bond" distances in the ground (singlet) and excited triplet states[7].

The infrared spectrum expected for N_2 dimers has been calculated, and compared with the limited available experimental data[8]. CARS spectra assigned to the CO_2 dimer have been reported[9], and it is suggested that a centrosymmetric structure is implied by the lack of coincidence of the Raman and infrared bands observed in the region associated with ν_1 of the monomer. The SO_2 dimer has been studied by mbers methods in the microwave and radiofrequency regions[10] and a tunnelling motion at 70 kHz interchanging non-equivalent SO_2 units identified. The intermolecular stretching force constant is found to be 0.026 mdyn/A!

A detailed study of the microwave spectra of several isotopic forms of N_2.HCN[11] gives a substitution structure consistent with a linear equilibrium configuration, with NN and CN triple bonds only slightly longer than in the separated molecules, but a significant lengthening of the CH bond. Complexes of OCS with HF[12] and HCl[13] are also effectively linear, with the hydrogen halide attached to oxygen.

Resolved hyperfine structure in $J=0\rightarrow 1$ lines of HF.HF, DF.DF, HF.DF and DF.HF, recorded using Fourier-transform microwave methods[14], has allowed the definition of an average equilibrium structure, which has the hydrogen-bond hydrogen atom only 7° away from the F..F axis but the terminal hydrogen some 60° off this axis. The two HF molecules in this structure interchange, and the planar tunnelling has been treated theoretically[15]. Microwave studies of HF complexes with H_2O[16] and D_2O[17] have been reported, and a more detailed understanding of the form of the bending potential, which has a double minimum, is emerging. A theoretical treatment of the H_2O dimer, based on a generalisation of the Internal Axis Method, has appeared[18]. FT microwave spectra of HI.HF have been reported[19]; this too seems to have a non-linear equilibrium structure. FTIR spectra of the CN stretching mode of HCN.HF at 0.004 cm^{-1} resolution have been analysed to give upper-state constants[20], assuming ground-state rotation constant and D_J values. FTIR and Raman bands assigned to dimers and trimers of HCN are reported both for cooled gas and free jet expansion samples, but rotationally resolved spectra are not yet available[21].

Several papers report the observation of complexes involving ammonia. The ammonia dimer appears not to have the expected $H_3N.H{-}NH_2$ hydrogen-bonded structure[22], but the observed mbers results do not make it clear what the true structure is, though it is apparently non-rigid[23], as is the complex with Ar. IR spectra of ammonia complexes with HCCH, CO_2, OCS, N_2O, HCN and Ar are also reported[23], and mbers microwave spectra of the N_2O complex have been analysed to show a structure with the ammonia nitrogen attached to the side of the N_2O molecule[24]. The $H_3N.H_2O$ complex, by contrast, is of the expected hydrogen-bonded form, $H_3N.H{-}OH$[25].

3 Diatomic Species

Table 1

Species	Ref.	Technique	State(s)	Information Derived
a) 2-electron species				
H_2	a	Reassignment of old and new B-F band FTIR line positions	F $^1\Sigma_g^+$ $(2p_\sigma)^2$	Term values and molecular constants
NaK	b	FT l.i.f.	X $^1\Sigma^+$, a $^3\Sigma^+$	Rot. anal.
Rb_2	c	l.i.f.	X $^1\Sigma_g^+$	Dunham coeffs., equilib. consts. incl. B_e
Cs_2	d	FT l.i.f.	D $^1\Sigma_u^+ \rightarrow$ (1) $^3\Sigma_g^-$	Rot. anal., many equilib. consts. incl. $r_e(1)$ = 554.25 pm
b) 4-electron species				
SnH^+	e	Emission	a $^3\Pi$ $\rightarrow$ X $^1\Sigma^+$	Rot. anal., many constants incl. $r_e(X)$ = 175.0(5) pm
He_2	f	Doppler-free optogalvanic spectra	b $^3\Pi_g$ - f $^3\Delta_u$	Spin-orbit and spin-spin couplings
c) 5-electron species				
SiH	g	IR (diode laser)	X $^2\Pi$	Anal. of Δv = 1 bands
GeH	h	IR and far-IR l.m.r.	X $^2\Pi$	Mol. parameters
d) 6-electron species				
OH^+	i	IR diff. freq. laser absorption	X $^3\Sigma^-$	Anal. of fundamental band, r_e = 102.893(1) pm
OH^+	j	Sub-mmw (1 THz) lines	X $^3\Sigma^-$	Mol. parameters
AsH	k	IR diode laser absorption	X $^3\Sigma^-$	Anal. of fundamental band, r_e = 152.315(5) pm
e) 7-electron species				
NH^-	l	IR colour-centre laser absorption	X $^2\Pi$	v = 0$\rightarrow$ 1 band analysed, equilib. consts.incl. r_e = 103.9 pm
OH	m	IR diode laser	X $^2\Pi$, v = 0,1	Reanalysis of pure rot. line positions; many mol. pars.
OH	n	Zeeman spectra	X $^2\Pi$	g-factors
OD	o	Far-IR l.m.r.	X $^2\Pi$ v=0,1,2,3	Reassignment and reanalysis of observed lines for higher v

Table 1 (cont)

Species	Ref.	Technique	State(s)	Information Derived
SD	p	l.m.r.	X $^2\Pi$	Equilib. consts.
f) 8-electron species				
NeH^+	q	Emission (FTIR)	X $^1\Sigma^+$	Mol. consts., incl.
NeD^+				r_e = 99.1195(21) pm
OH^-	r	IR colour-centre laser absorption	X, v=0→ 1	Vib. and rot. consts., r_e = 96.4313(48) pm
N_2^{++}	s	Emission	D $^1\Sigma_u^+$-X $^1\Sigma_g^+$	Rot. anal.
$^{15}N_2^{++}$	t	Emission	D - X	Rot. anal.
C_2	u	Doppler-limited dye laser excitation	d $^3\Pi_g$-a $^3\Pi_u$ v′=1,v″=0	Mol. parameters
C_2	v	l.i.f.	d - a	Mol. constants
C_2	w	IR colour-centre laser absorption	b $^3\Sigma_g^-$←a $^3\Pi_u$ v ← v+1, v=0,1,2	Improved constants
C_2	x	FTIR	b - a	Analysis of perturbations
C_2	y	Emission	C $^1\Pi_g$ - A$^1\Pi_u$	Rot. anal.
BN	z	Emission	A $^3\Pi$-X $^3\Pi$	Reanalysis, improved consts.
g) 9-electron species				
N_2^+	aa	l.i.f.	B $^2\Sigma_u^+$- X $^2\Sigma_g^+$	Spin-rot. consts.
N_2^+	bb	Emission	D′ $^2\Pi_g$- A $^2\Pi_u$	Rot. analysis
CO^+	cc	IR laser abs.	X $^2\Sigma^+$ v=0→ 1	Band consts.
C_2^-	dd	Emission	B $^2\Sigma_u^+$- X $^2\Sigma_g^+$	Consts. for A $^2\Pi_u$ state from perturbations
SiN	ee	Absorption	A $^2\Pi$ - X $^2\Sigma^+$	Rot. anal. of 2-0 band with microwave data
AlO	ff	Emission	D $^2\Sigma^+$,C $^2\Pi$-A $^2\Pi$	Rot. anal., B_e
AlO	gg	Emission	B $^2\Sigma^+$→X $^2\Sigma^+$	Rot. anal., mol. consts., RKR potential curves
TiN	hh	Laser excitation	A $^2\Pi$ - X $^2\Sigma$	Rot. anal., improved mol. constants
CaI,BaI	ii	Microwave	X $^2\Sigma^+$	Dunham coeffs., r_e = 282.8619(10)pm (CaI), 308.4798(11)pm (B
SrBr	jj	l.i.f.	B $^2\Sigma^+$- X $^2\Sigma^+$	Rot. anal.
SrBr,SrI	kk	Microwave	X $^2\Sigma^+$	Dunham coeffs., r_e = 273.5249 pm (SrBr), 297.3676 pm (SrI)
BaBr	ll	Microwave	X $^2\Sigma^+$	Dunham coeffs., r_e = 284.453(1) pm

Table 1 (cont)

Species	Ref.	Technique	State(s)	Information Derived
h) 10-electron species				
CF^+	mm	IR diode laser absorption	$X\ ^1\Sigma^+$ v=0→ 1	Band consts., r_e = 115.4272(35) pm
^{11}BF	nn	IR diode laser absorption	X, v=0→ 1	Upper-state consts., r_e = 126.280(5) pm
$^{13}C^{18}O$	oo	Emission (Herzberg bands)	$C\ ^1\Sigma^+ - A\ ^1\Pi$	Band consts., $r_e(A)$ = 123.38 pm
SiO	pp	FTIR in sunspots	X, v up to 10	Rot.anal., Dunham coeffs.
SiS	qq	Emission	$E\ ^1\Sigma^+ - X\ ^1\Sigma^+$	Rot. analysis, B_e
TiO	rr	Emission	$f\ ^1\Delta - a\ ^1\Delta$	Rot. analysis
ThO	ss	Emission	Ω (new state with $\Omega = 2$)	Rot. analysis, r_e = 185.6 pm
i) 11-electron species				
SiF	tt	Reanalysis of old optical spectra	$B\ ^2\Sigma^+ - A\ ^2\Sigma^+$	Improved mol. consts.
TiF	uu	Absorption	$^2\Delta, ^2\Phi - ^2\Delta$	Rot. analysis
PSe	vv	Emission	$A\ ^2\Pi - X\ ^2\Pi$	Mol. constants
MoN	ww	Emission	$^4\Pi \rightarrow X\ ^4\Sigma^-$	Rot. analysis
j) 12-electron species				
CuH	xx	FTIR emission	X, v=0,1,2	Improved mol. constants, r_e = 146.257 pm
$^{17}O_2$	yy	Mmwave	X, v=0	Ground-state constants
O_2	zz	(2+1)REMPI	$^1\Pi_g\ (3s_\sigma)$	Rot. analysis
O_2	aaa	Mmwave	$a\ ^1\Delta_g$	Improved mol. constants
O_2	bbb	Stimulated Raman	$X\ ^3\Sigma_g$	Many types of transitions
NF	ccc	Far-IR l.m.r.	$a\ ^1\Delta$ v=0,1,2	Mol. constants
PF	ddd	Microwave	$X\ ^3\Sigma^-$	Mol. constants, r_e = 158.9385(39) pm
PCl	eee	Microwave	$X\ ^3\Sigma^-$	Rot. constants
PCl	fff	IR diode laser absorption	X, v=0→ 1	Rot.anal., r_e = 201.4609(49) pm
SO	ggg	IR diode laser	$X\ ^3\Sigma^-$ v=0,1,2,3,4,5	Improved mol. consts.
UO	hhh	Two-colour REMPI	???	Lower-state consts., r″ = 183.7 pm
$^{130}Te_2$	iii	FTIR l.i.f.	???	Extension of analysis of vis. fluorescence

Table 1 (cont)

Species	Ref.	Technique	State(s)	Information Derived
k) 13-electron species				
MnS	jjj	l.i.f.	$A\ ^6\Sigma^+ - X\ ^6\Sigma^+$	Rot. analysis, $r_e(X)$ = 206.63(6) pm
l) 14-electron species				
ICl	kkk	J-selective p.l.s.	$A'\ ^3\Pi_2$	Mol. consts., $r_e(A')$ = 266.5(1) pm
I_2	lll	Optical-optical double resonance	$0_g^+\ (^3P_0)$	Mol. consts., r_e = 357.37 pm
I_2	mmm	Three-photon absorption	$F'\ (0_u^+)$	Upper-state consts., r_e = 347 pm
FeO	nnn	Emission (FTIR)	$?? - X\ ^5\Delta$	Improved g.s. consts., r_e = 161.64 pm
m) 17-electron species				
CuS	ooo	Laser excitation	$A\ ^2\Sigma - X\ ^2\Pi$	Rot. analysis
n) 18-electron species				
CuF	ppp	Microwave-optical double resonance	$b\ ^3\Pi$	Mol. constants

Notes: FT = Fourier Transform; l.i.f. = laser-induced fluorescence; l.m.r. = laser magnetic resonance; REMPI = Resonance-enhanced multi-photon ionisation; p.l.s. = polarisation-labelling spectroscopy.

References to Table 1

a) P.Senn, P.Quadrelli, K.Dressler and G.Herzberg, *J.Chem.Phys.*,1985,**83**,962.

b) A.J.Ross, C.Effantin, J.D'Incan and R.F.Barrow, *Mol.Phys.*,1985,**56**,903.

c) C.Amiot, P.Crozet and J.Vergés, *Chem.Phys.Letters*,1985,**121**,390.

d) C.Amiot and J.Vergés, *Chem.Phys.Letters*, 1985,**116**,273.

e) S.Yamaguchi, H.Obase, M.Tsuji and Y.Nishimura, *Chem.Phys.Letters*,1985,**119**,477.

f) K.Kawakita, K.Fukuda, Y.Adachi and C.Hirose, *J.Chem.Phys.*,1985,**82**,653.

g) P.B.Davies, N.A.Isaacs, S.A.Johnson and D.K.Russel, *J.Chem.Phys.*, 1985,**83**,2060.

h) J.M.Brown, K.M.Evenson and T.J.Sears, *J.Chem.Phys.*, 1985,**83**,3275.

i) M.W.Crofton, R.S.Altman, M.-F.Jagod and T.Oka, *J.Phys.Chem.*,1985,**89**,3614.

j) J.P.Bekooy, P.Verhoeve, W.L.Meerts and A.Dymanus, *J.Chem.Phys.*,1985,**82**,3868.

k) J.R.Anacona, P.B.Davies and S.A.Johnson, *Mol.Phys.*,1985,**56**,989.

l) D.M.Neumark, K.R.Lykke, T.Anderson and W.C.Lineberger, *J.Chem.Phys.*, 1985,**83**,4364.

References to Table 1 (cont)

m) B.Lemoine, M.Bogey and J.L.Destombes, *Chem.Phys.Letters*,1985,**117**,532.

n) W.M.van Herpen, W.L.Meerts and L.Veseth, *Chem.Phys.Letters*,1985,**120**,247.

o) J.M.Brown, J.E.Schubert, C.E.Brown, J.S.Geiger and D.R.Smith, *J.Mol.Spectrosc.*, 1985,**114**,185.

p) D.Zeitz, W.Bohle, J.Werner, A.Hinz and W.Urban, *Mol.Phys.*,1985,**54**,953.

q) R.S.Ram, P.F.Bernath and J.W.Brault, *J.Mol.Spectrosc.*,1985,**113**,451.

r) J.C.Owrutsky, N.H.Rosenbaum, L.M.Tack and R.J.Saykally, *J.Chem.Phys.*, 1985,**83**,5338.

s) D.Cossart, F.Launay, J.M.Robbe and G.Gandara, *J.Mol.Spectrosc.*,1985,**113**,142.

t) D.Cossart and F.Launay, *J.Mol.Spectrosc.*,1985,**113**,159.

u) T.Suzuki, S.Saito and E.Hirota, *J.Mol.Spectrosc.*,1985,**113**,399.

v) M.C.Curtis and P.J.Sarre, *J.Mol.Spectrosc.*,1985,**114**,427.

w) Wen-Bin Yan, R.F.Curl, A.J.Merer and P.G.Carrick, *J.Mol.Spectrosc.*, 1985,**112**,436.

x) F.Roux, G.Wannous, F.Michaud and J.Vergés, *J.Mol.Spectrosc.*,1985,**109**,334.

y) A.Antić-Jovanović, V.Bojović, D.S.Pesić, B.R.Vujisić, D.Rakotoarijimy and S.Weniger, *J.Mol.Spectrosc.*,1985,**110**,86.

z) H.Bredohl, I.Dubois, Y.Houbrechts and P.Nzohabonayo, *J.Mol.Spectrosc.*, 1985,**112**,430.

aa) S.D.Rosner, T.D.Gaily and R.A.Holt, *J.Mol.Spectrosc.*,1985,**109**,73.

bb) D.Cossart, C.Cossart-Magos, G.Gandara and J.M.Robbe, *J.Mol.Spectrosc.*, 1985,**109**,166.

cc) P.B.Davies and W.J.Rothwell, *J.Chem.Phys.*, 1985,**83**,5450.

dd) R.D.Mead, U.Hefter, P.A.Schulz and W.C.Lineberger, *J.Chem.Phys.*,1985,**82**,1723.

ee) S.C.Foster, K.G.Lubic and T.Amano, *J.Chem.Phys.*,1985,**82**,709.

ff) M.Singh and M.D.Saksena, *Can.J.Phys.*,1985,**63**,1162.

gg) J.A.Coxon and S.Naxakis, *J.Mol.Spectrosc.*, 1985,**111**,102.

hh) K.Brabaharan, J.A.Coxon and A.B.Yamashita, *Can.J.Phys.*,1985,**63**,997.

ii) T.Törring and K.Döbl, *Chem.Phys.Letters*,1985,**115**,328.

jj) J.O.Schröder and W.E.Ernst, *J.Mol.Spectrosc.*,1985,**112**,413.

kk) T.Törring, K.Döbl and G.Weiler, *Chem.Phys.Letters*,1985,**117**,539.

ll) W.E.Ernst, G.Weiler and T.Törring, *Chem.Phys.Letters*,1985,**121**,494.

mm) K.Kawaguchi and E.Hirota, *J.Chem.Phys.*,1985,**83**,1437.

nn) T.Nakanaga, H.Takeo, S.Kondo and C.Matsumura, *Chem.Phys.Letters*,1985,**114**,88.

oo) C.V.V.Prasad, S.P.Reddy and M.Sandys-Wunsch, *J.Mol.Spectrosc.*,1985,**114**,436.

pp) D.A.Glenar, A.R.Hill, D.E.Jennings and J.W.Brault, *J.Mol.Spectrosc.*, 1985,**111**,403.

qq) G.Lakshminarayana, B.J.Shetty and S.Gopal, *J.Mol.Spectrosc.*,1985,**112**,1.

rr) G.R.Brandes and D.C.Galehouse, *J.Mol.Spectrosc.*,1985,**109**,345.

References to Table 1 (cont)

ss) G.Edvinsson and A.Lagerqvist, *J.Mol.Spectrosc.*,1985,**113**,93.

tt) E.R.Comben and R.F.Barrow, *J.Mol.Spectrosc.*,1985,**109**,415.

uu) E.A.Shenyavskaya and V.M.Dubov, *J.Mol.Spectrosc.*,1985,**113**,85.

vv) S.N.Vempati, J.L.Ma and W.E.Jones, *J.Mol.Spectrosc.*, 1985,**110**,301.

ww) R.C.Carlson, J.K.Bates and T.M.Dunn, *J.Mol.Spectrosc.*,1985,**110**,215.

xx) R.S.Ram, P.F.Bernath and J.W.Brault, *J.Mol.Spectrosc.*,1985,**113**,269.

yy) G.Cazzoli and C.Degli Esposti, *Chem.Phys.Letters*, 1985,**113**,501.

zz) A.Sur, C.V.Ramana and S.D.Colson, *J.Chem.Phys.*,1985,**83**,904.

aaa) K.W.Hillig II, C.C.W.Chiu, W.G.Read and E.A.Cohen, *J.Mol.Spectrosc.*, 1985,**109**,205.

bbb) R.A.Hill, O.Owyoung and P.Esherick, *J.Mol.Spectrosc.*, 1985,**112**,233.

ccc) P.B.Davies, A.H.Ferguson, D.P.Stern and F.Temps, *J.Mol.Spectrosc.*, 1985,**113**,28.

ddd) S.Saito, Y.Endo and E.Hirota, *J.Chem.Phys.*, 1985,**82**,2947.

eee) T.Minowa, S.Saito and E.Hirota, *J.Chem.Phys.*, 1985,**83**,4939.

fff) H.Kanamori, C.Yamada, J.E.Butler, K.Kawaguchi and E.Hirota, *J.Chem.Phys.*, 1985,**83**,4945.

ggg) H.Kanamori, J.E.Butler, K.Kawaguchi, C.Yamada and E.Hirota, *J.Mol.Spectrosc.*, 1985,**113**,262.

hhh) M.C.Heaven, J.-P.Nicolai, S.J.Riley and E.K.Parkes, *Chem.Phys.Letters*, 1985,**119**,229.

iii) A.Topouzkhanian, O.Babaky, J.Verges, R.Willers and B.Wellegehausen, *J.Mol.Spectrosc.*, 1985,**113**,39.

jjj) M.Douay, B.Pinchemel and C.Dufour, *Can.J.Phys.*,1985,**63**,1381.

kkk) J.C.D.Brand, D.Bussieres and A.R.Hoy, *J.Mol.Spectrosc.*,1985,**113**,388.

lll) J.P.Perrot, A.J.Bouvier, A.bouvier, B.Femelat and J.Chevaleyre, *J.Mol.Spectrosc.*, 1985,**114**,60.

mmm) G.W.King and T.D.McLean, *Chem.Phys.Letters*, 1985,**121**,57.

nnn) A.W.Taylor, A.S.-C.Cheung and A.J.Merer, *J.Mol.Spectrosc.*,1985,**113**,487.

ooo) F.David, M.Douay and Y.Lefebvre, *J.Mol.Spectrosc.*, 1985,**112**,115.

ppp) T.C.Steimle, C.R.Brazier and T.M.Brown, *J.Mol.Spectrosc.*,1985,**110**,39.

4 Triatomic Molecules and Ions

A single line of H_2D^+ ($2_{21} \rightarrow 2_{20}$) has been observed near 156 GHz[26], and a new Hamiltonian for H_3^+ using curvilinear stretching co-ordinates has been used to analyse its spectra[27]. Laser photofragment spectroscopy has been applied to SiH_2^+, using dissociation due to the A $^2B_1$ Π ← X $^2A_1$ transition[28], the resulting rotation constants giving a ground-state structure with r SiH = 1.49(1) A, <HSiH =

119.0(5)°. Old and new data on the A $^2A_1$ - X $^2B_1$ transitions of ND_2 have been analysed to give molecular parameters for the 000 and 010 levels of the electronic ground state[29]. Absorption spectra of H_2O in the visible region (16500 - 25200 cm^{-1}) have been measured using FT techniques and 1174 lines assigned[30]. Line positions and intensities are reported[31] for three bands of $H_2{}^{18}O$ between 4400 and 6100 cm^{-1}. The pure rotation spectrum of H_2S between 1 and 1200 GHz has been calculated, and the 40 observed lines (out of 137 expected) fitted to a Padé-form Hamiltonian[32]. Far-infrared spectra of D_2S and HDS have been measured by FTIR methods with a resolution of 0.004 cm^{-1} and fitted to a conventional Hamiltonian with constants up to 10 th order[33]. The Q-branch of ν_1 of H_2S has been studied by CARS, with an effective resolution of 30 MHz[34] and FTIR spectra of the ν_2 bands of D_2S[35] and ^{33}S and ^{34}S H_2S[36] reported and analysed to give upper-state constants.

Fourier-transform spectra of laser-induced fluorescence in the 000-000 band of the A $^2\Pi$ - A $^2\Sigma^+$ transition of SrOH have been used to give improved molecular constants for both states[37]. The J=0→ 1 transition of $HN_2{}^+$ near 93 GHz has been studied at improved resolution (better than 0.1 MHz), which allows all expected hyperfine components to be assigned and correct values for the quadrupole coupling constants for the two different nitrogen atoms derived[38]. The Q-branch of the ν_2 band of $DN_2{}^+$ near 544 cm^{-1} has been resolved using a diode laser spectrometer[39], giving upper-state molecular constants, and similar results are reported for HCO^+, whose ν_2 band is near 828 cm^{-1},[40] and HCS^+, near 766 cm^{-1}.[41] Velocity-modulation laser spectroscopy of HCS^+ in the ν_1 region near 3141 cm^{-1} gives very precise upper- and lower-state constants[42]. Sub-Doppler spectra of the ν_3 band of HCN have been obtained using a molecular beam sample and a colour-centre IR laser[43]. Microwave-optical double resonance spectra of HNO in the A $^1A''$ (100) state allow the components of the dipole moment to be defined[44]. New observations of l.i.f. spectra of the A - X $^1A'$ transitions of HNO[45] and dye laser excitation spectra of the same transition of DNO[46] have been reported and analysed in terms of vibronic couplings and perturbations. Doppler-limited laser excitation[47] and l.i.f.[48] spectra of the A $^1A''$ - X $^1A'$ transitions of HSiF have been reported, and analysed to give molecular structures for both electronic states. Diode laser and laser magnetic resonance (l.m.r.) spectra of DO_2 in the ν_2 and ν_3 bands[49] complete the collection of vibrationally excited-state rotation constants required for the definition of an r_e structure for HO_2, and FTIR spectra of HOCl in the ν_1, ν_2, ν_3 and $2\nu_2$ bands give similar results[50] for this molecule. In both cases bond angles HOX are about 104°.

Nitrogen quadrupole coupling constants for the A $^2\Delta$ state of CCN have been derived from microwave-optical double resonance spectra[51]. The l-type doubling in the $A\ ^2\Sigma^+ - X\ ^2\Pi$ transitions of N_2O^+ has been fully analysed[52], and the ν_3 band of CO_2^+ has been studied by diode laser spectroscopy[53]. The anomalously low frequency (about 1423 cm^{-1}) is attributed to vibronic coupling with the low-lying A state. Laser excitation spectra of the $A\ ^2\Pi \rightarrow X\ ^2\Pi$ bands of OCS^+ [54] and laser fragmentation spectra of the $A \leftarrow X$ and $B\ ^2\Sigma \leftarrow X$ bands[55] have been analysed to give upper- and lower-state rotation constants. Microwave spectra of NCO in the v_2=0, 1 and 2 states have been observed[56] and molecular constants derived, while single lines in the ν_1 band (by laser magnetic resonance[57]) and the $A\ ^2\Sigma^+ - X\ ^2\Pi$ bands[58] (at sub-Doppler resolution) have been reported for the same species.

Very precise line positions in the ν_3 band of N_2O have been established using heterodyne frequency measurement[59], giving new standard frequencies, the bending bands of heavy isotopes of CO_2 have been studied by FTIR[60], and very weak lines of the $3\nu_2$ band of CO_2 have been identified by FTIR and assigned[61]. FTIR spectra of OCS in the ν_2[62] and ν_3[63] regions have been reported, giving new standards in the 490–560 cm^{-1} and 1975–2140 cm^{-1} regions respectively. Diode laser spectra of heavy isotopes of CS_2 in the ν_3 region[64] and of the $\nu_3-\nu_1$ band[65] of the normal species have been reported; in the latter case analysis of the new data with existing information gives many molecular constants. The near UV bands of $C^{34}S_2$ due to a $^3A_2 \leftarrow {}^1\Sigma_g$ transition have been analysed, giving rotation constants for several vibrational states and the isotopic shifts by comparison with data for the normal isotope[66]. Diode laser spectra of FCN allow resolution of the ν_1 bands of the normal and ^{13}C and ^{15}N isotopes and associated hot-bands[67]. Calculated distortion constants for various isotopic species of FCN in various vibrational states have been compared with experimental values[68].

Pure rotation transitions and far-infrared l.m.r. lines of PO_2 have been assigned[69] and used to derive molecular constants and an r_0 structure, with rPO = 146.65(41) pm and a bond angle of 135°. Millimeter-wave spectra of mono-^{18}O O_3 in the v_2=1 state have been observed, and analysed to give rotational and distortion constants up to the octic level[70], and IR and microwave data for $^{16}O_3$ in the ground state and the v_1=1 and v_3=1 states have been reported[71]. Sub-millimeter-wave spectra of SO_2 (400–1000 GHz) in the ground state and the v_2=1 state[72] and IR diode laser and IR-microwave double resonance spectra[73] of $S^{18}O_2$ have been used to derive molecular parameters. Laser-induced fluorescence spectra of the $A\ ^1A_2 - X\ ^1A_1$ bands of SO_2 in a supersonic jet are more easily analysed because of the removal of the congestion typical of spectra using static samples[74], and the vibronic origin has been unambiguously established. Zeeman splittings have been measured in the 320 nm bands of SO_2, allowing the magnetic moments of the $^1B_1$ and $^1A_2$ states involved to be deduced[75]. An a-type band assigned to ν_1 of S_2O has been studied between

1147 and 1183 cm^{-1} using a diode laser[76], and analysed giving ground-state and upper-state constants. A rotational analysis of the 000–040 band of the A $^1B_1 \rightarrow$ X $^1A_1$ emission of CF_2 has been reported[77], giving upper- and lower-state rotation constants. Microwave spectra of $^{18}O^{14}NF$ in vibrationally excited states have been assigned[78] and the resulting rotation constants used to define an r_e structure, with r NF = 151.658(25) pm, r NO = 113.155(23) pm, <ONF = 109.9220(72)°, and quadratic and cubic force field elements. Radio-frequency transitions of OF_2 within individual rotational levels in the presence of an external electric field have been studied by mbers methods, giving the dipole moment (0.30818(3) D), polarizability anisotropies and spin-rotation constants[79].

5 Tetra-atomic molecules and ions

More than 100 lines in the ν_3 band of CH_3^+ have been observed with a difference frequency laser[80]; the ion is planar, with r_0 CH = 109.151(9) pm. A single line, P(2,1), of the rotation-inversion spectrum of H_3O^+ has been observed at 307 192.41(5) MHz[81], close to the predicted position of 307 167(240) MHz. The ν_2 bands of H_3O^+[82] and D_3O^+[83] have been studied using diode laser spectrometers in the IR, and the equilibrium structure and potential function are well established (r_e OH = 97.5819 pm, inversion barrier = 672 cm^{-1} above equilibrium configuration). The ν_3 bands of H_3O^+ have also been studied[84]. FTIR spectra of ND_3 from 30–220 cm^{-1} have been analysed to give ground-state rotation-inversion parameters[85], the ν_2 bands of $^{15}NH_3$ have been studied using FTIR[86], and the NH stretching bands ν_1 and ν_3 of ^{14}N and $^{15}NH_3$ have been studied using FTIR and CARS[87], giving precise upper-state parameters. Transitions in the $2\nu_2$ and $2\nu_4$ bands of $^{15}NH_3$ with Δk = +2 or +3 have been analysed to give relative energies of different K states in the ground vibrational state[88], and FTIR spectra of the $\nu_1+\nu_2$ and $\nu_1+\nu_2-\nu_2$ bands of ^{14}N and $^{15}NH_3$ have been analysed to give upper-state parameters to eighth order[89]. Microwave lines between 9 and 300 GHz involving l-type doubling transitions in the v_4=1 state of AsH_3 have been observed; low-J transitions (J=1,2,3) can be analysed, but higher-J lines (J=6,7,8) are perturbed by Coriolis interactions with ν_2 levels[90].

FTIR spectra of formaldehyde in the 4000–5400 cm^{-1} region have been analysed to give rotation and distortion constants for the upper states of the $3\nu_2$, $2\nu_2+\nu_6$ and $\nu_3+\nu_5$ bands[91]. Millimeter-wave spectra of the protonated hydrogen cyanide $HCNH^+$ have been observed[92], giving B_0 and D_0 values in agreement with IR data for the same species. Microwave-optical double resonance spectra of thioformaldehyde[93] and pure microwave spectra of selenoformaldehyde[94] have also been reported. Selenium isotope data in the last study allows r CSe = 175.9

pm to be derived.

A rotational analysis of the $^2\Pi_i \rightarrow X\ ^2\Pi_i$ emission bands of HCCS has been presented[95], and laser excitation spectra of $HCCCl^+$ in the $A\ ^2\Pi \leftarrow X\ ^2\Pi$ bands have also been analysed[96], giving r_s structures for both electronic states from isotopic data. The ν_1 band of $HOCO^+$ has been studied using difference-frequency IR laser absorption[97,98], giving the rotation constants for the vibrationally excited state. More bands of HONO (ν_4, ν_5 and ν_6 of the trans form and ν_4 of the cis form) have been studied by FTIR at <0.01 cm^{-1} resolution[99], and a complete set of ground- and excited-state rotation and distortion constants is now available for the trans form.

Millimeter-wave spectra of C_3O have given very precise B_0 and D_0 values[100], and microwave data for ^{13}C and ^{18}O isotopic forms allow a substitution structure (r CC = 125.4, 130.6 pm, rCO = 115.0 pm) for this linear molecule to be defined[101]. A further study of the $A\ ^1A'' - X\ ^1A'$ bands of NCNO near 882 nm has appeared[102]. IR-microwave double resonance spectra of $^{10}BF_3$ using a tunable diode laser, which take advantage of the centrifugal-distortion induced dipole in the vibrational ground state, allow C_0 = 5161.40(5) MHz to be measured[103], and a similar study of the ^{11}B isotopic species gives the same value for C_0, though with a lower precision[104], from IR-microwave double resonance studies of the ν_3 band. The ν_3 band of NO_3 near 1492 cm^{-1} has been studied using a diode laser[105], giving r_0 NO = 124.0 pm. The ν_2 band of COF_2 near 960 cm^{-1} has been studied using FTIR and diode laser spectrometers, and the results analysed together with microwave data[106]. Microwave spectra of COClF, including isotopic species at natural abundance, allow a partial r_z structure to be determined (r CF = 133.429(7) pm, r CCl = 172.433(6) pm, assuming r CO = 117.3 pm, <OCCl = 127.5°, <OCF = 123.7°)[107]. Combined analysis of electron diffraction and microwave data gives an r_e structure for $COCl_2$ (r CO = 117.66(22) pm, r CCl = 173.65(12) pm, <ClCCl = 111.91(12)°)[108].

6 Penta-atomic molecules

Microwave spectra of $SiHD_3$, SiH_2D_2 and SiH_3D have been observed and an equilibrium SiH bond length of 147.41 pm deduced[109]. FTIR spectra of $^{70}GeH_3D$ have been analysed[110], leading to values for B, D and H constants including A_0 and D_K from perturbation-allowed transitions.

FTIR spectra of the 10-micron bands of CH_2=NH[111] and the 3-micron bands of the same species[112] have been analysed and the ν_1 band picked out from between the much stronger lines of HCN in the same region[113]; the rotation constants are now known for all first excited vibrational states. The ν_6 bands of three isotopic forms of GeH_3Cl have been analysed[114], giving a coherent set of upper-state rotation and distortion constants and isotopic frequencies.

Doppler-limited FTIR spectra (resolution 0.0054 cm^{-1}) of the ν_9 band of diazirine, H_2CN_2, have been analysed[115] and microwave spectra of the isomer cyanamide, H_2NCN, analysed using a semi-rigid bender model[116], giving a structure with a bent NCN unit (<NCN ~ 175°), a low-frequency, high-amplitude inversion of the NH_2 group (inversion barrier ~ 510(6) cm^{-1}) and significant variations of r NH, r NC and <HNH with the inversion angle. FTIR spectra of thioketene, $H_2C{=}C{=}S$, have been analysed[117], giving rotation constants and D_K in the ground state and the $v_7 = 1$ and $v_3 = v_8 = 1$ states. Microwave spectra of two conformers (*cis, cis* and *cis, trans*) of HNSNH have been identified and analysed[118], and partial structures derived.

The microwave spectra of several isotopic forms of HPF_2S have been reanalysed together with electron-diffraction and liquid-crystal n.m.r. data with appropriate vibrational corrections[119], giving an average structure that is more consistent with that of HPF_2Se than the original microwave-only structure. Mbers methods have been used to obtain very well resolved rotational spectra of HCCCN, from which the dipole moment, nitrogen quadrupole coupling constant and other parameters for ground and vibrationally excited states have been derived[120]. The 7.5 micron bands of HNO_3 (ν_3 and ν_4) have been studied by Stark modulation spectroscopy using a diode laser; this allows assignments to be made that FTIR spectra cannot provide because of the congestion in this region, and gives rotation and distortion constants for the $v_3 = 1$ state[121].

IR laser spectra in a supersonic jet sample allow assignments to be made for the ν_1 band of CF_3Cl[122]. The ν_4 band of CF_3I has been studied by FTIR methods[123] at 0.0005 cm^{-1} resolution, and FTIR spectra of CF_3I and CF_3Br in the ν_1 and ν_5 regions reported[124]. The rotational spectrum of OPF_3 has been reanalysed[125] with explicit inclusion of the $\Delta k = 3$ or 6 terms, giving an improved fit. Similar reanalysis gives no improvement for PH_3. The microwave spectrum of NCNCS has been analysed using a quasilinear model in which the reduced mass for the bend is allowed to vary with the bend angle[126].

7 Molecules with six or more atoms

In this section the organisation is by Group in the Periodic Table. The rotation constants for the expected isotopic modifications of B_2H_6 have been calculated from the r_z structure and force field deduced from the major study of the vibration-rotation spectra[127]. Microwave spectra of several isotopic modifications of B_2H_5Cl give an r_s structure, with a B..B distance of 177.6(3) pm and a terminal r BCl of 176.6(4) pm[128]. The ν_4 band of NH_2-BH_2 (BN stretch, near 1337 cm^{-1}) has been studied by FTIR methods and analysed to give ground-state rotation constants[129].

Analysis of the millimeter-wave spectrum of $CH_3C{\equiv}CSiH_3$ has given a value for the three-fold internal rotation barrier of about 3 cm^{-1} [130]. The inversion-rotation spectrum of 1,3-disilacyclobutane-1,1,3,3-d_4 in the microwave region includes strong Q branches due to the v = 0 → 1 transition with ν_0 about 43 GHz[131]. Microwave spectra of several isotopic forms of CH_3GeH_2CN give an r_s structure, including r Ge-CN = 192.7(4) pm, r Ge-CH_3 = 193.3(2) pm. The barrier to internal rotation of the methyl group is V_3 = 402(8) cm^{-1} [132].

Low-frequency motions of N_2H_4 are reanalysed, including the effects of tunnelling, using the double group $G_{16}^{(2)}$, which is claimed to result in a simpler treatment[133]. FTIR spectra of $CH_3C{\equiv}P$ and $CD_3C{\equiv}P$ have been reported[134] and the microwave spectrum of N≡C-C≡C-C≡P analysed[135] to give B_0. Rotation constants for both *gauche* and *trans* conformers of $CH_3CH_2PF_2$ have been reported, and partial structures derived with the aid of a diagnostic (constrained) least-squares fitting programme[136]. The microwave spectrum of the *trans* conformer of $ClCH_2POF_2$ has been analysed, and a structure derived in the same way[137]. FTIR spectra of the ν_3 band of PF_5 near 950 cm^{-1} have been analysed, including a hot-band, giving rotation and distortion constants for upper and lower states[138], while the bands near 1025 cm^{-1}, due to ν_5 and $2\nu_8$, have been studied by diode laser spectrometer using a jet-cooled sample, giving accurate band origins, rotation and distortion constants[139]. It has been shown that the parameter R_6 for a C_{4v} molecule such as XY_5 (square pyramidal) or XYF_5 can be used to constrain the refinement of the force field and allow the geometry to be defined more precisely; the method is illustrated using experimental data for XF_5, where X is Cl, Br or I, and SF_5X, where X is Cl or Br[140].

A massive paper summarising 10 years' work at a number of laboratories deals with diode laser IR spectra of the ν_3 band of UF_6 in supersonic jets; the rotation constant derived gives an r_0 bond length of 199.26(7) pm[141]. Diode laser spectra of the $\nu_1+\nu_3$ band of UF_6 near 1295 cm^{-1} have also been reported[142]; a static sample cooled to 226K was used.

References

1. M.D.Marshall, A.Charo, H.O.Leung and W.Klemperer, *J.Chem.Phys.*, 1985,**83**,4924.
2. B.J.Howard and A.S.Pine, *Chem.Phys.Letters*, 1985,**122**,1.
3. R.Viswanathan and T.R.Dyke, *J.Chem.Phys.*, 1985,**82**,1674.
4. H.S.Gutowsky, T.D.Klots, C.Chuang, C.A.Schmuttenmaer and T.Emilsson, *J.Chem.Phys.*, 1985,**83**,4817.
5. H.S.Gutowsky, T.D.Klots, C.Chuang, J.D.Keen, C.A.Schmuttenmaer and

T.Emilsson, *J.Amer.Chem.Soc.*, 1985,**107**,7174.
6. G.D.Hayman, J.Hodge, B.J.Howard, J.S.Muenter and T.R.Dyke, *Chem.Phys.Letters*, 1985,**118**,12.
7. M.-C.Duval, C.Jouvet and B.Soep, *Chem.Phys.Letters*, 1985,**119**,317.
8. G.Brocks and A. van der Avoird, *Mol.Phys.*, 1985,**55**,11.
9. G.A.Pubanz, M.Maroncelli and J.W.Nibler, *Chem.Phys.Letters*, 1985,**120**,313.
10. D.D.Nelson,Jr., G.T.Fraser and W.Klemperer, *J.Chem.Phys.*, 1985,**83**,945.
11. E.J.Goodwin and A.C.Legon, *J.Chem.Phys.*, 1985,**82**,4434.
12. A.C.Legon and L.C.Willoughby, *J.Mol.Struct.*, 1985,**131**,159.
13. E.J.Goodwin and A.C.Legon, *J.C.S.Faraday Trans.II*, 1985,1709.
14. H.S.Gutowsky, C.Chuang, J.D.Keen, T.D.Klots and T.Emilsson, *J.Chem.Phys.*, 1985,**83**,2070.
15. J.T.Hougen and N.Ohashi, *J.Mol.Spectrosc.*, 1985,**109**,134.
16. G.Cazzoli, P.G.Favero, D.G.Lister, A.C.Legon, D.J.Millen and Z.Kisiel, *Chem.Phys.Letters*, 1985,**117**,543.
17. Z.Kisiel, A.C.Legon and D.J.Millen, *J.Mol.Struct.*, 1985,**131**,201.
18. J.T.Hougen, *J.Mol.Spectrosc.*, 1985,**114**,395.
19. R.E.Bumgarner, C.D.Cogley and S.G.Kukolich, *Chem.Phys.Letters*, 1985,**122**,535.
20. B.A.Wofford, J.W.Bevan, W.B.Olson and W.J.Lafferty, *J.Chem.Phys.*, 1985,**83**,6188.
21. M.Maroncelli, G.A.Hopkins, J.W.Nibler and T.R.Dyke, *J.Chem.Phys.*, 1985,**83**,2129.
22. D.D.Nelson,Jr., G.T.Fraser and W.Klemperer, *J.Chem.Phys.*, 1985,**83**,6201.
23. G.T.Fraser, D.D.Nelson,Jr., A.Charo and W.Klemperer, *J.Chem.Phys.*, 1985,**82**,2535.
24. G.T.Fraser, D.D.Nelson,Jr., G.J.Gerfen and W.Klemperer, *J.Chem.Phys.*, 1985,**83**,5442.
25. P.Herbine and T.R.Dyke, *J.Chem.Phys.*, 1985,**83**,3768.
26. S.Saito, K.Kawaguchi and E.Hirota, *J.Chem.Phys.*, 1985,**82**,45.
27. V.Špirko, P.Jensen, P.R.Bunker and A.Čejchan, *J.Mol.Spectrosc.*, 1985,**112**,183.
28. M.C.Curtis, P.A.Jackson, P.J.Sarre and C.J.Whitham, *Mol.Phys.*, 1985,**56**,485.
29. R.E.Muenchausen, G.W.Hills, M.F.Merienne-Lafore, D.A.Ramsay, M.Vervloet and F.W.Birss, *J.Mol.Spectrosc.*, 1985,**112**,203.
30. C.Camy-Peyret, J.-M.Flaud, J.-Y.Mandin, J.-P.Chevillard, J.Brault, D.A.Ramsay, M.Vervloet and J.Chauville, *J.Mol.Spectrosc.*, 1985,**113**,208.
31. J.-P.Chevillard, J.-Y.Mandin, J.-M.Flaud and C.Camy-Peyret, *Can.J.Phys.*, 1985,**63**,1112.
32. A.V.Burenin, T.M.Fevral'skikh, A.A.Mel'nikov and S.M.Shapin, *J.Mol.Spectrosc.*, 1985,**109**,1.
33. C.Camy-Peyret, J.-M.Flaud and L.Lechuga-Fossat, *J.Mol.Spectrosc.*,

1985,**109**,300.

34. H.Frunder, R.Angstl, D.Illig, H.W.Schrotter, L.Lechuga-Fossat, J.-M.Flaud, C.Camy-Peyret and W.F.Murphy, *Can.J.Phys.*, 1985,**63**,1189.
35. J.R.Gillis, R.D.Blatherwick and F.S.Bonomo, *J.Mol.Spectrosc.*, 1985,**114**,228.
36. W.C.Lane, T.H.Edwards, J.R.Gillis, F.S.Bonomo and F.J.Murcray, *J.Mol.Spectrosc.*, 1985,**111**,320.
37. C.R.Brazier and P.F.Bernath, *J.Mol.Spectrosc.*, 1985,**114**,163.
38. G.Cazzoli, G.Corbelli, C.Degli Esposti and P.G.Favero, *Chem.Phys.Letters*, 1985,**118**,164.
39. T.J.Sears, *J.Chem.Phys.*, 1985,**82**,5757.
40. K.Kawaguchi, C.Yamada, S.Saito and E.Hirota, *J.Chem.Phys.*, 1985,**82**,1750.
41. P.B.Davies and W.J.Rothwell, *J.Chem.Phys.*, 1985,**83**,1496.
42. N.H.Rosenbaum, J.C.Owrutsky, L.M.Tack and R.J.Saykally, *J.Chem.Phys.*, 1985,**83**,4845.
43. E.K.Kyrö, P.Shoja-Chaghervand, M.Eliades, D.Danzeiser, S.G.Lieb and J.W.Bevan, *Can.J.Chem.*, 1985,**63**,1870.
44. K.Takagi, T.Suzuki, S.Saito and E.Hirota, *J.Chem.Phys.*, 1985,**83**,535.
45. R.N.Dixon and C.A.Rosser, *J.Mol.Spectrosc.*, 1985,**110**,262.
46. J.C.Petersen, *J.Mol.Spectrosc.*, 1985,**110**,277.
47. T.Suzuki, K.Hakuta, S.Saito and E.Hirota, *J.Chem.Phys.*, 1985,**82**,3580.
48. R.N.Dixon and N.G.Wright, *Chem.Phys.Letters*, 1985,**117**,280.
49. H.Uehara, K.Kawaguchi and E.Hirota, *J.Chem.Phys.*, 1985,**83**,5479.
50. C.M.Deeley and I.M.Mills, *J.Mol.Spectrosc.*, 1985,**114**,368.
51. T.Suzuki, S.Saito and E.Hirota, *J.Chem.Phys.*, 1985,**83**,6154.
52. M.Larzillière, S.Abed, M.Carré, M.L.Gaillard, J.Lermé and M.Broyer, *Chem.Phys.Letters*, 1985,**119**,55.
53. K.Kawaguchi, C.Yamada and E.Hirota, *J.Chem.Phys.*, 1985,**82**,1174.
54. M.Ochsner, M.Tsuji and J.P.Maier, *Chem.Phys.Letters*, 1985,**115**,373.
55. R.Kakoschke, U.Boesl, J.Hermann and E.W.Schlag, *Chem.Phys.Letters*, 1985,**119**,467.
56. K.Kawaguchi, S.Saito and E.Hirota, *Mol.Phys.*, 1985,**55**,341.
57. J.Werner, W.Seebass, M.Koch, R.F.Curl, W.Urban and J.M.Brown, *Mol.Phys.*, 1985,**56**,453.
58. R.N.Dixon, N.Little, M.J.Trenouth and C.M.Western, *Chem.Phys.Letters*, 1985,**121**,41.
59. J.S.Wells, A.Hinz and A.G.Maki, *J.Mol.Spectrosc.*, 1985,**114**,84.
60. K.Jolma, *J.Mol.Spectrosc.*, 1985,**111**,211.
61. D.C.Benner and C.P.Rinsland, *J.Mol.Spectrosc.*, 1985,**112**,18.
62. K.Jolma, V.-M.Horneman, J.Kauppinen and A.G.Maki, *J.Mol.Spectrosc.*, 1985,**113**,167.

63. N.Hunt, S.C.Foster, J.W.C.Johns and A.R.W.McKellar, *J.Mol.Spectrosc.*, 1985,**111**,42.
64. J.Lindenmayer and H.Jones, *J.Mol.Spectrosc.*, 1985,**110**,65.
65. G.Blanquet, E.Baeten, I.Cauuet, J.Walrand and C.P.Courtoy, *J.Mol.Spectrosc.*, 1985,**112**,55.
66. J.L.Hardwick and C.J.Seliskar, *J.Mol.Spectrosc.*, 1985,**113**,275.
67. H.Jones, J.Lindenmayer and M.Takami, *J.Mol.Spectrosc.*, 1985,**113**,339.
68. D.H.Whiffen, *J.Mol.Spectrosc.*, 1985,**114**,236.
69. K.Kawaguchi, S.saito, E.Hirota and N.Ohashi, *J.Chem.Phys.*, 1985,**82**,4893.
70. C.Chiu and E.A.Cohen, *J.Mol.Spectrosc.*, 1985,**109**,239.
71. H.M.Pickett, E.A.Cohen and J.S.Margolis, *J.Mol.Spectrosc.*, 1985,**110**,186.
72. P.A.Helminger and F.C.de Lucia, *J.Mol.Spectrosc.*, 1985,**111**,66.
73. J.Lindenmayer, H.Jones and V.Typke, *J.Mol.Spectrosc.*, 1985,**110**,357.
74. H.Watanabe, S.Tsuchita and S.Koda, *J.Mol.Spectrosc.*, 1985,**110**,136.
75. R.Kullmer and W.Demtröder, *J.Chem.Phys.*, 1985,**83**,2712.
76. J.Lindenmayer and H.Jones, *J.Mol.Spectrosc.*, 1985,**112**,71.
77. F.J.Comes and D.A.Ramsay, *J.Mol.Spectrosc.*, 1985,**113**,495.
78. C.Degli Esposti, G.Cazzoli and P.G.Favero, *J.Mol.Spectrosc.*, 1985,**109**,229.
79. R.L.DeLeon, D.Prichard and J.S.Muenter, *J.Chem.Phys.*, 1985,**83**,4962.
80. M.W.Crofton, W.A.Kreiner, M.-F.Jagod, B.D.Rehfuss and T.Oka, *J.Chem.Phys.*, 1985,**83**,3702.
81. G.M.Plummer, E.Herbst and F.C.de Lucia, *J.Chem.Phys.*, 1985,**83**,1428.
82. Di-Jia Liu, N.N.Haese and T.Oka, *J.Chem.Phys.*, 1985,**82**,5368.
83. T.J.Sears, P.R.Bunker, P.B.Davies, S.A.Johnson and V.Spirko, *J.Chem.Phys.*, 1985,**83**,2676.
84. M.H.Begemann and R.J.Saykally, *J.Chem.Phys.*, 1985,**82**,3570.
85. L.Fusina, G.Di Lonardo and J.W.C.Johns, *J.Mol.Spectrosc.*, 1985,**112**,211.
86. R.D'Cunha, Š.Urban, K.N.Rao, L.Henry and A.Valentin, *J.Mol.Spectrosc.*, 1985,**111**,352.
87. R.Angstl, H.Finsterhölzl, H.Frunder, D.Illig, D.Papousek, P.Pracna, K.N.Rao, H.W.Schrotter and Š.Urban, *J.Mol.Spectrosc.*, 1985,**114**,454.
88. Š.Urban, R.D'Cunha, K.N.Rao and D.Papousek, *J.Mol.Spectrosc.*, 1985,**111**,361.
89. Š.Urban, P.Misra and K.N.Rao, *J.Mol.Spectrosc.*, 1985,**114**,377.
90. F.Scappini, G.Di Lonardo, A.Guarnieri, A.F.Krupnov and V.P.Kazakov, *J.Mol.Spectrosc.*, 1985,**112**,377.
91. T.Tipton, J.-I.Choe, S.G.Kukolich and R.Hubbard, *J.Mol.Spectrosc.*, 1985,**114**,239.
92. M.Bogey, C.Demuynck and J.L.Destombes, *J.Chem.Phys.*, 1985,**83**,3703.
93. J.C.Petersen and D.A.Ramsay, *Chem.Phys.Letters*, 1985,**118**,31.
94. R.D.Brown, P.D.Godfrey and D.McNaughton, *Chem.Phys.Letters*, 1985,**118**,29.

95. B.Coquart, *Can.J.Phys.*, 1985,**63**,1362.
96. M.A.King, J.P.Maier and M.Ochsner, *J.Chem.Phys.*, 1985,**83**,3181.
97. T.Amano and K.Tanaka, *J.Chem.Phys.*, 1985,**82**,1045.
98. T.Amano and K.Tanaka, *J.Chem.Phys.*, 1985,**83**,3721.
99. C.M.Deeley, I.M.Mills, L.O.Halonen and J.Kauppinen, *Can.J.Phys.*, 1985,**63**,962.
100. T.B.Tang, H.Inouki, S.Saito, C.Yamada and E.Hirota, *Chem.Phys.Letters*, 1985,**116**,83.
101. R.D.Brown, P.D.Godfrey, P.S.Elmes, M.Rodler and L.M.Tack, *J.Amer.Chem.Soc.*, 1985,**107**,4112.
102. R.N.Dixon and P.Johnson, *J.Mol.Spectrosc.*, 1985,**114**,175.
103. S.Yamamoto, T.Nakanaga, H.Takeo, C.Matsumura, M.Takami and K.Kuchitsu, *Chem.Phys.Letters*, 1985,**122**,9.
104. S.Yamamoto, K.Kuchitsu, T.Nakanaga, H.Takeo, C.Matsumura and M.Takami, *J.Chem.Phys.*, 1985,**83**,1444.
105. T.Ishiwata, I.Tanaka, K.Kawaguchi and E.Hirota, *J.Chem.Phys.*, 1985,**82**,2196.
106. W.Lewis-Bevan, A.J.Merer, M.C.L.Gerry, P.B.Davies, A.J.Morton-Jones and P.A.Hamilton, *J.Mol.Spectrosc.*, 1985,**113**,458.
107. W.D.Anderson, M.C.L.Gerry and C.J.Marsden, *J.Mol.Spectrosc.*, 1985,**114**,70.
108. S.Yamamoto, M.Nakata and K.Kuchitsu, *J.Mol.Spectrosc.*, 1985,**112**,173.
109. K.Ohno, H.Matsuura, Y.Endo and E.Hirota, *J.Mol.Spectrosc.*, 1985,**111**,73.
110. R.Anttila and H.Bürger, *J.Mol.Spectrosc.*, 1985,**111**,201.
111. L.Halonen and G.Duxbury, *J.Chem.Phys.*, 1985,**83**,2078.
112. L.Halonen and G.Duxbury, *J.Chem.Phys.*, 1985,**83**,2091.
113. L.Halonen and G.Duxbury, *Chem.Phys.Letters*, 1985,**118**,246.
114. H.Bürger, P.Schulz and S.Cradock, *Z.Naturforsch.*, 1985,**40a**,383.
115. A.Gambi, M.Pedrali, M.Winnewisser and G.Guelachvili, *J.Mol.Spectrosc.*, 1985,**113**,250.
116. R.D.Brown, P.D.Godfrey and B.Kleibomer, *J.Mol.Spectrosc.*, 1985,**114**,257.
117. H.W.Kroto and D.McNaughton, *J.Mol.Spectrosc.*, 1985,**114**,473.
118. R.D.Suenram, F.J.Lovas and W.J.Stevens, *J.Mol.Spectrosc.*, 1985,**112**,482.
119. P.D.Blair, S.Cradock and D.W.H.Rankin, *J.Chem.Soc.Dalton Trans.*, 1985, 755.
120. R.L.DeLeon and J.S.Muenter, *J.Chem.Phys.*, 1985,**82**,1702.
121. C.R.Webster, R.D.May and M.R.Gunson, *Chem.Phys.Letters*, 1985,**121**,429.
122. P.B.Davies and A.J.Morton-Jones, *J.Chem.Soc.Faraday Trans.II*, 1985,1471.
123. P.K.Wahi, V.A.Job and V.B.Kartha, *J.Mol.Spectrosc.*, 1985,**114**,305.
124. H.Bürger, K.Burczyk, H.Hollenstein and M.Quack, *Mol.Phys.*, 1985,**55**,255.
125. G.Tarrago and M.Dang Nhu, *J.Mol.Spectrosc.*, 1985,**111**,425.
126. M.A.King, H.W.Kroto and B.M.Landsberg, *J.Mol.Spectrosc.*, 1985,**113**,1.
127. J.L.Duncan, *J.Mol.Spectrosc.*, 1985,**113**,63.

128. M.Sugie, H.Takeo and C.Matsumura, *J.Mol.Struct.*, 1985,**131**,225.

129. M.C.L.Gerry, W.Lewis-Bevan, A.J.Merer and N.P.C.Westwood, *J.Mol.Spectrosc.*, 1985,**110**,153.

130. J.Nakagawa, K.Yamada, M.Bester and G.Winnewisser, *J.Mol.Spectrsoc.*, 1985,**110**,74.

131. T.Kubota, K.Ueda, T.Tanaka and J.Laane, *J.Mol.Spectrosc.*, 1985,**114**,234.

132. J.R.Durig, A.B.Mohamad, G.M.Attia, Y.S.Li and S.Cradock, *J.Chem.Phys.*, 1985,**83**,10.

133. N.Ohashi and J.T.Hougen, *J.Mol.Spectrosc.*, 1985,**112**,384.

134. K.Ohno, H.Matsuura, D.McNaughton and H.W.Kroto, *J.Mol.Spectrosc.*, 1985,**111**,415.

135. M.C.Durrant, H.W.Kroto, D.McNaughton and J.F.Nixon, *J.Mol.Spectrosc.*, 1985,**109**,8.

136. P.Groner, J.S.Church, Y.S.Li and J.R.Durig, *J.Chem.Phys.*, 1985,**82**,3294.

137. B.J.van der Veken, P.Coppens, R.D.Johnson and J.R.Durig, *J.Chem.Phys.*, 1985,**83**,1517.

138. V.Dana, J.Bordé, L.Henry and A.Valentin, *J.Mol.Spectrosc.*, 1985,**114**,42.

139. S.Yamamoto, M.Takami and K.Kuchitsu, *J.Chem.Phys.*, 1985,**82**,3879.

140. P.Goulet, J.Chanussot and R.Jurek, *J.Mol.Struct.*, 1985,**131**,215.

141. J.P.Aldridge, E.G.Brock, H.Filip, H.Flicker, K.Fox, H.W.Galbraith, R.F.Holland, K.C.Kim, B.J.Krohn, D.W.Magnuson, W.B.Maier II, R.S.MacDowell, C.W.Patterson, W.B.Person, D.F.Smith and G.K.Werner, *J.Chem.Phys.*, 1985,**83**,34.

142. R.S.McDowell, M.J.Reisfeld, N.G.Nereson, B.J.Krohn and C.W.Patterson, *J.Mol.Spectrosc.*, 1985,**113**,243.

4
Characteristic Vibrations of Compounds of Main-group Elements

BY G. DAVIDSON

1 Group I

46 lines of the ν_2 band of H_3^+ have been assigned.[1]

Raman spectra have been obtained for three lithium-containing species isolated in Xe or Kr matrices. Bands were assignable to Li_2 and two forms of Li_3 - one fluxional, one rigid. The latter has ν_s at 353cm^{-1} and δ at 231cm^{-1}.[2]

Far-infrared spectra of solvated Li^+ cations in tetrahydrofuran solutions show only one band at ambient temperatures, e.g. 388 cm^{-1} (^{7}Li), 415cm^{-1} (^{6}Li) for LiCl at 298K.[3] A detailed assignment of internal and external modes was made from infrared and Raman spectra of $^7Li_2GeO_3$ and $^6Li_2GeO_3$ single crystals.[4] The Li-OH stretching and bending modes of a Li-H_2O complex in a Kr matrix were at *ca.* 900, 257cm^{-1} respectively (H/D and ^{6}Li/^{7}Li substitution were used).[5]

Raman spectra of aqueous $NaClO_4$ and $NaClO_4$/NaOH mixtures show that contact ion-pair formation occurs.[6] The F.T. infrared spectrum of bromosodalite shows bands at 200cm^{-1} (Na^+ cage vibrations), 107cm^{-1} (related to the Na-Br unit) and 294cm^{-1} (due to an Si-O-Al mode).[7]

$\nu_3(a_1)$, the K-O stretch, for KNO_2 in an Ar matrix is at 207 cm^{-1}.[8] The Raman spectrum of an aqueous KOH solution in the glassy state shows only one low-wavenumber band specific to the system, at 285cm^{-1}.[9]

2 Group II

Raman spectra of Al_2BeO_4 crystals gave an approximate assignment of vibrational modes.[10] Infrared spectra of $M(ROH)_6X_2$ and $M(ROH)_4X_2$, where M = Mg or Ca, X = Cl or Br, and R = Me or Et, show νMg-O *ca.* 375cm^{-1} (R = Et), 415, 430cm^{-1} (R = Me), νCa-O *ca.* 325cm^{-1} (R = Me or Et).[11] It has been possible to observe νCa-O for the $CaCl_2$/18-crown-6 complex in both infrared and Raman spectra.[12]

ν_3 (the Sr-O stretch) of SrOH is at 522cm^{-1} in the ground state ($\tilde{X}^2\Sigma^+$) and 544 cm^{-1} in the $\tilde{A}^2\Pi$ excited state. Other ground-state wavenumbers are : 361cm^{-1} (δSrOH, ν_2), with $2\nu_2$, $2\nu_3$, $3\nu_3$ at 727, 1048 and 1561cm^{-1} respectively.[13]

3 Group III

3.1 Boron.- The BH unit in boron-diffused silicon gives a band at 1873cm^{-1}.[14] F.T. infrared spectra of $HBCl_2$ and $DBCl_2$ gave values for all fundamentals except ν_3. These were used to calculate a general symmetry force field.[15] νBH is at 2410cm^{-1} in $[HB(O_2CH)_3]^-$, with νCH 2869cm^{-1}, νC=O 1699cm^{-1}, νC-O 1220cm^{-1}. In $[H_2B(O_2CH)_2]^-$, νBH are at 2360, 2335cm^{-1}, 1774cm^{-1} in the d_2 analogue, with νCH 2860cm^{-1}, νC=O 1700cm^{-1}, νC-O 1235cm^{-1}.[16]

Infrared and Raman spectra of $B_6X_6^{2-}$, where X = H, Cl, Br or I, were all assigned assuming O_h symmetry. The internal vibrations of the B_6 cage (750-1250cm^{-1}) are insensitive to the nature of X.[17] νBH for $B_{10}H_{10}CH^-$ salts depends very little on the cation size.[18] Infrared spectroscopy was used to characterise the new borane, $B_{12}H_{16}$, containing B_6 and B_8 units joined along a common B-B edge.[19]

The infrared spectrum of gaseous aminoborane, $H_2N{=}BH_2$, has been observed for the first time. 9 fundamentals above 800cm^{-1} were assigned - Table 1. The rotational fine structure of νB=N (ν_4) at 1337.474cm^{-1} was analysed.[20]

Table 1 Vibrational assignments for aminoborane, H_2NBH_2 (/cm^{-1})

a_1	$\nu_s NH_2$	3451	b_1	NH_2 o.o.p. wag	1005
	$\nu_s BH_2$	2495			
	$\delta_s NH_2$	1625	b_2	$\nu_{as} NH_2$	3533.8
	νBN	1337.474		$\nu_{as} BH_2$	2564
	$\delta_s BH_2$	1225		NH_2 rock	1131

Infrared and Raman spectra for $Me_3CB{\equiv}NCMe_3$ at natural boron isotopic abundance and enriched in ^{10}B showed that assignments could be made assuming a linear chain and C_{3v} symmetry. The predominantly νB≡N mode was at 2072cm^{-1} (^{10}B), 2009cm^{-1} (natural abundance $^{10}B/^{11}B$).[21] νBN in (1) was reassigned to 1039cm^{-1} (cf. an earlier value of 1271cm^{-1}).[22] Several ring modes were assigned from the infrared spectra of (2), where X = Me, Cl or Br.[23]

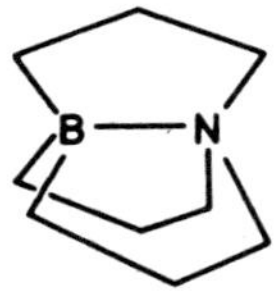

(1)

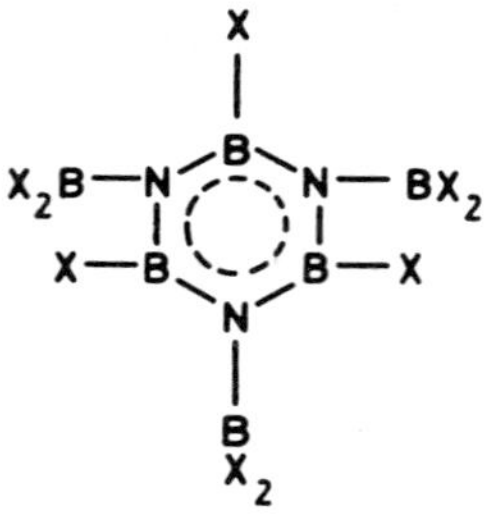

(2)

Isotopic ($^{10}B/^{11}B$) shifts in the infrared and Raman spectra of $B(OH)_4^-$ were shown to be normal, not anomalous as previously suggested.[24] Et_2B-O-Al(^iBu)-Al$^iBu_2$ is said to give νB-O-Al modes 640-670cm^{-1}, νAl-O-Al at 760-820cm^{-1}.[25] Infrared spectra were used to differentiate the cis and trans isomers of 2-isopropoxy-4,5- and -4,6-dimethyl-1,3,2-dioxaborinane.[26]

Infrared spectra of mixed-alkali diborate glasses $(Li,M)_2O.B_2O_3$, where M = Na or K, show non-linear shifts in νB-O and δBOB wavenumbers. There is, therefore, a mild mixed-alkali effect.[27] Raman studies show that boroxol ring concentrations in B_2O_3-GeO_2 and B_2O_3-GeO_2-SiO_2 glasses are dependent on temperature and concentration.[28] Similar data for Na_2O-B_2O_3-SiO_2 glasses containing AgX (X = Cl or Br) reveal the presence of SiO_4 tetrahedra, B_2O_6, B_4O_7 and boroxol ring groups.[29] Infrared spectra of Pb-Zn-borosilicate solder glass showed that BO_3 and BO_4 units were both present at high PbO concentrations,[30] as they were in PbO-B_2O_3-M_2O (M = Na, K) systems.[31] νB-S bands in $Ag_6B_{10}S_{16}$ are at 610, 640, 685, 735 and 760cm^{-1}.[32]

The infrared diode laser spectrum of ^{11}BF gave a vibrational wavenumber of 1378.685(1)cm^{-1}.[33] Infrared and Raman spectra of $MeAsH_2.BX_3$[34] and $Me_2AsH.BX_3$[35], where X = Cl, Br or I, can be assigned in terms of C_s molecular symmetry.

3.2 Aluminium.- There is infrared evidence (a band at 1710cm^{-1}) for the formation of aluminium hydride by the reaction of distilled water with aluminium powder activated by Ga, In or Sn.[36]

Raman data for $LiAlH_4$ and $LiAlD_4$, including data on a new α'-$LiAlD_4$ phase, and infrared and Raman data for crystalline $NaAlH_4$ gave partial vibrational assignments.[37,38] νAl-H bands for $Cp_4Ti_2AlH_3$ showed[39] that there are no terminal Al-H bonds, but $[(C_5Me_5)TiH_2AlH_2]_2$ gave $\nu Al\text{-}H_t$ at 1835cm^{-1}, $\nu Al\text{-}H_{br}$ 1600cm^{-1}.[40]

Infrared spectra of matrix-isolated monomers and dimers of MMe_3, where M = Al or Ga, and $AlEt_3$ were reported. The spectra of the monomeric aluminium compounds had not previously been reported. For MMe_3 the effective point symmetry was $\underline{D}_{3h}$.[41]

All infrared-active fundamentals were assigned for $AlCl_3.NH_3$ and $AlCl_3.ND_3$ vapours (staggered, $\underline{C}_{3v}$ symmetry). There was indirect evidence for an Al-N torsional mode, i.e. no free rotation about the Al-N bond.[42] The skeletal vibrational modes of AlX_3py_3 and $AlX_3(py\text{-}\underline{d}_5)_3$ (X = Cl, Br) show that these molecular adducts have <u>mer</u>-octahedral configurations.[43]

AlO has been isolated in an argon matrix at 18K. The X-state vibrational fundamental is at 975cm^{-1} (confirmed by ^{18}O substitution).[44] The Raman spectrum of Al + Al_2O_3 in an argon matrix contains a band at 472cm^{-1} assigned as the symmetric stretch of a linear, $\underline{D}_{\infty h}$, molecule, Al_2O; cf. Ga_2O, In_2O and Tl_2O, which are bent.[45] Al + O_2 in argon matrices give a cyclic dioxide initially, then a linear dioxide. Al_2O, AlO_2, $Al(O_2)_2$, Al_2O_2 and $Al(O_2)$ were all identified.[46] It was possible to deduce relationships between $\nu_sM\text{-}O$, νCO, the O-O distance etc. from Raman spectra of $M(acac)_3$ (M = Al, Ga or In).[47] νM-O modes were assigned, e.g. νAl-O 492cm^{-1}, in AlL_3, Si_2OL_6, $SnCl_2L_3$ and $SbCl_2L_3$, where L = $CHBr_2CO_2^-$.[48]

A νAlO_6 mode was assigned (580-590cm^{-1}) for $Al(IO_3)_2NO_3.6H_2O$, $Al(IO_3)_3.2HIO_3.6H_2O$ and their deuteriated forms.[49] The infrared spectrum of $BaO.Al_2O_3.7H_2O$ is consistent with the formulation $Ba_{\underline{n}}[Al_2(OH)_8]_{\underline{n}}.3\underline{n}H_2O$, with condensed (edge-sharing) AlO_6 groups.[50] The Al-O unit gives rise to most infrared features from 250-700cm^{-1} in crystalline $MAl(OH)_2CO_3$, where M = Na, K or NH_4. The data were analysed using the factor groups $\underline{D}_{2h}^{28}$ (M = Na, K) or $\underline{D}_{2h}^{17}$ (NH_4).[51] Infrared and/or Raman data were also reported for $CaO\text{-}Al_2O_3\text{-}B_2O_3$ glass (low B_2O_3 region),[52] silica-rich aluminosilicate glasses,[53] $NaAlSiO_4$, $KAlSi_2O_6$, $NaAlSi_3O_8$, $KAlSi_3O_8$ etc.,[54] gibbsite, böhmite, sapphire, ruby etc.,[55] and

$\underline{x}NaAlO_2.(100-\underline{x})GeO_2$ glasses.[56]

Raman spectra of hydrated zeolite A exchanged with Li^+, Na^+, K^+, Tl^+ or NH_4^+ show bands in 3 regions: (a) 300-500cm^{-1}, T-O (Al,Si) deformations, (b) 650-750cm^{-1}, νAl-O, (c) 900-1100cm^{-1}, νSi-O.[57] Raman bands characteristic of oxygen atoms in different environments were also assigned for zeolite, anion framework $Al_{12}Si_{12}O_{46}^{12-}$.[58]

An *ab initio* calculation has been made of the force field and infrared spectrum of AlF_2.[59] $(AlF)_2$ is observed when AlF is trapped in an inert-gas matrix. The infrared spectrum suggests a cyclic, D_{2h}, structure. Matrix reaction of this with O_2 produces $(OAlF)_2$, with an $FAl(\mu-O)_2AlF$, D_{2h}, structure.[60] Raman spectra of $AlCl_3$ solutions in methanol/water mixtures show a strong preference for H_2O coordination of Al^{3+}, but in $AlCl_3$-MeOH there is definite evidence for chloro complexes.[61]

3.3 Gallium.- Skeletal-mode assignments have been given for Ph(RS)GaX, where R = Et, ^{n}Pr, ^{i}Pr, Ph, CH_2Ph, X = Cl, Br or I,[62] and Me(RS)GaX, where R = Me, Et, ^{n}Pr, ^{i}Pr, Ph, CH_2Ph, X = Br or I.[63]

The vibrational spectra of $[Me_2NGaH_2]_2$ show that the dimeric unit is retained in both the crystalline state and in benzene solution.Quite detailed assignments were proposed, e.g. νGa-N at 458 and 498cm^{-1} in benzene solution.[64] Several assignments were proposed for $[Cl_2GaN(R)SiMe_3]_2$, where R = H or Me, e.g. νGaN 530, 502cm^{-1}, νGaCl 408, 375cm^{-1}.[65] νGa-N (480-500cm^{-1}) and νGa-P (280-305cm^{-1}) were assigned in $H_3N.GaX_3$ and $H_3P.GaX_3$, where X = Cl or Br. A normal-coordinate analysis shows very little inter-mode coupling for the NH_3 adduct but significant coupling between νGa-P and νGa-X in the PH_3 case.[66] νGa-P (coupled with PhP modes) is seen at 316, 260, 182cm^{-1} for $PhPH_2.GaCl_3$, $Ph_2PH.GaCl_3$ and $Ph_3P.GaCl_3$ respectively.[67]

Raman spectra of glasses in the $CaO-Ga_2O_3-B_2O_3$ system contained bands assignable to νGaO(of GaO_4) (*ca*.680 cm^{-1}) and Ga-O-Ga bending (*ca*.510cm^{-1}).[68] Some general skeletal-mode assignments were proposed for adducts of $I_2Ga(SMe)$.[69] The Raman spectra of $GaCl_3$ in KCl-$AlCl_3$ melts at 300°C show that $GaCl_4^-$ is present in basic melts (0.26 < pCl < 3.01). In acidic melts (pCl > 3.01) $Ga_2Cl_7^-$ and $GaAlCl_7^-$ are formed.[70]

3.4 Indium and Thallium.- (OEP)In(OAc) has νIn-O bands at 475 and 385cm^{-1}.[71] Raman spectra of aqueous $In(NO_3)_3$ solutions containing $NaNO_3$, HNO_3 or NaOH contained the following bands: $\nu_s InO_6$ of $In(H_2O)_6^{3+}$ 485cm^{-1}, νIn-O of a partly hydroxylated indium aquo-complex 406cm^{-1}, νIn-O of aquo complexes containing one or more coordinated NO_3^- groups 465,425cm^{-1}, $\nu In\text{-}ONO_2$ 270-290cm^{-1}.[72] Tentative assignments have been made to νMO_6 modes from the i.r. spectra of MVO_4, where M = In or Tl.[73] νIn-S lies in the range 260-370cm^{-1} for MIn_2S_4, where M = Ca, Sr or Ba.[74] Infrared spectra have been reported for the In(II) species $X(Y)InInY_2$, where X = Cl, Br or I, Y = Br or I, as the bis-(N,N,N',N'-tetramethylethanediamine) adducts.[75]

The vibrational spectrum of $PhTlF_2$ was assigned in terms of C_{2v} symmetry.[76] The Raman spectra of the reaction products of Tl_2O and SO_2 in an Ar matrix are not simply those of thallium sulphite and disulphite.[77] νTl-X and νTl-L modes have been assigned for the following: TlX_3L_2, where X_3 = Cl_3, $ClBr_2$, ClI_2, Br_2I, BrI_2, I_3, ClBrI, L = pyridine-N-oxide,[78] $TlCl_3L_2$, where L = pyridine-N-oxide or 4-chloropyridine-N-oxide, TlX_3L_2, where X = Cl or Br, L = pyridine-N-oxide or its 4-MeO, 4-Me, 4-CN or 4-NO_2 derivatives,[79] $TlX_3(Ph_3PO)_2$, where X = Cl, Br or I, $TlX_2Cl(Ph_3PO)_2$, where X = Br or I, $TlIBr_2(Ph_3PO)_2$, $TlI_2Br(Ph_3PO)_2$.[80] $TlLX_2$, where X = Cl, Br, L = *o*- or *m*-$C_2H_2B_{10}H_{10}$, have infrared and Raman spectra characteristic of an infinite-chain structure, with Tl_2X_2 four-membered rings.[81]

4 Group IV

4.1 Carbon.- Structural changes in various forms of carbon were followed by monitoring the characteristic Raman band at 1360 cm^{-1}.[82] Infrared absorption studies were used to follow the annealing of irradiated diamonds.[83] The Raman spectrum of diamond has been observed at pressures up to 27GPa. The fundamental phonon line shifts linearly with pressure and is an excellent pressure calibration standard.[84] M.O. calculations have been made of the predicted vibrational wavenumbers of a wide range of molecular ions, including CH_2^+, CH_3^+ and CH_5^+.[85]

Infrared and Raman spectra of $C(CN)_3^-$, $C(CN)_2(^{13}CN)^-$ and $C(CN)_2(C^{15}N)^-$ enabled a detailed vibrational assignment to be made, including the a_2' mode.[86] The carbene :$C(CF_3)_2$ has infrared bands at 1344, 1197, 1157 and 965cm^{-1} in an argon matrix at 12K.[87] Very high-resolution i.r. spectra gave very accurate values for ν_1[88] and ν_3[89] of HCN. FTIR and photoacoustic Raman spectroscopy of $(HCN)_2$ and $(HCN)_3$ in the gas phase gave νCN values for both species.[90] The infrared spectra of $HC^{14}N$ and $HC^{15}N$ in Ar, Kr and Xe matrices and of DCN in a Xe matrix gave assignments of fundamental wavenumbers and anharmonic constants. There was evidence for two trapping sites in Kr and Xe matrices.[91] Diode laser spectroscopy measurements on ν_1 (νC-F) of FCN, $FC^{15}N$ and $F^{13}CN$ showed that the band origins were at 1076.4921(2), 1063.6145(2) and 1070.9537(3)cm^{-1} respectively.[92]

Several papers reported high-resolution i.r. studies on methylene-imine, CH_2=NH.[93-5] The ν_9 band of H_2CN_2 was recorded using FTIR at Doppler-limited resolution. This CH_2 rocking mode is centred at 1124.944cm^{-1}.[96] The paramagnetic species CNO_2^{2-} is formed by the reaction of CO and NO at an MgO surface: νC=O 1624 cm^{-1}, δC=O 800cm^{-1}, ν_{as}CNO 1091cm^{-1}.[97]

The FTIR spectrum of CO in solid N_2 shows that the CO is trapped in a substitutional site.[98] A calculation has been made of the expected infrared spectrum of CO in an Ar matrix.[99] The ν_2 band of HCO^+ has been observed for the first time, using i.r. diode-laser spectroscopy, at 828.2305(9)cm^{-1}.[100] Raman spectra of aqueous solutions of Na^+ salts of $H^{12}C^{16}O_2^-$, $D^{12}C^{16}O_2^-$, $H^{13}C^{16}O_2^-$ and $H^{13}C^{18}O_2^-$ show that the formate ion has no plane of symmetry perpendicular to the ionic plane.[101] The expected vibrational wavenumbers have been calculated for protonated CO_2, i.e. CO_2H^+ and CO_2D^+.[102] The first high-resolution i.r. spectrum of the 2_0^1 band of COF_2, at 963cm^{-1}, has been reported.[103] $CO_2F_2^{2-}$ has been identified, as a triple ion with Tl^+, in Ar matrices (ν_{as}CO at 1382cm^{-1}).[104] Detailed assignments have been proposed from i.r. and Raman spectra of ClC(O)SCl[105] and CF_3OC(O)F.[106]

There have been several high-resolution i.r. studies of various isotopomers of CO_2[107-111] and of OCS.[112,113] Vibration-rotation bands were observed in the i.r. emission spectra of $^{12}C^{32}S$ and $^{12}C^{34}S$ in the $X'\Sigma^+$ state.[114] Mono- and dithiocarbonate anions CO_2S^{2-} and COS_2^{2-} were characterised, using ^{13}CO

and ^{18}O substitution, in an Ar matrix. CO_2S^{2-} has νCO at 1445, 1202 cm^{-1}, νCS at $603cm^{-1}$; COS_2^{2-} νCO at $1506cm^{-1}$, νCS at $606cm^{-1}$.[115] High-resolution i.r. studies have been carried out on CS_2 (several isotopic forms).[116,117] A detailed analysis was made of the gas-phase Raman spectrum of CS_2 in the $\nu_1/2\nu_2$ region.[118] The resonance Raman spectrum of CS_2 excited at 204 and 200nm suggests a bent, symmetrically stretched $^1B_2$ ($^1\Sigma_u^+$) upper state.[119] $C_{\underline{x}}S$ and related species show characteristic Raman bands at 1550, 1360 and $490cm^{-1}$.[120] A detailed assignment has been made of the i.r. and Raman spectra of CCl_3SBr ($\underline{C}_s$ symmetry, νCS $443cm^{-1}$, νSBr $325cm^{-1}$).[121]

Another series of high-resolution i.r. studies was concerned with CH_3F,[122] $CF_3{}^{35}Cl$,[123] CF_3Cl, CF_3Br (natural isotopic abundances),[124] $CF_3{}^{79}Br$, $^{12}CF_3I$, $^{13}CF_3I$,[125] and CF_3I.[126] Raman and far-infrared spectra of solid CCl_4 (monoclinic phase II) were consistent with the molecule having 16 formula units per unit cell.[127] The vibration/rotation i.r. spectrum (ν_3) of $C^{35}Cl_4$ gave a value for the ν_3 band centre of $798.7489(4)cm^{-1}$. An empirical anharmonic force field was used to analyse the $\nu_3/\nu_1+\nu_4$ Fermi resonance.[128]

<u>4.2 Silicon</u>.- SiH was identified by i.r. spectroscopy in a silane plasma discharge.[129] Si and H_2 in solid argon at 10K produce silylene, SiH_2. The fundamental wavenumbers are summarised in Table 2.[130] SiH modes in the i.r. spectrum of a hydrogenated

Table 2 Vibrational assignments for silylene ($/cm^{-1}$)

	SiH_2	SiHD	SiD_2
ν_1	1964.4	1973.3	1426.9
ν_2	994.8	854.3	719.8
ν_3	1973.3	1436.9	1439.1

silicon microcrystalline film showed the presence of SiH, SiH_2 and SiH_3 groups.[131] νSiH overtones to $\Delta\underline{v}$ = 5 were observed for SiH_4, Me_2SiH_2 and Me_3SiH, and to $\Delta\underline{v}$ = 3 for $MeSiH_3$.[132] Fundamental νSiH bands were reported for SiH_3CH_2Cl, $SiHD_2CH_2Cl$, SiH_3Et, $SiHD_2Et$, SiHDEtMe, $SiHEt_3$ etc.[133] High-resolution FTIR spectra of monoisotopic $H_3Si^{79}Br$ (850-1050cm^{-1}) gave accurate values of ν_2 (a_1) 929.75217(6)cm^{-1}, ν_5(e) 946.42569(4)cm^{-1}.[134] New i.r. and Raman data for gaseous $SiH_3CH{=}CH_2$ and $SiD_3CH{=}CH_2$ clarified previous assignments.[135]

I.r. bands due to :$SiMe_2$ were detected in a matrix as a result of u.v. irradiation of $Me_2Si(N_3)_2$.[136] Vibrational assignments were proposed for $CH_{\underline{n}}Cl_{3-\underline{n}}SiX_3$, where X = F or Cl, $\underline{n}$ = 1, 2 or 3.[137-9] Detailed vibrational assignments were given for Me_2SiCl_2, Me_3SiCl, $(Me_3Si)_2O$, $(Me_2SiO)_3$, $(Me_2SiO)_4$ etc.[140] and for $Me_2Si(OMe)_2$.[141] I.r. and Raman spectra of crystalline $SiPh_4$ showed that there is vibrational coupling between the phenyl groups.[142] The vibrational spectrum of (3) confirms the presence of a planar four-membered ring.[143] Detailed assignments were also

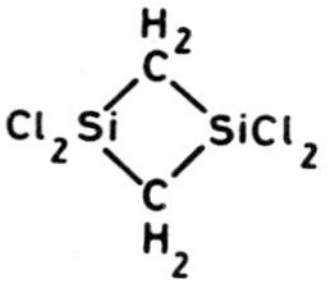

(3)

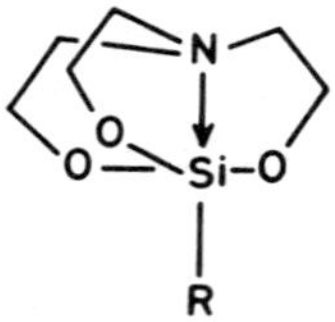

(4)

proposed for $(CR_3)_3SiOOSi(CR_3)_3$ (R = H or D),[144] $(SiMe_3)_2SiX_2$, $(SiMe_3)_2SiMeX$ (X = F, Cl, Br, I, OMe, Ph),[145] $Si_3(CH_3)_8$, $Si_3(CD_3)_8$ and several other isotopic analogues.[146]
I.r. bands were assigned for Si_2N_2O.[147] The Raman spectra of silatranes (4), where R = H, Me, vinyl or F, contain a νSiN band near 600cm^{-1}, confirming the presence of a transannular Si-N bond.[148] Raman spectra of β-sialons, $Si_{6-\underline{x}}Al_{\underline{x}}O_{\underline{x}}N_{8-\underline{x}}$, have νSiN above 800cm^{-1}, δNSiN and δSiNSi of β-Si_3N_4 400-750cm^{-1}.[149]

The photoproducts of SiH_4 + O_3 in an Ar matrix at 17K were shown to be SiO, H_2SiO, H(HO)SiO and $(HO)_2SiO$ (i.r.). HSiOH was formed in high yield by the reaction of SiH_4 and O atoms.[150] MNDO calculations gave predicted vibrational wavenumbers for SiO ($\underline{C}_{\infty v}$), Si_2O_2 ($\underline{D}_{2h}$) and Si_3O_3 ($\underline{D}_{3h}$), in only moderate agreement with experiment.[151] Calculated isotopic shifts for Si_2O_2 were in agreement with cyclic, $\underline{D}_{2h}$, geometry.[152] Several species with Si=O bonds were identified in the i.r. spectra of the reaction products of SiH_4 and O_3 in Ar matrices.[153] νSi-O in $Si(HPO_4)_2$ gives an i.r. band at 680cm^{-1}.[154]

There have been a large number of reports on the i.r. and/or Raman spectra of SiO_2- and silicate-containing species.[155-165] Vibrational assignments for $MSCl_3^-$, where M = Si or Ge, are summarised in Table 3.[166]

Table 3 Vibrational assignments for $MSCl_3^-$ ($/cm^{-1}$)

M =	Si	Ge
νMS	685	506
$\nu_{as}MCl_3$	492	371
$\nu_s MCl_3$	393	357
$\delta SMCl$	225	178
$\delta_s MCl_3$	220	163

Infrared spectra for amorphous Si-F alloys showed bands due to SiF, SiF_3, $(SiF)_2$ and $(SiF_2)_n$ chain units.[167] Matrix-isolation i.r. gave evidence for the formation of 1:1 complexes between SiF_4 or GeF_4 and $(CH_2)_nO$ (n = 2,3,4) and $(CH)_4O$.[168] SiF_5^- as the $^iPr_3NH^+$ salt gives vibrational modes characteristic of D_{3h} symmetry.[169] I.r. and Raman spectra of chlorinated amorphous silicon contain bands near 600cm^{-1} due to $\nu_{as}SiCl_2$, near 540cm^{-1} due to $\nu_s SiCl_2$ and $\nu SiCl$.[170]

4.3 Germanium.- High-resolution i.r. studies were made on GeH_4 in the ν_2[171] and ν_3[172] regions and on $GeH_3D(\nu_2)$ and $GeHD_3(\nu_1)$.[173] Complete vibrational assignments (using isotopic shifts, i.r. band shapes and Raman polarisation measurements) were proposed for CH_3GeH_2CN, CH_3GeD_2CN and CD_3GeH_2CN.[174] All active fundamentals (except the internal torsion) were assigned for $Me_3GeC{\equiv}CH$ and its perdeuterio analogue (C_{3v} symmetry).[175] νGe-C, νC-C and $\nu C{\equiv}C$ were assigned for $Et_3GeC{\equiv}CH$ and $Et_3GeC{\equiv}C{-}C{\equiv}CGeEt_3$.[176] Ge_n ring modes in $(Ph_2Ge)_n$, where n = 4,5 or 6, are in the range 140-330 cm^{-1}, strongly coupled with the mass-sensitive phenyl modes.[177] A Raman band at 274 cm^{-1} is seen for the gas-phase Ge_2 molecule. Other bands were due to a linear trimer (62, 198, 312cm^{-1}) and a tetrahedral tetramer (176, 250, 380cm^{-1}).[178]

Raman spectra of several germanates show that all contain only isolated GeO_4 tetrahedra.[179] νGe-O and νGe-S modes were assigned for (5), where X = O or S, and (6).[180] Raman spectra of glasses with the compositions GeS_3Br and $GeS_{3.25}Br_{0.75}$ contain bands due to Ge^{II}-Br.[181] I.r. and Raman spectra of $Na_4Ge(Se_2)_4$ show that

(5)

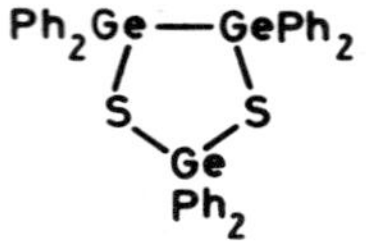

(6)

the symmetry of the anion is not greater than $\underline{C}_2$ and that there is extensive short-range order.[182,183]

4.4 Tin and Lead.- I.r. and Raman spectra of $MeSnCl_3$ can be assigned in terms of isolated, $\underline{C}_{3v}$, molecules.[184] Polarised i.r. spectra of $MeSnCl_3$, however, suggest that the molecules occupy sites of $\underline{C}_s$ symmetry in the crystals.[185] νSn-C, νSn-O, νSn-N and νSn-O-Sn were assigned in R_2SnL_2 and $R_2(L)SnOSn(L)R_2$, where L = N-phthaloyl derivatives of L-leucine, D-alanine or L-phenylalanine, R = Me, Et, ^nBu or $^nC_8H_{17}$.[186] Most skeletal modes were assigned fron i.r. and Raman spectra on $Ph_3MX.L$ adducts, where M = Sn or Pb, X = Cl, Br or I, L = dimethylene urea or 1,3-dimethyl-3,4,5,6-tetrahydro-2(1H)-pyrimidone.[187] Skeletal-mode assignments of $R_2Sn(TDT)$, where R = Me or Ph, H_2TDT = toluenedithiol-3,4, suggest 5-coordinate tin.[188] Me_2Sn(tetramethylenedithiocarbamate) shows νSn-C bands expected for a linear CSnC skeleton (ν_s 478cm^{-1} (Raman), ν_{as} 515cm^{-1}(i.r.)).[189] Skeletal modes were assigned in $Ph_3SnX(L)$, where L = (7), X = Cl, NCS, R = Me, ^tBu, $PhCH_2$, Cy etc.[190] $I(^tBu_2Sn)_4I$ has νSnSn at 119cm^{-1} and νSn-I at 167cm^{-1},

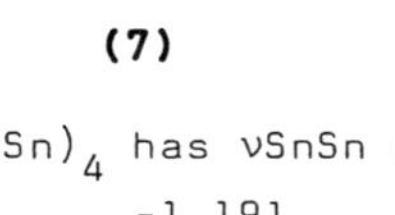

(7)

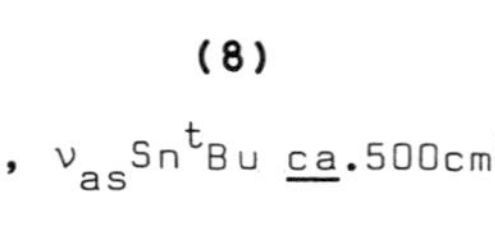

(8)

while $(^tBu_2Sn)_4$ has νSnSn at 125cm^{-1}, $\nu_{as}Sn^tBu$ ca.500cm^{-1} and ν_sSn^tBu ca.250cm^{-1}.[191]

Skeletal-mode assignments were made for 2,2'-azapyridine (AZP) complexes of Sn(IV): SnX_4(AZP), X = Cl, Br, $RSnX_3$(AZP), R = Me, ^tBu, ^noctyl, X = Cl, R_2SnX_2(AZP), R = Me, X = Cl, Br.[192] Transannular Sn...X bonding in (8), where X = O or S, leads to a strong, broad, i.r. band at 280-330cm^{-1}.[193] The infrared spectra of Ph_3SnO_2CR, where R = H, Me, Et, CMe_3, CF_3, CH_2Cl, etc., show that bulky R groups give monomeric molecules with unidentate carboxylates, while small R give polymeric chains with bridging, bidentate carboxylato groups.[194] $^nBuPh_2Sn(S_2CNMe_2)$ has νSn-S at 385cm^{-1} in the i.r. spectrum.[195]

νSn-Cl bands in $Cp_2Mo(Me_2SnCl_2)$ show the presence of terminal Sn-Cl and a secondary bridging bonded unit Sn-Cl...Sn.[196] A correlation has been found between νSn-Cl and the Mössbauer quadrupole splitting for $SnCl_4L_2$, where L = N,N-dimethylacetamide, DMF, DMS or tetrahydrothiophen.[197] Raman and far-infrared spectra of NEt_4^+ salts of $Ph_3SnX_2^-$, $Ph_2SnX_3^-$, Ph_3SnIX^-, $Ph_2SnX_2I^-$ (X = Cl or Br), $Ph_3SnClBr^-$ and $Ph_2SnCl_2Br^-$ gave skeletal vibrational assignments. ν_sSn-Ph were halogen dependent.[198]

$PbO_2(ReO_4)_2$ gives νPb-O *ca.*400cm^{-1}, while $Pb(OH)ReO_4$ gives an i.r. band at 775cm^{-1} assigned as νPb-OH.[199]

5 Group V

5.1 Nitrogen.- The fundamental mode of $a^1\ \Delta$ NH was detected.[200] Several vibrational/rotational lines of HN_2^+ were resolved using a tunable diode-laser, giving an accurate value for the ν_2 band origin.[201] The band origin for ν_6 of HNCO is at 656.287cm^{-1}.[202] I.r. spectra were obtained for matrix-isolated NH_2NO_2, $^{15}NH_2NO_2$, ND_2NO_2 and $^{15}ND_2NO_2$. All 12 fundamentals were assigned and a normal-coordinate analysis performed.[203] High-resolution FTIR spectra of $^{15}NH_3$ gave a value for the ν_2 band origin of 928.456592(19)cm^{-1}.[204] Temperature and pressure effects on Fermi resonance and hydrogen bonding in liquid NH_3 were followed by studying the ν_1 and $2\nu_4$ Raman bands.[205] Raman criteria were established for distinguishing between the $N_2H_5^+$ and $N_2H_6^{2+}$ cations.[206]

The fundamental of $^{14}N_2^-$ was seen at 1822cm^{-1} from the SERS of small potassium particles in a nitrogen matrix.[207] The predicted i.r. spectrum of the van der Waals dimer $(N_2)_2$ has been calculated.[208] FTIR spectra of matrix-isolated $^{14}N_2{}^{16}O$ and $^{15}N^{14}N^{16}O$ monomers, dimers and aggregates gave evidence on differential photolysis effects for these species.[209] I.r. and Raman spectra were reported and assigned for cis-F_3CNNCF_3, obtained for the first time.[210] Very precise measurements have been made of the ν_1 modes of N_3^- and NO_3^- in aqueous solutions. ν_1 values at infinite dilution were calculated to be 1342.7 ± 0.1 cm^{-1} (N_3^-), 1047.4 ± 0.1cm^{-1} (NO_3^-).[211]

Stimulated emission pumping and fluorescence dip spectroscopy characterised the metastable ($L'^2\Phi$) state of NO, ω_e = 1000.0(6) cm^{-1}, $\omega_e x_e$ = 10.90(64)cm^{-1}.[212] The Raman spectrum of $^{15}NO^+$ in β- and β"-alumina was used to obtain a harmonic wavenumber and cubic anharmonic force constant for NO^+.[213] The i.r. spectrum (ν_4 region near 1790cm^{-1}) of $(NO)_2$ has been obtained at temperatures down to 110K.[214] The molecule 'NO^+ $OTeF_5^-$' in the gaseous state exists as covalent molecules O=N-$OTeF_5$, with νN=O 1874cm^{-1}, νN-O 891cm^{-1}.[215] High-resolution FTIR spectra for cis- and trans-HONO, -DONO, -$HO^{15}NO$ and -$DO^{15}NO$ were used to obtain harmonic force fields.[216-8]

The i.r. spectrum of a dilute solution of NO_3^- in KBr at 4K showed significant fine structure to the ν_3 peak (main feature at 1385.58cm^{-1}). Other bands are at 1051.28cm^{-1} (ν_1), 825.60cm^{-1}(ν_2), and 723.90cm^{-1} (ν_4).[219] The $\nu_1(a_1)$ NO_3^- mode in aqueous KNO_3 was used to follow the relationship between solution structure, nucleation and growth. The free-ion wavenumber is at 1049cm^{-1}, with that due to a more ordered form at 1050.5cm^{-1}.[220] The degenerate NO stretch (ν_3) of the NO_3 radical has been studied in the gas phase at very high resolution. The results were consistent with $\underline{D}_{3h}$ symmetry and a $^2A_2'$ ground electronic state. The band origins are at 1492.3929(9)cm^{-1} ($^{14}NO_3$) and 1472.8331(13)cm^{-1} ($^{15}NO_3$).[221]

I.r. and Raman spectra of solid and liquid anhydrous HNO_3 revealed the following species: $(HNO_3)_x$ oligomers (chains), NO_2^+, $(NO_3.3HNO_3)^-$, two monohydrate forms $H_2\bar{O}H^+...ONO_2^-$ and $H_2O...HONO_2$, and free H_3O^+.[222] The Raman spectra of nitric acid solutions up to 250°C showed that νN-OH and δNO_2 of HNO_3 decreased as the temperature and concentration increased, showing reduced hydrogen bonding.[223] Concentrations of HNO_3, NO_3^- and NO_2^+ in solutions of HNO_3 in aqueous H_2SO_4 or $HClO_4$, and their dependence on medium composition, were monitored by Raman spectroscopy.[224]

5.2 Phosphorus.- A high-resolution analysis of the PH_3(g) i.r. spectrum (2153.697-2210.566cm^{-1}) has been reported.[225] I.r. and Raman spectra of $Me_2DP=S$, with data on the PH species, were used to give a detailed analysis and normal-coordinate calculation.[226] The Raman spectrum of gaseous phosphaethyne (HCP) had νCH and νC≡P bands at 3218, 1277cm^{-1} respectively, in good agreement with earlier i.r. data.[227]

(9)

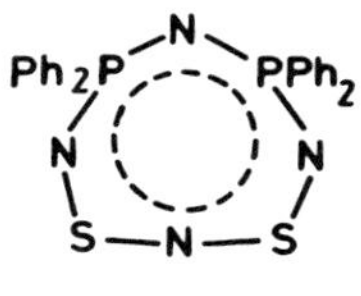

(10)

A normal-coordinate analysis for the A_1' modes of $(PNCl_2)_3$ has been made using $\underline{D}_{3h}$ symmetry.[228] Preresonance Raman spectra from (9) and (10) were reported. (10) gave evidence for excited-state distortion.[229] The Raman spectra of crystalline polyphosphides, MP_{15}, where M = Rb, K or Na, whose structures comprise infinite tubes of P atoms held together by M^+ and/or van der Waal's forces show bands below 140cm^{-1} due to intertube, 200-500cm^{-1} due to intratube vibrations.[230-1]

The dependence of νP=O on the nature of the substituents has been examined for 35 species O=PXYZ, where X,Y,Z = Me, OEt, NHEt, NMe_2 etc.[232] I.r. and Raman spectra of $MeSP(O)Cl_2$ show the presence of <u>gauche</u> and <u>trans</u> forms in liquid or in solutions (more <u>trans</u>). The solid state is all <u>trans</u>.[233] Similar data for $CH_3OP(O)Cl_2$ and $CD_3OP(O)Cl_2$ show that the annealed solid only contains the <u>trans</u> form.[234] A normal-coordinate analysis was reported for $RP(X)F_2$, where R = Me, Et or ^{t}Bu, X = O or S, and $EtP(Se)F_2$.[235] Detailed vibrational assignments were proposed for $(CH_3)_2PO(OCH_3)$, $(CH_3)_2PO(OCD_3)$, $(CD_3)_2PO(OCH_3)$, $(CD_3)_2PO(OCD_3)$ and $(CH_3)PO(O^{13}CH_3)$.[236]

The ν_1 mode of PO_4^{3-} in KH_2PO_4 was observed with ($\underline{yx}$) Raman scattering. It is weak in the $\underline{x}(\underline{yx})\underline{y}$ spectrum but very strong in the $\underline{z}(\underline{yx})\underline{y}$ scattering geometry.[237] A factor-group analysis of the PO_4^{3-} anion was made in $Sn^{II}HPO_4$ and $Pb^{II}HPO_4$.[238] Internal modes of the PO_4 group in vivianite, $Fe_3(PO_4)_2.8H_2O$, were identified in its Raman spectrum.[239] I.r. and Raman spectra of $K_3HP_2O_7.3H_2O$ and the anhydrous salt show the presence of $HP_2O_7^{3-}$ linked by hydrogen bonds to give infinite chains.[240] The i.r. and Raman spectra of Na_2O-SiO_2-P_2O_5 glasses show that the SiP_2O_7 glass structure is very different from the crystalline polymorphs, with no SiO_6 units present.[241] Assignments for the $P_4O_{12}^{4-}$ anion modes, from i.r. and Raman spectra of $M_2P_4O_{12}$, show that the effective symmetry is $\underline{S}_4$ (M = Co) or $\underline{D}_{2d}$ (Cu).[242] Some assignments were also made for the cyclic anion $P_{10}O_{30}^{10-}$, in polycrystalline $Ba_2Zn_3P_{10}O_{30}$.[243]

OCS and PX_3, where X = F or Cl, in an Ar matrix at 20K produce CO and SPX_3 on u.v. photolysis. They were identified by i.r. spectroscopy.[244] Raman and i.r. spectra of $AlPS_4$ were assigned on the basis of $\underline{D}_{2d}$ symmetry for the free molecule.[245] $P_2S_6^{2-}$, as K^+, Cs^+ or Tl^+ salts, has been shown to have a $\underline{D}_{2d}$ structure, like Al_2Cl_6.[246-7] The vibrational spectrum of $P_2S_6^{4-}$, as Cs^+ or Na^+ salts, could be assigned on the basis of perturbed $\underline{D}_{3d}$ symmetry. The predominantly P-P stretching mode was at 553 cm^{-1}.[248] I.r. spectra of P_4S_{10} and related species gave evidence of hydrolysis products.[249]

High-resolution i.r. data were found for PF_5 at low rotational and vibrational temperatures in the region of ν_3 [250] and $\nu_5/2\nu_8$.[251] The i.r. and Raman spectra of PF_6^- in $CsPF_6$ show that the site symmetry is less than $\underline{O}_h$ - a small effect, not revealed by X-ray diffraction.[252] Detailed vibrational assignments were given for PCl_5 as a melt or solution in benzene or CCl_4 and for $PCl_4^+PCl_6^-$ in the solid.[253] The structures of four compounds in the PCl_5-$SnCl_4$ system were elucidated by i.r. and Raman spectroscopy: $(PCl_4)_2SnCl_6$, $(PCl_4)_3(SnCl_6)(PCl_6)$, $(PCl_4)_2(Sn_2Cl_{10})$, and $(PCl_4)(SnCl_5)$. Isotopic splitting for ν_1 of PCl_4^+ was found in a number of cases.[254]

5.3 Arsenic.- νMC and νMS modes were assigned in (11), where M = As, R = H or Me, M = Sb, R = Me.[255] A detailed assignment of internal and external modes was made for Me_2AsOOH, Me_2AsOOD, $(CD_3)_2AsOOH$ and $(CD_3)_2AsOOD$. A phase change was detected at ca. 100K.[256] I.r. and Raman assignments for $CaAs_2O_6$ and $CaSb_2O_6$ were backed up by a normal-coordinate analysis.[257] I.r. and Raman spectra of crystalline and Raman spectra of molten PAs_3S_3 produced assignments for several modes and evidence for an equilibrium in the melt, involving PAs_3S_3, $P_2As_2S_3$ and α-As_4S_3.[258]

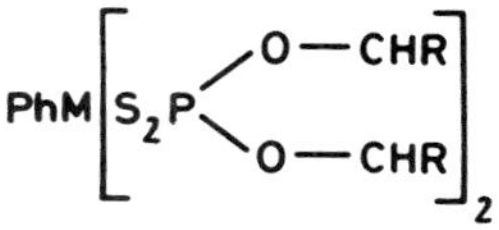

(11)

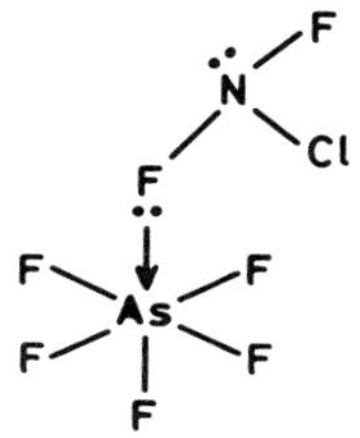

(12)

The Raman spectrum has been reported for a low-temperature 1:1 adduct between NF_2Cl and AsF_5. The data were only consistent with the formulation (12).[259] Some vibrational assignments for $AsClF_4$ were proposed from matrix i.r. and low-temperature Raman spectra - Table 4.[260] Some general assignments were given from the i.r. and Raman spectra of $(As_2SCl_5)^-$, $(As_2SCl_6)^{2-}$,[261] and $(PPh_4)_2(As_2Cl_8)$.[262]

Table 4 Some vibrational assignments for $AsClF_4$ ($/cm^{-1}$)

$\nu_{as}AsF_2$	789	$\nu_sAsF'_2$	620
ν_sAsF_2	745	$\nu AsCl$	463
$\nu_{as}AsF'_2$	720		

5.4. Antimony.- Raman spectra of antimony and bismuth clusters in rare-gas matrices showed that only dimers and tetramers were present. Resonance Raman scattering from dimers in excited electronic states was also seen.[263] $MeSb[N(SiMe_3)_2]_2$ has νSb-C at 499cm^{-1}, ν_sSbN_2 ca.400cm^{-1}. For $MeSbCl_2[N(SiMe_3)_2]_2$, νSb-C is at 539cm^{-1}, ν_sSbN_2 ca.375cm^{-1}, ν_sSbCl_2 258cm^{-1}, and $\nu_{as}SbCl_2$ 288cm^{-1}.[264] I.r. and Raman spectra for $(Me_3SiO)_3X$, where X = Sb, P=O or B, show that ν_sSb-O is sharply different from ν_sB-O or ν_sP-O, due to the interaction of Sb and O electron lone pairs.[265] $(Ph_2SbBrO)_2$ has $\nu Sb\text{-}O_{br}$ bands at 650 and 495cm^{-1}.[266] Bands due to E-O-E' (E,E' = Sb, Te) were assigned from the infrared spectra of $Sb_2Te_2O_9$ and Sb_2TeO_7.[267] (13), where X = O or Se, have νSbS_2 bands in the range 310-350cm^{-1} (infrared and Raman).[268]

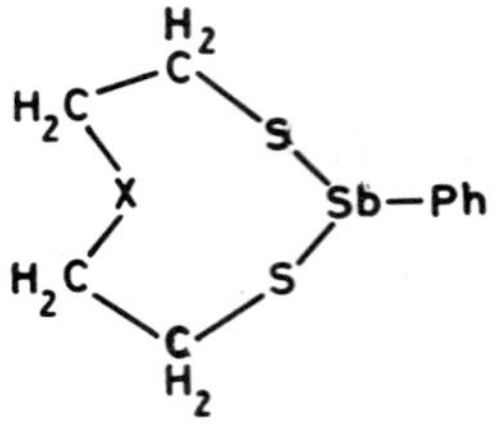

(13)

I.r. and Raman spectra of $SbCl_5L$, where L = benzonitrile or a number of substituted benzonitriles, are all consistent with effective $\underline{C}_{4v}$ symmetry for the $SbCl_5N$ unit.[269,270] νSbCl, with νSb-S contributions, lies in the range 250-320cm^{-1} in $SbCl_3$(L-L), where L-L = N,N'-disubstituted dithiomalonamides.[271] $Sb_2Br_8^{2-}$ as the PPh_4^+ salt has Sb-Br bands in the range 219-254cm^{-1}.[272]

5.5 Bismuth.- I.r. and Raman spectra of bismuth carbonates $(BiO)_2CO_3$ and $(BiO)_4(OH)_2CO_3$ have been reported. The vibrations of $(BiO)_n^{n+}$ give features in the range 350-600cm^{-1}.[273] νBiO modes were assigned from i.r. and Raman spectra of Bi_2MO_{20}, where M = Si or Ge.[274] (14), where X = O or S, R = Me or Ph, have νBiS_2

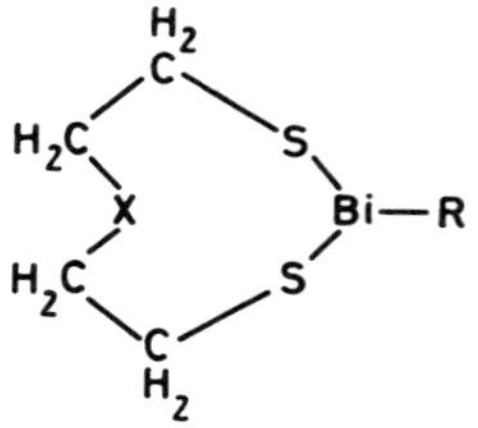

(14)

240-300cm^{-1} and νBi-Me 460-470cm^{-1}.[275] Vibrational spectra of $BiF_5(SbF_5)_n$, where n = 1.5, 2 or 3, are indicative of the presence of SbF_5 and BiF_5 units joined by cis-bridged fluorine atoms.[276]

6 Group VI

6.1 Oxygen.- The i.r. spectrum of $^3\Sigma^-OH^+$ shows that the band origin is at 2956.3724(29)cm^{-1}.[277] The i.r. spectrum of H_2O in liquid alkanes shows that there is no hydrogen bonding, and hence the presence of water aggregates is very unlikely.[278] A high-resolution i.r. spectrum of H_2O vapour has been reported, in the region of the $\nu_2 + 2\nu_3$ and $\nu_1 + \nu_2 + \nu_3$ bands.[279] Energy levels and intensities were calculated for the ν_1 and $2\nu_2$ i.r. bands of HDO.[280] High-resolution i.r. measurements on the ν_2 band of D_2O showed that the band centre was at 1178.3789 ±0.0005cm^{-1}.[281] Hydrogen bonding was studied using FTIR spectra of matrix-isolated $H_2O.CO$, $H_2O_2.CO$, $(H_2O)_2$ and $H_2O.H_2O_2$. Shifts in νOH increase along the series as written above.[282] Several groups have subjected H_3O^+ to high-resolution i.r. studies in various spectral ranges.[283-5]

A theoretical analysis of the i.r. spectra of liquid water and ice suggests that the large shifts in νOH on changing phase are due to anharmonicity rather than intermolecular coupling.[286] Infrared studies on vitrified liquid water suggest that amorphous forms of water made from the liquid and the vapour have different structures.[287] Reassignments of some isolated νOD wavenumbers in ice II have been proposed from Raman data.[288] Pressure and temperature dependences of Raman spectra were evaluated for ice I_h, II, III, V, VI and IX.[289] Inelastic neutron scattering data on polycrystalline ice I_h at 20K are in agreement with the presence of free H_2O molecules.[290] FTIR spectra were obtained for molecular complexes of H_2O, H_2O_2 and HOO$^\bullet$ with CO and CO_2, formed during photo-oxidation of H_2CO, and various isotopic variants. There was evidence for a cyclic dimer of HOO$^\bullet$ (15).[291]

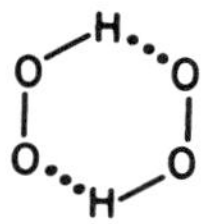

(15)

The ν_1 (O-O stretch) band of the O_2F radical in the gas phase has been observed for the first time; the band centre is at 1489cm^{-1}.[292] A complete i.r. absorption spectrum has been reported for gaseous FOOF, using a very long-pathlength cell. ν_1 is at 1210cm^{-1}, ν_2 630cm^{-1}, ν_3 ca.350cm^{-1}, ν_4 ca.200cm^{-1}, and ν_5 at 614cm^{-1}.[293]

6.2 Sulphur and Selenium Rings and Chains.- I.r. and Raman spectra were reported for the thiosulphate ion $(S_2{}^{16}O_{3-n}{}^{18}O_n)^{2-}$, where n = 0-3. The data were used for a normal-coordinate analysis.[294] The i.r. and Raman spectra of SrS_2, SrS_3-A and SrS_3-B show that both forms of SrS_3 contain distinct S_3^{2-} entities.[295] Raman spectra were used to identify the species in dilute solutions of sulphur in liquid ammonia. There was evidence for S_6^{2-} (Raman bands at 400, 440cm^{-1}).[296]

Fundamental vibrational wavenumbers of eleven possible six-membered ring systems $Se_{\underline{n}}S_{6-\underline{n}}$ and of all isomers of 1,2-Se_2S_5 and 1,2-Se_5S_2 were calculated using the Urey-Bradley force field. The calculated wavenumbers for SeS_5 and 1,2-Se_2S_5 were used to assign Raman spectra of these compounds.[297] A normal-coordinate analysis was performed on S_8, using Raman data collected at 10K.[298] $(PPh_4)_2S_8$ gives νSS bands at 500, 459, 431 and 390cm^{-1}, with a deformation mode near 282cm^{-1} from S_8^{2-}.[299] The i.r. and Raman spectra of S_9O and $S_{10}O$ show that both contain an exocyclic oxygen atom.[300]

6.3 Other Sulphur and Selenium Compounds.- The ν_2, ν_1 and ν_3 bands of $H_2{}^{32}S$, $H_2{}^{33}S$ and $H_2{}^{34}S$ were assigned from the FTIR spectra of H_2S (2150-2950cm^{-1}).[301] Rovibrational spectra of H_2S gas at low density were used to obtain polarisability constants for the molecule.[302] I.r. and Raman spectra suggested an assignment of internal modes for $H_2S_2I_2O_{14}^{4-}$ polyanion.[303]

Measurements of about 460 transitions of the ν_1 band of the unstable S_2O molecule gave a value for the band centre of 1166.454037(80)cm^{-1}.[304] Raman spectra of SO_2 and $S^{18}O_2$ over a wide temperature range reveal coupling between adjacent SO_2 molecules.[305] The band centre of the ν_1 band of $^{32}S^{18}O_2$ was found to be at 1101.16692(5)cm^{-1}.[306] I.r. spectra were obtained for SO_3 and its polymers in inert-gas matrices. The out-of-plane ν_2 mode shows considerable shifts with change of medium.[307] The vibrational spectra of FSO_3^- can be assigned assuming $\underline{C}_{3\underline{v}}$ symmetry (a_1 modes 1070, 740, 580cm^{-1}, e modes 1270, 480, 420cm^{-1}).[308] The Raman spectra of graphite fluorosulphate derivatives are consistent with a model involving charge-compensating back-donation from SO_3F^-.[309]

SO_4^{2-} modes were assigned from i.r. and Raman spectra of $LiRbSO_4$. There was evidence for interionic coupling between SO_4^{2-} ions.[310] A detailed vibrational assignment has been proposed for HSO_5^- in aqueous solution.[311]

Matrix i.r. spectra of 1:1 complexes between SF_4, SOF_2, SO_2F_2 and various amines or oxygen-containing bases all show a shift in νSF to lower wavenumbers compared to the parent fluoride or oxyfluoride. The spectra indicate the Lewis-acidity series: SF_4 > SOF_2 > SO_2F_2.[312] The Raman spectrum of $SCl_3^+AsF_6^-$ gave the following cation wavenumbers: ν_s 520cm^{-1}, ν_{as}531/547cm^{-1}, δ_s 285cm^{-1}, ν_{as} 217cm^{-1}.[313]

Raman spectra showed that SeO_2 solutions in aqueous HBr contained H_2SeO_3, $HSeO_2Br$, $SeOBr_3^-$, $SeBr_5^-$ and $SeBr_6^{2-}$.[314] Assignments were proposed for the hydrogen selenite ion, $HSeO_3^-$, on the basis of $\underline{C}_s$ symmetry.[315] Raman spectra of $RbHSeO_4$ gave evidence for a ferroelectric phase transition.[316] Reasonably complete vibrational assignments were proposed for CF_3SeO_3H, CF_3SeF_5 and CF_3SeF_4Cl.[317] The Raman spectra of SeX_2, where X = Cl or Br, in MeCN solution show that the dihalides are the dominant species but that Se_2Cl_2, $SeCl_4$ and Se_2Br_2 are also present.[318]

6.4 Tellurium.- νTe-C(alkyl) is at 530-540cm^{-1} in (16), where X = Cl, Br or I, R = Et, CH_2Ph, CH_2=CHCH$_2$, CH_2COPh etc.[319]

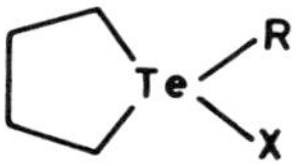

(16)

νTe-O (340-390cm^{-1}) was assigned in Ph_3Te[RCOCH=C(O)R], where R = CH_3, CF_3 or Ph, with unidentate β-diketonates.[320] $BiTeO_5$ gives i.r. bands characteristic of tetrahedral TeO_4 units.[321] The vibrational spectra of crystalline $(Ph_4As)_2TeCl_6$ show that in the orthorhombic form the anion is of $\underline{O}_h$ symmetry but that in the triclinic form the symmetry is lower.[322] I.r. and Raman spectra of orthorhombic Te_2Br gave evidence for low-wavenumber (ca.20cm^{-1}) phonons with very large polarisabilities.[323]

7 Group VII

Transition intensities of natural isotopic HF and HCl vapours were measured at Doppler-limited resolution.[324] Matrix-i.r. spectra of $(HF)_2$, (HF)(DF) and $(DF)_2$ gave the following assignments: $(HF)_2$ 3826cm^{-1}, (HF)(DF) 3831, 2808cm^{-1}, $(DF)_2$ 2803cm^{-1}.[325] I.r. and Raman spectra of liquid HF and of H_2F^+ were used to discuss their structures.[326] Vibrational spectra of KF.nHF (n = 1, 2, 3 or 4) were reported. For n = 3 or 4, bands due to $F_3H_2^-$ were seen. This species is also present in liquid HF solutions of KF.[327]

Co-deposition of HF and N_2 in argon matrices produces an N_2...HF complex with νHF 3881.4cm^{-1}, νNN 2332.1cm^{-1} (confirmed by $^{14}N^{15}N$ and $^{15}N_2$ substitution).[328] Hydrogen-bonded complexes of H_2X (X = S, Se) and HF were detected in Ar matrices at 12K. ν_sHF were at 3652, 3655cm^{-1} for H_2S...HF, H_2Se...HF respectively. HF...HSH was also seen (νHF 3799cm^{-1}), and there was evidence for H_2S...$(HF)_2$.[329] HX adducts (X = F, Cl or Br) with oxygen-containing heterocycles, e.g. $(CH_2)_nO$, where n = 2, 3, 4, or $(CH)_4O$, were characterised in low-temperature matrices. The νHX modes suggest that both HX.L and HX.2L species are formed.[330] Matrix-i.r. spectra of HX (X = F, Cl, Br) complexes with methyl-substituted cyclopropanes show that νHX is always shifted to lower wavenumbers.[331] I.r. spectra of 1:1, 1:2 and 1:3 adduct of RCN (R = Et, ^{i}Pr or ^{t}Bu) with HF, i.e. $RCN(HF)_n$, gave νHF assignments: (for R = ^{t}Bu) n = 1 3522cm^{-1}, n = 2 3630cm^{-1}, n = 3 3678 and 3667cm^{-1}.[332]

F-NCO has been identified in an Ar matrix, with i.r. bands at: 2172cm^{-1} (ν_{as}NCO), 861cm^{-1} (νNF), 695cm^{-1} (δNCO), 646cm^{-1} (δNCO) and 529cm^{-1} (δFNC). ^{15}N substitution was used to confirm the assignments.[333] I.r. spectra of matrix-isolated ClF with various oxygen bases gave evidence for the formation of 1:1 complexes *via* Cl to the O of the base.[334]

I.r. data showed that H_2S...X_2 (X = Cl or Br) complexes were formed in low-temperature Ar or N_2 matrices.[335] I.r. band strengths were measured for the fundamental and first overtone of HCl and DCl in liquid xenon solution.[336] Perturbation of νHX modes was studied in rare-gas hydrogen halide complexes (HX)Y, where X = Cl or Br, Y = Ar, Kr or Xe. The dissociation energy of (HCl)Xe was calculated to be 220cm^{-1}.[337]

A vibrational analysis of most of the $\tilde{A}^2\Sigma^+ \longrightarrow \tilde{X}^2\Pi$ and $\tilde{B}^2\Pi \longrightarrow \tilde{X}^2\Pi$ systems of XCN^+ (X = Cl, Br or I) gave vibrational wavenumbers of these cations in their $\tilde{X}^2\Pi$ states.[338] Several papers report i.r. and Raman data for XO_3^- anions (X = Cl, Br or I) with Sr^{2+} and Ba^{2+} counter-ions.[339-341]

Good agreement was achieved between observed and calculated depolarisation ratios of Q-branches in the continuous resonance Raman spectrum of $^{79}Br_2$.[342] I.r. and Raman data for $MF.2BrF_3$ and $MF.3BrF_3$, where M = Cs or Rb, are consistent with the formulations $M^+Br_2F_7^-$, $M^+Br_3F_{10}^-$ respectively.[343]

Iodosomethane, CH_3IO, has been characterised in an argon matrix at 17K. $\nu I-^{18}O$ is at 688.1cm^{-1}, $\nu I-^{16}O$ at 723.7cm^{-1}. CH_3IO_2 has $\nu_s IO_2$ at 795.9cm^{-1}, $\nu_{as} IO_2$ at 829.8cm^{-1}.[344] The splitting of νICl in ICl complexes with carbonyl bases is explained by the presence of two 1:1 stereoisomeric complexes (one linear, one bent).[345]

8 Group VIII

The FTIR emission spectrum of NeH^+ has been observed. The following values were obtained for (1,0) bands: $^{20}NeH^+$ 2677.8565(5)cm^{-1}, $^{22}NeH^+$ 2672.4960(10)cm^{-1}, $^{20}NeD^+$ 1984.5937(33)cm^{-1}.[346]

The Raman spectra of $Xe^{16}OF_5^-$ and $Xe^{18}OF_5^-$ are consistent with the presence of a stereochemically active lone pair. This gives a distorted octahedral arrangement of the five fluorines and one oxygen. The following assignments were proposed for the ^{16}O molecule: νXe=O 883cm^{-1}, $\nu_s XeF_4$(out of phase) 544cm^{-1}, $\nu_s XeF_4$ 524 cm^{-1}, $\nu_{as} XeF_4$ 473, 468, 435cm^{-1}, νXeF' 420cm^{-1}. Raman data were also given and assigned for $(XeOF_4)_3F^-$.[347]

References

1 J.K.G.Watson, S.C.Foster, A.R.W.McKellar, P.Bernath, T.Amano, F.S.Pan, M.W.Crofton, R.S.Altman and T.Oka, Can. J. Phys., 1984, 62, 1875.
2 M.Moskovits, W.Limm and T.Mejean, Jerusalem Symp. Quantum Chem. Biochem., 1984, 17, 437.
3 S.Chang, M.W.Severson and P.P.Schmidt, J.Phys. Chem., 1985, 89, 2892.
4 B.Orel, M.Klansjek, V.Moiseenko, M.Volnyanskii, N.Le Calvé, B.Pasquier and A.Novak, Phys. Stat. Solidi, B, 1985, 128, 53.
5 L.Manceron, A.Loutellier and J.P.Perchard, Chem. Phys., 1985, 92, 75.
6 A.G.Miller and J.W.Macklin, J. Phys. Chem., 1985, 89, 1193.
7 M.F.Miller, E.B.Bradley and L.T.Todd, Infrared Phys., 1985, 25, 531.
8 A.A.Belyaeva, V.N.Bukhmarina, Yu.B.Predtchenskii and L.D.Shcherba, Zh. Prikl. Spektrosk., 1984, 41, 938.
9 H.Kanno and J.Hiraishi, Bull. Chem. Soc. Japan, 1985, 58, 2701.
10 W.Jia, Q.Zheng, Y.Shang, Y.Wang, Z.Yao, S.He, H.Zhou and L.Ling, Kexue Tongbao (Foreign Lang. Ed.), 1985, 30, 452.
11 S.Halut-Desportes and E.Husson, Spectrochim. Acta, Part A, 1985, 41A, 661.
12 O.A.Raevskii, L.V.Golubev and V.E.Zubareva, Zh. Obshch. Khim. 1984, 54, 2764.
13 C.R.Brazier and P.F.Bernath, J. Mol. Spectrosc., 1985, 114, 163
14 Y.Du, Y.Zhang, G.Qin and S.Weng, Solid State Commun., 1985, 55, 501.
15 K.H.Lee, H.Takeo, S.Kondo and C.Matsumura, Bull. Chem. Soc. Japan, 1985, 58, 1772.
16 G.La Monica, G.A.Ardizzoia, F.Cariati, S.Cenini, and M.Pizzotti, Inorg. Chem., 1985, 24, 3920.

17 W.Preetz and J.Fritze, Z. Naturforsch.,Teil B, 1984, 39B, 1472.
18 S.F.Myasoedov, A.L.Popov, K.A.Solntsev and N.T.Kuznetsov, Russ. J. Inorg. Chem., 1985, 30, 54.
19 C.T.Brewer, R.G.Swisher, E.Sinn and R.N.Grimes, J. Amer. Chem. Soc., 1985, 107, 3558.
20 M.C.L.Gerry, W.Lewis-Bevan, A.J.Merer and N.P.C.Westwood, J. Mol. Spectrosc., 1985, 110, 153.
21 P.Klaeboe, D.Bougeard, B.Schrader, P.Paetzold and C. von Plotho, Spectrochim. Acta, Part A, 1985, 41A, 53.
22 K.J.Lee, P.D.Livant, M.L.McKee and S.D.Worley, J. Amer. Chem. Soc., 1985, 107, 5901.
23 H.Nöth, P.Otto and W.Storch, Chem. Ber., 1985, 118, 3020.
24 N.J.Camp bell, J.Flanagan and W.P.Griffith, J. Chem. Phys., 1985, 83, 3712.
25 L.Synoradzki, H.Bolesławski and J.Lewinski, J. Organometal. Chem., 1985, 284, 1.
26 V.V.Kuznetsov and A.I.Gren, Dopov. Akad. Nauk Ukr. RSR, Ser.B Geol., Khim., Biol. Nauki, 1984, 37. (Chem. Abs., 1985, 102 , 6592).
27 U.Selvaraj and K.J.Rao, Spectrochim. Acta, Part A,1984,40A, 1081.
28 I.N.Chakraborty and R.A.Condrate, Mater. Res. Soc. Symp. Proc., 1985, 41, 113.
29 Y.Jin, J.Shen and X.Chen, Guisuanyan Xuebao, 1985, 13, 164 (Chem. Abs., 1985, 103, 112505).
30 S.C.Chae and C.Y.Kim, Yoop Hakhoechi, 1984, 21, 349 (Chem. Abs., 1985, 103, 57895).
31 R.Oyamada, H.Koga and T.Koike, Yogyo Kyokaishi, 1985,93, 451 (Chem. Abs., 1985, 103, 95268).
32 B.Krebs and H.Diercks, Z. anorg. allg. Chem.,1984, 518, 101.
33 T.Nakanaga, H.Takeo, S.Kondo and C.Matsumura, Chem. Phys.Lett. 1985, 114, 88.
34 J.M.Chehayber and J.E.Drake, Spectrochim. Acta, Part A, 1985, 41A, 1197.
35 J.M.Chehayber and J.E.Drake, Spectrochim. Acta, Part A, 1985, 41A, 1205.
36 R.G.Sarmurzina and V.F.Vozdvizhenskii, Russ. J. Inorg. Chem., 1984, 29, 1224.
37 J.C.Bureau, B.Bonnetot, P.Claudy and H.Eddaoudi, Mater. Res. Bull., 1985, 20, 1147.
38 J.C.Bureau, J.P.Bastide, B.Bonnetot and H.Eddaoudi, Mater. Res. Bull., 1985, 20, 93.
39 E.B.Lobkovskii, G.L.Soloveichik, A.I.Sizov and B.M.Bulychev, J.Organometal. Chem., 1985, 280, 53.
40 V.K.Bel'skii, A.I.Sizov, B.M.Bulychev and G.L.Soloveichik, J.Organometal. Chem., 1985, 280, 67.
41 S.Kvisle and E.Rytter, Spectrochim. Acta, Part A, 1984, 40A, 939.
42 C.E.Sjøgren and E.Rytter, Spectrochim. Acta, Part A, 1985, 41A, 1277.
43 M.Dalibart and J.Derouaolt, Nouv. J. Chim., 1984, 8, 551.
44 S.J.Bares, M.Haak and J.W.Nibler, J. Chem. Phys., 1985, 82, 670.
45 I.V.Ovchinnikov, L.V.Serebrennikov and A.A.Mal'tsev, Zh. Fiz. Khim., 1985, 59, 1558.
46 L.V.Serebrennikov and A.A.Mal'tsev, Vestn. Mosk. Univ., Ser. 2: Khim., 1985, 26, 137.
47 B.A.Kolesov and I.K.Igumenov, Koord. Khim.,1985, 11, 485.

48 J.K.Puri and J.M.Miller, Inorg. Chim. Acta,1985, 101, 135.
49 T.G.Balicheva, G.A.Petrova and V.S.Kasperevich, Russ. J. Inorg. Chem., 1984, 29, 1435.
50 W.Gessner, D.Müller and H.Schubert, Z. anorg. allg. Chem., 1985, 524, 59.
51 C.J.Serna, J.V.Garcia-Ramos and M.J.Pena, Spectrochim. Acta, Part A, 1985, 41A, 697.
52 K.Fukumi, J.Fukunaga, N.Yoshida and M.Ihara, Yogyo Kyokaishi, 1984, 92, 680 (Chem. Abs., 1985, 102, 66200).
53 D.A.McKeown, F.L.Galeener and G.E.Brown, J. Non-Cryst. Solid 1984, 68, 361.
54 S.K.Sharma, J.A.Philpotts and D.W.Matson, J.Non-Cryst. Solids, 1985, 71, 403.
55 F.Pechar, Cryst. Res. Technol., 1985, 20, 239.
56 S.K.Sharma and D.W.Matson, J. Non-Cryst. Solids,1984,69,81.
57 P.K.Dutta and B.Del Barco, J. Phys. Chem., 1985, 89, 1861.
58 P.K.Dutta and B.Del Barco, J. Chem. Soc., Chem. Comm., 1985, 1297.
59 V.G.Solomonik and V.M.Ozerova, Zh. Fiz. Khim.,1984, 58,2610.
60 R.Ahlrichs, Z.Lin and H.Schnöckerl, Z. anorg. allg. Chem., 1984, 519, 155.
61 H.H.Emons, K.Pollmer, G.Kabisch and E.Kalman, Z. Phys. Chem., (Leipzig), 1985, 266, 553.
62 G.G.Hoffmann and H.Meixner, Z. anorg. allg. Chem., 1985, 523, 121.
63 G.G.Hoffmann, Z. Naturforsch., Teil B, 1985, 40B, 335.
64 P.L.Baxter, A.J.Downs, D.W.H.Rankin and H.E.Robertson, J. Chem. Soc., Dalton Trans., 1985, 807.
65 W.R.Nutt, J.A.Anderson, J.D.Odom, M.M.Williamson and B.H.Rubin, Inorg. Chem., 1985, 24, 159.
66 M.J.Taylor and S.Riethmiller, J. Raman Spectrosc.,1984,15,370
67 M.J.Taylor, D.S.Bohle and S.Riethmiller, J.Raman Spectrosc., 1984, 15, 393.
68 X.J.Zhao, J.Fukunaga, N.Yoshida and M.Ihara, Yogyo Kyokaishi, 1985, 93, 708 (Chem. Abs., 1985, 103, 223557).
69 G.G.Hoffmann and P.Resch, J. Organometal. Chem.,1985,295,137.
70 J.H.von Barner, Inorg. Chem., 1985, 24, 1686.
71 P.Cocolios, R.Guilard, D.Bayeul and C.Lecomte, Inorg. Chem., 1985, 24, 2058.
72 G.V.Kozhevnikova and G.Keresztury, Inorg. Chim. Acta, 1985, 98, 59.
73 E.J.Baran and M.E.Escobar, Spectrochim. Acta, Part A, 1985, 41A, 415.
74 B.N.Ivanov-Emin, V.I.Ivlieva, L.A.Filatenko, B.E.Zaitsev, G.Z.Kaziev and M.G.Sarabiya, Russ. J. Inorg. Chem., 1984, 29, 1120.
75 C.Peppe and D.G.Tuck, Can. J. Chem., 1984, 62, 2793.
76 B.P.Asthana and C.M.Pathak, Spectrochim. Acta, Part A, 1985, 41A, 595.
77 L.Peter and B.Meyer, Inorg. Chem., 1985, 24, 3071.
78 M.R.Bermejo, J.Irisarri and M.Gayoso, Synth. React. Inorg. Met.-Org. Chem., 1985, 15, 197.
79 A.Sanchez, J.Bravo, J.S.Casas, J.Sordo and M.Gayoso, An. Quim., Ser. B, 1983, 79, 160.
80 M.R.Bermejo, F.Botana and M.Gayoso, Rev. Port. Quim., 1984, 26, 113.

81 A.Ya.Usyatinskii, V.I.Bregadze, N.N.Godovikov, L.E.Vinogradova, L.A.Leites, A.I.Yanovskii and Yu.T.Struchkov, Izv. Akad. Nauk SSSR, Ser. Khim., 1984, 2009.

82 K.Tamai and M.Nakamizo, Kyushu Kogyo Gijutsu Shikensho, 1984, 2089 (Chem. Abs., 1985, 102, 148550).

83 G.S.Woods, Philos. Mag., B, 1984, 50, 673.

84 H.Boppart, J.van Straaten and I.F.Silvera, Phys. Rev.,B, 1985, 32, 1423.

85 D.J.DeFrees and A.D.McLean, J.Chem. Phys.,1985, 82, 333.

86 K.W.Hipps and A.T.Aplin, J.Phys. Chem., 1985, 89, 5459.

87 A.K.Mal'tsev, P.S.Zuev and O.M.Nefedov, Izv. Akad. Nauk SSSR, Ser. Khim., 1985, 957.

88 P.L.Varghese and R.K.Hanson, J. Quant. Spectrosc. Radiat. Transfer, 1984, 31, 545.

89 E.K.Lyro, P.Shoja-Chaghervand, M.Eliades, M.Danzeiser, S.G.Lieb and J.W.Bevan, Can. J. Chem., 1985, 63, 1870.

90 M.Maroncelli, R.A.Hopkins, J.W.Nibler and T.R.Dyke, J. Chem. Phys., 1985, 83, 2129.

91 A.D.Abbate and C.D.Moore, J. Chem. Phys., 1985, 82, 1255.

92 H.Jones, J.Lindenmayer and M.Takahi, J. Mol. Spectrosc., 1985, 113, 339.

93 L.Halonen and G.Duxbury, J.Chem. Phys., 1985, 83, 2091.

94 L.Halonen and G.Duxbury, J. Chem. Phys.,1985, 83, 2078.

95 L.Halonen and G.Duxbury, Chem. Phys. Lett.,1985, 118, 246.

96 A.Gambi, M.Pedrali, M.Winnewisser and G.Guelachvili, J. Mol. Spectrosc., 1985, 113, 250.

97 E.Garrone, E.Guglielminotti, A.Zecchina and E.Giamello, J. Chem. Soc. Farad. Trans.,I, 1984, 80, 2723.

98 B.Nelander, J. Phys. Chem., 1985, 89, 827.

99 K.Seiti, Bul. Shkencave Nat.,1984, 38, 49 (Chem. Abs., 1985, 103, 61804).

100 K.Kawaguchi, C.Yamada, S.Saito and E.Hirota, J. Chem. Phys., 1985, 82, 1750.

101 E.Spinner, Austral. J. Chem., 1985, 38, 47.

102 M.J.Frisch, H.F.Schaefer and J.S.Binkley, J.Phys. Chem., 1985, 89, 2192.

103 W.Lewis-Bevan, A.J.Merer, M.C.L.Gerry, P.B.Davies, A.J. Morton-Jone and P.A.Hamilton, J. Mol. Spectrosc.,1985,113,458

104 S.J.David and B.S.Ault, Inorg. Chem., 1985, 24, 1238.

105 C.O.Della Vedova, E.H.Cutin, E.L.Varetti and P.J.Aymonino, Can. J. Spectrosc., 1984, 29, 69.

106 E.H.Cutin, C.O.Della Vedova and P.J.Aymonino, An. Asoc. Quim. Argent., 1985, 73, 171.

107 K.Jolma, J. Mol. Spectrosc., 1985, 111, 211.

108 R.A.Toth, Appl. Opt., 1985, 24, 261.

109 M.O.Bulanin and P.V.Granskii, Opt. Spektrosk.,1984, 57, 771.

110 V.M.Devi, C.P.Rinsland and D.C.Brenner, Appl. Opt., 1984, 23, 4067.

111 D.C.Benner and C.P.Rinsland, J. Mol. Spectrosc., 1985, 112, 18.

112 K.Jolma, V.M.Horneman, J.Kauppinen and A.G.Maki, J. Mol. Spectrosc., 1985, 113, 167.

113 K.Tanaka, T.Tanaka and I.Suzuki, J. Chem. Phys.,1985,82,2835.

114 R.J.Winkel, S.P.Davis, R.Pecyner and J.W.Brault, Can. J. Phys., 1984, 62, 1414.

115 S.J.David and B.S.Ault, Inorg. Chem., 1985, 24, 1048.

116 G.Blanquet, E.Baeten, I.Cauuet, J.Walrand and C.P.Courtoy, J. Mol. Spectrosc., 1985, 112, 55.
117 E.Baeten, G.Blanquet, J.Walrand and C.P.Courtoy, Can. J. Phys., 1984, 62, 1286.
118 S,Montero, C.Domingo, F.Wetzel, H.Finsterhölzl and H.W. Schrötter, J. Raman Spectrosc., 1984, 15, 380.
119 R.A.Desiderio, D.P.Gerrity and B.S.Hudson, Chem. Phys. Lett., 1985, 115, 29.
120 C.H.Chang, R.A.Bayerlein and S.S.Chan, Carbon,1984, 22, 393.
121 M.R.Mazzieri, C.O. Della Vedova, E.H.Cutin, E.L.Varetti and P.J.Aymonino, J. Raman Spectrosc, 1985, 16, 242.
122 R.J.Lovell, S.J.Daunt, M.Dakhil, G.W.Halsey and W.E.Blass, J. Mol. Spectrosc., 1985, 110, 175.
123 H.Bürger and R.Grassow, Spectrochim. Acta, Part A,1984,40A, 1111.
124 M.K.Haj-Abdallah, H.Pascher, H.G.Häfele, A.Ruoff and H.Essig, Z. Naturforsch., Teil A, 1985, 40A, 233.
125 H.Bürger, K.Burczyk, H.Hollenstein and M.Quack, Mol. Phys., 1985, 55, 255.
126 K.Burczyk and H.Bürger, Spectrochim. Acta, Part A, 1984, 40A, 929.
127 A.Anderson, H.Basista, B.H.Torrie and W.Y.Zeng, Chem. Phys. Lett., 1985, 117, 240.
128 S.Yamamoto, M.Takami and K.Uchitsu, Reza Kagaku Kenkyu, 1984, 6, 170 (Chem. Abs., 1985, 102, 157134).
129 P.B.Davies, N.A.Isaacs, S.A.Johnson and D.K.Russell, J. Chem. Phys., 1985, 83, 2060.
130 L.Fredin, R.H.Hauge, Z.H.Kafafi and J.L.Margrave, J. Chem. Phys., 1985, 82, 3542.
131 T.Satoh and A.Hiraki, Jpn. J. Appl. Phys.,1985, 24, 491.
132 D.C.McKean, A.R.Morrisson and M.I.Kelly, Chem. Phys. Lett., 1984, 109, 347.
133 D.C.McKean, A.R.Morrisson, I.Torto and M.I.Kelly, Spectrochim. Acta, Part A, 1985, 41A, 25.
134 H.Bürger, H.Beckers and J.Kauppinen, J. Mol. Spectrosc., 1984, 108, 215.
135 V.F.Kalasinsky, S.E.Rodgers and J.A.S.Smith, Spectrochim. Acta, Part A, 1985, 41A, 155.
136 H.Vancik, G.Raabe, M.J.Michalczyk, R.West and J.Michl, J. Amer. Chem. Soc., 1985, 107, 4097.
137 S.V.Sin'ko, Yu.A.Pentin, G.M.Kuramshina, G.S.Gol'din and I.I.Baburina, Vestn. Mosk. Univ., Ser.2:Khim.,1985,26,280.
138 S.V.Sin'ko, Yu.A.Pentin, G.M.Kuramshina, G.S.Gol'din and I.I.Baburina, Vestn. Mosk.Univ.,Ser.2:Khim., 1985, 25, 548.
139 S.V.Sin'ko, Yu.A.Pentin, G.S.Gol'din, I.I.Baburina, G.M. Kuramshina and E.M.Protasov, Vestn. Mosk. Univ., Ser.2:Khim., 1985, 26, 131.
140 A.L.Smith and D.R.Anderson, Appl. Spectrosc.,1984, 38, 822.
141 T.F.Tenisheva and A.N.Lazarev, Zh. Priklad. Spektrosk., 1985, 43, 91.
142 G.Varsányi, B.Zelei, S.Dobos and M.Gál, Spectrochim. Acta, Part A, 1984, 40A, 529.
143 H.Schnöckel, L.Zhengyan, N.Auner, P.Bleckmann, and M. Hinrichsen, J. Mol. Struct., 1985, 127, 1.
144 I.S.Ignat'ev, A.V.Ganyushkin and A.N.Lazarev, Zh. Strukt. Khim., 1985, 26, 67.
145 K.Hassler, Spectrochim. Acta, Part A, 1985, 41A, 729.

146 K.Hassler, Spectrochim. Acta, Part A, 1984, 40A, 775.
147 M.I.Baraton, J.C.Labbe, P.Quintard and G.Roult, Mater. Res. Bull., 1985, 20, 1239.
148 H.Liu, X.Yang, Y.Wu, K.Lu and G.Wu, Guangpuxue Yu Guangpu Fenxi, 1985, 5, 28 (Chem. Abs., 1985, 103, 178317).
149 A.Takase and E.Tani, J. Mater. Sci. Lett., 1984, 3, 1058.
150 R.Withnall and L.Andrews, J.Phys. Chem.,1985, 89, 3261.
151 A.Reklat, Z. Chem., 1985, 25, 257.
152 V.V.Sliznev and V.G.Solomonik, Koord. Khim., 1984, 10, 1199.
153 R.Withnall and L.Andrews, J. Amer. Chem. Soc.,1985,107, 2567
154 A.N.Murashkevich, V.V.Pechkovskii, R.Ya.Mel'nikova, T.A. Barannikova and A.V.Chubarov, Russ. J. Inorg. Chem., 1984, 29, 1722.
155 S.C.Cerukuri, L.D.Pye, I.N.Chakraborty, R.A.Condrate, J.R. Ferraro, B.C.Cornilsen and K.Martin, Spectrosc. Lett., 1985, 18, 123.
156 W.J.McCracken, W.B.Person and L.L.Hench, J. Mater. Sci., 1985, 20, 3853.
157 W.Wojdowski, Acta Phys. Pol.,A, 1985, A67, 487.
158 S.V.Karpov, Ya.I.Ryskin, G.P.Stavitskaya and A.A.Shultin, Opt. Spekrosk., 1985, 58, 709.
159 D.W.Matson and S.K.Sharma, Geochim. Cosmochim. Acta, 1985, 49, 1913.
160 S.K.Sharma, D.W.Matson, J.A.Philpotts and T.L.Roush, J.Non-Cryst. Solids, 1984, 68, 99.
161 D.M.Krol and B.M.J.Smets, Phys. Chem. Glasses,1984,25, 119.
162 G.S.Henderson, G.M.Bancroft, W.E.Fleet and D.J.Rogers, Am. Mineral., 1985, 70, 946.
163 F.Gan, G.Huang and F.Lin, Guangxue Xuebao, 1984, 4, 339. (Chem. Abs., 1985, 102, 14273).
164 G.E.Walrafen, M.S.Hokmabadi, N.C.Holmes, W.J.Nellis and S.Henning, J. Chem. Phys., 1985, 82, 2472.
165 P.K.Dutta and D.C.Shieh, J. Raman Spectrosc., 1985, 16, 312.
166 U.Müller and V.Krug, Z.Naturforsch., Teil B,1985, 40B, 1015.
167 B.K.Agrawal and B.K.Ghosh, J. Phys.,C, 1985, 18, 3897.
168 B.S.Ault, J. Mol. Struct., 1985, 127, 357.
169 V.O.Gel'mbol'dt, L.A.Gavrilova, P.N.Yurkevich and A.A.Ennan, Russ. J. Inorg. Chem., 1985, 30, 910.
170 Z.Wang, H.Han, G.Li and X.Zhao, Bandaoti Xuebao, 1984, 5,478 (Chem. Abs., 1985, 102, 53225).
171 G.V.Vedeneeva, I.I.Zasavitskii, Yu.A.Kuritsyn, E.A.Segirev, O.N.Ulenikov, A.E.Cheglokov and A.P.Shotov, Opt. Spektrosk., 1985, 58, 571.
172 R.D.Schaeffer and R.W.Lovejoy, J. Mol. Spectrosc.,1985,113, 310.
173 K.Ohno, M.Kawamura and H.Matsuura, J. Sci. Hiroshima Univ., Ser. A: Phys, Chem., 1985, 49 (Chem. Abs.,1985, 103, 168961)
174 J.R.Durig, A.B.Mohammad, G.M.Attia, Y.S.Li and S.Cradock, J.Chem. Phys., 1985, 83, 10.
175 A.V.Belyakov, E.T.Bogoradovskii, V.S.Zavgorodnii, V.S. Nikitin, M.V.Polyakova and I.I.Baburina, Spectrochim. Acta, Part A, 1985, 41A, 1269.
176 K.Kamienska-Trela, H.Ilcewicz, H.Baranska and A.Labudzinska, Bull. Pol. Acad. Sci., Chem., 1984, 32, 143.
177 L.Ross and M.Dräger, Z. anorg. allg. Chem., 1984, 515, 141.
178 F.W.Froben and W.Schulze, Surf. Sci., 1985, 156, 765.

179 Yu.K.Voron'ko, A.B.Kudryavtsev, V.V.Osiko, A.A.Sobol and F.M.Spiridonov, Dokl. Akad. Nauk SSSR, 1985, 283, 1333.
180 M.Dräger and K.Häberle, J.Organometal. Chem., 1985, 280, 163.
181 L.Koudelka, M.Pisarčik and V.A.Ananichev, J.Mater.Sci. Lett., 1984, 3, 825.
182 G.Pfaff and A.Feltz, Z. Chem., 1984, 24, 446.
183 A.Feltz and G.Pfaff, J. Non-Cryst. Solids, 1985, 69, 425.
184 M.S.Soliman, M.A.Khattab and A.G.El-Kourashy, Bull. Soc. Chim. Belg., 1985, 94, 87.
185 J.M.Landry and J.E.Katon, J. Mol. Struct., 1984, 125, 253.
186 G.K.Sandhu, R.Gupta, S.S.Sandhu, R.V.Parish and K.Brown, J.Organometal. Chem., 1985, 279, 373.
187 C.T.Aitken and M.Onyszchuk, J.Organometal. Chem., 1985, 295, 149
188 K.Grätz, F.Huber, A.Silvestri, G.Alonzo and R.Barbieri, J.Organometal. Chem., 1985, 290, 41.
189 E,M,Holt, F.A.K.Nasser, A.Wilson and J.J.Zuckerman, Organometallics, 1985, 4, 2073.
190 L.H.Khoo and F.E.Smith, Polyhedron, 1985, 4, 447.
191 S.Adams and M.Dräger, J.Organometal. Chem., 1985, 288, 295.
192 E.Rivarola, A.Silvestri, G.Alonzo, R.Barbieri and R.H.Herber, Inorg. Chim. Acta, 1985, 99, 87.
193 M.Dräger, Z. anorg. allg. Chem., 1985, 527, 169.
194 M.Nadvornik, K.Handlr, J.Holeček and T.Janečkova, Sb. Ved. Pr. Vys. Sk. Chem.-Technol. Pardubice, 1984, 46, 45 (Chem. Abs., 1985, 103, 71427).
195 V.G.K.Das, C.Wei and E.Sinn, J.Organometal. Chem., 1985, 290, 291
196 V.K.Bel'skii, A.N.Protskii, B.M.Bulychev and G.L.Soloveichik, J.Organometal. Chem., 1985, 280, 45.
197 D.Tudela, V.Fernandez and J.D.Tornero, J. Chem. Soc., Dalton Trans., 1985, 1281
198 I.Wharf, R.Cuenca and M.Onyszchuk, Can. J. Spectrosc., 1984, 29, 31.
199 L.L.Zaitseva, A.A.Chizhov, A.A.Vashman and V.E.Samsonov, Russ. J. Iorg. Chem., 1985, 30, 826.
200 J.L.Hall, H.Adams, J.V.V.Kasper, R.F.Curl and F.K.Tittel, J.Opt. Soc. Am.,B, 1985, 2, 781.
201 T.J.Sears, J. Opt. Soc. Am.,B, 1985, 2, 786.
202 L.Fusina, M.Carlotti and B.Carli, Can. J. Phys., 1984, 62, 1452.
203 M.Nonella, R.P.Müller and J.R.Huber, J. Mol. Spectrosc., 1985, 112, 142.
204 R.D'Cunha, S.Urban, K.N.Rao, L.Henry and A.Valentin, J. Mol. Spectrosc., 1985, 111, 352.
205 M.Bradley, T.W.Zerda and J.Jonas, Spectrochim. Acta, Part A, 1984, 40A, 1117.
206 S.Miličev and J.Maček, Spectrochim. Acta, Part A, 1985, 41A, 651
207 W.Schulze, K.P.Charle and U.Kloss, Surf. Sci., 1985, 156, 822.
208 G.Brocks and A.van der Avoird, Mol. Phys., 1985, 55, 11.
209 J.R.Sodeau and R.Withnall, J. Phys. Chem., 1985, 89, 4484.
210 H.Bürger, G.Pawelke and H.Oberhammer, J. Mol. Struct., 1985, 128, 283.
211 K.J.Dean and G.R.Wilkinson, J. Raman Spectrosc., 1985, 16, 22.
212 M.R.Taherian, P.C.Cosby and T.G.Slanger, J. Chem. Phys., 1985, 83, 3878.
213 C.Myers and R.Frech, J. Solid State Chem., 1985, 59, 155.
214 P.Brechignac, S.De Benedictis, N.Halberstadt, B.J.Whittaker, and S.Avrillier, J. Chem. Phys., 1985, 83, 2064.

215 J.S.Thrasher and K.Seppelt, Z.anorg. allg. Chem.,1985,529,85.
216 C.M.Deeley, I.M.Mills, L.O.Halonen and J.Kauppinen, Can. J. Phys., 1985, 63, 962.
217 L.O.Halonen, C.M.Deeley, I.M.Mills and V.M.Horneman, Can. J. Phys., 1984, 62, 1300.
218 C.M.Deeley and I.M.Mills, Mol. Phys., 1985, 54, 23.
219 C.J.H.Schutte, J. Chem. Phys, 1985, 83, 448.
220 P.M.McMahon, K.A.Berglund and M.A.Larson, Process Technol. Proc., 1984, 2, 229.
221 T.Ishiwata, I.Tanaka, K.Kawaguchi and E.Hirota, J. Chem. Phys 1985, 82, 2196.
222 M.H.Herzog-Cance, A.Potier and J.Potier, Can. J. Chem., 1985, 63, 1492.
223 C.I.Ratcliffe and D.E.Irish, Can. J. Chem., 1985, 63,3521.
224 M.Sapoli, A.De Santis, N.C.Marziano, F.Pinna and A.Zingales, J.Phys. Chem., 1985, 89, 2864.
225 R.W.Lovejoy, R.D.Schaeffer, D.L.Frasco, C.C.Chackerian and R.W.Boese, J. Mol. Spectrosc., 1985, 109, 246.
226 I.A.Popova, M.A.Sarukhanov and Yu.Ya.Kharitonov, Russ. J. Inorg. Chem., 1985, 30, 937.
227 V.F.Kalasinsky and M.Pechsiri, J. Raman Spectrosc., 1985,16, 190.
228 I.Lipscitz and M.Family, Polym. Prepr.(Am. Chem. Soc., Div. Polym. Chem.), 1985, 26, 174.
229 Y.Y.Yang and J.I.Zink, Inorg. Chem., 1985, 24, 4012.
230 D.J.Olego, Phys. Rev.,B, 1985, 31, 2230.
231 J.C.Bollinger, G.Yvernault and T.Yvernault, Spectrochim. Acta, Part A, 1985, 41A, 399.
232 J.C.Bollinger, G.Yvernault and T.Yvernault, Spectrochim. Acta, Part A, 1985, 41A, 399.
233 E.A.Filipova, I.I.Vandyukova and R.R.Shagidullin, Zh. Strukt. Khim.,1985, 26, 55.
234 G.H.Pieters, B.J.van der Veken, A.J.Barnes, T.S.Little and J.R.Durig, J.Mol. Struct., 1984, 125, 243.
235 S.A.Katsyuba and R.R.Shagidullin, Zh. Priklad. Spektrosk., 1985, 42, 245.
236 B.J.van der Veken, R.L.Odeurs, M.A.Herman and J.R.Durig, Spectrochim. Acta, Part A, 1984, 40A, 563.
237 Y.Tominaga, M.Tokunaga and I.Tatsuzaki, Solid State Comm., 1985, 54, 979.
238 Ya.Mel'nikova, E.D.Dzyuba, V.V.Pechkovskii, I.E.Maloshonok and A.V.Chubarov, Russ. J. Inorg. Chem., 1984, 29, 1295.
239 B.Piriou and J.F.Poullen, J. Raman Spectrosc.,1984, 15, 343.
240 O.Sarr and L.Diop, Spectrochim. Acta, Part A, 1984, 40A,1011
241 I.N.Chakraborty and R.A.Condrate, Phys. Chem.Glasses, 1985, 26, 68.
242 V.Ramakrishnan, G.Aruldhas and A.Bigotto, Infrared Phys., 1985, 25, 665.
243 C.I.Cabello and E.J.Baran, Spectrochim. Acta, Part A, 1985, 41A, 1359.
244 M.Hawkins, M.J.Almond and A.J.Downs, J. Phys. Chem., 1985, 89, 3326.
245 C.Wibbelmann, B.N.Cyvin and S.J.Cyvin, Z. Naturforsch.,Teil A, 1985, 40A, 909.
246 W.Brockner, R.Becker, B.Eisenmann and H.Schäfer, Z. anorg. allg. Chem., 1985, 520, 51.
247 R.Becker and W.Brockner, Z. Naturforsch., Teil A, 1984, 39A, 1120.

248 S.J.Cyvin, B.N.Cyvin, C.Wibbelmann, R.Becker, W.Brockner and M.Parensen, Z. Naturforsch, Teil A,1985, 40A, 709.
249 J.J.Barieux and M.C.Demarcq, Rev. Chim. Min., 1984, 21,611.
250 M.Takami and H.Kuze, Reza Kagaku Kenkyu, 1984, 6, 167 (Chem. Abs., 1985, 102, 157133).
251 S.Yamamoto, M.Takami and K.Kuchitsu, J. Chem. Phys., 1985, 82, 3879.
252 R.B.English and A.M.Heyns, J. Crystallogr. Spectrosc. Res., 1984, 14, 531.
253 N.A.Chumaevskii, Russ. J. Inorg. Chem., 1984, 29, 1415.
254 J.Shamir, S.Luski, A.Bino, S.Cohen and D.Gibson, Inorg. Chem., 1985, 24, 2301.
255 R.K.Gupta, A.K.Rai, R.C.Mehrotra and V.K.Jain, Inorg. Chim. Acta, 1984, 88, 201.
256 M.Joannidou and F.Fillaux, J. Chim. Phys., 1984, 81, 397.
257 E.Husson, Y.Repelin and M.T.Vandenborre, Spectrochim. Acta, Part A, 1984, 40A, 1017.
258 W.Bues, M.Somer and W.Brockner, Z. anorg. allg. Chem., 1984, 516, 42.
259 K.Christe, W.W.Wilson, C.J.Schack and R.D.Wilson, Inorg. Chem 1985, 24, 303.
260 F.Claus, M.Glaser, V.Wölfel and R.Minkwitz, Z. anorg. allg. Chem., 1984, 517, 207.
261 A.T.Mohammed and U.Müller, Z. anorg. allg. Chem., 1985,523,45
262 A.T.Mohammed and U.Müller, Acta Crystallogr.,1985, C41,329.
263 B.Eberle, H.Sontag and R.Weber, Surf. Sci., 1985, 156, 751.
264 W.Kolondra, W.Schwarz and J.Weidlein, Z. Naturforsch., Teil B, 1985, 40B, 872.
265 V.S.Dernova, I.F.Kovalev, R.G.Kutlubaev, R.G.Mirskov, V.C.Chernova and M.G.Voronkov, Izv. Sib. Otd. Akad. Nauk SSSR Ser. Khim. Nauk, 1984, 88.
266 D.M.Wesolek, D.B.Sowerby and M.J.Begley, J. Organometal. Chem. 1985, 293, C5.
267 J.A.Alonso, A.Castro, A.Jerez, C.Pico and M.A.Veiga, J. Chem. Soc., Dalton Trans., 1985, 2225.
268 H.M.Hoffmann and M.Dräger, J. Organometal. Chem., 1985, 295,. 33.
269 A.Alvarez-Valdes, C.Gomez-Vaamonde, J.R.Masaguer, J.A. Garcia-Vazquez, Z.anorg. allg. Chem., 1985, 523, 227.
270 A.Alvarez-Valdes, J.R.Masaguer and J.A.Garcia-Vazquez, Spectrochim. Acta, Part A, 1984, 40A, 995.
271 J.M.Kisenyi, G.R.Willey and M.G.B.Drew, J. Chem. Soc., Dalton Trans., 1985, 1073.
272 A.T.Mohammed and U.Müller, Z. Naturforsch.,Teil B,1985, 40B, 562.
273 P.Taylor, S.Sunder and V.J.Lopata, Can. J. Chem.,1984,62,2863
274 W.Wojdowski, Phys. Stat. Solidi,B, 1985, 130, 121.
275 M.Dräger and B.M.Schmidt, J. Organometal. Chem.,1985,290,133
276 G.S.H.Chen, J.Passmore, P.Taylor, T.K.Whidden and P.S.White, J. Chem. Soc., Dalton Trans., 1985, 9.
277 M.W.Crofton, R.S.Altman, M.F.Jagod and T.Oka, J.Phys. Chem., 1985, 89, 3614.
278 M.P.Conrad and H.L.Strauss, Biophys. J.,1985, 48, 117.
279 O.K.Voitsekhovskaya, V.P.Kochanov, Yu.S.Makushkin, L.N. Sinitsa, A.M.Solodov, O.N.Sulakshina and V.N.Cherepanov, Opt. Spektrosk., 1985, 58, 1016.
280 A.Perrin, J.M.Flaud and C.Camy-Peyret, J. Mol. Spectrosc., 1985, 112, 153.

281 C.Camy-Peyret, J.M.Flaud, A.Mahmoudi, G.Guelachvili, and J.W.C.Johns, Int. J. Infrared Millimeter Waves, 1985, 6, 199.
282 T.-L.Tso and E.K.C.Lee, J.Phys. Chem., 1985, 89, 1612.
283 P.B.Davies, P.A.Hamilton and S.A.Johnston, J.Opt. Soc. Am,B, 1985, 2, 794.
284 D.J.Liu, N.N.Haese and T.Oka, J. Chem. Phys., 1985, 82, 5368.
285 M.H.Begemann and R.J.Saykally, J.Chem. Phys., 1985, 82, 3570.
286 J.R.Reimers and R.O.Watts, Chem. Phys., 1984, 91, 201.
287 E.Mayer, J.Phys. Chem., 1985, 89, 3474.
288 B.Minceva-Sukarova, W.F.Sherman and G.R.Wilkinson, Soectrochim. Acta, Part A, 1985, 41A, 315.
289 B.Minceva-Sukarova, W.F.Sherman and G.R.Wilkinson, J. Phys,C, 1984, 17, 5833.
290 C.Andreani, P.Bosi, F.Sacchetti and C.K.Loong, J. Chem. Phys. 1985, 83, 750.
291 T.-L.Tso and E.K.C.Lee, J.Phys. Chem., 1985, 89, 1618.
292 K.C.Kim and G.M.Campbell, Chem. Phys. Lett., 1985, 116, 236.
293 K.C.Kim and G.M.Campbell, J. Mol. Sruct., 1985, 129, 263.
294 S.Alvarez, V.Tabacik and J.Casabo, J. Mol. Struct., 1985, 130, 235.
295 H.D.Lutz, B.Oft and K.Wussow, Z. anorg. allg. Chem., 1985, 527, 118.
296 L.Bernard, J.P.Lelieur and G.Lepoutre, Nouv. J. Chim., 1985, 9, 199.
297 R.Laitinen, R.Steudel and E.-M.Strauss, J. Chem. Soc., Dalton Trans., 1985, 1869.
298 M.C.Chang, M.S.Jhon and H.Y.Kim, Bull. Korean Chem. Soc., 1985, 6, 29.
299 B.Czeske and K.Dehnicke, Z. Naturforsch., Teil B, 1985, 40B, 120
300 R.Steudel, T.Sandow and J.Steidel, Z.Naturforsch., Teil B, 1985, 40B, 594.
301 L.Lechuga-Fossat, J.M.Flaud, C.Camy-Peyret and J.W.C.Johns, Can. J. Phys., 1984, 62, 1889.
302 V.Mazzacurati, M.A.Ricci, G.Ruocco, G.Signorelli, M.Nardone, A.De Santis and M.Sampoli, Mol. Phys., 1985, 54, 1229.
303 C.I.Cabello and E.J.Baran, Monatsh., 1985, 116, 591.
304 J.Lindenmayer and H.Jones, J. Mol. Spectrosc., 1985, 112, 71.
305 M.H.Brooker and H.H.Eysel, J. Phys. Chem., 1984, 88, 6201.
306 J.Lindenmayer, H.Jones and V.Typke, J. Mol. Spectrosc., 1985, 110, 357.
307 V.E.Bondybey and J.H.English, J. Mol. Spectrosc., 1985, 109, 221
308 S.Mohan and M.K.A.Imam Khan, Ind. J. Phys,B, 1985, 59B, 288.
309 S.Karunanithy and F.Aubke, J. Chim. Phys., 1984, 81, 827.
310 V.Ramakrishnan, V.U.Nayar and G.Arudhas, Infrared Phys., 1985, 25, 607.
311 E.H.Appelman, L.J.Basile, H.Kim and J.R.Ferraro, Spectrochim. Acta, Part A, 1985, 41A, 1295.
312 C.S.Sass and B.S.Ault, J. Phys. Chem., 1985, 89, 1002.
313 R.Minkwitz, K.Jänichen, H.Prenzel and V.Wölfel, Z. Naturforsch., Teil B, 1985, 40B, 53.
314 J.Milne and P.Lahaie, Inorg. Chem., 1985, 24, 840.
315 Z.Mička, B.Kratochvíl, J.Podlahová and L.Niinistö, Coll. Czech. Chem. Comm., 1985, 50, 1368.
316 A.P.Fedotov, M.E.Smokotin and V.F.Shabanov, Fiz. Tverd. Tela, 1985, 27, 1886.
317 A.Haas and H.U.Weiler, Chem. Ber., 1985, 118, 943.
318 J.Milne, Polyhedron, 1985, 4, 65.
319 A.Z.Al-Rubaie, H.A.Y.Alshirayda, P.Granger and S.Chappelle, J.Organometal. Chem., 1985, 287, 321.

320 B.L.Khandelwal, A.K.Singh, H.B.Singh, K.M.Prasad, N.S. Bhandari and W.R.McWhinnie, J. Organometal. Chem.,1985,291, 185.
321 E.M.Spiridonov, L.A.Demina, V.A.Dolgikh, G.M.Antonyan,A.D. Rakceev, L.V.Bulgar, S.I.Lebedeva and T.N.Chvileva, Dokl. Akad. Nauk SSSR, 1984, 278, 199.
322 J.Milne, E.Philippot and M.Maurin, Rev. Chim. Min., 1984, 21, 749.
323 E.Anastassakis, J.S.Raptis and W.Richter, Phys. Stat. Solidi, B, 1985, 130, 161.
324 A.S.Pine, A.Fried and J.W.Elkins, J. Mol. Spectrosc, 1985, 109, 30.
325 R.D.Hunt and L.Andrews, J.Chem. Phys., 1985, 82, 4442.
326 P.V.Huong, Ann. Chim., 1984, 9, 649.
327 P.V.Huong and B.Desbat, Ann. Chim., 1984, 9, 781.
328 L.Andrews and S.R.Davis, J. Chem. Phys., 1985, 83, 4983.
329 R.T.Arlinghaus and L.Andrews, Inorg. Chem.,1985, 24, 1523.
330 B.S.Ault, J. Mol. Struct., 1985, 127, 343.
331 C.E.Truscott and B.S.Ault, J. Phys. Chem., 1985, 89, 1741.
332 S.R.Davis and L.Andrews, J. Mol. Spectrosc.,1985, 111, 219.
333 K.Gholivand, H.Willner, D.Bielefeldt and A.Haas, Z. Naturforsch., Teil B, 1984, 39B, 1211.
334 N.P.Machara and B.S.Ault, Inorg. Chem., 1985, 24, 4251.
335 U.P.Agarwal, A.J.Barnes and W.J.Orville-Thomas, Can. J. Chem. 1985, 63, 1705.
336 J.T.Knudtson and E.Weitz, J. Chem. Phys., 1985, 83, 927.
337 L.S.Bernstein and J.Wormhoudt, J.Chem. Phys., 1985, 82,4802.
338 J.Fulara, D.Klapstein, R.Kuhn and J.P.Maier, J. Phys. Chem., 1985, 89, 4213.
339 H.D.Lutz, W.Buchmeier, E.Alici and W.Eckers, Z. anorg. allg. Chem., 1985, 529, 46.
340 H.D.Lutz, H.Christian and W.Eckers, Spectrochim. Acta,Part A, 1985, 41A, 637.
341 W.Eckers and H.D.Lutz, Spectrochim. Acta, Part A,1985,41A, 1321.
342 P.Baierl and W.Kiefer, J.Raman Spectrosc., 1984, 15, 360.
343 L.Stein, J. Fluorine Chem., 1985, 27, 249.
344 M.Hawkins and L.Andrews, Inorg. Chem., 1985, 24, 3285.
345 M.Berthelot, M.Helbert and C.Laurence, Can. J. Chem., 1985, 63, 958.
346 R.S.Ram, P.F.Bernath and J.W.Brault, J. Mol. Spectrosc., 1985, 113, 451.
347 J.H.Holloway, V.Kaucic, D.Martin-Rovet, D.R.Russell, G.J.Schrobilgen and H.Selig, Inorg. Chem.,1985, 24, 678.

5
Vibrational Spectra of Transition-element Compounds

BY G. DAVIDSON

1 Detailed Studies

Sc, Ti and V react with H_2O in an Ar matrix at 15K to form non-linear HMOH. Vibrational assignments were made using $^{16}O/^{18}O$ and H/D shifts. Thus, for HScOH, νScH was at 1485.1cm^{-1} ($H_2{}^{16}O$), 1485.0cm^{-1}($H_2{}^{18}O$), 1070.0cm^{-1}(D_2O), νScO 715.8cm^{-1}($H_2{}^{16}O$), 690.4 cm^{-1}($H_2{}^{18}O$), 698.2cm^{-1}(D_2O).[1] Relationships between ν_sM-O, νCO and O-O intraligand distance were established from the Raman spectra of $M(acac)_3$, where M = Sc, V, Cr, Mn, Fe, Co, Rh or Ir. Dative π-bonding was thought to play only a small role in the total bonding.[2] $^{16}O/^{18}O$ isotopic shifts in the vibrational spectrum of VO_4^{3-} were shown to be normal, not anomalous as previously reported.[3]

Photolysis of M/H_2O adducts, where M = Cr, Mn, Fe, Co or Cu, in Ar matrices at 15K produces non-linear HMOH molecules. As in ref. 1, vibrational assignments were made based on isotopic shift data (including, for M = Cu, $^{63}Cu/^{65}Cu$).[4] Prolonged photo-oxidation of $M(CO)_6$, where M = Cr or W, in Ar matrices doped with $C^{16}O^{18}O$, gave (for M = Cr) i.r. bands at 964, 950 and 925cm^{-1} due to ν_3 of $Cr^{16}O_2$, $Cr^{16}O^{18}O$, $Cr^{18}O_2$ respectively and (for M = W) bands at 915, 890, 880 and 870cm^{-1} due to ν_3 of all the isotopic forms of WO_3.[5] Full vibrational assignments were obtained for CrF_6^- and CrF_6^{2-} and used in normal-coordinate analyses.[6] Matrix-isolation i.r. studies on $CrOF_4$ show (in N_2 matrices) that νCr=O(a_1) is at 1027.7cm^{-1} and νCr-F(e) at 746.3/741.6cm^{-1}. Chromium isotopic fine structure and relative intensities suggested a bond angle of about 106° for OCrF.[7]

H/D and $^{16}O/^{18}O$ isotopic studies were used to identify $\nu_{as}Fe_3O$ (600cm^{-1}) and δ_sFe_3O (300cm^{-1}) modes of $Fe_3O(OAc)_6py_3^{2+}$.[8] νCo-C_2 in $Co_4(CO)_{10}(\mu_4-\eta^2-C_2H_2)$ is at 619cm^{-1}(603cm^{-1} on ^{13}C isotopic substitution).[9]

Skeletal-mode assignments in $NiBr_2(H_4L)$, $NiBr(H_4L)_2$ and NiL_2^{6-}, where H_4L = ethylenediphosphinetetraacetic acid, were identified by $^{58}Ni/^{62}Ni$ isotopic shifts.[10] Similar assignments were made for Ni and Cu dithizonates,[11] and dithiocarbamates[12] used $^{58}Ni/^{62}Ni$ and $^{63}Cu/^{65}Cu$ isotopic shifts.

FTIR emission from CuH gave the following ω_e values: 1940.7462(41)cm^{-1} (^{63}CuH), 1940.2740(47)cm^{-1} (^{65}Cu).[13] I.r. and Raman spectra were obtained for $(Me_2S)AuX$ (X = Cl or Br), $(Me_2S)AuCl_3$ and their perdeuteriated derivatives, and skeletal modes were assigned. A normal-coordinate analysis suggested that the Me_2S ligand stabilised the Au-Cl bond trans to itself.[14]

νZn-N in $[Zn(NH_3)_4](ReO_4)_2$ was found by $^{14}N/^{15}N$ substitution.[15] I.r. spectra for $CdAl_{12}O_{19}$, $CdAl_4O_7$, $CdAl_2O_4$ and $CdGa_2O_4$ were interpreted using $^{106}Cd/^{116}Cd$ isotopic substitution.[16] A full assignment of vibrational modes was given from the Raman spectra of HgX_3Y^{2-}, where X ≠ Y = Cl, Br or I.[17]

A high-resolution i.r. study of ν_3 of $^{238}UF_6$ gave a band centre of 627.72387(11)cm^{-1}. The $^{235}U/^{238}U$ isotopic shift was found to be 0.60379 ± 0.00017cm^{-1}.[18]

Resonance Raman Spectra.- Resonance Raman spectra were reported for Ln_2O_3, where Ln = Eu, Dy or Tm.[19] The resonance Raman spectra of lanthanide acetylacetonate derivatives of α,β,γ,δ-tetraphenylporphyrinato show that the M-O stretch(sym.) is near 370cm^{-1}.[20]

Resonance Raman progressions were observed for Cr_2 isolated in solid Ar, Kr or Xe. In Xe the vibrational constants are ω_e 438cm^{-1} and $\omega_e x_e$ 14.5cm^{-1}. Another progression was assigned to a low-lying excited state of Cr_2.[21] The resonance Raman spectrum of (TTP)CrN, where TTP = tetra-p-tolylporphinato, shows no enhancemet of νCr≡N, but the Mn analogue gives a strongly enhanced νMn≡N mode.[22] Resonance Raman data for $Mo_2X_6(H_2O)_2^{2-}$, where X = Cl, Br or I, gave the following ground-state harmonic wavenumbers for the Mo-Mo stretch: 359.5cm^{-1}(Cl), 350.6cm^{-1}(Br) and 340.4cm^{-1}(I).[23]

Detailed excitation profiles were obtained for ν_1 and $2\nu_1$ bands from the resonance Raman spectra of MnO_4^{2-} isomorphously doped into K_2CrO_4 (excitation range 406.7-676.4nm). These show that the excited ($^2T_2$) state has Mn-O 0.035Å longer than in the ground (^2E) state.[24] Resonance Raman results were also given for MnO_4^- in $KClO_4$ at 5K [25] and ReS_4^-.[26] Time-resolved resonance Raman spectra for the emissive excited state of $Re_2Cl_8^{2-}$ showed that νRe-Re in the excited state was at the unexpectedly low wavenumber of 204cm^{-1}.[27]

Raman and resonance Raman spectra for $Fe(WS_4)_2^{3-}$, $Ag(MoS_4)_2^{3-}$ and a variety of related species gave useful information on the Fe (or Ag) $-MS_4$ delocalisation.[28]

The resonace Raman spectrum of Fe(salen)OC_6H_4-4-Me, a model for iron-tyrosinate proteins, gave a feature at $570cm^{-1}$ assigned to the Fe-O stretch, coupled to ring modes.[29] As in previous years, there have been a very large number of resonance Raman studies on iron-containing biomolecules.[30-40] A new development has been the use of picosecond time-resolved experiments to follow very rapid bonding processes in such molecules.[41,42]

I.r. and resonance Raman spectra of $CoX_2Pc(-1)$, where X = Cl or Br, and $CoX_2Pc(2-)$, where X = OH, F, Cl or Br, Pc = phthalocyaninato, were able to give values for $\nu_s CoX(a_{1g})$ and $\nu_{as} CoX(a_{2u})$ wavenumbers.[43]

In addition to a review of Raman and resonance Raman spectra of linear chain complexes (chiefly of Pd and Pt),[44] there have been several new reports for such systems: $[Pd(en)X_2]$-$[Pd(en)X_4]$, where X = Cl, Br, $[Pd(tn)Br_2][Pd(tn)Br_4]$, where tn = 1,3-diaminopropane,[45] $[Pt^{II}(en)_2][Pt^{IV}(en)_2X_2][Pt^{III}_2(H_2P_2O_5)_4X_2]$, where X = Br or I,[46] $[Pt(en)_2][Pt(en)_2Cl_2](ClO_4)_4$,[47] $[Pd^{II}(L-L)_2][Pt^{IV}(L-L)_2Cl_2](ClO_4)_4$, where L-L = en, 1,2- or 1,3-diaminopropane,[48] $K_4[Pt_2(pop)_4].2H_2O$, $K_4[Pt_2(pop)_4X_2].2H_2O$, $K_4[Pt_2(pop)_4X].\underline{n}H_2O$, where pop = $H_2P_2O_5^{2-}$, X = Cl, Br or I.[49]

Bands due mainly to νCu-S(Cys) and νCu-N(His) were identified near 400 and near $270cm^{-1}$ respectively in single-copper blue proteins from a variety of bacteria.[50]

Resonance Raman data for Ag_2 in rare-gas matrices were used to characterise trapping sites. νAg_2 lay in the range $190-203cm^{-1}$.[51]

2 Scandium, Yttrium, and the Lanthanoids

Earlier reference has been made to detailed studies on HScOH [1] and $Sc(acac)_3$,[2] with resonance Raman data for Ln_2O_3 (Ln = Eu, Dy or Tm).[19]

Addition of Sc_2O_3 to sodium silicate glasses gave new Raman bands assignable to Sc^{3+} ions coordinated by SiO_4^{4-} tetrahedra.[52]

The vibrational spectra of $\underline{c}$-type lanthanide oxides Ln_2O_3 (Ln = La, Gd, Sc) were assigned with the aid of a calculation of the spectrum using a polymer-chain approximation.[53] νM-O' was assigned for Ln(TPP)acac - the wavenumber was reduced by the effects of TPP coordination (TPP = tetraphenylporphinato).[54] The i.r. spectra of $Ln_2M_2O_7$ were assigned using $^{46}Ti/^{50}Ti$ and $^{98}Ru/^{104}Ru$ isotopic substitution (Ln = Gd, Lu, M = Ti, Ru).[55] I.r. and Raman data, with some assignments, were presented for

$^{40}CaLa_4Ti_4O_{15}$, $^{44}CaLa_4Ti_4O_{15}$, $SrLa_4Ti_4O_{15}$ and related species.[56] Normal-coordinate analyses were performed on $Ln_2Cu_2O_5$ (where Ln = Tb, Dy, Ho, Er, Tm, Yb, Lu, Y, Sc).[57] I.r. and Raman spectra of $NaLa(MoO_4)_2$ were interpreted on the basis of $\underline{C}_{4h}$ symmetry.[58] νLn-O in LnP_5O_{14}, where Ln= La, Ce, Pr, Nd, Sm, Eu, Gd, Tb, Dy, Ho, Y, Er, Tm, Yb or Lu, increased linearly with the atomic number of Ln.[59]

I.r. and Raman spectra of samples from the $LaCl_3$-$CuCl_2$ system show that branched, polymeric chains are formed. In the system $LaCl_3$-$CuCl_2$-KCl there was evidence for polymeric chains containing $CuCl_5$ ($\underline{C}_{4v}$) and $LaCl_6$ ($\underline{O}_h$).[60] I.r. spectra were obtained for GdI_3, HoI_3, LuI_3, $GdBr_3$ and $LuBr_3$ in inert-gas matrices.[61]

3 Titanium, Zirconium, and Hafnium

Reference has already been made to results on HTiOH,[1] $Ln_2Ti_2O_7$ (Ln = Gd or Lu),[55] and $CaLa_4Ti_4O_{15}$.[56]

There is Raman evidence for the existence of $Ti=O^{2+}$ in acidic aqueous solutions, i.e. νTi=O is at 975cm^{-1} in 0.08M $TiO(ClO_4)_2$ in 2M $HClO_4$.[62] $K[Ti(O_2)F_3].3H_2O$ has νTiO_2 at 530 and 610cm^{-1}.[63] The i.r. spectrum of K_2TiOF_4 contains a band at 810cm^{-1}, suggesting that there are infinite chains -Ti-O-Ti-O-Ti-O- in the structure.[64] ν_{as}Ti-O-Ti is at 820cm^{-1} in the i.r. spectrum of $[PcTi-O-TiPc]^{2+}$, where Pc = phthalocyaninato.[65]

ν_3 of gaseous TiF_4 has been observed in its i.r. spectrum at 772cm^{-1}, completing the vibrational assignment for this molecule.[66] Phase transitions in $MTiF_6.6H_2O$, where M = Mn or Zn, and their deuterio analogues were observed from changes in ν_s of TiF_6 with temperature.[67] νMCl modes were assigned as follows for trans-$MCl_2(dmpe)_2$: M = Ti 320cm^{-1}, V 312cm^{-1}, Cr 338cm^{-1}.[68] There is i.r. evidence for the dimerisation of $TiCl_4$ in argon matrices. The Ti_2Cl_8 seems to be a doubly bridged system of $\underline{C}_{2h}$ symmetry, with $\nu_{ax}TiCl_t$ (b_u) 544cm^{-1}, $\nu_{eq(as)}TiCl_t$ (a_u) 456cm^{-1}, and $\nu_{eq(s)}TiCl_t$ (b_u) 390cm^{-1}.[69] The Raman spectra of $A_2A'TiCl_6$, where A, A' = alkali-metal ions, at 77K show splitting of the t_{2g} mode due to static $\underline{D}_{4h}$ distortion.[70]

The FTIR spectrum of H_2 adsorbed on to ZrO_2 at room temperature contains bands at 1562 and 1371cm^{-1} due to νZrH, νZrHZr respectively, showing that dissociation of the H_2 has occurred.[71]

$M[ZrO(NCS)_3].2H_2O$, where M = Cs^+ or pyH^+, have i.r. bands at 916 and 615cm^{-1} assigned to two different bridged oxo-zirconium modes in an oligomeric structure.[72] Raman spectra of BaF_2-ZrF_4-FeF_2 glasses show that increasing the fraction of BaF_2 leads to an increase in the coordination number of Zr from 6 to 7.[73] The Raman spectra of melts containing $ZrCl_6^{2-}$, $HfCl_6^{2-}$, $NbCl_6^-$ and $TaCl_6^-$ are consistent with regular octahedral geometry for the anions. The symmetric stretch shifts to higher wavenumber as the size of the cation increases.[74] M-O stretches were assigned for ML_4, where M = Zr, Th or U, L = $CHBr_2CO_2^-$.[75]

Other species for which vibrational data have appeared are: Ga_2TiO_5,[76] $CpTiCl_2(NHR)$, where R = Et, ^{i}Pr, ^{t}Bu or Ph,[77] and $Cp_2Ti(1\text{-nor})Cl$, $Cp_2Ti(1\text{-nor})_2$, $CpTi(1\text{-nor})_3$, where 1-nor = 1-norbornyl.[78]

4 Vanadium, Niobium, and Tantalum

Earlier reference was made to HVOH,[1] $V(acac)_3$,[2] and VO_4^{3-}.[3]

Isotopic labelling experiments on $(C_5Me_5)_2V{\equiv}NPh$ show that the VN(nitrene) stretch is at 934cm^{-1} (923cm^{-1} for ^{15}N), correcting an earlier assignment.[79] The high wavenumber of the VN stretch in $Me_3Si\text{-}N{\equiv}VCl_3$ (1104cm^{-1}) was ascribed to coupling with the Si-N stretch.[80] νM=N modes were assigned in $X_3M{=}N\text{-}TeF_5$ (M = V, X = F or Cl) and $X_4M{=}N\text{-}TeF_5$ (M = Mo or W, X = F or Cl).[81] Skeletal-mode assignments for $[VCl_2(N_3S_2)]_\infty$, $[VCl_2(N_3S_2).NSCl]_2$ etc. include νVN_2V near 300cm^{-1}.[82]

νV=O in $VO(SO_3CF_3)_2$ (1025cm^{-1}) and $VO(SO_3CF_3)_3$ (1005cm^{-1}) show that there is no polymerisation via V=O.[83] $VO(RNCS_2)_2$, where R = di-Et-, di-Pr, di-iso-Pr etc., have νV=O 950-1000cm^{-1}, ν_{as}V-S 390-370cm^{-1}, and ν_sV-S 340-380cm^{-1}. A correlation was found between νV=O and the $b_2 \rightarrow e_\pi^*$ band.[84] A definitive vibrational assignment was proposed for $CrVO_4$ in terms of a site-symmetry analysis.[85] νVO_4 mode assignments for MVO_4 (M = In or Tl) suggest that the VO_4 groups are not completely isolated.[86,87]

I.r. data for $2MO.V_2O_5$, where M = Mg, Ca, Sr, Ba, Zn, Cd, Co, Cu or Mn, were used to find the VOV angle: it seemed that for M = Ca, Sr and Co VOV was linear in both crystalline and amorphous states, for M = Cd, Cu and Mn it is linear in both states, while for M = Mg or Zn VOV is bent in the crystal but linear in the amorphous state.[88] $V(O_2)_3^-$ has νVO_2 bands at 600 and 530cm^{-1}.[89] νV=O is at 980cm^{-1} in $VO(O_2)IDA^-$, where IDA = $C_4H_5NO_4^{2-}$.[90] νV=O wavenumbers were also assigned for $VOAsO_4.2H_2O$, $Li_{0.5}VOAsO_4.2H_2O$, and $LiVOAsO_4.2H_2O$.[91]

νVX modes were assigned in $Cp^*_2VX_2$, where Cp^* = C_5H_4Me, C_5HMe_4 or C_5Me_5, X_2 = Cl_2, Br_2, I_2, IBr or ICl.[92] νV-Cl modes in $[(N_2S_2)(VCl_5)_2]^{2-}$ are at 390, 355 and 331cm^{-1}, as expected for six-coordinate vanadium.[93]

Other vanadium species for which vibrational data were given are: V_2O_5,[94] $M[VO(O_2)_2].M_2SO_4$ (M = NH_4, Na or K),[95] $HV_2O_{11}^{3-}$,[96] $KVO(SO_4)_2$, $V_2O_3(SO_4)_2$,[97] $M[VO_2SeO_4].4H_2O$,[98] VO(acac)L, where L = py or substituted pyridine,[99] VOL_2 (L = dibenzoylmethane), $VOL_2(L')$ (L' = py or substituted pyridines),[100] $V_2O_5.\underline{n}H_2O$ xerogels,[101] $V_2O_5.\underline{x}H_2O$,[102] $M_2V_{12-\underline{x}}Mo_{\underline{x}}O_{\underline{y}}.\underline{n}H_2O$, where M = alkali metal, 0 < $\underline{x}$ < 3, 30.7 < $\underline{y}$ < 32.2, 0 < $\underline{n}$ < 11.2.[103]

νNb-H and νRh-H were identified in H_2-reduced rhodium/niobium oxide/silica catalysts.[104] The Raman and i.r. spectra of $NbF_5(O_2)^{2-}$ were assigned in terms of $\underline{C}_{2v}$ symmetry. The symmetric and antisymmetric Nb-O stretches were near 880, 900cm^{-1} respectively.[105] The i.r. spectra of $NbOX_3$, NbX_5 and NbX_5/Nb_2O_5 mixtures (X = Cl, Br or I) in Ar and N_2 matrices were assigned in terms of monomers and, for NbX_5, polymers.[106] νNb=O was at 940cm^{-1} in both $[NbOCl_4(OH_2)]^-$ and $[NbOCl_4(O_2PCl_2)]^{2-}$, with the νNb-O mode ca.240cm^{-1}.[107]

The i.r. spectra of $MW_5O_{18}S^{3-}$, where M = Nb or Ta, are very similar, except for νNb=S 510cm^{-1}, νTa=S 481cm^{-1}. The data were consistent with anion geometries of $\underline{C}_{4v}$, derived from $MW_5O_{19}^{3-}$ by replacement of one terminal M=O by M=S.[108] νM=S and νM-S were assigned in $M(S_2CNEt_2)_3S$, e.g. for M = Nb 498, 360cm^{-1} respectively.[109] Skeletal-mode assignments (not very specific) were given for $Nb_2X_4S_3$, where X = Cl or Br.[110] Earlier reference has been made to results on $NbCl_6^-$ and $TaCl_6^-$.[75]

νTaO is at 412cm^{-1} and $\nu TaCl_{br}$ at 326cm^{-1} in a number of salts containing $(Ta_6Cl_{12})(H_2O)_6^{3+}$.[111] I.r. data were reported for $MTaF_6$, where M = Li, Na, K, Rb or Cs.[112]

5 Chromium, Molybdenum, and Tungsten

A new metastable species of Cr_2 has been identified in matrices. This has ω_e at 78.6cm^{-1}, ω_ex_e = 0.4cm^{-1}.[113] Vibrational fine structure for the quadruply bonded dimers $M_2(mhp)_4$, where M = Cr, Mo, mhp = 2-methyl-6-hydoxypyridine, was analysed. For Cr no fine structure based on CrCr stretching was seen, but for M = Mo a vibrational progression was seen based on νMoMo of 380cm^{-1} (solid) or 344cm^{-1} (solution).[114] νCr-CO and νCr-CN were assigned in a vibrational study of $Cr(CO)_5(CNCOPh)$.[115] νCrC is at 1342 for $Br(OC)_4Cr{\equiv}CCH_3$ and 1299cm^{-1} in the CD_3 analogue. The shift on deuteriation shows that there is significant coupling between the CrC stretch and CH_3 modes.[116]

νCrN bands are at 490, 470 and 440cm^{-1}, with νCrCl at 330cm^{-1} in the i.r. spectrum of trans-$Cr(NH_3)_4(H_2O)Cl^{2+}$.[117] νCrP is at 285cm^{-1} in $[RCr(CO)_2(PR'_3)]_2$, where R = C_5H_4Me, R' = Ph, OMe or OEt.[118] νSb-M modes were assigned as follows for $MeC(CH_2Sb)_3M(CO)_5$ where M = Cr (228cm^{-1}), Mo (225cm^{-1}) or W (225cm^{-1}).[119]

The inelastic neutron scattering spectrum of chromous acid contains a band due to ν_sOCrO at 759cm^{-1}.[120] The Raman spectra of $CrO_4^{2-}.M^{2+}$, where M = Ni or Co, in a KBr crystal, show that there are two non-equivalent complexes, both with C_{3v} symmetry.[121] A definitive set of vibrational wavenumbers has been proposed for CrO_4^{2-} in a variety of crystalline compounds.[122] Skeletal-mode assignments were given for $[Cr_nFe_{3-n}(\mu_3\text{-}O)(glycine)_6(H_2O)_3].(NO_3)_7.xH_2O$, where n = 1, 2 or 3,[123] and for $[Cr_{3-n}Fe_nO(C_{14}H_{29}COO)_6(H_2O)_3].NO_3$, where n = 0-3.[124] ν_{as}Cr-O-Fe bands have been assigned for a range of μ-oxo-chromium iron porphyrin complexes (839-849cm^{-1}).[125] (TPP)CrOCr(TPP) has ν_{as}CrOCr at 860cm^{-1}, while νCr=O in OCr(TPP) is at 1020cm^{-1}.[126]

Solid CrF_4 and CrF_6 have i.r. spectra which show that they are fluorine-bridged polymers. Matrix-isolated CrF_4 and CrF_6, on the other hand, are simple tetrahedral and octahedral monomers, with strong i.r. bands at 784.3cm^{-1}, 763.2cm^{-1} respectively.[127]

MoMo stretching wavenumbers were assigned for a series of Mo_2-, Mo_3- and Mo_4-containing species, e.g. $Mo_4S_4(CN)_{12}^{8-}$ 228 cm^{-1}. This and other systems with a Mo_4S_4 core have νMoS modes at 372cm^{-1}(a_1), 363cm^{-1}(e), 341 and 313cm^{-1}(t_2).[128] The following νM-H assignments were made: $MoH(CN)_7^{4-}$ 1805cm^{-1}, $RhH(CN)_5^{3-}$ 1980cm^{-1}, $IrH(CN)_5^{3-}$ 2040cm^{-1}.[129]

The MoN stretches in $Mo(NO)(N_3)_3(terpy)$ include bands at 335 cm^{-1} (Mo-NO) and 315cm^{-1} (Mo-N_3).[130] νMo≡N in $MoN(N_3)_2Cl(terpy)$ is at 1005cm^{-1}.[131] Polymer-bound nitridomolybdenum(VI) complexes $MoNCl_3$(polystyrene-bound bipyridyl), and related species, give resonance Raman enhancement of a band at 1027cm^{-1} due to νM≡N.[132]

$\nu Mo{=}^{16}O$ and $\nu Mo{=}^{18}O$ were assigned in $MoO_2(S_2CNR_2)_2$ (R = Me, Et, ^nPr, ^iPr), $MoO(S_2CNR_2)_2$ (R = Me, Et ^nPr), $MoO[S_2P(OR)_2]$ (R = Me, Et). $\nu Mo{=}O$, $\nu MoOMo$ and $\delta MoOMo$ were also assigned for Mo_2O_3-$(S_2CNR_2)_4$ (R = Me, Et, ^nPr) and $Mo_2O_3[S_2P(OR)_2]_4$ (R = Me, Et), as well as $\nu Mo{=}O$, $\nu MoOMo$ and $\nu MoSMo$ in $Mo_2O_3S(S_2P^iPr_2)_4$ and $Mo_2O_3S(S_2CNEt_2)_2$. There was little change in the $MoO_2X_2^{2+}$ unit on going from $X_2 = O_2$ to OS to S_2.[133] $(Mo_6Cl_8)_2O_2$ has $\nu MoOMo$ and $\nu MoCl$ at 645cm^{-1} and 316cm^{-1} respectively.[134] $\nu Mo{=}O$ and $\nu Mo{-}O$ modes were assigned for <u>cis</u> and <u>trans</u> isomers of $L_2Mo_2O_4^{2+}$, where L = (1).[135] The i.r. spectra of the 7-coordinate complexes

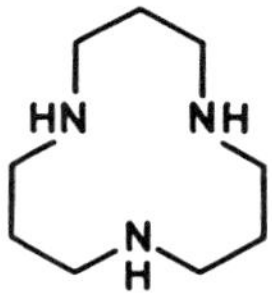

(1)

MoO_2LL', where L = en, 2-aminoethylpyridine, 2,2'-bipy etc., $L'H_2$ = salicylidene, show that <u>cis</u>-MoO_2 is present. Shifts in ν_sMoO_2 and $\nu_{as}MoO_2$ were related to the changing electron density at the Mo.[136] νMoO modes were assigned in <u>cis</u>-MoO_2dttd (885, 910cm^{-1}), MoO(dttd) (925cm^{-1}), and MoO(dpttd) (920cm^{-1}), where H_2dttd = 2,3,8,9-dibenzo-1,4,7,10-tetrathiadecane, H_2dpttd = 2,3,11,12-dibenzo-1,4,7,10,13-pentathiatridecane.[137] I.r. data for $(^tBu)_3SiOMoO_3^-$ and $Ph_3SiOMoO_3^-$ gave assignments for $\nu MoO_3(a_1)$, 900cm^{-1}, $\nu MoO_3(e)$, 880cm^{-1}, and $\nu MoOSi$, 330cm^{-1}.[138] The i.r. spectra of $Li_3M^{III}(MoO_4)_3$, where M^{III} = Al, In, Cr, Fe or Co, show strong coupling between the MoO_4 units.[139] νMoO bands in $Mo_2F_2(O^tBu)_4(PMe_3)_2$ are at 599, 540cm^{-1}, and in $Mo_4F_4(O^tBu)_8$ they are at 625, 555cm^{-1}. Both complexes have νMoF_t near 730cm^{-1} and the latter has νMoF_{br} at 510cm^{-1}.[140]

$Co(MoS_4)_2^{3-}$ has $\nu Mo{-}S_t$ at 481cm^{-1} and $\nu Mo{-}S_{br}$ at 466 and 445 cm^{-1}.[141] The following assignments were proposed for $Mo(S_2)Cl_3$: $\nu_{as}MoS_2$ 349cm^{-1}, ν_sMoS_2 335cm^{-1}, $\nu_{as}MoCl_2$ 313cm^{-1}, ν_sMoCl_2 291cm^{-1}, and $MoCl_2Mo$ 268, 235cm^{-1}. Skeletal-mode assignments were also given for $X_4Mo(\mu{-}S_2)_2MoX_4^{2-}$, where X = Cl, Br.[142] νMoS_t was at 494cm^{-1} and νMoS_2Mo at 462, 427cm^{-1} in $S_2MoS_2MoBr_2(NO)_2^-$.[143] $\nu Mo{-}S$ gives rise to more than one i.r. band for $[Mo(S_2CNEt_2)_4]_2$-Mo_6O_{19}, suggesting unsymmetrical Mo-dithiocarbamate bonding.[144]

The i.r. spectra of MOF_4, where M = Mo or W, contained the following bands: νMo=O 1035cm^{-1}, νMo-F 720cm^{-1}, νW=O 1055cm^{-1}, νW-F 698, 733cm^{-1}.[145] The i.r. spectrum of gaseous MoF_3Cl_2 shows that there are both terminal and bridging fluorine atoms present.[146] νMF bands in $M(CO)_2(S_2CNR_2)_2F^-$, where M = Mo or W, R = Me or Et, are all ca.415cm^{-1}, i.e. low for a terminal fluorine stretch.[147] The i.r. spectrum of a new, reactive form of β-$MoCl_2$ is very similar to that for the known form of this compound, and consistent with a tetranuclear structure.[148] Approximate νMCl_t and νMCl_2M mode assignments were given for $[MCl_4(PhC{\equiv}C\text{-}C{\equiv}CPh)]_2$, where M = Mo or W.[149] $[Mo_4Cl_8(PR_3)_2]_x$, where R = Et, ^{n}Pr, ^{n}Bu, or R_3 = Me_2Ph, all give very similar i.r. band patterns due to Mo-Cl stretching. Thus all appear to have the same structure.[150] α-$Mo_2Br_4(dppe)_2$, β-$Mo_2Br_4(dppe)_2$ and trans-$MoBr_2(dppe)_2$ have νMoBr modes as follows: 259,223cm^{-1}, 263cm^{-1}, 281cm^{-1}.[151]

νW-W is at 313cm^{-1} in $W_2(O_2C^tBu)_4$ and at 304cm^{-1} in $W_2(O_2CMe)_4$.[152] The Raman spectrum of μ-$HW_2(CO)_{10}^-$ was used to examine rotation of the proton around the W-W bond.[153] $Cl_3W(\mu\text{-}H)(\mu\text{-}Me_2S)_2WCl_2(Me_2S)$ has ν_sW-H-W at 1637cm^{-1}, ν_{as}W-H-W at 1245cm^{-1} (these are shifted to 1148, 889cm^{-1} respectively in the deuterio analogue).[154] νW≡C in $WCl_4(C^tBu)^-$ is at 986cm^{-1} (i.r.), 982cm^{-1}(Raman).[155] $WCl_5(IC{\equiv}CI)_2^-$ has νWC_2 bands in the range 908-830cm^{-1}, νWCl_4 333-280cm^{-1}, and νWCl_{trans} 240-217 cm^{-1}.[156]

The following assignments were proposed for $W_2NCl_{10}^-$: ν_{as}WNW 945cm^{-1}, δWNW 840cm^{-1}, νWCl_t 380,335cm^{-1}, νWCl_2W 290cm^{-1}.[157] $Cl_5WNSNWCl_5^{2-}$ has a νWN band at 1012/1020cm^{-1}, in the same region as for a terminal W≡N group.[158]

WO_2(2,3,4-trihydroxyazobenzene) gives rise to νWO_2 bands in the i.r. at 970 and 830cm^{-1}.[159] A factor-group analysis of the i.r. and Raman spectra of crystalline $W_2O_3(PO_4)_3$ suggests that the space group is $P2_1/m(C_{2h}^2)$.[160] νW=S bands lie in the range 525-550cm^{-1} in the i.r. spectra of $W_2S_4(Rdtc)_2$, where dtc = dithiocarbamate, R = morpholine, piperidine etc.[161]

A high-resolution i.r. spectrum covering the ν_3 band has been obtained for four isotopic species of WF_6 (^{182}W, ^{183}W, ^{184}W and ^{186}W).[162] $[WCl_4(Me_3SiC{\equiv}CSiMe_3)]_2$ has νWCl_t at 375, 361 and 340cm^{-1} and νWCl_2W at 313 and 280cm^{-1}.[163] νWBr modes were assigned in $[WBr_2(NO)_3]_2$ and $WBr_2(NO)_3L_2$, where L = PPh_3, L_2 = bipy. The dimer had νWBr_t at 225 and 203cm^{-1} and νWBr_{br} at 170cm^{-1}.[164]

Other reports of vibrational data for Cr, Mo and W compounds are as follows: $Cr(acac)_3$,[2] HCrOH,[4] CrO_2, WO_3,[5] CrF_6^-, CrF_6^{2-},[6] $CrOF_4$,[7] Cr_2 in matrices,[21] CrN(TPP),[22] $Mo_2X_6(H_2O)_2^{2-}$ (X = Cl, Br or I),[23] $NaLa(MoO_4)_2$,[58] trans-$CrCl_2(dmpe)_2$,[68] $X_4M{=}N{-}TeF_5$ (M = Mo, W, X = F,Cl),[81] $CrVO_4$,[85] $CrO_2(OOCR)_2$, CrO(OOCR), CrO(OOCR)py, where R = Me, Et, Pr or $CHCl_2$,[165] trans-MoH(SR)$(dppe)_2$ (dppe = $Ph_2PCH_2CH_2PPh_2$, R = ^{i}Pr, ^{t}Bu etc.),[166] XMO_4 (X = Ca, Pb, M = Mo, W),[167] $MnMoO_4.H_2O$, α-$MnMoO_4$, $Na_2Mn_2(MoO_4)_3.3H_2O$,[168] O=M(OR)(TPP) (R = H, Ph, M = Mo, W),[169] $LMoOX_2^+$ (X = Cl, Br or I, L = fac-coordinating 1,4,7-trimethyl-1,4,7-triazacyclononane),[170] $MoO(S_2CNR_2)_2(C_8H_{12})$, where R = Me, Et, C_8H_{12} = cyclo-octyne,[171] $NaCs_2[MO(CN)_5]$ (M = Mo, W),[172] $MoOXL_2$ (X = F, Cl, Br, I, HL = 4-morpholinodithiocarboxylic acid),[173] MoO_2L (L = tripodal or linear tetradentate N,S-donor ligands),[174] $K_2[MoO(O_2)_2(citrato)].\frac{1}{2}H_2O_2.3H_2O$,[175] $MoO_2Cl_2[(^iPrO)_2P(O)CH_2C(O)NEt_2]$,[176] $Zn_2Mo_3O_8$, $ScZnMo_3O_8$, $LiZn_2Mo_3O_8$, $Zn_3Mo_3O_8$,[177] $[syn\text{-}(S_2)OMo(\mu\text{-}S)_2MoS(S_2)]^{2-}$,[178] $Mo_2Cl_4(LL)_2$, where LL = dppm, dppe etc.,[179] WO_3,[180] $WO(OH)_2.2H_2O$,[181] $BaLa_2WO_7$.[182]

6 Manganese, Technetium, and Rhenium

νMnH in $MnH(CH_2{=}CH_2)(dmpe)_2$ is at 1725cm^{-1}.[183] νMn-C(a_{1g}) was assigned in $Mn(CNR)_6^+$ and $Mn(CNR)_6^{2+}$, where R = Me, Et, Pr, Bu, Cy. In each case the Mn^I species gives a higher-wavenumber band than the Mn^{II} analogue, e.g. for R = Me 292cm^{-1} (Mn^I), 270cm^{-1} (Mn^{II}).[184] νMn-C (ca.580cm^{-1}), νMn-S (ca.325cm^{-1}) and νMn-P (ca. 390cm^{-1}) were assigned for $[Mn(PR_3)X_2]_2CS_2$, where X = Cl, Br or I, PR_3 = P^nPr_3, P^nBu_3, $PPhMe_2$, $PPhEt_2$ or PPh^nPr_2.[185] $[MnX_2\{P(CH_2CH_2CN)_3\}]_2$ complexes (2), where X = Cl, Br, I or NCS, have had assignments made for νMnX_t, νMnXMn, νMnN and νMnP.[186]

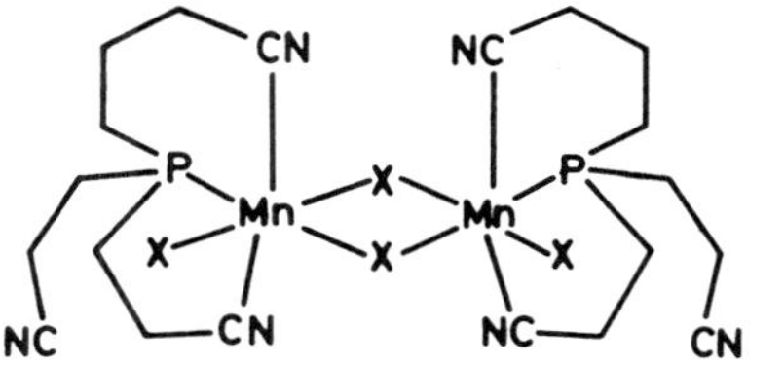

(2)

νM-O and νM-Cl modes were assigned for $MX_2(amide)_{\underline{n}}$, where M = Mn, Fe, Co, Ni, Cu, Cd, Ag, Pd, Pt, X = Cl or Br, amide = formamide, N-methylformamide or DMF, n = 1, 2 or 4.[187] Metal-oxygen stretching bands of $[(OC)_3M(Nb_2W_4O_{19})]^{3-}$, where M = Mn or Re, show that the $M(CO)_3^+$ units are bound to bridging oxygens in the cis-$Nb_2W_4O_{19}^{4-}$ structure.[188]

H_2 adsorbed on $^{99}Tc/Al_2O_3$ gives rise to a ν^{99}Tc-H band at 1810cm^{-1}.[189] MCl_4py_2, where M = Tc or Re, give 3 M-Cl stretching bands in the i.r. spectrum (300-360cm^{-1}) as well as two M-N stretches (220-265cm^{-1}), suggesting cis (C_{2v}) geometry.[190] Raman spectra of Tc(V) species ($TcOCl_4^-$ and $TcOCl_5^{2-}$) in 12M HCl and in CH_2Cl_2 solutions show that five-coordinate species are preferred in both cases, especially in aqueous media, where H_2O competes with Cl^- for the sixth coordination site.[191]

νRe-H(D) modes were assigned in $ReH(OR)(CO)(NO)(PPh_3)_2$, where R = Me or Et. For R = Me the bands were at 1871cm^{-1} (H), 1343cm^{-1} (D).[192] $(Me_4ReO)_2Mg(THF)_2$ has νRe-C at 540cm^{-1}, νRe=O at 960cm^{-1}, with νMg-O stretching modes at 320 and 370cm^{-1}.[193]

(3) has νRe=O as a very strong i.r. band at 935cm^{-1}. νReORe occur as weaker bands at 640cm^{-1} and 618cm^{-1}.[194] (4), on the other hand, only has νRe=O_t bands, at 812 and 792cm^{-1}, eliminating structures with Re-O-Re units.[195]

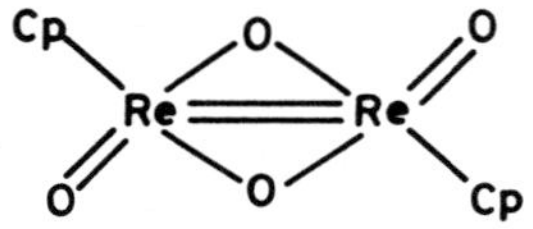

(3)

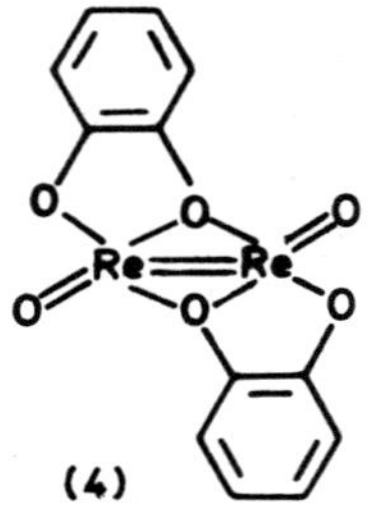

(4)

I.r. and Raman spectra of $Re_3Se_2Cl_5$ are consistent with the presence of an octahedral Re_6 cluster, with four μ_3-Se and four μ_3-Cl ligands, and six terminal chlorine atoms.[196]

The i.r. spectra of K_2ReF_6 at 10K and at room temperature are consistent with trigonal distortion of ReF_6^{2-}. Together with a vibrational analysis of the electronic spectrum, this gave a full assignment of the internal modes.[197]

Reports have also been made of some vibrational spectroscopic results for: $Mn(acac)_3$,[2] HMnOH,[4] MnN(TPP),[22] MnO_4^-,[24,25]

ReS_4^-,[26] $Re_2Cl_8^{2-}$,[27] $M(HQ)_2$, where M^{II} = Mn, Fe, Co or Zn, H_2Q = 1,2-benzoquinonedioxime,[198] $Re(NO)(SCN)_5^{2-}$,[199] $ReCl_4(NSCl)_2^-$.[200]

7 Iron, Ruthenium, and Osmium

The new species CpFeH has been detected in an argon matrix. νFeH gives infrared bands at 1749, 1745 and 1731cm^{-1} (1264, 1262 and 1257cm^{-1} in the deuterio analogue).[201] The i.r. spectrum of bis(cyano)phthalocyaninatoferrate(III) shows νFeC(a_{2u}) at 350cm^{-1}.[202] A normal-coordinate analysis of $Fe_4S_3(NO)_7^-$ shows that the skeletal modes are extensively delocalised.[203] Metal-ligand modes were identified in the new coordination complexes $MLCl_2$, where M = Fe or Cu, L =4,4'-didodecyl-2,2'-bipyridine or 4,4'-dioctadecyloxy-2,2'-bipyridine, using $^{54}Fe/^{56}Fe$ or $^{63}Cu/^{65}Cu$ isotopic shifts. The FeII complexes have a pseudo-octahedral structure (*D*$_3$ symmetry), while the CuII complexes are polymeric, with chlorine-bridged octahedral units.[204]

The vibrational spectra of two model compounds for haemerythrin, i.e. $[Fe_2O(OAc)_2(TACN)_2]^{2+}$ and $\{Fe_2O(OAc)_2[HB(pz)_3]_2\}$, show for the former ν_sFeOFe at 540cm^{-1} and for the latter ν_{as}FeOFe at 751cm^{-1}. These show the expected shifts on ^{18}O substitution. Some resonance Raman data were also given.[205] $Fe_2S_2(OAr)_4^{2-}$, where Ar = Ph or *p*-MeC$_6$H$_4$, have i.r. bands due to ν_{as}FeSFe at *ca.*400cm^{-1}, *ca.*345cm^{-1}, and a Raman band assigned as ν_sFeSFe at 383cm^{-1}. The terminal ligands also gave characteristic i.r. and Raman bands.[206]

The ruthenium-ruthenium stretch is at 250cm^{-1} for $[Ru(CO)_2(\mu\text{-}OOCEt)]_2$. νRuO is probably contributing to the bands at 292 and 204cm^{-1}, with νRuC(a_1) at 528cm^{-1}.[207] Assignments were proposed for νRuH in CpRu(L_2)H, where L_2 = $Ph_2P(CH_2)_nPPh_2$, *n* = 2 or 4, or (*R*)-$Ph_2PCH_2CHMePPh_2$ (1897-1908cm^{-1}).[208] νRuH bands appear at 1995 and 2040cm^{-1} in the i.r. spectrum of $CpRuH_3(PPh_3)$.[209] Assignments were also given for νRuH and νOsH in $MH(pyS)(CO)(PPh_3)_2$ and related species, where M = Ru or Os, pySH = pyridine-2-thiol.[210]

A detailed vibrational assignment of skeletal modes was derived from SERS data for $Ru(NH_3)_5L^{2+}$, $Os(NH_3)_5L^{2+}$ and $Os(NH_3)_5L^{3+}$, where L = py, pyrazine etc.[211]

The presence of only one νRu=O band (855cm^{-1}) in the infrared spectrum of *trans*-$Ru(TMC)O_2^{2+}$, where TMC = 1,4,8,11-tetramethyl-1,4,8,11-tetraazacyclotetradecane, confirms the *trans*-geometry.[212]

ν_{as}Ru=O is at 860cm^{-1} in Ru(tmea)$_2$O$_2^{2+}$, where (tmea)$_2$ = [Me$_2$N(CH$_2$)$_2$NMe$_2$]$_2$.[213] Quite detailed assignments (from i.r. and Raman spectra) were proposed for K$_4$M$_2$OCl$_{10}$.H$_2$O, where M = Ru or Os, backed up by a normal-coordinate analysis. ν_{as}M-O-M was at 873cm^{-1} (Ru), 850cm^{-1} (Os), with ν_sM-O-M at 253cm^{-1} (Ru) or 222cm^{-1} (Os). The selection rules for $\underline{D}_{4h}$ symmetry were not strictly obeyed in the solid-phase spectra.[214]

Ru$_2$ClL$_4$, where HL = 2-aminopyridine, has νRuCl at 311 cm^{-1}.[215] Raman spectra were recorded for Cs$_2$Ru$_2$Cl$_9$ in the temperature range 4-150K. All except one of the expected internal and external modes of Ru$_2$Cl$_9^{2-}$ were assigned.[216]

In OsCl$_5$(NCS)$^{2-}$, νOsN is at 285cm^{-1} (i.r.), 290cm^{-1} (Raman). The isomer OsCl$_5$(SCN)$^{2-}$ has νOsS at 265cm^{-1} (i.r.), 269cm^{-1} (Raman).[217] The OsO$_4^{2-}$ ion has been prepared, as Cs, Rb, K, Na or PPh$_4$ salts. ν_3 is near 840cm^{-1} and ν_4 near 250cm^{-1}. ν_3 is split in some cases, showing that the site symmetry is less than tetrahedral.[218] νOsO, νOs-OH and δOsO modes were assigned for OsO$_4$(OH)$_2^{2-}$, Os$_2$O$_8$(OH)$^-$ and OsO$_2$(OH)$_4^{2-}$.[219] Assignments of νOsO$_2$ and νOsX were made for OsO$_2$X$_2$L$_2$, where X = Cl or Br, L = SMe$_2$, SeMe$_2$: νOsO$_2$ 850cm^{-1}, νOsCl 310cm^{-1}, νOsBr 200cm^{-1}.[220] The complex Os$_2$O$_4$(OCMe$_2$CH$_2$N^{t}Bu)$_2$ has a Raman band at 960cm^{-1} due to νOs=O and an infrared band at 660cm^{-1} assigned as stretching mode of the Os$_2$O$_2$ unit.[221]

νOsCl modes were assigned in Os(NO)Cl$_2$(PPh$_3$)$_2$ (325cm^{-1}), Os(NO)ClBr(PPh$_3$)$_2$ (320cm^{-1}) and Os(NO)Cl(PPh$_3$)$_3$ (325cm^{-1}).[222] Osmium-chlorine stretches are seen in the ranges 300-320cm^{-1} and 260-285cm^{-1} in (η^6-$\underline{p}$-cymene)Os(L)Cl$_2$, where L = CO, CNCMe$_3$, Me$_2$SO etc.[223] Infrared and Raman spectra were reported and assigned for the new species OsX$_6^{3-}$ (X = Cl, Br or I) and for RhCl$_6^{2-}$, RuX$_6^{2-}$ (X = Cl or Br), and RuBr$_6^{3-}$. Thus, for OsBr$_6^{3-}$ $\nu_1(a_{1g})$ is at 201cm^{-1}, $\nu_2(e_g)$ 180cm^{-1}, $\nu_3(t_{1u})$ 200cm^{-1}, $\nu_4(t_{1u})$ 116cm^{-1}, and $\nu_5(t_{2g})$ 94cm^{-1}.[224]

Other iron- and ruthenium-containing species for which vibrational data have been reported are as follows: Fe(acac)$_3$,[2] HFeOH,[4] Fe$_3$O(OAc)$_6$py$_3$[FeCl$_4$],[8] Fe(WS$_4$)$_2^{3-}$, Fe(salen)OC$_6$H$_4$-4-Me,[29] a variety of iron-containing bio-molecules,[30-43] LnRu$_2$O$_7$, where Ln = Gd or Lu,[55] Fe(CHBr$_2$CO$_2$)$_3$,[75] Cr$_{\underline{n}}$Fe$_{3-\underline{n}}$(μ_3-O)(glycine)$_6$(H$_2$O)$_3^{7+}$ (where $\underline{n}$ = 1-3),[123] Cr$_{\underline{n}}$Fe$_{3-\underline{n}}$(μ_3-O)(C$_{14}$H$_{29}$COO)$_6$(H$_2$O)$_3^+$, where $\underline{n}$ = 1-3,[124] (TPP)CrOFe(TPP) and related systems,[125] FeX$_2$(amide)$_{\underline{n}}$, where X = Cl or Br,[187] $Fe(HQ)_2$, where H_2Q = 1,2-benzoquinone-dioxime,[198] and $M^{II}_3M^{III}_2WO_9$, where M^{II} = Ca, Sr or Ba, M^{III} = Fe, Co, In or Al.[225]

8 Cobalt, Rhodium, and Iridium

νCo-H-Co is in the range 1115-1140cm^{-1} in $[(C_5H_4CH_3)Co]_2HX_3$, where X = Cl, Br or I. νCoX_t is at 352cm^{-1} (Cl), 265cm^{-1} (Br), νCoX_{br} 314cm^{-1} (Cl), 240cm^{-1} (Br).[226] νMN modes were assigned for tetrazole (= TE) complexes $M(TE)_{1.5}Cl_2$, where M = Co, Ni or Cu, and $Cd(TE)_2Cl_2$.[227] νMN is in the range 410-450cm^{-1} in $Co(DH)_2(NO_2)A$ and $[Co(DH)_2(NO_2)X]^-$, where DH_2 = dimethylglyoxime, A = amine, X = Br, I, CN, NCS etc.[228]

νM-O and νM-X modes were assigned to i.r. bands in the ranges 369-394cm^{-1}, 300cm^{-1} (Cl) and 210cm^{-1} (Br) for $[MX(L)(EtOH)]_4$, where M = Co or Ni, X = Cl or Br, L = 3,5-dimethyl-1-oxymethyl-pyrazolate(1-).[229]

The rhodium-rhodium stretching modes were assigned as follows: for $Rh_2(O_2CMe)_4(L)_2$ L = PPh_3 289cm^{-1}, $AsPh_3$ 298cm^{-1}, $SbPh_3$ 306cm^{-1}.[230]

(5) has an i.r. band due to νRhH at 1950cm^{-1} (R_2P = (6)).[231]

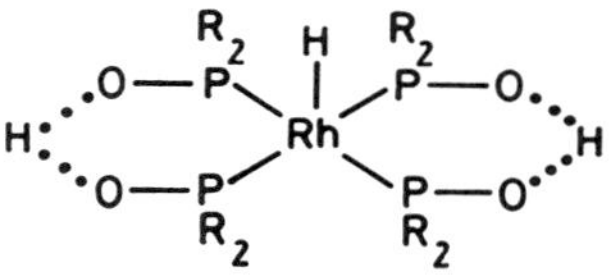

(5)

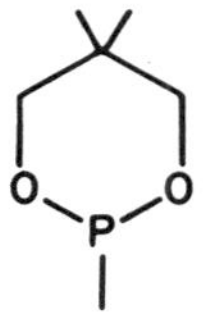

(6)

$\{Rh[(Ph_2P)_2NR]_2\}[cis\text{-}Rh(CO)Cl_2]$ has i.r. bands from the anion corresponding to a quadratic, planar structure, with symmetry $\underline{C}_{2v}$ (R = Ph or p-MeC_6H_4).[232]

νRhCl(bridging) bands are seen in the infrared spectrum of $RhCl_3(MTC)_2$, where MTC = N-methyl-o-ethylthiocarbamate. $Rh(MTC)X_3$, where X = Cl or Br, have an octahedral mer-configuration in the solid state and in solution.[233]

νIrH is at 2030cm^{-1} on H_2-reduced Ir/Al_2O_3 catalyst - confirmed by H/D substitution.[234] Ir-H stretching modes were also assigned for: $[IrH_2(COD)L]^+$ and $[IrH_2(COD)L'L'']^+$, where L is a bidentate sulphur ligand, L' is a unidentate sulphur ligand, and L" is a group IVB donor ligand;[235] Ir(chel)(L)(H)Cl, where chel =

$PPh_2(CH_2)_2SiMe_2$, L = COD, $PPh_2(CH_2)_2PPh_2$ or $PPh_2CH_2PPh_2$;[236] $(C_8H_{14})_2Ir$-μ-Cl_2-$\overline{IrH[CH_2CMe_2CH_2P}(CH_2CMe_3)_2][P(CH_2CMe_3)_3]$.[237]

$CsIrF_5Cl$ has been isolated for the first time. Its infrared and Raman spectra were assigned completely, on the basis of $\underline{C}_{4v}$ symmetry. The wavenumbers were shifted significantly to higher wavenumbers by comparison with the Ir(IV) analogue.[238] Infrared and Raman spectra were also assigned for $[IrCl_nBr_{6-n}]^{2-}$ and $[IrCl_nBr_{6-n}]^{3-}$, where $\underline{n}$ = 1-5, according to the point groups $\underline{C}_{4v}$ ($\underline{n}$ = 1,5), $\underline{D}_{4h}$ ($\underline{n}$= 2,4 -trans), $\underline{C}_{2v}$ ($\underline{n}$= 2,4 -cis) or $\underline{C}_{3v}$ ($\underline{n}$ = 3). Corresponding stretching wavenumbers are generally 10-25cm^{-1} higher in the Ir(IV) than in the Ir(III) complexes.[239]

Other cobalt, rhodium and iridium systems for which vibrational data were given are: $M(acac)_3$, where M = Co, Rh or Ir,[2] HCoOH,[4] $Co_4(CO)_{10}(\mu_4$-η^2-$C_2H_2)$,[9] $Co^{2+}.CrO_4^{2-}$,[121] $M_3^{II}Co_2WO_9$, where M^{II} = Ca, Sr or Ba.[225]

9 Nickel, Palladium, and Platinum

νNiC for 'on-top' CO adsorbed on nickel appears as several Raman bands in the range 420-509cm^{-1}. Bridged species give corresponding bands at 362 and 381cm^{-1}, with features thought to be NiNiC or OCNi deformations at 80and 185cm^{-1} respectively.[240] The following assignments were given for $(NC)_2NiS_5^{2-}$: νNi-C 440, 480 cm^{-1}, νNi-S 283cm^{-1}.[241] Detailed assignments of skeletal modes were reported from the i.r. spectra of nickel(II) and palladium (II) bis-salicylaldoximates.[242]

The nickel(II) complex of octaethylporphyrin $\underline{N}$-oxide gives rise to infrared bands showing the presence of the Ni-O-N unit.[243] $Ni(ClO_4)_2$ has νNi-O bands at 220cm^{-1} (infrared) and 274cm^{-1} (Raman).[244]

νPd-N in (7), where R = CO_2Et, and several related species, lies in the region 445-504cm^{-1}. νPd-C is probably between 370 and 415cm^{-1}.[245] νPd-Cl (300-320cm^{-1}) and νPd-P (ca.420cm^{-1}) were

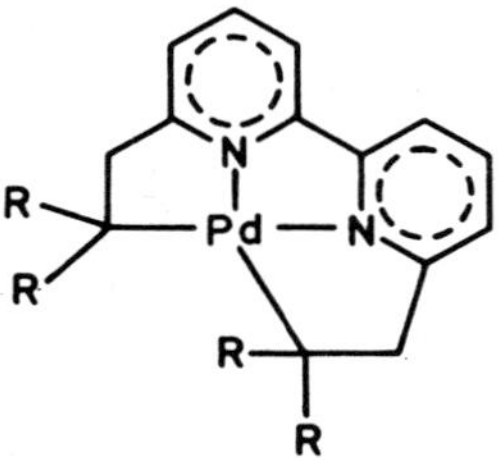

(7)

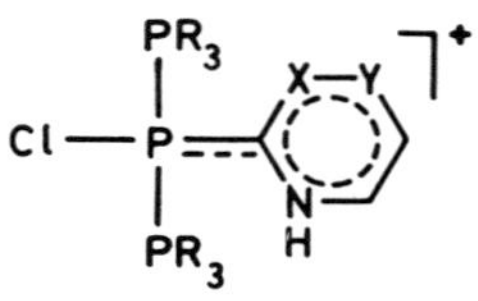

(8)

assigned for (8), where R_3 = Et_3 or Me_2Ph, X = N, Y = CH, or X = CH, Y = N.[246]

νPd-O was assigned to an i.r. band near 485cm^{-1} in Ba[Pd(OH)$_4$] and Ba[Pd(OD)$_4$]. δPd-OH lies between 755 and 1060cm^{-1} (650-850 cm^{-1} in the deuterio compound).[247] Stretching and deformation modes of the MO_4 skeleton were assigned from the i.r. and Raman spectra of M(acac)$_2$ and M(acac-d$_1$)$_2$, where M = Pd(II) or Cu(II).[248]

Vibrational spectral data were reported for a wide range of <u>cis</u>-PdX_2L_2 and <u>cis</u>-PtX_2L_2 species, as well as related Ni(II) and Rh(I) square-planar species (X = halide). Trends in M-L and M-X stretching wavenumbers were discussed, enabling the following spectrochemical <u>trans</u> influence series to be constructed: L = PR_3 > SbR_3 $\sim$ TeR_2 > AsR_3 > $SeMe_2$ > $AsPh_3$ > NH_3 $\sim$ SR_2.[249] νPd-X modes were also assigned for PdX_2L_2, where X = Cl or Br, L = $Ph_2P(CH_2)_2CN$, $Ph_2PCH_2CHCH_3CN$ etc. These were all consistent with <u>trans</u> coordination.[250] $Pd_2X_2(dmpm)_2$, where dmpm = bis(dimethylphosphino)methane, give Raman bands (in CH_2Cl_2 solution) due to νPdPd as follows: X = Cl 140cm^{-1}, X= Br 123cm^{-1}. In aqueous solution (X = Cl) νPdPd is at 140cm^{-1}, but there is also a band assigned to νPd-OH, at 351cm^{-1}.[251]

Pt(PPh$_3$)$_2$(SnCl$_3$)(H)(Cl)(OEt) has νPtH at 2120cm^{-1}, νPtCl at 300cm^{-1}, and νSnCl at 330cm^{-1}.[252] νPtH is at 2082cm^{-1} in <u>trans</u>-PtH[C(OMe)$_2$](PPh$_3$)$_2$ and related carbene complexes.[253]

Infrared spectra of Pt(CO)$_2$X$_2$, where X = Cl or Br (together with some Raman data) are consistent with planar, <u>cis</u> geometry in the crystal but a tetrahedral structure in the gas phase and in hydrocarbon solvents.[254,255]

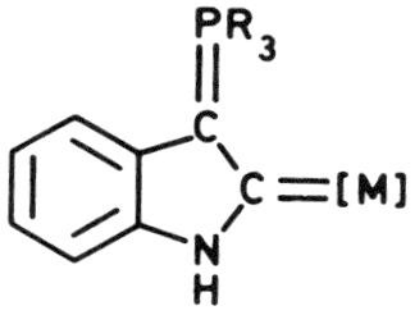

(9)

νPtCl was assigned in the 'carbene-ylide'complexes (9), where R_3 = Me_2Ph, M = <u>cis</u>-PtCl$_2$(PMe$_2$Ph) (288, 271cm^{-1}) or <u>trans</u>-PtCl(PR$'_3$)$_2^+$, where PR$'_3$ = PPh$_3$, PMePh$_2$ or PMe$_2$Ph, (<u>ca</u>.300

cm^{-1}).[256] Skeletal modes were assigned from the i.r. spectra of $(Et_4N)_2[PtCl_2(SnCl_3)_2]$.[257] $Pt_2(R_2SO)_2Cl_4$, where R = Me or Et, have $\nu Pt-Cl_t$ at 370cm^{-1} (Me) or 365cm^{-1} (Et) and $\nu Pt-Cl_{br}$ 270-290cm^{-1}. These data are consistent with a trans dinuclear structure, with two bridging chlorine atoms.[258] I.r. data were also recorded for $PtCl_4^{2-}$ and $PtCl_6^{2-}$ ions, as H_2tmen^{2+} salts, where H_2tmen = N,N,N',N'-tetramethylethylenediammonium.[259] Trans-$Pt(N_3)_4I_2^{2-}$ gives a resonance Raman spectrum with a progression in νPt-I (a_{1g}), for which ω = 138.5cm^{-1}.[260]

Other nickel, palladium and platinum species for which vibrational data were published are as follows: $NiBr_2(H_4L)$, $NiBr(H_4L)_2$, NiL_2^{6-}, where H_4L = ethylenediphosphinetetra-acetic acid;[10] mixed-valence Pd and Pt compounds (resonance Raman);[45-9] $Ni^{2+}.CrO_4^{2-}$;[121] $Ni(tetrazole)_{1.5}Cl_2$;[227] $Ni(HL)_2$, where HL = N,N'-diphenylthiorea;[261] $CpNiCl(PR_3)$, where R = Ph or Cy;[262] $Pd(C_6F_5)_2L_2$, where L = $HNEt_2$, NEt_3, cyclohexylamine, piperidine etc.;[263] $MX_2.2L$, where M = Pd, Pt, X = Cl, Br, L = CH_3NHOH;[264] $MM'(PR_3)_2Cl_4$, where M, M' = Pd or Pt, R_3 = Et_3, $^nPr_3$, $^nBu_3$, Ph_3, $PhMe_2$;[265] and $PtCl(C_6F_5)L$, $PtCl(p-HC_6F_4)L$, where L = phen or bipy.[266]

10 Copper, Silver, and Gold

νCuO bands were seen at 316, 288 and 257cm^{-1} in copper(II) di-2-ethylhexylphosphate at temperatures below 39°C, where the geometry is close to square planar. Above 39°C, when the geometry is nearly tetrahedral, the equivalent bands are at 330, 267 and 232cm^{-1}.[267] νCu-S modes in $Cu_3(S_4)_3^{3-}$ give Raman bands at 312, 287, 252 and 198cm^{-1}.[268] A similar mode is at 322cm^{-1} in the species $Cu_4[o-(SCH_2)_2C_6H_4]_3^{2-}$.[269] A band ascribed to νCuTe is seen at 180cm^{-1} in $(Et_2Te.CuCl)_n$, together with bands at 210 and 152cm^{-1} from the $CuCl_2Cu$ bridging unit.[270]

Infrared emission spectra were recorded for $(CuCl)_n$, where n = 1-4, and $(CuCl_2)_m$, where m = 1 or2. The following νCuCl assignments were made: n = 1 409cm^{-1}, 2 318cm^{-1}, 3 396cm^{-1}, 4 375cm^{-1}, m = 1 495cm^{-1}, 2 413cm^{-1}.[271] The i.r. spectra of MCl_2L, where M = Cu or Hg, L = 1-methyl-3,4-diphenylpyrazole, show the presence of both terminal and bridging M-Cl bonds, i.e they are both polymers.[272]

Three kinds of Ag_3 clusters were identified in Xe and Kr matrices by Raman spectroscopy.[273] Far-infrared spectra of metal atoms and clusters in silver zeolites A and Y show that the Ag-Ag stretches are in the range 125-200cm^{-1}. There is an Ag/cage vibration at 89cm^{-1} in zeolite Y.[274] The FTIR spectra of fully silver-ion-exchanged zeolite A show Ag-Ag skeletal modes in the range 50-175cm^{-1}. Data for $(Ag_3^x)_2$ clusters favour $\underline{D}_{2h}$ or $\underline{D}_{2d}$ geometry.[275]

SERS for pyridine adsorbed at silver electrodes shows a νAgN band at 221cm^{-1}.[276] Raman and infrared spectra of Ag_2HNSO_2NH and Ag_4NSO_2N, with data on the ^{15}N and deuterio derivatives, show that νAg-N is between 200 and 300cm^{-1}. In Ag_4NSO_2N νAgAg is believed to be at 288cm^{-1}.[277]

νAg-S bands were tentatively assigned for $Ag_2(S_6)_2^{2-}$ and $Ag(S_9)^-$.[278] The behaviour of νAgS at 190cm^{-1} in the Raman spectrum of Ag_3SbS_3 is related to the dynamics of the silver ions in the crystal.[279] The far-infrared spectra of the α- and β-phases of Ag_2HgI_4 show νAg-I at about 100cm^{-1} (α-phase) and νAg-I/νHg-I 80-120cm^{-1} (β-phase).[280]

The following assignments were proposed for $[Me_2Au(NHMe)]_2$: νAuC_2 564(i.r.)/567(Raman)cm^{-1}, 557cm^{-1}(Raman); νAuN 491 and 478cm^{-1} (i.r.); δAuC_2 282cm^{-1} (Raman).[281] νAuP (340-390cm^{-1}) and νAuBr (ca.220cm^{-1}) were seen in the i.r. spectra of $(AuBr)_2L$, where L = $Ph_2P(CH_2)_{\underline{n}}PPh_2$, $\underline{n}$ = 1,2,4,6,8,10,12 or 16. In $Au_2L_2^{2+}$, νAuP is between 340 and 370cm^{-1}.[282]

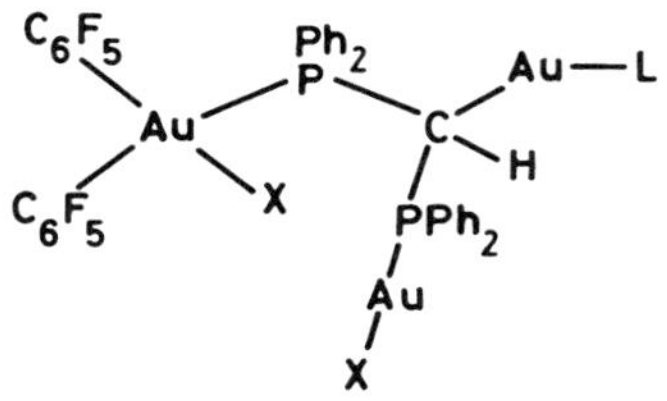

(10)

$AuCl_2(S_3N)$ has νAu-S at 370 and 326cm^{-1}, νAuCl 357 and 341cm^{-1} (in i.r. and Raman spectra).[283] The complex (10), where L = tetrahydrothiophen or $AsPh_3$, X = Cl, has νAuCl 330-338cm^{-1}; where L = $AsPh_3$, X = Br, νAuBr is at 230cm^{-1}. All have νAuC near 600cm^{-1}.[284] $Au^{III}Cl_3[CH(PPh_3)CO_2R]$ have νAuCl(trans to ylide) at 315cm^{-1}, $\nu_s AuCl_2$ 340cm^{-1}, $\nu_{as} AuCl_2$ 360cm^{-1}. In $Au^ICl[CH(PPh_3)CO_2R]$, νAuCl is at 330cm^{-1}.[285] In the complexes $Au(C_6F_5)X(phen)^+$, νAu-X modes are at 355cm^{-1} (X = Cl) or 245 cm^{-1} (Cl).[286]

Other reports of vibrational data for compounds of copper, silver and gold are as follows: HCuOH,[4] CuH,[13] $(Me_2S)AuX$, where X = Cl or Br, $(Me_2S)AuCl_3$,[14] $Ag(MoS_4)_2^{3-}$,[28] Ag_2 in matrices,[51] $LnCu_2O_5$ (Ln = most lanthanides),[57] $LaCl_3$-$CuCl_2$,[60] $Cu(tetrazole)_{1.5}Cl_2$,[227] $Cu(acac)_2$, $Cu(acac\text{-}d_1)_2$,[248] $Cu(en)_2^{2+}$, $Cu(trien)^{2+}$, $Cu(tetren)^{2+}$, $Cu(bipy)_2(NO_3)^+$,[287] MX_2L, where M = Cu or Zn, X = Cl or Br, L = di(2-benzothazolyl)sulphide,[288] $Cu_4OX_6L_4$, where X = Cl or Br, L = py, 2-, 3- or 4-methylpyridine,[289] $AuCl_3(NH_2CMe_3)$,[290] and $AuX_2X'L_n$, where L = isonicotinic acid or ethyl isonicotinate, X, X' = Cl, Br or I, *n* = 1,2, 3.[291]

11 Zinc, Cadmium, and Mercury

Infrared and Raman spectra of crystalline $(NPr_4)ZnCl_3$ are consistent with a chlorine-bridged structure. The symmetry is less than D_{2h}, and there is no centre of symmetry.[292] Phase transitions in single crystals of $(NH_4)_2ZnCl_4$ were followed by Raman spectroscopy. The room-temperature phase (III) was shown to have the space group C_s^2.[293] The vibrational modes of $Zn_2Br_6^{2-}$ were assigned from the infrared and Raman spectra of the NPr_4^+ salt. The anion was shown to have C_1 symmetry.[294]

Detailed skeletal-mode assignments, especially νHgX and νHgpy, were proposed for pyridine solutions of HgX_2, where X = Cl, Br or I, and for crystalline HgX_2py_2, where X = Br or I.[295] Some skeletal-mode assignments were also given for Hg(II) complexes of cysteine-containing oligopeptides.[296] νHgX were assigned for $(\eta^5\text{-}C_5H_4Me)(OC)_3MoHgX$, where X = Cl (266cm^{-1}) or SCN (240 cm^{-1}).[297] Infrared and Raman spectra of $Ph_3TeX.HgX'_2$ and $(Ph_3TeX)_2HgX'_2$, where X and X' = Cl, Br or I, indicate a weak Hg-Te interaction.[298] Principal component analysis of Raman spectra showed that there were three components in aqueous HCl solutions of $HgCl_2$, i.e. $HgCl_2$ (320cm^{-1}), $HgCl_3^-$ (290cm^{-1}), and $HgCl_4^{2-}$ (269cm^{-1}).[299] The following assignments were made for $(2abf)(HgBr_2)_2$, where 2abf = 2-aminobenzophenone: $\nu_{as}Hg^{(1)}Br_2$ 262cm^{-1}, $\nu_{as}Hg^{(2)}Br_2$ 218 cm^{-1}, $\nu_sHg^{(1)}Br_2$ 204cm^{-1}, $\nu_sHg^{(2)}Br_2$ 178cm^{-1}, and $\nu Hg^{(2)}N$ 250cm^{-1}.[300]

Vibrational data were also given for zinc, cadmium and mercury compounds as follows: $Zn(NH_3)_4(ReO_4)_2$,[15] $CdAl_{12}O_{19}$, $CdAl_4O_7$, $CdAl_2O_4$,[16] HgX_3Y^{2-}, where X, Y = Cl, Br or I,[17] $Cd(tetrazole)_2Cl_2$,[227] $HgCl_2L$, where L = 1-methyl-3,4-diphenylpyrazole,[272] ZnPcCl, where Pc = phthalocyaninato,[301] and β-HgI_2.[302]

12 The Actinoids

Earlier reference was made to studies on $^{238}UF_6$ [18] and ML_4, where M = U, Th, L = $CHBr_2COO^-$.[75]

Hydrogen adsorption on ThO_2 activated at 973K gave infrared bands at 1115 and 860cm^{-1} (with a shoulder at 910cm^{-1}) which were assigned to a partially reversibly adsorbed bridged species Th-H-Th.[303]

The oxygen isotope content of UO_2-containing species was determined from the intensities of the ν_3 bands of $^{16}OU^{16}O^{2+}$, $^{16}OU^{18}O^{2+}$ and $^{18}OU^{18}O^{2+}$ relative to the 678cm^{-1} band of acetate (in $NaUO_2(OAc)_3$) as an internal standard.[304] A normal-coordinate analysis has been performed on $K_3(UO_2)_2F_7.2H_2O$.[305]

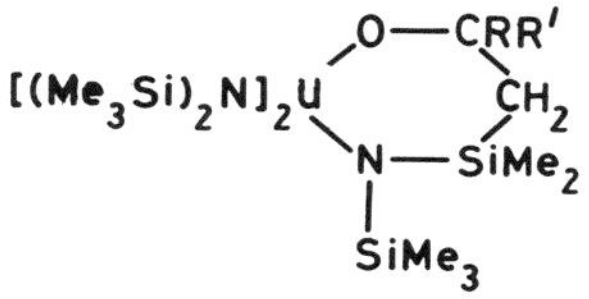

(11)

(11), where R, R' = H or various alkyls and aryls, all have νU-O near 480cm^{-1}.[306] The i.r. and Raman spectra of $UO_2(NCS)_2$-(X-N-Sal)$_2$ (X = NO_2, Cl, Br or OH) and $UO_2(NCS)_2$(X-N-Sal)$_3$ (X = OMe or Me) show that ν_s and ν_{as} UO_2 both decrease with X in the following series: X = NO_2 > Cl > Br > OH > Me > OMe.[307] The following assignments were proposed for polymeric $UO_2(PO_2F_2)_2$: ν_1UO_2 852cm^{-1}, ν_3UO_2 923cm^{-1} and ν_2UO_2 260 cm^{-1}.[308] νM-O (225-280cm^{-1}) and νM-N (632-682cm^{-1}) were assigned in UO_2^{2+} and Th^{4+} complexes of a variety of heterocyclic Schiff bases.[309] νU=O is in the range 885-900cm^{-1} and νU-F-U <u>ca</u>. 350cm^{-1} in several salts of $UO_2(O_2)F_2^{2-}$.[310]

A high-resolution study has been made of the i.r. spectrum of UF_6 in the region of the ν_3 band, using a tunable diode laser.[311] The $\nu_1+\nu_3$ combination band of $^{238}UF_6$ was studied at Doppler-limited resolution. There was little evidence for interaction of this level with other levels.[312]

There were brief mentions of vibrational spectroscopic data for the following species containing UO_2^{2+} groups: UO_2SO_4.2(ethylene glycol).H_2O;[313] UO_2X_2.mHMPA, where X = NO_3, $\frac{1}{2}SO_4$, m = 2, X = ClO_4, m = 4;[314] $(Ph_4As)_2UO_2X_4$, where X = Cl, NCS, NCSe, $ONC(CN)_2$, $N(CN)_2$;[315] $UO_2X_2L_2$, where X = Cl or NO_3, L = N,N'-tetramethylthiourea, N,N'-dimethylthiourea, monomethylthiourea etc.;[316] $UO_2XO_4(DMSO)_2.H_2O$, where X = S, Se or Cr;[317] $UO_2(DMPP)X_2$, where X = NCS, NO_3, DMPP = 3,5-dimethyl-1-(2'-pyridyl)pyrazole;[318] $UO_2L_2(NO_3)_2$, where L = 2,2'-bipy and related ligands, $UO_2L'(NO_3)_2$, where L' = 1,10-phen etc.;[319] $UO_2(H_2PO_4)_2.3H_2O$;[320] $(UO_2)_3(PO_4)_2.4H_2O$, $(UO_2)_{1.43}PO_4H_{0.14}$-(2 - 3.5)$H_2O$;[321] M[$UO_2(OAc)_4$], where M = Mg, Ba, Co or Zn.[322]

Electronic/vibrational spectra of $Cs_2NpO_2Cl_4$ and of $NpO_2Cl_4^{2-}$ in a $CsUO_2Cl_4$ matrix were analysed in terms of D_{4h} symmetry.[323]

References

1 J.W.Kauffman, R.H.Hauge and J.L.Margrave, J.Phys. Chem., 1985, 89, 3547.
2 B.A.Kolesov and I.K.Igumenov, Koord. Khim.,1985, 11, 485.
3 N.J.Campbell, J.Flanagan and W.P.Griffith, J. Chem. Phys., 1985, 83, 3712.
4 J.W.Kauffman, R.H.Hauge and J.L.Margrave, J. Phys. Chem., 1985, 89, 3541.
5 M.J.Almond, A.J.Downs and R.N.Perutz, Inorg. Chem., 1985, 24, 275.
6 R.Bougon, W.W.Wilson and K.O.Christe, Inorg. Chem., 1985, 24, 2286.
7 E.G.Hope, P.J.Jones, W.Levason, J.S.Ogden, M.Tajik and J.W.Turff, J. Chem. Soc., Dalton Trans., 1985, 529.
8 L.Montri and R.D.Cannon, Spectrchim. Acta, Part A, 1985, 41A, 643.
9 G.Gervasio, R.Rosetti and P.L.Stanghellini, Organometallics, 1985, 4, 1612.
10 J.Podlahová, L.Kavan, J.Šilha and J.Podlaha, Polyhedron, 1984, 3, 963.
11 D.Michalska and A.T.Kowal, Spectrochim. Acta, Part A, 1985, 41A , 1119.
12 H.O.Desseyn, A.C.Fabretti, F.Forghieri and C.Preti, Spectrochim. Acta, Part A, 1985, 41A, 1105.
13 R.S.Ram, P.F.Bernath and J.W.Brault, J. Mol. Spectrosc., 1985, 113, 269.
14 I.S.Butler, A.Neppel, K.R.Plowman and C.F.Shaw, J. Raman Spectrosc.,1984, 15, 310.
15 C.Tellez, Semina (Londrina, Braz.),1983, 4, 410.
16 N.V.Porotnikov, A.K.Vazhnov and K.I.Petrov, Russ. J. Inorg. Chem., 1985, 30, 341.
17 J.G.Contreras and G.V.Seguel, Spectrochim. Acta, Part A, 1985, 41A, 1113.

18 J.P.Aldridge, E.G.Brock, H.Filip, H.Flicker, K. Fox, H.W.Galbraith, R.F.Holland, K.C.Kim and B.J.Krohn, J. Chem. Phys., 1985, 83, 34.

19 L.A.Tucker, F.J.Carney, P.McMillan, S.H.Lin and L.Eyring Appl. Spectrosc., 1984, 38, 857.

20 Y.Zhao, G.Llu, G.Wang, and Y.Wang, Gaodeng Xuexiao Huaxue Xuebao, 1984, 5, 847 (Chem. Abs., 1985, 102, 69558).

21 M.Moskovits, W.Limm and T.Mejean, J. Phys. Chem.,1985,89,3886

22 C.Campocihara, J.A.Hofmann and D.F.Bocian, Inorg. Chem., 1985, 24, 449.

23 V.K.Ceylan, C.Sourisseau and J.V.Brencic, J. Raman Spectrosc. 1985, 16, 128.

24 R.J.H.Clark, T.J.Dines and J.M.Doherty, Inorg. Chem., 1985, 24, 2088.

25 A.Haav, K.Haller and L.Rebane, Eesti NSV Tead. Akad. Toim. Fuus., Mat., 1984, 33, 309 (Chem. Abs., 1985, 102, 35700).

26 B.Stewart and R.D.Peacock, Chem. Phys. Lett., 1985, 120, 307.

27 R.F.Dallinger, J. Amer. Chem. Soc., 1985, 107, 7202.

28 A.Müller and W.Hellmann, Spectrochim. Acta, Part A, 1985, 41A, 359.

29 J.W.Pyrz, A.L.Roe, L.J.Stern and L.Que, J. Amer. Chem. Soc., 1985, 107, 614.

30 C.M.Jones, C.R.Johnson, S.A.Asher and R.E.Shepherd, J. Amer. Chem. Soc., 1985, 107, 3772.

31 M.L.Mitchell, D.H.Campbell, T.G.Traylor and T.G.Spiro, Inorg. Chem., 1985, 24, 967.

32 A.J.Sitter, C.M.Reczek and J.Terner, Biochim. Biophys. Acta, 1985, 828, 229.

33 J.Terner, A.J.Sitter and C.M.Reczek, Biochim. Biophys. Acta, 1985, 828, 73.

34 S.Hashimoto, Y.Tatsuno and T.Kitagawa, Proc. Jpn. Acad., Ser. B, 1984, 60, 345.

35 H.Oshio, T.Ama, T.Watanabe and K.Nakamoto, Inorg. Chim. Acta, 1985, 96, 61.

36 M.Tsubaki and Y.Ichikawa, Biochim. Biophys. Acta,1985,827,268

37 J.M.Moulis, J.Meyer and M.Lutz, Biochem., 1984, 23, 6605.

38 E.A.Kerr, N.T.Yu and K.Gersonde, FEBS Lett.,1984, 178, 31.

39 S.Matsukama, K.Mawatari, Y.Yoneyama and T.Kitagawa, J. Amer. Chem. Soc., 1985, 107, 1108.

40 G.Smulevich and T.G.Spiro, J. Phys. Chem.,1985, 89, 5168.

41 J.M.Friedman, T.W.Scott, G.J.Fisanick, S.R.Simon, E.W. Findse, M.R.Ondrias and V.W.MacDonald, Science,1985,229,187.

42 E.W.Findsen, J.M.Friedman, M.R.Ondrias and S.R.Simon, Science, 1985, 229, 661

43 H.Homburg and W,Kalz, Z. Naturforsch., Teil B,1984, 39B,1490

44 R.J.H.Clark, Adv. Infrared Raman Spectrosc.,1984,11, 95.

45 R.J.H.Clark, V.B.Croud and M.Kurmoo, J. Chem. Soc., Dalton Trans., 1985, 815.

46 R.J.H.Clark and M.Kurmoo, J. Chem. Soc., Dalton Trans.,1985, 579.

47 M.Tanaka, S.Kurita, Y.Okada, T.Kojima and Y.Yamada , Chem. Phys., 1985, 96, 343.

48 R.J.H.Clark and V.B.Croud, Inorg. Chem., 1985, 24, 588.

49 M.Kurmoo and R.J.H.Clark, Inorg. Chem., 1985, 24, 4420.

50 D.F.Blair, G.W.Campbell, W.K.Cho, A.M.English, H.A.Fry, V.Lum, K.A.Norton, J.R.Schoonhover and S.J.Chan, J. Amer. Chem. Soc., 1985, 107, 5755.
51 P.S.Bechthold, U.Kettler and W.Kresser, Surf. Sci., 1985, 156, 875.
52 C.Nelson, D.R.Tallant and J.A.Shelnutt, J. Non-Cryst.Solids, 1984, 68, 87.
53 N.V.Porotnikov, O.I.Kondratov and K.I.Petrov, Russ. J. Inorg. Chem., 1985, 30, 931.
54 G.Liu, Y.Zhang, Y.Zhao, G.Wang and Y.Wang, Gaodeng Xuexao Huaxue Xuebao, 1984, 5, 713 (Chem. Abs., 1984,101,237537).
55 V.V.Fomichev, L.L.Kochergina, I.I.Olikov and K.I.Petrov, Russ. J. Inorg. Chem., 1985, 30, 344.
56 M.German and L.M.Kovba, Russ. J. Inorg. Chem.,1985, 30,176.
57 N.V.Porotnikov, K.I.Petrov, I.I.Olikov and B.U.Asanov, Russ. J. Inorg. Chem., 1984, 29, 1271.
58 V.Ramakrishnan and G.Aruldhas, Spectrochim. Acta, Part A, 1985, 41A, 1301.
59 S.Xi, S.Lan, G.Zheng, G.Hong and L.Li, Guangpuxue Yu Guangpu Fenxi, 1984, 4, 8 (Chem. Abs., 1985, 102, 14269).
60 N.P.Utorov, Yu.M.Bakshi, V.P.Bazov and A.I.Gel'bshtein, Koord. Khim., 1985, 11, 603.
61 N.S.Loktyushina and A.A.Mal'tsev, Zh. Fiz. Khim.,1984,58,2631
62 M.Graetzel and F.P.Rotzinger, Inorg. Chem., 1985, 24, 2320.
63 M.K.Chaudhuri and B.Das, Polyhedron, 1985, 4, 1449.
64 G.Pausewang and R.Schmidt, Z. anorg. allg. Chem.,1985,523,213
65 V.L.Goedken, G.Dessy, C.Ercolani, V.Fares and L.Gastaldi, Inorg. Chem., 1985, 24, 991.
66 T.C.DeVore and T.N.Gallaher, J. Chem. Phys.,1985, 82, 2512.
67 P.Choudhury, B.Ghosh, M.B.Patel and H.D.Bist, J. Raman Spectrosc., 1985, 16, 149.
68 G.S.Girolami, G.Wilkinson, A.M.R.Galas, M.Thornton-Pett and M.B.Hursthouse, J. Chem. Soc., Dalton Trans., 1985, 1339.
69 E.Rytter and S.Kvisle, Inorg. Chem., 1985, 24, 639.
70 R.Ameis, S.Kremer and D.Reinen, Inorg. Chem., 1985,24,2751.
71 T.Onishi, H.Abe, K.Maruya and K.Domen, J. Chem. Soc., Chem. Comm., 1985, 617.
72 M.Perec, Inorg. Chim. Acta, 1985, 103, 163.
73 T.Nishida, T.Nonaka and Y.Takashima, Bull. Chem. Soc. Japan, 1985, 58, 2255.
74 G.J.Kipouros, J.H.Flint and D.R.Sadoway, Inorg. Chem., 1985, 24, 3881.
75 J.K.Puri and J.M.Miller, Inorg. Chim.Acta,1985, 101, 135.
76 N.V.Porotnikov, V.G.Savenko and L.N.Margolin, Russ. J. Inorg. Chem., 1984, 29, 1408.
77 C.T.Jekel-Vroegop and J.H.Teuben, J. Organometal. Chem., 1985, 286, 309.
78 V.Dimitrov, J.Organometal. Chem., 1985, 282, 321.
79 J.H.Osborne and W.C.Trogler, Inorg. Chem., 1985, 24, 3098.
80 E.Schweda, K.D.Scherfist and K.Dehnicke, Z.anorg. allg. Chem., 1985, 528, 117.
81 R.Huppmann and K.Seppelt, Chem. Ber., 1985, 118, 457.
82 J.Hanich, M.Krestel, U.Müller, K.Dehnicke and D.Rehder, Z. Naturforsch., Teil B, 1984, 39B, 1686.
83 S.Singh, Amita, M.S.Gill and R.D.Verma, J. Fluorine Chem., 1985, 27, 133.
84 A.L.Doadrio-Villarejo, C.V.Ragel and A.Doadrio, An. R. Acad Farm., 1984, 50, 727.
85 E.J.Baran and M.C.Grasselli, An. Asoc. Quim. Argent., 1985, 73, 7.

86 E.J.Baran and M.E.Escobar, Spectrochim. Acta, Part A, 1985, 41A, 415.

87 B.Hajek, A.Muck and E.Santana, Coll.Czech. Chem. Comm., 1985, 50, 1465.

88 V.Dimitrov and Ya.Dimitriev, God. Vissh. Khim.-Tekhnol. Inst. Sofia,1983(publ. 1984),29, 376

89 M.K.Chaudhuri, S.K.Ghosh and N.S.Islam, Inorg. Chem., 1985, 24, 2706.

90 C.Djordjevic, S.A.Craig and E.Sinn, Inorg. Chem.,1985,24,1281

91 A.J.Jacobson, J.W.Johnson, J.F.Brody, J.C.Scanlon and J.T.Lewandowski, Inorg. Chem., 1985, 24, 1782.

92 M.Moran, J.R.Masaguer and V.Fernandez, J.Organometal. Chem., 1985, 291, 311.

93 J.Hanich, M.Krestel, U.Müller and K.Dehnicke, Z. anorg. allg. Chem., 1985, 522, 92.

94 P.Clauws, J.Broeckx and J.Vennik, Phys. Stat. Solidi, B, 1985, 131, 459.

95 M.K.Chaudhuri and N.S.Islam, Transition Met. Chem., 1985, 10, 333.

96 N.J.Campbell,J.Fllanagan, W.P.Griffith and A.C.Skapski, Transition Met. Chem., 1985, 10, 353.

97 M.P.Glazyrin, V.N.Krasil'nikov and A.A.Ivakin, Russ. J. Inorg. Chem., 1984, 29, 1783.

98 V.N.Krasil'nikov and A.A.Ivakin, Russ. J. Inorg. Chem., 1984, 29, 1613.

99 J.Martinez, A.Martinez and A.Doadrio, Thermochim. Acta, 1985, 87, 281.

100 J.Martinez, R.Lozano, A.Martinez and A.Doadrio, An. Quim., Ser. B, 1983, 79, 486.

101 Y.Repelin, E.Husson, L.Abello and G.Lucazeau, Spectrochim. Acta, Part A, 1985, 41A, 993.

102 L.Abello and G.Lucazeau, J. Chim. Phys., 1984, 81, 539.

103 G.S.Zakharova, V.L.Volkov and A.A.Ivakin, Russ. J. Inorg. Chem., 1985, 30, 823.

104 G.Gu, J.Liu, Y.Yang, D.Chen, J.Lin, K.Tsai and K.Guo, Wuli Huaxue Xuebao, 1985, 1, 177 (Chem. Abs.,1985, 103, 141229).

105 E.M.Nour, A.B.El-Sayed and F.Abdel-Rehim, Spectrochim. Acta, Part A, 1985, 41A, 865.

106 S.Nunziate-Cesaro, M.Maltese, M.Spoliti and B.Janis, Spectrochim. Acta, Part A, 1984, 40A, 579.

107 P.Klingelhöfer and U.Müller, Z. anorg. allg. Chem., 1984, 516, 85.

108 W.G.Klemperer and C.Schwartz, Inorg. Chem., 1985, 24,4459.

109 M.G.B.Drew, D.A.Rice and D.M.Williams, J. Chem. Soc., Dalton Trans., 1985, 1821.

110 M.G.B.Drew, D.A.Rice and D.M.Williams, J. Chem. Soc., Dalton Trans., 1985, 417.

111 N.Bričevič, Ž.Ružić-Toroš and B.Kojić-Prodić, J.Chem. Soc., Dalton Trans., 1985, 455.

112 A.I.Agulyanskii, V.Ya.Kuznetsov, L.A.Agulyanskaya and V.T.Kalinnikov, Koord. Khim., 1984, 10, 1512.

113 M.Moskovits, W.Limm and T.Mejean, J. Chem. Phys.,1985,82, 4875.

114 P.E.Fanwick, B.E.Burstein and G.B.Kaufmann, Inorg. Chem., 1985, 24, 1165.

115 P.Caillet, J.Y.Le Marouille and P.Lemaux, J. Organometal. Chem., 1985, 291, 191.

116 N.Q.Dao, M.Jovan, G.P.Fonseca, N.H.T.Hut and E.O.Fischer, J.Organometal. Chem., 1985, 287, 215.

117 J.Brencic, I.Leban and J.Zule, Z.anorg. allg. Chem., 1985, 521, 199.

118 R.M.Medusa, A.Alvarez-Valdes and J.R.Masaguer, J.Organometal. Chem., 1985, 294, 209.
119 J.Ellermann and A.Veit, J. Organometal. Chem.,1985, 290,307.
120 J.Tomkinson, A.D.Taylor, J.Howard, J.Eckert and J.A. Goldstone, J. Chem. Phys., 1985, 82, 1112.
121 Yu.P.Tsyashchenko and V.D.Danchuk, Ukr. Fiz. Zh. (Russ. Ed.), 1984, 29, 1745.
122 E.J.Baran and S.G.Manca, Spectrosc. Lett., 1984, 17, 827.
123 B.P.Straughan and O.M.Lam, Inorg. Chim. Acta, 1985, 98, 7.
124 Kh.M.Yakubov, G.L.Semenova, and M.M.Mansurov, Russ. J. Inorg. Chem., 1985, 30, 206.
125 D.J.Liston, B.J.Kennedy, K.S.Murray and B.O.West, Inorg. Chem., 1985, 24, 1561.
126 D.J.Liston and B.O.West, Inorg. Chem.,1985, 24, 1568.
127 F.G.Hope, P.J.Jones, W.Levason, J.S.Ogden, M.Tajik and J.W.Turff, J. Chem. Soc., Dalton Trans., 1985, 1443.
128 A.Müller, R.Jostes, W.Eltzner, C.Nie, E.Diemann, H.Bögge, M.Zimmermann, M.Dartmann, and U.Reinisch-Vogell, Inorg. Chem., 1985, 24, 2872.
129 M.J.Mockford and W.P.Griffith, J. Chem. Soc., Dalton Trans., 1985, 717.
130 J.Beck and J.Strähle, Z. Naturforsch., Teil B, 1985, 40B, 891.
131 J.Beck, E.Schweda and J.Strähle, Z. Naturforsch., Teil B, 1985, 40B, 1073.
132 N.Ueyama, H.Zaima, H.Okada and A,Nakamura, Inorg. Chim. Acta, 1984, 89, 19.
133 G.J-J.Chen, J.W.McDonald, D.C.Bravard and W.E.Newton Inorg. Chem., 1985, 24, 2327.
134 H.Schäfer, H.Plautz, H.-J,Abel and D.Lademann, Z. anorg. allg. Chem., 1985, 526, 168.
135 K.Wieghardt, M.Guttmann, P.Chaudhuri, W.Gebert, M.Minelli, C.G.Young and J.H.Enemark, Inorg. Chem., 1985, 24, 3151.
136 A.Syamal and M.R.Maurya, Transition Met. Chem., 1984, 10, 45
137 D.Sellmann and L.Zapf, Z. Naturforsch., Teil B, 1985, 40B, 368
138 W.G.Klemperer, V.V.Mainz, W.Shum and R.C.Wang, Inorg. Chem., 1985, 24, 1968.
139 M.A.Juri, M. del C. Viola and J.C.Pedregosa, Acta Sud Am. Quim., 1984, 4, 37.
140 M.H.Chisholm, D.L.Clark and J.C.Huffman, Polyhedron, 1985, 4, 1203.
141 W.H.Pan, D.C.Johnston, S.T.McKenna, R.R.Chianelli, T.R. Halbert, L.LHutchings and E.I.Stiefel, Inorg. Chim. Acta, 1985, 97, L17.
142 D.Fenske, B.Czeska, C.Schumacher, R.E.Schmidt and K.Dehnicke, Z. anorg. allg. Chem., 1985, 520, 7.
143 E.Herdtweck, C.Schumacher and K.Dehnicke, Z. anorg. allg. Chem., 1985, 526, 93.
144 K.S.Jasim, C.Chieh and T.C.W.Mak, J. Crystallogr. Spectrosc. Res., 1985, 15, 403.
145 V.D.Butskii, M.E.Ignatov and B.V.Golovanov, Russ. J. Inorg. Chem., 1985, 30, 455.
146 S.A.Ryabov, A.S.Alikhanyan, V.D.Butskii, O.G.Ellert and V.S.Pervov, Russ. J. Inprg. Chem., 1985, 30, 953.
147 S.J.N.Burgmayer and J.L.Templeton, Inorg. Chem., 1985, 24, 2224.
148 W.W.Beers and R.E.McCarley, Inorg. Chem., 1985, 24, 472.

149 K.Stahl, F.Weller and K.Dehnicke, Z. anorg. allg. Chem., 1984, 518, 175.
150 W.W.Beers and R.E.McCarley, Inorg. Chem., 1985, 24, 468.
151 P.A.Agaskar, F.A.Cotton, D.R.Derringer, G.L.Powell, D.R.Root and T.J.Smith, Inorg. Chem., 1985, 24, 2786.
152 D.J.Santure, J.C.Huffman and A.P.Sattelberger, Inorg. Chem., 1985, 24, 371.
153 N.V.Kharchevnikova and P.N.D'yachkov, Khim. Fiz.,1984, 3, 1507.
154 P.M.Boorman, K.J.Moynihan, V.D.Patel and J.F.Richardson, Inorg.Chem., 1985, 24, 2989.
155 J.Anhaus, Z.A.Siddiqui, H.W.Roesky and J.W.Bats, J. Chem. Soc., Dalton Trans., 1985, 2453.
156 K.Stahl, U.Müller and K.Dehnicke, Z. anorg. allg. Chem., 1985, 527, 7.
157 T.Godemeyer, A.Berg, H.-D.Gross, U.Müller and K.Dehnicke, Z. Naturforsch., Teil B, 1985, 40B, 999.
158 U.Kynast, W.Willing, U.Müller and K.Dehnicke, Z. anorg. allg. Chem., 1985, 529, 129.
159 D.G.Gambarov, A.G.Guseinov and A.M.Ayubova, Russ. J. Inorg. Chem., 1985, 30, 519.
160 C.Y.Kim and R.A.Condrate, J. Phys. Chem. Solids, 1984, 45, 1213.
161 R.Lozano, E.Alarcon, J.Roman and A.Doadrio, Polyhedron, 1984, 3, 1021.
162 M.Takami and H.Kuze, Reza Kagaku Kenkyu,1984, 6, 165 (Chem. Abs., 1985, 102, 140064).
163 E.Hey, F.Weller and K.Dehnicke, Z. anorg. allg. Chem., 1984, 514, 18.
164 A.Berg and K.Dehnicke, Z. Naturforsch., Teil B,1985, 40B, 842.
165 R.Kapoor, R.Sharma and P.Kapoor, Z. Naturforsch., Teil B, 1984, 39B, 1702.
166 C.Shortman and R.L.Richards, J.Organometal. Chem.,1985, 286, C3.
167 V.M.Nagiev, Sh.M.Efendiev and V.M.Borlakov, Phys. Stat. Solidi, B, 1984, 125, 467.
168 A.Clearfield, A.Moini and P.R.Rudolf, Inorg. Chem., 1985, 24, 4606.
169 T.N.Lomova, N.I.Volkova and B.D.Berezin, Russ. J. Inorg. Chem. 1985, 30, 352.
170 G.Backes-Dahmann, W.Herrmann, K.Wieghardt and J.Weiss, Inorg. Chem., 1985, 24, 485.
171 M.A.Bennett and I.W.Boyd, J. Organometal. Chem., 1985, 290, 165.
172 M.Dudek and A.Samotus, Transition Met. Chem.,1985, 10, 271.
173 S.Vasanthi, K.S.Nagaraja and M.Rudolph, Transition Met. Chem., 1984, 9, 382.
174 J.L.Corbin, K.F.Miller, M.Pariyadath, S.Wherland, A.E.Bruce, and E.I.Stiefel, Inorg. Chim. Acta, 1984, 90, 41.
175 J.Flanagan, W.P.Griffith, A.C.Skapski and R.W.Wiggins, Inorg. Chim. Acta, 1985, 96, L23.
176 S.M.Bowen, E.N.Duesler, D.J.McCabe and R.T.Paine, Inorg. Chem., 1985, 24, 1191.
177 C.C.Torardi and R.E.McCarley, Inorg. Chem., 1985, 24, 476.
178 X.Xin, N.L.Morris, G.B.Jameson and M.T.Pope, Inorg. Chem., 1985, 24, 3482.

179 N.F.Cole, D.R.Derringer, E.A.Fiore, D.J.Knoechel, R.K.Schmitt, and T.J.Smith, Inorg. Chem., 1985, 24, 1978.
180 S.Mohan and A.Nukunthan, Curr. Sci., 1985, 54, 858.
181 V.S.Khain, N.I.Zhomova and F.I.Andreev, Russ. J. Inorg. Chem. 1985, 30, 201.
182 V.I.Spitsyn, V.L.Balashov, L.N.Lykova and L.M.Kovba Dokl. Akad. Nauk SSSR, 1985, 282, 144.
183 G.S.Girolami, C.G.Howard, G.Wilkinson, H.M.Dawes, M.Thornton-Pett, M.Motevalli, and M.B.Hursthouse, J. Chem. Soc., Dalton Trans., 1985, 921.
184 R.M.Nielson and S.Wherland, Inorg. Chem., 1985, 24, 1803.
185 D.S.Barratt and C.A.McAuliffe, Inorg. Chim. Acta,1985, 97, 37.
186 D.S.Barratt, A.Hosseiny, C.A.McAuliffe and C.Stacey, J. Chem. Soc., Dalton Trans., 1985, 135.
187 D.B.Powell and A.Woollins, Spectrochim. Acta, Part A, 1985, 41A, 1023.
188 C.J.Besecker, V.W.Day, W.G.Klemperer and M.R.Thompson, Inorg. Chem., 1985, 24, 44.
189 K.Schmidt, Z. Chem., 1985, 25, 36.
190 O.Ya.Levanda, A.A.Oblova, A.F.Kuzina, L.I.Belyaeva and V.I.Spitsyn, Russ. J. Inorg. Chem.,1985, 30, 522.
191 R.W.Thomas, M.J.Heeg, R.C.Elder and E.Deutsch, Inorg. Chem., 1985, 24, 1472.
192 K.R.Grundy and K.N.Robertson, Inorg. Chem.,1985, 24, 3898.
193 P.Stavropoulos, P.G.Edwards, G.Wilkinson, M.Motevalli, K.M.A.Malik and M.B.Hursthouse, J. Chem. Soc., Dalton Trans., 1985, 2167.
194 W.A.Herrmann, R.Serrano, U.Küsthardt, E.Guggolz, B.Nuber and M.L.Ziegler, J. Organometal. Chem., 1985, 287, 329.
195 W.A.Herrmann, U.Küsthardt and E.Herdtweck, J.Organometal. Chem., 1985, 294, C33.
196 V.E.Fedorov, A.V.Mishchenko, B.A.Kolesov, S.P.Gubin, Yu.L.Slovokhotov and Yu.T.Struchkov, Izv. Akad. Nauk SSSR, Ser. Khim., 1984, 2159.
197 M.Bettinelli, L.Di Sipio, A.Pasquetto, G.Ingletto and A.Montenero, Inorg. Chim. Acta, 1985, 99, 37.
198 I.I.Kalinichenko, N.M.Titov and M.G.Ivanov, Russ. J. Inorg. Chem.,1984, 29, 1149.
199 A.Biswas, T.Mitra and S.Sen, Z. anorg. allg. Chem., 1985, 526, 179.
200 D.Fenske, A.Berg, F.Weller and K.Dehnicke, Z. anorg. allg. Chem., 1985, 527, 105.
201 D.W.Ball, Z.H.Kafafi, R.H.Hauge and J.L.Margrave, Inorg. Chem., 1985, 24, 3708.
202 W.Kalz, H.Homborg, H.Küppers, B.J.Kennedy and K.S.Murray, Z. Naturforsch., Teil B, 1984, 39B, 1478.
203 L.Zhang, Z.Liu, L.He, Y.Zheng and Y.Hu, Jiegou Huaxue, 1984, 3, 161 (Chem. Abs., 1985, 103, 112353).
204 A.T.Kowal and J.Skarzewski, Spectrochim. Acta, Part A, 1985, 41A, 563.
205 A.Spool, I.D.Williams and S.J.Lippard, Inorg. Chem., 1985, 24, 2156.
206 P.Beardwood and J.F.Gibson, J. Chem. Soc., Chem. Comm., 1985, 102.
207 G.Süss-Fink, G.Herrmann, P.Mrys, J.Ellermann and A.Veit, J.Organometal. Chem., 1985, 284, 263.

208 C.White and E.Cesarotti, J. Organometal. Chem., 1985, 287,123

209 G.J.Baird, S.G.Davies, S.D.Moon, S.J.Simpson and R.H.Jones, J.Chem. Soc., Dalton Trans., 1985, 1479.

210 P.Mura, B.C.Olby and S.D.Robinson, J. Chem. Soc., Dalton Trans., 1985, 2101.

211 S.Farquharson, P.A.Lay and M.J.Weaver, Spectrochim. Acta, Part A, 1984, 40A, 907.

212 C.-M.Che, K.-Y.Wong and C.-K.Poon, Inorg. Chem., 1985, 24, 1797.

213 T.C.W.Mak,C.-M.Che and K.-Y.Wong, J. Chem. Soc., Chem. Comm., 1985, 986.

214 I.A.Khartomik, A.B.Kovrikov, G.G.Novitskii D.S.Umreiko, T.M.Buslaeva and N.m.Sinitsyn, Russ. J. Inorg. Chem., 1984, 29, 1479.

215 A.R.Chakravarty, F.A.Cotton and D.A.Tocher, Inorg. Chem., 1985, 24, 172.

216 D.J.Lockwood, R.W.G.Syme and B.Briat, J.Phys.,C,1985,18,1721.

217 W.Preetz and U.Horns, Z. anorg. allg. Chem., 1984, 516,159.

218 W.Levason, M.Tajik and M.Webster, J.Chem. Soc., Dalton Trans. 1985, 1735.

219 M.C.Jewiss, W.Levason, M.Tajik, M.Webster and N.P.C.Walker, J. Chem. Soc., Dalton Trans.,1985, 199.

220 S.K.Harbron and W.Levason, J. Chem. Soc., Dalton Trans., 1985, 205.

221 W.P.Griffith, N.T.McManus, A.C.Skapski and A.J.Nielson, Inorg. Chim. Acta, 1985, 103, L5.

222 K.K.Pandey and S.R.Ahuja, Inorg. Chem.,1985, 24, 2855.

223 J.A.Cabeza and P.M.Maitlis, J.Chem. Soc., Dalton Trans., 1985, 573.

224 N.J.Campbell, V.A.Davis, W.P.Griffith and T.J.Townend, J. Chem. Soc., Dalton Trans., 1985, 1673.

225 M. del C.Viola, M.S.Augsburger, M.A.Juri and J.C.Pedregosa, An. Asoc. Quim. Argent., 1984, 72, 553.

226 S.Delgado, J.Macazaga, C.Moreno and J.R.Masaguer, J. Organometal. Chem., 1985, 289, 397.

227 I.M.Oglezneva and L.G.Lavrenova, Russ. J. Inorg. Chem., 1985, 30, 840.

228 J.Zsakó, J.Máthé, E.Kästner-Marian and C.Várhelyi, Acta Chim. Hung., 1985, 117, 323.

229 F.Paap, E.Bouwman, W.L.Driessen, R.A.G. de Graaff and J.Reedijk, J. Chem. Soc., Dalton Trans.,1985, 737.

230 R.J.H.Clark, A.J.Hempleman, H.M.Dawes, M.B.Hursthouse and C.D.Flint, J. Chem. Soc., Dalton Trans., 1985, 1775.

231 M.A.Bennett and T.R.B.Mitchell, J.Organometal. Chem., 1985, 295, 223.

232 J.Ellermann and G.Szucsányi, Z. anorg. allg. Chem., 1985, 520, 113.

233 G.Faraglia, F.Barbaro, R.Musumeci and L.Volponi, J. Coord. Chem., 1984, 13, 345.

234 W.Guo, C.Xu, Y.Zhou, Y.Chu and J.Liang, Guangpuxue Yu Guangpu Fenxi , 1985, 5, 31 (Chem. Abs., 1985, 103,184336).

235 J.C.Rodriguez, C.Claver and A.Ruiz, J. Organometal. Chem., 1985, 293, 115.

236 M.J.Auburn and S.J.Stobart, Inorg. Chem., 1985, 24, 318.

237 L.Dahlenburg and A.Yardimcioglu, J. Organometal. Chem., 1985, 291, 371.

238 W.Preetz and D.Tensfeldt, Z. anorg. allg. Chem., 1985, 522, 7.
239 W.Preetz and H.J.Steinebach, Z. Naturforsch., Teil B, 1985, 40B, 745.
240 H.A.Marzouk, K.A.Arunkumar and E.B.Bradley, Surf. Sci., 1984, 147, 477.
241 K.N.Udupa and S.Sarkar, J. Organometal. Chem., 1985, 284,C36
242 A.Bigotto and I.Pellegrini, Spectrochim. Acta, Part A, 1984, 40A, 979.
243 A.L.Balch, Y.W.Chan, M.Olmstead and M.W.Renner, J. Amer. Chem. Soc., 1985, 107, 2393.
244 J.L.Pascal, J.Potier, D.J.Jones, J.Roziere and A.Michalowicz, Inorg. Chem., 1985, 24, 238.
245 G.R.Newkome, W.E.Puckett, G.E.Kiefer, V.K.Gupta, F.R. Fronczek, D.C.Pantaleo, G.L.McClur, J.B.Simpson and W.A.Deutsch, Inorg. Chem.,1985, 24, 811.
246 B.Crociani, F.Dibianca, A.Giovenco and A.Scrivanti, J.Organometal. Chem., 1985, 291, 259.
247 B.N.Ivanov-Emin, L.P.Petrishcheva, B.E.Zaitsev, V.I. Ivlieva, A.S.Izmailovich and V.P.Dolganev, Russ. J. Inorg. Chem., 1984, 29, 1169.
248 B.Vlčková, B.Strauch and M.Horák, Coll. Czech. Chem. Comm., 1985, 50, 306.
249 J.Mink and P.L.Goggin, Kem. Kozl., 1984, 61, 275.
250 M.Habib, H.Trujillo, C.A.Alexander and B.N.Storhoff, Inorg. Chem., 1985, 24, 2344.
251 M.L.Kullberg, F.R.Lemke, D.R.Powell and C.P.Kubiak, Inorg. Chem., 1985, 24, 3589.
252 N.S.Akl and H.A.Tayim, J. Organometal. Chem., 1985, 297, 371.
253 R.A.Michelin, G.Facchin and R.Ros, J. Organometal. Chem., 1985, 279, C25.
254 S.M.Kim, V.Zh.Ushanov and O.V.Agagshkin, Izv. Akad. Nauk. Kaz. SSR, Ser. Khim., 1985, 26.
255 S.M.Kim and O.V.Agashkin, Izv. Akad. Nauk Kaz. SSR, Ser. Khim., 1984, 51.
256 G.Facchin, R.Campostrini and R.A.Michelin, J. Organometal. Chem,,1985, 294, C21.
257 E.N.Yurchenko, T.S.Khodashova, M.A.Porai-Koshits, V.P. Mel'nikova and V.F.Brattsev, Koord. Khim., 1985, 11, 359.
258 G.Annibale, M.Bonivento, L.Canovese, L.Cattalini, G.Michelon and M.L.Tobe, Inorg. Chem., 1985, 24, 797.
259 C.Bisi-Castellani, A.M.Manotti-Lanfredi, M.Tiripicchio, L.Maresca and G.Natile, Inorg. Chim. Acta, 1985, 90, 155.
260 W.Preetz and A.Grabowski, Z. anorg. allg. Chem., 1984, 518, 129.
261 T.V.Koksharova and A.I.Prisyazhnyuk, Russ. J. Inorg. Chem., 1984, 29, 1764.
262 E.Hernandez and P.Royo, J. Organometal. Chem.,1985, 291,387.
263 G.Garcia-Sanchez, G.Lopez Lopez and M.D. Santana Lario, An. Quim., Ser. B, 1983, 79, 214.
264 S.S.Bal'dman, S.D.Nasirdinov and Kh.T.Sharipov, Russ. J. Inorg. Chem., 1985, 30, 88.
265 H.C.Clark, G.Ferguson, V.K.Jain and M.Parvez, Inorg. Chem., 1985, 24, 1477.
266 G.B.Deacon and I.L.Grayson, J. Organometal. Chem., 1985, 292, 1.
267 E.S.Stoyanov, V.A.Mikhailov, I.N.Marov and V.K.Belyaeva Russ. J. Inorg. Chem., 1985, 30, 66.

268 A.Müller, F.W.Baumann, H.Bögge and K.Schmitz, Z. anorg. allg. Chem., 1985, 521, 89.
269 J.R.Nicholson, I.L.Abrahams, W.Clegg and C.D.Garner, Inorg. Chem., 1985, 24, 1092.
270 R.K.Chadha and J.E.Drake, J. Organometal. Chem.,1985, 286, 121.
271 M.Vanliere and T.C.DeVore, High. Temp. Sci.,1984, 18, 185.
272 S.P.Perlepes, D.Nicholls and M.R.Harrison, Inorg. Chim. Acta, 1985, 102, 61.
273 U.Kettler, P.B.Bechthold and W,Krasser, Surf. Sci., 1985, 156, 867.
274 M.D.Baker, G.A.Ozin and J.Godber, J. Phys. Chem., 1985, 89, 305.
275 M.D.Baker, J.Codber and G.A.Ozin, J. Phys. Chem., 1985, 89, 2299.
276 J.E.Pemberton, Chem. Phys. Lett., 1985, 115, 321.
277 A.Popitsch, J. Crystallogr. Spectrosc. Res.,1985, 15, 603.
278 A.Müller, E.Krickemeyer, M.Zimmerman, M.Römer, H.Bögge, M.Penk and K.Schmitz, Inorg. Chim. Acta, 1984, 90, L69.
279 P.J.S.Ewen, T.Han and W.Taylor,Solid State Commun., 1984, 52, 1041.
280 R.Sudharsanan and B.P.Clayman, Solid State Ionics, 1985, 15, 287.
281 U.Gräsle, W.Hiller and J.Strähle, Z. anorg. allg. Chem., 1985, 529, 29.
282 S.Al-Baker, W.E.Hill and C.A.McAuliffe, J. Chem. Soc., Dalton Trans., 1985, 2655.
283 J.D.Woollins, Polyhedron, 1984, 3, 1365.
284 R.Uson, A.Laguna, M.Laguna, B.R.Manzano, P.G.Jones and G.M.Sheldrick, J. Chem. Soc., Dalton Trans., 1985, 2417.
285 J.Vicente, M.T.Chicote, J.A.Cayuelas, J.Fernandez-Baeza, P.G.Jones, G.M.Sheldrick and P.Espinet, J.Chem. Soc., Dalton Trans., 1985, 1163.
286 R.Usón, A.Laguna, M.U. de la Orden , R.V.Parish and L.S.Moore, J. Organometal. Chem., 1985, 282, 145.
287 E.G.Bakalbassis, J.Mrozinski and C.A.Tsipis, Inorg. Chem., 1985, 24, 3548.
288 A.I.Prisyazhnyuk and V.A.Karpinchuk, Russ. J. Inorg. Chem., 1985, 30, 685.
289 E.Grazynska and W.Wojciechowski, Pol. J. Chem., 1984, 58,977
290 Y.Nagel and W.Beck, Z. anorg. allg. Chem., 1985, 529, 57.
291 M.A.S.Goher, Bull. Soc. Chim. Fr., 1985, 159.
292 J.G.Contreras, C.A.Lopez and G.V.Seguel, Spectrosc. Lett., 1985, 18, 71.
293 A.Agarwal, M.B.Patel, M.Pal and H.D.Bist, Spectrochim. Acta, Part A, 1984, 40A, 1063.
294 J.G.Contreras, C.A.Lopez and G.V.Seguel, Spectrosc. Lett., 1985, 18, 601.
295 I.Persson, M.Sandström, P.L.Goggin and A.Mosset, J. Chem. Soc. Dalton Trans., 1985, 1597.
296 N.Ueyama, M.Nakata and A.Nakamura, Bull. Chem. Soc. Japan, 1985, 58, 464.
297 M.Cano, R.Criado, E.Gutierrez-Puebla, A.Monge and M.P.Pardo, J.Organometal. Chem., 1985, 292, 375.
298 B.L.Khandelwal and S.K.Jain, Proc. Int. Conf. Org. Chem., Selenium, Tellurium,4th, 1983, 83.
299 P.B.Bentham, C.G.Romak and H.F.Shurvell, Can. J. Chem., 1985, 63, 2303.

300 I.M.Vezzosi, A.Albinati and F.Ganazzoli, Inorg. Chim. Acta, 1985, 96, 9.

301 M.Mossoyan-Deneux, D.Benlian, M.Pierrot, A.Fournel and J.P.Sorbier, Inorg. Chem., 1985, 24, 1878.

302 N.M.Belyi, A.V.Bobyr, I.S.Gorban, V.A.Gubanov and V.V.Frizel, Fiz. Tverd. Tela, 1985, 27, 2734.

303 J.Lamotte, J.C.Lavalley, V.Lorenzelli and E.Freund, J. Chem. Soc., Farad. Trans.,I, 1985, 81, 215.

304 S.A.Gaziev, N.A.Gorshkov, L.G.Mashrov and D.N.Suglobov, Radiokhim., 1984, 26, 316.

305 J.Heckly, Y.Repelin and N.Q.Dao, Spectrochim. Acta, Part A, 1985, 41A, 743.

306 A.Dormond, A. El Bouadili, A.Aaliti and C.Moise, J. Organometal. Chem., 1985, 288, C1.

307 W.I.Azeez and A.I.Abdulla, Inorg. Chim. Acta, 1985, 110, 15.

308 P.Vast and A.Semmoud, J. Fluorine Chem., 1985, 27, 47.

309 M.T.El-Haty and F.A.Adam, Bull. Soc. Chim. France,I,1984, 284.

310 M.N.Bhattacharjee, M.K.Chaudhuri and R.N.D.Purkayastha, J. Chem. Soc., Dalton Trans., 1985, 409.

311 M.Takami, T.Oyama, T.Watanabe, S.Namba and R.Nakane, Reza Kagaku Kenkyu, 1984, 6, 162(Chem. Abs., 1985, 102, 140063).

312 R.S.McDowell, M.J.Reisfeld, N.G.Nereson, B.J.Krohn and C.W.Patterson, J. Mol. Spectrosc., 1985, 113, 243.

313 L.B.Serezhkina, G.V.Vlasova and V.N.Serezhkin, Russ. J. Inorg. Chem., 1985, 30, 406.

314 V.V.Yakshin and N.L.Khokhlova, Koord. Khim., 1985, 11, 1088.

315 V.V.Skopenko, E.I.Ivanova, S.G.Andreeva and H.Keller, Dopov. Akad. Nauk Ukr. RSR, Ser. B, Geol. Khim., Biol. Nauki, 1985, 51 (Chem. Abs., 1985, 103, 63817).

316 A.O.Baghlaf, M.Ishaq, O.A.S.Ahmed and M.A.Al-Julani, Polyhedron, 1985, 4, 853.

317 V.N.Serezhkin and L.B.Serezhkina, Russ. J. Inorg. Chem., 1985, 30, 1156.

318 S.C.Nayar, V.Chakravortty and K.C.Dash, Ind. J. Chem. Sect. A 1984, 23A, 602.

319 I.S.Ahuja, Synth. React. Inorg. Met.-Org. Chem.,1985, 15, 93

320 R.Mercier, M.Pham Thi and P.Colomban,Solid State Ionics, 1985, 15, 113.

321 M.Pham Thi and P.Colomban, J. Less-Common Met., 1985, 108, 189.

322 K.K.Bhasin, S.Jindal, S.Singh and R.D.Verma, Synth. React. Inorg. Met.-Org Chem., 1985, 15, 733.

323 N.G.Gorshkov and L.G.Mashirov, Radiokhim., 1984, 26, 540.

6
Vibrational Spectra of Some Co-ordinated Ligands

BY G. DAVIDSON

1 Carbon, Silicon, and Tin Donors

The i.r. spectrum of $Li(C_2H_2)$ has been obtained in an Ar matrix. Isotopic studies ($^6Li/^7Li$, C_2H_2, C_2HD, C_2D_2 and $^{13}C_2H_2$) show that the structure is probably planar, with the Li bridging the π-system. νCC is at 1655 cm^{-1}, but the decrease was ascribed to shared electron density between C_2H_2 and Li, rather than Li-C_2H_2 π-back-bonding.[1]

Vibrational spectra were reported for MC_5H_5, where M = Li, Na or K, and the Raman spectra of LiC_5H_5, NaC_5H_5 in THF solution. From the ρCH wavenumbers it was shown that the polarity of the M-C_5H_5 increased on going from the solid to a THF solution. A weak Raman band (130-150cm^{-1}) for the Li and Na compounds in solution was assigned to an anion tilting motion in a tight ion pair $M^+C_5H_5^-$.[2]

The i.r. spectra of CpBeX and Cp_2Be in solution were compared. The data for Cp_2Be can best be explained by the formulation (1).[3] (2) and other complexes with this tin ligand have νSn-N in the range 365-405cm^{-1}.[4]

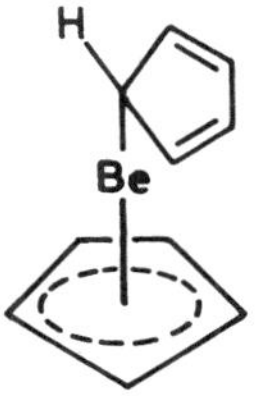

(1)

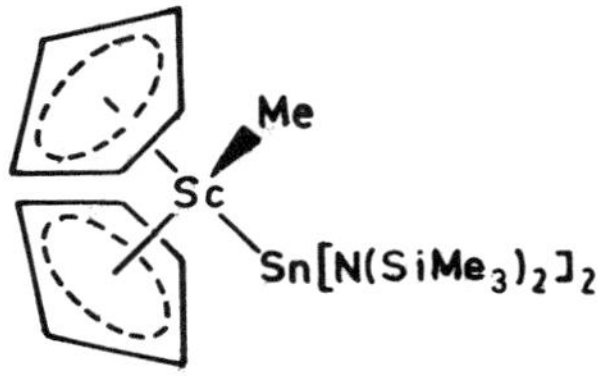

(2)

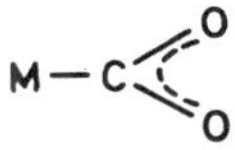

(3a)

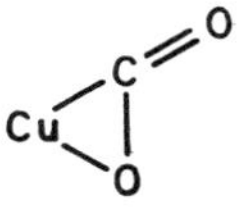

(3b)

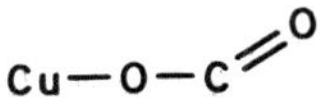

(3c)

νCO_2 modes for CO_2/transition-metal complexes in low-temperature matrices suggest that Ti, V and Cr complexes have a side-on structure, with angle ca.160°; Fe, Co and Ni give the structure (3a), while Cu gives either (3b or c).[5]

The disappearance of $\nu C{=}C$ and slight shift of $\nu C{=}O$ to lower wavenumbers, compared to the free ligand, show the coordination of V to C=C in Cp_2VL, where L = $CH_2{=}CHCHO$, $CH_3CH{=}CHCHO$ or $CH_2CHCOCH_3$.[6] Photolyses of MCp_2H_3 (M = Nb or Ta) or $MCp_2(H)CO$ produce 16-electron complexes MCp_2H and (in the latter case) small amounts of $MCp_2(CO)$. The reactions were followed by i.r. spectroscopy.[7]

'Isolated' methyl CH stretching wavenumbers (νCH^{is}) were reported for $M(CHD_2)(\eta^5\text{-}C_5H_5)(CO)_3$, where M = Cr, Mo or W. The C-H bond strengths decrease, Cr > Mo > W, reflecting a corresponding increase in the M-C bond strengths.[8] $\nu C{\equiv}N$ in $(OC)_5MOC{\equiv}N\text{-}N{=}CRR'$, where R,R' = H, Me, Et etc., M = Cr or W, are 40-50cm^{-1} higher than in the free ligand. Pt(II) analogues give a greater increase.[9] νCO in the ylide complexes (4) (X = CH, ER_2 = PPh_2, PMe_2, $P(NMe_2)_2$, $AsPh_2$; X = $C{=}PPh_3$,O, ER_2 = $AsPh_2$; X = NMe, ER_2 = $AsMe_2$; X = S, ER_2 = $AsPh_2$) show that the donor capacity of the ylides increases in the series: $H_2C{=}AsPh_2SAsPh_2 \sim H_2CAsPh_2OAsPh_2 \ll H_2C{=}AsMe_2NMeAsMe_2 \sim H_2C{=}AsPh_2CH_2AsPh_2 \ll H_2C{=}AsPh_2C({=}PPh_3)PPh_2$.[10]

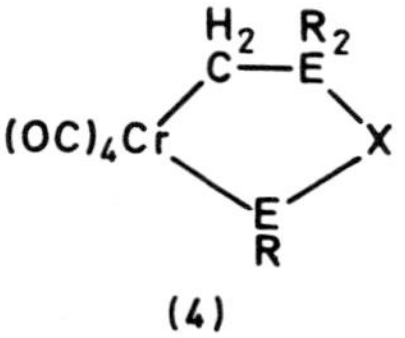

(4)

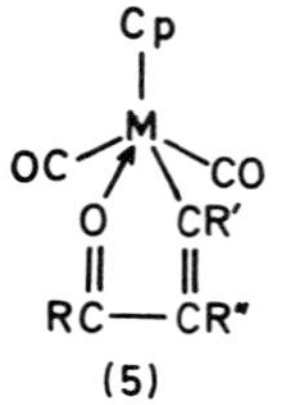

(5)

The complexes (5), where M = Cr, Cp = C_5H_5, C_5Me_5, R = Me, and a series of related Mo and W species all have $\nu C{=}O$ in the range 1510-1470cm^{-1}. This low value is a result of the C=O-M coordination.[11] Incoherent inelastic neutron scattering and far-infrared spectra were reported for $(C_7H_7)^+BF_4^-$ and $[(\eta^7\text{-}C_7H_7)M(CO)_3]^+BF_4^-$, where M = Cr, Mo or W. These revealed characteristic shifts in a number of tropylium modes on complexation.[12] $\nu C{=}O$ is near 1550cm^{-1} in (6), where M = Mo or W, showing the coordination of the ester oxygen to M.[13] $Mo_2(S_2CNEt_2)_3[\mu\text{-}CSC(S)S](\mu\text{-}S_3C_2NEt_2)$ has i.r. bands at 1038 and 995cm^{-1} due to the exo- and endocyclic modes of the -CSC(=S)s- ligand.[14] $[MCl_4(PhC{\equiv}C\text{-}C{\equiv}C\text{-}Ph)]_2$, where M = Mo or W, have νCC bands at 2135, 1703cm^{-1} (Mo), 2155, 1698cm^{-1}(W), showing that one C≡C is η^2-bonded to the metal, with one free.[15]

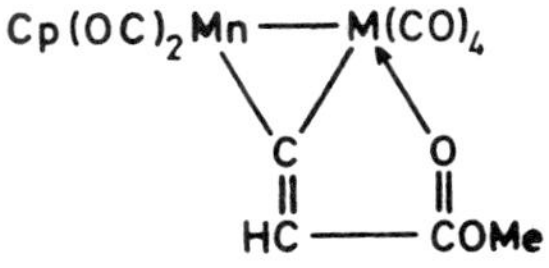

(6)

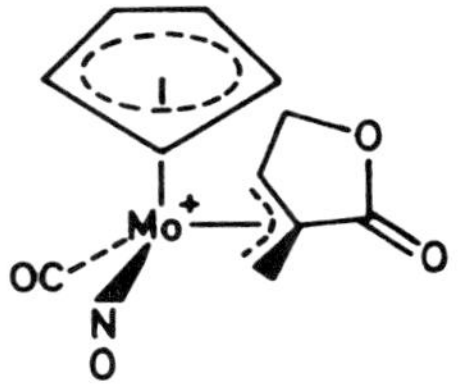

(7)

(7) has νC≡O and νC=O at higher wavenumber than in the neutral precursor dicarbonyl species, together with νNO at 1705cm^{-1}.[16] νC≡C in $MoO(S_2CNR_2)_2(C_8H_{12})$, where R = Me or Et, C_8H_{12} = cyclo-octyne, are near 1850cm^{-1}. Replacing the O by Cl_2 reduces this to about 1750cm^{-1}.[17] νC≡C for the cyclo-octyne complexes $M(S_2CNR_2)_2(C_8H_{12})\{(C_8H_{12})_2CO\}$, where R = Me or Et, are 20-30cm^{-1} lower for M = W than for M = Mo, showing the stronger bonding of C_8H_{12} to the third-row element.[18] νSnCl was assigned in the seven-coordinate complexes $(bipy)P(4\text{-}XC_6H_4)_3(OC)_2Mo(Cl)(SnCl_3)$.[19]

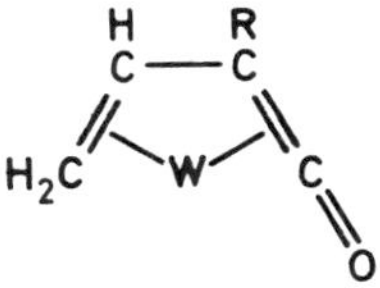

(8)

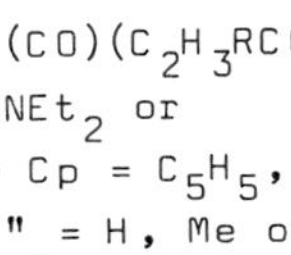

(9)

The coordinated vinylketene ligand (8) in $M(CO)(C_2H_3RCO)\text{-}(BB)_2$, where R = H or Me, M = Mo or W, BB = S_2CNEt_2 or $S_2P^iPr_2$, gives νCH at <u>ca</u>.3050cm^{-1}.[20] (9), where Cp = C_5H_5, C_5H_4Me or C_5Me_5, R = Me, Et, ^nPr or ^nBu, R,R',R" = H, Me or Ph, all have νC=O at rather low values (1595-1607cm^{-1}), showing reduced double-bond character.[21] (10) has $\nu C=O$ of the CO_2 unit of oxatungstcyclobutenone at $1645cm^{-1}$.[22] $[WCl_4(Me_3SiC{\equiv}CSiMe_3)]_2$ has νCC at $1621cm^{-1}$ ($2109cm^{-1}$ in the free ligand).[23] $WCl_4(IC{\equiv}CI)_2$ and $[WCl_5(IC{\equiv}CI)_2]^-$ have νCC of the sideways coordinated alkyne at 1619, $1590cm^{-1}$, 1640, $1605cm^{-1}$ respectively. The former has νCI at $687cm^{-1}$, compared to $720cm^{-1}$ in the free ligand.[24]

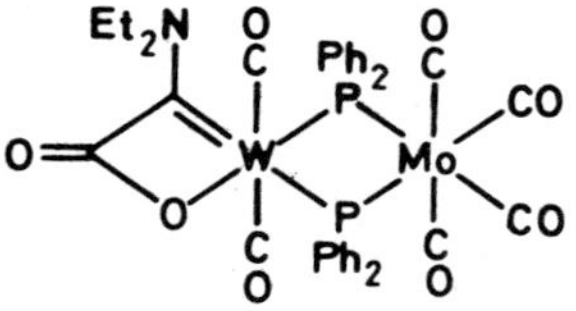

(10)

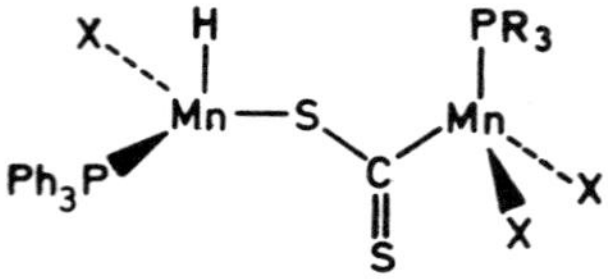

(11)

The coordinated CS_2 in (11), where X = Cl, Br or I, PR_3 = P^nPr_3, P^nBu_3, $PPhMe_2$, $PPhEt_2$ or PPh^nPr_2, gives i.r. bands due to $\nu C=S$ at $1110cm^{-1}$ and $\nu C-S$ in the range $800-830cm^{-1}$.[25] $\nu C=C$ bands in $Li_2[Mn(CH=CH_2)_4]$ and $Li_2[Mn(CH=CMe_2)_4]$ are at 1525 and $1555cm^{-1}$ respectively.[26] The coordinated ethene in $MnH(CH_2=CH_2)(dmpe)_2$ has νCH at 3020, $3005cm^{-1}$ and νCC at $1175cm^{-1}$.[27] $\nu C{\equiv}C$ modes gave weak i.r. bands near $2100cm^{-1}$ in $Mn(CO)_nL_{5-n}(C{\equiv}CR)$, where n = 3, L = dppe, dppm, phen, bipy, R = H, Ph or ^tBu, n = 2, L_{5-n} = (dppe or dppm)$(P(OR')_3)$.[28] The cyclobutadiene ligand in (12) has νCH at 2985, $2965cm^{-1}$, δCH_2 1462, $1371cm^{-1}$, and νCC $1535cm^{-1}$.[29] I.r. spectroscopy shows that reactions of (η^5-indenyl)$M(CO)_3$, where M = Mn or Re, with $AlCl_3.HCl$ or FSO_3H involve a metallotropic $\eta^5{\rightarrow}\eta^6$ rearrangement, with the metal carbonyl migrating from the 5- to the 6-membered ring of the aromatic ligand.[30]

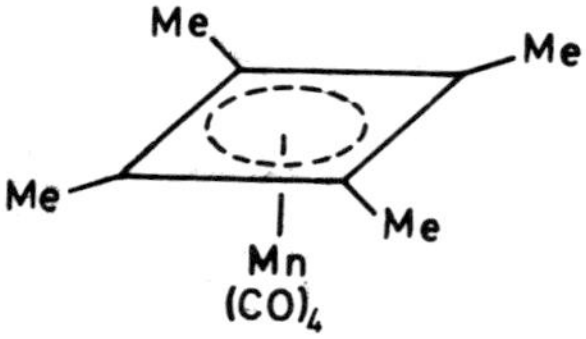

(12)

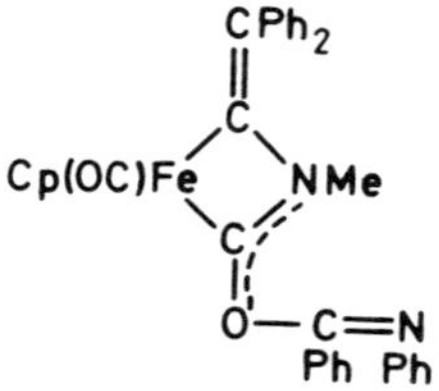

(13)

I.r. bands due to Fe=CH_2 were identified in Fe/CH_2N_2 systems in low-temperature matrices. The assignments were helped by ^{13}C substitution.[31] The formulation of (13) follows from the following assignments: $\nu C\overset{...}{=}O$ 1489cm^{-1}, $\nu C\equiv O$ 1925cm^{-1}, νC=N 1688cm^{-1} and νC-O 1242cm^{-1}.[32] (η^5-C_5H_5)FeH was detected in argon matrices, with νCH at 3106.5cm^{-1}, νCC at 1363cm^{-1}, together with several characteristic CH deformation bands.[33] Co-deposition of Fe, Co or Ni atoms with benzene or toluene at liquid-nitrogen temperatures gave evidence for the formation of 1:1 and 2:1 arene-metal π-complexes, with i.r. spectra analogous to that of $Cr(C_6H_6)_2$.[34]

The complex (14) has νC=O at 1549cm^{-1}, showing coordination to the Ru atom.[35] In (15), where Ar = Ph, 3,4-$Me_2C_6H_3$ etc., νC=C and νN=N appear as i.r. bands near 1550cm^{-1}.[36]

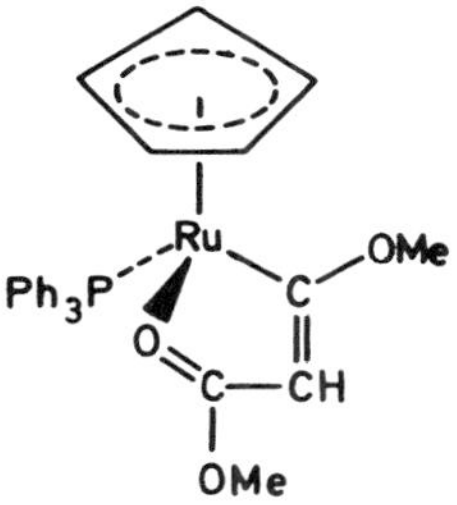

(14)

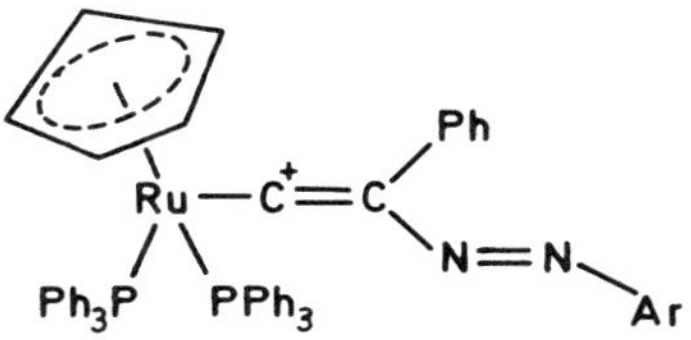

(15)

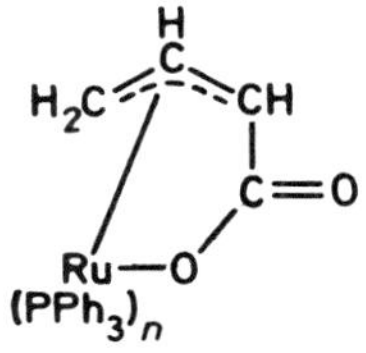

(16)

All νC=O bands in (16), where n = 2, PR_3 = PPh_3, P(C_6H_4-p-OMe)$_3$, n = 3, PR_3 = P(C_4H_4-p-Cl)$_3$, are at low wavenumbers (1535-1545cm^{-1}).[37] Ru(CO)[C_6H_3MeC(O)C_6H_4Me]XL_2, where X = Cl or I, L = PMe_2Ph or $AsMe_2$Ph, have νC=O(acyl) near 1590cm^{-1}, consistent with the presence of (17).[38]

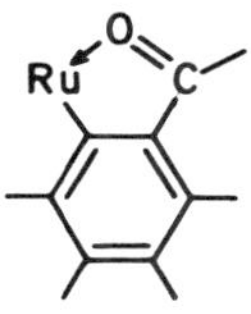

(17)

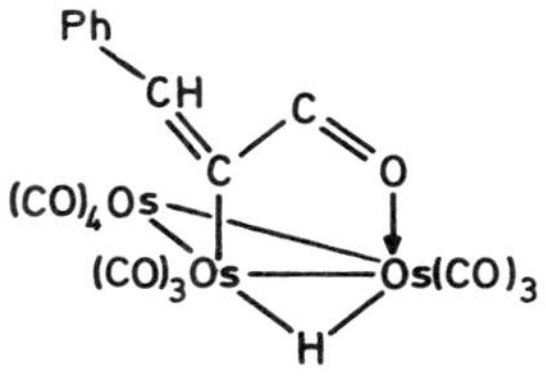

(18)

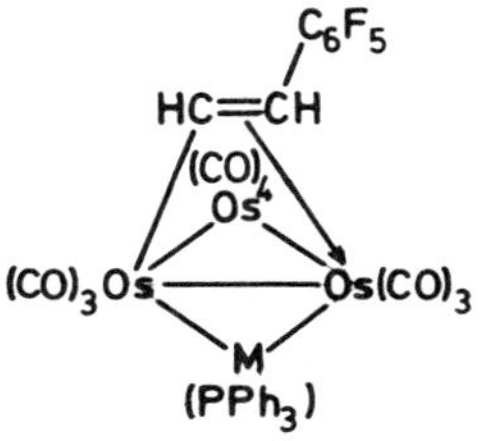

(19)

In (18) the νC=O band due to the coordinated carbonyl group is at $1515cm^{-1}$.[39] The νC=C bands in (19), where M = Cu, Ag or Au, were assigned as follows: (M = Cu) $1608cm^{-1}$, (Ag) $1602cm^{-1}$, (Au) $1618cm^{-1}$.[40]

Detailed ligand-mode assignments were made from the i.r. spectra of $Co_4(CO)_{10}(\mu_4\text{-}\eta^2\text{-HCCH})$ and its deuterio- and ^{13}C analogues. νCC was at very low wavenumbers, i.e. there is a great deal of π-back-bonding from the Co_4 unit. There is great similarity between the C_2H_2 modes here and for C_2H_2 adsorbed on metal surfaces.[41] νC≡C in RC≡CCo(salen)py, where salen = (20), are at $2160cm^{-1}$ (R = CF_3), $2120cm^{-1}$ (Ph).[42] (21) has νC=C at $1625cm^{-1}$ (M = Co) or $1646cm^{-1}$ (Rh).[43] νC=O is at $1750cm^{-1}$ and νC≡O at 2035 and 1980 cm^{-1} in (22).[44]

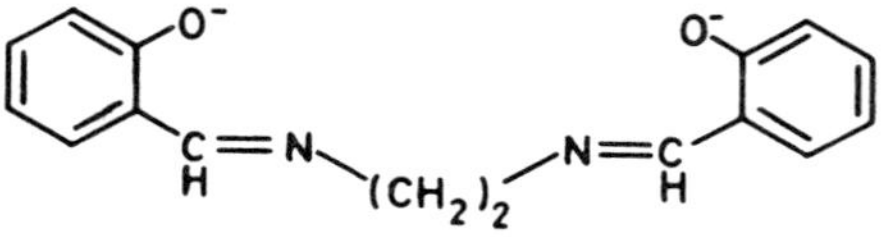

(20)

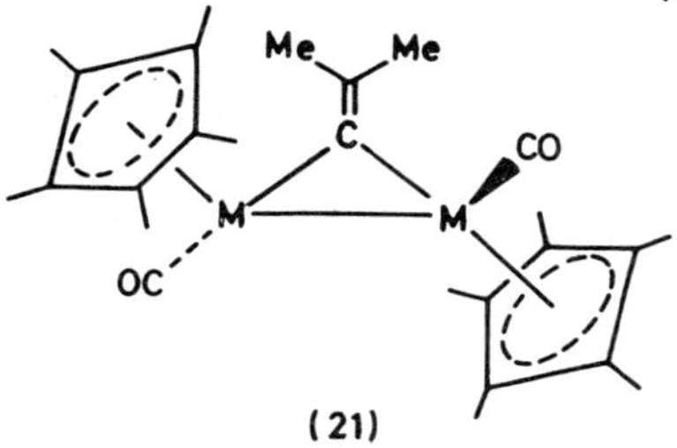

(21)

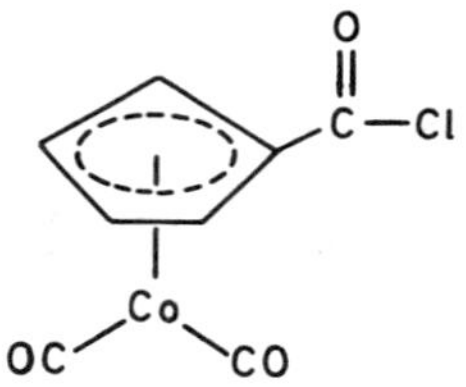

(22)

The Raman spectrum of C_2H_2 on Rh/Al_2O_3 showed that the dominant species at room temperature was $(\eta^2\text{-}\pi\text{-}C_2H_2)$, i.e. it is different from the species formed on single-crystal faces of rhodium.[45] νC=C wavenumbers were assigned from the i.r. spectra of trans-[RhCl(=C=CHR)(PR'$_3$)$_2$], where R' = ^{i}Pr, R = H, Me or Ph.[46] The C=C stretch in (23) is at $1646cm^{-1}$.[47] (2,6-^{t}Bu$_2$-4-Me-C_6H_2O)RhL, where L = $(CO)_2$, $(PPh_3)_2$, (CO)(PPh_3) or cyclo-C_8H_{12}, all have νC⩦C⩦O bands above $1500cm^{-1}$, i.e. much higher than νCO in the free phenol ($1255cm^{-1}$).[48] νC=O appears as a strong infrared band at $1510cm^{-1}$ in (24).[49] Analogous modes are at $1648cm^{-1}$ (L = PCy_3) or $1650cm^{-1}$ (PPh_3) in the species (L)Ni(CH_2CMeCMeCH_2COO), showing the presence of a unidentate carboxylate group.[50] νCC was assigned for (25), where R = H or Me and related species.[51]

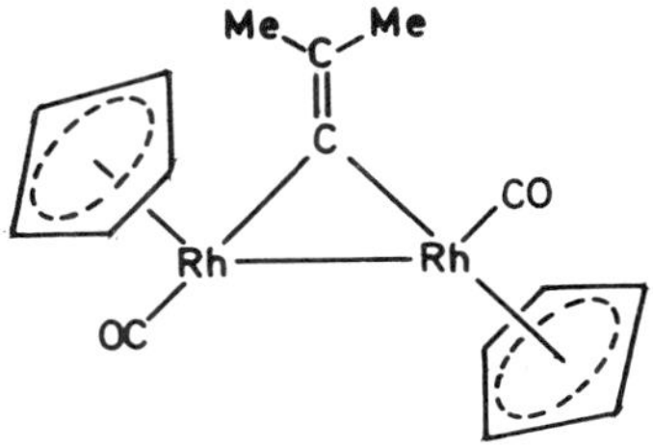

(23)

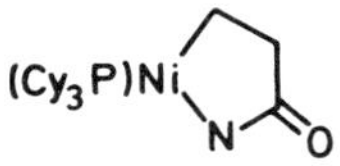

(24)

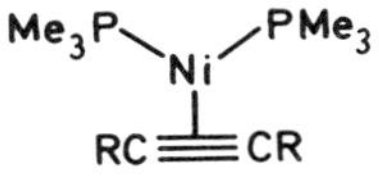

(25)

$[NiX(P^nBu_3)_2]_2(\mu PhC\equiv CR)$ have the following $\nu C\equiv C$ bands in their i.r. spectra: R = H, X = Cl or Br 1740cm^{-1}, R = Ph, X = Cl 1790 cm^{-1}.[52]

Surface ethylidyne complexes have been characterised for the first time on Pd/Al_2O_3 surfaces. The vibrational assignments are: δCH_3 1333cm^{-1}, νC-C 1088cm^{-1}, ν_sCH_3 2867cm^{-1}.[53] M(pnp)(CONHR) (where pnp = 2,6-bis(diphenylphosphinomethyl)pyridine; M = Pd, R = Ph, p-MeC_6H_4, p-$MeOC_6H_4$, C_6H_{11} or ^tBu; M = Pt, R = Ph) all have characteristic bands of carbamoyl ligands at 1580-1620cm^{-1}, with the NH stretch 3210-3400cm^{-1}.[54] (26), where R = Et or Ph, have νC=N about 20cm^{-1} lower than in the parent (27).[55] In (28) (where M = Pd, E = Se, P⁀P = Ph_2PCH_2-o-C_6H_4-CH_2PPh_2; M = Pt, E = S or Se, P⁀P = $Ph_2PCH_2CH_2PPh_2$) a band characterising the bridging CS_2 or CSSe ligand is seen near 900cm^{-1}.[56]

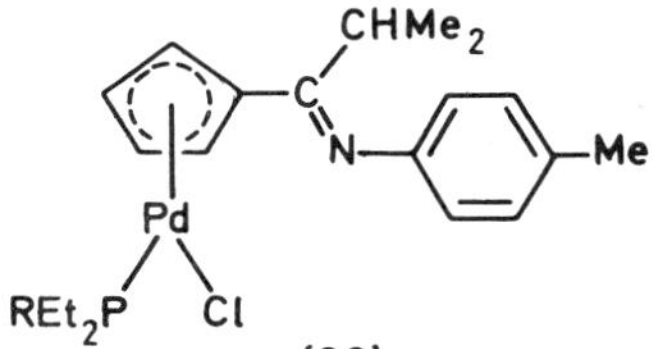

(26)

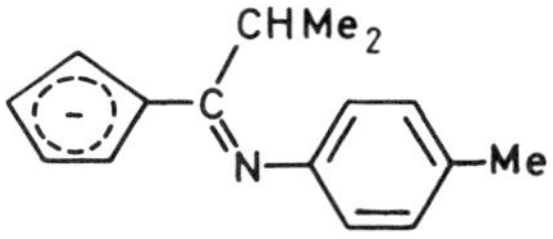

(27)

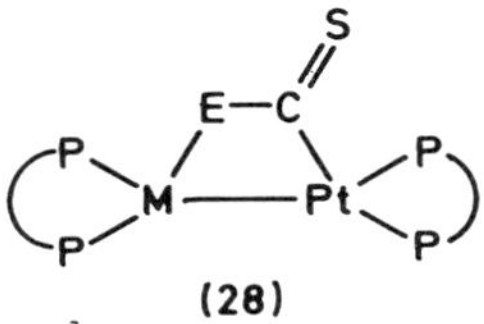

(28)

νNH is at 3344-3348cm^{-1} in (29), where PR_3 = PMe_2Ph, [M] = trans-$PtCl_2(PR'_3)_2^+$, PR'_3 = PPh_3, $PMePh_2$, PMe_2Ph. When [M] = cis-$PtCl_2(PMe_2Ph)_2$, however, it is at 3258cm^{-1}.[57] (30) has νC=N at

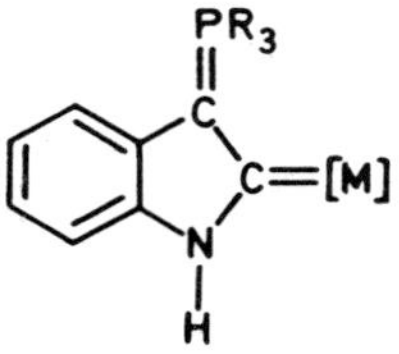

(29)

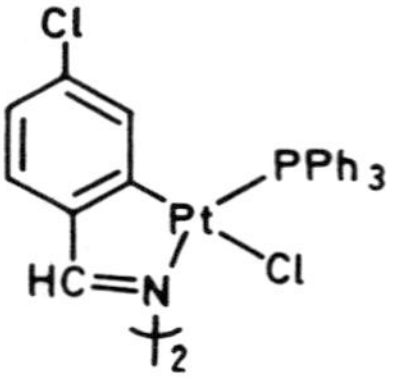

(30)

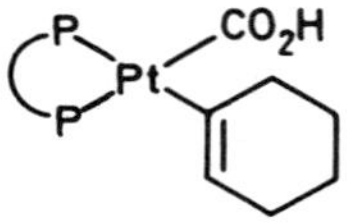

(31)

1600cm^{-1}, lowered as expected by the N-Pt coordination shown.[58] In (31), where P⌒P = $Ph_2P(CH_2)_nPPh_2$, n = 2,3 or 4, or cis-$Ph_2PCH=CHPPh_2$, νOH of the carboxylate group is at about 2670cm^{-1} in the solid (hydrogen bonding) but near 3440cm^{-1} (free monomer) in CH_2Cl_2 solution.[59] νC≡C is at 2110cm^{-1} and νC=O near 1700cm^{-1} in (32).[60] The complex (33) has νS=O at 1053cm^{-1}. Formation of an adduct by coordination of S=O to Cu lowers this to 966cm^{-1}.[61]

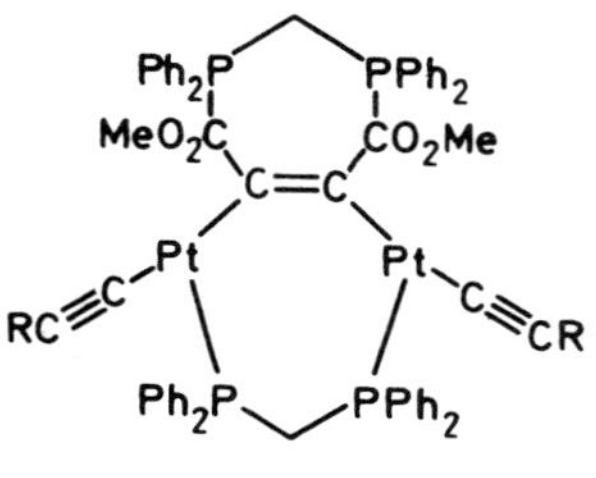

(32)

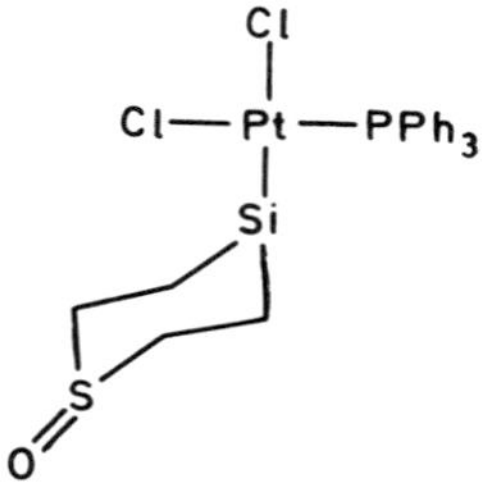

(33)

Assignmnents to νC=O were proposed for $[CuL(X)]^+ClO_4^-$, where X = C_2H_4, C_3H_6, C_6H_{10}, L = tmen, bipy etc. All were in the range 1515-1535cm^{-1}, but no mention was made of coupling with δCH_2. In $Cu(L)CO(ClO_4)$ there were bands characteristic of unidentate ClO_4 and νCO at 2110cm^{-1}.[62] $[Cu(PPh_3)(\mu_3\text{-}C{\equiv}CPh)]_4$ has a single νC≡C band at 2020cm^{-1}, confirming the equivalence of the phenylacetylide ligands.[63]

νCO in (34), where R = Me or Et, is at quite a low value. There must therefore be some back-donation from Au^I into the CO antibonding orbitals. Au^{III} analogues give νCO at much higher wavenumbers.[64]

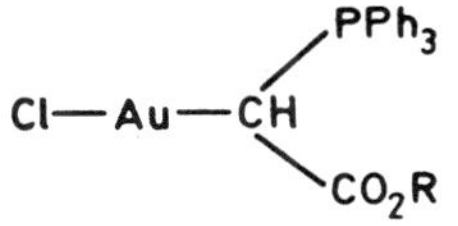

(34)

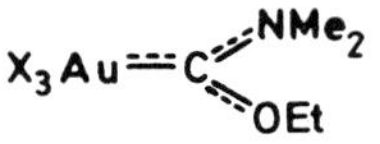

(35)

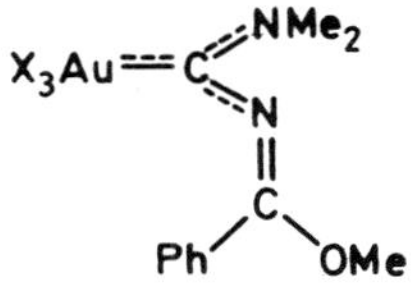

(36)

The complexes (35), where X = Cl, Br or I, have νC⋯N near 1610cm^{-1} and νC⋯O near 1310cm^{-1}. In (36) νC⋯N is near 1590 cm^{-1}, νC=N near 1660cm^{-1}.[65] νCC stretching modes decrease by 70-140cm^{-1} when alkynes or alkenes are adsorbed on gold electrodes. The bands were observed by SERS. It was suggested that π ⟶ s (σ-bond) overlap is more important than d→π* (π-bond) interaction.[66] Similar measurements for benzene and substituted benzenes adsorbed on gold electrodes gave evidence for π-adsorbate surface bonding.[67]

Ligand-mode assignments were proposed for the new species $(CF_3)_2Zn$ and $(C_6F_5)_2Zn$, as complexes with glyme, diglyme or pyridine, e.g. ν_s, $\nu_{as}CF_3$ were at 1150, 960cm^{-1} respectively in $(CF_3)_2Zn$.diglyme.[68]

I.r. and Raman spectra were reported, and some ligand modes assigned, for $Et_3GeC{\equiv}CH$, $Et_3GeC{\equiv}CGeEt_3$ and their ^{13}C-substituted analogues.[69] (37) gives νC=O assignments as follows: 1650cm^{-1} (R = Me), 1615cm^{-1} (OMe), 1658cm^{-1} (OEt).[70]

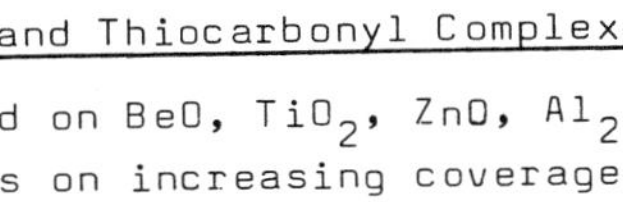

(37)

2 Carbonyl and Thiocarbonyl Complexes

^{12}CO and ^{13}CO adsorbed on BeO, TiO_2, ZnO, Al_2O_3 or SiO_2 show significant νCO shifts on increasing coverage, due to the effect on the metal electron-donating ability by adsorption at an adjacent site.[71]

νCO in $Cp_2M(CO)PMe_3$, where M = Ti, Zr or Hf, are much lower for M = Ti than for $Cp_2Ti(CO)_2$, showing the increased Ti-to-CO back-bonding in the mixed-ligand complexes. νCO decreases in the series Ti > Zr > Hf, again showing the trend to increased back-bonding.[72]

The i.r. spectra of $VCp(CO)_3EMe_2$, where E = S, Se, Te, show that νCO decreases in the order S > Se > Te.[73] Carbonyl stretches in (η^5-indenyl)$V(CO)_4$ are at 2022cm^{-1} (a_1), 1955, 1930 cm^{-1} (e). The splitting of the e mode shows the reduction in symmetry from $\underline{C}_{4v}$.[74] νCO modes were assigned for a wide range of $V(CO)_5PR_3^-$ complexes; all were consistent with approximate $\underline{C}_{4v}$ symmetry for the $V(CO)_5$ fragment.[75] A linear relationship was found between ^{51}V chemical shifts and σ/π-bonding parameters derived from νCO data for $V(CO)_5L^-$, where L = CNR, CN^-, NCR, NR_3, pyridines, $S(O)R_2$, $OCMe_2$, ER_2 (E = O, S, Se or Te).[76]

Carbonyl stretches for $M_2X_3(CO)_8^-$, where M = Nb or Ta, X = Cl, Br or I, show the bands expected for $\underline{C}_{4v}$ symmetry of the $M(CO)_4$ units, i.e. there is very little coupling across the M_2X_3 bridge.[77]

FTIR spectra of $Cr(CO)_2(CS)[(MeO)_3P]_3$ suggest mer-stereochemistry, with two carbonyls trans, not the expected fac-geometry.[78] I.r.(νCO region) was used to monitor the low-temperature matrix photochemistry of (η^4-diene)$Cr(CO)_4$ in Ar, N_2 or CO matrices.[79] I.r. and Raman data were used to follow the photolysis of $M(CO)_6$, where M = Cr or W, in the presence of CO_2 or N_2O. There was evidence, for example, for intermediates such as trans-$O_2M(CO)_4$.[80]

Fast time-resolved i.r. spectra show that the gas-phase photolysis of $Cr(CO)_6$ forms $Cr(CO)_5$. The first gas-phase i.r. spectrum of naked $Cr(CO)_5$ was obtained.[81] Convincing evidence was found for the formation of $Cr(CO)_5H_2$ in both H_2-doped liquid xenon[82] and H_2-saturated C_6H_{12}.[83] Transient i.r. spectra of $Cr(CO)_5.C_6H_{12}$ show that $Cr(CO)_5$ has $\underline{C}_{4v}$ geometry at room temperature, with a similar structure to that found in low-temperature matrices. There was evidence for $Cr(CO)_5(H_2O)$ in the presence of trace water.[84]

$M(CO)_5^-$, where M = Cr, Mo or W, give the carbonyl stretching bands expected for $\underline{D}_{3h}$ symmetry in the presence of large cations. Smaller cations lead to ion pairing, with dramatic effects on the νCO patterns.[85] νCO bands for fifty $M(CO)_5L$ complexes, where M = Cr, Mo or W, L = $R_2EE'R$ or $R_2EER'_2$ (R, R' = CH_3, CF_3, E = P, As, E' = S, Se or Te), show that the donor/acceptor ratio decreases as follows: EMe_3 > $Me_2EE(CF_3)_2$, $Me_2EE'CF_3$ > $(CF_3)_2EE'Me$ > $E(CF_3)_3$.[86] CO stretches for $M(CO)_5(R_2PTe)$, where M = Cr, Mo or W, R = ^tBu, show a symmetry less than $\underline{C}_{4v}$ for the $M(CO)_5$ unit.[87]

Absolute integrated i.r. intensities in CS_2 solutions for νCO, νCS and νCSe in $Cr(CO)_5(CX)$ and $(\eta^6\text{-}C_6H_6)Cr(CO)_2(CX)$, where X = O, S, Se, are consistent with the acceptor abilities CSe > CS > CO.[88] The bridging carbonyls in $CpFeM(CO)_6P(H)$-(N^iPr_2), where M = Cr or W, give i.r. bands close to $1810 cm^{-1}$.[89]

$(NMe_4)[Mo(CO)(CH_3SC_6H_4S)_3]$ has a very low carbonyl-stretching wavenumber ($1775 cm^{-1}$), showing that the electron density at the molybdenum atom must be very high.[90] (38), where P = PPh_3, has i.r. bands due to νCO_t at $1830 cm^{-1}$ (Mo), $1930 cm^{-1}$ (Ru), and to νCO_{br} at $1775 cm^{-1}$.[91] Cis-$Mo(CO)_4(4,4'\text{-}X_2\text{-bipy})$, where X = OMe, CMe_3, Me, CO_2Me or NO_2, give solution i.r. spectra showing preferential interaction between the solvent and equatorial CO ligands trans to the 4,4'-X_2-bipy.[92]

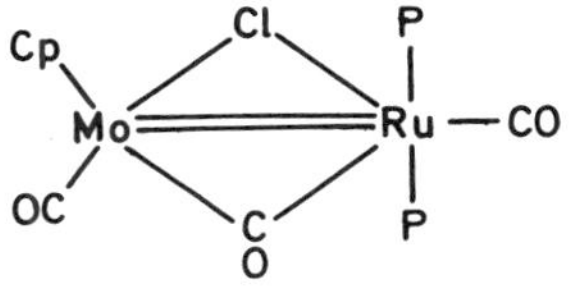

(38)

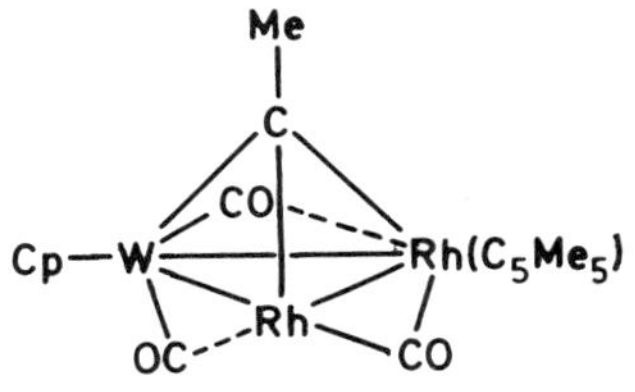

(39)

Semi-bridging and bridging carbonyls in (39) give νCO bands at 1860and $1795 cm^{-1}$.[93] I.r. (νCO), with ^{13}CO enrichment and energy-factored force-field calculations, were used to follow photochemical reactions of $W(\eta^1\text{-}C_3H_5)(\eta^5\text{-}C_5H_5)(CO)_3$ and $Mn(\eta^1\text{-}C_3H_5)(CO)_3$ in inert-gas matrices at 12K.[94] $WBr_3(CO)_4^-$ gives the carbonyl-stretching bands expected for $\underline{C}_{3v}$ symmetry. A very similar pattern was seen for $W_2Br_4(CO)_8$, in agreement with the observed crystal structure.[95] There was i.r. evidence for the formation of strong contact ion pairs in $\{NaW(CO)_5CH_2CH_2CH_2PPh_2\}$.[96]

Excited-state distortions of $W(CO)_5L$, where L = py, piperidine, as well as $Mo_2(CF_3CO_2)_4$ and $K_3Cr(CN)_5NO$ were calculated from emission spectra and preresonance Raman spectra, using the time-dependent theory of Raman spectroscopy.[97] I.r. spectra, in the carbonyl-stretching region, were used to follow the processes of CO and alkene loss on u.v. irradiation of $W(CO)_5$(alkene), where alkene = C_2H_4, C_3H_6 or 1-C_5H_{10}.[98]

Protonation of $Mn_2(\mu\text{-}\eta^2\text{-}CO)(CO)_4(dppm)_2$ forms $Mn_2H(\mu\text{-}\eta^2\text{-}CO)(CO)_4(dppm)_2^+$. This process is accompanied by destabilisation of the formally 4-electron-donating $\mu\text{-}\eta^2$-CO group, shown by a large increase in νCO, from 1648 to 1787 cm^{-1}.[99] $Mn_2(CO)_6(SPh)_3^-$ shows two terminal νCO only. Thus only the SPh groups take part in bridging.[100]

I.r. spectra (CO stretches) were used to monitor photochemical reactions of a number of complexes of the type $(OC)_5MM'(CO)_3$(α-di-imine), where M,M' = Mn or Re.[101-2] Four νCO bands were seen for $Mn_2(CO)_8[P_2(^tBuN)_2P_2(^tBuN)_2]$, showing local $\underline{C}_s$ symmetry for the cis-$Mn(CO)_4P_2$ fragment.[103]

Low-temperature FTIR and Raman spectra were reported for $Re_2(CO)_{10}$, together with ^{13}CO data. These were used to make detailed assignments, including Raman-active overtones and combinations for the first time. There was generally good agreement with $\underline{D}_{4d}$ selection rules, but there was some evidence for slight distortion in solution.[104]

The photochemistry of $CpM(CO)_2R$, where M = Fe or Ru, R = Me or Et, in frozen gas matrices and in solutions at -30°C was studied using i.r. spectroscopy.[105] Carbonyl-stretching modes were assigned from the i.r. spectra of $M(CO)_4L$ (where M = Fe, Ru or Os, L = EPh_3 (E = P, As, Sb), PMe_3, $P(OCH_2)_3CMe$; M = Ru or Os, L = $SbMe_3$). Axial/equatorial isomerism was shown to occur in solution, with the tendency to give the equatorial isomer being in the following series: Ru > Os > Fe; Sb > As > P; Ph > Me; $P(OCH_2)_3CMe$ > PMe_3, PPh_3.[106]

The transient species formed from the flash photolysis of $Fe(CO)_5$ in C_6D_6 was assigned from the i.r. spectrum as $Fe(CO)_4(C_6D_6)$.[107] I.r. spectra revealed the presence of $Fe(CO)_4$, $Fe_3(CO)_{12}$ and $Fe(CO)_4$(1-pentene) on the photolysis of $Fe(CO)_5$ on zeolite surfaces.[108] The interaction of $HFeCo_3(CO)_{12}$ with oxide supports gives a variety of surface metal carbonyl species, revealed by their i.r. spectra.[109] There was FTIR evidence for $HFe_4(C)(CO)_{12}^-$ after the interaction of $HFe_4(CH)(CO)_{12}$ with partially dehydroxylated alumina.[110]

The i.r. spectra of ^{13}CO-enriched $Ru(^{12}CO)_{5-x}(^{13}CO)_x$, where x = 1-5, in liquid xenon were reported and energy-factored force constants calculated.[111] νCO were only observed in the terminal region for the adducts $Ru_3(\mu\text{-dppm})_2(CO)_8(\gamma\text{-A})$, where A = AgO_2CCF_3, $Hg(O_2CCF_3)_2$ or HO_2CCF_3. Thus there is no shift of terminal to bridging CO on adduct formation.[112] Quantitative studies of νCO for $Ru_3(CO)_{12}$ adsorbed on hydrated Al_2O_3 gave evidence for the presence of several species.[113]

The carbonyl stretch is at 1902cm^{-1} in Os(MIX-DME)(CO)py, where MIX-DME = mesoporphyrin IX dimethyl ester. This is lower than in the Ru analogue, showing that $d_\pi(Os^{II}) \rightarrow \pi^*(CO)$ is greater than for Ru^{II}.[114]

Normal (non-enhanced) Raman spectra were reported for CO molecules adsorbed on Co at room temperature. There were several bands from linear adsorbed species, 2- and 3-fold chemisorbed species.[115] νCO modes from Co and Rh carbonyls immobilised on derivatised inorganic oxide surfaces were studied by Raman, FTIR and photo-acoustic FTIR.[116] The kinetics of CO substitution by $P(OPh)_3$ in $Rh(acac)(CO)_2$ were followed by i.r. spectroscopy.[117]

Carbonyl stretches for $[XHgCo(CO)_3L]_n$, where X = Cl, Br, n = 1 or 2, show that the tendency to form dimers decreases in the order: L = P^tBu_3 > $P(OMe)_3$ > $P(OEt)_3$ > $EtP(OEt)_2$ > $Et_2P(OEt)$ > $Et_2P(NEt_2)$ > PEt_3 > P^nBu_3 > $EtP(NEt_2)_2$ > $P(NEt_2)_3$.[118] The formulation of (40) as shown is revealed by the pattern of CO modes.[119]

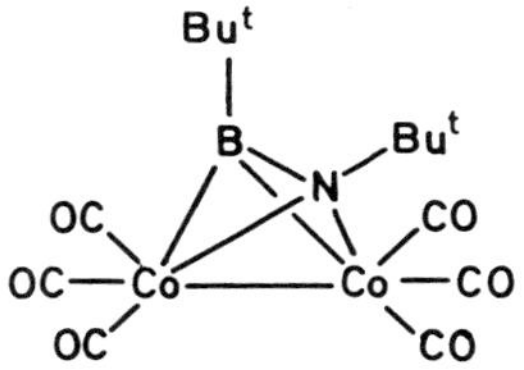

(40)

$[Co_{13}C_2(CO)_{24}]^{3-}$ gives the same pattern of carbonyl stretches as the corresponding tetra-anion, shifted ca.30cm^{-1}

to higher wavenumber because of the lower negative charge.[120] Evidence was found for a bridged carbonyl between Rh and the TiO_2 support when CO is adsorbed on a 0.5% Rh/TiO_2 film.[121] The presence of νCO at very low wavenumbers after adsorption of CO on Rh/SiO_2 containing Mn, Ti or Zr ions was explained by a model such as (41).[122]

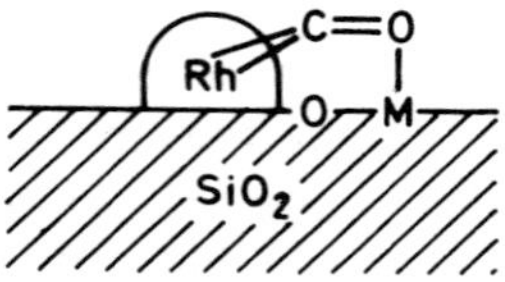

(41)

νCO values in Rh(β-diketone)(CO)L, where L = CO, PPh_3 or $P(OMe)_3$, show that increased electronegativity for substituents on the β-diketone leads to increases in νCO, especially for the dicarbonyl species.[123] The trans influence of nitrogenous bases was examined using νCO of trans-$Rh(CO)(PPh_3)_2L^+$, where L = py or substituted pyridine. A straight-line relationship was found between pK_a of L and νCO.[124] I.r. data (carbonyl stretches) for $Rh(CO)_2$acac and its ^{13}CO analogue were not consistent with the space group $P\bar{1}$ ($\bar{z}$ = 2) and indicate the absence of a centre of inversion.[125] (42), where the bridging ligand is dppm, has the following νCO modes: 2032cm^{-1} (unique CO_t), 2000, 1998cm^{-1} (remaining CO_t), 1872, 1869cm^{-1} (CO_{br}).[126] $K[Rh_6(CO)_{14}]$, in solution, has carbonyl stretches in both the terminal and bridging regions and appears to be isostructural with the cobalt analogue.[127]

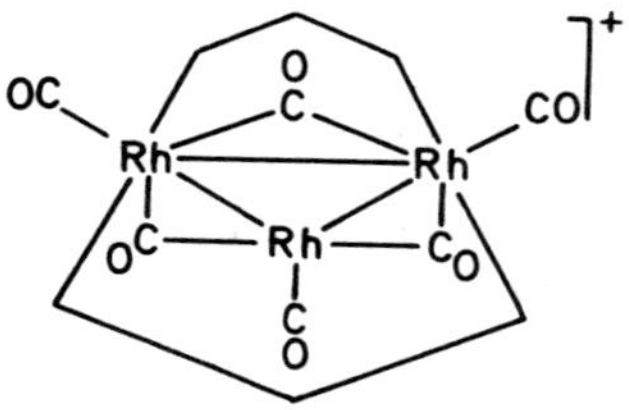

(42)

There are two types of CO adsorbed on a nickel catalyst, one linear (νCO 2060cm^{-1}) and one bridged (1960cm^{-1}).[128] I.r. spectra were reported for CO and NO adsorbed on polycrystalline NiO samples. There was evidence for dipole/dipole and chemical effects on the wavenumbers.[129]

CO adsorption on 5% Pd/Al_2O_3 was followed by i.r. spectroscopy (νCO). Bands shifted to higher wavenumber on increased coverage and on reduction of the surface by hydrogen.[130] The complexes cis-$M(C_6X_5)_2(CO)_2$, where M = Pd or Pt, X = F or Cl, were isolated for the first time. The carbonyl stretches at very high wavenumbers ($2186 cm^{-1}$ for Pd, X = F) show that there is negligible M-CO back-bonding.[131] Several linearly bound Pt-CO units were detected for CO adsorbed at 85K on a stepped Pt surface.[132] The vibrational spectra of $Pt(CO)_2X_2$, where X = Cl or Br, are consistent with a planar cis configuration in the crystalline state but a tetrahedral form in the gas phase or in hydrocarbon solutions.[133]

Linear, bridged and physisorbed CO molecules all gave characteristic CO stretches for CO on Cu(111) in the temperature range 7-77K.[134] CO chemisorption on reduced Cu/ZnO catalysts gives an i.r. band at $1580 cm^{-1}$ assigned to νCO for an unusual four-electron donor carbonyl species (43).[135] There is also evidence for two different Cu^{I}-CO complexes and a Cu^{II}-CO complex from the i.r. spectra of CO adsorbed on Cu/Zn oxide catalysts.[136] (44), where N⌒N = $PhCH{=}NCH_2CH_2N{=}CHPh$, gives νCO due to a bridging Cu^{I} carbonyl at $1958 cm^{-1}$.[137]

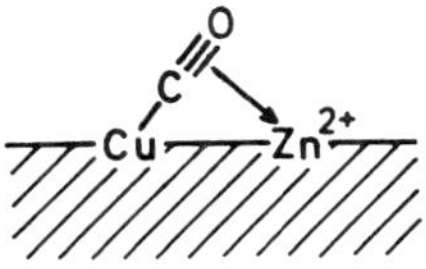

(43)

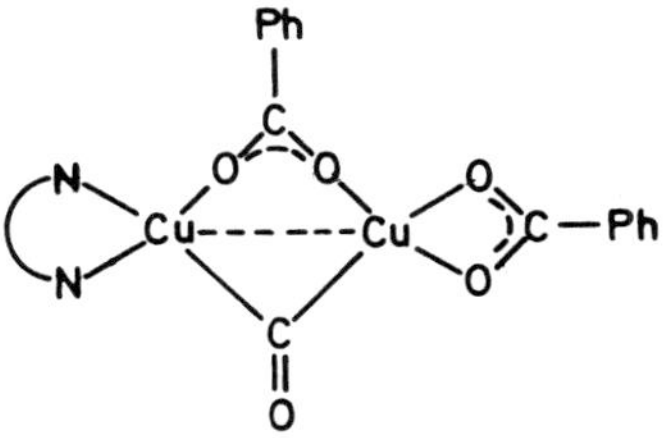

(44)

Physisorbed CO on SiO_2 surfaces containing isolated OH groups give νCO at 2158 and $2140 cm^{-1}$.[138]

3 Boron-containing Donors

(45) is formulated as shown since i.r. bands characteristic of unidentate HBH_3^- were seen.[139] $FeH(H_2BH_2)\{CH_3C(CH_2PPh_2)_3\}$ gives the expected features for a bidentate tetrahydroborate ligand.[140]

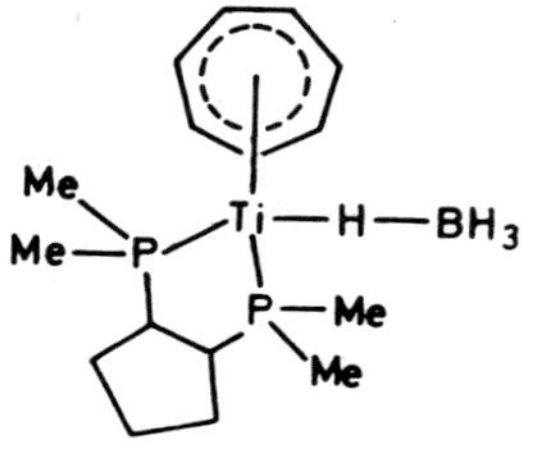

(45)

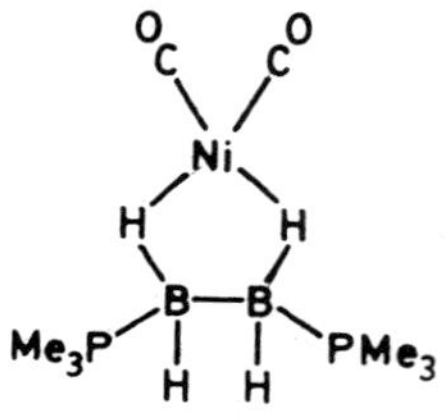

(46)

$Rh(Me_3B_3N_3Me_3)(LL')^+$, where LL' = diolefins or $(C_2H_4)_2$, have νBN of the coordinated hexamethylborazine at 1370-1400cm^{-1}, compared to the free ligand at 1407cm^{-1}.[141] (46) has νBH_t at 2322cm^{-1}, νBHNi at 1895cm^{-1} and νCO at 1933, 1909cm^{-1}. Bands below 1700cm^{-1} are very similar to those in the ZnII and CuI analogues.[142] The i.r. spectrum of $(Ph_3P)_2CuB_5H_8$ only has νBH above 2400cm^{-1}. Thus there is no feature which can be assigned to the bent Cu-H-B bridge.[143]

4 Nitrogen Donors

4.1 <u>Molecular Nitrogen, azido, and related complexes</u>.- <u>Trans</u>-$Cr(N_2)_2(dmpe)_2$ has only one νN_2 band in solution in the i.r. spectrum (1932cm^{-1}) and one in the Raman spectrum (2001cm^{-1}). The <u>cis</u> isomer has two νN_2 bands in the i.r., at 1920 and 1895 cm^{-1}.[144] $Mo_4O_8(OMe)_2(NNPh)_4^{2-}$ has i.r. bands at 1631, 1616, 1574 and 1522cm^{-1} characteristic of νNN for a <u>cis</u>-diazenido unit in $Mo(NNPh)_2^{2+}$.[145] $CpMo(PPh_3)(N_2C_6H_4Me$-<u>p</u>) and related species all have two i.r. bands in the range 1620-1550cm^{-1}, due to the arene-ring coupled NN stretch of a singly bent arenediazo ligand.[146] $MoCl_5(NSCl)^-$ gives the following i.r. features: νSCl <u>ca</u>.500cm^{-1}, ν_{as}MoNS 945cm^{-1}, ν_sMoNS 438cm^{-1} and δ_{as}NSCl 388cm^{-1}.[147]

<u>Cis</u>-$W(N_2)_2(Ph_2PCH_2PPh_2)_2$ gives two strong i.r. bands due to νNN, at 1935 and 2000cm^{-1}. <u>Trans</u>-$W(N_2)_2(Ph_2PCH{=}CHPPh_2)_2$ has

one strong νN_2 band, at 1965cm^{-1}, with a weak feature at 2022 cm^{-1}.[148] Similarly, cis-$W(N_2)_2(PMe_3)_4$ has two NN stretches, at 1980 and 1920cm^{-1} in the i.r., and $W(N_2)(PMe_3)_5$ has one, at 1905cm^{-1}.[149] Characteristic ligand modes were assigned for (47), including those for the bridging N_2S_2 unit at 840 and 825 cm^{-1}.[150]

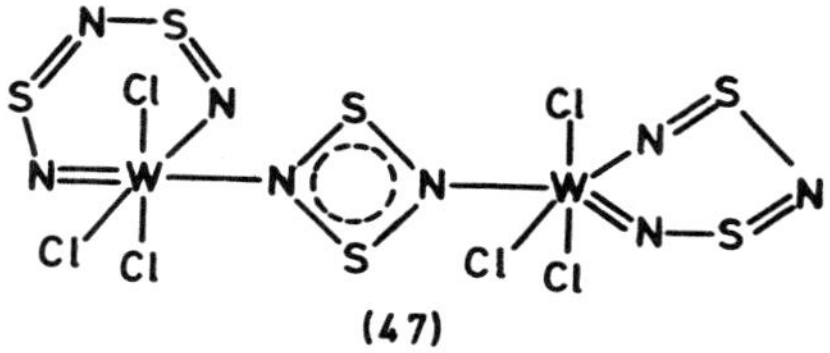

(47)

$Tc(PR_3)_2(CO)_2(ArN{\equiv}N{\equiv}NAr)$ (where R_3 = Me_2Ph, Ar = p-MeC_6H_4 or p-ClC_6H_4; R_3 = Ph_3, Ar = p-MeC_6H_4) give an i.r. band near 1260cm^{-1} characteristic of a chelating triazenido ligand.[151] νNN and νCN were seen in the ranges 2010-2030cm^{-1}, 2074-2120cm^{-1} respectively for mer-$ReCl(N_2)(CNR)[P(OMe)_3]_3$, where R = Me, Et, ^tBu or C_6H_4Me-4, and cis-$ReCl(N_2)(CNMe)(PPh_3)[P(OEt)_3]_2$. The lowest values were found for the last, where PPh_3 favours electron release from Re to the N_2 or CNR π-acceptors.[152] $ReCl_2(PPh_3)_2(NNHCOPh)(NHNHCOPh)$ has νNH 3150-3230cm^{-1}, νNN 1540-1600cm^{-1}.[153] νSCl of the NSCl ligand is at 450cm^{-1} in the i.r. spectrum of $ReCl_4(NSCl)_2^-$.[154]

Adsorbed N_2 on M-Al_2O_3-K, where M = Fe or Ru, gives i.r. bands at 2095, 2157cm^{-1} (Fe), 2050, 2169cm^{-1} (Ru), confirmed by ^{15}N substitution.[155] Matrix i.r. spectra of (48) show $\nu^{14}N_2$ at 2180cm^{-1}, $\nu^{14}N^{15}N$ at 2107cm^{-1}.[156]

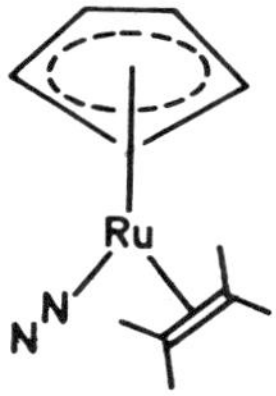

(48)

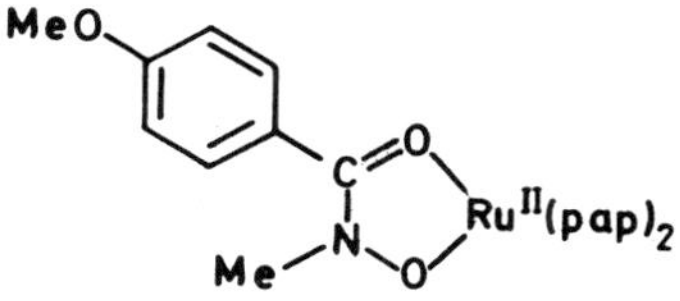

(49)

νN=N in complexes such as (49), where pap = 2-phenylazopyridine, are at $1315 cm^{-1}$, compared with $1425 cm^{-1}$ in free pap. Hence d(Ru)→π*(pap) back-bonding is significant.[157] Splitting of the NSCl ligand modes in $Ru(NSCl)_2ClX(PPh_3)_2$, where X = Cl, Br, CN or NCS, shows cis-coordination of the two NSCl ligands.[158]

$OsCl(COD)(N_2H_4)_3^+$, $Os(COD)(N_2H_4)_4^{2+}$ and $OsCl(COD)(NH_2NMe_2)_3^+$ all have νNN in the range $920-940 cm^{-1}$, characteristic of unidentate hydrazino ligands.[159]

Differences between i.r. and Raman wavenumbers of $\nu_{as}N_3$ in $Co(CN)_5N_3^{3-}$, as the K^+ salt, were interpreted in terms of a linear chain model of coupled anion oscillators.[160] The MLX_2 complexes, where M = Co or Ni, X = Cl, Br or I, L = $H_2NN=CMeCMe=NNH_2$, have i.r. spectra showing a polymeric structure, with each M coordinated by trans X,2 imino N from the same L and two amino N from different L.[161]

Triazenido ligand modes in Cp_2MY_2, where M = U or Th, Y = N,N'-di-p-tolyl-triazenide, show the unidentate coordination of Y.[162]

4.2 Amines and related ligands.- $CpTiCl_2(NHR)$, where R = Et, ^iPr, ^tBu or Ph, give bands in the i.r. spectra due to the amido ligands at ca. $3300 cm^{-1}$ (νNH), $1190 cm^{-1}$ (R = ^tBu), $1235 cm^{-1}$ (Ph) (νCN).[163] Some ligand-mode assignments were given from the i.r. spectra of trans-$Cr(NH_3)_4(H_2O)Cl^{2+}$, e.g. νNH at 3275, 3220 and $3145 cm^{-1}$.[164] Pyridine adsorbed on to porous Vycor glass with physisorbed $W(CO)_5$, from photolysis of $W(CO)_6$, forms adsorbed $W(CO)_5py$, with resonance Raman bands due to the py ligand at 1012 and $1042 cm^{-1}$.[165]

νNH_2 modes were assigned in $CpMn(CO)_2$(aniline), at $3325 cm^{-1}$ and $3280 cm^{-1}$, and in a series of complexes with substituted aniline derivatives. Oxidation to phenylaminyl complexes leads to an increase in νNH, to about $3340 cm^{-1}$.[166]

The resonance Raman spectra of singly reduced $Fe(bipy)_3^+$ shows that it contains two distinct chromophores.[167] $H_2Ru_3(NH)(CO)_9$ has νNH at $3362 cm^{-1}$ ($3353 cm^{-1}$ for ^{15}NO); in $HRu_3(NH_2)(CO)_{10}$ νNH bands are at 3405, $3345 cm^{-1}$ (3396, 3338 cm^{-1} for ^{15}N), and in $H_3Ru_4(NH_2)(CO)_{12}$ νNH bands are at 3367 and $3311 cm^{-1}$ (3358, $3304 cm^{-1}$ for ^{15}N).[168] The resonance Raman spectrum of $Ru(bipy)_3^{2+}$ adsorbed on porous Vycor glass shows that the ground state is similar to that in aqueous solution but that the electronic excited state of the adsorbed complex is very different from that in aqueous solution.[169] Multiple charge-transfer excited states of tris-(2,2'-bipyrimidine)ruthenium(II) were characterised by resonance Raman spectroscopy, e.g. $Ru^{III}(bpym)_2(bpym^-)^{2+}$.[170] The resonance Raman spectra of ground and low-lying excited states of $Ru(NH_3)_5(4\text{-acetylpyridine})^{2+}$ suggest that the charge-transfer state lies below the photo-active ligand field excited state, compared to the opposite order in the pyridine analogue.[171] The resonance Raman spectra of aqueous $Ru(bipy)_2(4\text{-acetylpyridine})_2^{2+}$ gave only ground-state modes for both ligands, due to the short lifetime of the MLCT state.[172] Detailed assignments of ligand modes were given using SERS for $Ru(NH_3)_5L^{2+}$, $Os(NH_3)_5L^{2+}$ and $Os(NH_3)_5L^{3+}$, where L = py, pyrazine etc.[173]

I.r., Raman and incoherent inelastic neutron scattering spectra show strong interionic vibrational coupling between a_{1g} modes in $[Co(NH_3)_6][MF_6]$, where M = Fe or Al.[174] The i.r. spectra of $ML(H_2O)_2$, where M = Co, Ni or Cu, H_2L = (50), show that the L is coordinated via the pyridine N and the amido N.[175]

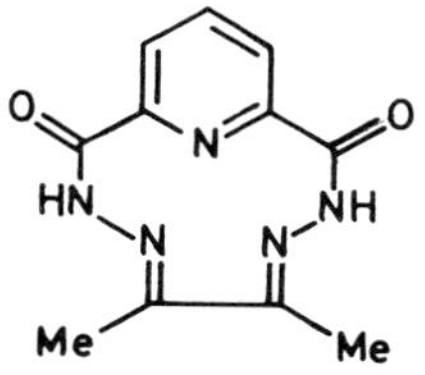

(50)

Interactions between $Cu(NH_3)_4^{2+}$ and zeolite Y can be investigated by examining Raman spectra in both the νNH and νCuN regions.[176] A surface-enhanced Raman scattering study on the structure of 2,2'-bipyridine adsorbed on a silver electrode showed at least 4 kinds of bipy molecule at the surface. One of these gave SERS very similar to $Ag^{I}(bipy)_2^{+}$.[177]

There is matrix i.r. evidence for the formation of 1:1 adducts between SiF_4 or GeF_4 and pyridine. Characteristic perturbation of pyridine modes was seen, suggesting strong interaction in the complexes.[178]

<u>4.3 Ligands containing >C=N< groups</u>.- $Cp_2Zr(N{=}CHMe)Cl$ has νC=N at $1700cm^{-1}$, i.e. $58cm^{-1}$ higher than in azomethine, HN=CHMe.[179] (51), where M = Cr or W, have νNH at $3280cm^{-1}$ and νC=N at 1572 cm^{-1} from the benzophenoneimine ligand.[180] SERS for $M^{II}(TPP)Cl$, where M = Cr, Mn or Fe, in the layered system CaF_2/roughened substrate/sample/Ag, show that they are converted to μ-oxo dimers.[181] A number of ligand modes were observed and assigned in the resonance Raman spectra of (TTP)MN, where M = Cr or Mn, TTP = tetra-p-tolylporphyrinato. The high-wavenumber $\nu C_{\alpha}C_m$ and $\nu C_{\beta}C_{\beta}$ modes are lower in the Cr than in the Mn complex, due to weakening of ring bonds caused by severe distortion of the macrocycle.[182]

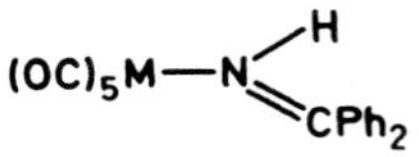

(51)

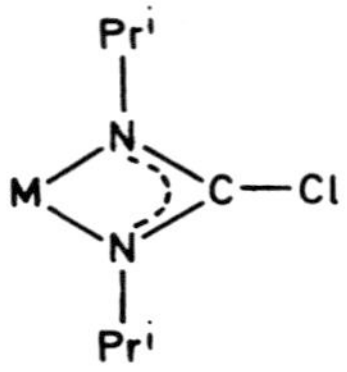

(52)

The i.r. spectra of $Cl_4M[^iPrNC(Cl)N^iPr]$, where M = Mo or Re, show that the C-Cl amidino ligand is coordinated as a symmetrical chelate (52).[183] Resonance Raman excitation profiles for $W(CO)_4L$, where L = R-N=CH-CH=N-R etc., were correlated with emission spectra for the complexes.[184] $(OC)_5W(N{=}CHPh)$, and substituted phenyl derivatives, have νN=C of the imine ligands in the range 1605-1640cm^{-1}, dependent on the electron-donating capacities of the phenyl substituent.[185]

$M(HQ)_2$, where M^{II} = Mn, Fe, Co or Zn, H_2Q = 1,2-benzoquinone-dioxime, have νC=N, νN-O and νOH bands consistent with the expected M-N coordination.[186] Re^VNPc, where Pc = phthalocyaninato, give ligand i.r. bands at $727cm^{-1}$ (γCH), 1070, 1127, $1169cm^{-1}$ (δCH of the 1,2-disubstituted benzene ring), and $1503cm^{-1}$ (νCC of benzene ring).[187]

Good agreement was found between observed (Raman) and calculated wavenumbers for the in-plane vibrational modes of iron phthalocyanine. Very extensive coupling between ligand modes was indicated, and hence there is no justification for assignments of specific wavenumbers to single internal coordinates.[188] I.r. and resonance Raman spectra were recorded for 20 Fe(TPP)LL' complexes to establish structure-sensitive bands. In the i.r.

spectra band I (1330-1350cm^{-1}) and band III (432-469cm^{-1}) are spin-state sensitive, while band II (790-806cm^{-1}) is oxidation-state sensitive. In the resonance Raman spectra, bands C (1498-1545cm^{-1}, anomalous polarisation) and D (1540-1565cm^{-1}, pol.) are spin-state sensitive, while band E (376-391cm^{-1}, pol.) is sensitive to spin and oxidation states.[189]

Resonance Raman data for $Fe(TPP)^-$ support a $\underline{d}^7$ Fe(I) formulation for the complex both at room and low temperatures.[190] Some ligand-mode assignments were proposed for (P)Fe(R), where R = Ph, C_6F_4H, C_6F_5, P = OEP, TPP, (m-Me)TPP or (p-Me)TPP.[191] Well resolved Raman spectra were obtained for Fe and Co tetra-sulphonated phthalocyanines adsorbed on silver electrodes, giving some tentative band assignments.[192] The electrochemical reduction of iron(III) tetrakis-(N-methyl-4-pyridiniumyl)porphine, Fe^{III}(TMPyP), was followed by resonance Raman spectroscopy. The reduction was shown to proceed in a high-spin state at both high and low pH values.[193] SERS was used to monitor the reduction of iron(III) protoporphyrin at a silver electrode. The enhancement pattern was similar to normal resonance Raman spectra, i.e. the molecular electronic states are not altered by interaction with the surface.[194]

Resonance Raman spectra at cryogenic temperatures of photodissociated haemoglobins show significant differences in peripheral haem substituent modes and the iron-histidine stretch, compared to the corresponding deoxygenated forms. The differences reflect differences in the ternary structure of the haem pocket between deoxyhaemoglobin and the CO-bound form.[195] Measurements were made of depolarisation ratio dispersion curves and excitation profiles for the 1375, 1583 and 1683cm^{-1} bands of oxyhaemoglobin. These gave detailed information on the actual symmetry of the haem group.[196] Differences in ternary and quaternary structures of fluoromethyl haemoglobin derivatives were probed by resonance Raman spectra excited in the u.v. region.[197]

The Raman spectra of haematoporphyrin derivatives (570-1640 cm^{-1}) in different solvents were shown to be very solvent dependent.[198] Raman spectra of oxidation- and spin-state marker bands of sperm-whale myoglobin (Mb) derivatives (oxyMb, deoxyMb, Fe^{III}Mb-CN) were recorded at different pH values. At all pH values above 6.5 the depolarisation ratios are independent of pH, consistent with the lack of a salt bridge between His(HC3)β and Asp(FG5)β (where Asp = aspartate).[199]

Resonance Raman spectra of horseradish peroxidase compound I free of photoreduction artifacts show changes in characteristic porphyrin modes, revealing loss of an electron from the a_{2u} π-orbital of the porphyrin ring, in comparison with compound II.[200] The resonance Raman spectra of iron chlorin complexes (models for the prosthetic groups of green haem-proteins), e.g. Fe^{III}(DC), where DC = deuteriochlorin IX dimethyl ester, showed effective symmetry of about C_2, compared to D_{4h} for metalloporphyrins.[201] Resonance Raman spectra were used to study haem-group structure in intermediates derived from haem-enzyme horseradish peroxidases.[202]

Assignments were made of oxidation and spin-state marker bands in the resonance Raman spectra of cytochrome oxidase in the resting, reduced, pulsed and oxygenated forms.[203]

Time-resolved resonance Raman spectroscopy was used to study the reduction of O_2 by both mixed-valence and reduced cytochrome oxidase. A photolabile intermediate was detected with wavenumbers characteristic of oxygenated haem was detected.[204] Similar experiments made it possible to monitor the structure of the haem group following photolysis of the CO complexes of haemoglobin, myoglobin and oxy-haemoglobin. For HbCO the porphyrin skeletal wavenumbers above 1450cm^{-1} (sensitive to core size) are lower than those of deoxyHb within 30ps of photolysis, and characteristic of a high-spin FeII haem complex containing a pair of weak-field axial ligands.[205] Nanosecond transient resonance Raman spectra were reported for the Fe^{II}-CO and Fe^{III}-NO photolysis products of horseradish peroxidase.[206]

Transient Raman spectra of photolysed carboxyhaemoglobins (CoHbA, CoHbA(Zurich)) show that the behaviour and geometry of the proximal side of the haem are very nearly the same in each case.[207] Characteristic ligand modes gave resonance Raman bands for a transient intermediate during the alkaline isomerisation reaction of ferricytochrome c.[208] The first picosecond, time-resolved resonance Raman spectrum has been reported for photolytic transients of myoglobin. These were used to discuss some aspects of ligand dynamics.[209]

Ligand-mode assignments in $Ni(LH)_2$, where H_2L = 1,2-hydroxylamino-oximes, are consistent with the presence of an NiN_4 planar chelate unit.[210]

I.r. and resonance Raman spectra were reported for Ni(2-FDP) and Ni(4-FDP), where FDP = formyldeuterioporphyrin IX dimethyl ester. All in-plane and many out-of-plane porphyrin modes and internal CHO modes were assigned. Formyl substitution at the 2- and 4-positions gave very similar spectra; hence there are very similar perturbations of the electronic structure of the macrocycle.[211]

Ligand modes were assigned for the symmetric chelating carbodi-imide complex $Cl_4P[^iPrNC(Cl)CN^iPr]$.[212]

<u>4.4 Cyanides, Isocyanides, and related ligands</u>.- $Cp_2V(\eta^1-C_3H_5)$-(2,6-xylylisocyanide) has νCN at 1965cm^{-1}, showing substantial π-back-donation to the isocyanide ligand.[213]

I.r. spectroscopy (77-293K) was used to study the reactive species in polyvinyl alcohol films. Thus $Cr(CN)_6^{3-}$ goes to $Cr(CN)_5^{2-}$ and $Cr(CN)_5(H_2O)^{2-}$.[214] Full infrared and Raman data were presented for $(OC)_5CrCNCOPh$, together with a force-field calculation. The results show that the benzoylisocyanide ligand is similar to CO in its interaction with Cr. There is extensive electron delocalisation in the C≡N-C=O system.[215] The following C≡N stretching assignments were reported for $K_8[Mo_4S_4(CN)_{12}].4H_2O$: 2093cm^{-1}($t_2$), 2100cm^{-1} ($a_1$), 2112cm^{-1} ($t_2$) and 2122cm^{-1} (e).[216] νCN (a_{1g}, e_g and t_{1u}) were assigned for $Mn(CNR)_6^+$ and $Mn(CNR)_6^{2+}$, where R = various alkyl and aryl groups. The wave-numbers were generally higher for Mn^{II} than for Mn^I, as expected.[217] $Re(SR)_3(NCR')_2$, where R' = Me or ^tBu, HSR = 2,4,6-tri-iso-propylbenzenethiol and related ligands, give a single νCN i.r. band, consistent with <u>trans</u>-$(NCR')_2$ ligands.[218]

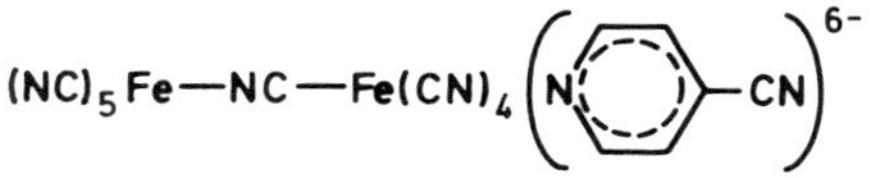

(53)

The i.r. spectrum of the new binuclear iron complex (53) contains a band due to νCN(bridging) at 2115cm^{-1}, with νCN_t at 2055cm^{-1}.[219] The i.r. spectra of $M(CNR)_6^{2+}$, where M = Ru or Os, R = $CHMe_2$, CMe_3, C_6H_{11}, all give νCN as a very strong band near 2200cm^{-1} (t_{1u} mode); the presence of a very i.r. feature near 2035cm^{-1} was explained by slight non-linearity of the M-C≡N-C units.[220]

Detailed assignments of CN have been proposed for $MH(CN)_5^{3-}$, where M = Rh or Ir (Table 1).[221] The complexes $Rh(L)(CO)(PPh_3)_2^+$,

Table 1 C≡N stretching wavenumbers for $MH(CN)_5^-$ (/cm^{-1})

M =	Rh	Ir
a_1	2147	2152
b_1	2127	2125
e	2119	2117

where L = CH_2=CHCN, CH_2=CMeCN, cis- and trans-MeCH=CHCN or CH_2=CHCH$_2$CN, all give increased νCN, but almost unshifted νC=C compared to the free ligands. Thus coordination has occurred via N and not the π-system of the olefinic group in L.[222] Exactly the same conclusion was reached from the i.r. spectra of $IrH_2(L)(CO)(PPh_3)_2^+$, where L = cis- or trans-MeCH=CHCN or CH_2=CHCH$_2$CN.[223]

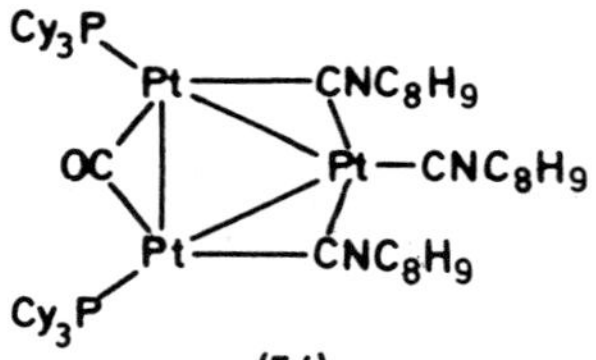

(54)

(54) has νCN_t at 2105cm^{-1}, νCN_{br} at 1730, 1695 and 1590cm^{-1}, νCO_{br} at 1815cm^{-1}.[224]

An increase, of 31cm^{-1}, in νCN on going from free benzonitrile to $[Cu(Cl_3CO)_2(PhCN)]_2$, shows the presence of a Cu-NC bond.[225] Evidence has been found that both gauche and trans conformers of succinonitrile on copper surfaces coordinate to the Cu via the π-system of one of the two C≡N groups.[226] SERS results for malonitrile at copper electrode surfaces point to exactly the same form of coordination here also.[227] SERS results for benzonitrile and benzyl cyanide in a silver sol show that the latter forms an Ag-NC bond but that the former is adsorbed 'face on' to the surface.[228]

Raman spectra were used to follow the complexation of $Hg(CN)_2$ by CN^- in liquid ammonia. The band assignments proposed are summarised in Table 2.[229]

Table 2 νCN assignments from $Hg(CN)_2$ + CN^- in liquid ammonia (/cm^{-1})

	ν_sCN	ν_{as}CN
$Hg(CN)_2(NH_3)_2$	2164	2161
$Hg(CN)_3(NH_3)^-$	2145	2148
$Hg(CN)_4^{2-}$	2139	2134

4.5 Nitrosyls.- Two forms of $Cp_2V(NO)I$ have been reported, with different νNO wavenumbers: 1590cm^{-1} and 1670cm^{-1}. That with lower νNO has either a bent VNO group or a linear VNO with a long N-O bond; this predominated at low temperatures. The higher νNO corresponds to a linear VNO, and this form is thought to have at least one Cp ligand with a hapticity less than 5.[230]

NO on reduced V_2O_5/Al_2O_3 catalysts gives νNO bands in the i.r. due to both mono- and dinitrosyls and to a nitrite species.[231]

νNO in CpCr(NO)(L)X (where L = PPh_3, X = Cl, Br or I; L = $P(OPh)_3$ or $P(OEt)_3$, X = I) occurs as a single sharp i.r. band 20-40cm^{-1} lower than in CpCr(CO)(NO)L. The order of decreasing νNO, CO > $P(OPh)_3$ > $P(OEt)_3$ > PPh_3, is consistent with the known electron-donating power of these ligands.[232] νCN and νNO both shift to higher wavenumbers on going from $Cr(NO)(CNR)_3(L-L)^+$ to the corresponding dications, as expected (R = Me, CMe_3, L-L = dppm, dppe). However, only one CN stretching band is seen for the dication, against 3 for the monocation. The reason for this is not known.[233]

The complex (55) has two strong i.r. bands due to νNO, i.e. the geometry is cis, as shown.[234] The same conclusion was drawn for the two nitrosyl groups in $S_2MoS_2MoBr_2(NO)_2^-$.[235] The νNO bands for $[WBr_2(NO)_3]_2$ correspond to C_s selection rules for for the structure (56).[236]

The Raman spectrum of $K_2[Fe(CN)_5NO]$ in the M→NO CT excited state shows νNO at 1835cm^{-1}, i.e. 105cm^{-1} lower than in the ground state (1940cm^{-1}). This shows that a change in geometry for FeNO from linear to bent has occurred.[237] $[(Me_5C_5)Fe(NO)]_2$ has νNO 37cm^{-1} lower than in the corresponding C_5H_5 complex.[238]

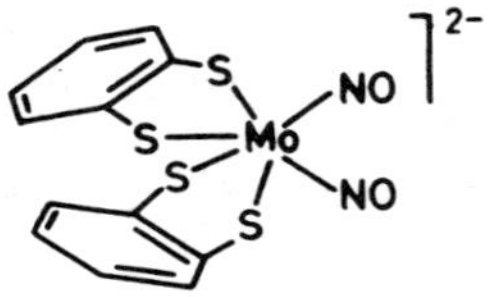

(55)

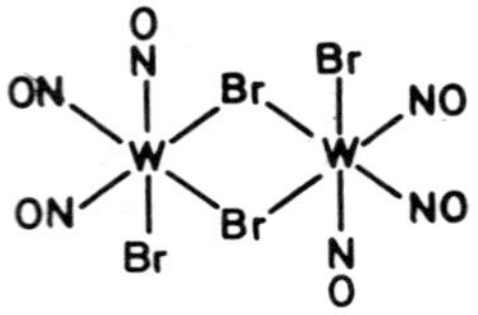

(56)

The NO stretching wavenumbers for (P)Fe(R)NO are listed in Table 3. These show that all are terminal nitrosyls, with values between those expected for linear $Fe-NO^+$ and bent $Fe-NO^-$.[239]

Table 3 Nitrosyl stretching wavenumbers for (P)Fe(R)NO ($/cm^{-1}$)

	P = OEP	TPP
R		
-	1670	1699
Me	1766	1789
^{n}Bu	-	1764
Ph	1795	1790

νNO values were assigned in a wide range of complexes (57), where R and R' are various alkyl groups, together with those for the <u>cis</u> analogues.[240] $HRu_3(CO)_{10}(\mu_2-NO)$ has νNO at $1550cm^{-1}$ and $Ru_3(CO)_{10}(\mu_2-NO)_2$ at 1524 and $1508cm^{-1}$.[241]

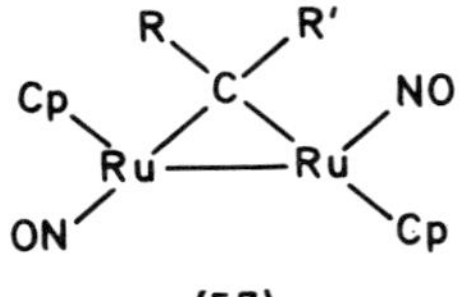

(57)

νNO (<u>ca</u>.$1800cm^{-1}$) in $Ir(O_2CR)_2(NO)(PPh_3)_2$, where R = CF_3 or C_2F_5, is at very different wavenumbers from that in $IrX_2(NO)(PPh_3)_2$, where X = halide. This suggests a different structure, and indeed the carboxylato complexes are essentially trigonal-bipyramidal, compared to the tetragonal-pyramidal form of the halo complexes.[242]

There is infrared evidence that Ni^{2+}-NO surface nitrosyls are formed by NO adsorption on pure NiO or NiO/MgO.[243] I.r. data also show that NO_2 adsorbed on Pt/SiO_2 dissociates to give Pt-NO species.[244] NO itself, when adsorbed on Pt/SiO_2, gives 3 distinct species, one of which is definitely linear and one bent.[245] I.r. studies were also made of NO chemisorbed at 298K on pre-reduced Pt/Al_2O_3, MoO_3/Al_2O_3 and $Pt/Al_2O_3/MoO_3$.[246]

5 Phosphorus and Antimony Donors

Complex (58) has νNP_3 bands at 871(i.r.)/873(Raman)cm^{-1} and 813(i.r.)/815(Raman)cm^{-1}.[247] νCO wavenumbers were assigned and Cotton-Kraihanzel force constants calculated for (59; R = CH_3, CF_3), (60), and related species.[248]

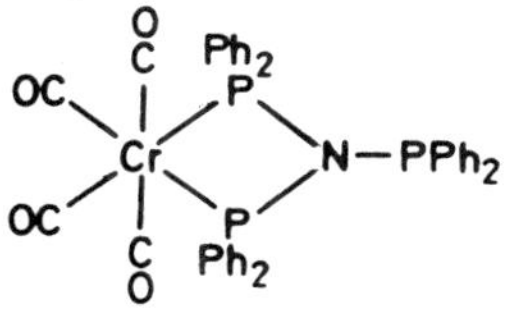

(58)

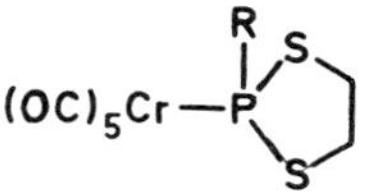

(59)

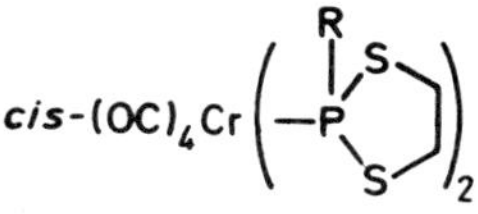

(60)

$[(OC)_5MPPh_2]_2O$, where M = Cr, Mo or W, all have ν_{as} and ν_sPOP more than 100cm^{-1} lower than in the free diphenylphosphinic acid anhydrides.[249] νPO_3 is at 868cm^{-1} in $\{RCr(CO)_2[P(OR')_3]\}_2$, where R = η^5-C_5H_4Me, R' = Me or Et, while δPO_3 contributes to a band at 497cm^{-1}.[250] A large number of ligand modes were assigned for the new complexes (61), where M = Cr, Mo or W. νSb_3 and δSb_3 modes were largely unshifted with respect to the free ligand values.[251]

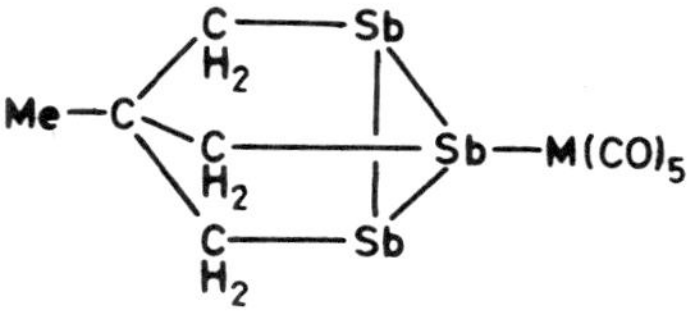

(61)

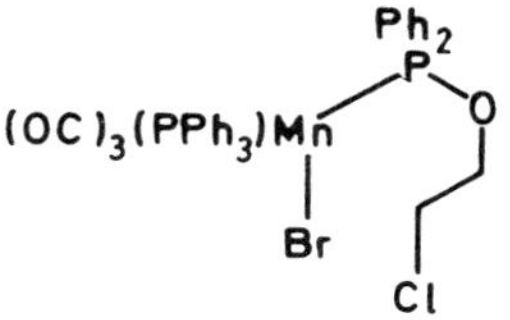

(62)

νPO in (62)is at $1031cm^{-1}$.[252] Axial substitution at the iron atom in (63) is revealed by the characteristic pattern of i.r. bands due to νCO.[253] (64), where Y = OMe or NMe_2, and related species give i.r. bands due to the PPh_3 ligands at 1090, 1430 and $1480cm^{-1}$.[254]

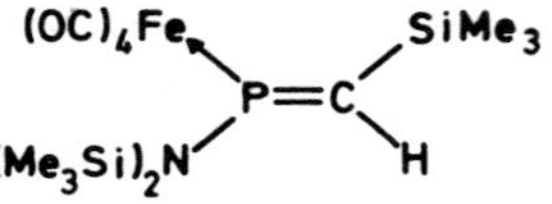
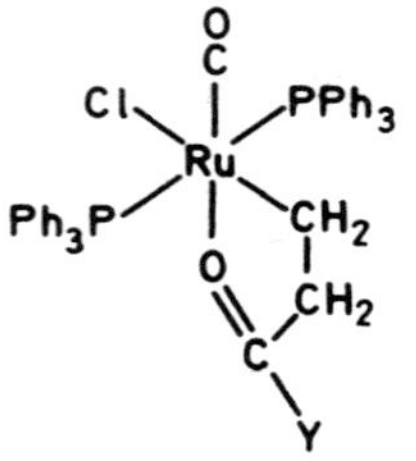
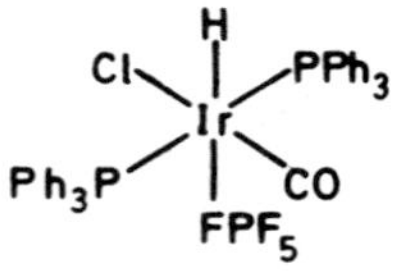

(63) (64) (65)

PPh_3 ligand modes were seen at 1435, 765, 745 and $700cm^{-1}$ in $[Rh(PPh_3)_3(MeCN)](BF_4)$.[255] Unidentate coordination of PF_6^- in (65) is revealed by the presence of four νPF bands (880, 810, 734 and $488cm^{-1}$).[256]

The formation of $Ag(AsPh_3)_3Cl$ gives only slight changes in i.r. bands due to the $AsPh_3$ ligand, compared to the free ligand.[257] The i.r. spectrum of ThP_3F_7 shows that it can be formulated as $Th_4F_4(\mu\text{-}L)_3$, where L = cyclic, tetradentate $(PF_2)_4^{4-}$.[258] $Cp_3U(PPh_2)$ gives the usual i.r. bands due to Cp and Ph groups, with νPC at $1435cm^{-1}$.[259]

Detailed assignments were proposed for PH_3 modes on H_3PGaX_3, where X = Cl or Br.[260] νPH and δPH modes in $Ph_nPH_{3-n}.GaCl_3$, where n = 1 or 2, are increased in wavenumber compared to the free-ligand values.[261]

6 Oxygen Donors

6.1 Molecular oxygen, peroxo, aquo, and related complexes.- H_2O modes were assigned for the $Li\text{-}H_2O$ complex in a krypton matrix. There were small wavenumber shifts but great changes in intensity compared to H_2O alone in a Kr matrix.[262] M (=Mg, Ca, Sr or Ba) and H_2O in argon matrices form $M.OH_2$ adducts. Complex formation is shown by shifts to lower wavenumber of the ν_2 H_2O bending mode.[263] I.r. spectra of $M(ROH)_6X_2$ and $M(ROH)_4X_2$, where M = Mg or Ca, R = Me or Et, X = Cl or Br, show that ROH modes above $800cm^{-1}$ are not much shifted by coordination but that ρOH modes move to lower wavenumbers.[264]

$K[Ti(O_2)F_3].3H_2O$ has νOO bands at 900 and 860cm^{-1} in both i.r. and Raman spectra, consistent with $\underline{C}_{2v}$ symmetry for the TiO_2 unit.[265] $M[VO(O_2)_2].M_2SO_4$, where M = NH_4, Na or K, contain peroxides bound to V^V in triangular, bidentate ($\underline{C}_{2v}$) fashion.[266] νOO of the coordinated O_2 is at 920cm^{-1} in $[VO(O_2)(IDA)]^-$, where IDA = $C_4H_5NO_4^{2-}$.[267] Equivalent modes are found near 850cm^{-1} for the Na^+ and K^+ salts of $V(O_2)_3^-$.[268] νO-O in $NbF_5(O_2)^{2-}$ is 955cm^{-1}.[269]

Inelastic neutron scattering of chromous acid, CrOHO, shows that ν_{as}OHO is at 2050cm^{-1} and γOHO at 1254cm^{-1}.[270] $[Cr_4(CO)_{12}(\mu_3\text{-}OH)_4]^{4-}$ has νOH at 3695cm^{-1} (2720cm^{-1} in the OD analogue), an unusually high value for μ_3-OH.[271] The Raman spectrum of $K_2[MoO(O_2)_2(citrato)].\frac{1}{2}H_2O_2.3H_2O$ contains νO-O of the molybdenum peroxo unit at 875cm^{-1}.[272]

Infrared and other spectroscopic data show that $[Re(CO)_3(OH)]_4$ on silica is hydrogen-bonded to surface OH groups.[273] Matrix-isolation i.r. of Fe(TPP) carbonyl complexes were reported, e.g. for $Fe(TPP)(CO)(O_2)$ there was evidence of σ-donation from CO to O_2.[274]

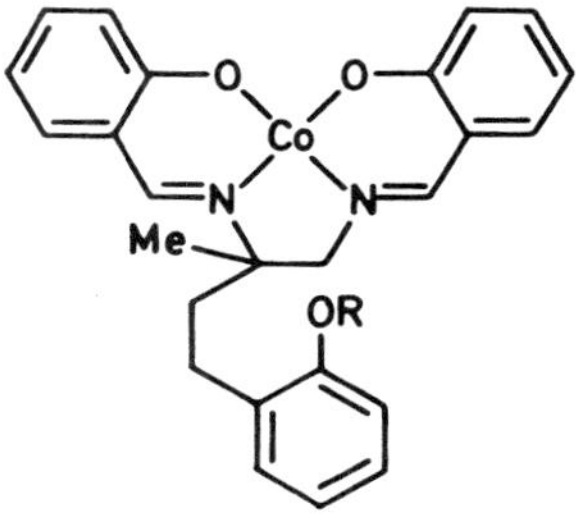

(66)

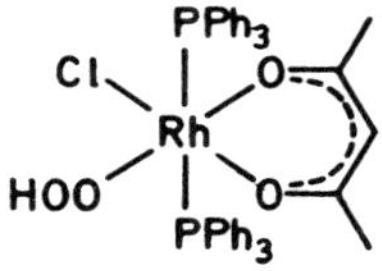

(67)

Resonance Raman spectra were reported for dioxygen adducts of (66), where R = H or Me. When R = Me, νOO is at 1148cm^{-1}, typical of six-coordinate $Co(salen)(base)O_2$-type complexes. When R = H, νOO is at 1095cm^{-1}, close to 'base-free' $Co(salen)O_2$. The results show that the pendant methoxyphenyl group is coordinated to the axial position, whereas the hydroxyphenyl group is not.[275] The low value of νO_2 in $Co(TPP)(SPh)O_2^-$ ($^{16}O_2$ 1122cm^{-1},

$^{18}O_2$ 1058cm^{-1}) is attributed to the presence of the lone pair of electrons on the thiolate ligand, which donate electron density to O_2 via π-overlap.[276] Resonance Raman spectra were reported for a large number of cobalt porphyrin complexes with nitrogenous base axial ligands, as dioxygen complexes. It is possible to get resonance enhancement of solvent bands whose wavenumbers match those of νO_2.[277] (67) has νOO of the hydroperoxo ligand at 813cm^{-1}. The assignment was confirmed by $^{18}O_2$ substitution (band shifted to 766cm^{-1}).[278]

νH_2O modes for $Cu(HCOO)_2.2H_2O$ crystals show that there are strong interactions between Cu and H_2O.[279] O_2 adsorbed on a silver electrode in an aqueous NaOH solution gives two Raman bands in solution: ca.700, ca.850cm^{-1}. Their relative intensities depend upon the electrode potential.[280]

A variety of salts containing $UO_2(O_2)F_2^{2-}$ have νO-O near 860 cm^{-1}.[281] νO-O in $K[BF_3(OOH)]$ appears as a weak infrared band at 837cm^{-1}.[282] νO_2 (peroxo) is at 860cm^{-1} in $B(O_2)F_3^{2-}$ and at 850 cm^{-1} in $B_2(O_2)_3F_2^{2-}$.[283]

$[^tBu_2Sn(OH)X]_2$ dimers form hydrogen bonds, -O-H...X, in crystals. For X = F these are strong, with νOH at 3060cm^{-1}; for X = Cl or Br, on the other hand, they are weak, with νOH at about 3450cm^{-1}, almost unshifted from the free-hydroxyl-group value.[284]

<u>6.2 Carboxylato and related complexes</u>.- A comprehensive analysis of the i.r. and Raman spectra of $M(CH_3COO)_2.4H_2O$, where M = Mg, Co or Ni, suggests Fermi resonance between acetato-group fundamentals and two-phonon modes associated with H_2O molecules.[285] Dibromoacetato complexes ($CHBr_2CO_2^-$ = L) ML_4 are shown to contain both chelating and bridging ligand groups, where M = Ti, Zr, Th or U. For ML_3 (M = Fe or Al), $SnCl_2L_2$ and $SbCl_2L_3$ only chelating ligands are present.[286]

I.r. spectra show the presence of both uni- and bidentate carboxylato groups in $CrO_2(OOCR)_2$, where R = Me, Et, Pr or $CHCl_2$.[287] Only bridging carboxylato groups were detected for $[Cr_2O(OOCR)_3]Cl$, where R = Me, Et or Pr.[288] Bands due to unidentate $CF_3CO_2^-$ ligands were seen, at 1720, 1660cm^{-1}, in the i.r. spectra of CH_2Cl_2 solutions of $Mo(CO)_2(PMe_3)_3(CF_3COO)_2$.[289] Several i.r. bands assignable to an η^2-carbonato group were seen for $CpMo(CO)_2(\eta^2-O_2CO)$.[290]

$W_2(O_2CR)_4$, where R = Me or ^tBu, have i.r. spectra in accord with bidentate carboxylate bridging, $\nu_{as}CO_2$ 1472cm^{-1} (^tBu), 1485cm^{-1} (Me).[291] $MnF_3(C_2O_4)^{2-}$ has oxalato i.r. bands due to bridging rather than chelating ligands: $\nu_{as}CO_2$, 1670cm^{-1}, ν_sCO_2, 1360, 1320cm^{-1}, δCO_2, 780, 750cm^{-1}.[292]

The infrared spectra of pentacoordinated Fe^{III} porphyrin carboxylates, (L)Fe(O_2CR), where L = OEP or TPP, R = Me, Et or Ph, show $\nu_{as}-\nu_sCO_2$ of 356-409cm^{-1}, i.e. the carboxylate groups are unidentate.[293] Quite detailed ligand-mode assignments were proposed for $[Ru(CO)_2(\mu\text{-OOCEt})]_2$, as dioxan or MeCN adducts, and $[Ru(CO)_2(OOCEt)]_n$. The ν_s and $\nu_{as}CO_2$ values were consistent with bidentate bridging carboxylates.[294]

There is i.r. evidence for the formation of acetato and acetyl species at the surface in the CO/H_2 reaction on SiO_2-supported Rh or Rh/Mn catalysts.[295] A novel mode of CO_2 coordination to Co, via a carbamato-like derivative, was inferred from the i.r. spectrum of Co(NO)$_2$(L)Cl, where L = (68).[296]

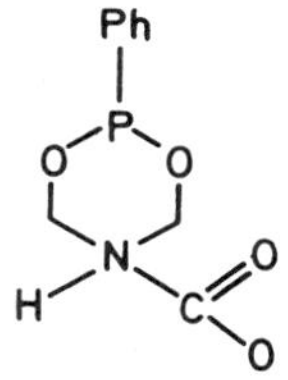

(68)

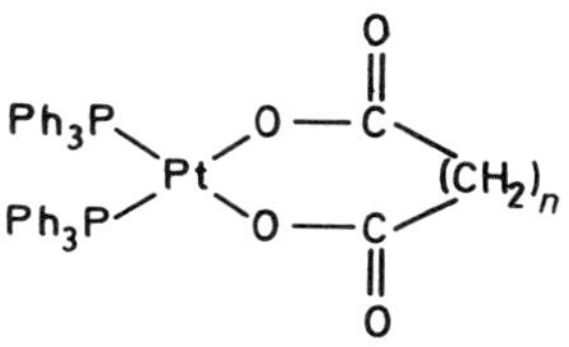

(69)

Ligand-mode assignments in the i.r. and Raman spectra of Pd^{II} and Cu^{II} bis-(2,4-pentanedionato) complexes show that the two ligands vibrate almost independently.[297] (69) has νC=O bands at 1770cm^{-1} (n = 1), 1740cm^{-1} (n = 2) or 1720cm^{-1} (n = 3).[298]

$(Ph_3P)_2Cu(O_2CH)$ has i.r. bands due to bidentate formato groups, while $(Ph_3P)_3Cu(O_2CH)$ and (triphos)Cu(O_2CH) contain unidentate formato groups.[299] The i.r. and Raman spectra of $Cu(HCOO)_2.2H_2O$ and its deuteriated analogue show the presence of two independent formate ions.[300] For $Cu(RCO_2)_2(L)$, where R = Me, Et, Pr, PhCH$_2$ or CH$_2$Cl, L = acetone, ½(dioxan) or THF, the carboxylate ligands are bridging bidentate, with L coordinated via the oxygen atom.[301]

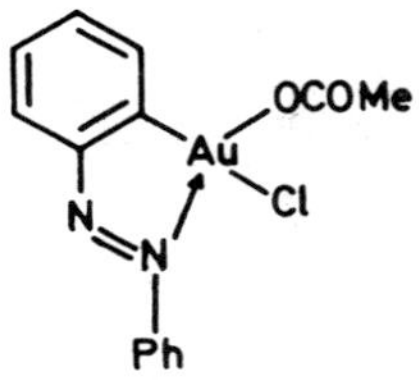

(70)

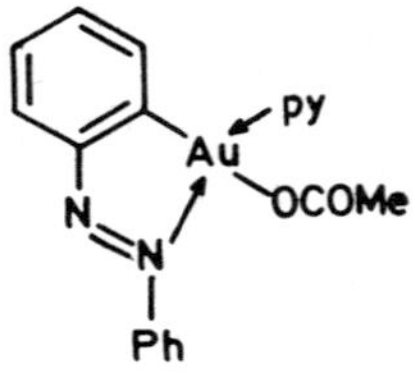

(71)

(70) and (71) have characteristic ν_{as} and $\nu_s CO_2$ bands for the acetato groups trans to N or Ph respectively. These bands are all seen in the complex with two acetato groups coordinated to the gold.[302]

$Cd(HCOO)_2(NH_3)_2$ is suggested to have unidentate formato groups on the basis of its i.r. spectrum.[303] I.r. bands due to carbonato ligands were identified in the $Co(NH_3)_6^{3+}$ salts of $M(CO_3)_3^{3-}$, where M = La, Sm, Gd or Ho. All the spectra were very complicated, showing the low effective symmetry of the ligands.[304] The i.r. spectra for $Ln(CF_3COO)_3.3H_2O$, where Ln = Sm, Eu, Er or Yb, show that they have dimeric structures, with bridging trifluoroacetato groups.[305]

Bidentate bridging carboxylato groups were also identified in $M[UO_2(OAc)_4]$ and $M[UO_2(O_2CCF_3)_4]$, where M = Mg, Ba, Co or Zn.[306] Infrared and Raman spectra of $UO_2(C_2O_4)(Bu_3PO)$ are consistent with a polymeric structure, containing quadridentate oxalato groups.[307]

I.r. spectra of $Me_2SiO_2Al_2(O^iPr)_4$ complexes with cinnamate, dihydrocinnamate or salicylate suggest that there is chelation of the carboxylato group to Al.[308] The Raman spectra of aqueous thallium(I) formate solutions showed the presence of bridging formato groups in concentrated solutions, as seen in the crystalline state.[309]

$\nu C{=}O$ bands were assigned for alkyltin iso-octylthioglycollates, $R_2Sn(IOTG)_2$, $R_2SnCl(IOTG)$ etc.[310] Carbonyl-stretching bands for $Ph_3Te[RCOCH{=}C(O)R']$, where R = CH_3, CF_3 or Ph, R' = CH_3, CF_3, Ph or 2-thienyl, show unidentate O-coordination of the β-diketonates.[311]

<u>6.3 Keto, alkoxy, ether and related complexes</u>.- νCOC modes were assigned in the i.r. spectra of crown-ether complexes of $LiClO_4$.[312] There is Raman evidence for the coordination of DMF to Li^+, Mg^{2+} and Al^{3+} in N,N-dimethylformamide/H_2O solutions.[313] Shifts in $\nu C=O$ indicated Mg-O coordination in $Mg(NO_3)_2.4CO(NH_2)_2$[314] and $Mg(NO_3)_2.6$(methylenediurea).[315] Ca-O coordination was detected similarly for $CaCl_2.4CO(NH_2)_2$.[316]

$\nu C=O$ modes in acetone complexes $Cp(CO)_2LML'^+$ (where M = Mo, L = PPh_3; M = W, L = PPh_3 or CO; L' = acetone) lie in the range 1640-1669cm^{-1}.[317] The carbonyl stretch due to the coordinated DMF in (TPP)Mn^{III}Br(DMF) is at 1658cm^{-1} (1675cm^{-1} in the free ligand).[318] The i.r. and Raman spectra of $MX_2(amide)_n$, where M = Mn, Fe, Co, Ni, Cu, Cd, Hg, Pd or Pt, amide = formamide, N-methylformamide (NMF) or DMF, X = Cl or Br, gave a complete assignment of ligand modes. All were O-coordinated, except for $NiCl_2(NMF)_4$, $NiCl_2(DMF)_2$ and $CuCl_2(DMF)_2$, where there is some evidence for N-coordination also.[319] Several diagnostic bands due to the OMe ligand were assigned for ReX(OMe)(CO)(NO)$(PPh_3)_2$, where X = H, D, Cl, Br or I.[320] The νCH of the bridging OMe group in $Rh_2(\mu\text{-OMe})(CO)_2(\mu\text{-dppm})_2$ is at 2803cm^{-1}.[321]

$Pd(C_6F_5)L_2(APPY)^+$, where L = PPh_3, P^nBu_3, or L_2 = bipy, APPY = $Ph_3PCHCOCH_3$, give $\nu C=O$ near 1520cm^{-1}, showing unprecedented Pd-O coordination of the keto-stabilised ylide. The neutral complex $Pd(C_6F_5)$(Cl)(tht)(APPY), where tht = tetrahydrothiophene, has $\nu C=O$ at 1655cm^{-1}, showing the normal Pd-C coordination.[322] The i.r. spectrum of $TmCl_2$.2THF gave assignments for some THF modes.[323] The i.r. spectra of $AlCl_3$.nTHF, where n = 1, 2 or 3, are consistent with Al-O coordination.[324]

<u>6.4 Ligands containing O-N or O-P bonds</u>.- I.r. spectra have been reported for $M[O_2P(OR)_2]$, where M = Li-Rb, R = 2-ethylhexyl, $M'[O_2P(OR)_2]_2$, where M' = Mg, Ca or Ba. When M = Li, the structure is a long-chain polymer, with bridging, bidentate P(O)O groups, M = Cs or Rb are monomers, with bidentate P(O)O, and M' = Mg is again a polymer.[325]

$Zr(Cp)_2Ph[ON(Ph)NO]$ gives i.r. bands at 1345, 1295, 1218 and 928cm^{-1}, typical of the $\overline{M[ON(R)NO]}$ chelate ring.[326] The $O_2PCl_2^-$ modes in $NbOCl_4(O_2PCl_2)^{2-}$ are consistent with unidentate coordination *via* one oxygen.[327]

Trans-$[Cr(cyclam)(NO_3)]NO_3.2H_2O$, where cyclam = 1,4,8,11-

tetra-azacyclotetradecane, has bands due to the coordinated nitrato group at 1515 and 1270cm^{-1}.[328] I.r. spectra of $Cr(O_2PF_2)_3.HPO_2F_2$, $M(O_2PF_2)_2.HPO_2F_2$, where M = Mn, Fe, Co or Ni, and 2 new forms of $Fe(O_2PF_2)_3$ all show the bands expected for bidentate bridging difluorophosphate ligands.[329] νC=O and νP=O bands for $MoO_2Cl_2[(^iPrO)_2P(O)CH_2C(O)NEt_2]$ are at lower wavenumbers than in the free ligand, i.e. both C=O and P=O are coordinated to the metal.[330]

The i.r. spectra of $M(pacO)(H_2O)_2$, where $pacO^{2-}$ = pyridine-2,6-dicarboxylate N-oxide ion, M = Mn, Co, Ni, Cu or Zn, do not show the expected decrease in νN-O (from the free ion) due to the O-coordination, which is found by X-ray diffraction. This may be due to a simultaneous increase because of the breakdown of the hydrogen bonding found in the free ligand.[331]

$Fe(salen)(NO_3)$ exists as two isomers, characterised by i.r. spectroscopy. One is a dimer, with unidentate nitrates, $[Fe(salen)ONO_2]_2$; the other is a monomer, with bidentate nitrate, $Fe(salen)O_2NO$. ^{15}N-substitution was used to identify NO_3 bands (salen = N,N'-ethylenebis(salicylideneaminato)).[332] I.r. bands indicative of unidentate phosphato (C_{3v} symmetry) were seen for $Rh(NH_3)_5PO_4$.[333]

Assignments were given for ν_s, $\nu_{as}NO_2$, δNO_2 and ρNO_2 for the trans-nitro complexes $Pd(NO_2)_2(PR_3)_2$, where R_3 = $MePh_2$, Me_2Ph, Ph_2 or Et_3; these were confirmed by ^{15}N substitution. All of the bands were consistent with Pd-N bonding, and the numbers of bands confirmed the trans geometry.[334] (72) has i.r. bands at 1375, 1315, 1275 and 910cm^{-1}, characteristic of the coordinated ON(Ph)NO ligand.[335]

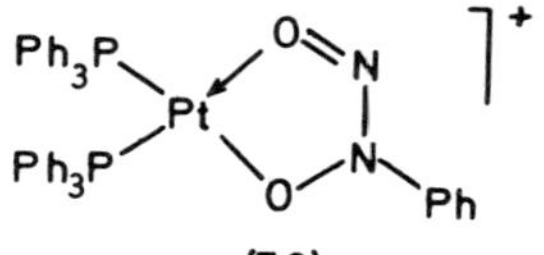

(72)

νP=O in $Eu(DIK)_2(NO_3)(Ph_3PO)_2$ is shifted to lower wavenumber compared to the free ligand, i.e. it is coordinated via O (DIK = various β-diketonate ligands).[336]

$U(NO_3)_4L_2$, where L = piperazine or substituted derivatives, have i.r. bands of chelating, bidentate nitrato only.[337] I.r. and Raman spectra of $UO_2(PO_2F_2)_2$ suggest a chain structure with OPO bridges (73).[338] The i.r. and Raman spectra of the adduct $P_2O_3Cl_4.2SbCl_5$ reveal the coordination of terminal O atoms of $P_2O_3Cl_4$ to Sb (74). ν_{as}POP, ν_sPOP are at 980, 440cm^{-1} respectively.[339]

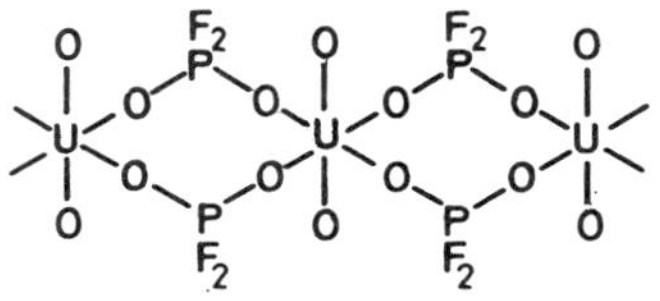

(73)

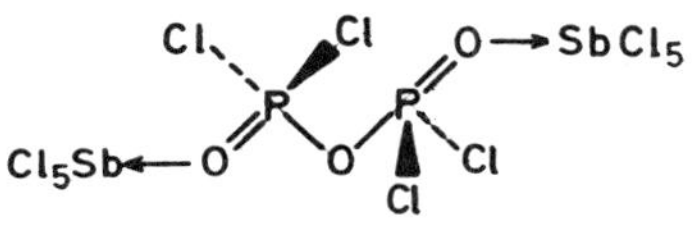

(74)

6.5 Ligands containing O-S, O-Se, or O-Te bonds.- I.r. data for $V(SO_3CF_3)_3$, $VO(SO_3CF_3)_2$ and $VO(SO_3CF_3)_3$ all suggest bridging bidentate bonding for SO_3CF_3.[340] SeO_4 modes were assigned for $VO_2SeO_4^-$: ν_1 825cm^{-1}, ν_3 885, 860cm^{-1}, ν_4 440, 420cm^{-1}.[341] Bands due to unidentate SO_4 ligands were assigned in $[Mo_6Cl_8](SO_4)_2$, at 1275, 1250, 1000, 645 and 615cm^{-1}.[342]

I.r. spectra were reported for $Ru(SO_3F)_3$ and $Ru(SO_3F)_5^-$, which have both uni- and bidentate ligands, and for $Ru(SO_3F)_6^{2-}$, which only contains unidentate ligands.[343] The i.r. spectra (SO_4^{2-} modes) of M_2CuSO_4, where M = Na, K, Rb, Cs or Tl, show that the nature of Cu-SO_4 coordination is dependent on M.[344]

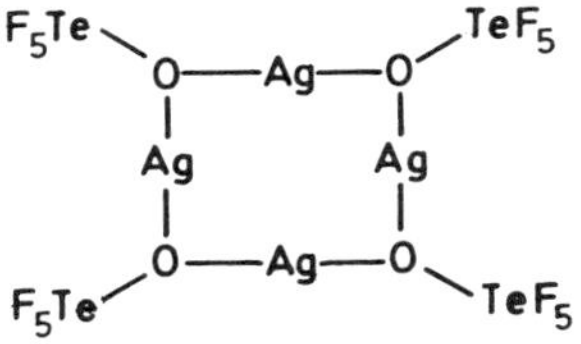

(75)

$Ag(OTeF_5)$ has νTeO at 815cm^{-1}(i.r.)/812, 845cm^{-1}(Raman), consistent with the structure (75).[345] The i.r. spectra of $UO_2XO_4(DMSO)_2.H_2O$, where X = S, Se or Cr, show bidentate, bridging coordination by the XO_4 units.[346]

6.6 Ligands containing O-Cl bonds.- FTIR spectra were obtained for the vapours above $NaClO_3$ and $LiClO_3$ at 360^0C. For $NaClO_3$ the most stable ion pair has terdentate geometry. A high percentage (more than 50%) of dimers were present in each case.[347] There was Raman evidence for ion-pair formation for alkali-metal perchlorates in $DMSO/H_2O$ solutions.[348]

The i.r. spectra of $Nb(ClO_4)_5$, $NbO(ClO_4)_3$ and NbO_2ClO_4 contain bands characteristic of both uni- and bidentate ClO_4 ligands. In $Nb(ClO_4)_6^-$ and $Nb(ClO_4)_7^{2-}$ there are only unidentate ligands.[349] Bands due to unidentate ClO_4 and free ClO_4^- ion were seen for $[RuL_3(ClO_4)]ClO_4$, where L = $Ph_2PCH_2PPh_2$, $Ph_2P(CH_2)_2$-PPh_2, $Ph_2As(CH_2)_2AsPh_2$ or $Ph_2P(CH_2)_2AsPh_2$.[350]

Anhydrous $M(ClO_4)_2$, where M = Co, Ni or Cu, have been prepared. The vibrational spectra were analysed in terms of tridentate perchlorato ligands for M = Co or Ni and two types of bidentate ligand for M = Cu.[351] Ligand-mode assignments were proposed for $CoL(ClO_4)_4.3H_2O$, where L = benzo-15-crown-5.[352] I.r. and Raman spectra were reported again for $Ni(ClO_4)_2$ (see ref.351).[353]

7 Sulphur Donors

The i.r. spectra of MX_4L, where M = Ti or Zr, X = Cl or Br, L = dithio-oxamide, all show M-S coordination for the dithio-oxamide.[354] I.r. data show bidentate dithiocarbamate ligands in $CpTi(S_2CNHR)Cl_2$ and $CpTi(S_2CNHR)Cl$, where R = $C_8H_5N_2S$, $C_9H_5N_2SCl_2$ or $C_9H_7N_2S$.[355] Similar results for $CpTi(S_2COR)Cl_2$, $CpTi(S_2COR)_2Cl$ and $CpTi(S_2COR)_3$, where R = Me, Et, Pr, Bu and C_5H_{11}, show that all are non-electrolytes with bidentate xanthato ligands.[356]

I.r. bands of $VO(RNCS_2)_2$, where R = Et_2, Pr_2, $^iPr_2$, Bu_2, $^iBu_2$, piperidine, pyrrolidine etc., were assigned to $\nu C{=}N$ ($1480\text{-}1515cm^{-1}$) and $\nu C\text{-}S$ ($1130\text{-}1150cm^{-1}$).[357] $\nu S\text{-}S$ lies in the range $580\text{-}600cm^{-1}$ in $Nb_2X_4S_3$, where X = Cl or Br, and their adducts with S-donors or MeCN.[358] $\nu S\text{-}S$ in $Mo(S_2)Cl_3$ is at 612 cm^{-1}, with very similar values in $[X_4Mo(\mu\text{-}S_2)_2MoX_4]^{2-}$, where X = Cl or Br.[359]

$MoCp(NO)[PS_2(OR)_2](S_2CNR'_2)$, where R = Et, ^iPr, R' = Me or Et, have νPS at $770\text{-}800cm^{-1}$ (i.e. unidentate dithiophosphorus ligand) and νCN at $1515\text{-}1580cm^{-1}$ (bidentate dithiocarbamato ligand).[360]

I.r. bands in the spectra of $MoOXL_2$, where X = F, Cl, Br or I, HL = 4-morpholino-dithiocarboxylic acid, are consistent with the dithiocarbamate acting as a univalent, bidentate ligand.[361] The νCN bands in $Mo(CO)_2(S_2CNR_2)_2X^-$, where X = F or N_3, R = Me or Et, are 15-17cm^{-1} lower than in $Mo(CO)_2(S_2CNR_2)_2$. This is due to an increase in the contribution from (76) rather than (77) in the ligands.[362]

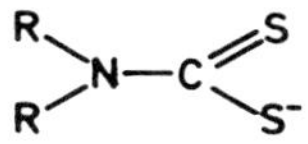

(76)

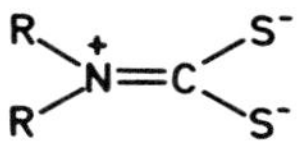

(77)

$Mo(S_2CNEt_2)_4^+$ does contain Mo^V - confirmed by the presence of a thioureide band at 1507cm^{-1}, intermediate between the values for known Mo^{IV} and Mo^{VI} species.[363] Splittings of ν_s and ν_{as}(CS) bands, near 1000cm^{-1}, 600cm^{-1} respectively, in $[Mo(S_2CNEt_2)_4]_2$-Mo_6O_{19} suggest a complex mode of bonding between the Mo and the dithiocarbamate.[364] $Mo_4(\mu_3\text{-}S)_4(\mu\text{-}S_2CNEt_2)_2(S_2CNEt_2)_4$ has only one νC=S i.r. band, showing bidentate coordination, but the presence of different types of ligand was revealed by the presence of two bands at 565, 551cm^{-1}.[365] νCO for $(OC)_5W[S{=}C(Ph)H]$ shows many bands, due to the presence of both η^1- and η^2-isomers, (78) and (79).[366]

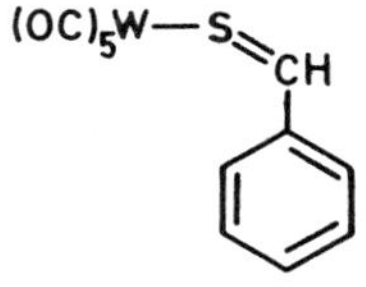

(78)

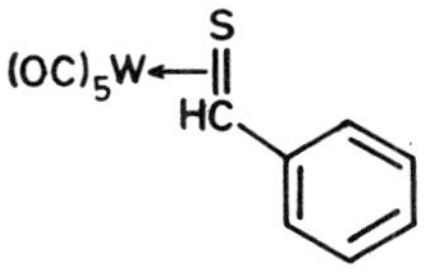

(79)

Xanthate and νCO modes were assigned for $SCo_3(CO)_7(S_2COMe)$.[367] $Co^{IV}(S_2CNRR')_3^+$, where R, R' = Me, Et, Ph or CH_2Ph, have νC⩵N about 20cm^{-1} higher than in the Co^{III} analogues.[368] The i.r. spectra of $HRhL_2.H_2O$, $H_3RhL_3.2H_2O$ etc., where H_2L = $HSCH_2CH(SH)$-CH_2SO_3Na, contain no νSH bands, showing coordination via both sulphur atoms.[369]

$(NC)_2NiS_5^{2-}$ has i.r. bands tentatively assigned to νSS at 460, 393 and 385cm^{-1}.[370] Solid $[O{=}PPh_2CH_2CMe(CH_2PPh_2)]Ni$-$(S_2CO)$ has νC=O from the chelating dithiocarbonato ligand. No carbon-sulphur stretching was detectable.[371]

A detailed assignment of ligand modes was proposed for ML_2, where M = Ni or Cu, L = (80; X = CH_2, O, S, NH, NMe, NPh). νC≕N was higher for the Ni than for the Cu complexes.[372]

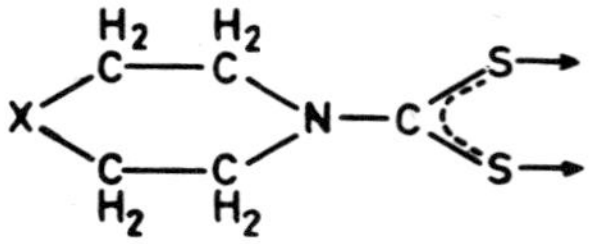

(80)

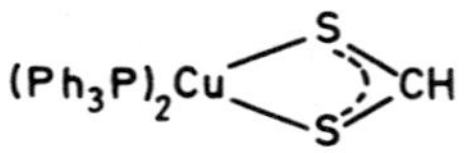

(81)

νS-S modes in $Cu_3(S_4)_3^{3-}$ give Raman features at 425, 435 and 469cm^{-1}.[373] (81) and (triphos)Cu(S_2CH), with bi- and unidentate dithioformates respectively, give very similar ligand modes, and not very different from those of free S_2CH^-. Hence the internal electronic distribution is almost unchanged on coordination.[374] $AuCl_3(S_3N)$ has νSN at 1052, 696cm^{-1}, δSN ca. 520cm^{-1}, and νSS at 476cm^{-1}.[375] $Au_4(dta)_4$, where dta = dithioacetato, has $\nu_{as}CS_2$ at 1160cm^{-1} and ν_sCS_2 at 899cm^{-1}, consistent with the multinuclear, completely bridged structure revealed by X-ray diffraction. Au(dtb), where dtb = dithiobenzoato, has $\nu_{as}CS_2$ at 1039cm^{-1}, ν_sCS_2 at 911cm^{-1} and δCS_2 at 656cm^{-1}, in agreement with a fully chelated dithiobenzoato ligand.[376]

There is a marked change in νY-H on coordination to form GaX_3.HYEt, where X = Br or I, Y = O, S, Se, showing that there is a strong interaction between the Ga and the chalcogenol.[377]

A number of ligand modes were assigned for $Ph_3GeS_2P(OMe)_2$ and $Ph_2Ge[S_2P(OMe)_2]_2$, showing that they are unidentate ligands.[378]

The i.r. spectra of $MeSb(S_2CNR_2)_2$, where R = Me or Et, R_2 = $-(CH_2)_2O(CH_2)_2-$, are all consistent with symmetrical bidentate coordination of the dithiocarbamates.[379] $SbCl_3$(L-L), where L-L = N,N'-disubstituted dithiomalonates RHNC(S)CH_2C(S)NHR (R = Et, ^{i}Pr, ^{n}Bu or Cy), have CN and CS shifted to higher and lower wavenumbers, respectively, compared to the free ligands. This is consistent with S,S-donation.[380]

8 Potentially Ambident Ligands

8.1 Cyanates, Thio- and Selenocyanates, and their Iso Analogues.- $M[ZrO(NCS)_3].2H_2O$, where M = Cs^+ or pyH^+, have NCS bands confirming Zr-NCS coordination.[381] V-NCS coordination was detected in the same way for $Cp^*V(NCS)_2$, where $Cp^* = C_5H_4Me$, C_5HMe_4 or C_5Me_5,[382] and W-NCS coordination in (82).[383]

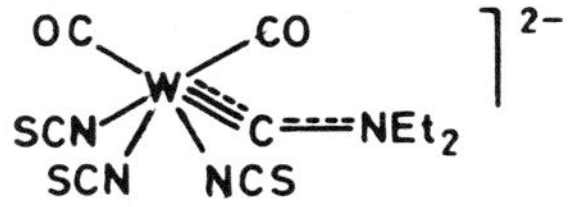

(82)

In the i.r. and Raman spectra of $OsCl_5(NCS)^{2-}$ and $OsCl_5(SCN)^{2-}$ νCN(S), νCS(N) and δNCS are at higher wavenumbers than νCN(N), νCS(S) and δSCN.[384] $W(CNO)(CO)_5^-$ has fulminate bands as follows: 2190cm^{-1} (2Xν_sCNO), 2110cm^{-1} (ν_{as}CNO) and 1087cm^{-1} (ν_sCNO). The complex isomerises in solution to $W(NCO)(CO)_5^-$, with ν_{as}NCO at 2235cm^{-1} and ν_sNCO at 1318cm^{-1}.[385]

νCN of NCS in $(C_5H_4Me)Co(CO)LL'$, where L = L' = NCS or L = NCS, L' = CN, show Co-NCS coordination in each case.[386] I.r. spectra of $L_2M(NCS)_2(NCSAg)_2$, where M = Co, Ni or Cu, L = urea, thiourea etc., all show that they have monomeric bridged structures.[387]

The i.r. spectra, chiefly νCN, of $Pd(SCN)_2L_2$, where L = PPh_3 or other arylphosphines, are consistent with Pd-S coordination.[388] There is i.r. evidence that $Pt(bipy)(SCN)_2$ has Pt-SCN coordination at room temperature, but on heating above 130°C converts it to the Pt-NCS form.[389]

Raman and i.r. spectra of species formed by CuSCN + SCN^- in soft donor solvents were reported. $Cu(NCS)_2^-$ has νCN at 2075cm^{-1} (i.r.)/2086cm^{-1}(Raman) and νCS at 791cm^{-1}. It is therefore N-bonded, i.e. Cu^+ is acting as a hard cation towards SCN^-.[390] The fulminato complex $Au(PPh_3)(CNO)$ has a characteristic i.r. band at 2156cm^{-1} (shifted to 2119cm^{-1} on ^{15}N substitution).[391]

The i.r. spectra of $MeSiH_2X$, where X = NCO, NCS or NCSe, show that all are Si-N-bonded.[392] The previously described 'F_5Se-NCO' has now been shown to be F_5Se-OCN, as it has νOCN bands at 2290cm^{-1} (as), 1104cm^{-1} (s).[393]

8.2 Ligands containing N and O atoms.- Infrared results for $Mg(IMP).5H_2O$, where $IMPH_2$ = inosine-5'-monophosphate acid, suggest Mg-N_7 coordination, together with indirect metal phosphate and metal carbonyl interaction.[394] The i.r. spectra of $M(acac)_2M'L$, where M = Mg or Zn, M' = Li, Na or K, HL = 1-nitroso-2-naphthol or 8-hydroxyquinoline, show that bridged and unbridged oxygen atoms are present.[395]

VOCl(L) and VO_2L_3, where H_2L = thio-Schiff bases from the condensation of salicylaldehyde, 2-hydroxyacetophenone etc., with S-(methyl or benzyl)dithiocarbazates, have i.r. spectra consistent with tridentate coordination via thiol S, azomethine N and phenolic O.[396] Re-investigation of the i.r. spectra of bis N-(2-mercaptoethyl)salicylideneaminato oxovanadium(IV) shows that the OH group of the ligand is coordinated to the metal, contrary to earlier reports.[397]

CrL_2Cl and MLCl, where M = Cr, Mn, Fe, Co, Ni or Cu, L^- = dihydrobis(succinimidyl)borate, have the L^- as a bidentate ligand via the O atoms of the succinimide ring system.[398] I.r. data for $[ML_2]X$ [where M = Cr, Co, X = Cl; M = Dy, Gd or Sm, X = NO_3; HL = o-HOC_6H_4N=CHC(O)R] show that the L is coordinated by the carbonyl O, the imino N and phenoxo O atoms.[399]

Ligand-mode assignments for Mn(II) complexes with cysteine, homocysteine and penicillamine all show coordination via NH_2 and COOH groups.[400]

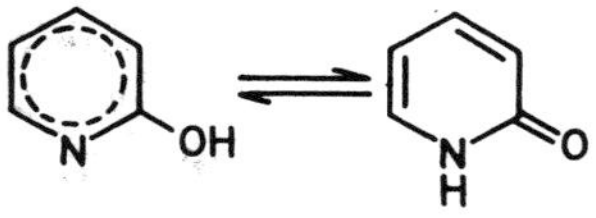

(83)

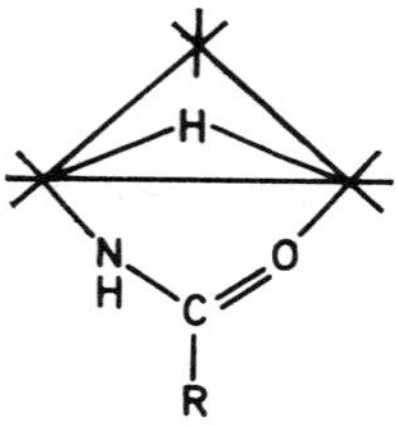

(84)

The i.r. band at 1660cm^{-1} in $Ru_2Cl(C_5H_4NO)_4(C_5H_5NO)$ shows the presence of the keto form of the 2-hydroxy-pyridine ligand (83).[401] νC=O(amide) in $Os_3(CO)_{10}(\mu$-H)(NHCOR), where R = H, Me, Ph, Et or Pr, always gives just one i.r. band, 1570-1597cm^{-1}, consistent with the presence of the unit (84).[402]

Assignments were proposed for nitro-ligand modes from the i.r. spectra of $Co(DH)_2(NO_2)A$ and $Co(DH)_2(NO_2)X^-$, where DH_2 = dimethylglyoxime, A = amine, X = I, Br, CN, NCS etc.[403] $M(guH)Cl_2$, where guH = guanine, M = Co, Cu or Zn, have i.r. spectra consistent with coordination of one or more ring nitrogens to M.[404] νC=N modes in all of the follwing are consistent with C=N-Co coordination: $CoL_2.2H_2O$, where HL = o-$HO_2CC_6H_4N=C(CH_3)R$, R = 2-furyl, or o-$HO_3SC_6H_4N$=CHC(O)R; $CoL'Cl_2.2H_2O$, where L' =

RC(=O)-CH=NC$_2$H$_4$N=CHC(=O)R; and CoL"$_2$ Cl$_2$, where L" = 2-Cl-4-O$_2$NC$_6$H$_3$N=CHC(=O)R.[405] I.r. spectra of ML$_2$, where HL = N-(dimethylaminomethyl)isatin β-thiosemicarbazone, M = Co, Ni, Cu, Zn or Hg, and FeL$_2$Cl show coordination via S, hydrazine N and ketone O.[406]

Shifts in νC=O and ring vibrations show that 1,8-naphthyridine-2,7-dicarboxylate coordinates by the ring N and carbonyl O to Ni in $Ni_2(2,7\text{-dc-1,8-napy})_3^{2-}$.[407] A detailed assignment of ligand modes was proposed for bis(salicylaldoximates) of Ni(II) and Pd(II).[408] (85) has νC=O at 1682cm^{-1}, consistent with a weak aldehyde C-H...Pt interaction.[409]

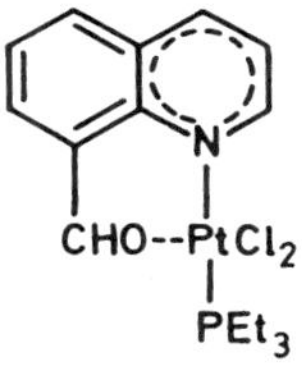

(85)

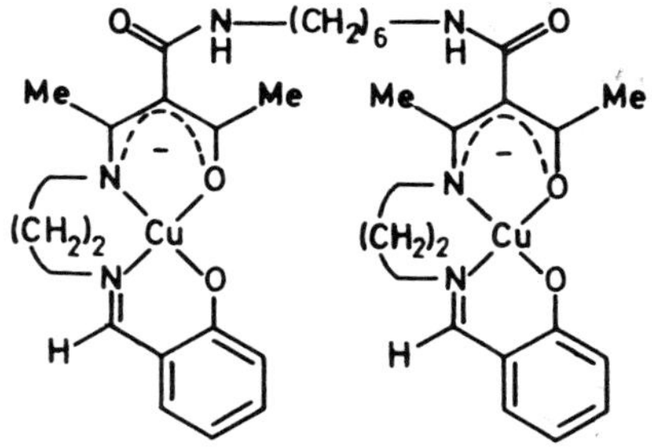

(86)

ν_sCO_2 and $\nu_{as}CO_2$ assignments for Cu(II) complexes of a variety of amino acids were used to assess the covalent character of the Cu-O bonds.[410] Dinuclear copper(II) complexes such as (86) have νC=O at 1650cm^{-1} and νNH at 3250cm^{-1}, from the substituted CONH group.[411]

In Ln(DPAAP)$_2$(NO$_3$), where DPAAP = 4-(2',4'-dihydroxyphenylazo)antipyrine, Ln = La, Pr, Nd, Sm, Gd, Tb, Dy, Ho or Y, the neutral bidentate ligand is attached by the carbonyl O and one of the azo N atoms.[412] The i.r. spectra of M$_2$L$_3$.6H$_2$O, where M = lanthanide, H$_2$L = phthalylsulphathiazole, show that the ligand is bidentate, via the carboxyl O and sulphonamido N atoms.[413]

The i.r. spectra of $UO_2L_3X_2$, where X = Cl, SCN, L = Schiff bases derived from furfural and $H_2NC_6H_4R$, R = H, *m*-Cl, *o*-, *m*- or *p*-Me etc., show coordination via the furan ring O and the azomethine N.[414] Ligand modes for UO_2^{2+} and Th^{4+} chelates with a variety of heterocyclic Schiff bases show that the ligands coordinate by the C=N nitrogen and phenolic oxygen.[415]

$SiCl_2L_2$, where HL = *p*-$H_2NSO_2C_6H_4N$=CHC_6H_3R-*o*-OH, and R = H, 3-, 4-, 5-CH_3 or 5-Cl, have i.r. spectra which show that L coordinates by the imino N and phenoxy O atoms.[416] Schiff-base coordination by two oxygen atoms was suggested from the i.r. of SnX_2.ML, where X = Cl, Br, NCS, M = Ni or Cu, H_2L = *o*-$HOC_6H_4(CR)$=NCH_2CH_2N=$C(R)C_6H_4OH$-*o* (R = H or Me).[417] In R_2SnL_2 and $R_2(L)SnOSn(L)R_2$, where L = *N*-phthaloyl derivatives of L-leucine, DL-alanine or L-phenylalanine, R = Me, Et, ^nBu or n-C_8H_{17}, there are bidentate carboxylates, with additional participation of the imido C=O in complex formation.[418]

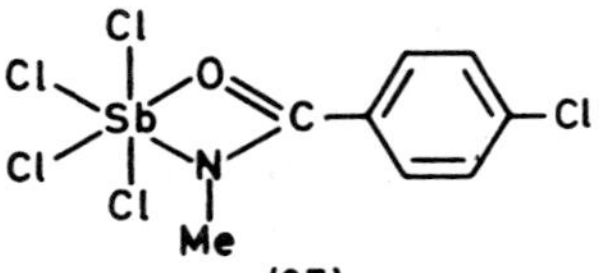

(87)

The i.r. spectrum of (87) shows the coordination of C=O to Sb, since νC=O is $90 cm^{-1}$ lower than in the free ligand.[419]

8.3 Ligands containing N and S atoms.- I.r. data show the coordination of dithiocarbazate via amino N and thiocarboxylato S atoms in $MOL_2.2H_2O$ (M = Zr or V) and $UO_2L_2.2H_2O$, where HL = dithiocarbazic acid.[420] νN_2S_2 is at $858 cm^{-1}$ and δN_2S_2 at 450 cm^{-1} in the new complex $(\mu\text{-}N_2S_2)(VCl_5)_2^{3-}$, with V-N coordination (88).[421]

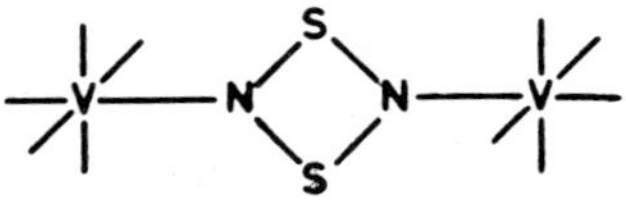

(88)

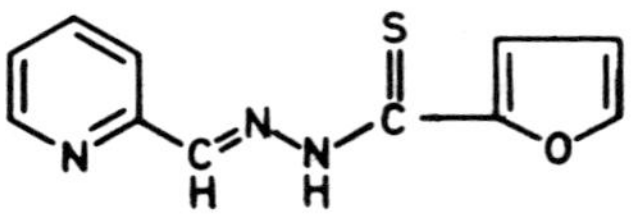

(89)

$M(pfth)_2$, where $pfth^-$ = (89), M = Mn, Fe, Co, Ni, Cu or Zn, involve coordination via azomethine and pyridine nitrogen atoms, as well as thiol S.[422] Characteristic ligand modes were assigned in (90) [423] and (91).[424]

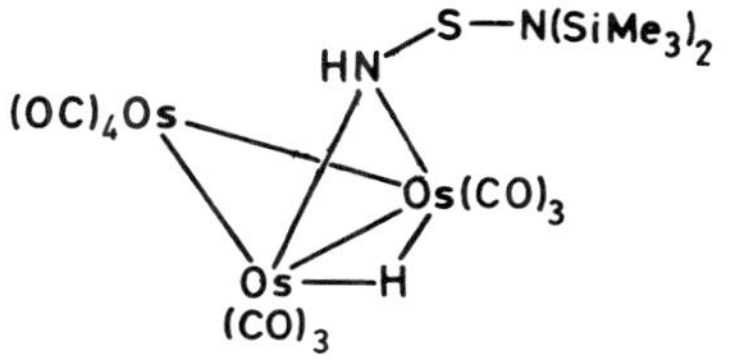

(90)

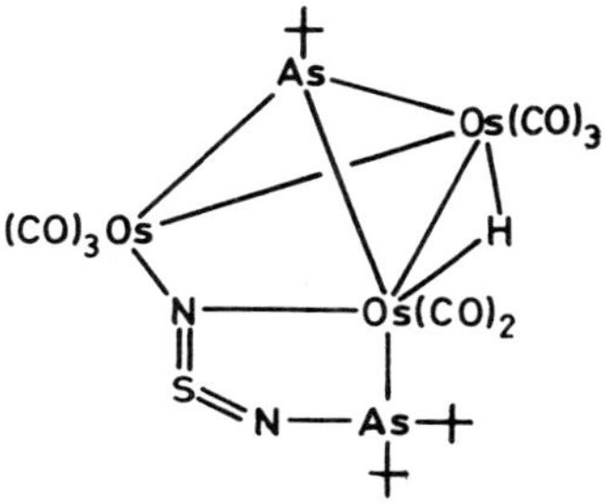

(91)

I.r. spectra of Co(II) complexes of 4,6-dimethylpyrimidine-2(1H)-thione (HL), i.e. $Co(HL)_2(H_2O)_2^{2+}$, $Co(HL)_2NO_3^+$, $Co(HL)_2SO_4$ and $Co(HL)_3^{2+}$, all reveal N,S-chelation via the non-protonated ring nitrogen and the exocyclic sulphur.[425] CoL_2^+, where HL = RCH=NNHC(S)NHPh, R = 5-methylpyrazol-3-yl, and related species, were shown to have ligand coordination by pyrazolyl N, azomethine N and S atoms.[426] Phosphazene N and thiocarbonyl S atoms are involved in coordination in $MLCl_2$, where L = RNHPPhNPhC(S)NHPh, RNHPRNRC(S)NHR, R = o-tolyl, M = Co, Ni, Cu or Zn, $AgL(NO_3)$ and $[FeLCl_2]Cl$.[427] Coordination of the ligand by the sulphur atom was inferred from i.r. and Raman results on ML_2Cl_2, where M = Co, Ni, Zn, Cd, L = 5-methyl-1-phenylhexahydro-1,3,5-triazine-2-thione.[428]

The i.r. spectrum of $Ni_6(MPDMA)_{12}$ and related species, where MPDMA = $^-S(CH_2)_3NMe_2$, shows that bridging mercaptoamine ligands are present, with S-coordination only. $Pd(MPDMA)_2Cl_2$, however, involves N- and S-coordination.[429] 1:1 and 1:4 allythiourea/Pd complexes only involve coordination by the nitrogen atom.[430] Ligand modes due to bridging $N_2S_2^{2-}$ were assigned in $[M(S_2N_2)(PPh_3)]_2$, where M = Pd or Pt.[431] Resonance Raman spectra of Pd(II), Pt(II), Cu(I), Cu(II) and Hg(II) dithizonate complexes identified vibrations associated with the chromophoric azo group. The Pd(II) and Pt(II) complexes are N,S-coordinated, with a trans, square-planar configuration; the Cu(I) complex contains a strong bond to S, and weak bonds to N, and a sulphur atom on an adjacent ligand, to give a polymeric structure.[432] I.r. data were used to suggest S,N,O-coordination in CuL [where H_2L = RC(O)CH_2C(R')=NR"; R' = Me, R = Me, PhNH etc.; R' = Ph, R = PhNH; R" = 5-mercapto-3-methyl-1,2,4-triazol-4-yl].[433]

SERS was used to study adsorption of thioacetamide[434] or N,N-dimethylthioformamide[435] on Cu or Ag electrodes. In both cases coordination to the metal occurred via S.

RHgL, where HL = 2-mercaptobenzothiazole, R = Me or Ph, have νC=N near 1560cm^{-1}. Some other tentative assignments were given.[436] The i.r. spectra of the thiosemicarbazide complexes Sb(CH_5N_3S)X_3, where X = Cl, Br or I, show that S and N (hydrazine terminal nitrogen) are coordinated to the Sb(III).[437]

8.4 Ligands containing S and O atoms.- $ZrCl_4$.9DMSO has a band at 463cm^{-1}, assigned as an S=O deformation, showing that Zr-S-coordination has taken place.[438] CpM(CO)$_3$(Ph_2SO)$^+$, where M = Mo or W, have νSO at 940cm^{-1} (M = Mo) or 910cm^{-1} (W), showing O-coordination by the sulphoxide ligand, as expected on steric grounds.[439]

The i.r. spectrum of the reaction product of Mn($PPhMe_2$)Br_2 and SO_2 shows that coordinated, bridging SO_2 is present (stoichiometry $MnLBr_2$.½SO_2).[440] Vibrational spectra were reported and assigned for Mn(SO_3F)$_3$, Cs_2[Mn(SO_3F)$_5$], M(CO)$_5SO_3F$ (M = Mn or Re) and Mn(CO)$_4SO_3F$. There were significant differences in SO_3F modes for uni- and bidentate coordination, exemplified by the Mn(CO)$_5$ and Mn(CO)$_4$ derivatives.[441]

νP=S (near 650cm^{-1}) and νP-S (near 540cm^{-1}) were assigned for Cp*Fe(CO)$_2$L, where Cp* = C_5H_5, C_5H_4Me or C_5Me_5, L = η^1-SP(S)(OR)$_2$, R = Et or ^{i}Pr.[442] νSO bands were seen at 1150 and 1220cm^{-1} in Ru(H_2O)(PPh_3)$_2$(tos)$_2$, where tos = p-toluene-sulphonato, i.e. the ligand is O-coordinated.[443] The complexes (η^6-p-cymene)Os(Me_2SO)XCl, where X = Cl or Me, have νSO at 1058cm^{-1} (Me) or 1120cm^{-1}. The latter is normally indicative of O-bonding, but the crystal structure shows that it is the sulphur atom which is bonded in both cases.[444]

The bridging SO_2 group in Ir_4(CO)$_9$(μ_2-CO)$_2$(μ_2-SO_2) has νSO bands at 1265 and 1096 cm^{-1}. This complex reacts with OMe^- to form $Ir_4(CO)_8(\mu_2\text{-}CO)_3(SO_2OMe)^-$, with an S-bonded methoxysulphonyl group: $\nu_{as}S{=}O$ 1227cm^{-1}, $\nu_sS{=}O$ 1090cm^{-1}, $\nu S\text{-}O$ 975cm^{-1} and δOSO 672cm^{-1}.[445] $Pt_2(R_2SO)_2Cl_4$ and trans-$Pt(R_2SO)_2Cl_2$, where R = Me or Et, all have an S-O stretching mode at 1130-1155cm^{-1}, diagnostic of terminally S-bonded sulphoxide.[446] The complex (92) has C=N at 1530cm^{-1}.[447]

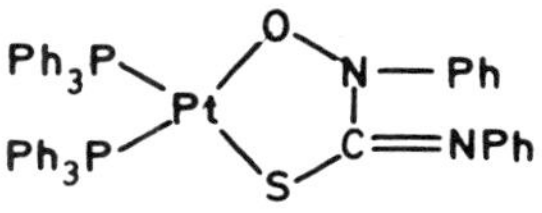

(92)

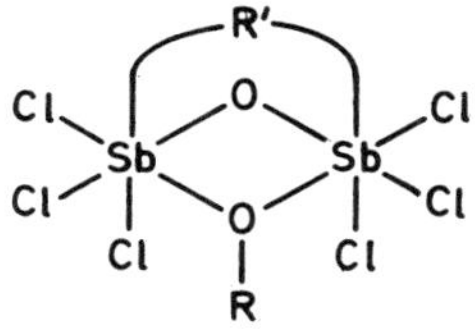

(93)

The i.r. spectra of the thiosulphates $In_2S_2O_3.2H_2O$, $In_2S_2O_3.InOH.2H_2O$ and $In_2S_2O_3.2InNO_3.2H_2O$ suggest In-O coordination in each case.[448] I.r. evidence points the presence of bidentate xanthate ligands in $GeO(S_2COR)_2.\underline{x}H_2O$.[449] I.r. and Raman data for (93), where R = H, Me or Et, R' = $MeSO_2$ or $MeSO_3$, all contain the expected bands for bridging methanesulphinato or methanesulphonato ligands.[450]

References

1 L.Manceron and L.Andrews, J. Amer. Chem. Soc., 1985, 107, 563.
2 I.A.Garbuzova, O.G.Garkusha, B.V.Lokshin, G.K.Borisov and T.S.Morozova, J. Organometal. Chem., 1985, 279, 327.
3 S.J.Pratten, M.K.Cooper and M.J.Aroney, Polyhedron, 1984, 3, 1347.
4 M.F.Lappert and P.P.Power, J. Chem. Soc., Dalton Trans., 1985, 51.
5 J.Mascetti and M.Tranquille, Surf. Sci., 1985, 156, 201.
6 M.Morán, J.J.Santos-García, J.R.Masaguer and V.Fernández, J.Organometal. Chem., 1985, 295, 327
7 R.F.G.Baynham, J.Chetwynd-Talbot, P.Grebenik, R.N.Perutz and M.H.A.Powell, J.Organometal. Chem., 1985, 284, 229.
8 D.C.McKean, G.P.McQuillan, A.R.Morrisson and I.Torto, J. Chem. Soc., Dalton Trans., 1985, 1207.
9 B.Weinberger, F.Degel and W.P.Fehlhammer, Chem. Ber., 1985, 118, 51.
10 L.Weber and D.Wewers, Organometallics ,1985, 4, 841.
11 H.G.Alt, H.E.Engelhardt, U.Thewalt and J.Riede, J. Organometal. Chem., 1985, 288, 165.
12 J.Howard and D.Graham, Spectrochim. Acta, Part A,1985, 41A, 815.
13 N.E.Kolobova, L.L.Ivanov, O.S.Zhvanko, A.S.Batsanov and Yu.T.Struchkov, J. Organometal. Chem., 1985, 279, 419.

14 S.J.N.Burgmayer and J.L.Templeton, Inorg. Chem., 1985, 24, 3939.
15 K.Stahl, F.Weller and K.Dehnicke, Z. anorg. allg. Chem., 1984, 518, 175.
16 C.Benamou and J.Benaim. J. Organometal. Chem., 1985, 280, 377
17 M.A.Bennett and I.W.Boyd, J. Organometal. Chem., 1985, 290, 165.
18 M.A.Bennett, I.W.Boyd, G.B.Robertson and W.A.Wickramasinghe, J.Organometal. Chem., 1985, 290, 181.
19 M.Panizo and M.Cano, J. Organometal. Chem., 1985, 287, 221.
20 J.L.Templeton, R.S.Herrick, C.A.Rusik, C.E.McKenna, J.W.McDonald and W.E.Newton, Inorg. Chem., 1985, 24, 1383.
21 H.G.Alt, J. Organometal. Chem., 1985, 288, 149.
22 E.O.Fischer, A.C.Filippou, H.G.Alt and U.Thewalt, Angew. Chem., Int. Ed. Engl., 1985, 24, 203.
23 E.Hey, F.Weller and K.Dehnicke, Z. anorg. allg. Chem., 1984, 514, 18.
24 K.Stahl, U.Müller and K.Dehnicke, Z. anorg. allg. Chem., 1985, 527, 7.
25 D.S.Barratt and C.A.McAuliffe, Inorg. Chim. Acta, 1985, 97,37
26 D.Gampe, K.Jacob and K.-H.Thiele, Z. anorg. allg. Chem., 1985, 526, 36.
27 G.S.Girolami, C.G.Howard, G.Wilkinson, H.M.Dawes, M. Thornto-Pett, M.Motevalli and M.B.Hursthouse, J. Chem. Soc., Dalton Trans., 1985, 921.
28 D.Miguel and V.Riera, J. Organometal. Chem., 1985, 293, 379.
29 K.Raab, M.Appel and W.Beck, J. Organometal. Chem., 1985, 290, C28.
30 M.G.Yezernitskaya, B.V.Lokshin, V.I.Zdanovich, I.A. Lobaova and N.E.Kolobova, J. Organometal. Chem., 1985,282,363
31 S.C.Chang, Z.H.Kafafi, R.H.Hauge, W.E.Billups and J.L.Margrave, J. Amer. Chem. Soc., 1985, 107. 1447.
32 W.P.Fehlhammer, P.Hirschmann, and A.Völkl, J. Organometal. Chem., 1985, 294, 251.
33 D.W.Ball, Z.H.Kafafi, R.H.Hauge and J.L.Margrave, Inorg. Chem., 1985, 24, 3708.
34 A.L.Shobert, I.C.Hisatsune and P.S.Skell, Spectrochim. Acta, Part A, 1984, 40A, 609.
35 M.I.Bruce, D.N.Duffy, M.G.Humphrey and A.G.Swincer, J. Organometal. Chem., 1985, 282, 383.
36 M.I.Bruce, C.Dean, D.N.Duffy, M.G.Humphrey and G.A. Koutsantonis, J. Organometal. Chem., 1985, 295, C40.
37 K.Sano, T.Yamamoto and A.Yamamoto, Z. Naturforsch., Teil B, 1985, 40B, 210.
38 Z.Danter, R.J.Mawby, C.D.Reynolds and D.R.Saunders, J. Chem. Soc., Dalton Trans., 1985, 1235.
39 A.J.Arce, Y. de Santis and A.J.Deeming, J. Orgnometal. Chem., 1985, 295, 365.
40 M.I.Bruce, E.Horn, J.G.Matisons and M.R.Shaw, J. Organometal. Chem., 1985, 286, 271.
41 G.Gervasio, R.Rossetti and P.L.Stanghellini, Organometallics, 1985, 4, 1612.
42 A.M. van den Bergen, R.L.Elliott, C.J.Lyons, K.P.Mackinnon and B.O.West, J. Organometal. Chem., 1985, 297, 361.
43 W.A.Herrmann, C.Weber, M.L.Ziegler and O.Serhadli, J. Organometal. Chem., 1985, 297, 245.
44 W.P.Hart, D.Shihna and M.D.Rausch, Inorg. Chem.,1985, 24, 111

45 W.L.Parker, A.R.Siedle and R.M.Hexter, J. Amer. Chem. Soc., 1985, 107, 264.

46 F.J.Garcia Alonso, A.Höhn, J.Wolf, H.Otto and H.Werner, Angew. Chem., Int. Ed. Engl., 1985, 24, 406.

47 W.A.Herrmann and C.Weber, J. Organometal. Chem.,1985, 282,C31

48 L.Dahlenburg and N.Höck, J. Organometal. Chem.,1985, 284, 129

49 H.Hoberg, K.Sümmermann and A.Milchereit, J. Organometal. Chem., 1985, 288, 237.

50 D.Walter, E.Dinjus, H.Görls, J.Sieler, O.Lindqvist and L.Andersen, J. Organometal. Chem.,1985, 286, 103.

51 K.R.Pörschke, R.Mynott, K.Angermund and C.Krüger, Z. Naturforsch., Teil B, 1985, 40B, 199.

52 E.Hernandez, I.Saez and P.Royo, J. Organometal. Chem., 1985, 293, 249.

53 T.P.Beebe, M.R.Albert and J.T.Yates, J. Catal.,1985, 96, 1.

54 G.Vasapollo, C.F.Nobile and A.Sacco, J. Organometal. Chem., 1985, 296, 435.

55 M.Onishi, K.Hiraki, Y.Ohama and A.Kurosaki, J. Organometal. Chem., 1985, 284, 403.

56 H.Ebner, H.Otto and H.Werner, Angew. Chem., Int. Ed. Engl., 1985, 24, 518.

57 G.Facchin, R.Campostrini and R.A.Michelin, J. Organometal. Chem., 1985, 294, C21.

58 R.M.Ceder and J.Sales, J. Organometal. Chem., 1985, 294, 389

59 M.A.Bennett and A.Rokicki, Organometallics,1985, 4, 180.

60 C.R.Langrick, P.G.Pringle and B.L.Shaw, J. Chem. Soc., Dalton Trans., 1985, 1015.

61 J. A.Davies, C.S.Hasselkus, C.N.Scimar, A.Sood and V.Uma, J. Chem. Soc. Dalton Trans., 1985, 209.

62 J.S.Thompson and R.M.Swiatek, Inorg. Chem., 1985, 24, 110.

63 L.Naldini, F.Demartin, M.Manassero, M.Sansoni, G.Rassu and M.A.Zoroddu, J. Organometal. Chem., 1985, 279, C42.

64 J.Vicente, M.T.Chicote, J.A.Cayvelas, J.Fernandez-Baeza, P.G.Jones, G.M.Sheldrick and P.Espinet, J. Chem. Soc., Dalton Trans., 1985, 1163.

65 E.O.Fischer and M.Böck, J. Organometal. Chem., 1985, 287,279

66 M.L.Patterson, and M.J.Weaver, J. Phys. Chem., 1985, 89,5046

67 P.Gao and M.J.Weaver, J. Phys. Chem., 1985, 89, 5040.

68 H.Lange and D.Naumann, J. Fluorine Chem., 1984, 26, 435.

69 K.Kamienska-Trela, H.Ilcewicz, H.Baranska and A.Labudzinska, Bull. Pol. Acad. Sci.,Chem.,1984, 32, 143.

70 T.S.Dory, J.J.Zuckermann and M.D.Rausch, J.Organometal. Chem., 1985, 281, C8.

71 A.A.Tsyganenko, L.A.Denisenko, S.M.Zverev and V.N.Filimonov, J. Catal., 1985, 94, 10.

72 L.B.Kool, M.D.Rausch, H.G.Alt, M.Hereberhold, B.Wolf and U.Thewalt, J. Organometal. Chem., 1985, 297, 159.

73 U.Belforte, F.Calderazzo and P.F.Zanazzi, Gazz. Chim. Ital., 1985, 115, 71.

74 R.M.Kowaleski, D.O.Kipp, K.J.Stauffer, P.N.Swepston and F.Basolo, Inorg. Chem., 1985, 24, 3750.

75 K.Ihmels and D.Rehder, Organometallics, 1985, 4, 1334.

76 K.Ihmels and D.Rehder, Organometallics, 1985, 4, 1340.

77 F.Calderazzo, M.Castellani, G.Pampaloni and P.F.Zanazzi, J. Chem. Soc., Dalton Trans., 1985, 1989.

78 P.H.Bird, A.A.Ismail and I.S.Butler, Inorg. Chem., 1985, 24, 2911.

79 W.Gerhartz, F.-W.Grevels, W.E.Klotzbücher, E.A.Körner von Gustorf and R.N.Perutz, Z.Naturforsch., Teil B, 1985, 40B,518
80 M.J.Almond, A.J.Downs and R.N.Perutz, Inorg. Chem., 1985, 24, 275.
81 T.A.Seder, S.P.Church, A.J.Oudekirk and E.Weitz, J. Amer. Chem. Soc., 1985, 107, 1432.
82 R.K.Upmacis, G.E.Gadd, M.Poliakoff, M.B.Simpson, J.J.Turner, R.Whyman and A.F.Simpson, J. Chem. Soc., Chem. Comm., 1985,27
83 S.P.Church, F.-W.Grevels, H.Hermann and K.Schaffner, J. Chem. Soc., Chem. Comm., 1985, 30.
84 S.P.Church, F.-W.Grevels, H.Hormann and K.Schaffner, Inorg. Chem., 1985, 24, 418.
85 J.M.Maher, R.P.Beatty and N.J.Cooper, Organometallics, 1985, 4, 1354.
86 J.Grobe and D.Levan, Z. anorg. allg. Chem., 1984, 518, 36.
87 N.Kuhn, H.Schumann and G.Wolmershäuser, J. Chem. Soc., Chem. Comm., 1985, 1595.
88 I.M.Baibich, A.M.English and I.S.Butler, Organometallics, 1984, 3, 1786.
89 R.B.King, W.K.Fu and E.M.Holt, Inorg. Chem., 1985, 24,3094.
90 D.Sellmann and L.Zapf, J. Organometal. Chem., 1985, 289, 57.
91 S.Sabo, B.Chaudret and D.Gervais, J. Organometal. Chem., 1985, 292, 411.
92 J.A.Connor and C.Overton, J. Organometal. Chem., 1985, 282, 349.
93 J.C.Jeffrey, C.Marsden and F.G.A.Stone, J. Chem. Soc., Dalton Trans., 1985, 1315.
94 R.B.Hitam, K.A.Mahmoud and A.J.Rest, J. Organometal. Chem., 1985, 291, 321.
95 F.A.Cotton, L.R.Falvello and J.H.Meadows, Inorg. Chem., 1985, 24, 514.
96 D.J.Darensbourg, R.Kudaroski and T.Delord, Organometallics, 1985, 4, 1094.
97 J.I.Zink, Coord. Chem. Rev., 1985, 64, 93.
98 K.R.Pope and M.S.Wrighton, Inorg. Chem., 1985, 24, 2792.
99 A.J.Deeming and S.Donovan-Mtunzi, Organometallics, 1985, 4, 693.
100 J.W.McDonald, Inorg. Chem., 1985, 24, 1734.
101 M.W.Kokkes, D.J.Stufkens and A.Oskam, Inorg. Chem., 1985, 24, 2934.
102 M.W.Kokkes, D.J.Stufkens and A.Oskam, Inorg. Chem., 1985, 24, 4411.
103 D.A.DuBois, E.N.Duesler and R.T.Paine, Inorg. Chem., 1985, 24, 3.
104 P.D.Harvey and I.S.Butler, Can. J. Chem., 1985, 63, 1510.
105 K.A.Mahmoud, A.J.Rest and H.G.Alt, J. Chem. Soc., Dalton Trans., 1985, 1365.
106 L.R.Martin, F.W.B.Einstein and R.K.Pomeroy, Inorg. Chem., 1985, 24, 2777.
107 S.P.Church, F.-W.Grevels, H.Hermann, J.M.Kelly, W.Klotz-bücher and K.Schaffner, J. Chem. Soc., Chem. Comm., 1985, 594.
108 S.L.Suib, A.Kostapapas, K.C.McMahon, J.C.Baxter and A.M.Winiecki, Inorg. Chem., 1985, 24, 858.
109 E.Guglielminotti, D.Osella and P.L.Stanghellini, J. Organometal. Chem., 1985, 281, 291.
110 M.A.Derzdzon, C.Tessier-Youngs, C.Woodcock, P.M.Blonsky, O.Leal, B.K.Teo, R.Burnall and D.F.Shriver, Inorg. Chem., 1985, 24, 2349.

111 M.F.Gregory, M.Poliakoff and J.J.Turner, J. Mol. Struct., 1985, 127, 247.
112 J.A.Ladd, H.Hope and A.L.Balch, Organometallics,1984, 3,1838
113 S.Dobos, I.Böszörményi, V.Silberer, L.Guczi and J.Mink, Inorg. Chim. Acta, 1985, 96, L13.
114 C.-M.Che, C.-K.Poon, W.-C.Chung and H.B.Gray, Inorg. Chem., 1985, 24, 1277.
115 H.A.Marzouk, E.B.Bradley and K.A.Arunkumar, Spectrosc. Lett., 1985, 18, 189.
116 S.I.Woo and C.G.Hill, J. Mol. Catal., 1985, 29, 209.
117 R.van Eldik, S.Aygen, H.Kelm, A.M.Trzeciak and J.J. Ziolkowski, Transition Met. Chem., 1985, 10, 167.
118 M.van Rentergem and G.P.van der Kelen, J. Mol. Struct., 1985, 127, 47.
119 P.Paetzold and K.Delpy, Chem. Ber., 1985, 118, 2552.
120 V.G.Albano, D.Brago, A.Fumagalli and S.Martinengo, J. Chem. Soc., Dalton Trans., 1985, 1137.
121 C.H.Dai and S.D.Worley, Chem. Phys. Lett., 1985, 114, 286.
122 M.Ichikawa and T.Fukushima, J. Phys. Chem., 1985, 89, 1564.
123 A.M.Trzeciak and J.J.Ziołkowski, Inorg. Chim. Acta, 1985, 96, 15.
124 C.Woods and C.Daffron, Inorg. Chim. Acta, 1985, 101, 13.
125 A.I.Rubailo, V.P.Selina and Yu.S.Varshavskii, Koord. Khim., 1984, 10, 1231.
126 S.P.Deraniyagala and K.R.Grundy, J. Chem. Soc., Dalton Trans. 1985, 1577.
127 S.Martinengo, A.Fumagalli and P.Chini, J. Organometal. Chem. 1985, 284, 275.
128 X.Li, Y.Fu and Z.Huang, Wuli Huaxue Xuebao, 1985, 1, 22 (Chem. Abs., 1985, 103, 43400).
129 E.E.Platero, S.Coluccia and A.Zecchina, Langmuir, 1985, 1,407
130 V.F.Vozdvizhenskii, V.M.Mishchenko and D.V.Sokol'skii, Dokl. Akad. Nauk SSSR, 1984, 279, 1153.
131 R.Usón, J.Forniés, M.Tomás and B.Menjón, Organometallics, 1985, 4, 1912.
132 B.E.Hayden, K.Kretzschmar, A.M.Bradshaw and B.G.Greenler, Surf. Sci., 1985, 149, 394.
133 S.M.Kim and O.V.Agashkin, Izv. Akad. Nauk Kaz. SSR, Ser.Khim. 1984, 51.
134 B.E.Hayden, K.Kretzshmar and A.M.Bradshwa, Surf. Sci., 1985, 155, 553.
135 G.Ghiotti, F.Boccuzzi and A.Chiorino, J. Chem. Soc., Chem. Comm., 1985, 1012.
136 G.Ghiotti, F.Boccuzzi and A.Chiorino, Stud. Surf. Sci. Catal., 1985, 21, 235.
137 A.Toth, C.Floriani, M.Pasquali, A.Chiesi-Villa, A.Gaetani-Manfredotti and C.Guasgini, Inorg. Chem., 1985, 24, 648.
138 T.P.Beebe, P.Gelin and J.T.Yates, Surf. Sci., 1985, 148,526.
139 C.E.Davies, I.M.Gardiner, J.C.Green, M.L.H.Green, N.J.Hazel, P.D.Grebenik, V.S.B.Mtetwa and K.Prout, J. Chem. Soc., Dalton Trans., 1985, 669.
140 C.A.Ghilardi, P.Innocenti, S.Midollini and A.Orlandini, J. Chem. Soc., Dalton Trans., 1985, 605.
141 M.Scotti, M.Valderrama, R.Ganz and H.Werner, J. Organometal. Chem., 1985, 286, 399.
142 S.A.Snow and G.Kodama, Inorg. Chem., 1985, 24, 795.
143 L.Barton and P.K.Rush, Inorg. Chem., 1985, 24, 3413.

144 J.E.Salt, G.S.Girolami, G.Wilkinson, M.Motevalli, M. Thornton-Pett and M.B.Hursthouse, J. Chem. Soc., Dalton Trans., 1985, 685.
145 T.-C.Hsieh and J.Zubieta, Inorg. Chem., 1985, 24, 1287.
146 F.J.Lalor and D.M.Condon, J. Organometal. Chem.,1985,284,327
147 U.Müller, P.Klingelhöfer, U.Kynast and K.Dehnicke, Z. anorg. allg. Chem., 1985, 520, 18.
148 M.Dadkhah, J.R.Dilworth, K.Fairman, C.T.Kan, R.L.Richards and D.L.Hughes, J. Chem. Soc., Dalton Trans., 1985, 1523.
149 E.Carmona, A.Galindo, M.L.Poveda and R.D.Rogers, Inorg. Chem., 1985, 24, 4033.
150 A.Berg, E.Conradi, U.Müller and K.Dehnicke, Z. anorg. allg. Chem., 1985, 529, 74.
151 A.Marchi, R.Rossi, A.Duatti, L.Magon, V.Bertolasi, V. Ferretti and G.Gilli, Inorg. Chem., 1985, 24, 4744.
152 M.F.N.N.Carvalho, A.J.L.Pombeiro, U.Schubert, O.Orama, C.J.Pickett and R.L.Richards, J. Chem. Soc., Dalton Trans., 1985, 2079.
153 T.Nicholson and J.Zubieta, J. Chem. Soc., Chem. Comm., 1985, 367.
154 D.Fenske, A.Berg, F.Weller and K.Dehnicke, Z. anorg. allg. Chem., 1985, 527, 105.
155 S.Liu, X.Zhang and J.Hu, Guangpuxue Yu Guangpu Fenxi, 1985, 5, 27 (Chem. Abs., 1985, 103, 184335).
156 D.M.Haddleton and R.N.Perutz, J. Chem. Soc., Chem. Comm., 1985, 1372.
157 P.Ghosh and A.Chakravorty, J. Chem. Soc., Dalton Trans., 1985, 361.
158 M.Gupta, R.F.N.Ashok, A.Mishra, V.B.S.Chauhan and U.C. Agarwala, J. Chem. Soc., Dalton Trans., 1985, 2449.
159 H.E.Oosthuizen, E.Singleton, J.S.Field and G.C.van Niekerk, J. Organometal. Chem., 18985, 279, 433.
160 O.E.Piro, Phys. Rev., B, 1985, 31, 1122.
161 H.C.Ram, R.S.Thakur and R.U.Sharma, J. Ind. Chem. Soc., 1984, 61, 627.
162 G.Paolucci, P.Zanella and A.Berton, J. Organometal. Chem., 1985, 295, 317.
163 C.T.Kekel-Vroegop and J.H.Teuben, J. Organometal. Chem., 1985, 286, 309.
164 J.V.Brenčič, I.Leban and J.Zule, Z. anorg. allg. Chem., 1985, 521, 199.
165 R.C.Simon, H.D.Gaffney and D.L.Morse, Inorg. Chem., 1985, 24, 2565.
166 D.Sellmann and J.Müller, J. Organometal. Chem., 1985,281,249
167 S.M.Angel, M.K.Dearmond, R.J.Donohoe and D.W.Wertz, J. Phys. Chem., 1985, 89, 282.
168 J.A.Smieja, R.E.Stevens, D.E.Fjare and W.L.Gladfelter, Inorg. Chem., 1985, 24, 3206.
169 S.Wei, S.Wolfgang, T.C.Strekas and H.D.Gaffney, J. Phys. Chem., 1985, 89, 974.
170 Y.C.Chung, N.Leventis, P.J.Wagner and G.E.Leroi, Inorg. Chem., 1985, 24, 1965.
171 Y.C.Chung, N.Leventis, P.J.Wagner and G.E.Leroi, J. Amer. Chem. Soc., 1985, 107, 1414.
172 Y.C.Chung, N.Leventis, P.J.Wagner and G.E.Leroi, J. Amer. Chem. Soc., 1985, 107, 1416.
173 S.Farquharson, P.A.Lee and M.J.Weaver, Spectrochim. Acta, Part A, 1984, 40A, 907.

174 U.A.Jayasooria and S.F.A.Kettle, J. Phys. Chem.,1985, 89, 944

175 V.B.Rana, S.P.Ratra, D.P.Singh and M.P.Teotia, Transition Met. Chem., 1985, 10, 115.

176 P.K.Dutta and R.E.Zaykoski, Inorg. Chem., 1985, 24, 3490.

177 M.Kim and K.Itoh, J. Electroanal. Chem., Interfacial Electrochem., 1985, 188, 137.

178 B.S.Ault, J. Mol. Struct., 1985, 129, 287.

179 W.Frömberg and G.Erker, J. Organometal. Chem., 1985,280,343.

180 H.Fischer and S.Zeuner, J. Organometal. Chem., 1985,286,201.

181 A.Takenaka, S.Takeuchi, Y.Kobayashi and K.Itoh, Surf. Sci., 1985, 158, 359.

182 C.Campochiara, J.A.Hofmann and D.F.Bocian, Inorg. Chem., 1985, 24, 449.

183 G.Rajca, W.Shwarz, J.Weidlein and K.Dehnicke, Z. anorg. allg. Chem., 1985, 522, 83.

184 P.C.Servaas, H.K.van Dijk, T.L.Snoeck, D.J.Stufkens and A.Oskam, Inorg. Chem., 1985, 24, 4494.

185 D.Czarkie and Y.Shvo, J. Organometal. Chem.,1985, 280, 123.

186 I.I.Kalinichenko, N.M.Titov and M.G.Ivanov, Russ. J. Inorg. Chem., 1984, 29, 1149.

187 A.Mrwa, S.Rummel and M.Starke, Z. Chem., 1985, 25, 186.

188 C.A.Melendres and V.A.Maroni, J. Raman Spectrosc., 1984, 15, 319.

189 H.Oshio, T.Ama, T.Watanabe, J.Kincaid and K.Nakamoto, Spectrochim. Acta, Part A, 1984, 40A, 863.

190 G.S.Srivatsa, D.T.Sawyer, N.J.Boldt and D.F.Bocian, Inorg. Chem., 1985, 24. 2123.

191 R.Guilard, B.Boisselier-Cocolios, A.Tabard, P.Cocolios and K.M.Kadish, Inorg. Chem., 1985, 24, 2509.

192 B.Simič-Glavaski, S.Začevič and E.Yeager, J. Amer. Chem. Soc. 1985, 107, 5625.

193 T.Koyama, M.Yamaga, M.Kim and K.Itoh, Inorg. Chem., 1985, 24, 4268.

194 L.A.Sanchez and T.G.Spiro, J. Phys. Chem., 1985, 89, 763.

195 D.L.Rousseau and M.R.Ondrias, Biophys. J., 1985, 47, 537.

196 R.Schweitzer-Stenner and W.Dreybrodt, J. Raman Spectrosc., 1985, 16, 111.

197 R.A.Copeland, S.Dasgupta and T.G.Spiro, J. Amer. Chem. Soc., 1985, 107, 3370.

198 S.H.Liu, H.L.Lee, J.H.Chen, W.J.Xu, Y.M.Ding, W.Zhang, J.R.Zhang and L.K.Hui, Proc. SPIE-Int. Soc. Opt. Eng., 1984(publ. 1985), 492, 129.

199 S.El-Naggar, W.Dreybrodt and R.Schweitzer-Stenner, Eur. Biophys. J., 1985, 12, 43.

200 W.A.Oertling and G.T.Babcock, J. Amer. Chem. Soc., 1985, 107, 6406.

201 L.A.Anderson, T.M.Loehr, C.K.Chang and A.G.Mauk, J. Amer. Chem. Soc., 1985, 107, 182.

202 H.E.van Wart and J.Zimmer, J. Amer. Chem. Soc., 1985, 107, 3379.

203 R.A.Copeland, A.Naqui, B.Chance and T.G.Spiro, FEBS Lett., 1985, 182, 375.

204 G.T.Babcock, J.M.Jean, L.N.Johnston, W.H.Woodruff and G.Palmer, J. Inorg. Biochem., 1985, 23, 243.

205 T.G.Spiro and J.Terner, Photochem. Photobiol. Proc. Int. Conf., 1983, 817.

206 G.Smulevich and T.G.Spiro, Biochim. Biophys. Acta, 1985, 830, 80.

207 T.W.Scott, J.M.Friedman and V.W.MacDonald, J. Amer. Chem. Soc., 1985, 107, 3702.
208 T.Uno, Y.Nishimura and M.Tsuboi, Biochem., 1984, 23, 6802.
209 E.W.Findsen, T.W.Scott, M.R.Chance, J.M.Friedman and M.R.Ondrias, J. Amer. Chem. Soc., 1985, 107, 3355.
210 V.N.Kirichenko, S.V.Larionov, I.A.Mikhailov, E.G. Bogushavskii, and L.B.Volodarskii, Russ. J. Inorg. Chem., 1984, 29, 1624.
211 D.L.Willems and D.F.Bocian, J. Phys. Chem., 1985, 89, 234.
212 W.Schwarz, G.Rajca and J.Weidlein, Z. anorg. allg. Chem., 1985, 525, 143.
213 J.Nieman and J.H.Teuben, J. Organometal. Chem., 1985, 287, 207.
214 P.E.Bloyce, R.H.Hooker, D.A.Lane and A.J.Rest,, J. Photochem., 1985, 28, 525.
215 P.Caillet, J.Y.Le Marouille and P.Lemaux, J. Organometal. Chem., 1985, 291, 191.
216 A.Müller, R.Jostes, W.Wltzner, C.Nie, E.Diemann, H.Bögge, M.Zimmermann, M.Dartmann and U.Reinsch-Vogell, Inorg. Chem., 1985, 24, 2872.
217 R.M.Nielson and S.Wherland, Inorg. Chem., 1985, 24, 1803.
218 P.J.Blower and J.R.Dilworth, J. Chem. Soc., Dalton Trans., 1985, 2305.
219 A.B.Altabef, S.A.Brandan and N.E.Katz, Polyhedron, 1985, 4, 227.
220 S.M.Tetrick and R.A.Walton, Inorg. Chem., 1985, 24, 3363.
221 M.M.Mockford and W.P.Griffith, J. Chem. Soc., Dalton Trans., 1985, 717.
222 M.K.Lee, I.B.Kim and C.S.Chin, J. Organometal. Chem., 1985, 290, 115.
223 S.H.Park, H.-K.Park and C.S.Chin, Inorg. Chem., 1985, 24, 1120.
224 C.E.Briaul, D.I.Gilmour, D.M.P.Mingos and R.W.M.Wardle, J. Chem. Soc., Dalton Trans., 1985, 1693.
225 M.Nakashima, M.Mikuriya and Y.Muto, Bull. Chem. Soc. Japan, 1985, 58, 968.
226 B.H.Loo, Y.G.Lee and D.O.Frazier, J. Phys. Chem.,1985,89, 4672.
227 B.H.Loo, Y.G.Lee and D.O.Frazier, Chem. Phys. Lett.,1985, 119, 312.
228 T.H.Joo, K.Kim, H.Kim and M.S.Kim, Chem. Phys. Lett., 1985, 117, 518.
229 G.J.Earl, P.Gans and J.B.Gill, J. Chem. Soc., Dalton Trans., 1985, 663.
230 F.Bottomley, J.Darkwa and P.S.White, J. Chem. Soc., Dalton Trans., 1985, 1435.
231 Z.Sobalík, V.Pour, L.A.Sokolova, O.V.Nevskaya and N.M. Popova, Coll. Czech. Chem. Comm., 1985, 50, 1259.
232 P.Legzdins and C.R.Nurse, Inorg. Chem., 1985, 24, 327.
233 W.R.Robinson, D.E.Wigley and R.A.Walton, Inorg. Chem., 1985, 24, 918.
234 D.Sellmann, L.Zapf, J.Keller and M.Moll, J. Organometal Chem. 1985, 289, 71.
235 E.Herdtweck, C.Schumacher and K.Dehnicke, Z. anorg. allg. Chem., 1985, 526, 93.
236 A.Berg and K.Dehnicke, Z. Naturforsch., Teil B, 1985,40B,842
237 Y.Y.Tang and J.I.Zink, J. Amer. Chem. Soc., 1985, 107,4799.
238 B.N.Diel, J. Organometal. Chem., 1985, 284, 257.

239 R.Guilard, G.Lagrange, A.Tabard, D.Lancon and K.M.Kadish, Inorg. Chem., 1985, 24, 3649.
240 W.A.Herrmann, M.Flöel, C.Weber, J.L.Hubbard abd A.Schäfer, J. Organometal. Chem., 1985, 286, 369.
241 J.P.Attard, B.F.G.Johnson, J.Lewis, J.M.Mace and P.R.Raithby, J. Chem. Soc., Dalton Trans. 1985, 1526.
242 A.Dobson, D.S.Moore, S.D.Robinson, A.M.R.Galas and M.B. Hursthouse, J. Chem. Soc., Dalton Trans., 1985, 611.
243 E.E.Platero, G.Spoto and A.Zecchina, J. Chem. Soc., Farad. Trans, I, 1985, 81, 1283.
244 B.A.Morrow, R.A.McFarlane and L.E.Moran, J. Phys. Chem., 1985, 89, 77.
245 B.A.Morrow, J.P.Chevrier and L.E.Moran, J. Catal., 1985, 91, 208.
246 H.C.Yao, W.G.Rothschild and H.S.Gandhi, Stud. Surf. Sci. Catal., 1984, 19, 71.
247 J.Ellermann and W.Wend, J. Organometal. Chem.,1985, 281,C29.
248 J.Grobe, D.Le Van and J.Szameitat, J. Organ metal. Chem., 1985, 289, 341.
249 C.Zeiher, J.Mohyla, I.-P.Lorenz and W.Hiller, J. Organometal. Chem., 1985, 286, 159.
250 R.M.Medina, A.Alvarez-Valdés and J.R.Maşaguer, J. Organometal. Chem., 1985, 294, 209.
251 J.Ellermann and A.Veit, J. Organometal. Chem., 1985, 290,307
252 E.Lindner and A.Brösamle, Chem. Ber., 1985, 118, 2134.
253 R.R.Ford, B.-L.Li, R.H.Neilson and R.J.Thoma, Inorg. Chem., 1985, 24, 1993.
254 K.Hiraki, N.Ochi, Y.Sasada, H.Hayashida, Y.Fuchita and S.Yamanaka, J. Chem. Soc., Dalton Trans., 1985, 873.
255 G.Pimblett, C.D.Garner and W.Clegg, J. Chem. Soc., Chem. Comm., 1985, 1977.
256 H.Bauer, U.Nagel and W.Beck, J. Organometal. Chem., 1985, 290, 219.
257 C.Pelizzi, G.Pelizzi and P.Tarasconi, J. Organometal. Chem., 1985, 281, 403.
258 V.D.Klimov, V.A.Legasov, A.V.Lebedev, S.S.Khoroshev, Yu.A. Teterin, A.S.Baev and N.S.Tolmacheva, Radiokhim., 1985, 27, 415.
259 G.Paolucci, G.Rossetto, P.Zanella and R.D.Fischer, J. Organometal. Chem., 1985, 284, 213.
260 M.J.Taylor and S.Riethmiller, J. Raman Spectrosc., 1984, 15, 370.
261 M.J.Taylor, D.S.Bohle and S.Riethmiller, J. Raman Spectrosc. 1984, 15, 393.
262 L.Manceron, A.Loutellier and J.P.Perchard, Chem. Phys., 1985, 92, 75.
263 J.W.Kauffman, R.H.Hauge and J.L.Margrave, High Temp. Sci., 1984, 18, 97.
264 S.Halut-Desportes and E.Musson, Spectrochim. Acta, Part A, 1985, 41A, 661.
265 M.K.Chaudhuri and B.Das, Polyhedron, 1985, 4, 1449.
266 M.K.Chaudhuri and N.S.Islam, Transition Met. Chem., 1985, 10 333.
267 C.Djordjevic, S.A.Craig and E.Sinn, Inorg. Chem.,1985,24,1281
268 M.K.Chaudhuri, S.K.Ghosh and N.S.Islam, Inorg. Chem., 1985, 24, 2708.
269 E.M.Nour, A.B.El-Sayed and F.Abdel-Rehim, Spectrochim. Acta, Part A, 1985, 41A, 865.
270 J.Tomkinson, A.D.Taylor, J.Howard, J.Eckert and J.A. Goldtsone, J. Chem. Phys., 1985, 82, 1112.

271 T.J.McNeese, T.E.Mueller, D.A.Wierda, D.J.Darensbourg and T. J.Delord, Inorg. Chem., 1985, 24, 3465.
272 J.Flanagan, W.P.Griffith, A.C.Skapski and R.W.Wiggins, Inorg. Chim. Acta, 1985, 96, L23.
273 P.S.Kirlin and B.C.Gates, Inorg. Chem.,1985, 24, 3914.
274 T.Kuroi, H.Oshio and K.Nakamoto, J. Phys. Chem., 1985, 89, 4087.
275 K.Bajdor, H.Oshio, K.Nakamoto, W.Kanda, H.Okawa and S.Kida, Inorg. Chim. Acta, 1985, 103, 63.
276 K.Nakamoto and H.Oshio, J. Amer. Chem. Soc.,1985, 107, 6518.
277 J.R.Kincaid, L.M.Proniewicz, K.Bajdor, A.Bruha and K. Nakamoto, J. Amer. Chem. Soc., 1985, 107, 6775.
278 H.Suzuki, S.Matsuura, Y.Moro-oka and T.Ikawa, J. Organometal. Chem., 1985, 286, 247.
279 A.M.Heyns, J. Mol. Struct., 1985, 127, 217.
280 N.Iwasaki, Y.Sasaki and Y.Nishina, Surf. Sci.,1985, 158,352.
281 M.N.Bhattacherjee, M.K.Chaudhuri and R.N.D.Purkayastha, J. Chem. Soc., Dalton Trans., 1985, 409.
282 G.P.Shchetinina, O.V.Brovkina, B.N.Chernyshov and E.G. Ippolitov, Russ. J. Inorg. Chem., 1985, 30, 460.
283 M.K.Chaudhuri and B.Das, Inorg. Chem., 1985, 24, 2580.
284 H.Puff, H.Hevendehl, K.Höfer, H.Reuter and W.Schuh, J. Organometal. Chem., 1985, 287, 163.
285 G.S.Raghuvanshi, D.P.Khandelwal and H.D.Bist, Spectrochim. Acta, Part A, 1985, 41A, 391.
286 J.K.Puri and J.M.Miller, Inorg. Chim. Acta, 1985, 101, 135.
287 R.Kapoor, R.Sharma and P.Kapoor, Z. Naturforsch., Teil B, 1984, 39B, 1702.
288 R.Kapoor, P.Kapoor and R.Sharma, Z. Naturforsch., Teil B, 1985, 40B, 247.
289 A.L.Beauchamp, F.Belanger-Gariepy and S.Arabi, Inorg. Chem., 1985, 24, 1860.
290 M.D.Curtis and K.R.Han, Inorg. Chem., 1985, 24, 378.
291 D.J.Santure, J.C.Huffman and A.P.Sattelberger, Inorg. Chem., 1985, 24, 371.
292 M.N.Bhattacherjee, M.K.Chaudhuri and R.N.P.Purkayastha, Inorg. Chem., 1985, 24, 447.
293 H.Oumus, C.Lecomte, J.Protas, P.Cocolios and R.Guilard, Polyhedron, 1984, 3, 651.
294 G.Süss-Fink, G.Herrmann, P.Morys, J.Ellermann and A.Veit, J. Organometal. Chem., 1985, 284, 263.
295 T.Fukushima, H.Arakawa and M.Ichikawa, J. Chem. Soc., Chem. Comm., 1985, 729.
296 M.Aresta, D.Ballivet-Tkachenko and M.C.Bonnet, J. Amer. Chem. Soc., 1985, 107, 2994.
297 B.Vlčková, B.Strauch and M.Horák, Coll. Czech. Chem. Comm., 1985, 50, 306.
298 O.J.Scherer, K.Hussong and G.Wolmershäuser, J. Organometal. Chem., 1985, 289, 215.
299 C.Bianchini, C.A.Ghilard, A.Meli, S.Midollini and A.Orlandini Inorg. Chem., 1985, 24, 924.
300 A.M.Heyns, J. Mol. Struct., 1985, 127, 9.
301 A.S.Tyagi and C.P.Srivastava, Proc. Ind. Nat. Sci. Acad., Part A, 1984, 50, 242.
302 J.Vicente, M.T.Chicote, M.D.Bermudez and M.Garcia-Garcia, J. Organometal. Chem., 1985, 295, 125.
303 A.I.Grigor'ev, N.V.Donchenko, K.M.Dunaeva and D.V.Debasov, Russ. J. Inorg. Chem., 1985, 30, 497.

304 M.Sakamoto, Nippon Kagaku Kaishi, 1985, 792 (Chem. Abs., 1985, 103, 29412).

305 V.Ya.Kavun, V.I.Sergienko, V.I.Kostin, B.N.Chernyshov and N.I.Steblevskaya, Russ. J. Inorg. Chem., 1985, 30, 190.

306 K.H.Bhasin, S.Jindal, S.Singh and R.D.Verma, Synth. React. Inorg. Met.-Org. Chem., 1985, 15, 733.

307 L.V.Kobets, T.A.Dik, G.F.Belyachits, D.S.Umreiko, I.A. Volodin, T.V.Zhavornkova and E.A.Filippov, Radiokhim., 1985, 27, 19.

308 R.Ramachandran, B.Singh, A.K.Narula, P.N.Kapoor, P.K.Gupta and P.N.Kapoor, Polyhedron, 1985, 4, 1007.

309 K.Ozutsumi, H.Ohtaki and A.Kusumegi, Bull. Chem. Soc. Japan, 1984, 57, 2612.

310 S.J.Blunden, L.A.Hobbs, P.J.Smith, A.G.Davies and S.B.Teo, J. Organometal. Chem., 1985, 282, 9.

311 B.L.Khandelwal, A.K.Singh, H.B.Singh, K.M.Prasad, N.S.Bhandari and W.R.McWhinnie, J. Organometal. Chem., 1985, 291, 185

312 S.He, J.Wu and Q.Zhang, Huaxue Xuebao, 1984, 42, 1183 (Chem. Abs., 1985, 102, 69532).

313 H.H.Emons, F.Birkeneder, K.Pollmer and G.Kabisch, Z. anorg. allg. Chem., 1985, 521, 231.

314 V.T.Orlova, O.E.Artemova, E.Ya.Makeeva, M.A.Shcherbanskii and I.N.Lepeshkov, Russ. J. Inorg. Chem., 1984, 29, 1688.

315 M.N.Nabiev, S.Usmanov, L.A.Abramova, B.M.Beglov, I.N. Lepeshkov, V.T.Orlova, S.Tukhtaev and N.N.Kadyeva, Russ. J. Inorg. Chem., 1985, 30, 131.

316 O.E.Artemova, I.N.Lepeshkov and V.T.Orlova, Russ. J. Inorg. Chem.,1984, 29, 1686.

317 K.Sünkel, G.Urban and W.Beck, J. Organometal. Chem., 1988, 290, 231.

318 C.L.Hill and M.M.Williamson, Inorg. Chem., 1985, 24, 2836.

319 D.B.Powell and A.Woollins, Spectrochim. Acta, Part A, 1985, 41A, 1023.

320 K.R.Grundy and K.N.Robertson, Inorg. Chem., 1985, 24,3898

321 S.P.Deraniyagala and K.R.Grundy,Inorg. Chem.,1985,24, 50.

322 R.Usón, J.Forniés, R.Navarra, P.Espinet and C.Mendívil, J. Organometal. Chem., 1985, 290, 125.

323 J.Zou and M.Tan, Yingyong Huaxue,1984, 1, 76 (Chem. Abs., 1985, 103, 114923).

324 K.A.Khaldoyanidi, V.N.Lubkova and I.I.Yakovlev, Russ. J. Inorg. Chem., 1985, 30, 154.

325 E.S.Stoyanov, V.I.Mikhailov and V.M.Popov, Koord. Khim., 1984, 10, 1619.

326 C.J.Jones, J.A.McCleverty and A.S.Rothin, J. Chem. Soc., Dalton Trans.,1985, 405.

327 P.Klingelhöfer and U.Müller, Z. anorg.allg. Chem.,1984, 516, 85.

328 N.A.P.Kane-Maguire, K.C.Wallace and D.B.Miller, Inorg. Chem. 1985, 24, 597.

329 M.F.A.Dove, R.C.Hibbert and N.Logan, J. Chem. Soc. Dalton Trans., 1985, 707.

330 S.M.Bowen, E.N.Duesler, D.J.McCabe and R.T.Paine, Inorg. Chem. 1985, 24, 1191.

331 L.C.Nathan, C.A.Doyle, A.M.Mooring, D.C.Zapien, S.K.Larsen and C.G.Pierpont, Inorg. Chem., 1985, 24, 2763.

332 J.C.Fanning, J.L.Resoe, G.C.Lickfield and M.E.Kotun, Inorg. Chem., 1985, 24, 2884.

333 M.Martinez and M.Ferrer, Inorg. Chem., 1985, 24, 792.
334 R.D.Feltham, G.Elbaze, R.Ortega, C.Eck and J.Dubrawski, Inorg. Chem., 1985, 24, 1503.
335 C.J.Jones, J.A.McCleverty and A.S.Rothin, J. Chem. Soc., Dalton Trans., 1985, 401.
336 N.I.Steblevskaya, V.E.Karasev and R.N.Shchelokov, Russ. J. Inorg. Chem., 1984, 29, 1274.
337 M.S.Manhas and A.K.Trikha, J. Chem. Soc., Dalton Trans., 1985, 1985.
338 P.Vast and A.Semmoud, J. Fluorine Chem., 1985, 27, 47.
339 A.W.Cooke, J.Pebler, F.Weller and K.Dehnicke, Z. anorg. allg. Chem., 1985, 524, 68.
340 S.Singh, Amita, M.S.Gill and R.D.Verma, J. Fluorine Chem., 1985, 27, 133.
341 V.N.Krasil'nikov and A.A.Ivakin, Russ. J. Inorg. Chem., 1984, 29, 1613.
342 H.Schäfer, H.Plautz, H.-J.Abel and D.Lademann, Z. anorg. allg. Chem., 1985, 526, 168.
343 P.C.Leung and F.Aubke, Can. J. Chem., 1984, 62, 2892.
344 B.Papánková, H.Langfelderová, P.Sivý and M.Serátor, Z. anorg. allg. Chem., 1985, 526, 213.
345 S.H.Strauss, M.D.Noirot and O.P.Anderson, Inorg. Chem., 1985, 24, 4307.
346 V.N.Serezhkin and L.B.Serezhkina, Russ. J. Inorg. Chem., 1985, 30, 1156.
347 G.Ritzhaupt, H.H.Richardson and J.P.Devlin, High Temp. Sci., 1985, 19, 163.
348 M.I.S.Sastry and S.Singh, Can. J. Chem., 1985, 63, 1351.
349 V.P.Babaeva and V.Ya.Rosolovskii, Russ. J. Inorg. Chem., 1984, 29, 1566.
350 M.H.T.Khan, H.C.Bajaj, M.R.H.Siddiqui, B.T.Khan, M.S.Reddy and K.V.Reddy, J. Chem. Soc., Dalton Trans., 1985, 2603.
351 J.L.Pascal, J.Potier and C.S.Zhang, J. Chem. Soc., Dalton Trans., 1985, 297.
352 L.Jin, X.Zeng, B.Chen, W.Zhu, J.Zhao and L.Huang, Zhonggao Xitu Xuebao, 1985, 3, 89 (Chem. Abs., 1985, 103, 170989).
353 J.L.Pascal, J.Potier, D.J.Jones, J.Rozière and A.Michalowicz, Inorg. Chem., 1985, 24, 238.
354 M.C.Martinez, M.T.Pereira, M.R.Bermejo and M.Gayoso, Acta Cienc. Compostelana, 1983, 20, 75.
355 O.P.Pandey, S.K.Sengupta and S.C.Tripathi, Monatsh., 1985, 116, 431.
356 O.P.Pandey, S.K.Sengupta and S.C.Tripathi, Polyhedron, 1984, 3, 695.
357 A.L.Doadrio-Villarejo, C.V.Ragel and A.Doadrio, An.R. Acad. Farm., 1984, 50, 727.
358 M.G.B.Drew, D.A.Rice and D.M.Williams, J. Chem. Soc., Dalton Trans., 1985, 417.
359 D.Fenske, B.Czeska, C.Schumacher, R.E.Schmidt and K.Dehnicke, Z. anorg. allg. Chem., 1985, 520, 7.
360 W.de Oliveira, J.-L.Migot, M.B.G.de Lima, J.Sala-Pala, J.Guerchais and J.-Y.Le Gall, J. Organometal Chem.,1985,284, 313.
361 S.Vasanthi, K.S.Nagaraja and M.R.Udupa, Transition Met. Chem. 1984, 9, 382.
362 S.J.N.Burgmayer and J.L.Templeton, Inorg. Chem.,1985,24,2224
363 K.S.Jasim, G.B.Umbach, C.Chieh and T.C.W.Mak, J. Crystallogr. Spectrosc. Res., 1985, 15, 271.

364 K.S.Jasim, C.Chieh and T.C.W.Mak, J. Crystallogr. Spectrosc. Res., 1985, 15, 403.
365 T.C.W.Mak, K.S.Jasim and C.Chieh, Inorg. Chem., 1985.24,1587
366 H.Fischer and S.Zeuner, Z. Naturforsch., Teil B, 1985, 40B, 954.
367 L.Marko, G.Gervasio, P.L.Stanghellini and G.Bor, Transition Met. Chem., 1985, 10, 344.
368 J.-P.Barbier, B.M.Ondo and R.P.Hugel, J. Chem. Soc., Dalton Trans., 1985, 597.
369 Kh.K.Ospanov, Yu.Ya.Kharitonov and M.Bigalieva, Koord. Khim., 1985, 11, 66.
370 K.N.Udupa and S.Sarkar, J. Organometal. Chem., 1985, 284,C36
371 C.Bianchini, C.A.Ghilardi, A.Meli and A.Orlandini, J.Organometal. Chem., 1985, 286, 259.
372 H.O.Desseyn, A.C.Fabretti, F.Forghieri and C.Preti, Spectrochim. Acta, Part A, 1985, 41A, 1105.
373 A.Müller, F.W.Baumann, H.Bögge and K.Schmitz, Z. anorg. allg. Chem., 1985, 521, 89.
374 C.Bianchini, C.A.Ghilardi, A.Meli, S.Midollini and A. Orlandini, Inorg. Chem., 1985, 24, 932.
375 J.D.Woollins, Polyhedron, 1984, 3, 1365.
376 B.Chiari, O.Piovesana, T.Tarantelli and P.F.Zanazzi, Inorg. Chem., 1985, 24, 366.
377 G.G.Hoffmann, Inorg. Chim. Acta, 1984, 90, L45.
378 R.K.Chadha, J.E.Drake and A.B.Sarkar, Inorg. Chem.,1985, 24, 3156.
379 M.Wieber, D.Wirth, J.Metter and C.Burschka, Z. anorg. allg. Chem., 1985, 520, 65.
380 J.M.Kisenyi, G.R.Willey and M.G.B.Drew, J. Chem. Soc., Dalton Trans., 1985, 1073.
381 M.Perec, Inorg. Chim. Acta, 1985, 103, 163.
382 M.Morán, J.R.Masaguer and V.Fernández, J. Organometal. Chem. 1985, 291, 311.
383 E.O.Fischer and D.Wittmann, J. Organometal. Chem.,1985,292, 245.
384 W.Preetz and U.Horns, Z. anorg. allg. Chem., 1984, 516, 159.
385 F.Beck, W.P.Fehlhammer and W.Beck, J. Organometal. Chem., 1985, 279, C22.
386 S.Delgado, J.Macazaga, C.Moreno and J.R.Masaguer, J.Organometal. Chem., 1985, 289, 397.
387 S.B.Sharma, T.N.Ojha, S.A.Khan and M.K.Singh, J. Ind. Chem. Soc., 1984, 61, 476.
388 V.K.Polovnyak, L.A.Tsytsyktueva and O.I.Rusetskii, Russ. J. Inorg. Chem., 1984, 29, 1535.
389 Yu.N.Kukushkin, L.V.Vrublevskaya, R.A.Vlasova, T.S. Isachkina, E.S.Postnikova and N.K.Sheleshkova, Russ. J. Inorg. Chem., 1985, 30, 224.
390 J.Rannou and M.Chabanel, Inorg. Chem., 1985, 24, 2319.
391 W.Bos, J.J.Bour, J.J.Steggerda and L.H.Pignolet, Inorg. Chem., 1985, 24, 4298.
392 S.K.Tandon, Proc. Natl. Acad. Sci. India, Sect. A,1984,519,65
393 K.Seppelt and H.Oberhammer, Inorg. Chem., 1985, 24, 1227.
394 H.A.Tajmir-Riari and T.Theophanides, Can. J. Chem., 1985, 63, 2065.
395 A.K.Banerjee, M.K.Ghosh, S.K.Sinha and S.K.Roy, J. Ind. Chem. Soc., 1985, 62, 272.

396 M.N.Mookerjee, R.V.Singh and J.P.Tandon, Acta Chim. Hung., 1984, 117, 89.
397 R.Seangprasertkij and T.L.Riechel, Inorg. Chem.,1985,24,1115
398 S.A.A.Zaidi, M.Jaria, S.Khan and Z.A.Siddiqi, Ind. J. Chem., Sect. A, 1985, 24A, 314.
399 P.Singh, V.Rani, G.C.Gupta and V.Singh, J. Ind. Chem. Soc., 1984, 61, 625.
400 A.Arlowska and W.Wojciechowski, Russ. J. Inorg. Chem., 1985, 30, 820.
401 A.R.Chakravarty, F.A.Cotton and D.A.Tocher, Inorg. Chem., 1985, 24, 172.
402 T.I.Odiaka, J. Organometal. Chem., 1985, 284, 95.
403 J.Zsakó, J.Máthé, E.Kästner-Marian and C.Várhelyi, Acta Chim. Hung., 1984, 117, 323.
404 C.M.Mikulski, L.Mattucci, L.Weiss and N.M.Karayannis, Inorg. Chim. Acta, 1985, 107, 81.
405 R.Dhakarey and G.Saxena, J. Chin. Chem. Soc., 1985, 32, 35.
406 R.P.Gupta and A.K.Srivastava, Synth. React. Inorg. Met.-Org. Chem., 1985, 15, 117.
407 H.Aghabozorg, R.C.Palenik and G.J.Palenik, Inorg. Chem., 1985, 24, 4214.
408 A.Bigotto and I.Pellegrini, Spectrochim. Acta, Part A, 1984, 40A, 979.
409 A.Albinati, C.G.Anklin and P.S.Pregosin, Inorg. Chim. Acta, 1984, 90, L37.
410 B.N.Misra and R.Kripal, Ind. J. Pure Appl. Phys., 1985, 23, 327.
411 N.Matsumoto, M.Asakawa and A.Ohyoshi, J. Chem. Soc., Dalton Trans., 1985, 107.
412 P.K.Radhakrishnan, J. Less-Common Met.,1985, 107, 45.
413 J.Wu, R.Deng and X.He, Lanzhou Daxue Xuebao, Ziran Kexueban, 1984, 20, 71 (Chem. Abs., 1985, 103, 31517).
414 C.R.Panda, S.C.Nayak, V.Chakravortty and K.C.Dasu, Ind. J. Chem. Sect. A, 1985, 24A, 141.
415 M.T.El-Haty and F.A.Adam, Bull. Soc. Chim. France, I,1984, 284.
416 N.S.Biradar, G.B.Krajagi, V.L.Roddabasanagoudar and T.M. Aminabhavi, Ind. J. Chem., Sect. A, 1985, 24A, 620.
417 B.K.Dwivedi and T.N.Srivastava, J. Ind. Chem. Soc.,1985, 62, 246.
418 G.K.Sandhu, R.Gupta, S.S.Sandhu, R.V.Parish and K.Brown, J. Organometal. Chem., 1985, 279, 373.
419 D.S.Morschheuser and H.P.Latscha, Z. anorg. allg. Chem., 1985, 525, 29.
420 R.K.Agarwal, G.Singh, V.Kapur and A.K.Srivastava, Curr. Sci., 1985, 54, 694.
421 J.Hanich, M.Krestel, U.Müller and K.Dehnicke, Z. anorg. allg. Chem., 1985, 522, 92.
422 N.K.Singh, S.Agrawal and R.C.Aggarwal, Polyhedron, 1984, 3, 1271.
423 G.Süss-Fink, W.Bühlmeyer, M.Herberhold, A.Gieren and T.Hübner, J. Organometal. Chem., 1985, 280, 129.
424 G.Süss-Fink, K.Guldner, M.Herberhold, A.Gieren and T.Hübner, J. Organometal. Chem., 1985, 279, 447.
425 R.Battistuzzi, Polyhedron, 1985, 4, 933.
426 N.Saha and N.Mukherjee, Synth. React. Inorg. Met.-Org.Chem., 1984, 14, 1151.

427 A.E.Arifien Pak. J. Sci. Ind. Res., 1984, 27, 127
428 O.S.Zakharova, D.D.Dobreva, L.A.Ignatova, V.V.Kravchenko and K.I.Petrov, Koord. Khim., 1984, 10, 1182.
429 H.Barrera, J.C.Bayon, J.Suades, C.German and J.P.Declerq, Polyhedron, 1984, 3, 969.
430 B.Morelli, Anal. Lett.,1984, 17, 2267.
431 R.Jones, P.F.Kelly, D.J.Williams and J.D.Woollins, J. Chem. Soc., Chem. Comm., 1985, 1325.
432 D.Michalska and A.T.Kowal, Spectrochim. Acta, Part A, 1985, 41A, 1119.
433 A.Syamal and M.A.B.Niazi, Synth. React. Inorg. Met.-Org. Chem. 1985, 15, 805.
434 B.H.Loo and Y.G.Lee, Chem. Phys. Lett., 1984, 112, 580.
435 B.H.Loo, Y.G.Lee and Z.Yazid, Chem. Phys. Lett.,1985,114,405
436 J.Bravo, J.S.Casas, M.V.Castano, M.Gayoso, Y.P.Mascarenhas, A.Sanchez, C.de O.P.Santos and J.Sord, Inorg. Chem., 1985, 24, 3435.
437 J.Maskowska and K.Oprządek, Polyhedron, 1985, 4, 23.
438 K.I.Arsenin, I.A.Sheka, L.A.Malinko, K.F.Karlysheva, L.V. Kolotylo, I.Ya.Pishchai and A.N.Antishko, Russ. J. Inorg. Chem., 1984, 29, 1283.
439 G.Urban, K.Sünkel and W.Beck, J. Organometal. Chem., 1985, 290, 329.
440 W.E.Hill, S.D.Worley, D.K.Paul and V.F.Newberry, Inorg. Chem., 1985, 24, 4429.
441 S.P.Mallela and F.Aubke, Inorg. Chem.,1985, 24, 2969.
442 M.Moran and I.Cuadrado, J. Organometal. Chem.,1985,295,353.
443 O.H.Bailey and A.Ludi, Inorg. Chem., 1985, 24, 2582.
444 J.Cabeza and P.M.Maitlis, J. Chem. Soc., Dalton Trans.,1985, 573.
445 D.Braga, R.Ros and R.Roulet, J. Organometal. Chem.,1985, 286, C8
446 G.Annibale, M.Bonivento, L.Canovese, L.Cattalini, G. Michelon and M.L.Tobe, Inorg. Chem., 1985, 24, 797.
447 S.Cenini, F.Porta, M.Pizzotti and C.Crotti, J. Chem. Soc., Dalton Trans., 1985, 163.
448 A.N.Red'kin, L.G.Dubovitskaya, V.A.Smirnov and V.S. Dimitiev, Russ. J. Inorg. Chem., 1984, 29, 1119.
449 I.I.Seifullina, A.I.Prisyazhnyuk and N.D.Kolker, Zh. Obshch. Khim., 1985, 55, 1122.
450 S.Blösl and A.Schmidt, Z. anorg. allg. Chem., 1985, 523, 99.

7
Mössbauer Spectroscopy

BY S. J. CLARK, J. D. DONALDSON AND S. M. GRIMES

1 Introduction

The format of this year's report is the same as that used last year. This introductory section provides details of books and review articles published during the review year. It is followed by sections that deal with the theoretical aspects of gamma-resonance spectroscopy and with advances in instrumentation and methodology. Sections 4, 5, and 6 contain detailed reviews on iron-57, tin-119, and other isotope Moessbauer effects respectively, while Section 7 reviews papers published on conversion electron and back-scattering gamma-resonance spectroscopy. The isotopes (energies in keV in parenthesis) that have been mentioned, some of them in review articles, during the year include: Fe-57(14.412 and 136.32), Ni-61(67.4), Zn-67(93.26), Ge-73(13.3), Ru-99(89.36), Sn-119(23.875), Sb-121(37.15), Te-125(35.46), I-127(57.6), I-129(27.72), La-139(165.9), Pr-141(145.4), Nd-145(67.3), Sm-149(35.8), Eu-151(21.64), Gd-155(86.54), Dy-161(80.7), Er-166(80.56), Tm-169(8.4), Yb-170(84.26), Yb-174(66.7), Ta-181(6.24), Ir-191(82.4), Ir-193(73.0), Au-197(77.35), Np-237(59.54), and U-238(44.7).

The eighth volume of the literature service 'Moessbauer Effect Data and Reference Journal'[1] was published in 1985 along with their handbooks on specialities in Moessbauer spectroscopy covering the topics tellurium-125, antimony-121, corrosion studies, and amorphous materials.

Books and Reviews.— A general review article on Moessbauer spectroscopy has been published in two parts[2]. The first part was concerned with the theory and uses of the iron-57 Moessbauer effect and dealt with the subject under the headings: fingerprint mode; chemical information; biological systems; structures, reaction intermediates and bonding. The second part dealt with information that can be obtained using isotopes other than Fe-57.

New possibilities for the application of gamma-resonance spectroscopy have been considered[3] and the topic has been reviewed with stress on the time-dependent hyperfine reactions involved with relaxation and diffusion[4].

A number of review publications have appeared since the last report that are concerned with the application of the Moessbauer effect to studies of a number of important systems including: the characterisation of Fischer-Tropsch catalysts[5]; minerals[6]; archaeological artifacts[7]; low-dimensional antiferromagnetic interactions in amorphous ferric fluorides[8]; the structures and magnetic properties of ternary Ba-Al and Th-Mn compounds[9]; iron atoms in biological systems[10]; silicate, phosphate and borate glasses[11]; ferrite materials[11]; and the structural role of Fe in a broad range of iron-containing glasses[12].

Moessbauer spectroscopy was included in a general paper on the uses of nuclear methods in materials science[13]. A number of reviews have also stressed the importance of using Moessbauer spectroscopy in conjunction with other techniques to obtain the maximum amount of information. Three examples of reviews of this type deal with (1) the complementary nature of the information obtained from Moessbauer and ESR studies on iron-containing proteins[14], (2) the use of EXAFS and Moessbauer data to analyse local structures in small particles of the heterogeneous catalyst systems Pt-Ir/Al_2O_3 and Co-Mo/Al_2O_3[15], and (3) the use of a variety of techniques including the Moessbauer effect to study the impurity levels in feldspars[16].

The application of gamma-resonance spectroscopy to metallurgy has been considered in a general article[17] and more specific topics reviewed have included the use of the Moessbauer effect in studies of iron and steel[18] and of gamma-Fe, metals containing hydrogen, and amorphous metals[19]. The uses of the technique as a structural probe in semi-conducting glasses have also been reviewed with some stress placed on the limitations of Moessbauer spectroscopy for this purpose[20]. Other general articles have described Moessbauer studies on hydrodesulphurisation catalysts[21] on inclusion compounds[22] and on goethite[23].

The uses of Moessbauer spectroscopy for low-temperature studies have been reviewed[24] as has the diffraction of Mössbauer gamma radiation by single crystals[25]. Other uses reviewed in the past year included applications in studies of phase transitions[26], of decorated defects in metals[27], of magnetic structures[28], and of impurity diffusion in metals[29]. The uses of Moessbauer isotopes as diamagnetic probes in solid-state chemistry experiments have also been considered in a general article[30].

Two recent review articles have been concerned with aspects of Moessbauer spectroscopy using actinide isotopes, viz the determination of *f*-electron structures[31] and studies of actinide intermetallics[32,33]. Results of Moessbauer studies have also been included in a book on the industrial uses of tin compounds[34] and in reviews on tin(II) compounds[35], mixed – valence ferrocenyl derivatives and tetranuclear iron clusters[36], and in the chemistry of the elements tin[37], tellurium[38], iron[39], and gold[40].

2 THEORETICAL

The Miedama and Van der Woude cellular model for the interpretation of Moessbauer isomer shifts in intermetallic compounds obtained from isotopes such as Fe-57, Ru-99, Eu-151, Gd-155, Ir-193, Pt-195, and Au-197 has been extended.[41] The value for the shifts for a given isotope in a series of intermetallic compounds is described in terms of a pair of constants that take account of interaction charge transfer and of interatomic *s*-*d*-electron conversion. Although the values of the constants are different from one isotope to another, a uniform pair of constants can be applied to all isotopes if the constants are scaled by taking account of the specific nuclear and electronic properties of each isotope.

The temperature-dependent double-time Green's function has been used[42] to calculate the anharmonic contributions to Moessbauer second-order Doppler shifts and the Debye-Waller factor for a cubic crystal containing a single-point pair defect. The effect of the presence of the defect is to broaden and slightly shift the resonance line. Even a weak anharmonicity is shown to produce a change in the time dependence of the correlation function and, as a consequence, adds temperature-dependent terms to the expression for the shift of the Moessbauer resonance line. The effects of the cubic and quartic anharmonicity on the second-order Doppler shift are also shown to become more important with increasing temperature. A linear response theory for surface Moessbauer atoms has also been developed to obtain information on the Debye-Waller factors for surface atoms[43].

A failure of the classical crystal-field model of gamma resonance spectroscopy has been claimed[44] because, in disagreement with experimental observations, it predicts an enhanced counting rate in a two-detector scheme. The authors favour a quantum electrodynamic model that agrees with non-enhanced counting rates.

The time dependence of Moessbauer gamma-radiation passing through an absorber of arbitrary thickness has been investigated on step-wise source modulation[45]. Expressions based on classical theory were derived for the transient pulse shapes, taking into account the finite rise time and the amplitude distribution of mechanical displacement in the gamma-radiation source. The predicted pulse shapes, voltages, and isomer shift dependences were found to be in good agreement with experimental data. The Stokes' polarisation method has been used in an analysis of Moessbauer radiation and data presented on polarisation and hyperfine parameters in the crystal field axis system for the M1 ($3/2^- \rightarrow 1/2^-$) and E2 ($2^+ \rightarrow 0^+$) radiations[46]. Ballet[47] has derived some useful expressions for the polarisation-dependent optical index of Fe-57 Moessbauer absorbers. The derivation of absorber transmissivities from the optical index is discussed along with the real rotation matrices that can be used to obtain wave-direction dependence and to simplify the treatment of texture and fraction anisotropy.

The change in the probabilities of internal conversion of M1 transitions at atomic nuclei when the conversion takes place in various ions was investigated[48]. The relative changes in the internal conversion coefficients and their deviations from the relative changes in the electron densities at the nucleus were calculated. The results of such calculations can be used for some Moessbauer isotopes.

A precise method of measurement of Moessbauer parameter changes in single symmetrical and nearly symmetrical resonance lines with arbitrary shapes has been developed[49]. The method, called the residual-area-analytical method, is a development of the then absorber approach for Lorentzian line-shapes and has the advantage of not requiring curve fitting or a knowledge of line shape. On test the results from use of the method were found to be in good agreement with experimental data. A method has also been developed for the calculation of Moessbauer line-shapes of hyperfine-split spectra using superoperator formalism to provide closed-form expressions for line-shape in the presence of magnetic dipole and electrical quadrupole interactions[50]. The authors claim that this procedure for obtaining line-shape has advantages over conventional methods in that averages over hyperfine field orientations may be calculated. The authors also demonstrate the use of their method for the case of Fe-57 with randomly orientated electrical field gradients in a uniform magnetic field. Two papers published in the review year dealt with the calculation of line shapes using the stochastic model[51] including the situation of a Moessbauer atom jumping in a tetrahedral site with a correlated electric field gradient reorientation and an $\underline{I}$ = 5/2 nuclear spin transition[52].

The problem associated with gamma-resonance line-shape broadening due to the diffusion of the Moessbauer atom in an arbitrary drift potential has been reduced to a solution of the appropriate Schroedinger equation[53]. The method used involved deriving analytical expressions for the spectra from diffusion analogues of the quantum oscillator and the Coulomb atom. The temperature dependence of the intensity of the unbroadened spectral component was found to be correlated with the arbitrary diffusion potential. The behaviour of the mean-square Moessbauer atom displacement with temperature derived from the solution was, however, shown to have non-trivial behaviour at each value of the wave vector of absorbed gamma quanta even in the simplest case of an anharmonic oscillator.

In many biological systems including whole cells, proteins, and membranes the gamma-resonance spectra obtained at temperatures above the melting point of their contained water are a superposition of the narrow lines normally expected from a solid sample and the broadened lines found for liquids. Nowik et al.[54] have calculated the expected Moessbauer line-shapes for harmonically bound nuclei in Brownian motion for the case of proteins at temperatures above the melting points of their contained water. The calculation is based on the assumption that the unfrozen conformational degrees of freedom of the protein macromolecules can be described in terms of damped harmonic oscillators acted on by random forces. The theory, in agreement with experimental data, predicts that the spectrum should be a superposition of a narrow and a broad-line resonance, that the total resonance absorption should decrease rapidly with increasing temperature, and that the Fe-57 spectra should be asymmetric doublets even when the Debye-Waller factor is isotropic. Outstanding agreement between calculated and observed spectra for these protein macromolecule systems is claimed.

A theory has been developed[55] for the time dependence of the recoilless radiation in Moessbauer spectroscopy to take account of coherent transient effects arising from phase modulation of the gamma-radiation. The time-dependent distortions affecting line-shape in conventional Moessbauer measurements were considered, and the time dependence of the oscillatory transient excited by a rapid sweep through the resonance was derived. The theory shows that transient experiments with a sinusoidal phase modulation should give good calibration of mechanical displacements and give more efficient separation of the source and absorber parameters than the conventional Moessbauer set-up. Linewidths and resonance line positions are obtained in the transient experiments in terms of a modulation frequency, which simplifies the calibration of the energy scale. When the phase of the incident radiation was switched rapidly, a simultaneous change was observed in the transmission intensity, and this change could be as large as eight times the maximum resonance absorption found in a conventional experiment. The use of the technique based on the theoretical calculations has been demonstrated for the 93 keV Moessbauer resonance line of Zn-67.

A theory of the Moessbauer effect in spectra of perturbed trinuclear exchange clusters has been developed[56] to explain the relationships between the structure of the exchange clusters and their Moessbauer spectra. The results, which are in good agreement with earlier experimental data, show that perturbations of the cluster geometry need not be reflected in the spectra because of a dependence of the quadrupole parameters in a relationship between the perturbation and cluster exchange interactions. In the presence of a strong magnetic field, however, the antisymmetric exchange can be suppressed and release static deformations that would be reflected in the Moessbauer spectrum.

Roy[57] has carried out a theoretical study of frequency moments and second-order Doppler shifts for dilute Fe-57 impurities in the high-temperature limit for a harmonic solid. He calculated the shifts from the Green's function of an impure crystal containing both mass disordering and force constant change terms in the crystal Hamiltonian and the frequency moments from McMillan ratios using standard Moessbauer recoil-free fractions at room temperatures. Roy and Kundu[58] have also reported their results of a theoretical study of stress-induced effects on Moessbauer impurities forming dislocations in some body-centred and face-centred-cubic metals.

A method of calculating changes in particle size from Moessbauer data has been described[59] and the use of the point-charge model to provide structural information based on iron and tin gamma-resonance data discussed.[60]

Two papers published during the year were concerned with the theory and use of grazing-incidence antireflection films for gamma-radiation[61,62]. Equations have also been developed[63] to describe the effects of passing Moessbauer gamma-radiation through crystals subjected to ultrasonic vibrations.

3 INSTRUMENTATION AND METHODOLOGY

Sakai and Sekizawa[64] have developed a new technique for obtaining Moessbauer spectra without using Doppler modulation. They obtained a spectrum from metallic iron using a new resonance technique involving magnetically modulated hyperfine-split energy levels of Fe-57 nuclei in a paramagnetic Pd metal source. Energy modulation was obtained by applying an external magnetic field of up to 22 kOe to the source cooled to 4.2K and varying this field slowly.

A Moessbauer spectrometer has been designed to enable the source to be placed close to the absorber surface and to obtain Moessbauer spectra from small sections of the absorber without the use of small aperture diaphragms[65]. The spectrometer design depends upon sinusoidal movement of the source by $\pm$ 10 mm s^{-1} at 420 Hz by means of an electromagnetically activated tuning fork to give a small amplitude (about 20 m). This method of source movement significantly decreases the activity required to obtain adequate percentage resonance dips. A Moessbauer system that avoids the use of an external wave form generator and provides a simultaneous display of 1024 channels has also been developed[66]. Two papers published during the review year dealt with microcomputer-controlled Moessbauer spectrometers[67,68]. The design, construction, and operation of a time-resolved Moessbauer spectrometer with a microcomputer for data recording and handling have also been described[69]. An article on a multiple micro-processor-based data-aquisition system for condensed-matter physics experiments included its use in Moessbauer spectroscopy[70].

A detailed procedure for the preparation of gamma-resonance sources with narrow unsplit emission lines has been described.[71] Co-57 sources with line-widths of 0.104-0.114 mm s^{-1} have been claimed from the procedure.

A laser interferometer for the absolute calibration of transducer velocities for Moessbauer effect studies has been developed[72] and a model of a distributed parameter electro-mechanical vibrator described[73]. Delayed-coincidence Moessbauer and quantuum beat measurements of a vibrating source have been obtained[74] using a Co-57 source vibrating at a frequency of about 10 MHz. The results were found to be in reasonable agreement with optical theory.

Conversion electron and Auger electrons emitted during the Moessbauer gamma event can be measured in a detector filled with helium containing a trace (0.1-1 wt%) impurity such as nitrogen and in which an electric field can be created. The conditions under which the signal-to-noise ratios and resolutions of the measured spectra can be improved are discussed[75]. Studies on resonance detection have also been carried out taking account of the weakening of outgoing secondary radiation from the converter[76].

A furnace system to enable Moessbauer measurements to be made at temperatures up to 1200K has been developed[77]. The system permits the study of samples at high temperatures in an external magnetic field of a few kG with a homogeneity of better than 1%. The results obtained with the furnace show that at a few degrees above the Curie temperature the small applied field induces a considerable hyperfine field in iron.

The upgrading of a microcomputer to provide a multiscaler system for Moessbauer spectroscopy has been described. The maximum number of channels that can be included is then limited only by the memory available[78].

A number of papers published during the review year have been concerned with methods of computer-fitting gamma-resonance spectra[79-85]. The programmes described include those designed to fit spectra with up to nineteen resonance peaks[80] and to take account of the effects of temperature and line overlap[85]. Two different methods for solving the transmission integral in Moessbauer spectroscopy have also been considered[81]. In one method the transmission integral is approximated by a single channel calculation while accounting for the nearby absorption cross-section by an approximate solution in the thin absorbed approximation. The second method is more elaborate and aims at producing a more exact solution with the inner part of transmission integral being convoluted numerically and the outer part by an asymptotic approximation. The finite impulse response operator method of deconvoluting spectra has been extended to deal with gamma-resonance spectra[84] and a computer algorithm for fitting the spectra using the Moore-Penrose generalised inverse A^+ method developed[83]. Two types of algorithm have also been suggested for fitting complex spectra that consist of overlapping six-line spectra from a distributed hyperfine field[84].

A coherent transient effect was found during a stopped-phase modulation experiment on a Co-57(Pd) gamma-source used by piezo-transformation.[86] The effect was expressed in terms of periodic pulses on passing the 14.4 Fe-57 keV radiation through an absorber. A method of pulse phase modulation for timing measurements in Moessbauer spectroscopy has also been devised[87]. A universal and relatively simple method was proposed for the numerical recognition of the time correlation for a system of interacting particles from their Moessbauer data[88]. The time dependence and Moessbauer spectra of resonance-filtered Fe-57 gamma-radiation has been studied by delayed-coincidence measurements and the coincidence spectra compared with the conventional Moessbauer spectra[89].

Moessbauer spectroscopy has been used in studies of various systems including liquids[90], glasses[91], semiconductors[92], coal ashes[93], and garnets[94].

The use of Moessbauer gamma-radiation to discriminate between elastic and inelastic scattering from various planes of single crystals of KCN has been described[95]. Both the Bragg reflection of Moessbauer Fe-57 radiation and the conversion-electron emission were measured precisely as functions of the angle and the data compared with the analagous functions for X-radiation[96]. The behaviour of the electromagnetic wave fields in the crystals is shown to be different for nuclear resonant and Rayleigh electron scattering. Suppression of the nuclear reactions was observed for the first time by measuring at the exact Bragg angle position and at a short distance from it. The degree of suppression of the electron yield for various nuclear transitions was determined and reported. A coincidence technique that efficiently reduces the detector background and increases the magnitude of the Moessbauer effect in a scattering experiment has been described[97]. A computer program has been written to analyse the scattering intensities used in radial distribution functions in the Rayleigh scattering of Moessbauer radiation[98]. The program follows that used for X-ray analysis of disordered systems and can provide information on the static and dynamic structures from the Fourier transform of the total and elastic scattering intensities.

Two papers that became available during the review year were concerned with laser spectroscopy of solids and its relationships with Moessbauer spectroscopy[99,100]. Very narrow optical homogeneous line-widths of the order of 1 kHz have been found for low-temperature zero-phonon transitions of dilute impurity ions such as Pr^{3+} in the LaF_3 lattice. Novel nonlinear optical-resonance techniques have been developed[99] to study these narrow lines. The techniques make use of ultra-stable phase-locked continuous-wave dye lasers, and the measurements can be made either in the frequency domain or the time domain. The authors claim that this technique effectively brings the advantages of Moessbauer effect studies into the optical region because the spectral line-widths are no longer limited by inhomogeneous strain broadening or by static local fields from neighbouring spins. The results obtained with Pr^{3+} in LaF_3 at 2 K are discussed in terms of a Monte line-broadening theory. Nuclear Raman spectroscopy is said to have several advantages over conventional Moessbauer spectroscopy including better resolution and a larger tuning range. In an experiment using 61.85 MHz of applied radio-frequency power, the first positive sideband of the least energetic fine-structure component of the transition and the first negative sideband of the most energetic component would overlap at the centre of the Fe-57 hyperfine spectrum in the absence of an isomer shift. If there is an isomer shift, both sidebands would be seen and they would be separated by twice the isomer shift. Because of the higher resolution of nuclear Raman spectroscopy the two absorption bands in a Fe-57 experiment can be seen by scanning the radio-frequency scales with the source at rest.

4 IRON-57

General Topics.— Nuclear Parameters and Metallic Iron. Systematic trends in the Fe-57 Moessbauer isomer shift have been examined[101]. It has long been known that isomer shift data provide information about the covalency of the iron-ligand bonds, that a correlation exists between isomer shift and the electronegativity of the ligand, and that a decrease in the iron coordination usually corresponds to a decrease in shift. The length of the iron-ligand bond is also thought to have some effect on the isomer shift. There are, however, many exceptions to these rules. The author examined the extent to which the second-order Doppler shift contributes to the isomer shift and compares published data for a large number of compounds containing (FeO_n) and (FeF_n) octahedra (Figure 1). His conclusions were three-fold. (1) The systematic decrease of isomer shift with the oxidation state of iron, the decrease, in general, of the shift with the coordination of iron, and the trend of a reduction in isomer shift with the electronegativity of the ligand are all verified. (2) There is no evidence for any correlation between the isomer shift and the length of the iron-ligand bonds. (3) Most of the anomalous isomer shift data may be accounted for by the concept of the antagonistic bond T-X (→Fe) (X = O or F, and T = any atom with a formal positive charge), which implies an increase in the ionic character of the Fe-X bond for an increase in the covalency of the T-X bond (and vice versa).

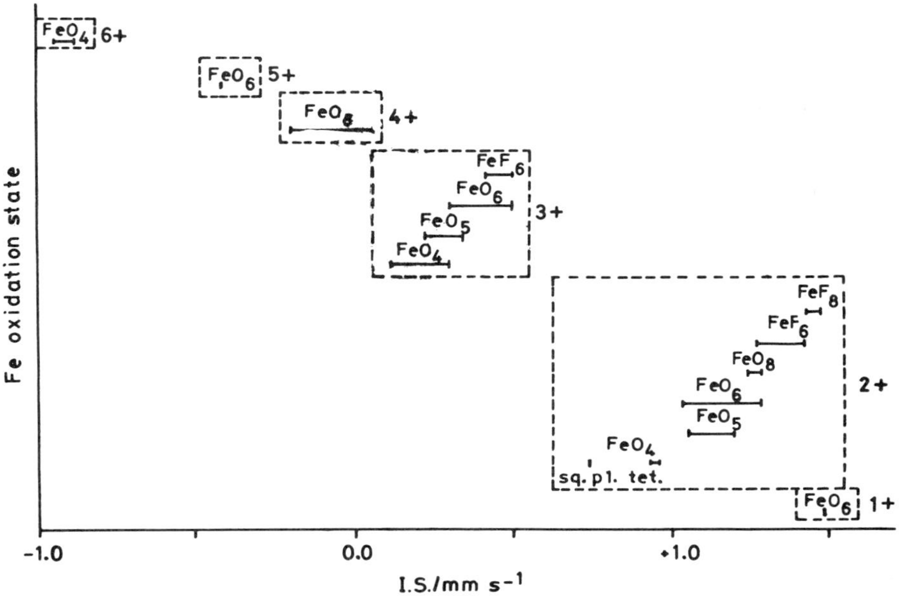

Figure 1 Room-temperature isomer shifts of various iron polyhedra (Based on data from reference 101)

Moessbauer spectroscopy has been used to examine magnetisation processes occurring near the surface of polycrystalline and single-crystal iron plates.[102] Two papers have been published on ultradispersed iron powders. An allotropic phase transition has been reported in an iron powder undergoing plastic deformation at room temperature and 2000 MPa pressure[103]. The particles (100 nm mean diameter) gave a paramagnetic singlet typical of the gamma-iron, together with a sextet identified with alpha-iron. The behaviour of a magnetic fluid containing 6.5 nm diameter iron particles has been studied using magnetic and Moessbauer measurements.[104] The 86K Moessbauer spectrum contained two lines which were identified with the superparamagnetic iron core and the FeO coating. Magnetic ordering in the FeO layer was only observed when the oxide coating was greater than 1 nm.

Iron-57 Impurity Studies, Polymers, and Ion Exchange. The dependence of the mixed electron-nuclear Moessbauer transitions on the external magnetic field has been studied for $Fe\text{-}57^{3+}$ ions doped into $Al(NO_3)_3.9H_2O$[105]. The molecular dynamics of lipids in phosphatidylcholine and phosphatidylcholine-cholesterol liposomes have been studied in the temperature range 80-350K using ferrocene, ferrocene carbaldehyde, and cholesterol ferrocene acetate as Moessbauer probes[106]. Smectic A, smectic B, and smectic C liquid crystals have been studied over a range of temperatures using Fe-57 Moessbauer spectroscopy[107].

Electrically conductive polymers formed by redox reactions between metal halides and polyacetylene and polypyrrole have been studied by gamma-resonance spectroscopy[108,109]. The Moessbauer Debye temperatures of $[CH(FeCl_4)_{\underline{y}}]_{\underline{x}}$ and $[CH(FeCl_4)_{\underline{y}}]_{\underline{x}}$ were found to be 89 and 120K respectively[108]. A similar study on polypyrrole revealed the formation of $FeCl_4^-$ species only[109]. A Moessbauer study of $FeCl_3$-doped poly(N-methyl-2,5- pyrrolene), however, revealed the formation of two distinct iron species which were identified with $FeCl_4^-$ and $FeCl_2.2H_2O$.[110] The glass transition in alpha,omega-dicarboxylatopolybutadiene neutralised with ferric ions has been studied by several techniques, including Moessbauer spectroscopy[111]. The temperature dependence of the recoil-free fraction deviated from Debye behaviour at 190K, the quasi-static glass transition, due to softening of the long-wavelength vibrational modes. The recoil-free fraction became unmeasureably small at the glass transition temperature ($\underline{T}_g$).

The state of iron in a Nafion perfluorinated cation-exchange membrane has been studied as a function of water content.[112] Three iron environments found in membranes containing less than 6% water were identified with isolated Fe^{3+} ions, ferric dimers, and clusters of Fe^{3+} ions. After exchange with K^+, a residue of iron remained which gave a spectrum similar to that of amorphous ferric hydroxide. The behaviour of Fe^{2+} in water-soaked Nafion 125 membranes has been studied at 90-250K[113]. The temperature dependences of the spectra were interpreted using a model based on overdamped harmonically bound Brownian motion. A Russian group have reported Moessbauer data on a series of acrylic acid-divinylbenzene cation-exchange copolymers[114-116] while other Soviet workers have examined Fe^{3+} sorbed onto an anion exchanger[117].

The effect of ferric ions on synthetic P zeolites has been studied[118]. Moessbauer spectra of the P_1 zeolite, where Fe is substituted for Al, showed that the zeolite has a higher thermal stability when the iron is tetrahedrally coordinated than when it is octahedrally coordinated. The Moessbauer spectra of iron cations in pentasil-type H zeolite have been compared with those observed in iron-impregnated H mordenite[119]. Very small alpha-Fe_2O_3 particles supported by zeolite-Y have been studied at a range of temperatures above and below the blocking temperature[120]. A new technique for calculating the average particle volume is described, which gave a value of 2.5 ± 0.4 nm for the mean diameter. The unusual Moessbauer spectra obtained for these particles are explained in terms of a high density of microcrystalline defects. Unequal magnetisation direction probabilities were significant and produced an anisotropy field estimated as 0.15.

Frozen Solutions and Matrix Isolation. The reductive decomposition of injections containing high-spin iron(III) complexes during preparation or storage has been studied[121]. Injections were quick-frozen and their iron(II) content determined from their Moessbauer spectra. A comparison of three model compounds and the correlation between the osmotic pressure of the solutions and their Fe(II) content was able to reveal the nature of the decomposition. Magnetically dilute frozen aqueous solutions containing Fe^{3+} have been studied at different temperatures.[122] Below $\underline{T}_g$ the spectra were affected by spin-lattice processes induced by acoustic phonons. Above this temperature, a rapid drop in $\underline{f}$, diffusional line-broadening, and an anomalous rise in spin-lattice relaxation frequency were all associated with a new form of spin-lattice relaxation which was in turn related to the molecular dynamics of the $[Fe(H_2O)_6]^{3+}$ complexes in the supercooled liquid.

Moessbauer measurements on iron clusters isolated in Ar and Xe matrices have been reported[123,124]. One study of 0.5 at.% iron embedded in solid Ar at 16K has identified Fe_3 as a non-magnetic component[124]. The relatively high abundance of Fe_3 compared with the larger magnetic clusters was attributed to the fast quenching effects of surface diffusion.

Emission Studies. Radioactive Co-57 implanted in frozen nitrogen, water, HCl, and xenon has been studied[125]. The results for ice showed stable arrangements of water molecules around the Co-57 atom and with a rearrangement above the amorphous-crystalline phase transitions. Results on the other matrices showed the existence of several different charge states produced by the decay and the possible formation of an Fe-N molecule. The emission spectra of $[^{57}Co(dipy)_3]Cl_2$ have also been recorded.[126] Moessbauer emission studies on ^{57}CoO doped with Li, Fe, In, and Ti have been described.[127] The electron recombination model proposed to explain the spectra takes into account the electronic character of the impurity defects. Gamma-resonance spectra of Co-57:La_2NiO_4 have been recorded at 297, 77 and 4.2K. Three non-equivalent positions for the $^{57}Co^{3+}$ impurity were identified and the existence of a uniquely charged Fe^{3+} daughter was discussed in terms of the oxide band structure.

Compounds of Iron.— High-spin Iron(II) Compounds. Electronic wave functions obtained from first-principles UHF calculations have been used to calculate isomer shifts for FeX_2 (X = Cl or Br), which were in good agreement with experimental data.[129] Hyperfine fields for these compounds were also calculated and found to be anisotropic about the molecular axes. The perpendicular and parallel field components were given as -199 and -193kG for $FeCl_3$

and -474 and -464kG for $FeBr_3$. The two-dimensional antiferromagnetic phases $K_2Co_{\underline{x}}Fe_{1-\underline{x}}F_4$ ($\underline{x}$ = 0.27, 0.55, 0.70)[130] and $K_2Fe_{\underline{x}}Zn_{1-\underline{x}}F_4$ (0.72 ⩽ $\underline{x}$ ⩽ 1)[131] have been studied with Moessbauer spectroscopy, and quasi-one-dimensional antiferromagnetic $RbFeCl_3$ has also received attention[132].

An analysis of experimentally observed distortions of impurity octahedral ferrous complexes has found a Jahn-Teller origin of the distortion[133]. A Moessbauer investigation of $[Fe(H_2O)_6]^{2+}$ in $M_{1-\underline{x}}Fe_{\underline{x}}SiF_6$ (M = Cu, $\underline{x}$ = 0.05) compared with data for isomorphic compounds (M = Ni, Zn, Co) suggested that the magnitude of this distortion and the sign of the e.f.g. are determined by the crystal field of the matrix. A linear decrease with temperature (77-550K) of the Moessbauer quadrupole splitting of iron in $FeCO_3$ has been observed and explained by orbit-lattice interactions.[134] Moessbauer data have been reported for the solid solutions $(Co_{1-\underline{x}}Fe_{\underline{x}})_3(PO_4)_2$, and a high-pressure modification of the $\underline{x}$ = 0.2 phase was found[135]. The products of the gamma-radiolysis of $Fe_{\underline{x}}Mg_{1-\underline{x}}(NH_4)_2(SO_4)_2.6H_2O$ have been studied[136]. The compound $Fe(O_2PF_2)_2.HPO_2$ has been prepared and characterised.[137] Two doublets with the same isomer shift (141 mm s^{-1}) but different splittings (2.66 and 2.22 mm s^{-1}) were associated with Fe(II) divided equally between two non-equivalent distorted octahedral sites.

The pyrazine (pyz) complexes $Fe(pyz)_2X_2$ (X = Cl, Br, I) have been prepared and characterised using IR, Raman, electronic, and Moessbauer spectroscopies[138]. These complexes have a polymeric pseudo-octahedral pyz-bridged structure containing six-coordinate high-spin Fe(II) in the FeN_4X_2 chromophore. Another polymeric compound, $[Fe(2,2\text{-bipyridine})oxalate_{1.1}]_{\underline{n}}$, has been studied at 295 and 4.2K and three iron states identified[139]. Moessbauer data for a series of iron(II) amino acid complexes have been reported[140]. The complexes prepared were: FeL_2 (L = Ala, PheGly, Ser, Gln, Trp, Met, MeCys, GlyGly), $FeL_2.2H_2O$ (L = Phe, Leu, Asp, Glu, His), Fe(CysCys), and $Fe(His)_2.H_2O$. Moessbauer data were also reported for an iron(II) inosine 5'-monophosphate complex[141]. The gamma-resonance spectra of two iron(II) complexes containing a binucleating clathrochelate ligand have been described[142]. Although both $[LFeMn]BF_4$ and

$[LFe_2]BF_4$ (where $H_3L = (1)$) contain high-spin iron(II), the temperature dependence of the spectra suggested that the second compound contains two inequivalent iron sites.

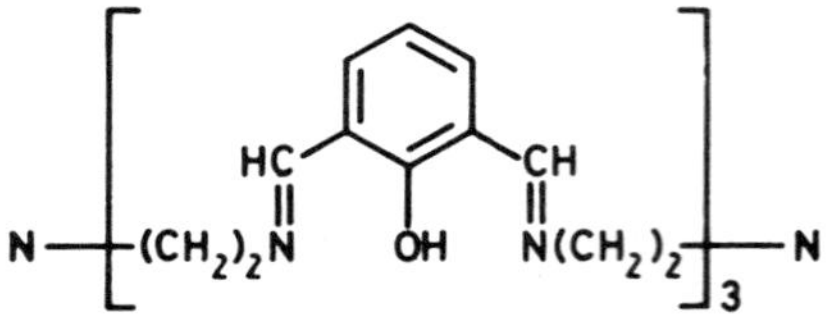

(1)

High-spin Iron(III) Compounds. Single-crystal susceptibility magnetisation and powder Moessbauer effect measurements have been reported for $[Me_3NH]FeCl_3.2H_2O$, a chainlike metamagnet[143]. Antiferromagnetic ordering was observed in both the susceptibility and the Moessbauer data around 3.2K. Gamma-resonance spectra of K_2FeF_5 recorded about the spin-flop phase transition have indicated that domains of antiferromagnetic and spin-flopped spins coexist in applied fields, B_{sf}, of 3.65 ± 0.4 T[144]. Systematic Moessbauer measurements of $A_2FeF_5.H_2O$ (A = K, Rb, NH_4) have been made at 78-300K[145]. While the quadrupole splittings of the K and Rb salts were found to be essentially temperature independent, that of the NH_4 salt was found to increase as the temperature was reduced below 240K, and this was taken to be evidence for a structural phase transition. The Moessbauer spectrum of ferri-magnetic $MnFeF_5.2H_2O$ has also been reported[146]. Moessbauer effect studies on NH_4FeF_6 have been carried out at low temperatures and in high magnetic fields and the spin correlation times were calculated[147]. The polarizable point-charge model has been used to interpret Moessbauer data on the e.f.g. at iron in alpha- and beta-$LiMnFeF_6$[148]. In the beta-phase the e.f.g. tensor lay along the *c*-axis, while in the alpha-phase the asymmetry parameter was large (about 0.5) and its *OZ* axis along the *a*-axis. Frequency-dependent susceptibility and Moessbauer experiments on the spin-glass $CsNiFeF_6$ have also been described[149].

The electric field gradient in the quasi-one-dimensional disordered compound $FeMgBO_4$ has been studied[150]. A probability distribution gave the mean values of $\Delta = +0.83$ mm s^{-1} at 30K, $\overline{\eta} = 0.58$, and the principal axis of the e.f.g. was found to lie in the ab-plane. Two new forms of $Fe(O_2PF_2)_3$ have been prepared and characterised[137]. Basic iron(III) sulphate has been the subject of three papers in the past year[151-153]. A broadening of the Moessbauer line-width of superparamagnetic particles of $Fe(OH)SO_4$ has been investigated and related to the behaviour of the internal magnetic field[151,152]. Linear-chain antiferromagnetism has been found and studied in the same compound above 111K and compared with that of $Fe(OH)CrO_4$[153].

Two trifluoroacetate compounds containing tetranuclear iron clusters $[Fe_4O_2(CF_3OOO)_x(H_2O)_y].2H_2O$ ($x = 8$, $y = 6$, and $x = 6$, $y = 3$) have been prepared and characterised[154,155]. The changes in Moessbauer lineshapes with paramagnetic relaxation time have been studied in a series of tris(beta-diketonato)-iron(III) complexes to determine the sign of the quadrupole splitting[156]. From the observed correlation it was shown that the sign should depend on the O-Fe-O bond angle. Other acetylacetonato-iron(III) complexes which have been studied are $Fe(acac)L_2X_2$ (L = pyridine, 3- and 4-picoline, X = Br, Cl)[157]

A series of monomeric Fe(III) complexes of the type FeLX (where L = a tetradentate S-alkylisothiosemicarbazide, X = Cl, Br, NCS) have been characterised using Moessbauer and magnetic methods[158]. While the isomer shift was found to be independent of X and the ligand substituents, the quadrupole splitting was affected by the nature of the substituents. The preparation of μ-oxoiron(III) phthalocyanine from a dihydroxoiron(III) precursor has provided the first definite proof of a pure high-spin $S = 5/2$ state in a phthalocyanine[159]. Moessbauer studies of iron(III)-reducing sugar complexes[160], of low-dimensional $[Fe(py)_3Cl_3.py]$[161], of iron(III) chloro complexes with N-arylglycosylamines[162], and of complexes with ligands containing the amide group[163] have been described. The rearrangement of ligands on the reaction of $[FeL_2Cl_2][FeCl_4]$ with 1,2,4-triazole, 4-allyl-1,2,4-triazole, and 4,4'-bipyridyl has been studied[164], and the Moessbauer spectra of mixed aquo(imidazole)iron(III) Schiff-base complexes[165] and of two forms of N,N'-ethylenebis(salicylidenaminato)(nitrato)-iron(III)[166] have been reported.

Intercalation Compounds Containing Iron. A Moessbauer relaxation study of FeOCl intercalation compounds has been described[167]. The results, obtained at temperatures of 100-300K, were interpreted in terms of an electron-hopping process which occurs within the timescale of the Moessbauer event. This relaxation followed a simple Arrhenius law, and the fraction of Fe(II) (0.10-0.13 per Fe(III) atom of the host matrix) was almost independent of both the nature of the guest species and the intercalation stoichiometry. Moessbauer spectra of graphite-iron intercalation compounds[168] and of the intercalate Fe_xTiS_2 have been recorded[169].

Mixed-Valence Compounds and Unusual Electronic States. Ternary BaF_2-ZrF_4-FeF_2 glasses have been studied to determine the local structure of ZrF_4-based glasses[170]. Moessbauer and Raman measurements revealed that the Fe^{2+} and Fe^{3+} ions are each surrounded by eight flourines and lie in interstitial sites in zigzag chains of ZrF_6 octahedra. A series of mixed-valence trinuclear iron acetate complexes have been prepared[171,172]. These complexes, $[Fe_3O(OAc)_6L_3]L$ (L = py or 4-ethylpyridine (L')), $[Fe_3O(OAc)_6L'_3]ClO_4$, $[Fe_3O(OAc)_6py_3]PhH$, and $[Fe_3O(OAc)_6L_3]$, were studied using Moessbauer, X-ray single crystal, and IR methods. The Moessbauer data showed that the rate of intramolecular electron transfer in the mixed-valence complex was dramatically affected by changing the solvate molecule. Unlike the solvent-free complex, the complexes containing the solvate exhibited temperature-dependent Moessbauer spectra. Bis(2,2'-bipyridine)dichloroiron(II) tetrachloroferrate(III), $[Fe(bipy)_2Cl_2][FeCl_4]$, has been studied by crystallographic and magnetic techniques and by zero- and high-field Moessbauer spectroscopy[173]. The Moessbauer and magnetic data suggest a complex combination of slow paramagnetic relaxation and weak magnetic order in the temperature range 10-1K. Fe-57 Moessbauer spectra of $(Fe(bipy)oxalate_{1.1})_n$ have also been recorded. The three doublets observed were assigned to Fe(II) $^5T_2$, Fe(II) $^1A_1$, and Fe(III) $^6A_1$ states, which maintain a nearly temperature-dependent ratio of spin ground states.

Several papers have been published on the mixed-valence biferrocenium cation (2) (where R = H, Bu, or CH_2Ph)[175,176]. X-ray structure determinations, together with DSC data and the effects of changing R or the anion (from I_3^- to Br_2I^-) on the temperature of the collapse of the two Moessbauer doublets into one, have indicated that lattice dynamics control the intramolecular

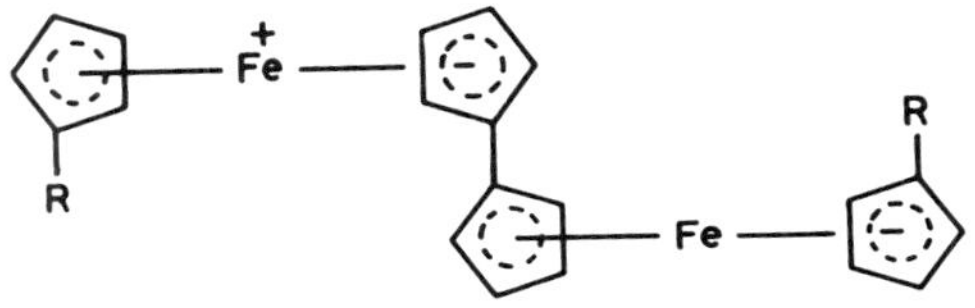

(2)

electron transfer in these compounds. The extent of delocalisation in (3-acetylfulvalene)(fulvalene)di-iron tri-iodide has been examined.[177] The Moessbauer and other data suggested a delocalised ionic structure unaffected by substitution at the 3-position. Polymer complexes of Fe(II) phthalocyanine, $(FePcL)_n.I_x$, have been prepared containing a bidentate bridging heterocyclic amine (L)[178]. The 50% iodinated complex with L = pyrazine was found to be a mixed-valence material containing low-spin Fe(II) and high-spin Fe(III) with isomer shifts of 0.18 and 0.37 mm s^{-1} and quadrupole splittings of 1.98 and 0.80 mm s^{-1} at 298 K respectively. Heteronuclear charge transfer between Co and Fe has led to the creation of an iron(IV) species in $Co(en)_3Fe(CN)_6$[179].

Spin-crossover Systems and Unusual Spin States. Quantitative light-induced spin-state trappings have been observed in the ferrous spin-crossover complexes $[Fe(phen)_2(NCS)_2]$ (phen = 1,10-phenanthroline)[180]. These Moessbauer data show high-spin states which have practically infinite lifetimes below 55 and 25 K respectively. Moessbauer and i.r. data have been presented which support the reformulation of the red complex $[Fe(phen)_2(NCS)_2]$ $.nH_2O$ as $[Fe(phen)_3]_2[Fe(NCS)]_4[NCS]_2.3H_2O$.[181] Variable-temperature Moessbauer data showed the presence of both high-spin and low-spin iron sites. The effects of pressure on the thermal hysteresis of the $^{5}\underline{T}_2 \rightleftarrows {}^{1}\underline{A}_1$ spin transition in $[Fe(phy)_2]X_2$ ($X = BF_4$, ClO_4, phy = 1,10-phenanthroline-2-carbaldehyde phenylhydrazone) have been studied[182]. The residual fractions were found to increase with pressure and the hysteresis width showed the linear dependence on pressure expected from Landau theory.

The spin-transition mechanisms in several spin-crossover Fe(III) complexes have been examined[183]. Differences in the spin-transition mechanisms were found to be reflected in the fraction of iron atoms that changed to the low-spin state at low temperatures, and the results were explained using a phenomenological model. The dynamics of spin-state interconversion on cooperativity have been studied in the complexes $[Fe(X\text{-}SalBzen)_2]Z$ (where X-SalBzen is the N_3O ligand derived from the Schiff-base condensation of N-benzylethylenediamine with X-substituted salicylaldehyde, X = 3-ethoxy or 3-allyl, Z = Cl, NO_3, or BPh_4)[184]. The complex $[Fe(3\text{-}allyl\text{-}SalBzen)_2]NO_3$ was shown to contain equal amounts of two crystallographically different cations, a low-spin ferric cation and a mixed-spin ferric cation which contained 67% low spin and 33% high spin at room temperature. The same workers have also studied the discontinuous spin-crossover complex [Fe(HL)L] and low-spin $[Fe(HL)L].H_2O$ (H_2L = pyruvic acid thiosemicarbazone)[185].

Relaxation-broadened Moessbauer spectra have been obtained for the trinuclear cluster complex $[Cu(Mesalen)]_2Fe(acac)(NO_3)_2$ (where $MesalenH_2$ = N,N'-bis(7-methylsalicilidenethylenediamine))[186,187]. Other techniques have identified the intermediate-spin, $\underline{S} = 3/2$, ground state. Moessbauer data obtained at liquid-helium temperatures were consistent with an effective relaxation rate which increases inversely with temperature. This was attributed to depopulation of an excited, slowly relaxing Kramers doublet with the simultaneous population of a more rapidly relaxing ground doublet.

Low-spin and Covalent Compounds. Room-temperature Moessbauer effect and variable-temperature magnetic-susceptibility measurements have been made on $Mg_4FeH_6Br_{3.5}Cl_{0.5}(THF)_8$.[188]

The spectral data, I.S. = +0.02 mm s^{-1} and zero splitting, were consistent with a $^1\underline{A}_1$, low-spin iron(II) ground state. The gamma-resonance spectrum of potassium hexacyanoferrate(II) trihydrate has been measured in an external magnetic field[189]. The single-line spectrum converted to a six-line spectrum within the magnetic field. The dehydration of the layered cyanide $Mg[Mg(hmta)_2Fe(CN)_6]_2.\underline{n}H_2O$ (hmta = hexamethylenetetramine) has been studied by Moessbauer spectroscopy[190]. The spectra of the hydrate corresponded to low-spin iron(II), while those of the anhydrous sample were typical of low-spin iron(III). An unusually large quadrupole splitting, of 1.85 mm s^{-1}, has been reported for anhydrous nitroprussic acid[191]. This value, which is large when compared with the values for the alkali-metal nitroprussides, was attributed to the intermolecular hydrogen bonding in the acid. A series of substituted pentacyanoferrate(II) complexes $Na_3[Fe(CN)_5L].\underline{n}H_2O$ (L = Me_2NH, Et_2NH, Et_3N, Pr_2NH, Bu_2NH, Bu_3N, Ph_2NH) have been prepared and studied with Moessbauer and i.r. spectroscopies.[191] All these complexes gave two-line spectra with quadrupole splittings of 0.73-1.92 mm s^{-1} and isomer shifts of -0.04-0.49 mm s^{-1} (relative to sodium nitroprusside). Partial isomer-shift quadrupole-splitting values were correlated with bonding characteristics of the ligands. The P-31 nmr and Fe-57 Moessbauer spectra of $Na_3[Fe(CN)_5L]$ (L = phosphine or phosphite) have been measured and correlated[193].

A Moessbauer effect study of tri-iron dodecacarbonyl, $Fe_3(CO)_{12}$, and the reduced carbide cluster compounds (PPN)-$[Fe_3(CO)_{10}CH]$, $(PPN)_2[Fe_3(CO)_9CCO]$, and (PPN)-$[Fe_2CO(CO)_9CCO]$ has been published (where PPN = bis(triphenylphosphine)nitrogen(1+)).[194] These compounds give two quadrupole doublets with the area ratio expected from their known crystal structures and have isomer shifts typical of diamagnetic Fe-CO clusters. Reduction to form the methylene and ketenylidene reduced the average isomer shift as the $\underline{s}$-electron density at the iron increased. These changes in data, relative to $Fe_3(CO)_{12}$, indicate that on reduction the added electrons were partly localised into orbitals associated with the metal sites. The thermal decomposition of $Fe_3(CO)_{12}$ adsorbed onto silica has been studied using several techniques, including Moessbauer spectroscopy.[165] The final products, particles of metallic iron and iron carbide and ferrous ions, gave a low Lamb-Moessbauer factor, which suggested that these particles were loosely bound to the silica. Fischer-Tropsch-like behaviour of these particles was indicated from the experimental results.

The Moessbauer spectra of a number of low-molecular-weight polymers containing ferrocene units have been studied at 80 - 300K.[196] A similar, non-linear, dependence of the absorption line area on temperature was seen in all the polymers. While an extended Debye model was unable to provide a satisfactory fit to the observed spectra, these data could be accounted for by assuming that each macromolecule vibrates within an anharmonic potential. Moessbauer spectroscopy has been used to study the dissolution of ferrocene in solid mixtures with human serum albumin, silica gel, and highly dispersed Teflon.[197] The gamma-resonance spectra of a series of [3]-ferrocenophane derivatives have been studied.[198] The Moessbauer spectra of ferrocenophanedione were recorded for the carbanion in frozen solution and had a quadrupole splitting similar to that of ferrocene. This was interpreted as indicating that the carbanion charge created a symmetric distribution about iron by repulsion of the $\underline{e}_{2g}$ electrons. Moessbauer data have also been reported for the dibutylferrocenium and iodotetramethyl [2]-ferrocenophanium tri-iodides [199] and for some mono- and diketone derivatives of [3]-, [4]-, and [5]-ferrocenophanes.[200]

Biological Systems and Related Compounds. A Chinese investigation of the Moessbauer spectra of iron porphyrin solutions has been published.[201] The range of values of isomer shift and quadrupole splitting were given as I.S. = 0.26-0.52, Q.S. = 0.49-1.02 mm s^{-1} for ferric iron and I.S. = 0.46-0.62 , Q.S. 1.06-1.84 mm s^{-1} for ferrous iron. The Moessbauer spectra of oxygenated frozen solutions of protoporphyrin IX iron(II), which contained 2-methylpiperidine or mercaptoethanol as the fifth iron ligand, have been compared with those of oxyhaemoglobin.[202] The results were discussed in relation to haemoprotein-oxygen binding. The molecular and electronic structures of (2,3,7,8,12,13,17,18-octaethylporphyrinato)iron(II) and its trans-7,8-dihydrate have been compared.[203] While these compounds share the $\underline{S}$ = 1 electronic ground state and have almost identical chemical shifts, their quadrupole splittings are different, 1.71 mm s^{-1} for the anhydrous compound and 2.55 mm s^{-1} for the hydrate. This difference was explained by the ring-induced rhombic magnetic anisotropy found in the hydrate by proton n.m.r. experiments. Moessbauer spectra of the iron porphyrins Fe(OEP)Cl, Fe[5-NO_2(OEP)]Cl, and Fe[(5,15-NO_2)$_2$(OEP)]Cl (OEP = octaethyl porphine) have been measured at 77 and 300K and in an external magnetic field of 5 kOe at room temperature[204]. Changes in spectral shape were explained by electron spin-spin relaxation, and the substituent effects of the nitro groups were discussed. Inter-heme spin coupling has been observed in $[Fe(OEP)(2\text{-}MeHIm)]^+$ dimers (where 2-MeHIm = 2-methylimidazole)[205,206]. Moessbauer spectra were recorded at 1.54-195K in a range of magnetic fields up to 6T, and magnetic susceptibility was also measured as a function of temperature. The low-temperature magnetic data could only be interpreted by assuming that spin-coupling occurred, and slow spin-relaxation at low temperatures allowed the observation of a small rhombic distortion, of $\underline{E}$ = 0.024 cm^{-1}, in low-field experiments. In other papers on porphyrins, electronic-structure calculations on model bis-(imidazole)iron porphyrins have been compared with Moessbauer data[207], and the gamma-resonance spectra of μ-oxo chromium-iron porphyrin compounds were reported[208].

The Moessbauer parameters of rat hemoglobin (Hb) have been found to vary with the donor animal and the strength of interaction between heterogeneous hemoglobin forms[209]. Natural oxidation of oxyHb resulted in a time-dependent reduction of quadrupole splitting in the temperature component, while chemical oxidation produced conformational changes in the protein which were also reflected in the Moessbauer spectra. Normal human foetal and adult hemoglobins have been investigated and differences in quadrupole splitting related to structural differences[210]. A model of the Moessbauer spectra of hemoglobins based on the presumed non-equivalence of iron(II) electronic structures in the alpha-, beta-, and gamma-subunits in tetramers was proposed. The met-aquo form of the monomeric insect hemoglobin CTT III has been studied before and after gamma-irradiation at 77K. Irradiation produced thermalised electrons which reduced the high-spin Fe(III) to low-spin Fe(II). The latter complex is a transition state which transformed into high-spin Fe(II) at $\underline{T}$ > 190K. This temperature corresponds to drastic changes in the temperature dependence of the Lamb-

Moessbauer factor. Both effects indicated a change in the intramolecular flexibility of the hemoglobin. Pocket dynamics of horse deoxyHb and metHb have been studied at 80-260K using ESR and Moessbauer spectroscopies[212]. A low-spin iron(II) complex was observed at temperatures down to 210K and was considered to be due to the coordination of N-histidine-E_7 as a sixth nitrogenous ligand[212]. The formation of a bis(histidine) complex by deoxyHb was accompanied by conformational fluctuations which affected the temperature dependence of the Lamb-Moessbauer factor.

The use of Fe-57 Moessbauer spectroscopy in the study of protein dynamics and, in particular, the motion of deoxymyoglobin, has been discussed[213]. The second-order Doppler shift of iron in deoxymyoglobin crystals has been determined in two series of experiments over the temperature range 5-280K[214]. In one the source and absorber were at the same temperature, in the other the source was at 298K. These experiments found a value for the Debye temperature of myoglobin, and the absence of any protein-specific motion from the second-order Doppler effect was noted. The hydrolysis of $FeLH_2O$ (where H_3L = N-[(2-(o-hydroxyphenyl)-glycino)ethyl]salicylidenimine), a model complex of the iron(III) environment in the transferrins, has been studied by Moessbauer spectroscopy[215]. A nitric oxide adduct of the binuclear Fe centre in deoxyhemerythrin from *Phascolopsis gouldii* has been studied[216] and Moessbauer data have been reported for covalently bound polyphospazene-heme complexes[217] and for the interaction of iron ions with L-DOPA melanin[218].

Moessbauer and ESR spectra of spinach ferredoxin and two model compounds (iron-mung bean trypsin inhibitor complexes) have been reported[219]. The data indicated that an antiferromagnetic couple existed between the two high-spin iron(II) in the [2Fe-2S] cluster. Active bovine heart aconitase, which contains a $[4Fe-4S]^{2+}$ cluster, and the inactive enzyme, which has a $[3Fe-4S]^{+}$ cluster, have been studied[220]. The enzyme was reacted with various substrates and frozen-solution Moessbauer spectroscopy used to identify reaction intermediates. A unique interpretation of the spectra was, however, not possible. Evidence has been obtained for a $[4Fe-4S]^{+}$ cluster with spin $S = 3/2$[221]. The protein Av2 (the azoferredoxin of the *Azobactor vinlandii* nitrogenase system) was studied by Moessbauer, ESR, and magnetic susceptibility measurements. Oxidised Av2 gave Moessbauer spectra typical of a diamagnetic $[4Fe\text{-}4S]^{2+}$ cluster, but when reacted with Mg-ATP or Mg-ADP a pronounced decline in quadrupole splitting was observed. These results, together with data from the other techniques, are the result of the heterogeneous nature of the reduced Av2, which contains two types of $[4Fe\text{-}4S]^{+}$ cluster with $\underline{S} = 1/2$ and $\underline{S} = 3/2$. Moessbauer and magnetic measurements have been used to analyse the active centre of hydrogenase from Desulfovibrio gigas[222]. The Moessbauer isomer shift of 0.38 mm s^{-1} and the quadrupole splitting of 0.87 mm s^{-1} with no magnetic splitting at 4K were in accord with a structure of $[4Fe\text{-}4S]^{3+}$ + $[4Fe\text{-}4S]^{2+}$. Other workers have studied a low-molecular-weight sulphite reductase from Desulfovibrio gigas.[223] The enzyme was found to contain a low-spin ferric siroheme and a $[4Fe\text{-}4S]^{2+}$ cluster. Moessbauer data have been reported for the iron-sulphur centres of milk xanthine oxidase[224] and for the mixed terminal ligand cubane compounds $Fe_4S_4(Et_2Dtc)_{\underline{n}}(X)_{4-\underline{n}}{}^{2-}$ (X = Cl, Br, I, RS, RO) were similar to those of the corresponding $[Fe_4S_4L_4]^{2-}$ cubane clusters.

The electronic structures of mononuclear Fe-S complexes with an $Fe^{II}S_4$ core and of binuclear Fe-Mo-S complexes containing the FeS_2Mo core have been calculated and the fine and hyperfine structure tensors were compared with Moessbauer spectral data.[227] For $[Fe(SPH)_4]$ and $[Fe(dts)_2]$ (H_2dts = 3,4-dimercapto-3-cyclobutene-1,2-dione), $\underline{D}$ is positive and $\underline{V}_{zz}$ negative, while for $[(SPh)_2FeS_2MoS_2]^{2-}$, $[S_5FeS_2MoS_2]$, and $[Cl_2FeS_2MoS_2]^{2-}$ $\underline{D}$ is negative and $\underline{V}_{zz}$ positive. In all these binuclear Fe-Mo-S complexes $\underline{V}_{zz}$ was aligned perpendicular to the Fe-Mo line. A single-crystal Moessbauer study of $[PhP][(PhCH_2)Me_3N][Cl_2FeS_2MoS_2]$ was used to confirm these results[228]. These cluster compounds are models for the active centres in nitrogenase and have been compared with the major iron components of nitrogenase[229]. Gamma-resonance spectra of a number of F-series nitrogen fixation model compounds of the type $(Et_4N)_2[Mo_{\underline{x}}Fe_{4-\underline{x}}S_4(SR)_4]$ have been reported[231]. The Moessbauer parameters provide evidence for a monocubic configuration in these compounds and were linear functions of $\underline{x}$, which implies isomorphous substitution of Mo and Fe. The MoFe-protein molybdoferredoxin, from Azobactor vinlandii, has been isolated and its Moessbauer spectra recorded at 125-175K[231,232]. Contrary to previous interpretations, four quadrupole doublets were fitted to four iron atoms in the ferrous doublet, two in the *S* doublet, 12 in the *D* doublet, and 16 in the *M* doublet. It was proposed that the Mo-Fe cofactor contains one Mo and eight Fe atoms. Mononuclear nonheme cofactor iron of putidamonooxin has been studied as a binary oxidised enzyme-substrate complex and as a ternary enzyme-substrate-NO complex[233]. From a comparison with the published Moessbauer spectra of similar compounds it was deduced that the iron was present with a coordination number greater than four. In the ternary complex the iron was in a five-coordinate, intermediate (S = 3/2) spin state.

Chromatophores from Ectothiorhodospira shaposhnikovii have been studied in different solvents using Rayleigh scattering of Moessbauer radiation and information on the membrane dynamics obtained[234]. Intermolecular dynamics of chromatophore membranes of Rhodopseudomonas sphaeroides have been determined by absorption Moessbauer spectroscopy after incorporation of Moessbauer probes into the membranes[235]. A compensation effect of environmental viscosity was observed. Iron-overloaded livers have been studied by Moessbauer spectroscopy. The spectra of normal liver samples, taken at room temperature, 77 and 4.2K, were typical of ferritin-hemosiderin compounds. A new iron resonance appeared in hemosiderosis, secondary overload, livers which had parameters typical of high-spin iron(III) and was characterised by super-paramagnetic behaviour which began above 77K. This component did not appear in cases of primary overload. Moessbauer absorption and emission studies have been used to determine the distribution and chemical states of Fe and Co in water hyacinth roots[237].

Oxide and Chalcogenide Compounds Containing Iron.— General and Hydroxides. Moessbauer spectroscopy has been used to study the temperature dependence of the hyperfine parameters of beta-FeOOH in paramagnetic chlorine[238]. External magnetic fields were applied to determine the sign and asymmetry of the e.f.g. at the Fe nuclei. Microcrystals of synthetic δ'-FeOOH have been studied using *X*-ray diffraction, electron microscopy, and the Moessbauer effect[239]. While the largest particles were ferro- or ferrimagnetic, the smallest particles had the very small magnetic moments typical of antiferromagnetic or spin-glass behaviour. The effects of metal cations on the disproportionation of ferrous hydroxide have been studied in the absence of oxygen and the Moessbauer spectra of the reaction products described[240]. SCF-X_α-SW-MO calculations have been used to explain differences between the Moessbauer spectra of FeOOH polymorphs and those of ferric oxides[241]. Replacement of oxygens in the Fe^{3+} coordination environment by hydroxyls was found to promote covalency between Fe^{3+} and O^{2-}. Spectral and physicochemical properties of submicron powders of goethite (alpha-FeOOH) and lepidocrocite (beta-FeOOH) have been related to their Moessbauer spectra.[242]

Wustite, Haematite, and Related Oxides. Moessbauer spectra of wustite, $Fe_{\underline{x}}O$ ($\underline{x}$ > 0.95), have been recorded[243]. At 298K the spectrum consisted of an asymmetric doublet which was fitted to a ferrous singlet, two ferrous doublets, and a ferric singlet. At 4.2K, however, a large number of unresolved lines were obtained. The room-temperature spectra were related to a model of the defect structure of wustite.

The Morin transition in natural haematite has been examined[244]. No specific-heat anomaly due to the Morin transition was found, although the expected changes were seen in the Moessbauer spectra. The results were consistent with the coexistence of two phases and a spread of about 40K over which the transitions take place. Submicron powders of haematite and maghemite have been studied.[242] The effects of small aluminium substitutions in haematite have been examined[245]. Characteristic Moessbauer temperatures and intrinsic isomer shifts were determined using the Debye approximation, and a maximum was found in both parameters at 4 mol.% Al substitution. A diamond anvil cell has been used in a Moessbauer effect and X-ray diffraction study of Fe_2O_3 at pressures up to 60 GPa[246]. Although two components were found in the high-pressure Moessbauer spectra, the X-ray data could not be explained by the corundum structure. A simple high-spin to low-spin transition could not, therefore, be considered for the 55 GPa phase transition in Fe_2O_3. Two independent studies have been published on Co-doped delta-Fe_2O_3. In the first, an emission study, it was found that some of the cobalt atoms diffused into the delta-Fe_2O_3 lattice after heat treatment at 300°C for one hour[247]. The second study discussed the coercive force and Moessbauer spectra in terms of the degree of penetration of Co^{2+} into the oxide lattice[248]. A Moessbauer and magnetisation study of finely dispersed antiferromagnetic grains of Fe_2O_3 in an Al_2O_3 matrix has been reported[249].

A Moessbauer emission study[250] has been made on the system $La_{1-\underline{x}}Sr_{\underline{x}}Co_{1-\underline{y}}Ti_{\underline{y}}O_3$ (for $\underline{x}$ = 0 or $\underline{y}$) over the temperature range 77-800K. No hyperfine splittings were observed even at 77K. Differences in behaviour between the two systems were explained by the generation of different valence and spin states of cobalt with increasing $\underline{x}$. Other systems which have been studied are Cr-doped nickel ferroaluminate[251], $SrO{\cdot}5.6Fe_2O_3$ doped with barium borate or kaolin[252], and Sc_2O_3-2 wt.% Fe_2O_3.[253]

Magnetite and Spinel-type Oxides. Ultrathin Fe_3O_4 films have been studied[254]. These films are superparamagnetic at low temperatures and paramagnetic at room temperature. The room-temperature spectra consisted of three doublets which were assigned to Fe^{3+} in the $\underline{A}$ site and $Fe^{2.5+}$ and Fe^{2+} in the $\underline{B}$ site. Moessbauer spectra of hydrous spinel iron oxide colloids and their evolution with thermal treatment have been interpreted by using a modified Weiss local field model[255]. Interparticle interactions were found to overcome the single-particle anisotropy energy. In another paper the spectral and physicochemical properties of submicron powders of Fe_3O_4 were related to its Moessbauer spectra.[242] A study of covalency effects and the pressure dependence of the Moessbauer isomer shifts for spinel-type ferrites has been described[256]. Although the cubic lattice parameter was the dominant influence on the pressure dependence of the isomer shift, consideration of covalency effects was also required to interpret the spectra. Electron localisation in ferrospinels with a mixed-valence tetrahedral sublattice has been investigated using Moessbauer spectroscopy[257]. The best interpretation of the spectra assumed completely localised electrons at room temperature with the presence of some, 4-17%, mixed-valence iron due to fast electron relaxations above 440K. Superexchange interactions in fluorosubstituted ferrites have also been studied[258].

Two papers have been published on Mo-bearing ferrites[259,260]. From XPS and Moessbauer spectra it was suggested that Fe_2MoO_4 can be described as $(Fe^{2+})_{tet}$ - $[Fe^{2+}Mo^{4+}]_{oct}O_4$. A Moessbauer study of $Fe_{3-x}M_xO_4$ (M = Al, Ga) has been described[261]. Measurements made at low temperatures using applied magnetic fields were used to study the cation distribution. The formation kinetics of mechanically activated $NiFe_2O_4$ have been studied as a function of the mechanical activation intensity[262]. The effects of oxygen loss on the crystallographic and magnetic properties of $CuFe_2O_4$ have been examined and the results of different heat treatments described[263]. Ferrimagnetic $Ga_xSn_yFe_{3-x-y}O_4$ has been studied as a function of the Ga^{3+} and Sn^{4+} content[264]. Cation distributions were determined at room temperature and 4.2K in an external magnetic field, and ferrous ions were found in A sites. The thermal decomposition of lithium ferrite, $Li_{0.5}Fe_{2.5}O_4$, has been studied[265] with the Moessbauer effect and the effects of zinc substitution reported.[265,266] Chromium-substituted lithium ferrites, $Li_{0.5}Fe_{2.5-x}Cr_xO_4$, have been studied for x = 0-2.[267,268] Six-line spectra were obtained for both A and B sites with chromium concentrations of up to x = 1.2, and a ferromagnetic relaxation occurred when x = 1.6. The decrease in internal magnetic field with increasing x was qualitatively explained by considering a dipolar field, zero-point spin deviation terms, and a supertransferred hyperfine field, while the relaxation effect was explained on the basis of domain wall displacements.

Manganese-zinc ferrites have been the subject of two investigations.[269-272] The effects of sintering and cation distributions were examined. Paramagnetic clusters found after annealing were attributed to non-random ordering caused by local charge compensation near cation vacancies[269,270]. In the other study, electron exchange between iron and manganese was studied as a function of composition[271] and the differences between two batches of different quality were examined[272]. The magnetic properties of $Zn_xMg_{1-x}Fe_2O_4$ have been studied at 77K[273] and 298K[274]. For $x \leqslant 0.6$, the 77K Moessbauer spectra showed two hyperfine fields which were assigned to iron in the A and B sites; relaxation behaviour was seen when x = 0.7, and a paramagnetic quadrupole doublet when $x \geqslant 0.8$. At room temperature relaxation behaviour began when $\underline{x} = 0.6$ and paramagnetism when $\underline{x} = 0.7$. This variation of hyperfine field with zinc concentration was explained in terms of $\underline{A}$-$\underline{B}$ and $\underline{B}$-$\underline{B}$ supertransferred hyperfine interactions. Cobalt-zinc[275], nickel-zinc[276], and cobalt-nickel-zinc ferrites[277] have also received attention.

Moessbauer effect studies have been described on the spinel systems $GaCoMnO_4$-$ZnFeMnO_4$[278], $Ga_{\underline{x}}Fe_{1-\underline{x}}(Ni,Cr)O_4$[279,280], and Al-[281] and Ga-doped[282] lithium ferrite. The influence of Ti^{4+} substitution on the Moessbauer parameters of zinc ferrites[283] and magnesium ferrite[284] has been investigated, and two papers were published on the magnetic properties of $Ge_{\underline{x}}Cu_{1-\underline{x}}Fe_2O_4$.[285,286] The antimony-substituted spinel ferrite $Zn_{\underline{x}}Ni_{5/3-\underline{x}}FeSb_{1/3}O_4$ has also been studied[287].

Other Oxides. Magnetic structure in the perovskite $SrLaFeO_4$ has been studied by Moessbauer spectroscopy[288]. The hyperfine field of the high-spin Fe(III) decreased as the temperature was raised from 80 to 320K and a value of 351K was obtained for the Néel temperature. A detailed investigation of the oxygen-deficient perovskite $(SrFe)_{3-\underline{y}}$ ($0.15 < \underline{y} < 0.25$) has been carried out using Moessbauer spectroscopy, $\underline{X}$-ray powder diffraction, and magnetic susceptibility measurements[289]. Extremely complex Moessbauer spectra were obtained at temperatures in the range 4.2-900K. Three chemically distinct types of iron were observed below 550K, but above 550K thermally activated electron transfer produced a single resonance. It was shown that ordering of oxygen vacancies occurred to produce layers of tetrahedrally coordinated ferric ions. It was tentatively suggested that in the idealised structure, $SrFeO_{2.75}$, electron delocalisation occurred in intervening layers which contained two types of octahedrally coordinated iron. Moessbauer spectra have been recorded for a number of garnets including: yttrium iron garnet (YIG) doped with Al,[290] YIG substituted with Ca, In, and Ge,[291] YIG containing Eu, Tm, Ca, and Ge,[292] YIG containing Gd, Ca, V, and In,[293] ion-implanted $(Eu,Sm,Lu,Ca)_3$-$(Fe,Ge)_5O_{12}$,[294] and the new garnets $Ln_3Fe_{5-\underline{t}}Ga_{\underline{t}}O_{12}$ (Ln = Nd, Pr)[295].

The possibility that the perovskite $BaFeO_3$ is formed in the early stages of the reaction kinetics of the hexagonal M-phase ferrite ($BaFe_{12}O_{19}$) has been examined[296]. The perovskite phase was formed by heating a mixture containing a 1:1 Ba:Fe ratio at low temperatures. X-ray, Moessbauer and magnetic methods have been used to study the effects of the gaseous medium on the phase composition, magnetic properties, and crystal structure of barium hexaferrites[297]. Differences between the Moessbauer spectra of W, M, and X types of barium hexaferrites have been compared[298]. It was suggested that the X-type consists of superimposed M and W types. Calcium-containing barium hexaferrite[299] and cadmium- and nickel-W-type hexaferrites[300] have also been studied by gamma-resonance spectroscopy.

Intervalence electron transfer between iron ions in $CaFe_3O_5$ has been studied. In the antiferromagnetic region below T_N (282K), the ferrous and ferric ions were well differentiated, but thermal electron transfer was observed in the paramagnetic region. Analysis of the data found that the electron transfer was limited to the trimer Fe^{3+}-OFe^{2+}-Fe^{3+} with a relaxation time of 180ns at 298K. The Moessbauer spectra of $BaFe_4Sn_2O_{11}$,[302] $SrFe_{12}O_{19}$,[303] and $BaCd_2Fe_{16}O_{27}$[304] have been recorded. Lithium insertion compounds of the high- and low-temperature insertion polymorphs of lithium stannoferrite have been examined[305] and Moessbauer emission spectra from Co-57-doped La_2NiO_4 recorded[306]. The quadrupole splitting distributions of the A- and B-site ferrous ions in iron titanate[307], the Moessbauer spectra of strontium iron vanadate[308] and of the bronzes theta-$(Fe_{1-y}Al_y)_xV_2O_5$[309] and Fe-57 doped $K_{0.3}MoO_3$[310] have been studied. Zinc iron manganese chromate[311] and the tungstates $MFe(WO_4)_2$ (M = Na, Li) have also received attention.

Inorganic Oxide Glasses Containing Iron. The ratio Fe^{3+}/total iron has been determined in a number of iron silicate glasses by wet chemical methods and Moessbauer spectroscopy[313]. Both techniques produced the same uncertainties in composition and were considered to give equally acceptable results. Another similar comparison found that the Moessbauer method overestimated the amount of Fe(II) by a factor of 1.2, and a modified instrument was developed to correct for this[314]. Moessbauer spectroscopy has been used to determine the redox equilibrium of iron and the structure of quenched melts in the system $Na_2Si_2O_5$-Fe_2O_3 to pressures of 40 kbar at 1400°C[315]. The fraction of the iron present as Fe^{3+} was found to decrease with increasing pressure and changed from being a network-former at 1 bar to a network-modifier at ⩾10 kbar. Fe^{2+} remained as a network-modifier in all quenched melts. These changes resulted in depolymerisation of the silicate melts. A similar study carried out at normal pressures has obtained a limit of 0.2 atoms of iron in tetrahedral sites per sodium atom[316]. Other workers have proposed a basic model of iron(III) in sodium silicate and sodium borosilicate glasses as a result of Moessbauer studies[317].

A general model has been proposed for the structural state of iron in silicate and aluminosilicate glasses in the systems Na_2O-Al_2O_3-SiO_2-Fe-O, CaO-Al_2O_3-SiO_2-Fe-O, and MgO-Al_2O_3-SiO_2-Fe-O.[318] For oxidised glasses with the ratio ferric iron/total iron < 0.5, the isomer shift for iron(III) lay in the range 0.22-0.33 mm s^{-1} at 298K, which corresponds to tetrahedrally coordinated iron(III). The quadrupole splittings of iron(III) in the sodium, calcium, and magnesium aluminosilicate glasses were 0.90, 1.3, and 1.55 mm s^{-1} respectively. For Fe(II) the Moessbauer parameters obtained were I.S. = 0.9-1.06 mm s^{-1}, Q.S. = 1.75-5.10 mm s^{-1} at 298K. The temperature dependences of these parameters suggested non-interacting ions, but the isomer shift values are intermediate between those normally attributed to tetrahedral and octahedral iron(II). An assignment of these shifts to Fe^{2+} in octahedral coordination was in agreement with other spectral studies. An amorphous region has been reported in the phase system Fe-Ca-Si-O and the magnetic properties reported[319]. The state of iron in some calcium magnesium silicate glasses has been investigated[320]. Initially both Fe(III) and Fe(II) were octahedrally coordinated, but after heat treatment the Fe(III) became tetrahedrally coordinated and the Fe(II) was found in a highly distorted octahedral environment.

Moessbauer spectra have been reported for iron-containing lithium borate[321,322] and borosilicate[323,324] glasses. The Moessbauer spectrum of a lithium borate glass with a 47.17% Li_2O content was characteristic of crystalline lithium ferrite[321]. It was found that when >2 mol.% of B_2O_3 was replaced with Fe_2O_3

most of the iron was present as Fe(III) in octahedral sites[323]. A correlated ESR and Moessbauer spectroscopic study of the glass system $2CaO$-$3B_2O_3$-$Al_2O_3$$x$$Fe_2O_3$ ($0.1 \leqslant x \leqslant 1$) has been reported[325]. When $x > 0.1$ iron(III) was found to occupy non-tetrahedral sites. An investigation of barium oxide-boron oxide-ferric oxide systems has been described[326]. Ferric iron was found in both four- and six-coordinate sites, and a value for the solubility limit of Fe_2O_3 was obtained. Mixed barium borate-barium vanadate glass systems have been studied[327] and it has been shown that iron in zinc borate glasses is present as iron(III) in tetrahedral sites[328].

The Moessbauer effect has been used to study iron-tellurite[329], strontium-iron tellurite[330], sodium-iron vanadate[331], and iron phosphate[332] glasses. Natural (pumice and obsidian) and synthetic glasses of granitic composition have been examined using the Moessbauer effect[333]. Iron(III) was found in tetrahedral sites when sufficient univalent cations (M^+) were available to balance the charges in both the $M^+Fe^{3+}O_2$ and the M^+AlO_2 complexes. In other phases, the distribution of iron between the tetrahedral and octahedral sites depended on the ratio of monovalent to divalent cations. The degree of polymerisation of these units was reflected in the quadrupole splittings. Increasing pressure reduced the geometric difference between the types of Fe^{2+}-O octahedra and produced a more even distribution of Fe^{2+} among these sites and therefore an ordering in the melts.

Minerals. A Libyan desert silicate glass has been examined with various techniques, including Moessbauer spectroscopy, and it was identified as a type of tektite formed by a meteoritic impact[334]. Moessbauer spectroscopy has been used to examine the valence and coordination of iron in other tektites and impactites[335] and in meteoritic hibonites[336]. Moessbauer spectroscopy has been used to discriminate pyroclastic deposits on the basis of their magnetite content[337]. Room-temperature spectra from seven types of volcanic ash were able to be differentiated by visual inspection. The transformation of magnetite to haematite via maghemite in Yakutian kimberlites has been studied[338].

Two papers have been published on chrysotile asbestos[339,340].

Several papers have appeared on sheet silicate minerals[341-344], including phlogopite[341,342], biotite[341,343], and 2:1:1 sheet silicates[344]. The Moessbauer spectra of laihunite, a mixed-valence iron olivine, have been studied[345], and it was identified as a component of the ferrifayalite type of mineral.[346] Synthetic fayalite has received attention[347], and Moessbauer data on several ferrispinels have been published[348]. Moessbauer studies on the calcium-containing minerals melilite[349], hedenbergite[350], and perovskite[351] were published. The nature of the iron sites and the effect of interlayer cations on the clay mineral nontronite have been studied[352,353]. Glauconites from Belgium and northern France have been investigated[354] and the stability of Ukrainian glaucanites to gamma-radiation studied[355]. Kaolins from France[356] and the Ukraine[357] have also been examined. Other minerals which have received attention are danalite[358], chromium spinellids[359], iron-nickel thiospinels[360], ilmenite[361], ankerite[362], natural and synthetic wolframites[363], and metavivianite[364].

Chalcogenides. A single-crystal Moessbauer effect study of naturally occurring FeS_2 has been made using linearly polarised gamma-rays.[365] Four equivalent, but differently orientated sites were examined, and it was found that: (i) the experimental peak area ratio was near unity in both polycrystalline and single-crystal absorbers, (ii) the Moessbauer *f*-fraction was isotropic, and (iii) the asymmetry parameter was zero. The distribution of iron in the lamellar compounds Fe_xZrX_2 (X = S, Se) has been studied[366]. Iron was distributed between both octahedral and tetrahedral voids of the van der Waal's gap in the host lattice with the octahedral sites preferred at low values of *x* and the iron predominantly occupying tetrahedral sites at high values of *x*. Moessbauer spectra of the chalcogenide spinels $Fe_{1-x}Co_xCrS_4$ have been studied for x = 0-0.8.[367] At 296K the isomer shift of the paramagnetic singlet decreased with increasing *x*. At 97K the compounds were ferromagnetic and, although the isomer shifts behaved in the same way, both quadrupole splitting and linewidth increased with *x*. These changes were explained by covalency effects and the contribution of indirect hyperfine interactions to the effective magnetic field at the iron nuclei.

Applications of Iron-57 Moessbauer Spectroscopy.— Catalysts. Highly dispersed iron particles consisting of 5 wt.% Fe on carbon black have been studied by in situ gamma-resonance spectroscopy at 295, 77, and 4K[368]. After hydrogen reduction at 740K 75-80% of the iron was reduced to alpha-iron with the remainder present as Fe^{2+}. The alpha-iron particles in this Fe/C-1 catalyst showed three effects characteristic of very small particles, viz. (i) large contributions to the spectra from superparamagnetic iron at 295 and 77K, (ii) the magnetic hyperfine field at the alpha-iron was enhanced by about 7 kOe at 4K, due to the influence of the demagnetising field on the particles, and (iii) the effective Debye temperatures of the catalysts were much smaller than those of bulk iron compounds. Carburization in syngas converted all the alpha-iron to iron carbides, while the Fe^{2+} was unaffected. Later analysis of the magnetic field dependence of the spectra obtained a value of 2.5 $\pm$ 0.2 nm for the average size of the particles[369]. Commercial, graphitised, and boron-doped carbon blacks used to support iron have been studied by in situ magnetisation and Moessbauer methods[370]. All these catalysts contained superparamagnetic iron with a particle size of 3.5-5.5 nm. The presence of a strong iron-carbon interaction could not, however, be confirmed. A modified model of relaxation for systems containing both a magnetic field and an electric field gradient has been used to explain the Moessbauer spectra of a small iron chi-carbide supported on high-surface-area carbon[371]. It was concluded that chi-carbide particles interact weakly with graphitic carbon, but very strongly with amorphous carbon. The formation of supported iron catalysts by decomposition of FePc[372] and the nature of iron oxide particles supported on Grafoil[373] have been studied.

The calcination and reduction of alpha-Fe_2O_3 supported on SiO_2, Al_2O_3, MgO, and carbon have been studied with the Moessbauer effect[374]. Reduction of the oxide to iron was increased by addition of nickel and zinc, but not possible on SiO_2 or MgO. Local temperatures of silica-supported Fischer-Tropsch catalysts have been studied by Moessbauer spectroscopy.[375] Temperatures were measured by monitoring the internal magnetic field of the iron carbide crystallites under reaction conditions. Rises of 13 and 19° respectively were observed at nominal bulk temperatures of 275 and 300°C when the gas flow to the catalyst was changed from helium to the syngas reaction mixture. *In situ* Moessbauer spectroscopy has been used to study the transformations of Fe/Al_2O_3 and Fe/ThO_2 catalysts under different stages of the calcination, reduction, carburisation, and Fischer-Tropsch synthesis reactions[376] and also structures of silica-supported FeM (M = Ru, Rh, Pd, Ir, Pt) catalysts[377]. Silica-supported iron nitride Fischer-Tropsch catalysts have also been studied.[378]

Iron-57 Moessbauer spectroscopy has been used to study the promoting effect of metals on the catalytic activity of clays in hydrometallisation processes[379]. Higher activity was associated with a higher Fe^{3+} concentration. Other catalytic processes which have been examined were the catalytic oxidation of phenols by Fe/Al_2O_3 or $Fe\text{-}Cu/Al_2O_3$[380], the oxidation of anthracene to anthraquinone[381], the photocatalytic conversion of nitrogen to ammonia by Fe_2O_3/TiO_2[382], and the photoassisted isomerisation of pent-1-ene by $Fe(CO)_5$ in zeolite[383].

Coal, Soils, and Sediments. A number of techniques have been used to investigate the low-temperature oxidation of a high-volatility and a low-volatility bituminous coal[384]. Moessbauer spectroscopy was the second most sensitive technique after Geiseler fluidity. A Moessbauer effect method has been designed for the measurement of the flow rate and metal contents of coal liquors containing ash with a known iron concentration. Two Chinese studies have used Moessbauer methods to determine the iron-bearing phases present in coal ash[386] and in coal and oil shale[387].

Moessbauer spectroscopy of soils from four loess profiles from Hesse has shown that the ratio of iron oxides in structural iron increases with increasing age[388]. This method was, however, ineffective where the age difference between samples was less than 100,000 years, and there were deviations observed due to topographical position. The chemical forms of iron in some Bulgarian soils[389] and in Danish Mesozoic sediments[390] have been studied. A possible relationship between the presence of ankerite and the formation of oil and gas in the sediments was identified[390]. The composition and origin of smectite from the metalliferous sediments of the Bauer Deep in the Pacific have been studied[391].

Cements and Ceramics. Addition of sodium carbonate to portland cement clinker changed the parameters of the Fe^{2+} and tetrahedral Fe^{3+} sites, while the octahedral Fe^{3+} disappeared[392]. The valence state and coordination of iron in high-iron cement clinkers have been studied[393]. Only ferric iron was found and had distorted octahedral and tetrahedral coordination. Moessbauer spectroscopy has been used to study corrosion reactions in arc-furnace refractories and to estimate the distribution of iron among the different components[394]. Colour effects in firing clay-power plant ash ceramic tiles have been investigated[395]. These colours were related to the local burning of residual ash and phase transformations of the iron oxides which initially made up 2.63% of the mixture. Iberian pottery from the 4th century B.C.[396] and Chinese ceramic glazes[397] have also been studied.

Other Applications. Billets formed by reduction of haematite ore in a rotary furnace have been studied[398] and silica-saturated lead-iron-oxygen-silica slags investigated[399]. Application of the Moessbauer effect to the non-destructive control of tool steel heat treatment has been considered[400], and the uses of gamma-resonance spectroscopy in the Egyptian and Qatari steel industries have been described[401]. Rust samples from mild-steel panels exposed to the atmosphere at 8 sites have been studied.[402] All the spectra showed superparamagnetic features at room temperature. The corrosion products were, therefore, identified from their Moessbauer spectra at 10K and by X-ray powder diffraction. Over 70% of all samples consisted of alpha-FeOOH and gamma-FeOOH, but only marine samples contained alpha-FeOOH and $Fe_{3-x}O_4$ (x = 0-0.33). The proportion of alpha-FeOOH increased with the SO_2 concentration at the test sites. Iron aerosols collected in remote, urban, and industrial areas have been examined[403]. Samples collected in remote areas gave spectra similar to those of clay minerals, while haematite and magnetite were predominant in the spectra of samples collected around industrial areas. Moessbauer spectroscopy has been used to study fossil bones from Egypt[404] and to evaluate the orientation of particles in magnetic recording tapes[405].

5 TIN

General Topics.— The change in the nuclear charge radius in going to the ground state of Sn-119 from its 23.87 keV excited state has been measured by internal-conversion methods[406]. The method used involved implanting radioactive Sb-119 into the host matrices $CaSnO_3$, beta-Sn, Pt, Y, Au, and Pb and measuring both the Moessbauer emission spectra and the conversion-electron spectra. The value of $\Delta R/R$ obtained from the correlation between the Moessbauer shift data and the intensity ratios of the Q-shell to N_1-shell conversion electrons was $0.87 \pm 0.25 \times 10^{-4}$ for a uniform charge distribution of $R = 1.2 \times A^{1/3}$ fm. The contact density and the number of 5s-electrons on beta-Sn have been calculated as 72.3 ± 4.1 a.u. and 1.87 ± 0.13 respectively from the Moessbauer shift data[407]. Dirac-Slater self-consistent field calculations have also been carried out on some tin ions, and the results have been used in a simple model to evaluate effective valence electron populations and hence the chemical-isomer shifts[408].

Isomer-shift and linewidth measurements were carried out on tin-foil and annealed tin-foil samples[409]. The effect of pressure on the parameters of the foil causes only a small overall change in the shift but a substantial broadening of the resonance line by, for example, 20% at 3.09×10^6 N m^{-2} and by 29% at 4.63×10^6 N m^{-2}. A theoretical model based on classical dispersion theory has been developed to deal with the problem of the influence of the hyperfine structure on time-dependent Moessbauer spectra. An experiment using a $BaSnO_3$ source and an SnO_2 absorber was used to provide experimental data for comparison with the suggestions of the model[410].

A number of papers published during the review year dealt with the use of the Sn-119 Moessbauer effect in studies of the applications of tin-containing materials[411-420]. The technique has been used[411] to study the changes in three organotin stabilisers in a PVC matrix irradiated with up to 20 Mrad of gamma radiation. The final product of prolonged irradiation is always $SnCl_4$ with the rate of degradation being dependent both on the stabiliser and on the extent of thermomechanical damage to the polymer on processing. Moessbauer spectroscopy was also used to determine the final oxidation state of tin in Ir-Sn-alumina catalysts after hydrogen activation[412]. The results showed that, while SnO_2 itself is reduced to Sn at 673K on treatment with hydrogen, an SnO_2-Al_2O_3 mixture is reduced only to tin(II) and that the presence of In in the catalyst promotes reduction of Sn(IV) to Sn at lower temperatures with the formation of a Sn-In alloy.

The products of the reaction of $SnCl_4$ with soda-lime glass have been studied by gamma-resonance spectroscopy[413] as have the effects of the presence of Sn(II) on the mechanical properties of sodium silicate glasses[418]. Two papers published during the year were concerned with studies on the use of tin phosphates for silk weighting[414,415]. The areas under the Moessbauer resonance peaks for weighted silk samples were found to reach equilibrium after two treatments with tin(IV) chloride-disodium phosphate mixtures. A method of controlling the weighting process using Moessbauer parameters has been devised as a result of these studies.

Moessbauer effect studies were used to determine the chemical changes in the tin species introduced during hydroliquefaction of Victorian brown coal. The tin was found to be present in the coal as hydrous tin(IV) oxide, which is dehydrated at hydroliquefaction temperatures and reduced by H_2 to SnO and beta-Sn[416].

Moessbauer spectroscopic results have shown that organotin toxicants added to neoprene-based elastomeric marine antifouling coatings undergo considerable structural changes on incorporation into the elastomer. The spectra obtained from elastomers containing 2.6 weight % of toxicant such as the tributyltin oxide stearate and carbonate derivatives are the same within experimental error, having shifts of 1.41 to 1.44 mm s^{-1} and splittings of 2.78-2.84 mm s^{-1}. These data can be compared with shifts of 1.25 and 1.45 mm s^{-1} and splittings of 1.53 and 3.65 mm s^{-1} for the oxide and stearate respectively. The authors suggest that the toxicants react with the elastomer to form tributyltin chloride even though this compound has Moessbauer parameters (shift = 1.52 mm s^{-1}, splitting = 3.42 mm s^{-1}) that are not consistent with the data for the toxicants in the elastomer[417].

The Moessbauer emission spectra of tin in suspension in ethyl alcohol have been used to study the electron-scavenging effects of the addition of various solutes[419] and the mechanism of the sorption of chloroplatinum(II) complexes into tin(II) oxides determined[420]. The results suggest that the sorption of the Pt complexes is not simple and that it involves ion exchange, oxidation-reduction reactions, and complexing.

The Sn-119 gamma-resonance spectra have been recorded for the cluster Zintl anions $[Sn_5]^{2-}$, $[Sn_9]^{4-}$, $[TlSn_8]^{5-}$ and $[Sn_2Bi_2]^{2-}$, $[SnSe_4]^{4-}$, $[SnTe_4]^{4-}$, and $[SnSe_{3-\underline{x}}Te_{\underline{x}}]_{\underline{z}}^{2\underline{z}-}$ and the data interpreted in terms of known crystal structures[421]. The Moessbauer parameters for solid derivatives of these ions are in Table 1. The derivatives of $[Sn_5]^{2-}$ and $[Sn_9]^{4-}$ and $[TlSn_8]^{5-}$ give spectra with overlapping single-line spectra, while that of $[Sn_2Bi_2]^{2-}$ is a quadrupole doublet with shift and splitting parameters of 2.75 and 1.07 mm s^{-1} respectively. The data for $[SnSl_{3-\underline{x}}Te_{\underline{x}}]_{\underline{z}}^{2\underline{z}-}$ ($\underline{x}$ = 0-3, $\underline{z}$ = 1 or 2) are consistent with distorted tetrahedral tin environments.

Table 1 Moessbauer data for cluster Zintl phases

Zintl compound	Temp./ K	Shift/ mm s^{-1}	Splitting/ mm s^{-1}	Line width/ mm s^{-1}	% of form
$[Na(crypt)]_2[Sn_5]$	77	2.32	...	0.98	79.1
		3.00	...	0.98	20.9
	137	2.31	...	0.89	83.3
		2.85	...	0.89	16.7
	190	2.31	...	0.89	84.5
		2.79	...	0.89	15.5
$[Na(crypt)]_4[Sn_9]$	4.2	2.17	...	1.90	...
	77	2.25	...	0.80	41.6
		2.67	...	0.80	16.8
		2.98	...	0.80	41.6
	165	2.33	...	0.81	50.0
		2.92	...	0.81	50.0
Na_2TlSn_8	77	2.19	...	1.35	35.8
		2.63	...	1.35	28.4
		3.05	...	1.35	35.8
$Na_5(cn)_2TlSn_8$	77	2.17	...	1.00	43.5
		2.67	...	1.00	13.0
		3.02	...	1.00	43.5
$[K(crypt)]_2[Sn_2Bi_2]$.en	77	2.75	1.07	1.06	...
$[Na(crypt)]_4[SnSe_4]$	77	1.47	0.62	0.95	...
$[Na(crypt)]_4[SnTe_4]$	77	1.76	0.63	1.03	...

Moessbauer studies have been carried out on Sn-119 impurity atoms in a number of systems. The effects of radiation damage in Nb_3Sn, SnS_2, and Sn films caused by Sn-119 have been considered in a review[422], and data on Sn-119 in As_2S_3 films have shown that irreversible changes occur in the film during photoirradiation and annealing processes[423]. The hyperfine interactions at Sn sites in samples of Sn-doped In_2O_3 containing 0.025, 1, and 5 at.% Sn have been measured. The Sn enters the lattice as Sn(IV) to give a shift of 0.22 mm s^{-1} and a splitting of 0.60 mm s^{-1}.[425] The effects of the implantation of $[Sn\text{-}119]^+$ ions and of the radioactive ions $[In\text{-}119]^+$, $[Sn\text{-}119m]^+$, $[Sb\text{-}119]^+$, and $[Te\text{-}119m]^+$ into silicon carbide have been investigated by both Moessbauer emission and conversion-electron spectroscopy[425]. Gamma-resonance emission and perturbed angular correlation (PAC) techniques were combined in a study of impurity defects in ion-implanted gallium arsenide and phosphide lattices. Moessbauer data suggested that the prominent defects in room-temperature implants are impurity-vacancy complexes, and PAC data confirmed that the line-shape of the Moessbauer resonance line should be interpreted in terms of an unresolved quadrupole splitting[426]. The spectra of Sn-119 nuclei doped into the antimonides FeSb, CoSb, and NiSb have also been obtained and those for the FeSb system interpreted in terms of the presence of Sn in the trigonal-bipyramidal sites of an NiAs-type lattice[427].

Radioactive In-119 ions decaying to the 23.9 keV Moessbauer isotope have been used to study impurity-vacancy complexes in In-119-doped iron metal. Evidence was found for impurities in substitutional sites and for two impurity defect sites with magnetic fields at the tin nuclei of 10 and 20% respectively greater than that at the substitutional Sn atoms[428]. The hyperfine field at Sn-119 impurity atoms in GdMnSi was found to be due to short-range interactions and to have a value of -27.6 T at 5K.[429] Moessbauer spectroscopy has also been used to study the magnetic hyperfine field at Sn-119 atoms substituted at Al sites in $NdAl_2$[430].

Moessbauer spectroscopy was used to study the behaviour of Co-57 and Sn-119 impurities in the magnetic superconductor phase Y_9Co_7. The shape and asymmetry of the quadrupole spectra were interpreted in terms of the three non-equivalent Co sites in the structure[431]. Both Fe-57 and Sn-119 isotopes were also used to study constitutional and thermally created defects in beta-PdIn, but the experiments with Sn-119 only gave an integral measure of the total amount of defects[432]. The hyperfine interactions of Sn-119 impurity atoms replacing Al in the ferromagnetic RAl_3 phases (R = Pr, Nd, Sn, Tb, Ho, and Dy) were investigated by gamma-resonance spectroscopy at 5K[432]. A considerable difference was found between the dependences of the hyperfine fields in the spin of the rare-earth atoms present between R = Pr, Nd, and Sn and R = Tb, Ho, and Dy. This difference was interpreted in terms of large orbital contributions to the hyperfine fields at Sn impurities with, for the heavier R atoms (R = Tb, Ho, and Dy), an additional spin contribution.

The Moessbauer spectra have been measured for an Au-2 atom % Sn alloy annealed in air at temperatures between 673 and 1073 K. SnO_2 was identified as the only internally oxidised phase, and the process of oxidation was interpreted on the basis of changes occurring in the Moessbauer spectra. Oxide formation occurred at the surfaces, and there was no preferential oxidation along the Grace boundary. When the alloys were annealed in hydrogen at 1123K the oxide was reduced to Sn, which then dissolved in the Au matrix.

A similar study of internal oxidation has been carried out on Ag-Sn alloys over the temperature range 200-850°. The product formed at temperatures above 500° consists of agglomerates of SnO_2, while that found below 300° gave an isomer shift of 0.29 mm s^{-1} from $CaSnO_3$ and a splitting of 0.32 mm s^{-1}. It was suggested that the lower-temperature product could be in the form of Sn atom-oxide complexes in the Ag matrix. The effect of cold working on the parameters of the oxidised Sn was also studied, and it was found that the as-rolled samples have the same Moessbauer parameters as annealed samples but at lower temperatures[435]. The effects of particle sizes on the oxidation of the Ag-Sn alloys were also studied[436]. The value of the shift was used as an indication of the particle size of the oxides formed. The oxidation was found to be more rapid than that predicted theoretically, but the deviations from the theoretical predictions did decrease with decreasing oxide particle size.

When Nb_3Sn samples are irradiated with neutrons the lattice parameters increase, and this has been shown to be reflected in a change in Moessbauer shift[437]. Gamma-resonance studies of $FeSn_2$ obtained below the Néel temperature of 378K show that a second magnetic transition occurs at 93K[438].

Sn-119 Moessbauer studies on alloy phases described during the review year include those on splat-cooled Sn-Ge-Cu alloy,[439] on the Heusler alloys Pd_2TbSn,[440] Pd_2DySn,[440] and Pd_2MSn (M = Tb-Yb),[441] on $LaRh_{1.1}Sn_4$,[442] on $LaNi_{4.7}Sn_{0.3}$,[443] on Cu-Zn-Sn alloys,[444] and on Sn-containing Al-Mg-Si alloys.[445] A comparison of the data for $LaNi_{4.7}Sn_{0.3}$ and its hydride showed that there is no interaction between Sn and H in this system.[443]

Four papers published in the last year dealt with the uses of the Moessbauer effect in studies of electrodeposited tin[445-449]. <u>In situ</u> gamma-resonance studies were used to characterise the nature of the passivation film formed in a tin borate solution at -1180 mV against a standard calomel electrode[446]. The film at more negative potentials (-1180 to -780 mV) is a mixture of hydrous tin(II) and tin(IV) oxides, while at less negative potentials it consists only of hydrous tin(IV) oxide. Moessbauer spectroscopy has been used to study the interface formation between electrodeposited Sn and the substrate metals Ag[447], Pt[447], Ni[448], and Al[448]. The Co-Sn alloys deposited from an alkaline sulphate solution were studied using both the Sn-119 absorption and Co-57 emission effects[449]. The data show that Sn dissolved in Co and CoSn are the components obtained from a fresh plating solution but, as the solution becomes exhausted, gamma-Co_3Sn_2 is formed. During the electroplating process a precipitate is formed in the electrolyte which is richer in Sn(II), in contrast to the electrolyte solution which is rich in Sn(IV).

<u>Inorganic Tin(II) Compounds</u>.—Moessbauer spectroscopy was used to study the disproportionation reaction of SnO_2 at temperatures between 200 and 75° and over time spans of 0.5 to 100 hours.[450] The reaction starts at about 250°C giving Sn and Sn_3O_4 as products. In the range 250-425° the authors claim that the

products are Sn and Sn_3O_4 but that above 425° the Sn_3O_4 decomposes slowly to give Sn and SnO_2 until at over 550° the products are the thermodynamically stable Sn and SnO_2.

The polarographic oxidation and reduction behaviour of $SnSO_4$ solutions was studied in the presence of various diphosphonic acids. Sn-119 data obtained on a 1-hydroxyethane-1,1-diphosphonic acid-Sn sample oxidised with various quantities of pertechnetate ions confirmed that the proces involved a 4-electron transfer[451]. The formation of a Sn(II) 1-hydroxyethylidine-1,1-diphosphonic acid (EHDP) complex from a 0.01M solution of $SnCl_2$ in a 0.1M solution of the acid has also been proved by the Sn-119 Moessbauer effect.[452,453] In the absence of EHDP the spectrum of a frozen aqueous solution of $SnCl_2$ has shift = 3.55 mm s^{-1} and splitting = 1.12 mm s^{-1}. The spectrum of fresh solutions containing EHDP has shift = 3.30 mm s^{-1} and splitting = 1.94 mm s^{-1}. When the solutions are allowed to stand, a singlet due to a tin(IV) oxidation product also appears in the spectrum. Potassium [hydrogen(bis-maleato)stannate(II)] is triclinic and its structure consists of layers of discrete $[Sn(CHCOO:CHCOO)(CHCOO:CHCOOH)]_2^{2-}$ moieties with K ions located in holes between the layers. In the discrete unit are two different maleate groups, an anisobidentate maleate ligand and a terminal unidentate monoprotomaleate ligand. The Sn atom lies in a trigonal-pyramidal environment with three Sn-O distances, to the maleate and monoprotomaleate ligands, of 219.9 and 221.2 pm respectively. These bonds lengths are consistent with the Sn-119 Moessbauer data. The preference for tin to exist in a trigonal-pyramidal environment, even in the presence of potential chelating ligands[454], is clearly shown in the structure of this complex Sn(II) maleate.

X-ray diffraction and Moessbauer effect studies have been used to show that Sn_3S_4 and Sn_2S_3 can be prepared by mechano-chemical methods.[455] The effect of the presence of an inert additive in the suppression of explosive character in the synthesis of the sulphides mechanochemically was also studied. The initiation of a high-temperature reaction by the heat produced in the mechanochemical explosion reaction was also observed for the first time. The Moessbauer parameters for Sn in $Pb_{1-x}Sn_xS$ and

$Pb_{1-\underline{x}}Sn_{\underline{x}}Se$ solid solutions show[456] that tin atoms are donor centres with charge states determined by acceptor impurities in the solid solutions. The two states of the Sn donor atoms in the solid solutions were identified as Sn^{2+} and Sn^{4+} and the two-electron exchange between them was observed in $Pb_{1-\underline{x}}Sn_{\underline{x}}Se$ solid solution doped with Na and Tl acceptor impurities. The phase transitions in SnTe have been studied by Moessbauer spectroscopy. Two transitions were identified, viz a structural ferroelectric transition at 90K and an unidentified first-order transition at 140 K.[457]

Birchall et al.[458] obtained the Sn-119 Moessbauer parameters of a tin(II) fluoride sample sealed under vacuum in a Teflon container at temperatures of 298, 378, 443, and 473K. They found no significant changes in the parameters over the temperature range studied other than the expected steady fall in the recoil-free fraction as the temperature increases. In contrast to their earlier work, they found no discontinuities in the graphs of shift splitting and recoil-free fraction. When the experiment was repeated in air, some hydrolysis of the SnF_2 took place to give a product with isomer-shift and quadrupole-splitting parameters of 3.04 mm s^{-1} from $CaSnO_3$ and 2.20 mm s^{-1} respectively. They now claim that the parameters previously assigned to beta, gamma, and liquid tin(II) fluoride are erroneous and that the spectra from which the parameters were obtained were due to the hydrolysis product. Tin(II) fluoride reacts with AsF_5 and SbF_5 in a 2:1 ratio to give $Sn_2F_3[AsF_6]$ and $Sn_2F_3[SbF_6]$, and reaction of $Sn_2F.[MF_6]$ (M = As or Sb) with SnF_2 in liquid SO_2 also produces $Sn_2F_3[MF_6]$. The Sn-119 Moessbauer data for the $Sn_2F_3[MF_6]$ compounds were reported and compared with those for SnF_2, $SnF[MF_6]$, and $Sn(SbF_6)_2$.

Sn-119 Moessbauer spectroscopy was used to study the state of Sn atoms in the borosilicate matrix of a porous glass after densification at 1173K. The porous glass was obtained by leaching a DV-1 glass in 3M HCL and was impregnated with a saturated solution of $SnCl_2$ in HCL at 298K. The formation of chemical bonds between Sn atoms and monomeric SiO_2 in the volume of pores was proved, but there was no evidence for the formation of intermediate SnO and SnO_2 during thermal treatment[460].

Stannane reacts in fluorosulphonic acid at -90°C to produce SnH_3^+ which was characterised by H-1, C-13, and Sn-119 n.m.r. Hydrogen evolution studies along with Sn-119 n.m.r. and Sn-119 Moessbauer spectroscopy were used to show that SnH_3^+ decomposes at higher temperatures to give a Sn(II) species with Moessbauer shift and splitting parameters (4.33 and 0.58 mm s^{-1} at 77K) that are similar to those of $Sn(SO_3F)_2$ (4.18 and 0.68 mm s^{-1}). The authors suggested that the most likely course of the reaction was the two-stage process:[461]

$$SnH_4 + HSO_3F \longrightarrow SnH_3^+ + SO_3F^- + H_2$$

$$SnH_3^+ + 5HSO_3F \longrightarrow Sn^{2+} + 2H_2SO_4 + 2SO_2 + 4HF + SO_3F^-$$

Evidence for the formation of H_2 in the first stage was provided by proton n.m.r. and, although there was no direct evidence for the presence of SO_2 in the second stage, the pale blue colour of the solutions was taken to be indicative of polycations of sulphur arising from the reaction of SO_2 and HSO_3F in the presence of H_2.

A variable-temperature gamma-resonance study of the Chevrel phase $SnMo_6S_8$ showed that there was a sudden change in the shift, the Debye-Waller factor, and the ratio of intensities of the components of the quadrupole splitting at about 80K. Since X-ray diffraction data do not show any phase changes at this temperature the anomaly has been attributed to an electronic rearrangement at the tin atom[462]. Electronic structure calculations have also been carried out for clusters representing the $SnMo_6S_8$ lattice with and without oxygen doping. The effects of the presence of oxygen on the Moessbauer parameters were determined from theoretical considerations and the results compared with experimental parameters[463]. The shift does not change much when a sulphur atom in the tin environment is substituted with an oxygen, but there is a significant change in splitting because of the sensitivity of the e.f.g. to the Sn-O distance. An increase in the Sn-O bond distance of 10 pm reduces the splitting by about 20%.

The following tin-containing materials with Moessbauer resonance lines in the Sn(II) region of the spectrum were mentioned earlier in this chapter: oxide catalysts[412], glasses[413], hydrous oxide passivation films[446], and electroplating sludges[449].

Inorganic Tin(IV) Compounds.— Gamma-resonance spectroscopy has been used to study the formation of the tin(IV) oxide ore cassiterite in several tin-mining areas[464]. The relationships between the values of the quadrupole splitting parameters and the unit cell volumes of minerals and between the intensity ratios of the quadrupole split peaks and the unit-cell dimensions were determined. The authors concluded that both the presence of trace elements in the cassiterite lattice and their concentrations have an influence on both the unit cell and Moessbauer parameters of the minerals.

Both Sn-119 and Fe-57 spectra have been used to study the oxide phases $BaFe_4Sn_2O_{11}$[302] and $LiFeSnO_4$.[305] Variable-temperature data on $BaFe_4Sn_2O_{11}$ show that magnetic ordering starts at 77K and is complete by 4.2K and that a magnetic hyperfine field of 45 kG is transferred to the Sn atoms. Sn-119 data on the high- and low-temperature polymorphs of $LiFeSnO_4$ show that there are subtle differences in the Sn-O bonding and in the distortion of the SnO_6 polyhedra in the two forms. Moessbauer spectra have been used to obtain information on the structures of the lilac ceramic pigments obtained by sintering of 0.5 and 5% $K_2Cr_2O_3$ with SnO_2. The presence of a weak quadrupole splitting in the spectra was taken to suggest the incorporation of Cr^{3+} in the SnO_2 lattice[465]. X-ray diffraction, electron spin resonance spectra, and Sn-119 Moessbauer spectra have been used to characterise tin-molybdenum oxides prepared by calcination of the precipitates obtained from mixed Sn-Mo solutions[466]. The calcination process involves the initial dehydration of amorphous gels to give, at low temperatures, highly disordered rutile-type structures capable of accommodating high concentrations of Mo. Under moderate thermal treatment the Mo segregates from the growing rutile phase to produce MoO_3, which volatilises off at higher temperatures. Prolonged calcination at high temperatures gives products that can be described as solid solutions containing low levels of Mo(V) in a rutile SnO_2 lattice. These processes are reflected in the Sn-119 Moessbauer data obtained in the heat-treated products. The spectra of products obtained at 300-400°C and those calcined at higher temperatures are best fitted to two lines characteristic of quadrupole split Sn(IV) resonance lines. The splittings for the poorly crystalline solids obtained at 300°C had values in the range 0.54-0.61 mm s^{-1} and were independent of the nominal Mo contents, while the splittings for samples calcined at higher temperatures were similar to those obtained for samples of SnO_2 prepared under identical conditions.

The Moessbauer shifts for the singlet resonance lines of $SnTe_3O_8$ and $Sn(SeO_3)_2$ have values of about -0.07 mm s^{-1}, which means that the Sn atoms must be present in these structures in an octahedral coordination similar to that found in SnO_2 but with slightly longer Sn-O bonds. The spectrum of the basic selenite $2SnO_2.3Sn(SeO_3)_2$ is also a singlet with a shift of +0.03 mm s^{-1}, but this compound has a much lower recoil-free fraction than both $SnTe_3O_8$ and $Sn(SeO_3)_2$.[467]

Tin(IV) chloride solutions trapped in a porous glass were studied by Sn-119 Moessbauer spectroscopy at ambient temperatures and at 80K. The data suggested that for aqueous acidic solution the $SnCl_4$ was not in contact with the glass surface[468]. A Moessbauer study has also been carried out on polyacetylene doped with seleted metal halides including $SnCl_4$. A lattice temperature of 95K was obtained for $[CH(SnCl_5)_{\underline{y}}]_{\underline{x}}$ from the temperature dependence of the recoil-free fraction of the tin chloride-doped polymer[108]. The authors also reported that the ternary compound $[CH(SnCl_5)_{\underline{y}}(FeCl_4)_{\underline{x}}]_{\underline{x}}$ could be obtained from polyacetylene doped with $SnCl_4$ and $FeCl_3$ but that the Moessbauer parameters are not significantly different from those of $[CH(SnCl_5)_{\underline{y}}]_{\underline{x}}$. Polypyrrole can be reduced in thf solution and then reoxidised with $SnCl_4$. The only Sn-containing species in the reaction products was identified from its Moessbauer spectrum as a tin(IV) chloride complex[109]. The Moessbauer data for the complexes $SnX_4.2L$ (X = Cl, Br, or I, L = $C_6H_5NH_2$ or $(CH_3)_2SO$) have been used along with tin X-ray emission spectroscopic results to study the ways in which tin uses its bonding electrons in complex formation[469].

Moessbauer spectra have been reported[470] for the cis-trans isomers of $SnCl_4L_2$ [where L = N,N-dimethylacetamide (dma), N,N-dimethylformamide (dmf), dimethylsulphoxide (dmso), and tetrahydrothiophen (tht)] and for the cis-trans isomer of $SnBr_4(dmf)_2$. The spectra obtained are shown in Figure 2. There are clear differences in the spectra of the isomers of $SnBr_4(dmf)_2$, $SnCl_4(dmf)_2$, and $SnCl_4(dma)_2$ in that the spectra for the trans isomers show resolvable quadrupole splittings. The other spectra can be fitted to doublets, but the values (Table 2) of the quadrupole splittings in some cases are too small to be significant in view of the natural linewidth for the Sn-119 isotopes. The authors also show that there is a relationship between the quadrupole splitting values for a wide range of $SnCl_4.L_2$ complexes and the infrared Sn-Cl stretching frequencies. The decrease in value of the Sn-X stretching and an increase in the Sn-L stretching frequency in going from the cis to the trans isomer in the complexes with dmf, dmso, and dmf were taken to indicate differences in the bond strengths of the tin to the halide and ligand-donor atoms, which may, in part, account for the fact that the ratios of the quadrupole splittings for the cis and trans isomers in these complexes are below the theoretically predicted value of 2. The other factor that could be affecting the ratio would be a deviation for regular geometry in one or both of the isotopes of each pair.

Table 2 Moessbauer parameters of cis and trans complexes of tin(IV) halides

Compound	δ/mm s^{-1}	Δ/mm s^{-1}	$\Delta_{trans}/\Delta_{trans}$
trans-$SnCl_4(tht)_2$	0.72	0.35	1.5
cis-$SnCl_4(tht)_2$	0.70	0.24	
trans-$SnCl_4(dmso)_2$	0.41	0.57	1.4
cis-$SnCl_4(dmso)_2$	0.40	0.41	
trans-$SnCl_4(dmf)_2$	0.38	0.73	1.4
cis-$SnCl_4(dmf)_2$	0.39	0.53	
trans-$SnBr_4(dmf)_2$	0.66	0.83	1.9
cis-$SnBr_4(dmf)_2$	0.66	0.44	
trans-$SnCl_4(dma)_2$	0.38	0.78	1.7
cis-$SnCl_4(dma)_2$	0.38	0.45	

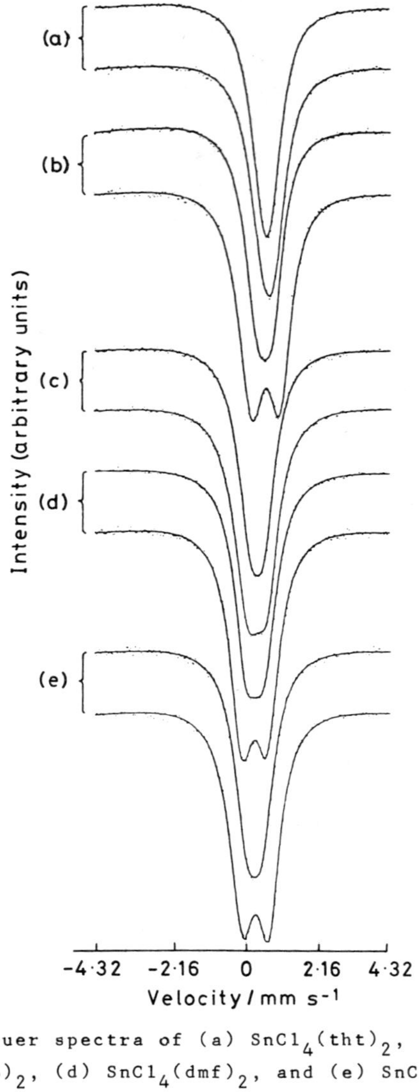

Figure 2 Moessbauer spectra of (a) $SnCl_4(tht)_2$, (b) $SnBr_4(dmf)_2$, (c) $SnCl_4(dmso)_2$, (d) $SnCl_4(dmf)_2$, and (e) $SnCl_4(dma)_2$. The upper spectrum of each pair corresponds to the cis isomer. (Reproduced by permission from J. Chem. Soc., Dalton Trans., 1985, 1281)

The interaction of $SnCl_4$ and $SnBr_4$ with a series of cyclic and aromatic ethers in solution and in the solid state was determined by a combination of Moessbauer spectroscopy and infrared and n.m.r. data. The data for nine $SnCl_4L_2$ complexes were found to be consistent with octahedral tin geometry and <u>trans</u>-coordinated ligands in the solid state[471]. Variable-temperature Moessbauer spectroscopy was used along with other techniques to study a range of new complexes of tin(IV) halides with 2,2'-azopyridine. All of the new complexes have 1:1 stoichiometry with tin in distorted octahedral sites with four bonds to halogen atoms, one to a ligand azo-group N atom, and one to a ligand ring nitrogen.[472]

The possible use of the Sn(IV) complex of potentially tridentate dianionic molecules (L) as extractants for tin was studied. As part of the study, compounds of the type SnL_2 and $SnCl_2L_2$ were prepared and characterised by Moessbauer spectroscopy. The best extraction was obtained with L = 2-(2'-hydroxyphenyl)-8-quinoliriol [473].

The tin(IV)-cyanide derivations $SnCl_2CN(Me_3SiCN)$ and solvated $SnCl_2XCN$ compounds (X = Br, I) have been prepared and characterised by Sn-119 Moessbauer spectroscopy[474]. The quadrupole splitting of $SnCl_3CN(Me_3SiCN)$ is essentially temperature independent in the range 77-291K, but the isomer shift decreases linearly with temperature at the rate of 3.04×10^{-4} mm s^{-1} K^{-1} because of second-order Doppler shifts. The slope of the $d(\ln \underline{A})/d\underline{T}$ graph is -1.0×10^{-2} K^{-1}, and this is in the range expected for a polymeric tin-containing structure.

Two papers[465,476] published during the review year dealt with compounds containing Sn-transition-metal bonds. A study of the temperature dependence of the recoil-free fractions in the range 77-300K was carried out on complexes of tin halides with platinum-group metals. It was suggested that the temperature dependence of the fraction depended upon the strength of the bonds to Sn and so the normal vibrations of the Sn atom vibrations could be calculated. The change in the temperature dependences between the complexes was explained in terms of an increase in the number of low-frequency vibrations in the order Br < Cl < F.[475] Sn-119

Moessbauer data were used[476] to confirm the presence of the stannandiyl moiety in the compound (3), which reacts with thf to give (4).

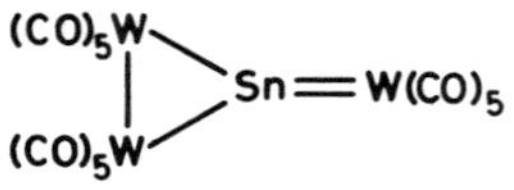

(3)

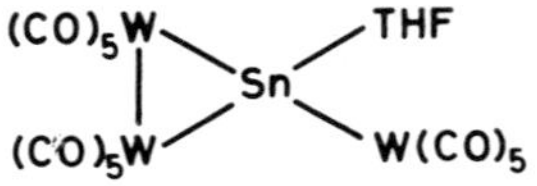

(4)

The following tin-containing materials with Moessbauer resonance lines in the Sn(IV) region of the spectrum were mentioned earlier in this chapter: oxide,[410] indium tin oxide,[424] chloride,[411] sulphide,[422] and tin in glass surfaces,[413] in hydroxide passivation film,[446] in electroplating sludges,[449] in alloy oxidation products,[434-436] and in substituted diphosphonic acid solutions.[453]

Organotin(IV) Compounds.— The Moessbauer data for trimethyl- and triethyl-organotin oxide compounds are listed in a recently published part of 'Gmelin Handbook of Inorganic Chemistry'[477].

The methylstannanes $SnMe_{4-n}H_n$ (n = 1-3) react with fluorosulphuric acid at -90°C to produce $SnMe_{3-n}H^+$ species. Sn-119 Moessbauer data were used to show that these species decomposed at higher temperatures[461]. The spectra for the decomposed sample from solutions of $SnMeH_2^+$ show a tin(II) resonance with shift and splitting data of 4.42 and 0.6 mm s^{-1} that are similar to $Sn(SO_3F)_2$ and a tin(IV) resonance with shift and splitting data of 1.86 and 5.22 mm s^{-1} respectively that are similar to those of $SnMe_2(SO_3F)_2$, suggesting the following decomposition reaction:

$$2SnMeH_2^{+} + 4HSO_3F \longrightarrow SnMe_2^{2+} + Sn^{2+} + H_2SO_4 + SO_2 + 2HF + 2H_2 + 2SO_3F^-$$

The spectra obtained from heated solutions of $SnMe_2H_2$ and $SnMe_3H$ in fluorosulphuric acid contain one resonance - a Sn(IV) spectrum with parameters similar to those of $SnMe_2(SO_3F)_2$.

The gamma-resonance data were included in a study of the spectroscopic and thermal-decomposition behaviour of the glycylglycine complexes $RSnCl_3(H_2glygly)$.[478] The isomer shifts show the expected trends in that they have lower values for the complexes than for the parent monoorganotin trichlorides. These shifts are also lower than those of the corresponding diorganotin dihalide complexes, and this is consistent with the expected increase in the use of tin s-electrons when chloride is replaced by an alkyl group. The shift data for the diorganotin complexes $R_2Cl_2(H_2glygly)$ (R = Me, n-Bu, n-Oct, and Ph) suggest that the bonding between the tin atom and the glygly ligand is relatively weak. P-31, C-13 and proton n.m.r. spectra have been used along with Sn-119 Moessbauer data in studies on mono-, di-, and tri-organotin(IV) dialkylthiophosphates[479]. The n.m.r. data are consistent with tetrahedral tin sites in triorganotin(IV) compounds such as $Me_3Sn[SSP(OEt)_2]$, although the Moessbauer spectra do suggest five- rather than four-coordinated tin. The Moessbauer isomer-shift data suggest weak Sn-S bonding to the ligand.

The structures of the alkyltin iso-octylthioglycollates (IOTG) $R_2SnCl_{\underline{n}}(IOTG)_{2-\underline{n}}$ ($\underline{n}$ = 0, 1) and $RSn_{\underline{n}}(IOTG)_{3-\underline{n}}$ ($\underline{n}$ = 0-2) have been investigated with n.m.r. and some of the results compared with gamma-resonance data[480]. Isooctylthioglycollate and isocretyl maleate derivatives were also included in a Moessbauer study of the degration of organotin-stabilised PVC[411]. The Moessbauer effect was used to show that the organic groups in diorganotin halide adducts with 2,2'-azopyridine are in trans-positions in a six - coordinated Sn site[472].

The temperature dependence of the recoil-free fraction for thirteen phenyl- and eleven cyclohexyl-tin compounds has been investigated[481]. The compounds chosen either had known crystal structures or had structures that could easily be inferred from other data. The phenyl compounds that were known to consist of non-interacting monomers all had large values for the temperature dependence of the recoil-free fraction [(1.37 - 2.8) x $10^{-2}K^{-1}$] that reflected the relative vibration freedom of the tin atoms in each case. The phenyltin compounds that were known to be polymeric fell into two groups with temperature dependences of about 1.1 x $10^{-2}K^{-1}$ and 1.8 x $10^{-2}K^{-1}$ respectively. The values for the second group fell in the range normally associated with non-polymeric systems. In order to interpret the data, the authors defined the polymers in terms of three layers of complexity: the primary structure, is related to the monomer from which the polymer is built, its coordination number, geometry, stereochemistry, etc.; the secondary structure, is used to describe the dimensionality of the polymer, one-, two-dimensional, chain, sheet, etc.; and the term tertiary structure, which covers the three-dimensional disposition of secondary structure. While the primary and secondary structures play a major role in defining the characteristics of the polymer, they do not alone necessitate a restriction in the vibrational motion of the atom incorporated into the polymer, as evidenced by the data for the second group of phenyltin polymers. The Moessbauer variable-temperature role of the polymer's tertiary structure has therefore to be considered more fully. Four classes of tertiary structure were identified, viz (1) rod-like structures with linear Sn-X-Sn linkages, (2) zig-zag structures with bent Sn-X-Sn linkages where X is a simple group such as a halide or hydroxide, (3) zig-zag structures with bent Sn-X-Sn linkages where the X group is larger and forms a more flexible S-shaped bridging link, and (4) structures with very flexible X groups that can form a helical polymer. In terms of the variable-temperature Moessbauer data, the lattice rigidity decreases (a) with decreasing strength of the Sn-X bridging bond and (b) with increasing displacement of the bridging mass away from the Sn...Sn vector, i.e. in the order rod-like structures, zig-zag structures, S-shaped structures and helical structures. Compounds with the latter two structures show the highest values for the temperature dependence of the recoil-free fraction, while the most rigid of the structures, the rod-like structures, have the lowest values. The variable-temperature Moessbauer data for the cyclohexyltin compounds were interpreted in a similar fashion to those of the phenyltin compounds in that monomers [such as $Sn(C_6H_{11})_3Br$] and weakly polymeric compounds [such as $Sn(C_6H_{11})_3(O_2CMe)$] have the highest values of temperature dependence of f while rod-like polymeric materials [such as $Sn(C_6H_{11})_3Cl$] have low values.

A number of new complexes of diorganotin(IV) compounds with nucleosides have been prepared and characterised using a number of spectroscopic techniques including Moessbauer spectroscopy[482]. The gamma-resonance parameters were also discussed in a study of the antitumor activity of di- and tri-organotin complexes with amino acids[483].

The crystal structure of dimethyltin(IV) bispyrrolidinecarbodithioate has been determined. The Me-Sn-Me angle in the structure was found to be 137.3° in contrast to that of 123.5° predicted from the Moessbauer parameters of this compound[484].

Sixteen compounds of general formula R_2SnL_2 and $R_2(L)SnO$-$Sn(L)R_2$ (where L is an N-acetylamino acid) have been prepared and characterised. The data for the R_2SnL_2 have been interpreted in terms of the presence of distorted trans-octahedral tin sites, while those for the $R_2(L)SnOSn(L)R_2$ compounds were said to be consistent with 5-coordinated tin sites with bridging oxygen atoms and nearly linear C-Sn-C groups.[485]

The chemical isomer-shift data for the compounds R_3SnX (R = alkyl, phenyl, X = F, Cl, OH, OR^1, CN, NCS, NCO, or Ng) have been correlated with partial atomic charges on the tin. A unique shift-partial atomic charge relationship was obtained for all five coordinated R_3SnL species[486]. Partial quadrupole splitting values were used to suggest the coordination of the tin atoms in a series of 22 organotin compounds $(R_{n-3}R'_3Sn)_2X$[489]. Harrison et al. have characterised two trimethyltindipeptide methyl ester complexes[488], and the Sn-119 Moessbauer effect has been used to help in the description of the bonding in organotin(IV) esters of tetramethylenedithiocarbamic acid[489]. Eleven new complexes of Ph_3SnCl with N-alkylsalicylideneimines have been prepared and characterised with infrared, n.m.r., and Moessbauer spectroscopic data[490]. The data are consistent with trigonal-bipyramidal tin environment (5) in which the Sn is bonded to the ligand through the phenolic oxygen atoms.

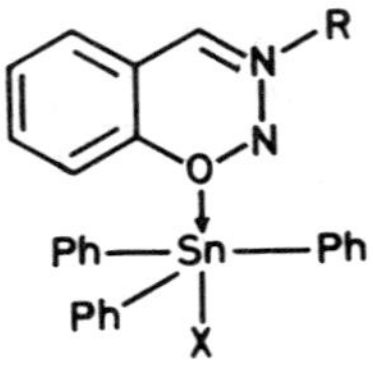

(5)

Both Sn-119 Moessbauer and Sn-119 n.m.r. data have been collected for a series of sterically hindered tetraorganotin(IV) derivatives[491]. The most unusual spectra obtained were for tetra-adamant-1-yltin, $Sn(ad)_4$ (see Figure 3).

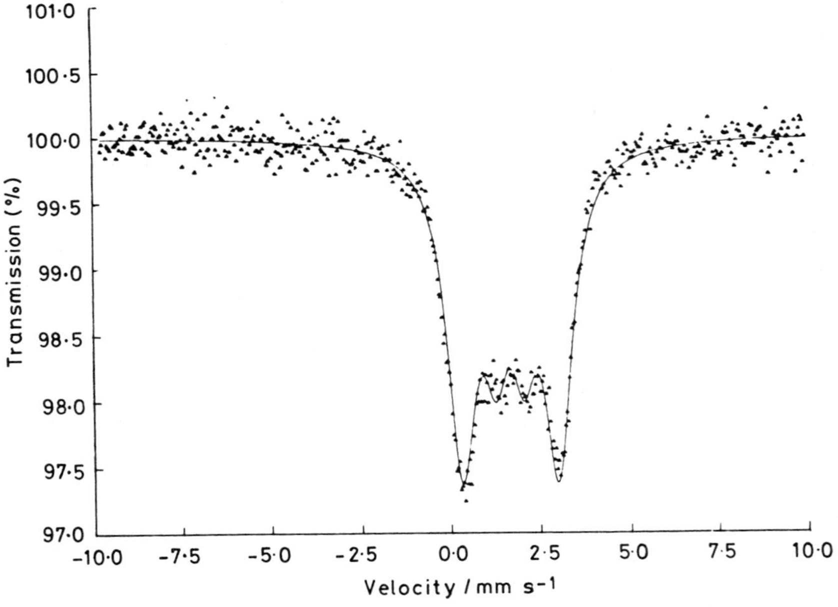

Figure 3 Sn-119 Moessbauer spectrum of $Sn(ad)_4$ at 80K (Reproduced by permission from J. Chem. Soc., Dalton Trans., 1985, 169)

At 80K the spectra show evidence for two tin sites, one with a large quadrupole splitting (2.7 mm s^{-1}) and one with a small quadrupole splitting (0.76 mm s^{-1}). The variable-temperature data also show the two sites over the temperature range studied. From the ratio of the resonance lines the site with the larger splitting must account for two-thirds of the Sn atoms in $Sn(ad)_4$. Molecular models showed that four (ad) groups around a tin atom with normal Sn-C distances would result in an extremely sterically hindered structure, and it is likely that at least one Sn-C bond would be longer than the others (6) or that an effectively three-coordinated tin structure would be formed with the fourth (ad) group balancing the charge in the lattice (7) or that a five-coordinated structure would be formed (8). Site (6), the

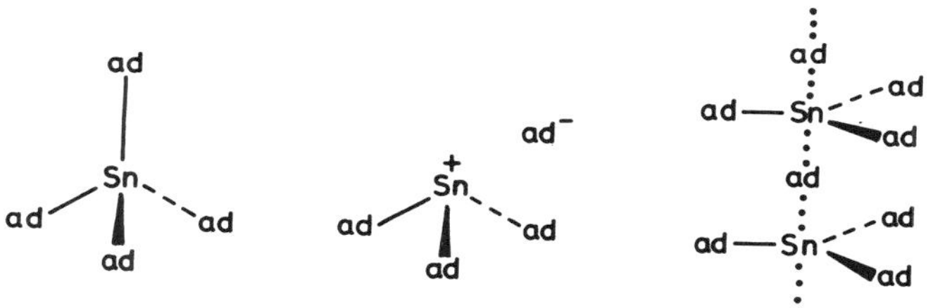

(6) (7) (8)

distorted tetrahedral site, could account for the resonance with the low quadrupole splitting, whilst the five-or three-coordinated sites could account for the resonance with the larger splitting. The isomer shifts for both of the sites in $Sn(ad)_4$ are about 1.6 mm s^{-1} and are the largest yet recorded for SnR_4 compounds. The high shifts were said to be due to the electron-donating properties of the (ad) groups, allowing more s-electron density to remain on the tin atoms.

The Moessbauer spectrum at 80K for quench-frozen $SnPh_4$ has also been obtained[491]. The spectrum consists of a singlet with a large linewidth which narrows on annealing at 193K to give a narrow line. The authors suggested that the most likely explanation of this effect is that defects caused in the structure due to cracking and internal stress could generate the wide line and that annealing would remove the defects. Many other compounds of the type SnR_3R' that contain potentially sterically hindering groups bonded to tin also have broad singlet resonances that can be interpreted in the same way as the data for $SnPh_4$.

6 OTHER ELEMENTS

This section reviews the data for elements other than iron or tin. In each of the three main subsections (main-group elements, transition-metal elements, and lanthanide and actinide elements) the isotopes are discussed in order of increasing atomic number.

Main-Group Elements.— Antimony (Sb-121). The Sb-121 Moessbauer and Raman spectroscopic data for a number of antimony compounds are presented and discussed in a thesis[492].

Variable-temperature gamma-resonance data in the temperature region 4.5 - 150K for polycrystalline antimony were used[493] to obtain a Debye temperature of 180K for the element. The isomer shift for Sb at 100K was measured as -11.4 mm s^{-1} from a $BaSnO_3$ source cooled to 100K, and no quadrupole broadening of the single resonance line was detected. Variable-temperature Moessbauer data have also been obtained for InSb over the range 11-160K and for 3-methylpyridinium tetrabromoantimonate(III), 2-chloropyridinium tetrabromoantimonate(III), and Me_3SbCl_2 over the range 11-100K. There is no evidence from the data of large anharmonicity effects or temperature dependence of the characteristic Moessbauer temperature for any of the compounds over the temperature ranges studied. The Debye temperature determined for InSb using the thin absorber approximation was 160 $\pm$ 5K.[494]

The Moessbauer effect has been used to study the electronic configurations of Sb present both as a component of and as an added impurity to III-V and II-VI semiconductors. The impurity effects were studied by measuring the emission spectra from implanted Xe-121, which decays via Te-121 to Sb-121[495]. Antimony Moessbauer spectra have also been used to study the vitreous products obtained in the glass-forming region of the $Tl_2S-Sb_2S_3$ system, that is over the composition range 60-95 mole % Sb_2S_3. The Moessbauer spectra were interpreted in the light of the measured glass transition temperatures[496]. In another study of antimony sulphides, Moessbauer spectroscopy was used to determine the oxidation state of the Sb atoms present[497].

Moessbauer spectra have been obtained for the antimony(III)

catecholatohalides $(C_6H_4O_2)SbX$, and the data are said to be consistent with a trigonal-bipyramidal distribution of electron pairs around Sb.[498] A crystal-structure determination of $[SbL(phen)_2]BPh_4$ (L = catechol, phen = 1,10-phenanthroline) showed that the antimony atom is part of a complex catecholato-antimony(III) unit chelated by two phenanthioline ligands. The three ligands coordinated to the Sb are arranged so that space is left for a sterically active non-bonding electron pair. The Moessbauer parameters of $[SbL(phen)_2]BPh_4$ are discussed in terms of the crystal structures[499].

The infrared spectra and Moessbauer parameters of $P_2O_3 \cdot 2SbCl_5$ are consistent with complex formation between $SbCl_5$ molecules and the terminal O atom of the P_2O_3 group[500]. The Sb-121 data for the Sb(V) compounds SbO_2F, $SbOFCl_2$, $Sb_5O_7Cl_{11}$, and $[SbCl_4(OPMe_2)]$ have all been shown to be consistent with the presence of octahedrally coordinated antimony(V)[501].

The synthesis of $SbCl_5$-doped polyacetylene has been described[502]. The doped polymer samples containing molar compositions of 0.005-0.125 $SbCl_5$ were characterised by a number of techniques including Moessbauer spectroscopy. The gamma-resonance data for samples doped with less than 0.005 moles of $SbCl_5$ unequivocally demonstrate only the presence of $SbCl_3$, while samples doped with >0.07 moles $SbCl_5$ show resonance lines from both Sb(III) and Sb(V)[502].

<u>Tellurium (Te-125)</u>. A value of $\Delta \underline{R}/\underline{R}$ of 0.76×10^{-4} was obtained for the 35.4 keV Te-125 isotope from emission and conversion-electron studies on I-125 implanted in the metal matrices Cu, Sn, Pt, Te, Au, and Zn[503]. A uniform charge distribution of $1.2 \times A^{1/3}$ fm was assumed.

The phase transitions in TeO_2 with pressure were studied on single-crystal and polycrystalline samples over the pressure range 12-50 kbar. A second-order transition from tetragonal to orthorhombic was identified at about 9 kbar, and the material becomes brittle at about 12 kbar.[504]

Moessbauer effect studies have been carried out on nucleogenic Te-125 monomers formed by the decay of Te-125m embedded in solid Ar[505]. The Te-125 spectrum consisted of two components: (1) a quadrupole split component with shift and splitting values of 0.35 mm s^{-1} (from ZnTe) and 9.2 mm s^{-1} respectively and (2) a singlet with shift 0.15 mm s^{-1}. The quadrupole sub-spectrum was attributed to a Te(0) species and the singlet to a Te(-1) species. A novel single-line absorber, Mg_3TeO_6, with a high recoil-free fraction was developed for Te-125 emission experiments. The data obtained were used, along with relativistic Dirac-Slater calculations, to produce a new isomer-shift calibration curve for Te-125. The calibration consisted of a non-linear relationship between shift and the number of p-holes in the closed-shell $5s^25p^6$ configuration. The change in the mean-squared charge radius $\Delta(r^2)$ for Te-125 calculated from the calibration was 3.4 x 10^{-3} fm^2.

Iodine (I-127 and I-129). Moessbauer effect studies were carried out on the nucleogenic I-129 monomers formed by the decay of Te-129m in solid Ar[505]. The spectrum from this dilute rare-gas matrix-isolated ion-implanted sample was a single quadrupole split resonance peak with a shift 0.75 mm s^{-1} from a ZnTe source and a splitting -685(20)HMz that was attributed to an I(0) species. A new single-line absorber, Na_5IO_6, with a high recoil-free fraction for Te-129m emission studies was developed. The change in the mean-squared charge radius $\Delta(r^2)$ obtained for I-129 from the calibration data for isomer shifts was 19.9 x $10^{-3}fm^2$.

The I-127 Moessbauer data for a series of CuI complexes listed have been obtained.[506] The most striking feature of the spectra is that none of them displays any resolvable quadrupole coupling. In fact, the linewidths of the spectra obtained by assuming a zero quadrupole splitting in one fitting procedure are comparable with that found for CuI in which the tetrahedral coordination of the Cu should mean negligible quadrupole coupling. The maximum possible values for the coupling constants (eQV_{zz}) for the CuI complexes are shown in Table 3 and are in the range 410-690 MHz. These values can be compared with the coupling constants of up to 3,000 MHz found for the covalent interactions of iodine bonded to main-group elements in non-cubic configurations. The authors, therefore concluded that for the Cu(I)-iodine-containing compounds either there is very little distortion of the electronic environment from the spherically symmetrical valence configuration of the purely ionic iodide ion and that the Cu(2)-I bond is ionic or that in the Cu(I)-I bonding electrons are withdrawn nearly equally from the three 5*p*-orbitals of iodine. The authors favoured the former explanation but they did not consider the possibility that any covalent interaction between Cu(I) and I could involve molecular orbitals populated essentially by Cu electrons.

Table 3 I-127 Moessbauer parameters for complexes of CuI, assuming (i) single lines, (ii) quadrupole split sites

Complex		Isomer shift/ $mm\ s^{-1}$ from KI	$e^{127}Q_g V_{zz}$/ MHz
$[CuI(PPh_3)_3]$	(i)	-0.12(6)	
	(ii)	-0.03(6)	-470(60)
$[(CuI)_2(PPh_3)_3]$	(i)	-0.10(6)	
	(ii)	-0.04(6)	-450(50)
$[\{CuI(PPh_3)\}_4]$	(i)	-0.12(3)	
	(ii)	-0.65(9)	-620(70)
		+0.42(12)	690(60)
$[\{CuI(PH_2Ph)_2\}_2]$	(i)	-0.06(6)	
	(ii)	+0.01(6)	-410(60)
$[\{(CuI)_2(S_2Ph_2)\}_n]$		-0.09(3)	
CuI		-0.03(6)	

Resonance Raman and I-129 Moessbauer spectroscopy were used in a study of I-doped polythienylenes prepared by polymerisation of dibromothiophenes. The iodine species identified in the polymers were polyiodides consisting of I_3^- and I_2 units, and this means that the I must have been reduced by taking electrons from the polymer in the doping process. The result was said to be consistent with an increase in the electrical conductivity of the polymer on doping it with iodine[507].

The hyperfine interactions in the Te-V system were followed in V_3Te_4, V_5Te_8, and VTe_2 by the Te-129 Moessbauer emission. A transferred magnetic hyperfine field at the daughter I-129 sites in $V_{2.6}Te_4$, which has a Néel temperature at 52K, was observed at 16K. The isomer shift decreases with decreasing V content and the ionicity calculated was said to be in agreement with earlier results on Cr-Te and similar systems[508].

Transition-Metal Elements.— The isotopes Ru-99, Ir-193, and Au-197 were included in a study of intermetallic compounds designed to provide an interpretation of Moessbauer isomer shifts[41].

Nickel (Ni-61). Both the Ni-61 and the Fe-57 Moessbauer effects were used to study a partially ordered sample of Ni_3Fe. The calculated occupation probabilities of Ni neighbours were used to estimate the changes in the magnetic hyperfine field at Ni due to the presence of one Fe atom in the first coordination sphere.[509]

Zinc (Zn-67). Ikonan et al.[55] used the 93 keV Zn-67 isotope in an investigation of coherent transient effects arising from phase modulation of the gamma-radiation. In studies of Zn-67 in ZnO transient experiments with Zn-67 mechanical displacements of about 10^{-13}m were resolved and, as an extension of the study, the recoil-free fractions of both the Ga(67):ZnO source and the Zn-67-enriched ZnO were determined as 0.0212(16) and 0.0213(18) respectively.

The 93.3 keV Moessbauer resonance in Zn-67 involves a gamma transition between a 5/2 ground state and a 1/2 excited state. Since the excited state will not split under the interaction of an electric-field gradient tensor, the quadrupolar hyperfine spectrum is comparatively simple and easily interpreted. The spin of the ground state, on the other hand, is high enough to allow a separation of the field gradient V_{zz} from the asymmetry parameter in cases where the symmetry is lower than axial. Furthermore, the resonance linewidth is extremely narrow on account of the rather long half-life of the 93.3 keV excited state. Thus field gradients can be measured with an accuracy characteristic for microwave resonance methods.

Since ZnF_2 is the most ionic zinc compound, the point-charge model would be expected to provide an appropriate description of the hyperfine parameters. Potzel and Kalvius[510] have, however, shown that it is not the case because neither the sign of the e.f.g. nor the asymmetry parameter are correctly predicted by the point-charge model. They used the spectrum of (Zn-67)F_2 to show how much more elaborate calculations are required to explain the experimental results. The spectrum of ZnF_2 consists of three lines at the positions -86.07, -56.25, and -40.23 μm s^{-1} from a (Ga-67)ZnO source. During the experiment a single crystal of ZnF_2 was also aligned such that the *c*-axis of its rutile unit cell was parallel to the direction of the gamma-rays. In this orientation one of the quadrupole lines is not observed and the spectrum consists of two lines separated by 4.80 μm s^{-1}. The isomer shift for ZnF_2 obtained from the spectrum was -6.4 ± 4 μm s^{-1}, and the second-order shift arising from the different Debye temperatures of ZnO and ZnF_2 was 3.5 μm s^{-1}. The Debye temperature obtained for ZnF_2 was 294 ± 13K. To determine the e.f.g. Potzel and Kalvius considered the energy eigenvalues for an I = 5/2 state, and from the ratio of the energy separations $(E_{\pm 5/2} - E_{\pm 3/2})/(E_{\pm 3/2} - E_{\pm 1/2})$ the asymmetry parameter was calculated as 0.24 ± 0.05 by solving the corresponding equation numerically. Using this value for the asymmetry parameter, the quadrupole frequency $e^2V_{zz}Q$ was determined as +100.5±1.5 μm s^{-1}, from which the V_{zz} value of +(2.09 ± 0.23) x 10^{17} V/cm^2 was derived.

Ruthenium (Ru-99). The Ru-99 isotope has been used as a probe atom in niobium hydrides, and the results showed that the electron density at ruthenium decreased with increasing x in NbH_x (x = 0.7-1.0).[511] The Ru-99 Moessbauer effect was also used to study the properties of single crystals grown by chemical vapour transport in the system TiO_2-RuO_2. The results showed that all Ru oxidation states from V+ to III+ were present in the solid solutions obtained[512].

Tantalum (Ta-181). The Ta-181 Moessbauer spectra were measured in the same sample of TaS_2 before and after the intercalation of hydrated Na in a study of the charge-transfer induced change arising from the intercalation[513]. The intercalation compound $Na_{0.33}(H_2O)_2TaS_2$ has a single Moessbauer resonance line with a higher shift than the parent TaS_2, as would be expected because of the lower valency of Ta in the Na-containing compounds.

The quadrupole interaction in TaS_2 is smaller after intercalation, in agreement with perturbed angular correlation data. The 2H-$TaSe_2$ phase has also been studied[514] using the Moessbauer effect and, because of the high resolution possible with the Ta-181 resonance, it was found that the splitting between the $\pm5/2 \rightarrow \pm3/3$ and $\pm7/2 \rightarrow \pm5/2$ energy levels could be resolved. In another study of selenides and sulphides, the 6.2 keV Ta-181 resonance was used to study the derivatives Tl_3TaX_4 (X = S or Se). The isomer shifts for $TlTaS_4$ and $TlTaSe_4$ from Ta metals were found to be -10.16 and +3.85 mm s^{-1} respectively, values which are about 25 mm s^{-1} lower than the values for the shifts in the corresponding Cu_3TaE_4 series[515].

Iridium (Ir-193). The Ir-193 Moessbauer effect was used in two studies of its parent nucleus ^{193}Os.[516,517] An (Os-193). $Os_{0.08}Fe_{0.92}$ source was used in one of the studies[516] of the effect of the orientation of the Os-193 ground state on the substate populations of the 73 keV $1/2^+$ Moessbauer state. The different Moessbauer spectra obtained at high and low temperatures arising from different populations are shown in Figure 4.

The gamma-ray distribution coefficients combined with a quadrupole alignment measurement from the literature Q (Os-193) lead to a value for the ground state of +0.48(6)b. The magnetic moment of the ground state of Os-193($3/2^-$) has also been reported as -0.75(3)μN.

The Ir-193 isotope has been used as a probe atom in niobium hydrides, and the results obtained showed that the electron density at Ir decreases with increasing x in NbH_x (x = 0.7-1.0).[511]

Gold (Au-197). Gamma-resonance measurements carried out on bi- and tri-nuclear Au(I) complexes containing Ph_3PAu^+ units bridged with a halogen or an oxygen atom showed that the spectra consisted of one doublet and that the parameters were almost identical to those obtained for monomeric Au(I) compounds[518].

The first dithiocarboxylate derivatives of gold have been obtained.[519] The compounds have the compositions $Au(CH_3CS_2)$,

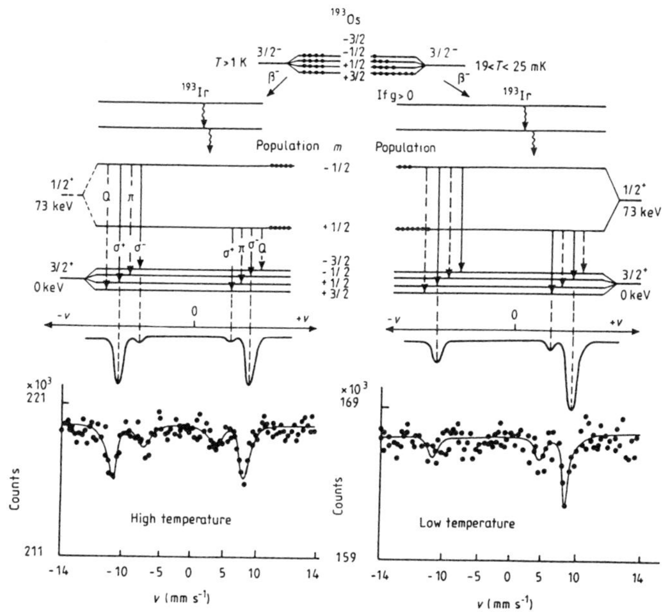

Figure 4 Calculated Ir-193 Moessbauer spectra assuming different populations of the parent Os-193 levels compared to experimental spectra obtained with a polarised source at high and low temperatures. At high temperatures the $\frac{1}{2}-\frac{1}{2}$> and the $\frac{1}{2}+\frac{1}{2}$> substate populations are equal and the spectrum is symmetrical about zero velocity. At low temperatures the substate populations are not equal and the spectrum is not symmetrical
(Reproduced by permission from J. Phys. G, 1985, 11, 287)

$Au(PhCS_2)$, and $Au(PhCS_2)(PH_2CCS_2)$. The Au-197 parameters for $Au(CH_3CS_2)$ are I.S. = 1.40 mm s^{-1} (relative to Au in Pt metal) and Q.S. = 6.13 mm s^{-1}. The structure of $Au(CH_3CS_2)$ consists (9) of a tetramer of gold atoms bridged in pairs by CH_3CS_2 groups.

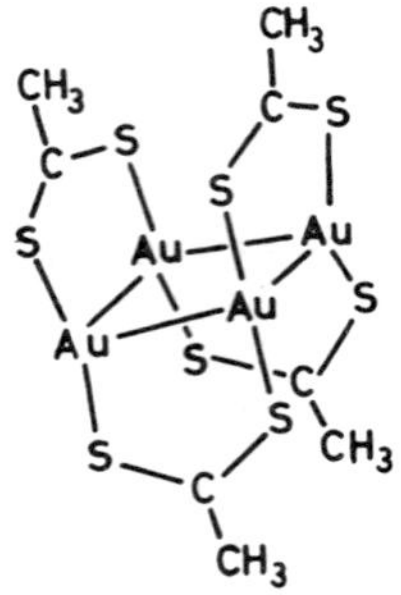

(9)

The Au-Au distances, average 301.3 pm, are long compared with those of 288.4 pm found in the metal, but it is still surprising that the Moessbauer parameters provide no evidence of direct interaction between the Au atoms of the tetramer.

Addition of a bidentate ligand (L) [where L = 1,10-phenanthroline or o-phenylenebis(dimethylarsine)] to solutions of $Au(C_6F_5)X_2$. tetrahydrothiophene (X = Cl or Br) leads to products that could contain five-coordinated Au(III). The Moessbauer spectra of the products, however, were shown to be consistent with the formation of square-planer Au(III) complexes with only weak interactions in one axial position[520].

Au-193 has been used as a probe atom in NbH_x, and it has been shown that the electron density at the Au increases with x over the range x = 0.7-1.0[511].

The Moessbauer spectra have been obtained for Au-197 in CsAu, which has the CsCl-type structure, at 4.2K and at pressures of 1 bar and 27, 30, and 40 kbar[521]. Evidence was found for a phase change to a structure of lower symmetry at a pressure just below 27 kbar. The 1 bar spectrum for CsAu showed the presence of a small amount of an impurity Au site, but the main singlet splits into a quadrupole doublet and shifts to lower energy to give the parameters I.S. = 6.95 mm s^{-1}, Q.S. = 1.51 mm s^{-1}. The presence of a phase transition at about 27 kbar is also indicated by an increase in the Debye temperature from 72K at 1 bar to 79K at 27 bar. The changes in the gold Moessbauer data were taken by the authors to suggest greater covalency in the bonding in the high-temperature phase.

Lanthanide and Actinide Elements.—During the past year papers have been published on Moessbauer effect studies using isotopes of europium, gadolinium, dysprosium, erbium, thulium, ytterbium, and neptunium. The last year has seen a large number of papers published on the Eu-151 Moessbauer spectra of europium 1:2:2 compounds.

Europium (Eu-151). Line-broadening due to paramagnetic relaxation has been noted in Moessbauer spectra of Eu^{2+} in frozen glassy solutions[522]. This broadening was used as an indicator of the degree of dispersion of europium. Two papers have used the Eu-151 Moessbauer effect to identify the valence states of europium in the europium-doped phosphors $ABeF_3$[523] and $AF\text{-}AlF_3$[524] (A = Na, K, Rb, Cs). Both systems contained Eu^{3+} in addition to Eu^{2+}. Structural, magnetic, and Moessbauer data have been reported for the europium dihalides EuX_2 (X = Cl, Br, I).[525] A linear correlation between the Moessbauer isomer shift and the saturation hyperfine field was found and interpreted in terms of changes in bond ionicity which result in increasing 6*s*-electron and spin densities with decreasing ionic character.[525] Samples of the non-stoichiometric mixed-valence compound $EuBr_{3-x}$ have been examined by Moessbauer and other methods[526]. These results, together with those from an earlier study on $EuCl_3$, were consistent with a thermally activated mixed-valence process in which a 4*f*-electron is promoted into the 5*d*6*s*-conduction band for a characteristic time of about 10^{-6} s at 78K.

The magnetoplumite-type aluminates $La_{1-x}EuMgAl_{11}O_{19}$ ($0 < x \leqslant 1$) have been studied by Moessbauer spectroscopy.[527,528] These phases were prepared by thermal treatment of amorphous coprecipitates, and the influence of the growth process on the Eu valence state and environment was examined. The hyperfine structure of Eu^{3+} in the $x = 0.3$ phase was analysed and compared with the results of an ESR study on the $\underline{x}$ = 0.02 phase. Eu-151 Moessbauer spectroscopy has also been used to investigate the phosphor $Y_{2-\underline{x}}Eu_{\underline{x}}WO_6$[529]. The spectra were interpreted in terms of the substitution of europium into three non-equivalent yttrium sites. Small differences in the covalency for the three sites explained the different isomer shifts, and the relative occupancy of one of the sites was strongly dependent on $\underline{x}$.

The unusual nature of $EuRh_3B_2$ has been noted by three groups of researchers[530-532]. A magnetic and Moessbauer effect study has determined that the ferromagnetic properties of nearly stoichiometric $EuRh_3B_2$ originate from the ordering of the spins of Eu^{2+} perpendicular to the hexagonal $\underline{c}$-axis[530]. The value of the Eu-151 quadrupole splitting, 48 mm s^{-1}, was larger than any previously reported for any europium system and was attributed to the very short Eu-Eu separation. Structural, magnetic, Moessbauer effect, and L_{III} absorption-edge measurements have been used to follow changes in the valence state and magnetic behaviour in $Eu_{\underline{x}}Rh_3B_2$ ($0.6 \leqslant \underline{x} \leqslant 1$).[531] The Eu valency changed from a predominance of Eu^{2+} for $\underline{x}$ = 1.0 to a combination of divalent and mixed-valence states at lower values of $\underline{x}$. A quadrupole interaction of $\underline{e}^2\underline{qQ}$ = 770 MHz was reported. The Moessbauer linewidth of $EuPd_6B_4$ has been found to reach a maximum when its isomer shift varies most strongly with temperature[533]. This anomalous line-broadening was associated with a distribution of the local europium valency and a suitable model devised.

Detailed Moessbauer and L_{III} $\underline{X}$-ray absorption-edge experiments on $EuPd_2Si_2$ have been reported.[534,535,537] A quantitative correlation of these two methods has been made over the range of the thermally induced valence change of 0.51 electrons[534]. The resultant relationship between Eu-151 isomer shift and the L_{III}-determined valency was used to provide an estimate of the valence states of other europium 1:2:2 compounds and to emphasise the extent of the coexistance of magnetic ordering and valence mixing in these compounds. Another study has examined $EuPd_2Si_2$ with the same techniques over the temperature range 1.7-300K.[535] Data obtained by the two techniques from identical samples was used to discuss the valence/isomer-shift calibration problem, and the behaviour of the satellite line observed in the Moessbauer spectra was studied. Other workers have described a study of diluted europium in $Y_xCu_2Si_2$ (x = 0.03-0.1) and compared the data with those of the concentrated (x = 1) phase[537]. Six techniques, including gamma-resonance spectroscopy, were used and only slight differences were found between the concentrated and dilute materials. The dominant feature of both systems was a very strong variation of the europium valency with temperature from about 2.8 at 4K to around 2.45 at 800K. A new evaluation of the temperature dependence of the Moessbauer spectra of $EuPd_2Si_2$ and $EuCu_2Si_2$ has been presented which combined the Moessbauer effect results with X-ray diffraction data and chemical analyses to provide new information on the local valence behaviour in these materials.[537] The Moessbauer linewidth of $EuPd_2Si_2$ has been related to the isomer shift as described for $EuPd_6B_4$[533]. Moessbauer effect, magnetic susceptibility, X-ray diffraction, and L_{III} absorption-edge measurements have been made on the series $Eu(Pd_{1-x}Au_x)_2Si_2$ (x = 0.1)[538,539]. This system orders antiferromagnetically for $x \geqslant 0.18$. When $x > 0.4$ a $4f^7$-ground state is found for the Eu^{2+} ion. When $x < 0.4$, however, the magnetic data suggested a pure, divalent $4f^7$-ground state while the other techniques suggested unstable valence behaviour. This was explained by assuming a partial extension of the $4f$-shell radius which preserves the localised character of the 4f-electrons.

Extensive magnetic and Moessbauer studies have been reported on EuM_2Ge_2 (M = Rh, Ru)[540,541] and $Eu(Pd,Rh,Ru)Si_{2-z}Ge_z$.[542,543] $EuRh_2Ge_2$ behaved similarly to the silicide with two magnetic phase transitions, corresponding to the antiferromagnetic ordering of the local rare-earth moments and the itinerant electron ordering of the rhodium sublattice[540]. Studies of $EuRu_2Ge_2$ over the temperature range 4.1-600K have obtained the unusually large value e.f.g. of $A_2\langle r^2\rangle$= 425K with eqQ = -440 MHz[541]. This compound is also the first-reported example of an anisotropic recoil-free fraction in a metallic rare-earth compound. The Debye temperatures obtained parallel and perpendicular to the c-axis were 450 and 180K respectively. Comparison of the magnetic properties and Eu-151 hyperfine interactions of $Eu^{2+}Rh_2Ge_2$, $Eu^{2+}Ru_2$-Ge_2, $Eu^{2+}Rh_2Si_2$, and $Eu^{3+}Rh_2Si_2$ with other similar rare-earth systems has been studied, and the crystal-field effects are also significant. The properties of $EuPd_{2-x-y}Rh_xRu_ySi_{2-z}Ge_z$ were studied in terms of the valency changes, and it was found that the local chemical environment was the dominant influence in determining the europium valency[542,543].

Anomalous behaviour of the Moessbauer resonance linewidth in the mixed-valence compound $EuNi_2P_2$ has been described in a series of papers.[532,533,544,545] A broad maximum in linewidth was observed around 40K which was explained by an inhomogeneous width of the interconfigurational excitation energy in addition to the homogeneous fluctuation width[533]. Whereas the average valency in other europium-based materials of this type has tended to approach 3 as $T \rightarrow 0$, this system had an average valency of 2.5 at 1.4K.[544,545] The strongly mixed-valent state of europium in this compound has also been demonstrated by the variation of the mean valency with pressures up to 90 kbar at 4.2K[546]. Contradictory evidence has been obtained for the ground state of europium in $EuPd_2P_2$[547, 548]. While Moessbauer and magnetisation measurements suggested a divalent $4f^7$-state, the L_{III} absorption-edge spectrum was that expected for a mixed-valency material. Possible reasons suggested for this were: the existence of a partly extended 4f-radius, significant 4f-covalent bonding, and configuration-interaction effects on the L_{III}-edge spectra.

The atomic cell model of Miedema and Van de Woude has been used to clarify the redistribution of electronic charge when two elements form a compound or alloy. A series of papers have described the application of this model to Eu-base intermetallics and used two quantities derived from this model, the sign and magnitude of the enthalpy of formation, to calculate the Moessbauer isomer shift[549,550]. The magnetic properties of intermetallic compounds in the series Eu-Ga and Eu-Sn were investigated.[550] These compounds followed the Curie-Weiss law in the paramagnetic region and showed antiferromagnetic ordering at low temperatures. Metamagnetic behaviour was observed in Eu_3Ga_8 and in the low-temperature phase of $EuGa_2$ together with a volume contraction associated with compound formation. The values of the Eu-151 isomer shift could be accurately predicted using the cellular model after correction for this volume contraction.

Gadolinium (Gd-155). The cellular model has been applied to the Gd-155 Moessbauer isotope[549,551,552]. This model was applied to Gd-155 in Eu-base intermetallics[549], and volume effects on the isomer shift of intermetallic compounds containing gadolinium were observed[551]. This model was also applied to a range of ternary compounds containing gadolinium[552]. Changes in the magnetic properties and Gd-155 hyperfine field have been studied in GdM compounds (M = Rh, Pd, Cu, Ag, and Au) when these materials were charged with hydrogen gas[553]. Internal hyperfine fields of up to 35 T were observed. Gd-155 Moessbauer data have been described for amorphous $Gd_{20}Y_{30}Ag_{50}$[554]. A distribution of quadrupole splittings was observed in the paramagnetic state with magnetic ordering appearing below 20K. A hyperfine field of about 10 T was present at 4.2K.

$Gd_2Fe_{14}B$ has been the subject of two studies.[555,556] Single-crystal Moessbauer data have shown that the main axis of the e.f.g. lies in different directions at the two Gd sites[556]. Other workers have reported large crystal-electric-fields (CEF) at the Gd sites in $GdCo_3B_2$[557] and the Moessbauer spectra of GdS[558]. The Chevrel-phase chalcogenides $GdMo_6X_8$ (X = S or Se) have been studied with the Gd-155 Moessbauer effect[559]. Quadrupole interactions of 545 and 470 MHz were reported for the sulphide and selenide respectively. These results suggested the presence of a large $\underline{B}_2^0\underline{I}_2^0$ term in the CEF Hamiltonian in all rare-earth Chevrel-phase sulphides and selenides and indicated that previous results based on a cubic Hamiltonian are not valid.

Dysprosium (Dy-161). Dysprosium-161 Moessbauer spectra of $Dy_2Fe_{14}B$ have been recorded over the temperature range 4.2-300K[560]. The spectra were analysed in terms of two subspectra associated with the two crystallographic Dy sites, both having a crystal-field split ground state of -15/2. The temperature variation of the hyperfine field and quadrupole splitting and bulk magnetisation measurements were used to derive values of the magnetic coupling constants between the iron moments ($\underline{J}_{FeFe}$) and between the dysprosium and iron ($\underline{J}_{DyFe}$). An investigation of $Dy_2Fe_{14}BH_{\underline{x}}$ has included 4.2K Dy-151 Moessbauer data[561]. Moessbauer and magnetic studies have shown the measured hyperfine magnetic field in $DyCo_3B_2$ to be less than that of the Dy^{3+} free-ion value[557].

The hyperfine interaction parameters at the Dy-161 and Fe-57 nuclei in $Dy_2Co_yFe_{17-y}$ ($y \leqslant 17$) and $Dy_{2-y}Y_yFe_{17}$ ($y \leqslant 2$) have been determined at 4.2K[562]. Changes on the hyperfine parameters were determined as a function of *y* in each case and the results explained by a *d*-electron model. In the case of the yttrium compound, the excess of the magnetic hyperfine field at the Dy-161 nuclei and the reduced field at the Fe-57 nuclei was explained as a result of the presence of the magnetically polarised 5*d*- (or 4*d*-) electrons in these phases. Experimental determinations of the Dy-161 Moessbauer parameters of amorphous dysprosium alloys have been compared with theoretical calculations made using a quadratic crystal-field model[563]. The good agreement obtained showed that higher-order crystal-field terms are negligible in this type of alloy. Crystalline and amorphous DyAg has been studied using Moessbauer spectroscopy and muon spin relaxation[564]. The effects of lowering of local symmetry were significant in the electronic ground state of Dy^{3+} and the fluctuation rate of the local magnetic moment. Moessbauer spectroscopy proved less sensitive than muon spin relaxation of high-spin fluctuation rates.

Erbium (Er-166). Moessbauer and magnetic measurements have been carried out on the hexagonal boride $ErFe_4B$.[565] The Er-166 Moessbauer studies found a nearly free hyperfine field of 8545 kOe at the erbium sites which indicated that the exchange interaction in $ErFe_4B$ is much stronger than the crystal-field interaction.

Thulium (Tm-169). A nonmagnetic Γ_1 singlet ground state has been identified from the Tm-169 Moessbauer spectra of TmB_{12}[566]. The same authors have reported the spectra of cubic $TmCu_5$, in which a quadrupole splitting was observed at temperatures as low as 2.3K[567]. This was explained as indicating the presence of a dynamic lattice distortion.

Ytterbium (Yb-170, Yb-174). A neutron diffraction and Yb-170 Moessbauer investigation of the perovskite phase $TbTiO_{2.95}$ has been described[568]. The Moessbauer data gave a value of 2.0 μB for the Yb moment and 42K for the Curie temperature. Below T_c

the Yb sublattice magnetisation followed a Brillouin function, and a distribution of hyperfine fields observed at low temperature was attributed to a random distribution of defects around the Yb sites. A slow-relaxation Moessbauer emission spectrum has been obtained at 0.11K for paramagnetic Yb^{3+} in Pd.[569] The two lowest crystal-field states were quasi-degenerate, with a separation of 2.5K, and the magnetic hyperfine spectrum was very sensitive to crystal field broadening away from cubic symmetry, which caused a static broadening of the lines. The e.f.g. on Yb in $YbPd_2Si_2$, $YbCu_2Si_2$, and YbCuAl have been recorded and interpreted using the interconfiguration fluctuation model[570]. Clear evidence of the persistence of crystal-field effects in the intermediate valence (4+) state was found, and Yb^{4+} was observed in Yb diluted into $TmPd_2Si_2$.

A single paper has been published which described the use of the Yb-174 Moessbauer effect[571]. The amorphous alloys $Yb_{80}X_{20}$ (X = Cu, Ag, Au, Pd, Bi) were studied, and the distribution of quadrupole splittings was compared with a calculated charge distribution corresponding to dense random packing of ions. While most of the alloys gave spectra consistent with this model, the data for $Yb_{80}Bi_{20}$ showed clear evidence of short-range order which was attributed to the greater covalency of bismuth.

Neptunium (Np-237). A study of barium fluoride-neptunium fluoride system has been made using X-ray diffraction and Np-237 Moessbauer spectroscopy[572]. A Moessbauer study of NpO_2 has been made in external magnetic fields of up to 8 T and in the temperature range 1.5-160K[573]. The oxide remained paramagnetic well below 25K. Changes in the specific heat, magnetic susceptibility, and Moessbauer spectra were therefore attributed to a structural transition which was described as an inhomogeneous, compressional, expansional, internal distortion of the oxygen sublattice. A series of oxides with the general formula $A_2(NpO_2)_2V_2O_8$ (A = K, Rb, Tl) have been prepared[574]. The Moessbauer spectra indicated strong quadrupolar interactions and isomer-shift values among the most positive found for hexavalent neptunium compounds.

The isostructural oxide chalcogenides NpOS and NpOSe have been studied using Np-237 Moessbauer spectroscopy[575]. These compounds have the PbClF structure and contained only Np(IV). Only the selenide was magnetically ordered at 4.2K. The same authors have reported 4.2K Moessbauer data for Np_2O_2Te[576]. This compound was magnetic and had an isomer shift of +26.3 mm s^{-1} (relative to $NpAl_2$) which was consistent with Np(III).

The tetragonal intermetallic phase $NpCu_4Al_8$ has been studied by neutron diffraction (at 2-300K) and Moessbauer spectroscopic (at 4.2-77K) methods[577]. A Moessbauer isomer shift of +14.3 mm s^{-1} (relative to $NpAl_2$) was obtained which suggested the presence of the non-Kramers Np^{3+} ($^5I_4$) ion. A magnetic splitting was observed at temperatures below 45K and was fully resolved at 4.2K with $\underline{B}_{eff}$ = 330 T. Neutron diffraction, however, found no evidence for magnetic order at any temperature. These data were explained in terms of paramagnetic relaxation involving a low-lying doublet state followed by a singlet state.

7 Backscatter Conversion-Electron Moessbauer Spectroscopy

There appears to have been an increase in papers describing the developments and application of the CEMS technique published during the past year. As in previous years, this section presents a summary of the theoretical developments in methodolgy and instrumentation followed by descriptions of the recent applications of the technique in studies of iron-, tin-, europium-, and, for the first time, tellurium-containing materials.

Four reviews have appeared in the literature over the past year dealing with the applications of conversion-electron Moessbauer spectroscopy. The application of CEMS[578] for characterisation of surfaces modified using directed energy processing techniques has been reviewed. In another review[579] examples are given for the analysis of electrodeposited Zn-Ni and Zn-Fe alloy layers on steel sheets using a combination of conversion-electron Moessbauer, X-ray photoelectron, and Auger-electron spectroscopic techniques. Some new developments in CEMS and its recent applications in a study of ion-implanted materials have been presented by Sawicki[580]. The author has stated that on-beam CEMS measurements are now possible at temperatures from 4 to 1200K in ultrahigh vacuum and that the application of Fe-57 resonance is rapidly expanding and experiments with Sn-119, Eu-151, and Au-197 are being carried out. Ogale et al.[581] have also reviewed the recent methods in the field of ion bombardment of material surfaces. In the review the use of novel characterisation concepts to explore the ion-beam-induced phenomena has been emphasised and in this context the importance of conversion-electron Moessbauer spectroscopy discussed. Also included in the review are examples of the use of CEMS to study ion implantation, ion-beam mixing, and corrosion of ion-bombardment surfaces.

The effect of gamma-magnetic resonance on conversion-electron polarisation has been considered[582] by the density matrix method. Analytical equations for Moessbauer spectral parameters such as intensity, area, and linewidth have been derived[583] by analysing the spectra recorded in the 2 -geometry of backscattering using various kinds of secondary emission: X-ray, resonantly scattered Moessbauer, conversion electron, and Auger electron. The effect of surface roughness[584] on the scattering and energy loss of internal conversion electrons has been studied by Monte Carlo simulation. The initial electron energy was 7.3 keV corresponding to the k-conversion electron in Fe-57 CEMS and DCEMS. Roughness was modelled by a surface oscillation with amplitudes and wavelengths in the range 1-10^3 nm. Total transmission, angular distributions, energy-loss profiles, and transmission into restricted energy and angular intervals were simulated by assuming Fe as scattering material. Conditions for the use of conventional weight functions in Fe-57 CEMS and DCEMS have also been obtained. In another paper Liljequist and Ismail[585] have determined the simultaneous energy loss and angular distribution of 7.3 and 13.6 keV internal conversion electrons as a function of their starting depth in an Fe sample by Monte Carlo simulation. A complete mapping of the Fe-57 DCEMS K- and L-conversion-electron weight functions has been obtained at different electron-emission angles relative to the absorber surface normal and at different electron energies >6.3 keV. Detection of electrons at large angles leads to a drastic contraction of the weight function toward the surface.

Weight functions for internal conversion Moessbauer effect, computed from Monte Carlo simulation of electron transport, have been tested for scaling properties[586]. Energy scaling is found to hold with high accuracy for a wide range of materials and electron energies, and mass scaling from Fe to other materials is adequate except for low-atomic-number materials. Meyer et al.[587] have carried out zero-field Moessbauer measurements between 1.9 and 30K and at room temperature on a Au-3 at.% Fe sample. A quite well defined change of regime was observed at 19.5K well above the measured cusp temperature (16.4K) for the sample. In the temperature region below 19.5K the spectra have been described in terms of a static inhomogeneously broadened field distribution, and at higher temperatures the spectra were compatible with relaxation broadening. It has been shown[588] that the combination of a grazing incidence total external reflection geometry with conversion-electron detection produces surface-sensitive Moessbauer spectra which are selective for the first few nanometres of the solid. This approach has been shown to be a simply implemented method of obtaining Moessbauer spectra of surfaces.

It has been shown[589] that the surface selectivity of resonance-electron Moessbauer spectroscopy with a proportional counter can be improved by detecting the sum signals of K-conversion and K-Auger electrons. The probing depth has been estimated from the range of K-Auger electrons to be 120nm in the case of Fe-57 Moessbauer measurements. Furthermore, depth-selective Moessbauer spectroscopy for two layers has been made possible by observing the above-mentioned spectrum and the conventional resonance-electron Moessbauer spectrum, the probing depths of which are 120nm and 300 nm respectively. In order to record several Moessbauer spectra simultaneously, as in the case of depth-selective Moessbauer spectroscopy, a microcomputer equipped with interface circuits has been used. A method to study the composition and depth distribution of phases present in surface layers of 0-50nm has been developed[590]. The method is based on the analysis of the Moessbauer spectra recorded by means of conversion electrons emerging at approximately the initial energy at different angles relative to the layer surface. Another paper reports[591] an experimental investigation of the depth and surface selectivity in Fe-57 DCEMS performed using a high-resolution electron-energy analyser combined with a Moessbauer spectrometer. DCEM spectra were recorded in ultrahigh vacuum at different electron energies E from 6.5 to 7.3 keV (with 2.7% energy resolution) and for the first time at different electron-emission angles, theta, relative to the absorber surface normal (from theta = 10° to 72°) using a test absorber consisting of a thin alpha-Fe layer on a thick stainless-steel substrate. The alpha-Fe and stainless-steel Moessbauer spectra areas were found to be in very good agreement with theoretical calculated values. It has been confirmed that the surface signal is enhanced by measuring at glancing angles relative to the absorber surface. Surprisingly, two papers have appeared in the literature of Gurachevskii et al.[592, 593] which describe a method used to detect low-energy conversion and Auger electrons which are involved in the transition of the resonance-excited Fe-57 nucleus to the ground state. A photoelectron multiplier was used as a detection of light pulses from microdischarges in rarified air, caused by electrons emitted from the surface of a sample under irradiation.

A thin proportional counter for scattering Moessbauer spectrometry (CEMS and X-ray Moessbauer spectrometry) has been designed[594] and its working characteristics studied to find the optimal condition for the effective detection of the resonant scattered X-ray. With the newly designed proportional counter, scattering Moessbauer spectra of steels were measured and the magnetism and chemical composition of the steel surface layers analysed. The construction of a simple gas-flow proportional counter suitable for operation between 100 and 400K without the need for an evacuated cryogenic system has been described[595]. Different gas mixtures were studied over the temperature range and He/5%CO was found to be most suitable at 100K. The system was tested using standard foils to obtain the optimum operating conditions. Low-concentration Fe-based samples were also studied to show the importance of investigating surface phenomena at different temperatures. In another paper[596] the working principle, features, and construction of a beta-resonance detector used for backscattering Moessbauer spectroscopy have been described along with examples for stainless-steel and Fe plate samples. A conversion-electron Moessbauer spectrometer which allows measurements to be made at low temperature(80K) has been fabricated[597] using a ceramic electron multiplier. It was used to discriminate the CEM spectrum of beta-FeOOH (T_N = 273K) from gamma-FeOOH (T_N = 50K), which were produced on steel surfaces as corrosion products following exposure of steel in an air-HCl-H_2O atmosphere. The instrumentation for the combination of a Moessbauer spectrometer with a hollow-core beta spectrometer has been described[598] in detail. The conversion-electron spectra obtained were found to be unsatisfactory for the measurements of charge and spin densities of the 4*s*-electrons.

Iron.—The favourable properties of the Fe-57 Moessbauer gamma-radiation and its associated conversion electrons result in most of the applications of the CEMS technique over the past year being concerned with the iron-57 effect. These applications are now discussed under the headings 'films and implantation studies', 'steels and corrosion products', and 'chemical reactions'.

Films and Implantation Studies. Hyperfine interaction parameters near clean and Ag-covered Fe(110) surfaces were measured[599] for the first time by means of *in situ* conversion-electron Moessbauer spectroscopy. The increase in the quadrupole splitting reflected the reduction in symmetry in the topmost monolayer, and the magnetic hyperfine fields provided information on magnetic order near the surface with monolayer resolution. An additive hyperfine field perturbation model has been used[600] to analyse the CEMS of composition-modulated Fe-V films. The theoretical prediction of a reduced moment at the interface was verified and the antiferromagnetic moment on the V atoms near the interface supported, but the predicted oscillatory moment of the inner Fe layers was discounted. Magnetisation processes near the surface of polycrystalline and single-crystal Fe plates were studied by Moessbauer emission and CEM spectroscopy and by the magneto-optical Kerr effect[601]. The Moessbauer methods were used to probe depths of 5nm-20 m. The values of H obtained were found to be consistent with the break in the bulk magnetisation curve. CEMS and EXAFS were used[602] to study emersed electrode investigations of passivated iron films.

CEMS has also been used to study carbon deposition on iron[603]. Carbon deposits were formed on Fe foils by the catalytic decomposition of a Me_2CO-CO_2 mixture and then studied by conversion-electron Moessbauer spectroscopy. The results indicated that wustite (FeO) catalysed the C deposition process and cementite

(Fe_3C) played a role in forming carbon filaments by disseminating the wustite. The surface technique has been used by Tomashevskii et al.[604] for the determination of the carburised layer structure in Fe. Tests were made on Armco Fe saturated with C-14, and the distribution of C-14 in Fe was determined by layer-by-layer removal. The C diffusion coefficient value ($\underline{D} = 2.2 \times 10^{-7}$ cm^2/s) was calculated for the tested carburisation conditions, and information about the structural state of the diffusion zone was obtained from analysis of the Moessbauer spectra.

A number of papers have appeared reporting the use of conversion-electron Moessbauer spectroscopy to study ion-beam-induced atomic mixing at iron-metal interfaces[605-608]. Two papers deal with Fe-Al interfaces[605,606]. In the first paper[605] the authors used both CEM and transmission Moessbauer spectroscopies so that the structural state of nearly equiatomic FeAl multilayers before and after Xe^{3+} ion-beam mixing could be studied. The as-ion-beam-mixed material gave rise to a Moessbauer spectrum exhibiting a doublet. From the CEMS-TEM study of annealing (300-750K) after ion-beam mixing the well known B2 superstructure could be detected at temperatures as low as 470K. In the second paper[606] both ion-beam-induced atomic mixing and the effect of thermally activated transformations at the Fe-Al interface were studied. The CEMS spectra of the as-deposited sample showed that not all the Fe-57 atoms in the interface region see the environment as in alpha-Fe but have one or more Al neighbours. The interface layers were transformed on bombardment with 100 keV Ar^+ ions at a dose of 10^{16} $ions/cm^2$ into a random metastable alloy having an average composition of $Fe_{55}Al_{45}$. Results of the annealing studies on this system have also been described. In addition to these studies Ogale et al.[607,608] have investigated the ion-beam-induced and subsequent thermal transformations at the Fe-Si and Fe-Ge interfaces. In the former[607], the ion-beam-mixed sample showed significant changes in the values of hyperfine interaction parameters subsequent to vacuum annealing treatment at 450K for one hour while precipitation of alpha-Fe and Fe_3Si were obtained on annealing at 700K. In the paper dealing with the iron-germanium interface, the ion-beam-mixed samples exhibited the presence of the magnetic FeGe phase along with Fe atoms situated in interstitial positions having a tetrahedral environment about Ge.

The as-deposited as well as ion-beam-mixed composites showed significant structural relaxations on annealing at 350K. Similarities and differences between this interface and those of Fe-Al and Fe-Si were then considered. Both Fe-57 and Sn-119 conversion-electron Moessbauer spectrometry were applied to the investigation of the compositional variations of the Fe-Sn intermetallic alloy produced at the surface of 90nm thick Sn-electro-deposited steel by thermal treatment[609]. The fully alloyed layer of $FeSn_2$, FeSn + $FeSn_2$, FeSn, FeSn + Fe_3SnC, and Fe_3SnC was formed by the thermal treatment for 30 minutes at 300, 400, 500, 600, and 700K respectively, and the thickness of the Fe-Sn layer was observed to increase with increasing temperature.

The CEM spectra have been recorded[610] for Fe-40%at. Al implanted with Al^{2+} or Fe^{2+} ions and for Fe-50%at. Al implanted with Ar^{3+} ions. The variation of the long-distance order parameter as a function of the magnetic hyperfine splitting was given for Al^{2+}-implanted Fe-40%at. Al, and no magnetic hyperfine splitting was observed for implantation in Fe-50%at. Al.

Steels and Corrosion Products. In a review[611] of recent developments in conversion-electron Moessbauer spectroscopy for surface analysis Ujihira has described the application of the technique to the following: the characterisation of corrosion products or phosphated coatings produced on steel, the layer-by-layer analysis of the hardened steel surface, the estimation of intermetallic compounds formed between Fe-Zn and Fe-Sn interfaces, and studies of surface deposition of Fe compounds by the crystallisation of amorphous Fe.

Results have been presented[612] on the anisotropy of the Moessbauer effect on the surface of the steel 45 sample with crystallographic texture of the surface resulting from grinding. Two Moessbauer spectroscopic methods[613] were developed for the study of the magnetic texture of steel sheets, which were based on recording the conversion electrons or the characteristic emission accompanying internal conversion of gamma-quanta. For electrical-technology steel the former method showed a nearly isotropic magnetic texture whereas the latter one indicated the magnetisation of domains predominantly along the sheet rolling direction. The phase structure of laser-modified surface layers of alloys[614] has also been studied by CEMS.

The corrosion products[615] of mild steel in an acid medium have been identified using CEMS. In another paper by Rao et al.[616] the results of CEMS characterisation of boride phases before and after pulsed laser treatment at 3 J/cm^2 have been presented. A pack cementation was used to produce 10-45 thick borided layers. The CEMS spectrum showed different boride phases present on the steel, such as Fe_2B and FeB, and, as the thickness of the boride layer increased, the FeB was increasingly formed over an underlying Fe_2B layer of laser.

Chemical Reactions. In situ conversion-electron Moessbauer studies have been carried out on alpha- and gamma-Fe_2O_3 gas-sensor materials[617]. Information on the surface states of the oxides was obtained at 400°C (alpha phase) and at 250°C (gamma phase). The gas-sensing activity depended on a number of factors including the amount of remaining sulphate ion + the microstructure, and a small amount of Fe(II) species generated through the reduction of alpha-Fe_2O_3 gas sensor was confirmed to be due to the reduction of gamma-Fe_2O_3 to $Fe_{3-x}O_4$. Thin films of Fe_2O_3 deposited onto transparent conducting glasses by a spray pyrolysis technique were characterised by CEMS[618]. The growth of thin Fe_2O_3 films was found to be dependent upon the substrate material used, viz tin(IV) oxide, indium oxide, and zinc oxide coated glasses. The CEMS spectra of $CoFe_{2-x}Al_xO_4$ (x = 0.1, 0.3, 0.6, 0.8, and 1.0) were measured[619] at room temperature and the cation distributions for all samples derived and compared with those distributions for powdered samples from normal transmission measurements. The fraction of Co^{2+} ions at the tetrahedral lattice sites was found to increase linearly with the Al content. In another paper[620] the CEMS results on the cation distribution in $Co^{2+}Fe_{2-x}Co^{3+}O_4$ ($0 < x < 2$) showed a shift of Co^{2+} from octahedral to tetrahedral sites when Co^{3+} was substituted in $CoFe_2O_4$. This change resulted in enhanced optical absorption, enhanced magneto-optical effects and a lower Curie temperature.

Single-crystal garnet films of composition $(YSmCaLu)_3$-$(FeGe)_{5O12}$ were implanted 621 with 100 keV Ne^+ ions with doses of 2 x $10^{14}Ne^{+}/cm^{2}$ and 6 x $10^{14}Ne^{+}/cm^{2}$. After annealing at temperatures from 600 to 1000K the films were investigated by CEMS at room temperature. The effects of ion implantation could be recovered completely by annealing at 1000K. Both CEMS and X-ray double-axis spectroscopy[622] were used to investigate the magnetic and structural properties respectively of the thin surface layer that forms on Ca-doped YIG films as a result of annealing in a hydrogen atmosphere. In this study the stresses in both the surface layer and bulk of the film were determined.

Finally, Eymery has reviewed[623] the principle of Moessbauer backscattering spectroscopy and in doing so has demonstrated the value of the technique in the study of the mineralogy of samples of sulphides, carbonates, oxides, phosphates, silicates, and meteorite minerals.

Other Elements.—In principle, the CEMS technique can be used for any of the Moessbauer isotopes. However, as in previous years, in addition to those reporting applications of Fe-57 Moessbauer effect only a few papers report the use of other isotopes. Two papers were published in the past year describing studies involving the use of Sn-119; one paper reports the use of the Eu-151 isotope, and for the first time a paper has appeared detailing a CEM study involving the Te-125 isotope.

The implantation behaviour of stable $^{119}Sn^{+}$ ions and radioactive $^{119}In^{+}$, $^{119m}Sn^{+}$, $^{119}Sb^{+}$, and $^{119m}Te^{+}$ ions in SiC has been investigated[425] by conversion-electron Moessbauer spectroscopy on the 24 keV transition of Sn-119 and by Moessbauer emission spectroscopy on the 24 keV gamma-radiation emitted by the Sn-119 daughter after the decays of the radioactive isotopes. The Moessbauer spectra could be decomposed in most cases into two groups of lines, one originating from Sn-119 atoms on substitutional Si sites, the other from various Sn-vacancy complexes distinguished by their Moessbauer parameters. Annealing experiments revealed a strong dependence of the structure of the defects and the formation and annealing kinetics on the chemical nature of the impurities. Sn-119 conversion-electron Moessbauer spectroscopy has also been applied at 4.2, 80, and 300K to study the alloying effects due to high dose (2 x 10^{16} ions/cm^{2})

implantation of $^{119}Sn^+$ ions into Al. The nature of the local states of Sn and their thermal evolution after stepwise annealing at <400° has been discussed. The three main components of the spectra obtained were ascribed to substitutional Sn, to Sn in highly compressed positions, and to SnO_2.

The Eu-151 Moessbauer spectra of surface-oxidised europium(II) sulphate have been measured[625] using gamma-ray transmission and conversion-electron techniques. Comparison of the spectra demonstrated that the change in the oxidation state of trace amounts of Eu induced by surface reactions could be characterised by the CEM technique. The photochemical reduction of Eu(III) oxalate to an Eu(II) species after irradiation with a low-pressure Hg lamp was also demonstrated using CEM spectroscopy.

Radioactive I-125 was implanted into six metal matrices (Cu, Zn, Sn, Pt, Au, and Te) by means of an electromagnetic isotope separator, and the Te-125 internal conversion-electron Moessbauer spectra were recorded using an Fe-free beta spectrometer[503]. From the Moessbauer data the relative change of the nuclear charge radius in the 35.46 keV transition of Te-125 was found to be $\underline{R}/\underline{R}$ = 0.76 x 10^{-4} with the assumption of a uniform charge distribution of $\underline{R}$ = 1.2 x $A^{1/3}$ fm.

(References begin overleaf)

References

1. 'Moessbauer Effect Reference and Data Journal' ed. J.G. Stevens, V.E. Stevens, R.W. White and J.L. Gibson, Moessbauer Effect Data Center, University of North Carolina, Ashville, N.C. 28814, 1985, Vol. 8.
2. R.V. Parish, Chem. in Britain, 1985, 21, 546 and 740.
3. K. Burger, Magy. Kem. Lapja, 1985, 40, 309.
4. K. Ruebenbauer, Adv. Colloid Interface Sci., 1985 23, 257.
5. L.N. Mulay and T. Pannaparayil, ACS Symp. Ser., 1985, 288, 498.
6. A.G. Maddock, Chem. Bonding Spectrosc. Miner. Chem., 1985, 141.
7. K.S. Venkateswarlu and V. Ramshesh, J. Archaeol. Chem., 1984, 2, 41.
8. P.B. Fabrichnyi, Inorg. Solid Fluorides, 1985, 519.
9. W. Suski, Physica B+C, 1985, 130, 195.
10. K. Tomala, Zesz. Nauk. Uniw. Jagiellon, Pr. Biol. Mol., 1984, 12, 75.
11. H. Sano, Seramikkusu, 1985, 20, 530.
12. M.D. Dyar, Am. Mineral, 1985, 70, 304.
13. M.J. Fluss, Mater. Sci. Forum, 1984, 2, 1.
14. B.H. Huynh and T.A. Kent, Adv. Inorg. Biochem., 1984, 6, 163.
15. G.L. Haller, Appl. Surf. Sci., 1985, 20, 351.
16. G. Lehmann, NATO ASI Ser., Ser.C, 1984, 137, 121.
17. A. Vertes, Magy. Alum., 1985, 22, 52.
18. Y. Ujihira, Bunseki, 1985, 231.
19. U. Gonser and H.G. Wagner, Hyperfine Interact., 1985, 26, 769.
20. P. Boolchand, Phys. Prop. Amorphous Mater., 1985, 221.
21. N. Burriesci, M. Petrera and A. Gennaro, Chim. Ind., 1985, 67, 494.
22. J.E.D. Davies, Inclusion Compd., 1984, 3, 37.
23. W. Wolski, Farbe Lack, 1985, 91, 184.
24. B. Kolk, S. Afr. J. Sci., 1985, 81, 662.
25. E.V. Zolotoyabko and E.M. Iolin, Latv. PSR Zinat. Akad. Vestis, 1985,76.
26. R. Vijayaraghavan and L.C. Gupta, Mater. Sci. Forum, 1985, 3, 365.
27. P. Jena, Mater. Sci. Forum, 1984, 2, 27.
28. W.M. Reiff, NATO ASI Ser., Ser. C., 1985, 140, 355.
29. G. Vogl and W. Petry, Festkoerperprobleme, 1985, 25, 655.
30. P.B. Fabrichnyi, Zh. Vses. Khim. O-va. im. D.I. Mendeleeva, 1985, 30, 143.
31. G.M. Kalvius, Hyperfine Interact., 1985, 26, 793.
32. G.M. Kalvius, W. Potzel, J. Moser, F.J. Litterst, L. Asch, J. Zaenkert, U. Potzel, A. Kratzer and M. Wunsch, Physica B+C, 1985, 130, 393.
33. G.M. Kalvius, W. Potzel and J. Moser, INIS Atomindex, 1985, 16, No. 16:047430.
34. S.J. Blunden, P.A. Cusack, and R. Hill, 'The Industrial Uses of Tin Chemicals', The Royal Society of Chemistry, London, 1985.
35. S.W. Ng and J.J. Zuckerman, Ad. Inorg. Chem. Radiochem., 1985, 29, 297.
36. W.E. Geiger and N.G. Connelly, Ad. Organometal.Chem., 1986, 24, 87.
37. P.G. Harrison, Annu. Rep. Prog. Chem., Sect. A, 1984, 81, 95.
38. F.J. Berry, Annu. Rep. Prog. Chem., Sect. A, 1984, 81, 137.
39. B.W. Fitzsimmons, Annu. Rep. Prog. Chem., Sect. A, 1984, 81, 225.
40. J. Silver, Annu. Rep. Prog. Chem., Sect. A, 1984, 81, 301.
41. J.W.C. De Vries, R.C. Thiel and K.H.J. Buschow, J. Phys. F. Met Phys., 1985, 15, 2403.
42. D.N. Sahu, Acta. Phys. Pol., A., 1985, 68, 781.
43. E.L. Albuquerque and C.A. Dos Santos, Phys. Status Solidi, B, 1985, 132, K5.
44. J. Javanainen, P. Helisto, E. Ikonen and T. Katila, Phys. Rev. Lett., 1985, 55, 2063.
45. E. Realo, M. Haas and J. Jogi, Zh. Eksp. Teor. Fiz,1985, 88, 1864.
46. A.L. Sharma, Alimuddin and K.R. Rama, Nuovo Cimento Soc. Ital. Fis., D, 1985, 5D, 147.
47. O. Ballet, Hyperfine Interact., 1985, 23, 133.
48. K. Makariunas and E. Makariuniene, Liet. Fiz. Rinkinys, 1985, 25, 111.

49. E. Ratner, Nucl. Instrum. Methods Phys. Res., Sect. B, 1984, 233, 518.
50. N. Blaes, H. Fischer and U. Gonser, Nucl. Instrum. Methods Phys. Res., Section B, 1985, 9, 201.
51. P. Moretti, Lett. Nuovo Cimento Soc. Ital. Fis., 1985, 43, 155.
52. V.D. Gorobchenko and F.J. Litterst, Phys. Status Solidi B, 1985, 127, 351.
53. V.E. Sedov, Phys. Status Solidi B, 1985, 131, 299.
54. I. Nowik, E.R. Bauminger, S.G. Cohen and S. Ofer, Phys. Rev. A, 1985, 31, 2291.
55. E. Ikonen, P. Helisto, T. Katila and K. Riski, Phys. Rev. A: Gen. Phys., 1985, 32, 2298.
56. A.L. Shames, V.E. Fainzil'berg, M.I. Belinskii and B.S. Tsukerblat, Nek. Vopr. Fiz. Neravnovesnykh Protsessov Poluprovodn. Dielektr., 1984, 32.
57. S.K. Roy, Pramana, 1985, 24, 773.
58. S.K. Roy and N. Kundu, Pramana, 1985, 25, 319.
59. T. Peev and A. Vertes, Acta Chim. Hung., 1985, 119, 57.
60. Y. Wang, Huaxue Xuebao, 1985, 43, 313.
61. J.P. Hannon, G.T. Trammell, M. Mueller, E. Gerdau, R. Rueffer and H. Winkler, Phys. Rev. B: Condens. Matter, 1985, 32, 6363.
62. J.P. Hannon, G.T. Trammell, M. Mueller, E. Gerdau, R. Rueffer and H. Winkler, Phys. Rev. B: Condens. Matter, 1985, 32, 6374.
63. Baryshevskii, J. Phys. C, 1985, 18, 191.
64. N. Sakai and H. Sekizawa, J. Phys. Soc. Jpn., 1985, 54, 474.
65. S.M. Cheremisin, I.V. Matveev and A.Y. Dudkin, Prib. Tekh. Eksp., 1985, 6, 35.
66. E. Wallingford and C. Bordeleau, Rev. Sci. Instrum., 1985, 56, 1253.
67. E.V. Bil'dyukevich, V.L. Gurachevskii, Y.A. Litvinovich, M. Mashlan, O.V. Misevich, A.L. Kholmetskii and V.A. Chudakov, Prib. Tekh. Eksp., 1985, 6, 62.
68. J.F. Harmon, Gov. Rep. Announce Index, 1985, 85, 123.
69. K.F. Purcell and M.P. Edwards, Rev. Sci. Instrum., 1985, 56, 108.
70. M.K. Sanyal, V.C. Sahni and G.P. Das, Phys. News (Bombay), 1985, 16, 53.
71. W. Zhang, Z. Wang, A. Shong, D. Su, T. Fan and G. Su, He Huaxue Yu Fangshe Huaxue, 1985, 7, 189.
72. Y. Xia, R. Liu, S. Wang, Z. Cao, Q. Chen, L. Shen, Z. Gan, W. Mao, Y. Li and S. Jiang, Nanjing Daxue Xuebao, Ziran Kexue, 1984, 1, 59.
73. S. Ormandzhiev and T. Semova, Bulg. J. Phys., 1985, 12, 110.
74. G.L. Zhang, E.H. Du Marchie van Voorthuysen and H. De Waard, Phys. Lett. A, 1985, 111, 145.
75. V.S. Vartanov and B.G. Zemskov, USSR Pat. 1984 Su 1,069,538.
76. V.V. Morozov, Zh. Tekh. Fiz., 1985, 55, 683.
77. B. Kolk, A. Bleloch, D.B. Hall, Y. Zheng and K.E. Patton-Hall, Rev. Sci. Instrum., 1985, 56, 1597.
78. J.B. Mueller, Nucl. Instrum. Methods Phys. Res., Sect. A, 1985, 235, 142.
79. Y. Zhang and X. Qu, He Dianzixue Yu Tance Jishu, 1985, 5, 266.
80. D.S. Zakhar'in, F.K. Chibirova and S.I. Reiman, Prib. Tekh. Eksp., 1985, 4, 49.
81. P. Jernberg, Atomindex, 1985, 16, 012558.
82. K. Shimokoshi, T. Kanzaki and R.N. Jones, Appl. Spectrosc., 1985, 39, 1014.
83. M. Jin, Jilin Daxue Ziran Kexue Xuebao, 1985, (1), 71.
84. M. Jin, Jilin Daxue Ziran Kexue Xuebao, 1985, (2), 65.
85. R.T. Eakin, Diss. Abstr. Int. B., 1985, 45, 2198.
86. E. Realo, K. Rebane, M. Haas and J. Eigi, Pis'ma Zh. Eksp. Teor. Fiz., 1984, 40, 477.
87. Y.V. Shvyd'ko, Phys. Lett. A, 1985, 113, 273.
88. L. Tsankov, J. Phys. D, 1985, 18, 1645.
89. L. Liu, W. Zhou, L. Xue and M. Hu, Hejishu, 1984, (6), 38.
90. A.R. Arakelyan, E.M. Arutyunyan and M.M. Slavinskii, Izv. Akad. Nauk Arm. SSR, Fiz., 1985, 20, 278.

91. D.Massiot, J. Non-Cryst. Solids, 1985, 69, 371.
92. E. Antonicik, Proc. Int. Conf. Phys. Semicond., 17th, 1984, 1017.
93. P.H. Kydd, U.S. Pat., US4,491,955.
94. G. Balestrino and S. Geller, J. Magn. Magn. Mater., 1985, 49, 225.
95. F. Wittmann and J. Jex, Solid State commun., 1985, 53, 407.
96. G.V. Smirnov and A.I. Chumakov, Zh. Eksp. Teor. Fiz., 1985, 89, 1169.
97. J.J. Bara and B.F. Bogacz, Nucl. Instrum. Methods Phys. Res., Sect. A, 1985, 238, 469.
98. A. Deriu, J. Appl. Crystallogr., 1985, 18, 122.
99. R.G. Brewer and R.G. de Voe, NATO Conf. Ser., 1985, 9, 171.
100. S.S. Wagel, B.D. de Paola, and C.B. Collins, Proc. Int. Conf. Lasers, 1984, 257.
101. F. Menil, J. Phys. Chem. Solids, 1985, 46, 763.
102. H. de Waard, J.A. Sawicki, B.D. Sawicka, F. Pleiter, and K. Franz, J. Magn. Magn. Mater., 1985, 49, 50.
103. V.I. Novikov, P.F. Relushko, L.I. Trusov, Yu.A. Lopukhov, Fiz. Tverd. Tela (Leningrad), 1984, 26, 3447.
104. N. Ayoub, M.A. Kobeissi, R.W. Chantrell, K. O'Grady, and J. Popplewell, J. Phys. F, 1985, 15, 2229.
105. A.M. Afanas'ev, M.A. Chuev, W. Zinn, Ch. Sauer, and A. Holzwarth, Zh. Eksp. Teor. Fiz., 1985, 89, 182.
106. V.Ya. Rochev, E.F. Makarov, O.P. Kevdin, L.B. Kukushkina, and G.N. Kosova, Biol. Membr., 1985, 2, 95.
107. R.P. Marande, Diss. Abstr. Int. B, 1985, 45, 3268.
108. Z. Kucharski, M. Jozefowicz, J. Suwalski, A. Pron, and S. Lefrant, Mol. Cryst. Liq. Cryst., 1985, 117, 133.
109. A. Pron, Z. Kucharski, C. Budrowski, M. Zagorska, S. Krichene, J. Suwalski, G. Dehe. and S. Lefrant, J. Chem. Phys., 1985, 83, 5923.
110. T. Yamamoto and H. Sakai, J. Mater. Sci. Lett., 1985, 4, 916.
111. A. Meagher, G. Smyth, V.J. McBrierty, J.M.D. Coey, and M. Pineri, J. Chem. Phys., 1985, 83, 2552.
112. A. Meagher, B. Rodmacq, J.M.D. Coey, and M. Pineri, NATO ASI Ser., Ser. E, 1985, 98, 332.
113. E.R. Bauminger, I. Nowik, S. Ofer, and C. Heitner-Wirguin, Polymer, 1985, 26 , 1829.
114. A.S. Plachinda, D.B. Poblinkov, B.N. Laskorin, E.F. Makarov, A.A. Logvinenko, V.V. Krylova, and A.I. Zorina, Koord. Khim., 1984, 10, 1669.
115. A.S. Plachinda, B.N. Laskorin, D.B. Poblinkov, E.F. Makarov, N.G. Zhukova, V.V. Krylova, A.A. Logvinenko, A.Yu. Permogorov, and V.I. Khromov, Dokl. Akad. Nauk SSSR, 1985, 280, 938.
116. A.S. Plachinda, D.B. Poblinkov, B.N. Laskorin, E.F. Makarov, N.G. Zhukova, V.V. Krylova, A.A. Logvinenko, A.Yu. Permogorov, and V.I. Khromov, Izv. Akad. Nauk. SSSR, Ser. Khim., 1985, 1740.
117. V.L. Gutsanu, K.I. Turte, R.A. Stukan, and V.A. Gafiichuk, Zh. Fiz. Khim., 1985, 59, 693.
118. Yu.I. Gogolyuk, E.V. Pol'shin, and V.V. Mazykin, Izv. Akad. Nauk SSSR, Neorg. Mater., 1985, 21, 71.
119. D. Barb, N.M. Grecu, D. Tarina, J. Blum, and N. Goidea, Rev. Roum. Phys., 1984, 29, 647.
120. D.G. Rancourt, S.R. Julian, and J.M. Daniels, J. Magn. Magn. Mater., 1985, 49, 305.
121. I. Zay, A. Vertes, G.T. Nagy, M. Suba, and K. Burger, J. Radioanal. Nucl. Chem., 1985, 88, 343.
122. S. Moerup and J.E. Knudsen, J. Phys. C, 1985, 18, 2943.
123. P.A. Montano, G.K. Shenoy, T.I. Morrison, and W. Schulze, Springer Proc. Phys., 1984, 2, 231.
124. H. Micklitz, M. Pasternak, and J.P. Sanchez, Solid State Commun., 1985, 55, 667.

125. S. Bukshoan, I. Dezsi, H. Pattyn, M. van Rossum, V. Verbiest, G. Langouche, and R. Coussement, Radiat. Eff., 1985, 89, 215.
126. A.L. Kulikov and M.I. Afanasov, Vestn. Mosk. Univ., Ser. 2: Khim., 1984, 25, 610.
127. J. Fontcuberta, X. Obradors, J. Rodriguez, and J. Tejada, J. Phys. Chem. Solids, 1985, 46, 301; 305.
128. J. Fontcuberta and J.B. Goodenough, J. Solid State Chem.,1985, 56, 116.
129. K.C. Mishra, K.J. Duff, P. Kelires, S.K. Mishra, and T.P. Das, Phys. Rev. B, 1985, 32, 58.
130. W.A.H.M. Vlak, D.J. Dikken, A.F.M. Arts, and H.W. de Wijn, Phys. Rev. B, 1985, 32, 4496.
131. D.J. Dikken, A.F.M. Arts, H.W. de Wijn, W.A.H.M. Vlak, and E. Frikkee, Phys. Rev. B, 1985, 32, 5600.
132. H.H.A. Smit, H.J.M. de Groot, R.C. Thiel, and L.J. de Jongh, Solid State Commun., 1985, 53, 573.
133. V.G. Ksenofontov, A.Ta. Laptienko, I.V. Ruban, B. Ya. Sukharevsky, and A.V. Pustovit, Solid State Commun., 1985, 53, 9.
134. K.K.P.Srivatava, Phys. Rev. B, 1985, 32, 3282.
135. A.G. Nord., G. Aaberg, H. Annersten, T. Ericsson, and T. Stefanidis, Chem. Scr., 1985, 25, 189.
136. N. Sakai, F. Sakai, and H. Sekizawa, Bull. Chem. Soc, Jpn., 1985, 58, 574.
137. M.F.A. Dove, R.C. Hibbert, and N. Logan, J. Chem. Soc., Dalton Trans., 1985, 707.
138. K. Rachlewicz, J. Pietrzyk, and K. Drabent, Polyhedron, 1985, 4, 1777.
139. M. Krause, C. Michalk, M. Benedix, and J. Suwalski, Phys. Status Solidi A, 1985, 90, 657.
140. B.W. FitzSimmons, A. Hume, L.F. Larkworthy, M.H. Turnbull, and A. Yavan, Inorg. Chim. Acta, 1985, 106, 109.
141. E. Molins, A. Labarta, X. Tejada, and V. Moreno, Z. Phys. B, 1985, 59, 419.
142. M.D. Timken, W.A. Marritt, D.N. Hendrickson, R.A. Gagne, and E. Sinn, Inorg. Chem., 1985, 24, 4202.
143. C.P. Landee, R.E. Greeney, W.M. Reiff, J.H. Zhang, J. Chalupa, and M.A. Novotny, J. Appl. Phys. , 1985, 57, 3343.
144. Q.A. Pankhurst, C.E. Johnson, and M.F. Thomas, J. Phys. C, 1985, 18, 3249.
145. C.S.M. Partiti, A. Piccini, and H.R. Rechenberg, Solid State Commun., 1985, 56, 687.
146. E.R. Jones, jr., C. Van Hine, T. Datta, L. Cathey, and D.G. Karraker, Inorg. Chem., 1985, 24, 3888.
147. J. Pebler, J. Solid State Chem., 1985, 56 , 58.
148. J. Teillet and G. Courbion, J. Phys. Chem. Solids, 1985, 46, 559.
149. C. Pappa, J. Hammann, G. Jehanno, and C. Jacobi, J. Phys. C, 1985, 18, 2817.
150. Q.A. Pankhurst, M.F. Thomas, and B.M. Wanklyn, J. Phys. C, 1985, 18, 1255.
151. P.C. Morais, A.L. Tronconi, and K.S. Neto, J. Appl. Phys., 1985, 57, 1291.
152. P.C. Morais and K.S. Neto, J. Appl. Phys., 1985, 58, 2080.
153. M.G. Townsend, G. Longworth, and J.E. Dutrizac, Phys. Rev. B, 1985, 31, 2966.
154. R.A. Stukan, V.I. Ponomarev, V.I. Nifontov, K. Turte, and L.O. Atovmyan, Zh. Strukt. Khim., 1985, 26, 62.
155. K.I. Turte, S.A. Bobkova, B. Ya. Kuyavskaya, I.N. Ivlevba, V.I. Ponomarev, and M.E. Veksel'man, Koord. Khim., 1985, 11, 1106.
156. S. Yamauchi, Y. Sakai, and T. Tominaga, Bull. Chem. Soc. Jpn., 1985, 58, 442.

157. S. Wadja and W. Kielbus, Proc. Sch.-Symp. Inorg. Biochem. Mol. Biophys., 1985, 237.
158. S.G. Shova, M.A. Yampol'skaya, B.G. Zemskov, N.V. Gerbeleu, K.I. Turte, and I.V. Ivleva, Zh. Neorg. Khim., 1985, 30, 2309.
159. B.J. Kennedy, K.S. Murray, P.R. Zwack, H. Homberg, and W. Kalz, Inorg. Chem., 1985, 24, 3302.
160. S. Wolowiec and K. Drabent, J. Radioanal. Nucl. Chem., 1985, 95, 1.
161. J. Janicki, M. Jaruszczyk, R. Krzyminiewski, J. Pietrzak, J. Suwalski, and I. Sledzinska, Acta Phys. Pol., 1985, A68, 69.
162. N.I. Kaletina, Koord. Khim., 1985, 11, 219.
163. V. Ravindar, P. Lingaiah, M. Vithal, and R. Jagannathan, Ind. J. Chem., 1985, 24A, 485.
164. J.G. Vos, Inorg. Chim. Acta, 1985, 99, 117.
165. B.J. Kennedy, G. Brain, E. Horn, K.S. Murray, and M.R. Snow, Inorg. Chem., 1985, 24, 1647.
166. J.C. Fanning, J.L. Resce, G.C. Lickfield, and M.E. Kotun, Inorg. Chem., 1985, 24, 2884.
167. R.H. Herber and H. Eckert, Phys. Rev. B, 1985, 31, 34.
168. A.T. Shuvaev, B.Yu. Khel'mer, T.A. Lyubeznova, V.L. Kraizman, A.S. Mirmil'shtein, L.D. Kvacheva, Yu.N. Novikov, and M.E. Vol'pinas, Izv. Akad. Nauk SSSR, Ser. Fiz., 1985, 49, 1471.
169. M. Inoue, M. Matsumoto, H. Negishi, and H. Sakai, J. Magn. Magn. Mater., 1985, 53, 131.
170. T. Nishida, T. Nonaka, and T. Takashima, Bull. Chem. Soc. Jpn., 1985, 58, 2255.
171. S.M. Oh, T. Kambara, D.N. Hendrickson, M. Norai, K. Kaji, S.E. Woehler, and R.J. Wittebort, J. Am. Chem. Soc., 1985, 107, 5540.
172. S.M. Oh, D.N. Hendrickson, K.L. Hassett, and R.E. Davies, J. Am. Chem. Soc., 1985, 107, 8009.
173. E.H. Witten, W.M. Reiff, K. Lazar, B.W. Sullivan, and B.M. Foxman, Inorg. Chem., 1985, 24, 4585.
174. M. Bendix, R. Bendix, C. Michalk, and J. Suwalski, Inorg. Chim. Acta, 1985, 96, Ll.
175. M.J. Cohn, T.Y. Dong, D.N. Hendrickson, S.J. Geib, and A.L. Rheingold, J. Chem. Soc., Chem. Commun., 1985, 1095.
176. T.Y. Dong, D.N. Hendrickson, K. Iwai, M.J. Cohn, S.J. Geib, A.L. Rheingold, H. Sano, I. Motoyama, and S. Nakashima, J. Am. Chem. Soc., 1985, 107, 7996.
177. M.F. Moore, and D.N. Hendrickson, Inorg. Chem., 1985, 24, 1236.
178. H.-H. Wei and H.-L. Shyu, Polyhedron, 1985, 4, 979.
179. S. Das, S. Ganguli, and M. Bhattacharya, J. Phys. Chem. Solids, 1985, 46, 143.
180. S. Decurtins, P.Guetlich, C.P. Koehler, and H. Spiering, J. Chem. Soc., Solid State Commun., 1985, 430.
181. S.Savage, J.L. Zhou, and P.G. Maddock, J. Chem. Soc., Dalton Trans., 1985, 991.
182. E. Koenig, G. Ritter, J. Waigel, and H.A. Goodwin, J. Chem. Phys., 1985, 83, 3055.
183. Y. Maeda, N. Tsutsumi, and Y. Takashima, J. Radioanal. Nucl. Chem., 1985, 93, 253.
184 M.D. Timken, D.N. Hendrickson, and E. Sinn, Inorg. Chem., 1985, 24, 3947.
185. M.D. Timken, S.R. Wilson, and D.N. Hendrickson, Inorg. Chem., 1985, 24, 3450.
186. I. Morgenstern-Badarau and H.H. Wickman, J. Chem. Soc., Solid State Commun., 1985, 176.
187. I. Morgenstern-Badarau and H.H. Wickman, Inorg. Chem., 1985, 24, 1889.
188. R.O. Moyer, jr., R. Lindsay, S. Suib, R.P. Zerger, J. Tanaka, and S.G. Gibbons, Inorg. Chem., 1985, 24, 3890.

189. X. Wang, J. Niu, H. Chen, X. Xin, A. Dai, Y. Zhang, Kexue Tongbao (For. Lang. Ed.), 1985, 30, 279.
190. S. Das, S. Ganguli, and M. Bhattacharya, J. Phys. C, 1985, 18, 3809.
191. D.V. Parwate and A.N. Garg, Ind. J. Chem., 1985, 24A, 414.
192. D.V. Parwate and A.N. Garg, Z. Naturforsch., B, 1985, 40B, 1495.
193. H. Inoue M. Sasagawa, and E.Fluck, Z. Naturforsch., B, 1985, 40B, 22.
194. C.G. Benson, G.J. Long, J.W. Kolis, and D.F. Schriver, J. Am. Chem. Soc., 1985, 107, 5297.
195. D. Rojas, P. Bussiere, J.A. Dalmon, A. Cholin, J.M. Basset, and D. Oliver, Surf. Sci., 1985, 156, 516.
196. J.S. Brooks, C.M. Care, S. Plimley, and C.G. Corfield, Hyperfine Interact., 1984, 20, 151.
197. I.G. Gusakovskaya, S.I. Pirumova, and N.S. Ovanesyan, Zh. Fiz. Khim., 1985, 59, 2495.
198. B. Lukas, C.W. Patterson, R.M.G. Roberts, and J. Silver, J. Organomet. Chem., 1985, 286, 209.
199. K. Sato, S. Nakashima, M. Watanabe, I. Motoyama, and H. Sano, Nippon Kagaku Kaishi, 1985, 580.
200. S. Toma, E. Solcaniova, and A.G. Nagy, J. Organomet. Chem., 1985, 288, 331.
201. X. Ding, Y. Hsia, R. Liu, S. Wang, S. Shi, L. Chen, S. Jian, T. Huang, and S. Lian, Nanjing Daxue Xuebao, Ziran Kexue, 1985, 21, 253.
202. B. Lukas and J. Silver, Inorg. Chim. Acta, 1985, 106, 219.
203. S.H. Strauss, M.E. Silver, K.M. Long, R.G. Thompson, R.A. Hudgens, K. Spartalian, and J.A. Ibers, J. Am. Chem. Soc., 1985, 107, 4207.
204. X. Yang, L. Gong, and Y. Gu, Bopuxue Zazhi, 1985, 2, 7.
205. G.P. Gupta, G. Lang, W.R. Scheidt, D.K. Geiger, and C.A. Reed, J. Chem. Phys., 1985, 83, 5945.
206. W.R. Scheidt, D.K. Geiger, Y.J. Lee, and C.A. Reed, J. Am. Chem. Soc., 1985, 107, 5693.
207. V.I. Kleskov, B.N. Burykin, and R.E. Garibov, Teor. Eksp. Khim., 1985, 21, 146.
208. D.J. Liston, B.J. Kennedy, K.S. Murray, and B.O. West, Inorg. Chem., 1985, 24, 1561.
209. N.P. Didenko, V.I. Zelentsov, N.S. Fedorova, V.A. Cha, and E.M. Chuprikova, Deposited. Doc., 1984, VINITI 4974.
210. M.I. Oshtrakh and V.A. Semenkin, Mol. Biol. (Moscow), 1985, 19, 1310.
211. V.E. Prusakov, R.A. Stukan, R.M. Davidov, and K. Gersonde, FEBS Lett., 1985, 186, 158.
212. A. Levy and J.M. Rifkind, Biochemistry, 1985, 24, 6050.
213. F. Parak, Struct. Motion: Membr., Nucleic Acids proteins, Proc. Int. Symp. Struct. Dyn. Membr., Nucleic Acids Proteins 1984, (pub. 1985), 243.
214. L. Reinisch, J. Heidemeier, and F. Parak, Eur. Biophys. J., 1985, 12, 167.
215. C.J. Carrano, K. Spartalian, G.V.N.A. Rao, V.L. Pecoraro, and M. Sundaralingam, J. Am. Chem. Soc., 1985, 107, 1651.
216. J.M. Nocek, D.M. Kurtz, jr., J.T. Sage, P.G. Debrunner, M.J. Maroney, and L.Que, jr., J. Am. Chem. Soc., 1985, 107, 3382.
217. H.R. Allcock, T.X. Neenan, and B. Boso, Inorg. Chem., 1985, 24, 2656.
218. V.A. Lapina, A.E. Dontsov, and M.A. Ostrovskii, Koord. Khim., 1985, 11, 1234.
219. N. Zhu, Y. Zhang, Y. Chen., and D. Zhu, Kexue Tongbao (For. Lang. Ed.), 1984, 29, 1685.
220. T.A. Ken, M.H. Emptage, H. Merkle, M.C. Kennedy, H. Beinert, and E. Muenck, J. Biol. Chem., 1985, 260, 6871.

221. P.A. Lindahl, E.P. Day, T.A. Kent, W.H. Orme-Johnson, and E. Muenck, J. Biol. Chem., 1985, 260, 11160.
222. T. Yagi, K. Kimura, and H. Inokuchi, J. Biochem., 1985, 97, 181.
223. B.H. Huynh, L. Kang, D.V. DerVartanian, H.D. Peck, jr., and J. LeGail, J. Biol. Chem., 1984, 259, 15373.
224. R. Hille, W.R. Hagen, and W.R. Dunham, J. Biol. Chem., 1985, 260, 10569.
225. M.G. Kanatzidis, D. Coucouvanis, A. Simopoulos, A. Kostikas, and V. Papaefthymiou, J. Am. Chem. Soc., 1985, 107, 4925.
226. M.G. Kanatzidis, W.R. Hagen, W.R. Dunham, R.K. Lester, and D. Coucouvanis, J. Am. Chem. Soc., 1985, 107, 953.
227. A.X. Trautwein, E. Bill, R. Blaes, S. Lauer, H. Winkler, and A. Kostikas, J. Chem. Phys., 1985, 82, 3584.
228. H. Winkler, E. Bill, S. Lauer, U. Lauer, A.X. Trautwein, A. Kostikas, V. Papaefthymiou, H. Boegge, A. Mueller, et al, J. Chem. Phys., 1985, 82, 3594.
229. S. Simopoulos, A. Kostikas, and V. Papaefthymiou, Comm. Eur. Communities, [Rep] EUR, 1985, EUR 9686.
230. Y. Wang, Heijishu, 1984, 37.
231. W.R. Dunham, W.R. Hagen, A. Braaksma, H. Haaker, S. Gheller, W.E. Newton, and B. Smith, Nitrogen Fixation Res. Prog., Proc. Int. Symp., 6th., 1985, 591.
232. W.R. Dunham, W.R. Hagen, A. Braaksma, H.J. Grande, Eur. J. Biochem., 1985, 146, 497.
233. E. Bill, F.H. Bernhardt, A.X. Trautwein, and H. Winkler, Eur. J. Biochem., 1985, 147, 177.
234. Yu.F. Krupyanskii, D. Bade, I.V. Sharkevich, N.Ya. Uspenskaya, A.A. Kononenko, I.P. Suzdalev, F. Parak, V.I. Goldanskii, R.L. Moessbauer, and A.B. Rubin, Eur. Biophys. J., 1985, 12, 107.
235. K.V. Shaitin, I.N. Nikolaev, A.A. Kononenko, and A.B. Rubin, Biofizika, 1985, 30, 249.
236. J.N. Rimbert, F. Dumas, C. Kellershohn, R. Girot, and P. Brissot, Biochimie, 1985, 67, 663.
237. S. Ambe, F. Ambe, and T. Nozaki, Int. J. Appl. Radiat. Isot., 1985, 36, 7.
238. D.G. Chambaere, E. de Grave, R.L. Vanleerberghe, and R.E. Vanleerberghe, Hyperfine Interact., 1984, 20, 249.
239. M.B. Madsen, S. Moerup, C.J.W. Koch, and O.K. Borggaard, Surf. Sci., 1985, 156, 328.
240. J. Zhang, Y. Xia, S. Wang, R. Liu, and S. Qi, Kexue Tongbao (For. Lang. Ed.), 1985, 30, 1175.
241. D.M. Sherman, Phys. Chem. Miner., 1985, 12, 311.
242. R.V. Morris, H.V. Lauer, jr., C.A. Lawson, E.K. Gibson, jr., G.A. Nace, and C. Stewart, J. Geophys. Res. B, 1985, 90, 3126.
243. C.A. McCammon and D.C. Price, Phys. Chem. Miner., 1985, 12, 250.
244. K.D. Jayasuriya, A.M. Stewart, and S.J. Campbell, J. Phys. Chem. Solids, 1985, 46, 625.
245. E. de Grave, A.E. Verbeeck, and D.G. Chambaere, Phys. Lett. A, 1985, 107A, 181.
246. T. Suzuki, T. Yagi, S. Akimoto, A. Ito, and S. Morimoto, Solid State Phys. Pressure: Recent Adv. Anvil Devices, 1985, 149.
247. H. Mehner, S. Nagy, H.J. Schierhorn, and A. Vertes, J. Radioanal. Nucl. Chem., 1985, 94, 210.
248. Y. Xia, R. Liu, S. Wang, C. Xia, and D. Shen, Wuli Xuebao, 1985, 34, 1185.
249. J.L. Dormann, J.R. Cui, and C. Sella, J. Appl. Phys., 1985, 57, 4283.
250. V. Kumar, R.M. Singru, O.Parkash, and D. Bahadur, Phys. Status Solidi B, 1985, 128, 223.

251. T.D. Bakuma and E.A. Zhurakovskii, Izv. Akad. Nauk. SSSR, Neorg. Mater., 1984, 20, 2021.
252. V.P.Pashchenko, D.Ya. Serebro, S.M. Lisityn, A.K. Prokopenko, O.A. Kostyagina, L.Ya. Selivanova, and G.M. Ryabova, Izv. Akad. Nauk SSSR, Neorg. Mater., 1985, 21, 1415.
253. F.K. Egorov, G.A. Teterin, E.M. Menchuk, S.P. Ekimov, T.M. Shkol'nikova, and E.G. Semin, Ukr. Khim. Zh. (Russ. Ed.), 1985, 51, 1139.
254. T.Terashima and Y. Bando, J. Phys. Soc. Jpn., 1985, 54, 3920.
255. E. Tronc and D. Bonin, J. Phys. Lett., 1985, 46, 437.
256. V.I. Nikolaev, V.S. Rusakov, and N.I. Christyakova, Phys. Status Solidi A, 1985, 91, K139.
257. R. Gerardin, A. Ramdani, C. Gleizer, B. Gillot, and B. Durand, J. Solid State Chem., 1985, 57, 215.
258. Sh.Sh. Bahkirov, A. Liberman, and A.M. Khasanov, Fiz. Tverd. Tela (Leningrad), 1985, 26, 3465.
259. H. Furukawa, T. Kanzaki, and T. Katsura, J. Electron Spectrosc. Relat. Phenom., 1985, 36, 1.
260. H. Furukawa, T. Kanzaki, and T. Katsura, J. Chem. Soc., Dalton Trans., 1985, 1713.
261. M. Sosenberg, P. Deppe, H.U. Janssen, V.A.M. Brabers, and F.S. Li, J. Appl. Phys., 1985, 57, 3740.
262. Yu.T. Pavlukhin, N.I. Ens, and Ya.Ya. Medikov, Izv. Sib. Otd. Akad. Nauk. SSSR, Ser. Khim. Nauk, 1985, 24.
263. B. Hannover, M. Lenglet, R. Chopova, and J.C. Tellier, Mater. Chem. Phys., 1985, 13, 449.
264. K. Melzer, J. Suwalski, Z. Kucharski, and C. Gera, Acta Phys. Pol. A, 1985, A68, 433.
265. A.Gruskova, Acta Phys. Slovaca, 1985, 35, 256.
266. X. Yang, Y. Gu, H. Tang, R. He, P. He, Wuli Xuebao, 1985, 34, 1133.
267. N.K. Gill and R.K. Puri, Ind. J. Pure Appl. Phys., 1985, 23, 71.
268. N.K. Gill and R.K. Puri, Spectrochim. Acta, 1985, 41A, 1005.
269. C. Michalk and J. Suwalski, Hermsdorfer Tech. Mitt., 1984, 24, 2106.
270. C. Michalk and J. Suwalski, Acta Phys. Pol. A, 1985, A68, 441.
271. V.P. Pashchenko, A.V. Kopaev, G.T. Bovkina, A.K. Prokopenko, and I.F. Klochai, Izv. Akad. Nauk SSSR, Neorg. Mater., 1985, 21, 1773.
272. V.P. Pashchenko, G.T. Bovkina, V.V. Klimov, G.F. Afonina, and L.N. Sharanevich, Poroshk. Metall. (Kiev), 1985, 52.
273. R.G. Kulkarni and H.H. Joshi, Solid State Commum., 1985, 53, 1005.
274. H.H. Joshi, S.N. Rao, and R.G. Kulkarni, J. Mater. Sci. Lett., 1985, 4, 1249.
275. K. Muraleedharan, J.K. Srivastava, V.R. Marathe, and R. Vjayaraghavan, J. Phys. C, 1985, 18, 5355.
276. T. Peev, T. Mitev, T. Dimova, and L. Peneva, God. Vissh. Khim.-Tekhnol. Inst., Sofia 1983, (Pub. 1984), 21, 238.
277. S. Besenicar and D. Hanzel, J. Phys., Colloq., 1985, 169.
278. R. Ramachandran and V.S. Darshane, Proc.- Ind. Acad. Sci., Chem. Sci., 1985, 95, 297.
279. K. Muraleedharan, J.K. Srivastava, V.R. Marathe, R. Vjayaraghavan, J.A. Kulkarni, and V.S. Darahane, Proc. Nucl. Phys. Solid State Phys. Symp. 1982, (Pub. 1984), 25C, 117.
280. K. Muraleedharan, J.K. Srivastava, V.R. Marathe, R. Vjayaraghavan, J.A. Kulkarni, and V.S. Darahane, Solid State Commun., 1985, 55, 363.
281. S.K. Kulshreshtha and G. Ritter, J. Mater. Sci., 1985, 20, 821.
282. S.K. Kulshreshtha and G. Ritter, J. Mater. Sci., 1985, 20, 3926.
283. C.P. Rana, R. Sharma, S. Phanjoubam, and J.S. Baijal, J. Less-Common Met., 1985, 109, L15.
284. R.A. Brand, H. Georges-Gilbert, J. Hubsch, and J.A. Heller, J. Phys. F, 1985, 15, 1987.

285. G.J. Baldha, S.N. Rao, and R.G. Kulkarni, J. Mater. Sci. Lett., 1985, 4, 1536.
286. G.J. Balkdha and R.G. Kulkarni, J. Mater. Sci., 1985, 20, 3675.
287. M.K. El-Nimr, H.A. Saleh, and M.K. Fayek, Appl. Phys. A, 1985, A38, 67.
288. F. Abbattista, D. Mazza, M. Vallino, A. Delunas, and V. Maxia, Lett. Nuova Cimento Soc. Ital. Fis., 1985, 43, 223.
289. T.C. Gibb, J. Chem. Soc., Dalton Trans., 1985, 1455.
290. C. Michalk and W. Thiel, Phys. Status Solidi A, 1985, 90, 325.
291. S. Li and G. Li, Kexue Tongbao (For. Lang. Ed.), 1985, 30, 593.
292. Sh.Sh. Bashkirov, N.G. Ivoilov, and E.S. Romanov, Fiz. Tverd. Tela (Leningrad), 1985, 27, 2853.
293. P. Zhang, Q.Ling, S. Guo, G. Li, Y. Zhang, and L. Zhang, Yingyong, Kexue Xuebao, 1985, 3, 169.
294. S.Ye. Yurchenko, Yu.V. Baldokhin, N.F. Shukin, G.M. Fedichkin, and I.S. Smirnov, Acta Phys. Pol., A, 1985, A68, 317.
295. I.S. Lyubutin, B.V. Mill, R.I. Chalabov, and A.V. Butashin, J. Phys. Chem. Solids, 1985, 46, 363.
296. S. Garcia and E. Altshuler, Phys. Status Solidi A, 1985, 89, 427.
297. V.P. Pashchenko, P.P. Kirichok, G.M. Ryabova, N.P. Kapustin, I.F. Klochai, and A.M. Nesterov, Poroshk. Metall. (Kiev), 1985, 94.
298. P.Liu, H. Jiao, Y. Zhang, H. Lu, Y. Du, and T. Wang, Wuli Xuebao, 1985, 34, 129.
299. M.N. Shipko, L.M. Letyuk, and L.A. Bashkirov, Ukr. Khim. Zh. (Russ. Ed.), 1985, 30, 1518.
300. M.K. Fayek and M.K. Elnimr, Ind. J. Pure Appl. Phys., 1985, 23, 145.
301. R. Gerardin, E. Millon, J.F. Brice, O. Evrard, and G. Le Caer, J. Phys. Chem. Solids, 1985, 46, 1163.
302. T. Birchall, C. Hallett, and D. Hanzel, Solid State Commun., 1985, 56, 77.
303. A. Buzas, I. Nagy-Czako, and A. Vertes, Acta Chim. Hung., 1984, 117, 401.
304. N. Suarez, J.L. Sanchez, and S. Garcia, Phys. Status Solidi A, 1985, 90, K171.
305. M. Greenblatt, E. Wang, H. Eckert, N. Kimura, R. Herber, and J.V. Waszczak, Inorg. Chem., 1985, 24, 1661.
306. J. Fontcuberta and J.B. Goodenough, J. Solid State Chem., 1985, 56, 116.
307. R. Vanleerberghe and R.E. Vanleerberghe, Hyperfine Interact., 1985, 23, 75.
308. J. Janicki, A. Maryanowska, J. Pietrzak, and J. Suwalski, Acta Phys. Pol., A, 1985, A68, 425.
309. J.J. Bara, B.F. Bogacz, M. Pekala, and A. Polaczek, J. Solid State Chem., 1985, 58, 143.
310. J.Y. Veuillen, R. Chevalier, S. Sallomon, J. Dumas, J. Marcus, and C. Schlenker, Lect. Notes Phys., 1985, 217, 129.
311. B.A. Mulla, and V.S. Darshane, Ind. J. Chem., 1985, 24A, 359.
312. A.S. Kamzin, Fiz. Tverd. Tela (Leningrad), 1985, 27, 1573.
313. B.O. Mysen, I.S.E. Charmichael, and D. Virgo, Contrib. Mineral. Petrol., 1985, 90, 101.
314. D.S. Goldman and D.E. Bewley, J. Am. Ceram. Soc., 1985, 68, 691.
315. B.O. Mysen and D. Virgo, Phys. Chem. Miner., 1985, 12, 191.
316. T. Pagnier, J.L. Souquet, and M. Ribes, Mater. Res. Bull., 1985, 20, 225.
317. Y. Hsia, R, Liu, S. Wang, Y. Cheng, C. Xu, and S. Pan, Microstruct. Charact. Mater. Non-Microsc. Tech., Proc. Risoe Int. Symp. Metall. Mater. Sci. 5th, 1984, 307.
318. D. Virgo and B.O. Mysen, Phys. Chem. Miner., 1985, 12, 65.
319. M. Sugimoto, T. Takahashi, K. Naya, N. Hiratsuka, and Y. Soeno, Nippon Oyo Jiki Gakkaishi, 1985, 9, 167.

320. V.V. Galitskaya, Ya.A. Fedorovskii, A.B. Klyukin, A.I. Ovchinnikov, and A.D. Mokrushin, Fiz. Khim. Stekla, 1984, 10, 724.
321. A.G. Mostafa, A.M. Sanad, I. Kashif, and A.A. El-Saghier, J. Mater. Sci. Lett., 1985, 4, 571.
322. T. Nishida and Y. Takashima, Chem. Lett., 1985, 1271.
323. A.M. Nassar, S.S. Gomaa, S.M. Salman, and F. Mostafa, Cent. Glass Ceram. Res. Inst. Bull., 1983, 30, 62.
324. A.M. Nassar, S.M. Salman, S.S. Gomaa, and F. Mostafa, Cent. Glass Ceram. Res. Inst. Bull., 1983, 30, 106.
325. C.S. Sundandana and R. Jagannathan, Solid State Commun, 1985, 53, 985.
326. N. Kishore, T.K. Bansal, R. Kamal, and R.G Mendiratta, J. Non-Cryst. Solids, 1985, 69, 213.
327. T.K. Bansal, N. Kishore, and R.G Mendiratta, Phys. Chem. Glasses, 1985, 26, 94.
328. R.V. Anavekar, N. Devaraj, and J. Ramakrishna, Curr. Sci., 1985, 54, 1063.
329. A.A. Bahgat, E.E. Shaisha, A.I. Sabry, and N.A. Eissa, Phys. Status Solidi A, 1985, 90, K25.
330. E.E. Shaisha, A.A. Bahgat, A.I. Sabry, and N.A. Eissa, Phys. Chem. Glasses, 1985, 26, 91.
331. D.K. Kanchan, R.K. Pura, and R.G. Mendiratta, Phys. Chem. Glasses, 1985, 26, 217.
332. S. Courant, M. Duclot, T. Pagnier, and M. Ribes, Solid State Ionics, 1985, 15, 147.
333. B. Spiering and F.A. Siefert, Contrib. Mineral. Petrol., 1985, 90, 63.
334. H.A. Sallam, N.A. Eissa, I. Dezsi, and D.L. Nagy, Acta Phys. Hung., 1985, 57, 125.
335. N.N. Korotaeva, A.V. Polosin, and T.V. Malysheva, Geokhimiya, 1985, 899.
336. R.G. Burns and V.M. Burns, J. Geophys. Res., B, 1984, 89, C313.
337. J.M. Daniels, H.Y. Lam, D.G. Rancourt, J.A. Westgate, and D. York, Earth Planet. Sci. Lett., 1985, 73, 430.
338. A.V. Podgaetskii, D.D. Kotel'nikov, Iu.B. Voitkovskii, I.P. Ilupi, and N.N. Zinchuk, Dokl. Akad. Nauk SSSR, 1985, 282, 1238.
339. J.K. de Waele, M.J. Luys, E.F. Vasant, and F.C. Adams, J. Trace Microprobe Tech., 1984, 2, 87.
340. V.F. Krutikov, and F.M. Bulatov, Mineral. Zh., 1985, 7, 78.
341. A.I. Bakhtin, F.M. Bulatov, and I.S. Guziev, Geokhimiya, 1985, 1380.
342. Sh.Sh. Bashkirov, F.M. Bulatov, and E.Kh. Ivoilova, Dokl. Akad. Nauk SSSR, 1985, 282, 343.
343. M.G. Townsend and G. Longworth, Phys. Chem. Miner., 1985, 12, 141.
344. O. Ballet, J.M.D. Coey, and K.J. Burke, Phys. Chem. Miner., 1985, 12,370.
345. X. Kan and J.M.D Coey, Am. Mineral., 1985, 70, 576.
346. M.W. Schaeffer, Am. Mineral., 1985, 70, 729.
347. N.N. Razumovskaya, A.D. Vaskevich, Yu.B. Voitovskii, Izv. Akad. Nauk SSSR, Neorg. Mater., 1985, 21, 1383.
348. G.S. Momdzhi, Izv. Vyssh. Uchebn. Zaved., Geol. Razved., 1985, 26.
349. M. Akasaka, and H. Ohashi, Phys. Chem. Miner., 1985, 12, 13.
350. J.M.D Coey and S. Ghose, Solid State Commun., 1985, 53, 143.
351. I.J. Muir, J.B. Metson, and G.M. Bancroft, Can. Mineral., 1985, 22, 689.
352. J.H. Johnson and C.M. Cardile, Clays Clay Miner., 1985, 33, 21.
353. C.M. Cardile and J.H. Johnson, Clays Clay Miner., 1985, 33, 295.
354. E. De Grave, J. Vandenbruwaene, and E. Elewaut, Clay Miner., 1985, 20, 171.
355. V.P. Ivanitskii, N.I. Buchinskaya, M.A. Plastinina, and D.I. Shvets, Mineral. Zh., 1985, 7, 54.

356. T. Moya, L. Fournes, P. Hagenmuller, L. Kolditz, G. Blumenthal, and G. Wegener, Z. Chem., 1985, 25, 343.
357. L.S. Sonkin, V.V. Mank, E.V. Pol'shin, and B.F. Gorbachev, Dokl. Akad. Nauk SSSR, 1985, 284, 459.
358. Y. Maeda, Y. Takashima, and K. Ishida, Bull. Chem. Soc. Jpn., 1985, 58, 1047.
359. M. Rang and Z. Qiu, Chanchun Dizhi Xueyuan Xuebao, 1985, 29.
360. D.J. Vaughan and J.R. Craig, Am. Mineral., 1985, 70, 1036.
361. V.A. Reznichenko, I.A. Maksutov, T.A. Pryakhina, A.I. Vorobeichik, V.F. Podzerko, and I.A. Karyazin, Izv. Akad. Nauk SSSR, Met., 1985, 34.
362. E. de Grave and R. Vochten, Phys. Chem. Miner., 1985, 12, 108.
363. R. Guillen and J.R. Regnard, Phys. Chem. Miner., 1985, 12, 246.
364. K.A. Rodgers and J.H. Johnston, Neues Jahrb. Mineral., Monatsh., 1985, 539.
365. R. Garg, E. Galvano da Silva, Y.S. Liu, and V.K. Garg, Solid State Commun., 1985, 54, 305.
366. M. Buhannic, P. Colombet, and M. Danot, Nouv. J. Chim., 1985, 9, 519.
367. Sh.Sh. Bashkirov, R.K. Gubaidullin, Sh.R. Safin, and V.N. Zaritskii, Fiz. Tverd. Tela (Leningrad), 1985, 27, 730.
368. J.W. Niemantsverdriet, A.M. van der Kraan, W.N. Delgass, and M.A. Vannice, J. Phys. Chem., 1985, 89, Z67.
369. P.H. Christensen, S. Moerup, and J.W. Niemantsverdriet, J. Phys. Chem., 1985, 89, 4898.
370. L.N. Mulay, A.V.P. Rao, P.L. Walker, jr., K.J. Yoon, and M.A. Vannice, Carbon, 1985, 23, 501.
371. S.C. Lin and J. Phillips, J. Appl. Phys., 1985, 58, 1943.
372. D.A. Scherson, C.A. Fierro, D. Tryk, S.L. Gupta, E.B. Yeager, J. Eldridge, and R.W. Hoffman, J. Electroanal. Chem. Interfacial Electrochem., 1985, 184, 419.
373. J. Phillips, Y. Chen, and J.A. Dumescic, ACS Symp. Ser., 1985, 288, 518.
374. H.H. Wei and K.H. Lin, J. Chin. Chem. Soc. (Taipei), 1984, 31, 329, 335.
375. R.J. Matyi, J.B. Butt, and L.H. Schwartz, J. Catal., 1985, 91, 185.
376. P.P. Vaishnava, P.I. Ktorides, P.A. Montano, K.J. Mbadcam, and G.A. Nelson, J. Catal., 1985, 96, 301.
377. J.W. Niemantsverdriet, J.A.C. Van Kaam, C.F.J. Flipse, and A.M. Van der Kraan, J. Catal., 1985, 96, 58.
378. E. Yeh, N.K. Jaggi, J.B. Butt, and L.H. Schwartz, J. Catal., 1985, 91, 231.
379. M. Rosa-Brussin, H. Constant, R. Iraldi, E. Jaimes, and F. Gonzalez-Jimenez, Acta Cient. Venez., 1984, 35, 332.
380. N. Al-Hayek, J.P. Eymery, and M. Dore, Water Res., 1985, 19, 657.
381. A. Spinzi, E. Burzo, and M. Rogalski, React. Kinet. Catal. Lett., 1985, 28, 181.
382. D. Cordischi, N. Burriesci, F. D'Alba, M. Petrera, G. Polizzotti, and M. Schiavello, J. Solid State Chem., 1985, 56, 182.
383. S.L. Suib, A. Kostapapas, K.C. McMahon, J.C. Baxter, and A.M. Winiecki, Inorg. Chem., 1985, 24, 858.
384. G.P. Huffman, F.E. Huggins, G.R. Dumyre, A.J. Pignocco, M.C. Lin, Fuel, 1985, 64, 849.
385. P.H. Kidd, UK Pat. Appl. GB 2,144,214.
386. N. Yang, Y. Xia, G. Chen, and R. Liu, Ranliao Huaxue Xuebao, 1984, 12, 375.
387. R. Hong, Z. Gan, and Y. Gu, Ranliao Huaxue Xuebao, 1985, 13, 90.
388. J. Ensling, P. Guetlich, and A. Semmel, Geol. Jahrb. Hessen, 1984, 112, 67.
389. I. Georgieva and S. Mirchev, Pochvozn. Agrokhim., 1984, 19, 72.

390. S. Moerup, J. Franck, J. Van Wonterghem, H. Roy-Poulsen and L. Larsen, Fuel, 1985, 65, 528.
391. T.G. Cole, Geochim. Cosmochim. Acta., 1985, 49, 221.
392. N.A. Eissa and M.Y. Hassan, J. Mater Sci. Lett., 1985, 4, 37.
393. X. Feng, P. Yan, Y. Xia, R. Liu and S. Wang, Guisuanyan Xuebao, 1985, 13, 153.
394. C. Saragovi-Badler and C. Puglisi, Ceram. Int., 1984, 10, 105; 11.
395. M. Kulbekov, S.T. Suleimenov, S. Zh. Saibulatov, I.M. De and E.F. Galyutina, Vestn. Akad. Nauk Kaz. SSR 1985, 25.
396. J.R. Gancedo, M. Garcia, A. Hernandez-Laguna, C. Ruix-Garcia and J. Palomares, Archeometry, 1985, 27, 75.
397. R. Hong, D. Deng and Y. Gu, Guisuanyan Xuebao, 1985, 13, 375.
398. J. Chen, C. Wang and G. Wang, Jinshu Xuebao, 1985, 21, B66.
399. M.J. Hollit, J.D. Cashion and L.J. Brown, Metall. Slags, Fluxes, Int. Symp., Proc., 2nd., 1984, 961.
400. L.A. Alekseev and A.I. Shamov, Zavod. Lab., 1985, 51, 33.
401. N.A. Eissa, INIS Atomindex, 1985, 16, Abstr. No. 16:019061.
402. A.K. Singh, T. Ericsson, L. Haeggstroem, and J. Gullman, Corros. Sci., 1985, 25, 931.
403. D. Raj and S. Kumar, J. Radioanal. Nucl. Chem., 1985, 94, 171.
404. T.Z.A. Abou, S.M. Higgy, M.Y. Hassan, and N.A. Ali, Egypt. J. Phys., 1985, 11, 55.
405. S. Kitahata, M. Kishimoto, and M. Amemiya, J. Appl. Phys., 1985, 57, 3931.
406. H. Muramatsu, T. Miura, H. Nakahara, M. Fujioka, E. Tanaka, and A. Hashizume, Hyperfine Interact., 1984, 20, 305.
407. H. Muramatsu, T. Miura, H. Nakahara, and M. Fujioka, Z. Phys. A, 1985, 321, 699.
408. J. Blomquist, Sci. Tech. Aerosp. Rep., 1985, 23, Abstr. No. N85-31973.
409. B.K. Das, A.K. Tewari, and D.L. Bhattacharya, Phys. Status Solidi B, 1985, 132, 23.
410. D. Balabanski, E. Vapirev, and P. Kamenov, J. Phys., 1985, 46, 1387.
411. J.S. Brooks, D.W. Allen, and J. Unwin, Polym. Degrad. Stab., 1985, 10, 79.
412. R. Frety, M. Guenin, P. Bussiere, and Y.L. Lam, Mater. Sci. Monogr., 1985, 28, 1055.
413. A.S. Sanyal and J. Mukerji, Phys. Chem. Glasses, 1985, 26, 135.
414. T. Busova and D. Khristov, Textilveredlung, 1985, 20, 93.
415. S. Winterthur and H. Rohner, Textilveredlung, 1985, 20, 95.
416. P.S. Cook, J.D. Cashion and P.J. Cassidy, Fuel, 1985, 64, 1121.
417. D.W. Allen, S. Bailey, J.S. Brooks, and B.F. Taylor, Chem. Ind., 1985, 826, 513.
418. L.V. Kurbanova and P.P. Seregin, Fiz. Khim. Stekla, 1985, 11, 244.
419. S. Ichiba and T. Sakamoto, Bull. Chem. Soc. Jpn., 1985, 58, 1323.
420. M.M. Kolosova, S.A. Simanova, E.N. Yurchenko, E.S. Boichinova, and V.I. Kuznetsov, Zh. Neorg. Khim., 1985, 30, 1493.
421. T. Birchall, R.C. Burns, L.A. Devereux and G.J. Schrobilgen, Inorg. Chem., 1985, 24, 890.
422. R.H. Herber, Mater. Sci. Forum., 1984, 2, 49.
423. V.M. Lyubin and L.N. Seregina, Fiz. Khim. Stekla., 1985, 11, 626.
424. R.C. Mercader, F.H. Sanchez, L.A. Mendoza-Zelis, L. Terminiello, A.G. Bibiloni, C.P. Massolo, J. Desimoni and A.R. Lopez-Garcia, Hyperfine Interact., 1985, 23, 211.
425. J.W. Petersen, G. Weyer, H. Loft Nielsen, S. Damgaard, W.J. Choyke, and H. Andreasen, Hyperfine Interact., 1985, 23, 17.
426. K. Bonde-Nielsen, H. Grann, H. Haas, F.T. Pedersen, and G. Weyer, Int. Conf. Defects Semicond., Proc., 13th., 1984, 1065.
427. E.A. Vasil'ev, V.A. Virchenko, T.M. Tkachenko, and A.V. Shablovskii, Dokl.Akad. Nauk BSSR, 1985, 29, 985.
428. G. Weyer, F.T. Pedersen, and H. Grann, Nucl. Instrum. Methods Phys. Res., Sect. B, 1985, 7, 103.

429. N.N. Delyagin, V.I. Krylov, G.T. Mudzhiri, V.I. Nesterov, and S.I. Reiman, Phys. Status Solidi B, 1985, 131, 555.
430. S.I. Reiman and N.I. Rokhlov, Fiz. Tverd. Tela (Leningrad), 1985, 27, 1587.
431. A. Kolodziejczyk and J. Zukrowski, J. Phys. F, 1985, 15, 121.
432. H. Hahn, M. Ghafari, R. Dieudonne, and U. Gonser, Scr. Metall., 1985, 19, 615.
433. N.N. Delyagin, V.I. Krylov, N.I. Moreva, G.T. Mudzhiri, V.I. Nesterov, and S.I. Reiman, Zh. Eksp. Teor. Fiz., 1985, 88, 300.
434. I. Tanaka, S. Nasu, and F.E. Fujita, Nippon Kinzoku Gakkaishi, 1985, 49, 701.
435. F.H. Sanchez, R.C. Mercader, A.F. Pasquevich, A.G. Bibiloni, and A. Lopez-Garcia, Hyperfine Interact., 1984, 20, 295.
436. F.H. Sanchez, R.C. Mercader, A.F. Pasquevich, A.G. Bibiloni, and A.R. Lopez-Garcia, Phys. Status Solidi A, 1985, 89, 17.
437. I.Y. Bezotosnyi, N.N. Degtyarenko, V.F. Elesin, I.V. Zakharchenko, V.P. Martin, and I.N. Nikolaev, Radiats. Eff. Met. Splavakh, Mater. Vses. Soveshch., 3rd., 1983, 65.
438. G. Le Caer, B. Malaman, G. Venturini, D. Fruchart , and B. Roques, J. Phys. F, 1985, 15, 1813.
439. H. Sirkin, B. Arcondo, E. Duhalde, E. Nassif, and G. Quintana, Z. Naturforsch., A, 1985, A40, 548.
440. A.M. Umarji, S.K. Malik, and G.K. Shenoy, Solid State Commun., 1985, 53, 1029.
441. S.K. Malik, A.M. Umarji, and G.K. Shenoy, Phys. Rev. B, 1985, 31, 6971.
442. J. Olejniczak, A. Rojek, and C. Sulkowski, Phys. Status Solidi B, 1985, 130, 73.
443. F.W. Oliver, W. Morgan, E.C. Hammond, S. Wood, and L. May, J. Appl. Phys., 1985, 57, 3250.
444. J.E. Frackowiak and D. Ostrowski, Conf. Appl. Crystallogr., 1984, 11, 286.
445. A. Vertes, S. Nagy, T. Turmezey, and A. Griger, Magy. Alum., 1985, 22, 181.
446. M.L. Varsanyi, J. Jaen, A. Vertes, and L. Kiss, Electrochim. Acta, 1985, 30, 529.
447. H. Mehner and A. Vertes, J. Radioanal. Nucl. Chem., 1985, 96, 353.
448. H. Mehner and A. Vertes, J. Radioanal. Nucl. Chem., 1985, 89, 153.
449. A. Vertes, J. Jaen, and M.L. Varsanyi, J. Radioanal. Nucl. Chem., 1985, 90, 383.
450. F. Gauzzi, B. Verdini, A. Maddalena, and G. Principi, Inorg. Chim. Acta, 1985, 104, 1.
451. R.A.M.J. Slaessens, J.G.M. Van der Linden, J.E.J. Schmitz, and N.H.M. De Jong, Inorg. Chim. Acta, 1985, 109, 123.
452. N. Alberola and R. Point, Int. J. Appl. Radiat. Isot., 1985, 36, 150.
453. N. Alberola, Polyhedron, 1985, 4, 1853.
454. J.D. Donaldson, S.M. Grimes, A. Nicolaides, and P.J. Smith, Polyhedron, 1985, 4, 391.
455. K.G. Tschakarov, V. Rusanov, and G. Gospodinov, J. Solid State Chem., 1985, 59, 265.
456. F.S. Nasredinov, L.V. Prokof'eva, P.P. Seregin, S.V. Zarubo, A. Ermolaev, and A.N. Kurmantaev, Phys. Status Solidi B, 1985, 130, 727.
457. S.S. Bashkirov, I.A. Dobryakov, A.B. Liberman, and S.S. Tsarevskii, Kristallografiya, 1985, 30, 1016.
458. T. Birchall, G. Denes, K. Ruebenbauer, and J. Pannetier, J. Chem. Soc., Dalton Trans., 1985, 1275.
459. T. Birchall, J.E. Vekris, B. Frlec, D. Gantar, and D. Hanzel, J. Fluorine Chem., 1985, 27, 61.
460. L.V. Kurbanova, Y.P. Tarlakov and V.P. Podkhalyuzin, Fiz. Khim. Stekla, 1985, 11, 120.
461. T. Birchall and V. Manivannen, J. Chem. Soc., Dalton Trans., 1985, 2671.

462. I.A. Evstyukhina, Y.M. Devyatko, and O.V. Sokolova, Zh. Tekh. Fiz., 1984, 54, 2380.
463. D. Guenzburger, D.E. Ellis, P.A. Montano, G.K. Shenoy, S.K. Malik, D.G. Hinks, P. Vaishnava, and C.W. Kimball, Phys. Rev. B, 1985, 32, 4398.
464. J. Du, J. Deng, D. Yao, L. Zen, R. Chen, Z. Yu, and Y. Zheng, Zhongshan Daxue Xuebao, Ziran Kexueban, 1985, 3, 104.
465. J. Rotsche, U. Reichel, R. Schroeder, and C. Pietzsch, Silikattechnik, 1985, 36, 316.
466. F.J. Berry and C. Hallett, J. Chem. Soc., Dalton Trans., 1985, 45.
467. K. Chakurov, G. Gospodinov, and V. Rusanov, Bulg. J. Phys., 1984, 11, 601.
468. K. Burger, A. Vertes, M. Suba, and I. Dekany, Inorg. Chim. Acta, 1985, 100, 17.
469. L.M. Dautov, B.D. Thurkavaev, E.R. Ishkenov, and N.K. Ergard, Fiz. Tverd. Tela (Leningrad), 1985, 27, 3150.
470. D. Tudela, V. Fernandez, and J.D. Tornero, J. Chem. Soc., Dalton Trans., 1985, 1281.
471. S.J. Blunden, D. Searle and P.J. Smith, Inorg. Chim. Acta, 1985, 98, 185.
472. E. Rivarola, A. Silvestri, G. Alonzo, R. Barbieri, and R.H. Herber, Inorg. Chim. Acta, 1985, 99, 87.
473. E. Uhlemann, H. Reichmann and H. Mehner, Anal. Chim. Acta 1985, 170, 319.
474. D. Tudela, V. Fernandez and J.D. Tornero, Inorg. Chem., 1985, 24, 3892.
475. E.N. Yurchenko, V.I. Kuznetsov, and E.B. Burgina, Teor. Eksp. Khim., 1985, 21, 194.
476. O. Scheidsteger, G. Huttner, K. Dehnicke, and J. Pebler, Angew. Chem., 1985, 97, 434.
477. H. Schumann and I. Schumann, Gmelin Handbook of Inorganic Chemistry, 8th Edition. Sn-Organic Compounds, Part II, Trimethyltin- and Triethyltin-oxygen Compounds, Springer, Berlin, 1985.
478. L. Pellerito, M.T. Lo Giudice, G.C. Stocco, J.D. Donaldson, S.M. Grimes, Polyhedron, 1985, 4, 747.
479. H.C. Clark, V.K. Jain, R.C. Mehrotra, B.P. Singh, G. Srivastavam, and T. Birchall, J. Organomet. Chem., 1985, 279, 385.
480. S.J. Blunden, L.A. Hobbs and P.J. Smith, J. Organomet. Chem., 1985, 282, 9.
481. K.C. Molloy and K. Quill, J. Chem. Soc., Dalton Trans., 1985, 1417.
482. G. Ruisi, M.T. Lo Giudice, and L. Pellerito, Inorg. Chim. Acta, 1984, 93, 161.
483. F. Huber, G. Roge, L. Carl, G. Atassi, F. Spreafico, S. Filippeschi, R. Barbieri, A. Silvestri, E. Rivarola, G. Ruisi, F. Di Bianca, and G. Alonzo, J. Chem. Soc., Dalton Trans., 1985, 523.
484. T.P. Lockhart, W.F. Manders, and E.O. Schlemper, J. Am. Chem. Soc., 1985, 107, 7451.
485. G.K. Sandhu, R. Gupta, S.S. Sandhu, and R.V. Parish, Polyhedron, 1985, 4, 81.
486. R. Barbieri, A. Silvestri, G. Ruisi, and G. Alonzo, Inorg. Chim. Acta, 1985, 97, 113.
487. M. Gielen, K. Jurkschat, B. Mahieu, and D. Apers, J. Organomet. Chem., 1985, 286, 145.
488. P.G. Harrison and N.W. Sharpe, Inorg. Chim. Acta, 1985, 108, 7.
489. E.M. Holt, F.A.K. Nasser, A. Wilson Jr., and J.J. Zuckerman, Organometallics, 1985, 4, 2073.
490. L.E. Khoo and F.E. Smith, Polyhedron, 1985, 4, 447.
491. C.S. Frampton, R.M.G. Roberts, J. Silver, J.F. Warmsley, and B. Yavari, J. Chem. Soc., Dalton Trans., 1985, 169.
492. S.W. Hedges, Diss. Abstr. Int. B, 1985, 45, 3239.
493. P. Jernberg, L. Haeggstroem, and T. Sundqvist, Phys. Scr., 1985, 31, 215.
494. S.W. Hedges and L.H. Bowen, Mater. Sci. Forum, 1984, 2, 65.
495. G. Weyer, H. Andreasen, and H. De Waard, Phys. Status Solidi B, 1985, 132, 219.

496. J. Olivier-Fourcade, J.C. Jumas, N. Rey, E. Philippot, and M. Maurin, J. Solid State Chem., 1985, 59, 174.
497. B. Kalman, N.V. Zsuzsanna, V. Attila, and M.J. Afanaszov, Magy. Kem. Foly., 1985, 91, 356.
498. G. Alonzo, N. Bertazzi, and F. Huber, Can. J. Spectrosc., 1985, 30, 21.
499. F. Huber, H. Preut, G. Alonzo, and N. Bertazzi, Inorg. Chim. Acta, 1985, 102, 181.
500. A.W. Cooke, J. Pebler, F. Weller, and K. Dehnicke, Z. Anorg. Allg. Chem., 1985, 524, 68.
501. J. Pebler and K. Dehnicke, Z. Anorg. Allg. Chem., 1984, 517, 103.
502. L. Soderholm, C. Mathis, B. Francois, and J.M. Friedt, Synth. Met., 1985, 10, 261.
503. T. Miura, Y. Hatsukawa, K. Endo, H. Nakahara, M. Fujioka, and E. Tanaka, Genshikaku Kenkyu, 1985, 30, 43.
504. A.A. Kornilova, E.V. Kapitanov, R.M. Kuz'min, and A.A. Opalenko, Vestn. Mosk. Univ., Ser. 3: Fiz., Astron., 1985, 26, 87.
505. M. Van der Heyden, M. Pasternak, and G. Langouche, J. Phys. Chem. Solids, 1985, 46, 1221.
506. R.J. Batchelor and T. Birchall, J. Chem. Soc., Dalton Trans., 1985, 1727.
507. H. Sakai, M. Mizota, Y. Maeda, T. Yamamoto, and A. Yamamoto, Bull. Chem. Soc. Jpn., 1985, 58, 926.
508. T. Ohtani and H. Sakai, Solid State Commun., 1986, 57, 81.
509. M. Likasiak, T. Zemcik, J. Suwalski, and J. Piekoszewski, Acta Phys. Pol. A, 1985, 68, 79.
510. W. Potzel and G.M. Kalvius, Phys. Lett. A, 1985, 110, 165.
511. R. Wordel and F.E. Wagner, J. Less-Common Met., 1984, 101, 427.
512. P. Triggs, Helv. Phys. Acta, 1985, 58, 657.
513. D. Salomon, A. Lerf, W. Biberacher, T. Butz, and S. Saibene, Chem. Phys. Lett., 1985, 119, 238.
514. M. Eibschutz, D. Salomon, and F.J. Di Salvo, Phys. Lett. A, 1985, 113, 277.
515. K. Zitter, J. Schmand, and R. Schoellhorn, Mater. Res. Bull., 1985, 20, 787.
516. I. Berkes, B. Hlimi, G. Marest, J. Sau, E.H. Sayouty, and K. Heyde, J. Phys. G, 1985, 11, 287.
517. I. Berkes, B. Hlimi, G. Marest, J. Sau, and E.H. Sayouty, Hyperfine Interact., 1985, 22, 591.
518. Y. Uchida, M. Katada, H. Sano, H. Sakai, and Y. Maeda, J. Radioanal. Nucl. Chem., 1985, 94, 215.
519. B. Chiari, O. Piovesana, T. Tarantelli, and P.F. Zanazzi, Inorg. Chem., 1985, 24, 366.
520. R. Uson, A. Laguna, M.U. De la Orden, R.V. Parish, and L.S. Moore, J. Organomet. Chem., 1985, 282, 145.
521. J. Stanek, S.S. Hafner, and F. Hensel, Phys. Rev. B, 1985, 32, 3129.
522. S. Yamauchi, Y. Minai, and T. Tominaga, J. Radioanal. Nucl. Chem., 1985, 93, 237.
523. C. Shi, F. Ambe, and K. Takahashi, Nippon Kagaku Kaishi, 1985, 152.
524. C. Shi, K. Takahashi, and F. Ambe, Zhongguo Xitu Xuebao, 1985, 3, 53.
525. J.P. Sanchex, J.M. Friedt, H. Barnighuasen, and A.J. Van Duyneveldt, Inorg. Chem., 1985, 34, 405.
526. S.J. Lyle and W.A. Westall, J. Less-Common Met., 1985, 106, 109.
527. P. Thuery, F. Laville, E. Tronc, A.M. Lejus, and D. Vivien, Rev. Chim. Miner., 1985, 22, 216.
528. E. Tronc, D. Saber, A.M. Lejus, and D. Vivien, J. Less-Common Met., 1985, 111, 321.
529. H.M. Van Noort and T.J.A. Popma, Solid State Commun., 1985, 55, 77.
530. S.A. Shaheen, M. Abd-Elmeguid, H. Micklitz, J.S. Schilling, P. Kalvins, and R.N. Shelton, Phys. Rev. Lett., 1985, 55, 312.

531. S.K. Malik, G.K. Shenoy, S.M. Heald, and J.M. Tranquada, Phys. Rev. Lett., 1985, 55, 316.
532. R. Vijayaraghavan, J. Magn. Magn. Mater., 1985, 47-48, 561.
533. I. Nowik, E.V. Sampathkumaran, and G. Kaindl, Solid State Commun., 1985, 55, 721.
534. E. Kemly, M. Croft, V. Murgai, L.C. Gupta, C. Godart, R.D. Parks, and C.U. Segre, J. Magn. Magn. Mater., 1985, 47-48, 403.
535. G. Wortman, K.H. Frank, E.V. Sampathkumaran, B. Perscheid, G. Schmeister, and G. Kaindl, J. Magn. Magn. Mater., 1985, 49, 325.
536. B. Bittins, K. Keulerz, A. Scherzberg, J.P. Sanchez, W. Boksch, H.F. Braun, J. Roehler, H. Schneider, P. Weidner, and D. Wohlleben, Z. Phys. B, 1985, 62, 21.
537. M. Abd-Elmeguid, C. Sauer, and W. Zinn, J. Phys. C, 1985, 18, 345.
538. M. Abd-Elmeguid, C. Sauer, U. Koebler, W. Zinn, J. Roehler, and K. Keulerz, J. Magn. Magn. Mater., 1985, 47-48, 417.
539. M. Abd-Elmeguid, C. Sauer, U. Koebler, and W. Zinn, Z. Phys. B, 1985, 60, 239.
540. I. Felner and I Nowik, J. Phys. Chem. Solids, 1985, 46, 681.
541. I. Nowik and I. Felner, Physica B+C, 1985, 130, 433.
542. I. Felner and I Nowik, Acta Phys. Pol. A, 1985, A68, 97.
543. I. Felner and I Nowik, J. Magn. Magn. Mater., 1985, 47-48, 420.
544. R. Nagarajan, G.K. Shenoy, L.C. Gupta, and E.V. Sampathkumaran, J. Magn. Magn. Mater., 1985, 47-48, 413.
545. R. Nagarajan, G.K. Shenoy, L.C. Gupta, and E.V. Sampathkumaran, Phys. Rev. B, 1985, 32, 2846.
546. B. Perscheid, E.V. Sampathkumaran, and G. Kaindl, J. Magn. Magn. Mater., 1985, 47-48, 410.
547. E.V. Sampathkumaran, B. Perscheid, W. Krone, and G. Kaindl, J. Magn. Magn. Mater., 1985, 47-48, 407.
548. E.V. Sampathkumaran, G. Kaindl, W. Krone, B. Perscheid, and R. Vijayaraghavan, Phys. Rev. Lett., 1985, 54, 1067.
549. J.W.C. De Vries, INIS Atomindex, 1985, 16, Abstr. No. 16:040963.
550. J.W.C. De Vries, R.C. Thiel, and K.H.J. Buschow, Physica B+C, 1985, 128, 265, 309.
551. J.W.C. De Vries, R.C. Thiel, and K.H.J. Buschow, J. Phys. F, 1985, 15, 1413.
552. J.W.C. De Vries, R.C. Thiel, and K.H.J. Buschow, J. Less-Common Met., 1985, 111, 313.
553. J.W.C. De Vries, R.C. Thiel, and K.H.J. Buschow, Physica B+C, 1985, 128, 183.
554. J.L. Oddou, C. Jeandey, G. Czjzek, V. Oestreich, H. Schmidt, INIS Atomindex, 1985, 16, Abstr. No. 16:044987.
555. K.H.J. Buschow, R.C. Thiel, and J.W.C. De Vries, J. Phys. F, 1985, 15, L93.
556. M. Boge, J.M.D. Coey, G. Czjzek, D. Givord, C. Jeandey, H.S. Li, and J.L. Oddou, Solid State Commun., 1985, 55, 295.
557. S.K. Malik, A.M. Umarji, and G.K. Shenoy, J. Appl. Phys., 1985, 57, 3252.
558. J.P. Sanchez, J.M. Friedt, K. Tomala, and F. Holtzberg, Phys. Lett. A, 1985, 111A, 83.
559. D.R. Noakes, G.K. Shenoy, and D.G. Hinks, Phys. Rev. B, 1985, 31, 5712.
560. P.C.M. Gubbens, A.M. Van der Kraan, and K.H.J. Buschow, Phys. Status Solidi B, 1985, 130, 575.
561. L.P. Ferreira, R. Guillen, P. Vulliet, A. Yaouanc, D. Fruchart, P. Wolfers, P. L'Heritier, R. Fruchart, J. Magn. Magn. Mater., 1985, 53, 145.
562. J. Pszczola, J. Zukrowski, K. Krop, J. Suwalski, Z. Kucharski, and M. Lukasiak, Physica B+C, 1985, 130, 436, 446.
563. M. Maurer and J.M. Friedt, Rapidly Quenched Met., Proc. Int. Conf., 5th. 1984, Pub. 1985, 2, 1173.
564. G.M. Kalvius, F.J. Litterst, L. Asch, and J. Chappert, Hyperfine Interact., 1985, 25, 709.

565. P.P. Vaishnava, C.W. Kimball, A.M. Umarji, S.K. Malik, and G.K. Shenoy, J. Magn. Magn. Mater., 1985, 49, 286.
566. P.C.M. Gubbens, A.M. Van der Kraan, and K.H.J. Buschow, Physica B+C, 1985, 130, 412.
567. P.C.M. Gubbens, A.M. Van der Kraan, and K.H.J. Buschow, Physica B+C, 1985, 130, 415.
568. J.E. Greedan, L. Sonderholm, and J.M. Friedt, J. Solid State Chem., 1985, 59, 242.
569. P. Bonville and R. Conte, J. Phys. (Les Ulis, Fr.), 1985, 46, 1553.
570. P. Bonville and J.A. Hodges, J. Magn. Magn. Mater., 1985, 47-48, 152.
571. D. Weschenfelder, Energy Res. Abstr., 1985, 10, Abstr. No. 7094.
572. H. Abazli, J. Jove, and A. Coussin, J. Less-Common Met., 1985, 108, 123.
573. J.M. Friedt, F.J. Litterst, and J. Rebizant, Phys. Rev. B, 1985, 32, 257.
574. A. Tabuteau, H.X. Yang, J. Jove, T. Thevenin, and M. Pages, Mater. Res. Bull., 1985, 20, 595.
575. T. Thevenin, J. Jove, and M. Pages, Mater. Res. Bull., 1985, 20, 723.
576. T. Thevenin, J. Jove, and M. Pages, Mater. Res. Bull., 1985, 20, 1075.
577. J. Gal, H. Pinto, S. Fredo, M. Melamud, H. Shaked, R. Caciuffo, F.J. Litterst, L. Asch, W. Potzel, and G.M. Kalvius, J. Magn. Magn. Mater., 1985, 50, L123.
578. S.B. Ogale, S.V. Ghaisas, S.M. Kanetkar, and V.G. Bide, Proc. Indian Natl. Acad., Part A, 1985, 51, 211.
579. Y. Fukuda, Y. Tsuchiya, Y. Shima, and M. Terasaka, Kinzoku Hyomen Gijutsu, 1985, 36, 494.
580. J.A. Sawicki, Mater. Sci. Eng., 1985, 69, 501.
581. S.B. Ogale, S.V. Ghaisas, S.M. Kanetkar, and V.G. Bide, Proc. Indian Natl. Acad., Chem Sci., 1985, 95, 207.
582. A.V. Mitin and E.A. Popov, Phys. Status Solidi B, 1985, 132, 523.
583. V.V. Morovoz, Zh. Tekh. Fiz., 1985, 55, 2196.
584. D. Liljequist and M. Ismail, Nucl. Instrum. Methods Phys. Res., 1985, A239, 273.
585. D. Liljequist and M. Ismail, Phys. Rev. B, 1985, 31, 4131.
586. F. Salvat, R. Mayol, J.D. Martinez, and J. Parallada, Nucl. Instrum. Methods Phys. Res., 1985, B6, 547.
587. C. Meyer, F. Hartmann-Boutron, Y. Gros, and L.A. Campbell, J. Magn. Magn. Mater., 1985, 46, 254.
588. J.C. Frost, B.C.C. Cowie, S.N. Chapman, and J.F. Marshall, Appl. Phys. Lett., 1985, 47, 581.
589. M. Takafuchi and T. Kobayashi, Nucl. Instrum. Methods Phys. Res., 1985, B9, 223.
590. P. Topalov and A. Proikova, Nucl. Instrum. Methods Phys. Res., 1985, A236, 142.
591. D. Liljequist, M. Ismail, K. Saneyoshi, K. Debusmann, W. Keune, R.A. Brand, and W. Kaiuka, Phys. Rev. B, 1985, 31, 4137.
592. V.L. Gurachevskii, M. Mashlan, O.V. Misevich, A.L. Kholmetskii, and V.A. Chudakov, Prib. Tekh. Eksp., 1985, 38.
593. V.L. Gurachevskii, M. Mashlan, O.V. Misevich, A.L. Kholmetskii, and V.A. Chudakov, Dokl. Akad. Nauk BSSR, 1985, 29, 329.
594. Y. Ujihira, Bunseki Kagaku, 1985, 34, T5.
595. D. Cook and E. Agyekum, Nucl. Instrum. Methods Phys. Res., 1985, B12, 515.
596. Y. Zhang, J. Zhu, X. Wang, C. Wang, He Dianzixue Yu Tance Jishu, 1985, 5, 214.
597. A. Handa, J. Kobayashi, and Y. Ujihira, 1985, 20, 581.
598. K. Hisatake, T. Toriyama, H. Miyasaka, K. Ueoka and Y. Fujita, Genshikaku Kenkyu, 1985, 30(5), 65.
599. J. Korecki and U. Gradmann, Phys. Rev. Lett., 1985, 55, 2491.
600. N.K. Jaggi, L.H. Schwartz, H.K. Wang, and J.B. Ketterson, J. Magn. Magn. Mater., 1985, 49, 1.

601. H. de Waard, J.A. Sawicki, B.D. Sawicka, F. Pleiter, and K. Frentz, J. Magn. Magn. Mater., 1985, 49, 50.
602. M.E. Kordesch, Diss. Abstr. B, 1985, 45, 2593.
603. I. Stewart, M.J. Tricker, and J.A. Cairns, J. Catal., 1985, 94, 360.
604. N.A. Tomashevskii and V.F. Mazanko, Metallofizika, 1985, 7, 118.
605. C. Jaouen, J.P. Eymery, E.L. Mathe, and J. Delafond, Mater. Sci. Eng., 1985, 69, 483.
606. V.P. Godbole, S.M. Chaudhari, S.V. Gaisas, S.M. Kanetkar, S.B. Ogale, and V.G. Bhide, Phys. Rev. B, 1985, 31, 5703.
607. S.B. Ogale, R. Joshee, V.P. Godbole, S.M. Kanetkar, and V.G. Bhide, J. Appl. Phys., 1985, 57, 2915.
608. J. Patankar, Y.V. Bhandarkar, S.M. Kanetkar, S.B. Ogale, and V.G. Bhide, Nucl. Instrum. Methods Phys. Res., 1985, B7-8, 720.
609. M. Fujinami, K. Yoshida, Y. Ujihira, and M. Terasaka, J. Mater. Sci., 1985, 4099.
610. A. Fnidiki and J. Eymery, Scr. Metall., 1985, 19, 847.
611. Y. Ujihara, Rev. Anal. Chem., 1985, 8, 125.
612. V.L. Gurachevskii, M. Mashlon, A.L. Kholmetskii, V.A. Chudakov, G.P. Shpen'kov, Dokl. Akad. Nauk BSSR, 1985, 29, 917.
613. N.S. Kolpakov, V.A. Kondratenko, V.P. Mel'nichuk, V.Yu. Miloserdin. K.E. Nilov, Yu.V. Petrikin, and S.V. Yakutik, Izv. Akad. Nauk SSSR, Ser. Fiz., 1985, 49, 1656.
614. A.G. Grogor'yants, A.N. Safanov, Yu.V. Baldokhin, T.V. Gulyaeva, V.M. Tarasenko, and E.V. Kauts, Elektron. Obrab. Mater., 1984, 26.
615. K. Choudhury, D. Das, C.K. Majumdar, K.R.P.M. Rao, and M. Adhikari, Phys. Status. Solidi A, 1985, 91, 29.
616. K.R.P.M. Rao, D.K. Sood, A.K. Jain, and J. Subrahmanyam, Proc. Nucl. Phys. Solid State Phys. Symp. 1982, (Publ. 1984), 25, 523.
617. M. Fujinami and Y. Ujihara, J. Mater. Sci., 1985, 20, 1859.
618. K. Nomura, Y. Ujihara, K. Itoh, and K. Honda, Thin Solid Films, 1985, 128, 225.
619. H.M. Van Noort, J.W.D. Martens, and W.L. Peeters, Mater. Sci. Bull., 1985, 20, 41.
620. J.W.D. Martens, W.L. Peeters, H.M. Van Noort, and M. Erman, J. Phys. Chem. Solids, 1985, 46, 411.
621. X.Z. Zhou, R.J. Pollard, and A.H. Morrish, IEEE Trans. Magn., 1985, MAG-21, 1703.
622. G. Balestrino, S. Lagomarsino, and A. Tucciarrone, Thin Solid Films, 1985, 125, 263.
623. J.P. Eymery, Bull. Mineral., 1985, 108, 125.

8
Gas-phase Molecular Structures Determined by Electron Diffraction

BY D. W. H. RANKIN AND H. E. ROBERTSON

1 Introduction

This review covers all structures of gaseous inorganic molecules determined by electron diffraction and published in 1985. We hope that we have succeeded in finding all the relevant papers, and as usual we have taken a catholic view of the term 'inorganic'. The breadth of material described here is very great: there are compounds with as few as 2 or as many as 96 atoms, and the majority of papers describe more than just electron diffraction work. There is preparative chemistry, vibrational, rotational, and nuclear magnetic resonance spectroscopy, and there are many different types of theoretical treatment of structure. So there is great diversity in both subject and technique of study, and although the total number of structures described each year seems to be more or less constant the quality of the work is undoubtedly rising. As time goes by we tend to become more critical as reviewers, but we find that there are fewer papers causing us to raise our eyebrows than there were a few years ago. Of particular note this year is the increased number of structures of organometallic compounds, as most of these must be included with the more difficult subjects of study.

One paper which does not report new structural data but which analyses results for many compounds[1] will undoubtedly prove to be useful to structural inorganic chemists. A proposed modification to the Schomaker-Stevenson rule gives a much better agreement of experimental and calculated distances for very polar bonds. The new equation is:

$$\underline{r}(A-B) = \underline{r}_A + \underline{r}_B - 0.085|\underline{x}_A - \underline{x}_B|^{1.4}$$

Here $\underline{r}_A$ and $\underline{r}_B$ are covalent radii for atoms A and B, and $\underline{x}_A$ and $\underline{x}_B$ are their Pauling electronegativities.

The allocation of compounds to the various sections of this report is inevitably somewhat arbitrary, so the following list is provided to speed up the search for any particular compound.

Section 2, Groups I to III: NaCl, KCl, RbCl, CsCl, and dimers, $KAlCl_4$, $RbAlCl_4$, $CsAlCl_4$, $Mg(C_5H_5)$(neo-pentyl), BF_3, $(GaH_2NMe_2)_2$, $GaMe_3(OMe_2)$.
Section 3, Group IV: CCl_3F, CCl_2O, cyclo-$SiF_2C_4H_8$, SiH_3(cyclo-C_5H_{10}), GeH_3(cyclo-C_4H_8), $CH_3COOGeH_3$, $Me_3SnC{\equiv}CH$, $[Sn(OBu^t{}_2)]_2$, $(Me_2SiNMe)_2$, Me_3SiNCO, $LiCH(SiMe_3)_2$, $Ge[CH(SiMe_3)_2]_2$, $M[N(SiMe_3)_2]_n$ (n=2, M=Cd, Hg; n=3, M=Sc, Ce, Pr), $PH(OSiMe)_2$.
Section 4, Group V: N_2O, cis-$CF_3N{=}NCF_3$, perfluoro-(1,4-dimethyl-2,3-diazacyclohepta-1,3-diene), 3,3-bis(azidomethyl)-oxetane, $CF_3N{=}CHF$, $CF_3N{=}CHCl$, cyclo-C_4F_7NO, o-$C_6H_4Cl(NO_2)$, N-chloropyrollidine, $Me_3N.Br_2$, PCl_2F_3, PF_2HS, $PBu^t{}_3O$, $PBu^t{}_3NH$.
Section 5, Group VI: $SBu^t{}_2$, 3-chloro-propylene sulphide, ClSCOCl, $(MeSO_2CH_2)_2$, CF_3SF_5, $Se(CF_3)_2F_2$, SeF_5OCN, $TeCl_2$.
Section 6, Transition elements: VCl_2, $CrCl_2$, and $CoBr_2$, all with their dimers, $CdBr_2$, WCl_3, W_2Cl_6, NbF_5, TaF_5, $ReOCl_4$, $Zn(C_5Me_5)_2$, Mn(neo-pentyl)$_2$, $Cr(NO)_4$, $Mn(CO)(NO)_3$, $Fe(CO)_2(NO)_2$, $Co(CO)_3(NO)$, $Cr(CO)_5(CF_3NC)$, Cu(acetate)$_2$, Cr(acetate)$_2$, Er(dipivaloylmethanato)$_3$.

2 Compounds of Elements in Main Groups I, II, and III

Any first-year chemistry students should be able to describe the structure of crystalline sodium chloride, although they might have more difficulty with caesium chloride. But try asking them what they might expect to find in gaseous sodium chloride, and you will get a wide range of answers! The structures of four gaseous alkali-metal halides, the chlorides of sodium, potassium, rubidium, and caesium, have now been studied by electron diffraction,[2] at temperatures ranging from 840 to 960 K. At these temperatures the vapour pressures are around 0.01 torr, and in each case the vapour consisted of a mixture of monomeric and dimeric molecules. The proportion of dimer (defined as number of dimer molecules relative to the total number of monomers and dimers) was found to be between 9 and 18%, and all the dimer molecules had planar, rhombic structures, in which the bonds were longer than in the corresponding monomers but shorter than in the crystalline phase. The angles at the alkali-metal atoms in the dimers were smallest for the largest metal atoms. The most important parameters are listed in Table 1.

Table 1 Parameters (r_a) for alkali-metal chlorides

Parameter	NaCl	KCl	RbCl	CsCl
r(M-Cl)monomer/Å	2.388(8)	2.703(8)	2.817(4)	2.940(12)
r(M-Cl)dimer/Å	2.584(34)	2.950(54)	3.008(22)	3.017(32)
r(M-Cl)crystal/Å	2.814	3.139	3.285	3.560
<ClMCl/°	101.4(24)	96.0(24)	88.2(20)	83.6(1.3)
Proportion of dimer/%	16.6(66)	95.5(56)	12.4(34)	17.6(9.2)

Electron diffraction data for the tetrachloroaluminates of potassium, rubidium, and caesium have been reanalysed,[3] to investigate the possible presence of dimers or of aluminium trichloride. Failure to allow for such species could well lead to erroneous structures, but fortunately it was found that there could be at the most 10% aluminium trichloride present, and even less of dimeric molecules $(MAlCl_4)_2$, and that the original conclusions about the structures of the $MAlCl_4$ monomers were still valid. The molecules have double MClAl bridges and overall C_{2v} symmetry, and it was assumed that the $AlCl_4$ units had regular tetrahedral configurations. The Al-Cl bondlengths (r_g) in the three compounds are 2.153(6), 2.151(6), and 2.149(9) Å, and the K-Cl, Rb-Cl, and Cs-Cl distances are 2.98(5), 3.16(3), and 3.31(7) Å respectively.

Organometallic compounds, involving both main-group and transition metals, feature prominently in this report. Neopentyl(cyclopentadienyl)magnesium has been studied at 360 K,[4] and it was shown that the cyclopentadienyl group is η^5-bonded, with assumed C_{5v} local symmetry, and the Mg-C and C-C distances

(r_a) are 2.328(7) and 1.426(4) Å respectively. The neopentyl group was assumed to have T_d symmetry, with one hydrogen atom of the neopentane molecule replaced by magnesium, and it was also assumed that the substituted carbon atom lay on the C_5 axis of the MgC_5H_5 unit. The Mg-C bond length was then found to be 2.12(2) Å and the C-C bond lengths to be 1.532(5) Å. The angle MgCC refined to 126(4)°, this large value reflecting the strong interactions between the tertiary butyl and cyclopentadienyl-magnesium substituents on the linking carbon atom.

This magnesium compound is one of the more difficult subjects for study by electron diffraction, and many assumptions must be made in the data analysis. At the other extreme, a study of boron trifluoride[5] shows how much information can be deduced about a simple molecule, using both electron diffraction and spectroscopic data - vibrational frequencies, isotopic shifts, Coriolis constants and centrifugal distortion constants. Analysis of all these data simultaneously gave a value of 1.31074(8) Å for r_e^h(B-F), and the harmonic force constants F_{11}(A'), F_{22}(A_2"), and F_{33}, F_{34} and F_{44}(E') were found to be 8.768(27), 0.2884(3), 6.929(46), -0.453(17), and 0.5044(28) x 10^2 N m^{-1} respectively - remarkably precise values.

The structures of two gallium compounds were reported in 1985. Dimethylamidogallane[6] exists as a dimer in the gas phase at room temperature, and not as a monomer, as had previously been thought. The overall symmetry is effectively D_{2h}, and the planar Ga_2N_2 skeleton is almost perfectly square, with r_a(Ga-N) 2.027(4) Å and <GaNGa 90.6(8)°. The Ga-H bond length is 1.49(4) Å, at the short end of the known range; the dimethylamido groups have r(N-C) 1.463(13) Å and <CNC 109.6(16)°. Vibrational spectra indicate that the dimeric structure is retained both in benzene solution and in the crystalline state.

The adduct of trimethylgallium and dimethyl ether[7] has a structure very similar to that of the analogous aluminium compound. The most interesting feature is the almost planar arrangement of the three bonds to oxygen, with the angles COC and GaOC being 111.3(16)° and 124.1° (uncertainty not specified), so that the sum of the angles at oxygen is 359.5°. The non-planarity is indicated by a parameter defined as the angle between the Ga-O bond and the bisector of the COC angle. This refined to -7(11)°, the negative value indicating that the C-O bonds tend to eclipse two of the Ga-C bonds. Other parameters are: $\underline{r}_a$(Ga-C) 1.965(4), $\underline{r}$(Ga-O) 2.045(21), $\underline{r}$(C-O) 1.431(5) Å, and <CGaO 103.5(18)°. The last angle corresponds to a CGaC angle of 114.8°, so the effect of adduct formation has been to deform the planar structure of trimethylgallium, but nowhere near enough to give the gallium atom regular tetrahedral coordination.

3 Compounds of Elements in Main Group IV

In a study of trichlorofluoromethane, CCl_3F,[8] electron diffraction data were analysed to give an $\underline{r}_g$ structure with $\underline{r}$(C-Cl) 1.7634(11), $\underline{r}$(C-F) 1.343(5) Å, and <ClCCl 110.5(2)°. The refined amplitudes of vibration were then taken, together with vibrational frequencies, Coriolis coupling constants, and a centrifugal distortion constant, and from these data a set of quadratic force constants was derived. These force constants were then used to calculate vibrational corrections, so that a combined structural analysis of the electron diffraction data and ground-state rotation constants could be undertaken. In the resulting zero-point average structure ($\underline{r}_{av}$) the C-F and C-Cl bond lengths were 1.342(3) and 1.7609(10) Å respectively, and the ClCCl angle was 110.5(1)°.

Phosgene, CCl_2O, has been the subject of an even more elaborate study.[9] Rotation constants were obtained for eight different isotopic species in the ground state and for one of these species in the first excited state of each of the six vibrational modes. Analysis of electron diffraction intensities gave the $\underline{r}_z$ structure, and then changes in the average structure for each of the vibrationally excited states were deduced, using the rotation constants (Table 2). The third-order anharmonic constants were obtained by analysing these structural changes, and the equilibrium structure could then be deduced: $\underline{r}_e$(C-Cl) 1.7365(12), $\underline{r}_e$(C=O) 1.1766(22) Å, and $<_e$ClCCl 111.91(12)°.

Table 2 Structural changes for vibrational excited states of CCl_2O

Mode	$\delta \underline{r}$(C=O)/10^{-4}/Å	$\delta \underline{r}$(C-Cl)/10^{-4}/Å	δ<ClCCl/10^{-2}/°
ν_1	+64(30)	+7(13)	+15(13)
ν_2	0(8)	+37(3)	-4(3)
ν_3	0(10)	+6(1)	+16(1)
ν_4	-19(7)	0(10)	+1(10)
ν_5	0(13)	+50(5)	-12(5)
ν_6	0(13)	+15(5)	-15(5)

Various models can be used to describe five-membered ring compounds. For 1,1-difluorosilacyclopentane,[10] a twisted form with $\underline{C}_2$ symmetry and an envelope form with $\underline{C}_s$ symmetry were considered first, but then a mixture of the two (30 envelope:70 twist) gave a much better fit to the data. This suggested that the barrier to pseudorotation was low, so a dynamic model was used subsequently. With this model the potential function was found to have a minimum for a $\underline{C}_2$ structure with the dihedral angle C(1)C(2)C(3)C(4) 56.0(32)°, and the barrier to pseudorotation, which coincided with an envelope form, was 9(4) kJ mol^{-1} above the minimum. At the potential maximum the dihedral angle must be zero, and it was assumed that the envelope flap angle (between the CCCC and SiCC planes) was 125°. Under these conditions the following parameters ($\underline{r}_g$) were obtained: $\underline{r}$(C-C) (mean) 1.553(15), $\underline{r}$(Si-C) 1.853(3), $\underline{r}$(Si-F) 1.582(6) Å, <CSiF 113.4(3), <CCC 106.0(10), and <SiCC 101.3(13)°. These figures imply that <CSiC is 98.5(20)° and <FSiF is 105(2)°.

Molecules of cyclopentylsilane also undergo pseudorotational motion.[11] The most stable conformation is an envelope form with $\underline{C}_s$ symmetry, in which the SiH_3 group occupies an equatorial position; the axial form is about 3.2 kJ mol^{-1} less stable. A dynamic model was used to describe the ring puckering motion. The mean C-C bond length ($\underline{r}_g$) was found to be 1.548(2) Å, $\underline{r}$(Si-C) was 1.874(4) Å, and the angles SiCC were 113.1(6)°.

Cyclobutylgermane exists in two conformations at 280-290 K.[12] In the more abundant form [77(3)%] the germyl group occupies an equatorial position and the ring puckering angle [defined as the angle between the planes C(1)C(2)C(4) and C(2)C(3)C(4)] is 25.3(31)°, while in the second conformer the germyl group is in an axial position and the ring puckering angle is 20.4(36)°. Both forms have $\underline{C}_s$ symmetry. The relative abundances indicate that the energy difference between the conformations is 3.1(1) kJ mol^{-1}. The equivalent energy difference for cyclobutylsilane is only 0.8(4) kJ mol^{-1}, and as electronegative substituents tend to stabilise equatorial isomers of cyclobutyl derivatives it is suggested that germanium may be more electronegative than silicon. Other geometrical parameters ($\underline{r}_a$) include: $\underline{r}$(C-C) 1.557(3) and 1.558, $\underline{r}$(Ge-C) 1.948(4) Å, and <C(2)C(1)C(4) 89.6(7)°. The angles between the Ge-C bond and the plane C(1)C(2)C(4) for the equatorial and axial conformers are 131.2(11) and 126.4(19)° respectively.

Molecules of germyl acetate adopt a cis-planar conformation in the gas phase.[13] This involves the Ge-O bond eclipsing the C=O bond and leads to a very short non-bonded Ge...O distance, 2.839(12) Å, much less than the sum of van der Waals radii, which is 3.6 Å. The corresponding distance is short in other silyl and germyl esters and thioesters, but this seems to be the most extreme example encountered so far. The Ge-O bond length, 1.830(4) Å, is unusually long, but other parameters are much as would be expected for esters of this type: $\underline{r}_a$(C-O) 1.357(5), $\underline{r}$(C=O) 1.229(5), $\underline{r}$(C-C) 1.549(9) Å, <GeOC 113.0(5), <OCO 124.0(6), and <CCO 110.3(5)°.

In a new analysis of the structure of trimethylstannylacetylene[14] a force field has been used to calculate vibrational amplitudes and perpendicular amplitude corrections and thereby to eliminate many of the problems of correlations between parameters. Using a dynamic model to allow for the large amplitude torsional motions of the methyl groups, the following parameters ($\underline{r}_g$, $<_\alpha$) were obtained: $\underline{r}$(Sn-C_{methyl}) 2.141(10), $\underline{r}$(Sn-$C_{ethynyl}$) 2.126(25), $\underline{r}$(C≡C) 1.233(18) Å, <C_{Me}Sn$C_{ethynyl}$ 104.7(20)°. The errors quoted here are three times the standard deviations. The new results make it clear that Sn-$C_{ethynyl}$ bonds in trimethylstannyl alkynes are consistently longer, by some 0.03 to 0.04 Å, than in tri- or tetra-alkynyl stannanes. This is believed to result from the greater electronegativity of the ethynyl group, compared with that of the methyl group.

Gas-phase data for bis(t-butoxy)tin at 385 K are consistent with the presence of dimers, although the possibility that small percentages of monomers or other species are also present cannot be excluded.[15] The two tin atoms are linked by two bridging butoxy groups, with r_a(Sn-O) in the planar ring 2.16(1) Å and <OSnO 76(2)°. The C-O bonds of the bridging groups lie very nearly in the plane of the ring, the sum of angles at oxygen being 358(2)°. The terminal butoxy groups lie one on either side of the ring, with the Sn-O bonds [1.97(2) Å long] almost perpendicular to the Sn...Sn vector and the butyl groups occupying positions as far from the ring as possible. The SnOC angles are 132(2)°, which is unusually wide. In the same note there is a report of the crystal structure of the analogous germanium compound, but with the butyl groups replaced by (tri-t-butyl)methyl groups. This extremely crowded molecule is, not surprisingly, monomeric.

Once again there are many structures of compounds containing methyl-substituted silyl groups to report. Hexamethylcyclodisilazane[16] has a planar Si_2N_2 ring, with r_g(Si-N) 1.736(4) Å and <NSiN 86.4(3)°. The N-C bonds, which are 1.460(5) Å long, also lie in the ring plane, or very nearly so. In fact slight puckering (around 6°) was observed, but it was concluded that in the equilibrium structure the $Si_2N_2C_2$ group of atoms is planar. The Si-C bond length was given as 1.864(4) Å and the NSiC angle as 115.4(5)°. This corresponds to a CSiC angle of 107.9(4)°.

The structure of trimethylsilylisocyanate has been redetermined.[17] In the r_a structure the SiNCO group is markedly non-linear, with <SiNC 156.9(30), <NCO 165.8(36)°, and the N=C bond eclipses one Si-C bond, and the Me_3Si group is also tilted by 5.7(24)° away from the CO group, so that its C_3 axis no longer coincides with the Si-N bond. These deviations from overall C_{3v} symmetry for the C_3SiNCO group were shown to be compatible with a pseudolinear structure similar to that of SiH_3NCO. The Si-N bondlength was reported to be 1.740(4) Å, between the values of

1.69 and 1.76 Å found in earlier studies by microwave spectroscopy and electron diffraction. Other bondlengths are $\underline{r}$(Si-C) 1.864(2), $\underline{r}$(N=C) 1.202(16), and $\underline{r}$(C=O) 1.176(10) Å, and the CSiC angle is 108.8(25)°.

The structure of bis(trimethylsilyl)methyl lithium, $LiCH[Si(CH_3)_3]_2$, has been studied in the gas phase at 413 K.[18] This is the first organic alkali-metal derivative to be studied in this way. The data were consistent with the presence of monomers only, or at least a very high percentage of monomers. It was assumed that the $C(SiMe_3)_2$ groups had local $\underline{C}_2$ symmetry and that the central Si-C bonds were longer than the terminal ones by 0.015 Å. The Li-C distance ($\underline{r}_a$) is 2.03(6) Å, and this first reported value for a mono-coordinate lithium-to-carbon bondlength should be compared with Be-C and B-C distances of 1.80 and 1.58 Å in $Be(CH_3)_2$ and $B(CH_3)_3$. The central Si-C distance is given as 1.892(2) Å, and important angles are: <SiCSi 119.0(15), $<C_{central}SiC$ 111.0(12), and <SiCLi 115(2)°. All these parameters show effects of steric crowding, which is also important in $Ge[CH(SiMe_3)_2]_2$.[19] This compound is monomeric in the gas phase at 430 K, with $\underline{r}_a$(Ge-C) 2.04 Å (exceptionally long, normally around 1.95 Å) and <CGeC 107(2)° (10° larger than in the analogous tin compound). In this case the mean Si-C bondlength is given as 1.89(1) Å and the angles GeCSi and SiCSi as 111(1) and 113(1)° respectively. The angles at carbon are therefore smaller than in the lithium compound, but there is a restriction on their increase, because the two bulky alkyl groups can interact with each other. This germanium compound is dimeric in the crystalline phase, with three-coordinate but non-planar germanium atoms, while the lithium compound is polymeric, with two-coordinate lithium atoms.

Structures of five compounds containing bis(trimethylsilyl)amido groups have been reported. These are the bis-amides of cadmium[20] and mercury [21] and the tris-amides of scandium,[22] cerium,[23] and praseodymium.[23] The cadmium and mercury compounds, studied at 375 and 390 K, have linear N-metal-N arrangements, planar coordination at the nitrogen atoms, and staggered Si_2NMNSi_2 skeletons, with $\underline{D}_{2d}$ symmetry. In the scandium compound at 450 K the ScN_3 group of atoms is essentially planar, and each

$ScNSi_2$ group is also planar, but the latter groups are twisted 55(2)° to the central plane, giving the skeleton overall $\underline{D}_3$ symmetry. In contrast, the cerium and praseodymium compounds (studied at 435 and 430 K) have skeletons with $\underline{C}_3$ symmetry, and the central MN_3 units are pyramidal, with NCeN and NPrN angles of 112(3) and 113(3)°. In these compounds the planar $MNSi_2$ groups are twisted, by 42(5) (Ce) and 47(5)° (Pr), from the positions in which the skeletons would have $\underline{C}_{3v}$ symmetry, and in each case the two MNSi angles are no longer equal. Other parameters for all these compounds are given in Table 3.

Table 3 Parameters ($\underline{r}_a$) for some bis(trimethylsilyl)amido compounds

Compound	$Cd(N)_2$	$Hg(N)_2$	$Sc(N)_3$	$Ce(N)_3$	$Pr(N)_3$
$\underline{r}$(M-N)/Å	2.03(2)	2.01(2)	2.02(3)	2.33(4)	2.31(4)
$\underline{r}$(Si-N)/Å	1.727(7)	1.732(9)	1.715(8)	1.696(15)	1.705(15)
$\underline{r}$(Si-C)/Å	1.881(3)	1.883(6)	1.881(4)	1.887(8)	1.890(8)
<MNSi/°	114.0(6)	116.0(10)	116.0(10)	{108(2) 123(2)	{108(2) 123(2)
<NSiC/°	113.6(12)	111.8(12)	110.7(10)	117(2)	116(2)

The structure of gaseous bis(trimethylsilyl)hypophosphite, $PH(OSiMe_3)_2$, has been determined.[24] The conformation of this compound presents considerable problems, as dihedral angles about two P-O and two Si-O bonds must be defined, and there may of course be more than one conformer present in the gas. Two different conformations fitted the data reasonably well, and in both of them the Si-O, P-O, and Si-C distances were 1.637(2), 1.657(2) (assumed to be 0.02 Å greater than P-O), and 1.905(1) Å respectively, the OPO angle was 99 or 100(1)° and the OSiC angles were 109.1 or 110.5(5)°. In the structure that fitted the data best one SiOP angle was 113.2(15)° and the other was 141.5(11)°. Both trimethylsilyl groups were in positions such that the O-Si bond was almost anti with respect to the P-H bond.

4 Compounds of Elements in Main Group V

The thermally averaged structure of nitrous oxide, N_2O, has been measured at several temperatures in the range 293-932 K.[25] Force field calculations based on the variational wave-function method and on the moment method indicate that the R_a N-N distance should increase from ca. 1.1309 to 1.1350 Å and the N-O distance from 1.1918 to 1.1930 Å over this temperature range, while the non-bonded N...O distance should remain more or less constant, at about 2.3182 Å. These changes have now been observed experimentally, and the amplitudes of vibration also vary in a predictable manner. It was noted that the stretch-bend interaction force constant is of particular importance in describing the temperature dependence of the structural parameters.

The cis isomer of hexafluoroazomethane, $CF_3N{=}NCF_3$, has now been prepared and its structure determined.[26] The CF_3 groups, which are staggered with respect to the N=N bond, interact strongly with each other, and the consequent strain is partially released by widening of the NNC angles, from 113.4° in the trans isomer to 122.1(8)° in the cis form, lengthening of the C-N bonds by 0.03 Å to 1.494(5) Å (r_g), and tilting of the local C_3 axes of the CF_3 groups by 5.6(12)°, away from each other. The other parameters, including the N=N bond length, are not very different from those of the other isomer: r(N=N) 1.229(8), r(C-F) 1.324(3) Å, <FCF 109.2(4)°.

Electron diffraction data for perfluoro-(1,4-dimethyl-2,3-diazacyclohepta-1,3-diene)[27] at 308 K are consistent with a structure having C_2 symmetry and a dihedral angle C=N-N=C of 52° but planar N-N=C-C fragments. Principal parameters (r_g) are: r(N-N) 1.349(19), r(N=C) 1.257(11), r(C-C)(mean) 1.536(3), r(C-F)(mean) 1.340(2) Å, <NNC 121.9(7), <NCC(ring) 124.0(6), <NCC(CF_3) 117.8(7), <CCF(CF_3) 111.1(3) and <FCF(mean,CF_2) 107.8(6)°. The CF_3 groups were found to be orientated with one C-F bond eclipsing a C=N bond.

The central four-membered ring of 3,3-bis(azidomethyl)-oxetane is reported to be planar[28], with r_g(C-C) 1.547(3) and r(C-O) 1.458(7) Å and <COC 94.0(16)°. The azidomethyl substituents were assumed to have linear N_3 groups, and the two N-N distances were found to be 1.240(2) and 1.142(2) Å, with the terminal bonds shorter than the inner ones. In the substituent groups C-N distance was 1.470(9) Å, the C-C distance the same as in the ring (assumed), the NNC and NCC angles 119.2(10) and 111.4(12)° respectively, and the angle between the two C(ring)-C(substituent) bonds 108.9(23)°.

Both CF_3N=CHF and CF_3N=CHCl exist as two isomers, the form with the halogen atom trans to the CF_3 group (the E isomer) being more stable. The ratio of isomers has been shown by NMR spectroscopy to be 6.7:1 for the chloride and 15.4:1 for the fluoride, and electron diffraction data are consistent with this, although they do not give the ratios so precisely.[29] Analysis based on sets of parameters which were assumed to apply to both isomers, but with increased CNC and NCCl angles for the Z isomer, gave the following results (r_a) for the chloride: r(C-F) 1.332(4), r(N-C) 1.427(6), r(N=C) 1.277(11), r(C-Cl) 1.707(5) Å, <CNC 116.3(8), <CNCl 119.5(9)°, tilt of CF_3 group 4.6(5)° away from the nearby C-H bond. For the fluoride: r(C-F)(CF_3) 1.332(4), r(N-C) 1.414(7), r(N=C) 1.277(7), r(C-F)(CHF) 1.312(10) Å, <CNC 117.3(9) and <NCF(CHF) 117.7(14)°; the CF_3 group is tilted 4.8(7)° away from the C-H bond.

The structure of heptafluoro(nitroso)cyclobutane[30] is very similar to that of perfluoro-cyclobutane itself, with r(C-C) 1.573(5) and r(C-F) 1.331(3) Å, <FCF 109.1(4)°, and the ring puckering angle (between two CCC planes) 23.0(18)°. It was not possible to tell whether the nitroso group occupied an axial or an equatorial position, but it was clear that the group itself adopted an exo orientation, so that the N-O bond eclipsed the adjacent C-F bond. The N=O bondlength was not refined but was fixed at 1.20 Å. The C-N bondlength refined to 1.521(15) Å and the angle CNO to 111.9(22)°.

In ortho-chloronitrobenzene[31] the plane of the nitro group is twisted 33.9(4)° from the plane of the rest of the atoms, and the other parameters describing this substituent are similar to those in related compounds: r_g(N=O) 1.226(2), r(C-N) 1.462(12) Å, and <ONO 123.6(10)°. Both substituents are slightly bent away from each other, with angles C(2)C(1)N 123.5(10)° and C(1)C(2)Cl 120.8(11)°. The C-Cl bondlength is 1.721(3) Å. The ring is also distorted from regular hexagonal symmetry, although all C-C distances were assumed to be equal [1.387(2) Å]: the angles C(2)C(1)C(6), C(1)C(2)C(3), and C(2)C(3)C(4) are 121.4(12), 121.7(7), and 117.2(10)° respectively. As in other substituted benzenes, the angles between carbon atoms at the positions of electronegative substituents are greater than 120°.

The conformations and geometric structures of *N*-chloro- and *N*-methyl-pyrrolidine have been studied by electron diffraction and by *ab initio* calculations.[32] For the chloro compound the axial envelope conformation is found experimentally to be more stable than the equatorial form [ΔG = 3(1) kJ mol^{-1}], but the *ab initio* calculations indicate a very small preference for the equatorial conformer. Experimental parameters (r_a) include: r(C-N) 1.476(5), r(C-C) 1.535(3) [C(2)-C(3)] and 1.553(3) [C(3)-C(4)], r(N-Cl) 1.736(3) Å, <CNC 108.0(8) and <CNCl 110.5(7)°. The angle between the planes CCCC and CNC is 44.7(13)°.

The adduct of trimethylamine and dibromine is not very strongly bound, and in a study of its structure in the gas phase[33] it was found that it was 48(7)% dissociated. The N-Br distance (r_a) was quite long, 1.98(3) Å, and the Br-Br bond, 2.29(1) Å, no longer than in free dibromine. The angle between these two bonds was found to be 112(3)°, which is interesting, because the I_2 and ICl adducts in the solid phase involve linear N-I-halogen coordination. The N-Br bond did not coincide with the (assumed) C_3 axis of the NMe_3 group but was tilted 9(2)° towards the other bromine atom. All this evidence is consistent with a very weak N-Br interaction, so it is particularly surprising to find that the C-N bonds have lengthened from 1.46 Å in free trimethylamine to 1.67(2) Å while the CNC angles have decreased from 111 to 104(2)°.

A reliable modern structure for PCl_2F_3 provides a welcome addition to the literature.[34] At least one modern inorganic chemistry textbook perpetuates the story that the chlorine atoms occupy axial positions! This new work shows quite clearly that they are indeed equatorial, with $\underline{r}_g$(P-Cl) 2.004(2) Å, although a model with a very small amount of a form with one chlorine axial and one equatorial (mole fraction 0.019±0.044) and some hydrolysis product, $PF_3O.2HCl$ (mole fraction 0.054±0.017) gave a slightly better fit to the data. As in other fluorophosphoranes, the axial P-F bonds are longer than the equatorial ones: the bondlengths are 1.593(4) and 1.546(9) Å. Distortions from a regular trigonal-bipyramidal structure are very small, and the angles $F_{ax}PF_{eq}$ and ClPCl are 89.3(3) and 122.0(5)° respectively.

Rotation constants obtained by microwave spectroscopy and dipolar coupling constants observed in n.m.r. spectra for solutions in a liquid-crystal solvent have been combined with electron diffraction data to give a complete structure for PF_2HS.[35] Refinements based on various combinations of data are presented, and the most precise results are based on simultaneous analysis of all the available data. The P=S, P-F, and P-H distances ($\underline{r}_\alpha^{\,o}$) are 1.875(3), 1.546(2), and 1.419(13) Å, and the angles FPF, FPS, and HPS are 99.1(1), 117.2(2), and 116.4(6)° respectively. Using electron diffraction data alone, the parameters for the heavy atom group PF_2S are close to those given above, but the parameters defining the position of the hydrogen atom could not be refined satisfactorily.

The bulky substituents in tri(t-butyl)phosphine oxide and tri(t-butyl)phosphine imide confer unusual thermal and chemical stability on these compounds, and studies of their structures[36] now show how the butyl groups protect the oxygen atom and imide group from attack. In both compounds the steric interactions between the butyl groups are reduced by widening of the angles CPC from around 100° in most phosphine oxides and imides to 109-113°, lengthening of P-C bonds from 1.83 to 1.89-1.92 Å, and tilting of the local $\underline{C}_3$ axes of the butyl groups by about 3° away from each other. Important parameters ($\underline{r}_a$) for PBu_3O are: $\underline{r}$(P=O) 1.590(12), $\underline{r}$(P-C) 1.888(6), $\underline{r}$(C-C) 1.519(3) Å, <CPC 112.9(5) and <CCC 109.1(4)°. Parameters for PBu_3NH include: $\underline{r}$(P=N) 1.652(11), $\underline{r}$(P-C) 1.913(6), $\underline{r}$(C-C) 1.532(2) Å, <CPC 109.4(7) and <CCC 107.8(5)°. The P=O and P=N bonds are exceptionally long, so much

so that they should really be considered as single bonds and the compounds as $PBu_3^+-O^-$ and $PBu_3^+-NH^-$. There is no steric requirement for this elongation, which must be regarded as electronic in origin.

5 Compounds of Elements in Main Group VI

Steric strain between the t-butyl groups in di(t-butyl)sulphide[37] is released by twisting of the groups 12(8)° away from the staggered conformation, giving $\underline{C}_2$ overall symmetry, tilting of the local $\underline{C}_3$ axis of these groups by 7(2)° away from each other, lengthening of the S-C bonds by 0.04 Å from the normal length in an unstrained sulphide to 1.854(5) Å, and by widening of the CSC angle from the normal value of around 100° to 113.2(12)°. The parameters for the butyl groups [$\underline{r}_g$(C-C) 1.539(3) Å, $<_\alpha$CCC 109.6(5)°] are unexceptional.

There are two different gauche conformers of 3-chloropropylene sulphide present in the gas phase at 308 K.[38] In the more abundant conformer [82(8)%] the dihedral angle SCCCl is −129.6(48)°, and it is 75.7° in the other form. Other parameters were assumed to be the same for both forms, and the following values ($\underline{r}_a$) were obtained: $\underline{r}$(C-C)(ring) 1.492(23), $\underline{r}$(C-C)(chain) 1.516(24), $\underline{r}$(C-S) 1.822(13) (two different ring bonds, assumed equal), $\underline{r}$(C-Cl) 1.798(23) Å, <CCC 144.1(13), <CCS 114.3(13), <CCCl 109.8(12), and <CSC 48.3(7)°.

In chlorocarbonylsulphenylchloride, ClSCOCl,[39] the major conformer has the chlorine atoms <u>anti</u> with respect to one another, although a small amount of a second form [6.5(99)% at 308 K] may also be present. The bondlengths and angles are: $\underline{r}$(C=O) 1.183(5), $\underline{r}$(C-Cl) 1.749(8), $\underline{r}$(C-S) 1.791(9), $\underline{r}$(S-Cl) 2.010(4) Å, <SCO 126.9(20), <SCCl 106.0(2), and <CSCl 100.6(4)°.

In an attempt to compare gas- and solid-phase structures, 1,2-bis(methylsulphonyl)ethane, $CH_3SO_2CH_2CH_2SO_2CH_3$, has been studied at 478 K.[40] This molecule was particularly difficult to study by electron diffraction, and it was necessary to fix several geometrical parameters, particularly those relating to the valence angles for the CH_3SO_2 groups, at values consistent with those found for related compounds. With these constraints the bondlengths (r_g) for S=O, C-C, S-C(terminal), and S-C(central) bonds refined to 1.443(3), 1.561(12), 1.772(25), and 1.808(25) Å respectively and the CCS and CSC angles to 106.4(21) and 100.4(20)°. The main change from crystal to gas phase is lengthening of the S-C bonds. The preferred model had an inversion centre, and the conformation was similar to that found in the crystal. The SCCS dihedral angle was therefore 180°, and the CSCC dihedral angles were close to 90°, so that each of the two terminal CSC planes was essentially perpendicular to the plane of the central heavy atoms.

The coordination at sulphur in (pentafluorothio)-trifluoromethane[41] is slightly distorted from that of a regular octahedron and closely resembles that in $SClF_5$. The S-F(equatorial, *i.e.* cis to the CF_3 group) bonds [r_a, 1.572(2) Å] are marginally longer than the S-F(axial) bond [1.562(7) Å], and the angle $F_{ax}SF_{eq}$ is 89.5(2)°. The S-C and C-F distances are 1.887(8) and 1.319(2) Å respectively, and the FCF angle is 108.7(5)°. The compound was also studied by microwave spectroscopy, but combined analysis of electron diffraction data and rotation constants was not attempted, because the large amplitude torsional vibration makes calculation of reliable vibrational correction terms impossible.

In bis(trifluoromethyl)selenium difluoride[42] the CF_3 groups occupy equatorial positions of a framework derived from a trigonal bipyramid. The overall symmetry is C_2, with the CF_3 groups twisted 19.8(11)° away from the positions in which one C-F bond of each group is anti with respect to an Se-C bond. The CF_3 groups do not appear to be significantly tilted, the refined value for the tilt angle being 0.7(21)°, with the local C_3 axes of the groups inclined away from each other, no longer coinciding with the C-Se bonds. The Se-C bondlength (r_a) is 2.022(8) Å, about 0.08 Å longer than in $Se(CH_3)_2$ and 0.04 Å longer than in $Se(CF_3)_2$, thus reflecting the changes noted for S-C bonds. The Se-F bonds [length 1.827(5) Å] are unusually long, compared with 1.771 Å for the axial bonds in SeF_4, for example, and again the trend is consistent with what has been observed for analogous sulphur compounds. However, the CSeC angle, 118.7(17)°, should be compared with the corresponding CSC angle of 97.3° in

$(CF_3)_2SF_2$. Other parameters of importance are: $\underline{r}$(C-F) 1.314(3) Å, <FCF 110.1(4)°, and <FSeF 157.8(35)°. This last angle indicates considerable displacement of the axial fluorines from their positions in a regular trigonal-bipyramidal structure.

It is now believed that the compound previously formulated as pentafluoroselenium isocyanate, SeF_5NCO, is in fact the cyanate, SeF_5OCN.[43] Its structure, compared with those of analogous sulphur and tellurium compounds, was anomalous, but ^{77}Se, ^{14}N, and ^{15}N n.m.r. data and vibrational spectra have been recorded, and various structures, including the nitrile oxide, SeF_5CNO, and fulminate, SeF_5ONC, have been considered. The evidence is now consistently in favour of the cyanate structure, although why this is the favoured form for selenium, but not sulphur or tellurium, is not clear. The principle structural parameters are: $\underline{r}$(Se-O) 1.794(6), $\underline{r}$(O-C) 1.257(10), $\underline{r}$(C≡N) 1.181(10) Å, <SeOC 116.8(8), and <OCN 172(3)°.

Tellurium dichloride is an interesting but elusive compound. It is not thermodynamically stable in the solid phase, although there is a eutectic point in the phase diagram of Te-$TeCl_4$ at 470 K, corresponding to a tellurium:chloride ratio which is very close to 1:2. It is known that $TeCl_2$ molecules are present in the liquid phase under these conditions and that $TeCl_2$ is also the dominant species in the vapour above the eutectic mixture. An electron diffraction study of the vapour obtained from a mixture of this composition at 480 K showed that it consisted of essentially pure $TeCl_2$.[44] The distance $\underline{r}_a$(Te-Cl) was found to be 2.329(3) Å and the ClTeCl angle to be 97.0(6)°. Both these parameters are exactly as expected, and the series of structures for OCl_2, SCl_2, $SeCl_2$, and $TeCl_2$ provide an excellent example of simple structural trends down a main group of the periodic table.

6 Compounds of Transition Elements

There have been several new studies of metal dihalides, and for most of these systems, which must be studied at elevated temperatures, there are uncertainties about the bond angle at the metal atom and about the presence of dimers as well as monomers. Vanadium dichloride at 1330 K[45] is 91(2)% monomer, with $\underline{r}_g$(V-Cl)

2.172(6) Å and <ClVCl 91(2)°. In the small amount of dimer the molecules have a puckered structure with a double chloride bridge, with r(V-Cl) 2.341(16) for the bridge bonds and <ClVCl 98.1(13)°. The length of the terminal V-Cl bonds in the dimer, which has C_{2v} symmetry and equatorial terminal bonds, could not be refined, and results are given for two refinements, with this distance equal to, or 0.03 Å greater than, the bondlength in the monomer. The same strategy was used for chromium dichloride, studied at 1170 K,[45] but this time the other parameters were slightly different in the two refinements. For the monomer [72(5)% abundant] r(Cr-Cl) is 2.207(10)/2.200(11) Å and <ClCrCl is 109.8(24)/111.8(25)°, while for the dimer r(Cr-Cl)(bridge) is 2.353(10)/2.357(10) Å and <ClCrCl is 82.0(7)/82.8(8)°.

Cobalt dibromide at 910 K is 95% monomeric,[46] with r_g(Co-Br) 2.241(5) and r(Br...Br) 4.387(9) Å. The latter distance is 0.095 Å less than twice the bondlength, and this difference is believed to be a shrinkage effect, caused by a large amplitude bending vibration. The data enable the vibrational frequency for the bending mode to be estimated at 68(8) cm^{-1}. The small amount of dimer is shown to be considerably puckered, with a C_{2v} symmetry structure analogous to those of V_2Cl_4 and Cr_2Cl_4. The bridge Co-Br distance is 2.429(14) Å. Cadmium bromide has also been investigated,[47] at 940-1020 K. In this case only monomers were detected, but again a shrinkage effect was observed, with r_g(Cd-Br) 2.394(2) Å and r(Br...Br) 4.743(4) Å. These are averages of values obtained in several independent experiments.

The structure of gaseous tungsten trichloride is described in two papers. At 800 K the molecules are essentially all monomers,[48] with a T-shaped structure derived from a trigonal bipyramid, in which the two axial bonds [r_a, 2.322(9) Å] are longer than the equatorial bond [2.168(9) Å]. The angle $Cl_{ax}WCl_{eq}$ is 89(1)°, and the $Cl_{ax}WCl_{ax}$ angle is given as 150(5)°, which indicates that there is considerable distortion from the idealised configuration. At 670 K the gas consists of 75% dimers and 25% monomers.[49] The dimer molecules have double chlorine bridges and D_{2h} symmetry. The terminal and bridging W-Cl distances (r_g) are 2.211(5) and 2.323(6) Å respectively, and the angles ClWCl are 119(4)° for the angle between two terminal bonds and 94.7(13)° for two bridging bonds.

Niobium pentafluoride has been investigated at 610 and 850 K and tantalum pentafluoride at 610 K.[50] In both cases trigonal-bipyramidal models fitted the data better than square-pyramidal models, and it was also shown that the axial bonds were longer than the equatorial bonds. The Nb-F bondlengths were found to be 1.905(2) and 1.829(1) Å at 610 K and 1.894(2) and 1.823(1) Å at 850 K, while the Ta-F bondlengths are 1.886(5) and 1.842(3) Å.

A square-pyramidal structure has been observed for tetrachloro-oxo-rhenium(VI), $ReCl_4O$, at 340 K.[51] The Re-Cl and Re=O bondlengths are 2.270(5) and 1.663(9) Å respectively, and the angle OReCl is 105.5(15)°. This places the rhenium atom well above the plane of the four chlorine atoms, a structural characteristic which has been observed in related compounds.

In bis(pentamethylcyclopentadienyl)zinc[52] at 400 K there is one η^5 ring with r_a(Zn-C) 2.30(2) Å and one σ-bonded ring with r(Zn-C) 2.09(5) Å. The angle between the plane of this second ring and the Zn-C bond is only 87(3)°, which seems to indicate that the interactions between the metal atom and the ring must involve more than just one carbon atom. The rings are therefore effectively parallel to each other. As many assumptions had to be made so that some geometrical parameters could be refined, detailed interpretation of the significance of the structure must wait until more information is available.

The radial distribution curve for bis(neo-pentyl)manganese at 410 K is consistent with the presence of monomers only.[53] Assuming C_2 symmetry, that the coordination at manganese was linear, and that the quaternary carbon atoms had a regular tetrahedral configuration, the Mn-C and C-C bondlengths (r_a) refined to 2.104(6) and 1.552(3) Å respectively and the MnCC angle to 120(1)°. The two t-butyl groups were almost anti with respect to one another, with the dihedral angle CC(Mn)CC 160(30)°.

Two interesting papers describe the structures of $Cr(NO)_4$, $Mn(CO)(NO)_3$, $Fe(CO)_2(NO)_2$,[54] and $Co(CO)_3(NO)$,[55] all of which are isoelectronic with $Ni(CO)_4$. All the compounds were studied at around room temperature, and in each case the M-Y-O groups were linear or very nearly so, in the r_g structure, so that T_d, C_{3v}, or C_{2v} symmetry could be assumed, as appropriate. Important details of the four new structures are given in Table 4. The bonds M-N and M-C both get shorter as the atomic number of M increases, although the total bond order of the metal bonds changes very little. The pattern of structural changes is consistent with the NO ligand being a much better electron acceptor than CO.

The structure of the trifluoromethylisocyanide ligand when coordinated to a metal should be substantially different from that of the free molecule, and the changes have been revealed by a study of $Cr(CO)_5(CF_3NC)$ in the gas phase at 310 K.[56] The N=C bond is lengthened from 1.171 to 1.220(16) Å, while the C=N-C angle has changed from 180 to 142(4)°, both these parameters indicating strong back π-donation from the metal atom to the C=N bond. Other parameters (r_a) include: r(C=O) 1.140(3), r(C-F) 1.338(11), r(N-C) 1.40(3), r(Cr-C)(average) 1.918(3) Å, <FCF 108.2(11)°. The $Cr(CO)_5$ group was assumed to have C_{4v} local symmetry, and the angle $C_{ax}CrC_{eq}$ refined to 87.2(14)°.

Table 4 Geometrical parameters in metal carbonyl nitrosyl compounds

Compound	$Cr(NO)_4$	$Mn(CO)(NO)_3$	$Fe(CO)_2(NO)_2$	$Co(CO)_3(NO)$
r(M-N)/Å	1.763(2)	1.717(2)	1.688(3)	1.671(6)
r(M-C)/Å		1.947(7)	1.883(3)	1.843(3)
r(N=O)/Å	1.171(2)	1.167(2)	1.171(4)	1.180(4)
r(C=O)/Å		1.145(6)	1.140(4)	1.136(4)
<NMN/°	109.5	112.5(43)	114.2(62)	
<NMC/°		106.2(49)	108.4(13)	107.7(14)
<CMC/°			110.2(25)	111.2(13)

Copper(I) acetate is dimeric with a planar skeleton at 490 K.[57] Assuming $\underline{D}_{2h}$ symmetry, the following bondlengths ($\underline{r}_g$) were obtained: Cu-Cu 2.491(3), Cu-O 1.868(2), C-O 1.270(2), and C-C 1.506(3) Å. The angle OCuO is 172.5(1)°. The metal-metal bond is short in this example, but in dichromium(II) tetra-acetate it is dramatically shorter.[58] This compound, which has four acetate groups bridging the metal-metal bonded core ($\underline{D}_{4h}$ symmetry), has a formal quadruple bond, but attempts to measure the bondlength, which was expected to be very short, have hitherto been frustrated. Either additional ligands coordinate to the metal in the vacant axial sites, or the oxygen atoms of the molecule itself coordinate to the chromium, as happens with the anhydrous compound in the crystalline phase. Using a quartz nozzle at 470 K, the uncoordinated compound has now been studied. The Cr≡Cr distance ($\underline{r}_a$) is 1.966(14) Å, more than 0.3 Å shorter than in the crystal. Other parameters are: $\underline{r}$(Cr-O) 2.014(8), $\underline{r}$(C-O) 1.267(5), $\underline{r}$(C-C) 1.525(16) Å, and <CrCrO 94.5(3)°.

We end this chapter with the largest molecule, tris(dipivaloylmethanato)erbium(III).[59] This molecule has 96 atoms and about 3500 atom pairs to contribute to the radial distribution curve. At 450 K the structure is similar to that in the solid phase, with $\underline{C}_3$ overall symmetry. The basic form is that of a trigonal prism, but each ErO_2 plane is twisted 12.9(13)° away from the position it would occupy if the symmetry was $\underline{D}_{3h}$, and there is also folding about each of the three local O...O axes, by 22.2(15)°. The ligands were assumed to have $\underline{C}_{2v}$ local symmetry and the butyl groups to have $\underline{C}_{3v}$ symmetry. Under these conditions the following parameters ($\underline{r}_a$) were obtained: $\underline{r}$(Er-O) 2.218(5), $\underline{r}$(C-O) 1.274(4), $\underline{r}$(C-C)(ring) 1.406(6) Å, and <OErO (within one ligand) 75.0(5)°. The C-C distance within the t-butyl groups is 1.550(6) Å.

References

1. R. Blom and A. Haaland, J. Mol. Struct., 1985, 128, 21.
2. R.J. Mawhorter, M. Fink and J.G. Hartley, J. Chem. Phys., 1985, 83, 4418.
3. V.V. Ugarov, Yu. Sh. Kalaichev and A.I. Kolesnikov, Zh. Strukt. Khim., 1985, 26(3), 47.

4. R.A. Andersen, R. Blom, A. Haaland, B.E.R. Schilling and H.V. Volden, Acta Chem. Scand., Ser. A, 1985, 39, 563.
5. A.G. Gershikov and V.P. Spiridonov, Vestnik Moskovskogo Universiteta, Khimija, 1985, 40(1), 15.
6. P.L. Baxter, A.J. Downs, D.W.H. Rankin and H.E. Robertson, J. Chem. Soc., Dalton Trans., 1985, 807.
7. V.P. Novikov, A.V. Golubinskii, V.S. Mastryukov, L.V. Vilkov, L.M. Golubinskaya and V.I. Bregadze, Zh. Strukt. Khim., 1985, 26(1), 34.
8. S. Konaka, M. Takeuchi and M. Kimura, J. Mol. Struct., 1985, 131, 317.
9. S. Yamamoto, M. Nakata and K. Kuchitsu, J. Mol. Spectrosc., 1985, 112, 173.
10. Q. Shen and M. Dakkouri, J. Mol. Struct., 1985, 130, 283.
11. Q. Shen and M. Dakkouri, J. Mol. Struct., 1985, 131, 327.
12. M. Dakkouri, J. Mol. Struct., 1985, 130, 289.
13. E.A.V. Ebsworth, C.M. Huntley and D.W.H. Rankin, J. Organometal. Chem., 1985, 281, 63.
14. A.V. Belyakov, L.S. Khaikin, L.V. Vilkov, G.M. Apal'kova, V.S. Nikitin, E.T. Bogoradovskii and V.S. Zavgorodnii, J. Mol. Struct., 1985, 129, 17.
15. T. Fjeldberg, P.B. Hitchcock, M.F. Lappert, S.J. Smith and A.J. Thorne, J. Chem. Soc., Chem. Commun., 1985, 939.
16. E. Gergő, G. Schultz and I. Hargittai, J. Organometal. Chem., 1985, 292, 343.
17. S. Cradock, C.M. Huntley and J.R. Durig, J. Mol. Struct., 1985, 127, 319.
18. T. Fjeldberg, M.F. Lappert and A.J. Thorne, J. Mol. Struct., 1985, 127, 95.
19. T. Fjeldberg, A. Haaland, B.E.R. Schilling, H.V. Volden, M.F. Lappert and A.J. Thorne, J. Organometal. Chem., 1985, 280, C43.
20. E.C. Alyea, K.J. Fisher and T. Fjeldberg, J. Mol. Struct., 1985, 127, 325.
21. E.C. Alyea, K.J. Fisher and T. Fjeldberg, J. Mol. Struct., 1985, 130, 263.
22. T. Fjeldberg and R.A. Andersen, J. Mol. Struct., 1985, 128, 49.
23. T. Fjeldberg and R.A. Andersen, J. Mol. Struct., 1985, 129, 93.
24. S.A. Shaidulin, O.A. Litvinov, V.A. Naumov, T. Ya. Stepanova and G.V. Romanov, Zh. Strukt. Khim., 1985, 26(2), 175.
25. B.R. Miller and M. Fink, J. Chem. Phys., 1985, 83, 939.
26. H. Bürger, G. Pawelke and H. Oberhammer, J. Mol. Struct., 1985, 128, 283.
27. B. Beagley and R.G. Pritchard, J. Mol. Struct., 1985, 130, 55.
28. C. George, A.H. Lowrey and J. Karle, J. Mol. Struct., 1985, 127, 107.
29. D. Lentz and H. Oberhammer, Inorg. Chem., 1985, 24, 4665.
30. H.M. Marsden, H. Oberhammer and J.M. Shreeve, Inorg. Chem., 1985, 24, 4756.
31. O.G. Batyukhova, N.I. Sadova, L.V. Vilkov and Yu.A. Pankrushev, Zh. Strukt. Khim., 1985, 26(5), 175.
32. G. Pfafferott, H. Oberhammer and J.E. Boggs, J. Am. Chem. Soc., 1985, 107, 2309.

33. S. Shibata and J. Iwata, J. Chem. Soc., Perkin Trans. II, 1985, 9.
34. R.J. French, K. Hedberg, J.M. Shreeve and K.D. Gupta, Inorg. Chem., 1985, 24, 2774.
35. P.D. Blair, S. Cradock and D.W.H. Rankin, J. Chem. Soc., Dalton Trans., 1985, 755.
36. D.W.H. Rankin, H.E. Robertson, R. Seip, H. Schmidbaur and G. Blaschke, J. Chem. Soc., Dalton Trans., 1985, 827.
37. A. Tsuboyama, S. Konaka and M. Kimura, J. Mol. Struct., 1985, 127, 77.
38. Q. Shen and K. Hagen, J. Mol. Struct., 1985, 131, 309.
39. Q. Shen and K. Hagen, J. Mol. Struct., 1985, 128, 41.
40. J. Brunvoll and I. Hargittai, Acta Chem. Scand., Ser. A, 1985, 39, 667.
41. C.J. Marsden, D. Christen and H. Oberhammer, J. Mol. Struct., 1985, 131, 299.
42. P.L. Baxter, A.J. Downs, A.M. Forster, M.J. Goode, D.W.H. Rankin and H.E. Robertson, J. Chem. Soc., Dalton Trans., 1985, 941.
43. K. Seppelt and H. Oberhammer, Inorg. Chem., 1985, 24, 1227.
44. L. Fernholt, A. Haaland, H.V. Volden and R. Kniep, J. Mol. Struct., 1985, 128, 29.
45. M. Hargittai, O.V. Dorofeeva and J. Tremmel, Inorg. Chem., 1985, 24, 3963.
46. M. Hargittai, O.V. Dorofeeva and J. Tremmel, Inorg. Chem., 1985, 24, 245.
47. V.M. Petrov, A.N. Utkin, G.V. Girichev and A.A. Ivanov, Zhur. Strukt. Khim., 1985, 26(2), 52.
48. Yu. S. Ezhov and S.A. Komarov, Zh. Strukt. Khim., 1985, 26(5), 178.
49. Yu. S. Ezhov and S.A. Komarov, Zh. Strukt. Khim., 1985, 26(6), 60.
50. V.N. Petrova, G.V. Girichev, V.M. Petrov and V.K. Goncharuk, Zhur. Strukt. Khim., 1985, 26(2), 56.
51. K. Hagen, R.J. Hobson, D.A. Rice and N. Turp, J. Mol. Struct., 1985, 128, 33.
52. R. Blom, A. Haaland and J. Weidlein, J. Chem. Soc., Chem. Commun., 1985, 266.
53. R.A. Andersen, A. Haaland, K. Rypdal and H.V. Volden, J. Chem. Soc., Chem. Commun., 1985, 1807.
54. L. Hedberg, K. Hedberg, S.K. Satija and B.I. Swanson, Inorg. Chem., 1985, 24, 2766.
55. K. Hedberg, L. Hedberg, K. Hagen, R.R. Ryan and L.H. Jones, Inorg. Chem., 1985, 24, 2771.
56. H. Oberhammer and D. Lentz, Inorg. Chem., 1985, 24, 1271.
57. K. Iijima, T. Itoh and S. Shibata, J. Chem. Soc., Dalton Trans., 1985, 2555.
58. S.N. Ketkar and M. Fink, J. Am. Chem. Soc., 1985, 107, 338.
59. S. Shibata, K. Iijima and S. Kimura, J. Mol. Struct., 1985, 131, 113.